AF370782

Prospects and Challenges of Bioeconomy Sustainability Assessment

Prospects and Challenges of Bioeconomy Sustainability Assessment

Editors

Idiano D'Adamo
Massimo Gastaldi

MDPI • Basel • Beijing • Wuhan • Barcelona • Belgrade • Manchester • Tokyo • Cluj • Tianjin

Editors
Idiano D'Adamo
Sapienza University of Rome
Rome
Italy

Massimo Gastaldi
University of L'Aquila
L'Aquila
Italy

Editorial Office
MDPI
St. Alban-Anlage 66
4052 Basel, Switzerland

This is a reprint of articles from the Special Issue published online in the open access journal *Sustainability* (ISSN 2071-1050) (available at: https://www.mdpi.com/journal/sustainability/special_issues/Bioecon_Sust).

For citation purposes, cite each article independently as indicated on the article page online and as indicated below:

LastName, A.A.; LastName, B.B.; LastName, C.C. Article Title. *Journal Name* **Year**, *Volume Number*, Page Range.

ISBN 978-3-0365-8296-2 (Hbk)
ISBN 978-3-0365-8297-9 (PDF)

Cover image courtesy of Idiano D'Adamo

Contents

About the Editors

Idiano D'Adamo

Idiano D'Adamo is an Associate Professor of Management Engineering at the Department of Computer, Control, and Management Engineering of Sapienza University of Rome, where he teaches Business Management, Economics of Technology and Management and Economics and Management of Energy Sources and Services. Idiano was among the world's top 2% scientists for three consecutive years.

He received his Master of Science in Management Engineering in 2008 and a Ph.D. in Electrical and Information Engineering in 2012, both from the University of L'Aquila. He has worked at the University of Sheffield, the National Research Council of Italy, Politecnico di Milano, University of L'Aquila and Unitelma Sapienza—University of Rome. In August 2015, he obtained the Elsevier Atlas Award with a work published in *Renewable and Sustainable Energy Reviews*. He collaborates continuously with several journals in a variety of roles: Editor in Chief (*Sustainability*), Subject Editor (*Sustainable Production and Consumption*), Associate Editor (*Environment, Development and Sustainability, Global Journal of Flexible Systems Management, Frontiers in Sustainability*), Editorial Board Member (e.g., *Scientific Reports, Sustainable Operations and Computers*), and is a reviewer for about 100 journals indexed in Scopus.

During his academic career, Idiano D'Adamo has published 130 papers in the Scopus database, reaching an H-index of 43. He has participated in scientific research projects (PRIN2022 "Protecting the Environment: Advances in Circular Economy (PEACE)", PNRR-PE "Made-in-Italy circolare e sostenibile", Horizon 2020 "Star ProBio", Life "Force of the Future"), has collaborated with relevant national institutes (MATTM, CNBBSV, SVIMEZ) and has written for major national newspapers (Formiche, il Messaggero). His current research interests are bioeconomy, circular economy, green energy, sustainability and waste management.

Massimo Gastaldi

Massimo Gastaldi is now Full Professor in Managerial Engineering at the University of L'Aquila, Department of Industrial and Information Engineering and Economics where he teaches Industrial Economics and Analysis of Financial Systems. Since 1993, he has worked for the National Research Council (C.N.R.) and at University of Rome "Tor Vergata". He was visiting Professor at the California State University, San Marcos, USA, and at Colorado University, Builder, USA. He is a responsible manager of private and public research projects. He was editor in chief of the *International Journal of Management and Network Economics*. He collaborates with several journals as Guest Editor (*Global Journal of Flexible System Management, Frontier in Sustainability and Sustainability*) and as a reviewer for several journals indexed in Scopus. Current research issues: cleaner production, sustainability, and circular economy. He is author of 121 papers published on refereed books, journals, and conference proceedings in the Scopus database with 2700 citations (H-index 32).

Preface to "Prospects and Challenges of Bioeconomy Sustainability Assessment"

This Special Issue, "Prospects and Challenges of Bioeconomy Sustainability Assessment", aims to propose the benefits of virtuous use of human and physical resources to achieve sustainable development goals.

Great changes happen with teamwork, harmony and attention to the next generation. The cover image depicts the building of the University of L'Aquila, where, between October 2002 and February 2008, Idiano was a student of engineering management and Massimo was his professor. Giving young people confidence, listening to them and embarking on a path of improvement leads us to look at the future with more serenity. We need, as Pope Francesco says, "a new kind of education, one that allows us to overcome the current globalization of indifference and the culture of waste".

Students have not only rights, but also duties. They are asked to use part of their time to research solutions for current problems and research what aspects of professional life make them proud of what they do. At the same time, professors are not only asked to transfer their background, but to ensure listening time toward students to hear their doubts and curiosities.

A long-lasting professor–student relationship can create a sustainable society.

This Special Issue was made possible by the effort, talent and support of the Managing Editor, Mr. Nikola Milic.

Idiano D'Adamo and Massimo Gastaldi
Editors

 *sustainability*

Editorial

Perspectives and Challenges on Sustainability: Drivers, Opportunities and Policy Implications in Universities

Idiano D'Adamo [1,*] and Massimo Gastaldi [2]

1 Department of Computer, Control and Management Engineering, Sapienza University of Rome, Via Ariosto 25, 00185 Rome, Italy
2 Department of Industrial Engineering, Information and Economics, University of L'Aquila, Via G. Gronchi 18, 67100 L'Aquila, Italy
* Correspondence: idiano.dadamo@uniroma1.it

Citation: D'Adamo, I.; Gastaldi, M. Perspectives and Challenges on Sustainability: Drivers, Opportunities and Policy Implications in Universities. *Sustainability* **2023**, *15*, 3564. https://doi.org/10.3390/su15043564

Received: 5 February 2023
Accepted: 7 February 2023
Published: 15 February 2023

Sustainability calls for contributions from all countries in the evaluation of all its components: nations and regions should invest in research and development, prioritizing the use of green and circular resources [1]. Furthermore, stakeholder engagement, collaboration amongst countries and the inclusion of both young talents and experienced professionals are required to spur a process towards a sustainable transition [2]. We submit that in order to define flexible, transparent and robust strategies for sustainability in the higher education sector, stakeholders must be involved [3] and such involvement should guarantee that interests must be composed to let the "sustainable hand" emerge. In fact, "sustainability does not belong to the selfish. It is a way of life, an approach to pass on to future generations that their dreams can come true" [4]. Some studies have underlined the importance of considering human resource management practices and related socio-economic and psychological supports to foster competitiveness in higher education institutions (HEIs) [5]. For decades, sustainability researchers have been adamant that HEIs should be transformed into more sustainable and inclusive contexts. HEIs have the potential to contribute significantly to the United Nations General Assembly's 2015 Sustainable Development Goals (SDGs) and 2030 Agenda [6]. Historically, sustainability discourse has tended to focus on environmental and economic dimensions, while giving less emphasis to the social dimension. However, the COVID-19 pandemic underlined the importance of the social dimension, particularly with regard to student support during stressful periods [7]. More specifically, the pandemic highlighted critical issues in the education system, indicating an overall lack of resilience. The shift to online learning revealed that some professors were unfamiliar with information technology and some students did not have the resources to participate online; additionally, many students reported problems concentrating and struggled to follow the new educational model [8]. These challenges limited the extent to which SDG 4 could be achieved, as inclusive and accessible education for all did not materialize [9]. Indeed, SDGs are typically addressed more broadly, with SDG 4 and SDG 9 receiving the greatest and least attention, respectively [10]. Of course, the relevance of individual SDGs changes according to the priorities set by each country under consideration [11]. The issue of sustainability is well established, yet the pandemic period has underscored the urgency of finding sustainable solutions. As part of that challenge, HEIs are clearly called upon to provide responses to the various SDGs. An essential requirement is stakeholder engagement.

Research has highlighted that HEIs promote sustainability through the two main channels: institutional initiatives and campus operations [12] (e.g., concrete, "virtuous" projects [13] that are able to realize a zero-carbon solution [14]). Arguably, despite increased awareness of the SDGs, the majority of HEIs have not yet fully embraced sustainability in the curriculum; nor have they created appropriate learning environments. This is because, over the last decades, HEIs around the world have prioritized the high employability of their graduates [15]. We submit that it is time to rethink educational strategy given that sustainable issues have assumed a central position in business and government agendas [16]. In this way, a gap emerges that the literature needs to consider.

Some authors have proposed a tetrahedron framework, with students at the center and vertices representing student competence, teaching methodology, professors and alliances [17]. This framework aligns with the literature, which highlights that students may play a decisive role in the achievement of sustainability goals [18]. Whereas college students have been shown to present a moderate degree of knowledge about sustainability, their behaviors have been found to be insufficient to promote a sustainable transition [19]. Furthermore, some analyses have measured the impact of a university course on students' knowledge, showing an increase in sustainable attitudes and identifying two pillars of future civil society: sustainable education and youth confidence [20]. Young scholars have the opportunity to collaborate to address sustainability issues [21,22] and hone their green skills [23] to foster transformative innovation [24]. We submit that, in order to analyze whether HEIs are capable of equipping their students and scholars accordingly, new methodologies must be developed to assess sustainability learning practices [25,26]. It emerges that the goal of HEIs should be to create value for students, who should be given more attention.

Research has also highlighted the need to identify indicators to measure the impact of initiatives, emphasizing the need for policies with quantifiable effects [27]. Some policy suggestions call for the inclusion of sustainable topics in all professional curricula, as it has emerged that these topics are sometimes only loosely associated with relevant subjects, such as waste recycling and social inclusion [19]. Problems related to environmental pollution, resource scarcity and social inequalities highlight the need not only for trained graduates, but also graduates capable of putting their theoretical knowledge into practice. Encouraging results have emerged from approaches in which students propose projects to solve real problems [28]. Some analyses have shown that, after taking courses on sustainability topics, students tend to increase their focus on bio-products and bio-energy issues and develop a greater interest in careers in the bioeconomy [29]. The multidisciplinary nature of the course should respond to the need to strengthen and improve courses to meet current challenges by providing solutions. Moreover, the market itself may require new professional skills.

In this direction, encouraging results have emerged from approaches in which students propose projects to solve real problems [28]. One example of a student-centered model is project-based learning (PjBL), whereby students independently create their own learning experiences [30]. PjBL aims at enhancing students' critical thinking and ability to solve complex contextual problems. In particular, it emphasizes the need to stimulate emotional learning. The goal is for students to engage and interact with diverse communities, acquire information from multiple sources, enter into collaborative patterns and develop team-building strategies [31]. Additionally, the literature emphasizes the "living lab" model, which, unlike a traditional lab, operates in a real-world context, putting stakeholders at the center [32]. This approach will prompt many professors to review their knowledge transfer habits and encourage more collaboration with students in order to monitor the progress of project work.

A recent report released by carbon consultancy firm EcoAct aims to establish the level of emissions released by UK universities, and suggests a pathway to zero emissions that would include a variety of initiatives. These include an online portal where different emissions data are collected, so that universities can get a better idea of the emissions of the businesses they work with; an incentive system for graduates to offer their sustainability skills to the industry for placements and internships; and suggesting that students do not travel back home during vacations [33]. In their sustainability initiatives, we argue that HEIs should strike the right balance between internationalization [34] and connection with local realities [35]. In fact, some studies have highlighted the relevance of environmental education centers in adhering to the guiding principles of sustainable development while attending to local contexts [36]. Furthermore, the assessment and mapping of ecosystem services may improve knowledge co-production for sustainable territorial management [37]. Indeed, institutional investors have been found to play an important role in the development of clean energy infrastructure, entrepreneurship, poverty reduction, corporate social responsibility and firm development [38]. Initiatives involving different universities should concern all countries, in order to prevent a hyper-concentration of certain countries over others [39]. Sustainability is therefore called upon to define the point of harmony between the needs of internalization and the development of local realities.

In fact, the world is increasingly interconnected, and internationalization is needed to learn about other cultures and to compete and grow amidst initiatives based in other countries. The Erasmus initiative has already fostered this perspective in students, and universities, more generally, may become places where knowledge ecosystems can be realized. Universities must be able to attract scholars from around the world to find solutions to problems and improve social welfare. They must also reward achievement and ensure the free exchange of knowledge. This may activate competitive advantages, with an eye to the balance of ecosystems. Training meetings should be encouraged, where participants can compare ideas from the academic, public and business spheres. In this context, attention should also be paid to student projects, which could benefit from being directly tested by a panel of stakeholders, with rewards offered for the most compelling projects and ideas.

This Special Issue, titled "Prospects and Challenges of Bioeconomy Sustainability Assessment," suggests the need for multiple approaches (i.e., quantitative, qualitative, hybrid) applied across different sectors of sustainability, to illustrate the real advantage of sustainability. HEIs may play an essential role in sustainability, as key agents in the education of future leaders who will contribute to the dissemination of SDG principles. Academia has long recognized the importance of this role, and several studies have illustrated different approaches to promoting sustainability within HEIs, as well as visions that require immediate action. The sustainability challenge depicts ecosystems that are no longer in balance, due to human actions that have gone beyond that which is sustainable, resulting in objective environmental change. Importantly, some emerging phenomena may limit the extent to which current problems can be addressed. One example is "sustainable washing," which describes the tendency to label oneself a sustainability expert despite lacking the necessary qualifications. To avoid sustainable washing, researchers must take care not to present themselves as experts in sustainability unless they have produced appropriate scientific papers on the topic or carried out sustainability projects. In addition, graduate courses should not be considered sustainable simply by virtue of the inclusion of symbolic words in the course title, if there is no accompanying real change in the teaching content.

The literature also indicates a need to develop interdisciplinary pathways to encourage group work that incorporates different perspectives and knowledge bases, as well as a mixture of resources and skills. Such group work may aim to propose solutions for everyday challenges, with a forward-looking perspective. In this vein, professors should challenge themselves to value the different perspectives of young people, giving them time to grow and establish relationships as they work to attain a degree. Collaborations may also involve the launch of start-ups, as innovation and sustainability march in the same direction. While these models already exist, it is not always clear what governs them or how young people can access them. Simplification and economic resources are needed to encourage youth in this direction. The aim of collaborations should be to identify and perceive new needs, select appropriate market segments and implement blue ocean strategies, where possible. However, there must be a clear awareness of the risk component concerning real markets, particularly with respect to global giants.

There are many drivers and opportunities related to the new role of universities. Their goal should be to govern the sustainability challenges and put students at the heart of the process, bestowing on them not only rights but also duties. Essentially, students must develop an ability to connect theoretical knowledge with practical experience, and acquire tacit and complex knowledge. Such knowledge takes a significant amount of time to acquire and, as such, requires patience. Universities face multiple challenges. Suggestions for meeting these challenges may arise from evolutionary economics, which holds that successful organizations replicate successful organizational routines and abandon unsuccessful ones (e.g., looking solely at the enhancement of local resources), as well as the ecology of organizations, which holds that inertia is limited by the mechanism of selection (e.g., if curricula are not adjusted, enrolment will decrease). Moreover, the political context is also relevant. The use of public funding needs to be monitored, as unproductive public spending is unsustainable due to passing the interest on debt to future generations. However, it is necessary to invest in research, skills and the creation of hubs of excellence that can make universities protagonists within ecosystems.

To pursue the goal of sustainability, universities should lead cultural change by listening more to students, increasing their sense of responsibility to others and directing their assignments to impact the greatest number of people. At the same time, policymakers should propose solutions that accrue benefits to many, and not the few. Now is the time to build community, overcome selfishness and reduce stress. The needs of local territories must be met, but through sustainable and resilient solutions at the national level, to prevent a weakening of the system as a whole. Sustainable choices do not always require compromise, but choices must be forward-looking, even at the expense of a lack of consensus in the short term. The knowledge triangle (i.e., education, research, innovation) must be linked with the mission of service to society.

Author Contributions: Conceptualization, I.D. and M.G.; writing—original draft preparation, I.D. and M.G.; writing—review and editing, I.D. and M.G.; supervision, I.D. All authors have read and agreed to the published version of the manuscript.

Funding: This research received no external funding.

Conflicts of Interest: The authors declare no conflict of interest.

References

1. D'Adamo, I.; Gastaldi, M.; Morone, P.; Rosa, P.; Sassanelli, C.; Settembre-Blundo, D.; Shen, Y. Bioeconomy of Sustainability: Drivers, Opportunities and Policy Implications. *Sustainability* **2022**, *14*, 200. [CrossRef]
2. D'Adamo, I.; Gastaldi, M. Sustainable Development Goals: A Regional Overview Based on Multi-Criteria Decision Analysis. *Sustainability* **2022**, *14*, 9779. [CrossRef]
3. Miquelajauregui, Y.; Bojórquez-Tapia, L.A.; Eakin, H.; Gómez-Priego, P.; Pedroza-Páez, D. Challenges and opportunities for universities in building adaptive capacities for sustainability: Lessons from Mexico, Central America and the Caribbean. *Clim. Policy* **2022**, *22*, 637–651. [CrossRef]
4. D'Adamo, I.; Gastaldi, M.; Morone, P. Economic sustainable development goals: Assessments and perspectives in Europe. *J. Clean. Prod.* **2022**, *354*, 131730. [CrossRef]
5. Mohiuddin, M.; Hosseini, E.; Faradonbeh, S.B.; Sabokro, M. Achieving Human Resource Management Sustainability in Universities. *Int. J. Environ. Res. Public Health* **2022**, *19*, 928. [CrossRef] [PubMed]
6. Serafini, P.G.; de Moura, J.M.; de Almeida, M.R.; de Rezende, J.F.D. Sustainable Development Goals in Higher Education Institutions: A systematic literature review. *J. Clean. Prod.* **2022**, *370*, 133473. [CrossRef]
7. Gamage, K.A.A.; Munguia, N.; Velazquez, L. Happy Sustainability: A Future Quest for More Sustainable Universities. *Soc. Sci.* **2022**, *11*, 24. [CrossRef]
8. Maican, M.-A.; Cocoradă, E. Online Foreign Language Learning in Higher Education and Its Correlates during the COVID-19 Pandemic. *Sustainability* **2021**, *13*, 781. [CrossRef]
9. Crawford, J.; Cifuentes-Faura, J. Sustainability in Higher Education during the COVID-19 Pandemic: A Systematic Review. *Sustainability* **2022**, *14*, 1879. [CrossRef]
10. Alcántara-Rubio, L.; Valderrama-Hernández, R.; Solís-Espallargas, C.; Ruiz-Morales, J. The implementation of the SDGs in universities: A systematic review. *Environ. Educ. Res.* **2022**, *28*, 1585–1615. [CrossRef]
11. Alvarez-Risco, A.; Del-Aguila-Arcentales, S.; Rosen, M.A.; García-Ibarra, V.; Maycotte-Felkel, S.; Martínez-Toro, G.M. Expectations and Interests of University Students in COVID-19 Times about Sustainable Development Goals: Evidence from Colombia, Ecuador, Mexico, and Peru. *Sustainability* **2021**, *13*, 3306. [CrossRef]
12. Lambrechts, W.; Van Liedekerke, L.; Van Petegem, P. Higher education for sustainable development in Flanders: Balancing between normative and transformative approaches. *Environ. Educ. Res.* **2018**, *24*, 1284–1300. [CrossRef]

13. de Souza Silva, J.L.; Barbosa de Melo, K.; dos Santos, K.V.; Yoiti Sakô, E.; Kitayama da Silva, M.; Soeiro Moreira, H.; Bolognesi Archilli, G.; Ito Cypriano, J.G.; Campos, R.E.; Pereira da Silva, L.C.; et al. Case study of photovoltaic power plants in a model of sustainable university in Brazil. *Renew. Energy* **2022**, *196*, 247–260. [CrossRef]

14. Osorio, A.M.; Úsuga, L.F.; Vásquez, R.E.; Nieto-Londoño, C.; Rinaudo, M.E.; Martínez, J.A.; Leal Filho, W. Towards Carbon Neutrality in Higher Education Institutions: Case of Two Private Universities in Colombia. *Sustainability* **2022**, *14*, 1774. [CrossRef]

15. Ramakrishna, S.; Jose, R. Addressing sustainability gaps. *Sci. Total Environ.* **2022**, *806*, 151208. [CrossRef]

16. France, J.; Milovanovic, J.; Shealy, T.; Godwin, A. Engineering students' agency beliefs and career goals to engage in sustainable development: Differences between first-year students and seniors. *Int. J. Sustain. High. Educ.* **2022**, *23*, 1580–1603. [CrossRef]

17. Zamora-Polo, F.; Sánchez-Martín, J. Teaching for a Better World. Sustainability and Sustainable Development Goals in the Construction of a Change-Maker University. *Sustainability* **2019**, *11*, 4224. [CrossRef]

18. Wright, C.; Ritter, L.J.; Wisse Gonzales, C. Cultivating a Collaborative Culture for Ensuring Sustainable Development Goals in Higher Education: An Integrative Case Study. *Sustainability* **2022**, *14*, 1273. [CrossRef]

19. Wendlandt Amézaga, T.R.; Camarena, J.L.; Celaya Figueroa, R.; Garduño Realivazquez, K.A. Measuring sustainable development knowledge, attitudes, and behaviors: Evidence from university students in Mexico. *Environ. Dev. Sustain.* **2022**, *24*, 765–788. [CrossRef]

20. Biancardi, A.; Colasante, A.; D'Adamo, I. Sustainable education and youth confidence as pillars of future civil society. *Sci. Rep.* **2023**, *13*, 955. [CrossRef]

21. Dieu, H.D.T.; Kim, O.D.T.; Bich, H.N.V. Sustainable Development of Collaborative Problem Solving Competency for Technical Students through Experiential Learning (A Case Study in Planning Skills Subject at Ho Chi Minh city University of Technology and Education). In Proceedings of the 2018 4th International Conference on Green Technology and Sustainable Development (GTSD), Ho Chi Minh City, Vietnam, 23–24 November 2018; pp. 505–510.

22. Hernandez-Aguilera, J.N.; Anderson, W.; Bridges, A.L.; Fernandez, M.P.; Hansen, W.D.; Maurer, M.L.; Ilboudo Nébié, E.K.; Stock, A. Supporting interdisciplinary careers for sustainability. *Nat. Sustain.* **2021**, *4*, 374–375. [CrossRef]

23. Sady, M.; Żak, A.; Rzepka, K. The Role of Universities in Sustainability-Oriented Competencies Development: Insights from an Empirical Study on Polish Universities. *Adm. Sci.* **2019**, *9*, 62. [CrossRef]

24. El-Jardali, F.; Ataya, N.; Fadlallah, R. Changing roles of universities in the era of SDGs: Rising up to the global challenge through institutionalising partnerships with governments and communities. *Health Res. Policy Syst.* **2018**, *16*, 38. [CrossRef]

25. Albareda-Tiana, S.; Vidal-Raméntol, S.; Fernández-Morilla, M. Implementing the sustainable development goals at University level. *Int. J. Sustain. High. Educ.* **2018**, *19*, 473–497. [CrossRef]

26. Kioupi, V.; Voulvoulis, N. The Contribution of Higher Education to Sustainability: The Development and Assessment of Sustainability Competences in a University Case Study. *Educ. Sci.* **2022**, *12*, 406. [CrossRef]

27. Qian, H.; Ye, M.; Liu, J.; Gao, D. Evaluation of and Policy Measures for the Sustainable Development of National Experimental Teaching Demonstration Centers in Chinese Universities and Colleges. *SAGE Open* **2022**, *12*, 21582440211068516. [CrossRef]

28. Affolderbach, J. Translating green economy concepts into practice: Ideas pitches as learning tools for sustainability education. *J. Geogr. High. Educ.* **2022**, *46*, 43–60. [CrossRef]

29. McAlexander, S.L.; McCance, K.; Blanchard, M.R.; Venditti, R.A. Investigating the Experiences, Beliefs, and Career Intentions of Historically Underrepresented Science and Engineering Undergraduates Engaged in an Academic and Internship Program. *Sustainability* **2022**, *14*, 1486. [CrossRef]

30. Fletcher, G.J.O.; Simpson, J.A.; Thomas, G. Ideals, perceptions, and evaluations in early relationship development. *J. Pers. Soc. Psychol.* **2000**, *79*, 933–940. [CrossRef] [PubMed]

31. Paristiowati, M.; Rahmawati, Y.; Fitriani, E.; Satrio, J.A.; Putri Hasibuan, N.A. Developing Preservice Chemistry Teachers’ Engagement with Sustainability Education through an Online Project-Based Learning Summer Course Program. *Sustainability* **2022**, *14*, 1783.

32. Purcell, W.M.; Henriksen, H.; Spengler, J.D. Universities as the engine of transformational sustainability toward delivering the sustainable development goals. *Int. J. Sustain. High. Educ.* **2019**, *20*, 1343–1357. [CrossRef]

33. Williams, T. What is the Real Carbon Footprint of Universities? Available online: https://www.timeshighereducation.com/depth/what-real-carbon-footprint-universities (accessed on 23 January 2023).

34. Ramaswamy, M.; Marciniuk, D.D.; Csonka, V.; Colò, L.; Saso, L. Reimagining Internationalization in Higher Education Through the United Nations Sustainable Development Goals for the Betterment of Society. *J. Stud. Int. Educ.* **2021**, *25*, 388–406. [CrossRef]

35. Berchin, I.I.; de Aguiar Dutra, A.R.; Guerra, J.B.S.O. de A. How do higher education institutions promote sustainable development? A literature review. *Sustain. Dev.* **2021**, *29*, 1204–1222. [CrossRef]

36. Eliades, F.; Doula, M.K.; Papamichael, I.; Vardopoulos, I.; Voukkali, I.; Zorpas, A.A. Carving out a Niche in the Sustainability Confluence for Environmental Education Centers in Cyprus and Greece. *Sustainability* **2022**, *14*, 8368. [CrossRef]

37. González-García, A.; Aguado, M.; Solascasas, P.; Palomo, I.; González, J.A.; García-Llorente, M.; Hevia, V.; Olmo, R.M.; López-Santiago, C.A.; Benayas, J.; et al. Co-producing an ecosystem services-based plan for sustainable university campuses. *Landsc. Urban Plan.* **2023**, *230*, 104630. [CrossRef]

38. Khan, P.A.; Johl, S.K.; Akhtar, S.; Asif, M.; Salameh, A.A.; Kanesan, T. Open Innovation of Institutional Investors and Higher Education System in Creating Open Approach for SDG-4 Quality Education: A Conceptual Review. *J. Open Innov. Technol. Mark. Complex.* **2022**, *8*, 49. [CrossRef]
39. Arnaldo Valdés, R.M.; Gómez Comendador, V.F. European Universities Initiative: How Universities May Contribute to a More Sustainable Society. *Sustainability* **2022**, *14*, 471. [CrossRef]

Article

Sustainable Development Goals: A Regional Overview Based on Multi-Criteria Decision Analysis

Idiano D'Adamo [1,*] and **Massimo Gastaldi** [2]

1   Department of Computer, Control and Management Engineering, Sapienza University of Rome, Via Ariosto 25, 00185 Rome, Italy

2   Department of Industrial Engineering, Information and Economics, University of L'Aquila, Via G. Gronchi 18, 67100 L'Aquila, Italy

*   Correspondence: idiano.dadamo@uniroma1.it

Abstract: The Sustainable Development Goals (SDGs) have the ambitious goal of protecting the planet, eradicating poverty and providing peace and prosperity for all citizens. The challenge is certainly very ambitious and it is necessary to monitor progress toward these SDGs over time. This work is based on the multi-criteria decision analysis and aims to build a framework that can be replicated. A necessary condition for this aim is that the data are available and that they are as recent as possible. This work is based on 28 targets with data mainly from 2019 to 2020 and related to Italian regions. The results show that Trentino Alto Adige and Valle d'Aosta have the best performance and, in general, the northern territory has several realities that perform positively toward the SDGs. Important results are also present at the level of central Italy (in particular Marche and Toscana), while at the southern level the situation is not flourishing, with the sole exception of Abruzzo. The policy implications thus drive the need for targeted green investments for southern regions, projects that nationally promote the "green, bio and circular Made in Italy" brand that can enhance territorial distinctiveness, and the necessary collaboration among regions to be poles of excellence based on available resources and skills.

Keywords: Italy; indicators; multi-criteria decision analysis; policy implications; Sustainable Development Goals

Citation: D'Adamo, I.; Gastaldi, M. Sustainable Development Goals: A Regional Overview Based on Multi-Criteria Decision Analysis. *Sustainability* **2022**, *14*, 9779. https://doi.org/10.3390/su14159779

Academic Editor: Antonis A. Zorpas

Received: 8 July 2022
Accepted: 7 August 2022
Published: 8 August 2022

Publisher's Note: MDPI stays neutral with regard to jurisdictional claims in published maps and institutional affiliations.

1. Introduction

The 2030 Agenda for Sustainable Development, adopted by all member states of the United Nations in 2015, aims to achieve peace and prosperity across the world. These aspects are even more amplified by a global disruption brought about by the COVID-19 pandemic that has resulted in the deaths of many people, but has also changed the socio-economic system.

The 17 SDGs require urgent action by all countries—developed and developing—through a global partnership [1]. The theme is very broad, and therefore needs to be declined in all areas. There is a demand for an interdisciplinary approach that can combine artificial intelligence with the water–energy–food nexus [2], that frames the contribution of sustainability from a technological perspective [3], that identifies the role of eco-systems innovation [4], how public administrations implement sustainable practices [5], circular approaches [6], businesses that optimize resources by following principles of competitiveness and value-added [7], responsible consumer behavior [8], and the contribution of culture [9] and tourism [10] towards human welfare. These are all actions that should lead to a reduction in greenhouse gas emissions [11] using new approaches based on the circular economy and bioeconomy [12,13]. New approaches that rely on the important role of sustainable approaches [14], the contribution of new generations through training [15], and the role of stakeholder engagement and renewable communities thus assume a key role [16]. In this context, society demands to adopt a model based on the sustainable hand [17]. In addition, it is important to emphasize the strategic role of human resources on how certain characteristics, such as capability, motivation and opportunity, impact performance [18].

The topic of the SDGs has acquired a key role in the literature [19,20]. The approach toward the SDGs calls for developing new approaches [21], but the problem remains difficult to define given that the goals address different scales and each SDG has multiple aspects to consider [22]. All territories, regardless of their size, are called upon to achieve the SDGs [23]. However, each territory has its own peculiarities and characteristics, so it becomes important to identify and adopt a National Strategy for Sustainable Development [24]. Following this perspective, several approaches are considered, since the areas of analysis are very broad [7,25,26].

The role of indicators within the sustainability assessment literature is very important. The purpose of an indicator is to provide summary information, to gather different perspectives, and to identify outputs and relationships useful to stakeholders. Similarly, they play a role in increasing the awareness and responsibility of territories toward achieving the SDGs. However, the complexity of countries' paths to sustainable development cannot be fully understood by resorting to a single multifunctional ranking indicator [27]. An overall indicator can be based on partial indicators, but then it is necessary to aggregate the different information [28]. Analyses are conducted at the local level [29] and at the global level [30]. Particular emphasis is placed on Europe aiming for the ambitious goal of being climate neutral [31,32]. Some authors highlight the need for active input not only from member states, but also at the central level of European institutions [33]. In this direction, however, the European Green Deal calls for great efforts, and therefore needs policies that know how to have models implemented that actually follow sustainable principles.

Sustainable goals are as cross-cutting as they are specific, making it necessary to adopt methods that can aggregate them [34,35]. The literature shows that ranking among different alternatives is likely to increase both awareness and accountability toward achieving the SDGs [36,37].

Multi-criteria decision analysis (MCDA) seems to be a suitable approach to provide real support to public decision makers in policy making [38]. MCDA is an established method in the literature that can assess sustainable performance and can be used to aggregate contributions from different sustainable indicators. MCDA based on data obtained from the main databases can be aggregated with each other in order to provide an aggregate indicator and ranking. The level of accuracy of the result increases as the number of available data increases, but also depending on the objective to be constructed. This research is based on a strand of research that aims to provide results and critical interpretations of values to frame the performance of certain territorial realities. The method has been applied in Italy at the level of individual regions [39] or at the level of individual cities [40]. In addition, the methodological approach has been used to measure the economic performance of European countries by demonstrating that gross domestic product cannot interpret the economic sustainability of a country [41]. This research follows this strand of analysis, in which the gap of providing new recent data, assessing spatial performance and identifying new methodological steps is filled. In addition, some works give particular attention to the role of targets toward the SDGs [42,43]. In this way, the focus of this work is duplicative. From a methodological point of view, it aims to provide a sustainable framework that can be useful for policy makers, but, in general, for all citizens to control and monitor the progress of their territory towards the SDGs. From a managerial point of view, the ranking analysis and decomposition of the final result according to sustainability dimension and geographic macro-area is useful to identify practices, actions and policies to be implemented.

2. Materials and Methods

The MCDA method is very useful in decision-making processes because it has the advantage of synthesizing a large amount of data. The decision maker can use the synthesized elements calculated in order to make a choice among different alternatives. The method has the advantage of being easy to use, well established in the literature, and used in various fields, including those related to sustainability [44,45]. MCDA is based on the identification of suitable criteria to achieve the objective. The main disadvantage is that it does not consider dynamic aspects and does not provide direct assessments of interactions between variables.

MCDA is a methodology that is part of the field of operations research and is defined as a method that considers conflicting criteria within a decision-making process to identify the best among alternative solutions [46]. The method can have characteristics of objectivity when the criteria are measurable, allowing a fair comparison between the same alternatives [47,48]. The output is a synthesis of multiple data, which, only through decomposition can be understood whether they are convergent or divergent. The literature shows how countries can be considered alternatives. In fact, if comparing products and technologies, the expected result of MCDA is to see, for example, the most profitable project or the one with the best quality. Instead, when analyzing countries, a ranking can be proposed to define the strengths and weaknesses of individual performances and to understand possible relationships [39,49,50].

Section 1 showed how the MCDA applies to European countries, which are then identified as alternatives. In this paper, the analysis focuses on Italian regions. The highest value identifies the region that best performs from a sustainability perspective [39]. The criteria selected depend primarily on data availability. Other papers have used an approach based on European or national statistical data [51,52], while in this paper we chose to use data released by ASviS (Italian Alliance for Sustainable Development) [42,53]. The rationale is due to the use of the most recent data and the active role that ASviS plays in the country.

Specifically, the latest report shows values for 28 Targets attributable to 16 of the 17 SDGs [54]—Table 1.

Table 1. List of criteria.

SDG	Target	Year	Unit
SDG 1	Target 1.2—By 2030, reduce the number of people at risk of poverty or social exclusion by 20% compared with 2019	2019	%
SDG 2	Target 2.4 (a)—By 2030, reduce the use of distributed fertilizer in agriculture by 20% compared to 2020	2020	quintals per ha
SDG 2	Target 2.4 (b)—By 2030, reach the share of 25% of UAA invested by organic crops	2019	%
SDG 3	Target 3.4—By 2025, reduce the probability of dying from noncommunicable diseases by 25% compared to 2013	2018	%
SDG 3	Target 3.6—By 2030, halve road traffic injuries compared to 2019	2020	per 10,000 population
SDG 4	Target 4.1 (a)—By 2030, reduce the number of students who do not reach the sufficient level of numerical proficiency (18–19 years old) below the 15% quota	2021	%
SDG 4	Target 4.1 (b)—By 2030, reduce the number of students who do not reach the sufficient level of literacy proficiency (18–19 years old) below the 15% quota	2021	%
SDG 4	Target 4.1 (c)—By 2030, reduce early exit from education and training (18–24 years old) below the 9% rate	2020	%
SDG 4	Target 4.3—By 2030, reach the 50% share of college graduates (30–34 years old)	2020	%
SDG 5	Target 5.5—By 2030, halve the gender employment gap compared with 2020	2020	females/males * 100
SDG 6	Target 6.3—By 2027 ensure high or good ecological quality status for all surface water bodies	2015	%

Table 1. *Cont.*

SDG	Target	Year	Unit
SDG 6	Target 6.4—By 2030, achieve 90% efficiency share of drinking water distribution networks	2018	%
SDG 7	Target 7.2—By 2030, achieve 40% share of energy from renewable sources.	2018	%
SDG 7	Target 7.3—By 2030, reduce gross final energy consumption by 14.4% compared to 2019	2019	ktoe per 10,000 population
SDG 8	Target 8.5—By 2030 to reach 78% employment rate (20–64 years old)	2020	%
SDG 8	Target 8.6—By 2030, reduce the share of NEETs to below 9% (15–29 years old)	2020	%
SDG 9	Target 9.5—By 2030, reach the share of 3% of GDP devoted to research and development	2019	%
SDG 9	Target 9.c—By 2026, ensure all households have Gigabit network coverage	2019	%
SDG 10	Target 10.4—By 2030, reduce the disposable income inequality index to the levels observed in the best of European countries	2018	s80/s20
SDG 11	Target 11.2—By 2030, increase the number of seat-km per inhabitant offered by public transport by 26% compared to 2004	2019	places-Km per inhabitant
SDG 11	Target 11.6—By 2030, reduce exceedances of the PM10 limit to below 3 days per year	2019	days
SDG 12	Target 12.4—By 2030, reduce the share of municipal waste generated per capita by 27% compared to 2003	2019	kg/inhab.* year
SDG 13	Target 13.2—By 2030, reduce CO2 and other climate-changing gas emissions by 55% from 1990 levels	2017	ton CO2 equivalent per capita
SDG 14	Target 14.5—By 2030, reach 30% share of marine protected areas.	2019	%
SDG 15	Target 15.3—By 2050, bring the increase in annual land consumption to zero	2020	ha per 100,000 population
SDG 15	Target 15.5—By 2030, reach 30% share of protected land areas	2019	%
SDG 16	Target 16.3—By 2030, reduce overcrowding in correctional institutions to zero	2020	%
SDG 16	Target 16.7—By 2030, reduce the average duration of civil proceedings to the levels observed in the best of the Italian regions	2020	days

It is shown that most of the targets (22 out of 28) have a time referring to 2019. The division of the 28 targets among the different SDGs is not fair: there are in fact four associated with SDG 4, one with SDG 1, SDG 5, SDG 10, SDG 12, SDG 13, and SDG 14, and two with the remaining nine SDGs. SDG 17 does not present any criteria. The choice of targets used matches all those that are reported in the ASviS report. We also want to emphasize that the latest available data are proposed.

All values are compared with each other and it emerges that they are homogeneous, so no changes need to be made in order to make the more populous regions comparable with the less populous ones. The aggregate sustainability indicator is based on the product between a value and a weight associated with the different criteria and calculated for all alternatives.

There are twenty-one alternatives, as all nineteen regions are considered, while the two provinces Bolzano and Trento are proposed for the Trentino Alto Adige region. We want to underline that in the result figures proposed in Section 3, the average value for the Trentino Alto Adige region will be proposed as an average value between two provinces. The values associated with the different criteria follow the normalized 0–1 method [41]: the least performing value among all twenty-one alternatives is assigned the value 0, while the best performing one is assigned the value 1. Finally, an intermediate value is calculated for all the remaining alternatives by interpolation—Table 2A,B. It should be specified that the 14.5 target was calculated only for ten regions (Liguria, Friuli, Toscana, Lazio, Abruzzo, Campania, Puglia, Calabria, Sicilia and Sardegna), while for the others the relevant target and associated SDG were not counted.

Table 2. List of values. (**A**) IT = Italy; A1 = Piemonte; A2 = Valle d'Aosta; A3 = Liguria; A4 = Lombardia; A5 = Province Bolzano; A6 = Province Trento; A7 = Veneto; A8 = Friuli Venezia Giulia; A9 = Emilia Romagna; A10 = Toscana; * = Estimated. (**B**) A11 = Umbria; A12 = Marche; A13 = Lazio; A14 = Abruzzo; A15 = Molise; A16 = Campania; A17 = Puglia; A18 = Basilicata; A19 = Calabria; A20 = Sicilia; A21 = Sardegna.

(A)											
Target	IT	A1	A2	A3	A4	A5	A6	A7	A8	A9	A10
1.2	0.579	0.793	1.000	0.772	0.805	0.918	0.858	0.928	0.863	0.822	0.745
2.4 (a)	0.702	0.619	1.000	0.571	0.000	0.988	0.976	0.298	0.452	0.321	0.857
2.4 (b)	0.338	0.000	0.029	0.199	0.019	0.013	0.003	0.029	0.006	0.325	0.527
3.4	0.478	0.435	0.565	0.457	0.587	0.783	1.000	0.674	0.543	0.674	0.652
3.6	0.681	0.799	0.821	0.000	0.688	0.516	0.842	0.667	0.717	0.405	0.337
4.1 (a)	0.410	0.726	0.739	0.491	0.860	0.549	1.000	0.653	0.664	0.649	0.528
4.1 (b)	0.420	0.708	0.783	0.542	0.816	0.466	1.000	0.656	0.605	0.613	0.478
4.1 (c)	0.548	0.643	0.678	0.757	0.652	0.452	1.000	0.774	0.948	0.878	0.670
4.3	0.579	0.642	0.585	0.572	0.893	0.503	0.962	0.723	0.836	0.893	0.604
5.5	0.535	0.765	1.000	0.756	0.731	0.822	0.867	0.598	0.686	0.796	0.782
6.3	0.384	0.503	0.874	0.707	0.260	0.933	0.826	0.340	0.388	0.248	0.298
6.4	0.406	0.585	1.000	0.448	0.770	0.857	0.648	0.439	0.296	0.728	0.382
7.2	0.146	0.130	1.000	0.000	0.071	0.735	0.469	0.117	0.146	0.037	0.119
7.3	0.653	0.364	0.000	0.734	0.266	0.231	0.254	0.231	0.104	0.058	0.474
8.5	0.555	0.756	0.854	0.692	0.838	1.000	0.863	0.808	0.841	0.896	0.811
8.6	0.566	0.705	0.825	0.693	0.801	1.000	0.912	0.908	0.952	0.861	0.817
9.5	0.051	0.092	0.000	0.051	0.041	0.015	1.000	0.046	0.062	0.082	0.056
9.c	0.674	0.697	0.127	0.984	0.636	0.000	0.737	0.376	0.434	0.592	0.516
10.4	0.520	0.720	1.000	0.760	0.720	0.760	0.940	0.940	0.900	0.820	0.780
11.2	0.388	0.428	0.000	0.349	1.000	0.289	0.329	0.463	0.342	0.210	0.238
11.6	0.434 *	0.000	0.928	0.964	0.133	0.988	0.964	0.145	0.711	0.277	0.711
12.4	0.518 *	0.547	0.191	0.421	0.595	0.540	0.460	0.560	0.534	0.000	0.165
13.2	0.647	0.471	0.435	0.600	0.494	0.671	0.671	0.553	0.247	0.365	0.671
14.5	0.370	0.000	0.000	0.130	0.000	0.000	0.000	0.000	0.109	0.000	1.000
15.3	0.663	0.585	0.539	1.000	0.720	0.684	0.808	0.389	0.834	0.622	0.813
15.5	0.331	0.203	0.436	0.124	0.150	0.556	0.853	0.128	0.192	0.094	0.165
16.3	0.416	0.398	0.555	0.161	0.109	0.317	1.000	0.180	0.025	0.426	0.451
16.7	0.605	0.934	1.000	0.866	0.824	0.937	0.951	0.756	0.949	0.853	0.708

Table 2. *Cont.*

	(B)										
Target	**A11**	**A12**	**A13**	**A14**	**A15**	**A16**	**A17**	**A18**	**A19**	**A20**	**A21**
1.2	0.875	0.736	0.599	0.594	0.279	0.000	0.296	0.361	0.238	0.024	0.519
2.4 (a)	0.833	0.774	0.786	0.810	0.905	0.786	0.845	0.976	0.881	0.940	0.940
2.4 (b)	0.277	0.543	0.576	0.196	0.029	0.251	0.495	0.505	1.000	0.659	0.158
3.4	0.630	0.739	0.435	0.500	0.326	0.000	0.543	0.435	0.413	0.283	0.435
3.6	0.681	0.452	0.505	0.760	0.968	0.993	0.581	0.932	1.000	0.763	0.875
4.1 (a)	0.384	0.409	0.315	0.229	0.543	0.000	0.071	0.256	0.052	0.052	0.190
4.1 (b)	0.447	0.460	0.410	0.290	0.503	0.000	0.101	0.265	0.014	0.145	0.236
4.1 (c)	0.713	0.835	0.652	0.991	0.939	0.183	0.330	0.809	0.243	0.000	0.643
4.3	0.881	0.780	1.000	0.811	0.409	0.157	0.075	0.384	0.132	0.000	0.409
5.5	0.762	0.686	0.618	0.388	0.382	0.000	0.031	0.159	0.014	0.037	0.640
6.3	0.000	0.344	0.342	0.346	0.010	0.301	0.011	0.025	1.000	0.571	0.507
6.4	0.030	0.648	0.075	0.000	0.299	0.301	0.313	0.313	0.319	0.152	0.131
7.2	0.209	0.133	0.012	0.245	0.413	0.113	0.113	0.530	0.439	0.064	0.213
7.3	0.272	0.723	0.682	0.607	0.717	1.000	0.642	0.694	0.960	0.988	0.740
8.5	0.729	0.747	0.619	0.527	0.396	0.003	0.171	0.311	0.000	0.003	0.341
8.6	0.749	0.781	0.602	0.669	0.367	0.120	0.323	0.446	0.116	0.000	0.454
9.5	0.026	0.031	0.072	0.031	0.036	0.041	0.015	0.005	0.005	0.015	0.021
9.c	0.392	0.115	1.000	0.268	0.033	0.840	0.455	0.178	0.150	0.577	0.225
10.4	0.920	0.860	0.580	0.800	0.620	0.160	0.500	0.860	0.600	0.000	0.500
11.2	0.112	0.152	0.559	0.180	0.010	0.121	0.126	0.051	0.096	0.103	0.269
11.6	0.614	0.602	0.181	0.855	1.000	0.566	0.843	0.940	0.771	0.747	0.361
12.4	0.476	0.450	0.502	0.657	0.955	0.693	0.634	1.000	0.854	0.689	0.683
13.2	0.553	0.835	0.671	0.776	0.376	1.000	0.294	0.471	0.753	0.659	0.000
14.5	0.000	0.000	0.109	0.261	0.000	0.435	0.217	0.000	0.196	0.478	0.413
15.3	0.829	0.617	0.725	0.124	0.000	0.922	0.466	0.332	0.881	0.689	0.306
15.5	0.218	0.297	0.410	1.000	0.000	0.906	0.455	0.665	0.571	0.342	0.083
16.3	0.559	0.507	0.320	0.507	0.013	0.411	0.000	0.517	0.673	0.695	0.894
16.7	0.556	0.752	0.602	0.688	0.583	0.332	0.369	0.000	0.122	0.347	0.509

Two distinct methods are proposed for weights [39]:

- Equal weights among SDGs (EWG) scenario;
- Equal weights among indicators (EWI) scenario.

In the EWG scenario, the different targets are aggregated within the reference SDG. As a result, SDGs with only one indicator will be given no change (e.g., SDG 1), while those with two indicators (e.g., SDG 2) will be given an average value in which the targets will be given equal importance. For SDG 4, the same principle applies where an average value is proposed for the four targets. The EWI scenario, on the other hand, assumes that all targets are given equal relevance. Thus, the main difference between the two scenarios is that EWI allows all individual contributions to be assessed without performing any intermediate operation on the final MCDA value. In contrast, in the EWG scenario, equal relevance is preferred to be given to the individual SDG, regardless of the number of targets in it.

The reference scenario will be the EWG scenario and the final value will be broken down according to the three dimensions of sustainability: environmental (SDGs 6, 13–15), economic (SDGs 7–9, 11 and 12) and social (SDGs 1–5, 10 and 16) [35,55]. Furthermore, starting from this composition, it will be assumed that the value of the three dimensions of the SDGs is not the same. It is considered that the involvement of experts could be a subjective element to detect the incidence of individual SDGs, and therefore, in order to identify a more objective method, it was chosen to consider what has been proposed in the literature. The Scopus database was used (access date on 5 July 2022) and within the Article title, Abstract and Keywords the number in which these word combinations were present was detected:

- Environmental Sustainability with 113,518 items (44%);
- Economic Sustainability with 75,134 items (29%);
- Social Sustainability with 71,339 items (26%).

The literature review found that placing equal weights on indicators is a widespread and accepted practice [56]. In particular, the logical idea is to assign equal adequacy to individual indicators [57]. This approach is therefore chosen in several papers, and other analyses show it as the preferred criterion [28,58], but point out that alternative scenarios can also be developed [40]. In particular, some authors prefer to opt for a different weight [59].

In addition, the analysis will also be conducted at the territorial level by identifying three macro-areas:

- North—Valle d'Aosta, Piemonte, Lombardia, Liguria, Trentino Alto Adige, Veneto, Friuli Venezia Giulia and Emilia Romagna.
- Center—Toscana, Umbria, Marche and Lazio.
- South—Abruzzo, Molise, Campania, Puglia, Basilicata, Calabria, Sardegna and Sicilia.

3. Results

The results of this work allow for the mapping of Italian regions based solely on the targets considered; they are not exhaustive of an overall assessment of the SDGs. In fact, the number of criteria considered is not comparable to that of other studies [39], but it is crucial as it allows new information to be provided to decision makers. The sustainability challenge does not admit delay and can move in the wake of pragmatism. For the purposes of the EWG scenario, it should be pointed out that after the first intermediate phase in which targets were normalized in the 0–1 range and aggregated within the relevant SDG, a second normalization phase was applied. The values were restored to the maximum value of 1, which was not reached if within the specific SDG, all targets were rewarded the same alternative. The sustainability score (output of MCDA) is shown in Figures 1 and 2 for the EWG and EWI scenarios, respectively.

The results show that the Trentino Alto Adige region excels, with the Province of Trento having a leading position in both rankings (0.827 and 0.785 in the EWG and EWI scenarios). Its value detects a very significant performance toward the maximum value of 1 and its gap is quite significant compared to the alternatives that follow it in the rankings: 0.134 compared to the Province of Bolzano in the EWG scenario and 0.157 compared to Valle d'Aosta in the EWI scenario. These two alternatives trade relative position in the two scenarios in third place with a small gap from the relative second. Thus, the first picture that emerges from these results is the important figure of the northern regions. In particular, the Province of Trento is first in several Targets (3.4, 4.1, 9.5 and 16.3), the Province of Bolzano in Targets 8.5 and 8.6 and Valle d'Aosta in other Targets (1.2, 2.4a, 5.5, 6.4, 7.2, 10.4 and 16.7). Maximum performance is achieved in these targets. However, even these virtuous regions have some aspects that need to be improved. Target 2.4b has a low value in all three territories; the Province of Bolzano has a low value for the Target 9.5c and Valle d'Aosta for the Targets 7.3, 9.5 and 11.2.

Provincia Trento	0.827
Provincia Bolzano	0.693
Valle d'Aosta	0.686
Marche	0.617
Toscana	0.611
Liguria	0.579
Lombardia	0.579
Piemonte	0.569
Umbria	0.569
Abruzzo	0.554
Veneto	0.545
Italy	0.544
Friuli Venezia Giulia	0.539
Lazio	0.532
Basilicata	0.528
Emilia Romagna	0.518
Calabria	0.515
Sardegna	0.462
Molise	0.459
Campania	0.441
Sicilia	0.411
Puglia	0.381

Figure 1. Sustainability Score in EWG scenario.

Provincia Trento	0.785
Valle d'Aosta	0.628
Provincia Bolzano	0.612
Marche	0.556
Toscana	0.548
Lombardia	0.536
Liguria	0.529
Piemonte	0.528
Friuli Venezia Giulia	0.514
Umbria	0.508
Abruzzo	0.504
Emilia Romagna	0.502
Lazio	0.498
Veneto	0.495
Italy	0.494
Basilicata	0.460
Calabria	0.446
Sardegna	0.418
Molise	0.411
Campania	0.380
Sicilia	0.358
Puglia	0.333

Figure 2. Sustainability Score in EWI scenario.

Scrolling down the ranking, the relevant performance of other northern regions, such as Liguria, Lombardia, Piemonte and Veneto, is confirmed, which are always above the national average (0.544 in the EWG scenario and 0.494 in the EWI scenario). We want to underline the particular performance of Friuli Venezia Giulia and Emilia Romagna, which are above average only in the EWI scenario.

However, the fourth place in both scenarios belongs to a central region (Marche). It does not top in any of the targets considered, but shows significant performance in Targets 4.1c, 10.4 and 13.2, while it shows weak performance in Target 9.5. Marche is followed by Toscana in the ranking, and Umbria, among the central regions, appears above the national average. Again, Lazio appears above the national average only in the EWI scenario.

Regarding the southern regions, the certainly interesting figure is that of Abruzzo, the only region among the eight to have a value above the national average in both scenarios. Important performances are achieved in several Targets (2.4a, 4.1c, 4.3, 10.4 and 11.6) and the highest performance in Target 15.5. In contrast, weak performances are achieved in Targets 6.4, 9.5, and 15.3. The interesting fact is this region prevails over other northern regions (Veneto, Emilia Romagna) and the remaining central region (Lazio). In the EWG scenario it also precedes Friuli. However, it should be remembered that this analysis refers only to the analysis involving the 28 targets identified by ASviS.

As mentioned, all other southern regions occupy the last positions in the ranking and, in particular, the weakest results are recorded by the three most populous regions in this macro-area: Campania, Sicilia and Puglia. The results showed that the value of Target 9.5 was low for several alternatives. In order to give an answer to this aspect, it is useful to apply the concentration indicator for each target. The excellent performance of the Province of Trento and the strong gap with all other alternatives leads to a very low concentration indicator. A similar situation occurs for Target 14.5, which we said is valid for only a few alternatives and with a better performance than Toscana. In contrast, the results tend to be very similar for Targets 2.4a and 10.4—Figure 3.

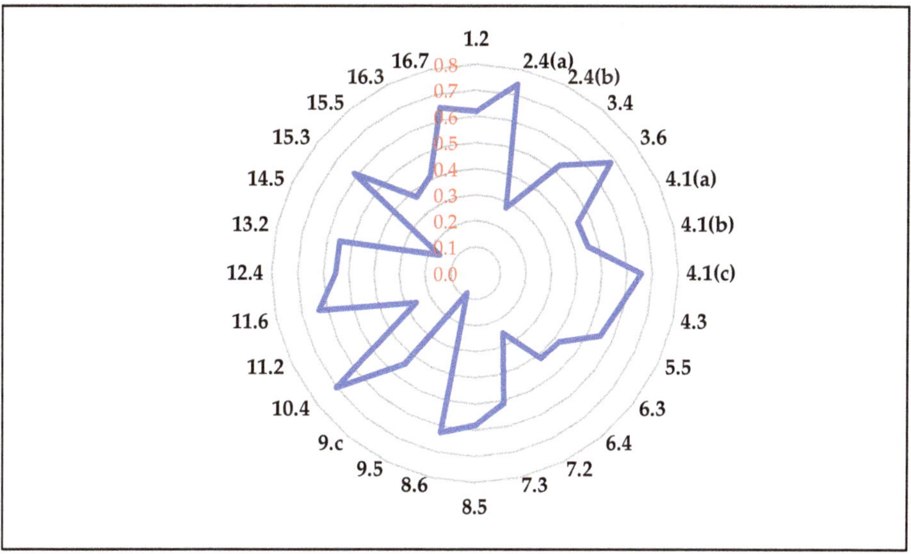

Figure 3. Indicator concentration. All normalized values are reported in red colors for each target.

A further analysis that can be conducted is on the basis of the sustainability dimension. Again, it is worth noting that the results obtained refer only to the targets proposed by ASviS, but they provide an interesting overview—Figure 4.

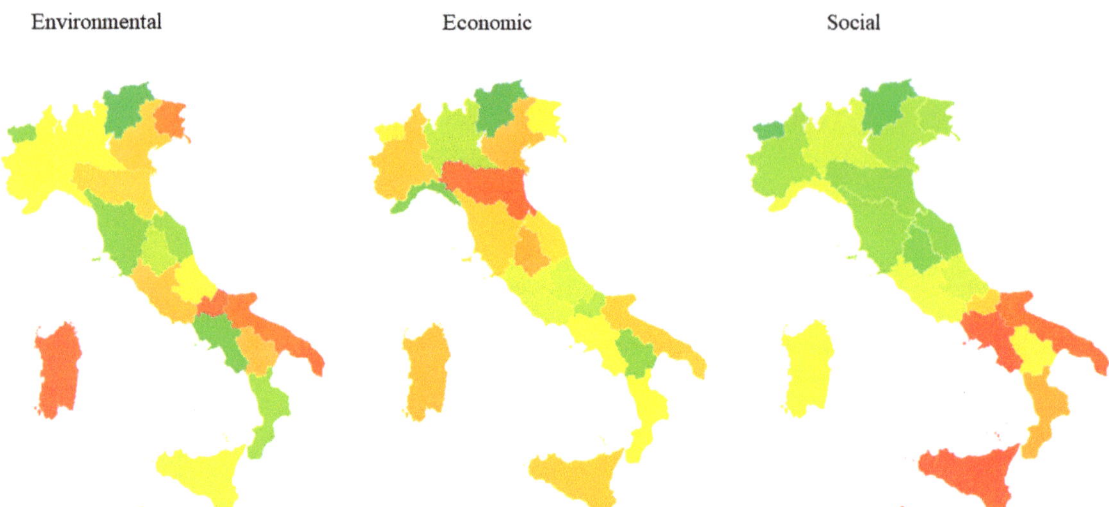

Figure 4. Sustainability Score in EWG scenario—Environmental, economic and social group.

The Province of Trento is a leader in all dimensions considered. Together with the Province of Bolzano (second in the environmental side and third in the economic side) and Lombardia, these are the only territorial realities that are above the national average in all dimensions. Valle d'Aosta (second in the social), Toscana, Marche and Umbria are regions that only in the economic dimension show a lower performance than the national average. Liguria (second in the economic) and Abruzzo present a mirror assessment relative to the environmental dimension in which they perform worse than the national average. Finally, Puglia is the only region that in no dimension performs better than the national average.

The proposed results thus showed that within territorial areas regions' behaviors vary, but they show a gap that particularly affects the South. Unlike other studies, even in the environmental dimension it does not tend to perform optimally—Figure 5. However, it should be pointed out that these results are strongly affected by the small number of targets: thirteen related to the social dimension, nine to the economic dimension and six to the environmental dimension.

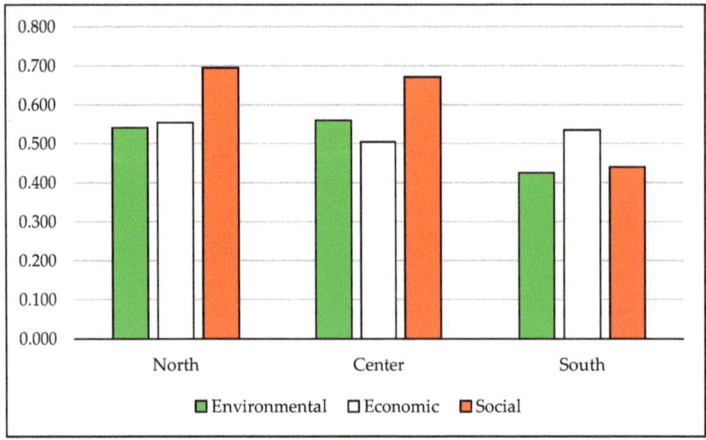

Figure 5. Sustainability Score in EWG scenario—A comparison among macro-areas.

The northern regions excel with a value of 0.615, which is slightly higher than the 0.582 associated with the central regions. In contrast, the difference with the southern regions appears to be much more significant, with a value of 0.469. These results can be explained by the reasons seen for the individual regions, since these are values that are obtained simply by aggregating the regions' data associated with the relevant macro-areas. The south has a significant gap in both the social dimension (in which northern Italy excels) and the environmental dimension (in which central Italy excels). The particular figure, however, is for the economic dimension in which the variations are much less significant and see the North excel, followed by the South. This result is affected by the weak performance of Emilia Romagna, which presents the worst value in terms of Target 12.4.

Finally, the last analysis that can be conducted is one in which the values proposed in Section 2 are always considered, but in this alternative scenario, a different weight is given to the criteria. Specifically, a different weight is not given to each individual target or SDG, but to its dimension—Figure 6.

Provincia Trento	0.809
Provincia Bolzano	0.712
Valle d'Aosta	0.660
Toscana	0.608
Marche	0.607
Liguria	0.560
Lombardia	0.556
Calabria	0.544
Piemonte	0.537
Abruzzo	0.528
Italy	0.522
Campania	0.517
Umbria	0.500
Lazio	0.499
Veneto	0.498
Basilicata	0.495
Friuli Venezia Giulia	0.478
Emilia Romagna	0.472
Sicilia	0.449
Sardegna	0.402
Molise	0.382
Puglia	0.368

Figure 6. Sustainability Score in EWG scenario with different weights.

The results see the Province of Trento still in first place with 0.809, followed by the Province of Bolzano with 0.712. Compared to the baseline scenario, it occurs that Calabria exceeds the national average figure (0.522), while Umbria and Veneto take the reverse path. This result evidently depends on environmental performance, which, in the weighted average, accounts for 44 percent. There is also a significant increase in Campania, which has a better environmental performance than Calabria, which is next in the economic dimension, but closes the ranking in the social dimension.

4. Conclusions

The theme of sustainability has now become a term that is used in all contexts in order to identify positive action towards civil society and in the protection of eco-systems. However, our society is far from applying sustainable hand concepts, as there is an interest, often economic, that overrides the other dimensions. The SDGs are a remarkably relevant benchmark, but not all targets can always be quantified. In such scenarios, it becomes essential to identify benchmarks and make them as objective as possible.

Moreover, the sustainability challenge can only be met if all countries move toward the same goal. The world is a puzzle in which if the largest emitting components do not implement responsible behaviors, inevitably the other components of the puzzle will be harmed as well. In this context, policy makes the choices and should move toward a shared goal that is useful for the future. However, it is up to all of us citizens to make a contribution, big or small it does not matter, but we should be protagonists of a participatory model. We need to promote best practices, demand reforms and choose responsible consumption patterns. Inevitably, businesses, to compete in the marketplace, will also have to adapt. This is particularly so at this time when both the COVID-19 pandemic and the conflict in Ukraine have brought attention to the relevance of having resources locally. Within this framework, the bioeconomy plays a key role, as it enables the use of resources that potentially move toward sustainability.

This work proposes an easily replicable method of analysis in which, regardless of the number of criteria available, it is possible to provide a sketch of an area. A limitation of the work is certainly the small number of the targets, but the main methodological contribution is to provide a framework that can be adopted. The framework, based on MCDA, consists of the following steps: (i) finding recent sustainable data for each alternative; (ii) homogenization among these data; (iii) identification of weights for these criteria by choosing alternative methods; and (iv) decomposition analysis of the final result according to its components (e.g., sustainability dimension and definition of a geographic macro-area). Another limitation of this paper is that it proposes an alternative scenario based on the weights of the three dimensions of sustainability, but values obtained with unequally weighted weights can also be proposed in order to see how the results change. In addition, if you make the assignment of weights through experts, the results may change depending on the categories of stakeholders.

Such work, however, also provides managerial insights. Although the analyses cannot be extended to the overall level of sustainability, they allow for, at the level of the targets analyzed, the observation of how performance in the Italian territory varies widely. Trentino Alto Adige (with the provinces of Trento and Bolzano) and Valle d'Aosta have very relevant results and, in general, most of the northern regions move in that direction. The central regions also show positive results and, in particular, Marche and Toscana propose high values of sustainability score. Finally, among the southern ones, only Abruzzo stands out. It thus emerges that very populous regions, such as Emilia Romagna, Lazio, Campania, Puglia and Sicilia, require interventions in the targets examined in this paper. This also prompts reflection on whether current city patterns, in which a depopulation of inland areas to densely populated areas takes place, is a positive aspect for sustainability. This work has not explored this issue in depth, but the question needs to be asked if we are trying to understand whether we want a decentralized or centralized society. Perhaps the solution lies in how resources are distributed differently in various territories, without being concentrated. However, research has the task of investigating these issues.

The result that emerges from this work is the need to optimize the resources at one's disposal, to invest in the availability of these data that are strategic for making decisions, and to identify policies that can move more resources. The target on research and development indicates that many territories can invest in this aspect, and likewise the green and circular resources need to be enhanced. Sometimes, the social aspect is neglected because it is more difficult to identify. However, the key turning point could be precisely in citizen involvement, collaboration between territorial realities, and the ability to match young talent with more experienced figures. Sustainability is not a trend, and the natural disasters of recent times define the urgency of implementing policies that promote sustainable actions and practices. A sustainability that should be demonstrated with numbers by moving from innovative and resilient ideas.

The results of this work show that there are opportunities for sustainability that have not yet been well explored, highlighting a weak performance of southern regions; European and Italian policies place special emphasis on this. However, public money should be spent only when it produces a benefit for the community, otherwise it generates a debt to be borne by future generations. The political choice can, therefore, be to reward a "green, bio and circular Made in Italy" in which the sad European record of NEET (not [engaged] in education, employment or training) in Italy is overcome and young people are allowed to make their contribution to this society. We need to put resources and skills at the center of the sustainable agenda, each with its own profile and taking advantage of the benefits associated with interdisciplinary. Investment is the best medicine for an economy that risks periods of stagnation, and through the development of the various excellences that characterize Italian territories, this Mediterranean country can play a decisive role in European economic recovery.

Author Contributions: Conceptualization, I.D. and M.G.; methodology, I.D. and M.G.; data curation, I.D. and M.G.; writing—original draft preparation, I.D. and M.G.; writing—review and editing, I.D. and M.G.; supervision, I.D. All authors have read and agreed to the published version of the manuscript.

Funding: This research received no external funding.

Institutional Review Board Statement: Not applicable.

Informed Consent Statement: Not applicable.

Data Availability Statement: Not applicable.

Acknowledgments: We are grateful to Yichen Shen for her passion and expertise in handling this special issue.

Conflicts of Interest: The authors declare no conflict of interest.

References

1. Halkos, G.; Gkampoura, E.-C. Where do we stand on the 17 Sustainable Development Goals? An overview on progress. *Econ. Anal. Policy* **2021**, *70*, 94–122. [CrossRef]
2. Di Vaio, A.; Varriale, L.; Di Gregorio, A.; Adomako, S. Corporate social performance and non-financial reporting in the cruise industry: Paving the way towards UN Agenda 2030. *Corp. Soc. Responsib. Environ. Manag.* **2022**, in press. [CrossRef]
3. Vacchi, M.; Siligardi, C.; Demaria, F.; Cedillo-González, E.I.; González-Sánchez, R.; Settembre-Blundo, D. Technological Sustainability or Sustainable Technology? A Multidimensional Vision of Sustainability in Manufacturing. *Sustainability* **2021**, *13*, 9942. [CrossRef]
4. Ben Amara, D.; Chen, H. Evidence for the Mediating Effects of Eco-Innovation and the Impact of Driving Factors on Sustainable Business Growth of Agribusiness. *Glob. J. Flex. Syst. Manag.* **2021**, *22*, 251–266. [CrossRef]
5. Liu, J.; Ma, Y.; Appolloni, A.; Cheng, W. How external stakeholders drive the green public procurement practice? An organizational learning perspective. *J. Public Procure.* **2021**, *21*, 138–166. [CrossRef]
6. Ippolito, N.M.; Amato, A.; Innocenzi, V.; Ferella, F.; Zueva, S.; Beolchini, F.; Vegliò, F. Integrating life cycle assessment and life cycle costing of fluorescent spent lamps recycling by hydrometallurgical processes aimed at the rare earths recovery. *J. Environ. Chem. Eng.* **2022**, *10*, 107064. [CrossRef]
7. Taddei, E.; Sassanelli, C.; Rosa, P.; Terzi, S. Circular supply chains in the era of Industry 4.0: A systematic literature review. *Comput. Ind. Eng.* **2022**, *170*, 108268. [CrossRef]
8. Moustairas, I.; Vardopoulos, I.; Kavouras, S.; Salvati, L.; Zorpas, A.A. Exploring factors that affect public acceptance of establishing an urban environmental education and recycling center. *Sustain. Chem. Pharm.* **2022**, *25*, 100605. [CrossRef]
9. Duxbury, N.; Kangas, A.; De Beukelaer, C. Cultural policies for sustainable development: Four strategic paths. *Int. J. Cult. Policy* **2017**, *23*, 214–230. [CrossRef]
10. Grelaud, M.; Ziveri, P. The generation of marine litter in Mediterranean island beaches as an effect of tourism and its mitigation. *Sci. Rep.* **2020**, *10*, 20326. [CrossRef]
11. Cucchiella, F.; Gastaldi, M.; Miliacca, M. The management of greenhouse gas emissions and its effects on firm performance. *J. Clean. Prod.* **2017**, *167*, 1387–1400. [CrossRef]
12. Dwivedi, A.; Moktadir, M.A.; Chiappetta Jabbour, C.J.; de Carvalho, D.E. Integrating the circular economy and industry 4.0 for sustainable development: Implications for responsible footwear production in a big data-driven world. *Technol. Forecast. Soc. Chang.* **2022**, *175*, 121335. [CrossRef]

13. Morone, P.; Imbert, E. Food waste and social acceptance of a circular bioeconomy: The role of stakeholders. *Curr. Opin. Green Sustain. Chem.* **2020**, *23*, 55–60. [CrossRef]

14. Backhouse, M.; Lühmann, M.; Tittor, A. Global Inequalities in the Bioeconomy: Thinking Continuity and Change in View of the Global Soy Complex. *Sustainability* **2022**, *14*, 5481. [CrossRef]

15. Eliades, F.; Doula, M.K.; Papamichael, I.; Vardopoulos, I.; Voukkali, I.; Zorpas, A.A. Carving out a Niche in the Sustainability Confluence for Environmental Education Centers in Cyprus and Greece. *Sustainability* **2022**, *14*, 8368. [CrossRef]

16. D'Adamo, I.; Sassanelli, C. Biomethane Community: A Research Agenda towards Sustainability. *Sustainability* **2022**, *14*, 4735. [CrossRef]

17. D'Adamo, I.; Gastaldi, M.; Morone, P.; Rosa, P.; Sassanelli, C.; Settembre-blundo, D.; Shen, Y. Bioeconomy of Sustainability: Drivers, Opportunities and Policy Implications. *Sustainability* **2022**, *14*, 200. [CrossRef]

18. Li, S.; Jia, R.; Seufert, J.H.; Hu, W.; Luo, J. The impact of ability-, motivation- and opportunity-enhancing strategic human resource management on performance: The mediating roles of emotional capability and intellectual capital. *Asia Pacific J. Hum. Resour.* **2022**, *60*, 453–478. [CrossRef]

19. Voukkali, I.; Zorpas, A.A. Evaluation of urban metabolism assessment methods through SWOT analysis and analytical hierocracy process. *Sci. Total Environ.* **2021**, *807*, 150700. [CrossRef]

20. Sarker, M.R.; Moktadir, M.A.; Santibanez-Gonzalez, E.D.R. Social Sustainability Challenges Towards Flexible Supply Chain Management: Post-COVID-19 Perspective. *Glob. J. Flex. Syst. Manag.* **2021**, *22*, 199–218. [CrossRef]

21. Gusmão Caiado, R.G.; Leal Filho, W.; Quelhas, O.L.G.; Luiz de Mattos Nascimento, D.; Ávila, L.V. A literature-based review on potentials and constraints in the implementation of the sustainable development goals. *J. Clean. Prod.* **2018**, *198*, 1276–1288. [CrossRef]

22. Allen, C.; Metternicht, G.; Wiedmann, T. Prioritising SDG targets: Assessing baselines, gaps and interlinkages. *Sustain. Sci.* **2019**, *14*, 421–438. [CrossRef]

23. Yigitcanlar, T.; Kamruzzaman, M. Does smart city policy lead to sustainability of cities? *Land Use Policy* **2018**, *73*, 49–58. [CrossRef]

24. Mallick, S.K.; Das, P.; Maity, B.; Rudra, S.; Pramanik, M.; Pradhan, B.; Sahana, M. Understanding future urban growth, urban resilience and sustainable development of small cities using prediction-adaptation-resilience (PAR) approach. *Sustain. Cities Soc.* **2021**, *74*, 103196. [CrossRef]

25. Loizia, P.; Voukkali, I.; Zorpas, A.A.; Navarro Pedreño, J.; Chatziparaskeva, G.; Inglezakis, V.J.; Vardopoulos, I.; Doula, M. Measuring the level of environmental performance in insular areas, through key performed indicators, in the framework of waste strategy development. *Sci. Total Environ.* **2021**, *753*, 141974. [CrossRef]

26. Krishankumar, R.; Mishra, A.R.; Ravichandran, K.S.; Peng, X.; Zavadskas, E.K.; Cavallaro, F.; Mardani, A. A Group Decision Framework for Renewable Energy Source Selection under Interval-Valued Probabilistic linguistic Term Set. *Energies* **2020**, *13*, 986. [CrossRef]

27. Sciarra, C.; Chiarotti, G.; Ridolfi, L.; Laio, F. A network approach to rank countries chasing sustainable development. *Sci. Rep.* **2021**, *11*, 15441. [CrossRef]

28. Marti, L.; Puertas, R. Assessment of sustainability using a synthetic index. *Environ. Impact Assess. Rev.* **2020**, *84*, 106375. [CrossRef]

29. Xu, Z.; Chau, S.N.; Chen, X.; Zhang, J.; Li, Y.; Dietz, T.; Wang, J.; Winkler, J.A.; Fan, F.; Huang, B.; et al. Assessing progress towards sustainable development over space and time. *Nature* **2020**, *577*, 74–78. [CrossRef]

30. Miola, A.; Schiltz, F. Measuring sustainable development goals performance: How to monitor policy action in the 2030 Agenda implementation? *Ecol. Econ.* **2019**, *164*, 106373. [CrossRef]

31. Hametner, M.; Kostetckaia, M. Frontrunners and laggards: How fast are the EU member states progressing towards the sustainable development goals? *Ecol. Econ.* **2020**, *177*, 106775. [CrossRef]

32. Kostetckaia, M.; Hametner, M. How Sustainable Development Goals interlinkages influence European Union countries' progress towards the 2030 Agenda. *Sustain. Dev.* **2022**, in press. [CrossRef]

33. Ionescu, G.H.; Firoiu, D.; Tănasie, A.; Sorin, T.; Pîrvu, R.; Manta, A. Assessing the Achievement of the SDG Targets for Health and Well-Being at EU Level by 2030. *Sustainability* **2020**, *12*, 5829. [CrossRef]

34. Giannetti, B.F.; Agostinho, F.; Almeida, C.M.V.B.; Liu, G.; Contreras, L.E.V.; Vandecasteele, C.; Coscieme, L.; Sutton, P.; Poveda, C. Insights on the United Nations Sustainable Development Goals scope: Are they aligned with a 'strong' sustainable development? *J. Clean. Prod.* **2020**, *252*, 119574. [CrossRef]

35. Costanza, R.; Daly, L.; Fioramonti, L.; Giovannini, E.; Kubiszewski, I.; Mortensen, L.F.; Pickett, K.E.; Ragnarsdottir, K.V.; De Vogli, R.; Wilkinson, R. Modelling and measuring sustainable wellbeing in connection with the UN Sustainable Development Goals. *Ecol. Econ.* **2016**, *130*, 350–355. [CrossRef]

36. García López, J.; Sisto, R.; Benayas, J.; de Juanes, Á.; Lumbreras, J.; Mataix, C. Assessment of the Results and Methodology of the Sustainable Development Index for Spanish Cities. *Sustainability* **2021**, *13*, 6487. [CrossRef]

37. Cheng, D.; Xue, Q.; Hubacek, K.; Fan, J.; Shan, Y.; Zhou, Y.; Coffman, D.M.; Managi, S.; Zhang, X. Inclusive wealth index measuring sustainable development potentials for Chinese cities. *Glob. Environ. Chang.* **2022**, *72*, 102417. [CrossRef]

38. Ricciolini, E.; Rocchi, L.; Cardinali, M.; Paolotti, L.; Ruiz, F.; Cabello, J.M.; Boggia, A. Assessing Progress Towards SDGs Implementation Using Multiple Reference Point Based Multicriteria Methods: The Case Study of the European Countries. *Soc. Indic. Res.* **2022**, *162*, 1233–1260. [CrossRef]

39. D'Adamo, I.; Gastaldi, M.; Imbriani, C.; Morone, P. Assessing regional performance for the Sustainable Development Goals in Italy. *Sci. Rep.* **2021**, *11*, 24117. [CrossRef]
40. D'Adamo, I.; Gastaldi, M.; Ioppolo, G.; Morone, P. An analysis of Sustainable Development Goals in Italian cities: Performance measurements and policy implications. *Land Use Policy* **2022**, *120*, 106278. [CrossRef]
41. D'Adamo, I.; Gastaldi, M.; Morone, P. Economic sustainable development goals: Assessments and perspectives in Europe. *J. Clean. Prod.* **2022**, *354*, 131730. [CrossRef]
42. Richiedei, A.; Pezzagno, M. Territorializing and Monitoring of Sustainable Development Goals in Italy: An Overview. *Sustainability* **2022**, *14*, 3056. [CrossRef]
43. Zeng, Y.; Runting, R.K.; Watson, J.E.M.; Carrasco, L.R. Telecoupled environmental impacts are an obstacle to meeting the sustainable development goals. *Sustain. Dev.* **2022**, *30*, 76–82. [CrossRef]
44. Jamwal, A.; Agrawal, R.; Sharma, M.; Kumar, V. Review on multi-criteria decision analysis in sustainable manufacturing decision making. *Int. J. Sustain. Eng.* **2021**, *14*, 202–225. [CrossRef]
45. Radmehr, A.; Bozorg-Haddad, O.; Loáiciga, H.A. Integrated strategic planning and multi-criteria decision-making framework with its application to agricultural water management. *Sci. Rep.* **2022**, *12*, 8406. [CrossRef] [PubMed]
46. Han, Q.; Zhu, Y.; Ke, G.Y.; Hipel, K.W. An ordinal classification of brownfield remediation projects in China for the allocation of government funding. *Land Use Policy* **2018**, *77*, 220–230. [CrossRef]
47. Nesticò, A.; Elia, C.; Naddeo, V. Sustainability of urban regeneration projects: Novel selection model based on analytic network process and zero-one goal programming. *Land Use Policy* **2020**, *99*, 104831. [CrossRef]
48. Sironen, S.; Primmer, E.; Leskinen, P.; Similä, J.; Punttila, P. Context sensitive policy instruments: A multi-criteria decision analysis for safeguarding forest habitats in Southwestern Finland. *Land Use Policy* **2020**, *92*, 104460. [CrossRef]
49. Vlachokostas, C.; Michailidou, A.V.; Achillas, C. Multi-Criteria Decision Analysis towards promoting Waste-to-Energy Management Strategies: A critical review. *Renew. Sustain. Energy Rev.* **2021**, *138*, 110563. [CrossRef]
50. Phillis, A.; Grigoroudis, E.; Kouikoglou, V.S. Assessing national energy sustainability using multiple criteria decision analysis. *Int. J. Sustain. Dev. World Ecol.* **2021**, *28*, 18–35. [CrossRef]
51. Streimikiene, D.; Siksnelyte, I.; Zavadskas, E.K.; Cavallaro, F. The Impact of Greening Tax Systems on Sustainable Energy Development in the Baltic States. *Energies* **2018**, *11*, 1193. [CrossRef]
52. Resce, G.; Schiltz, F. Sustainable Development in Europe: A Multicriteria Decision Analysis. *Rev. Income Wealth* **2021**, *67*, 509–529. [CrossRef]
53. Mazziotta, M.; Pareto, A. Measuring Well-Being Over Time: The Adjusted Mazziotta–Pareto Index Versus Other Non-compensatory Indices. *Soc. Indic. Res.* **2018**, *136*, 967–976. [CrossRef]
54. AsviS Report Territories 2021. Available online: https://asvis.it/rapporto-territori-2021/ (accessed on 28 June 2022).
55. Kettunen, M.; Boywer, C.; Vaculova, L.; Charveriat, C. *Sustainable Development Goals and the EU: Uncovering the Nexus between External and Internal Policies*; Institute of European Environmental Policy: Brussels, Belgium, 2018.
56. Gan, X.; Fernandez, I.C.; Guo, J.; Wilson, M.; Zhao, Y.; Zhou, B.; Wu, J. When to use what: Methods for weighting and aggregating sustainability indicators. *Ecol. Indic.* **2017**, *81*, 491–502. [CrossRef]
57. Roszkowska, E.; Filipowicz-Chomko, M. Measuring sustainable development in the education area using multi-criteria methods: A case study. *Cent. Eur. J. Oper. Res.* **2020**, *28*, 1219–1241. [CrossRef]
58. Sachs, J.; Schmidt-Traub, G.; Kroll, C.; Lafortune, G.; Fuller, G. *SDG Index and Dashboards Report 2018*; Bertelsmann Stiftung and Sustainable Development Solutions Network (SDSN): New York, NY, USA, 2018.
59. Guijarro, F.; Poyatos, J.A. Designing a Sustainable Development Goal Index through a Goal Programming Model: The Case of EU-28 Countries. *Sustainability* **2018**, *10*, 3167. [CrossRef]

 *sustainability*

Article

Biomethane Community: A Research Agenda towards Sustainability

Idiano D'Adamo [1,*] and Claudio Sassanelli [2]

1 Department of Computer, Control and Management Engineering, Sapienza University of Rome, Via Ariosto 25, 00185 Rome, Italy
2 Department of Mechanics, Mathematics and Management, Politecnico di Bari, Via Orabona 4, 70125 Bari, Italy; claudio.sassanelli@poliba.it
* Correspondence: idiano.dadamo@uniroma1.it

Abstract: The bioeconomy is an effective solution to align with the sustainability agenda and to meet the pressing calls for action from Cop26 on a global scale. The topic of the circular bioeconomy has gained a key role in the literature, while the theme of energy community is a basic form of social aggregation among stakeholders. This work focuses on biomethane and proposes a framework based on several criteria that are evaluated using a hybrid Analytic Hierarchy Process (AHP) and 10-point scale methodology. The results show that regulation and energy community are considered the two most relevant categories. The overall ranking of criteria sees the stakeholders' engagement as the most important, followed by more significant subsidies for small- and medium-sized plants and the principle of self-sufficiency applied at the inter-regional level. Subsequently, the Italian Adriatic corridor composed of four MMAP (Marche, Molise, Abruzzo, and Puglia) regions is considered as a case study in order to evaluate the possible environmental (854 thousand $\frac{\text{tons } CO_2eq}{\text{year}}$) and economic (from 49 million EUR to 405 million EUR in function of plant size) benefits associated with potential biomethane production of 681.6 million m^3. It is found that the biomethane community is an enabler of sustainability and this strategy can be used for sharing different natural resources.

Keywords: AHP; bioeconomy; biomethane; energy community; Italy; point scale; stakeholders' engagement; sustainability

Citation: D'Adamo, I.; Sassanelli, C. Biomethane Community: A Research Agenda towards Sustainability. *Sustainability* **2022**, *14*, 4735. https:// doi.org/10.3390/su14084735

Academic Editor: Muhammad Abdul Qyyum

Received: 15 March 2022
Accepted: 13 April 2022
Published: 15 April 2022

Publisher's Note: MDPI stays neutral with regard to jurisdictional claims in published maps and institutional affiliations.

1. Introduction

In a world constantly affected by wars, geopolitical risks, and a constantly surging need to modify production models to cope with resource depletion and scarcity, sustainability [1] is becoming the main challenge worldwide, involving different types of stakeholders in society. Sustainability is "development that meets the needs of the current generation without compromising the ability of future generations to meet their own needs" [2]. Sustainability is the balance point between economic prosperity, environmental improvement, and social equity, famously known as the three dimensions. The Triple Bottom Line captures the spectrum of values that organizations must embrace to be competitive given the increasing weight they are gaining [3].

Sustainable development [4] and the Circular Economy (CE) [5] are the dominant concepts suggesting new business models [6] that are able to reach the Sustainable Development Goals (SDGs) [7]. The CE proposes the criteria of narrowing, slowing, and closing resource loops [8] to promote sustainable development [9]. For this aim, the concept of waste is proposed as an added value [10,11].

Europe, with its strategy of the European Green Deal, aims to be the first climate-neutral continent by 2050 [12]. Consequently, the European Commission explicitly seeks to achieve a just transition to a low-carbon energy system. The topic is very much felt in the literature and is becoming increasingly relevant. Some authors highlight three different perspectives: "energy justice occurring within community energy initiatives, between initiatives and related actors, and beyond initiatives" [13]. Energy communities need government tools to develop, and a review of the topic highlighted four categories: (i) community planning and capacity; (ii) environmental protection; (iii) grid access; and (iv) payment-based [14]. For this scope, it is crucial to enhance the integration of the prosumer [15] but also to clarify the role of the stakeholders [16]. These concepts can be studied on individual projects [17] or can identify suitable indicators to measure performance [18].

In addition, the recent war in Ukraine has highlighted energy problems of national self-sufficiency for some countries, and society is no always able to reduce and limit the impact of the main industries on energy consumption (manufacturing [19], transportation [20], and both civil and industrial buildings [21]) through digital technologies adoption. In the current situation, to avoid a placebo effect, this is no longer sufficient, and a breakthrough is needed. The greater energy dependence on foreign countries and the greater dependence on speculative aspects related to the different actors involved in the chain of sale of the energy sources ask for urgent actions in the safe restoration and valorization of waste and biomass to produce valuable and competitive materials and energy [22]. The bioeconomy plays a key role in trying to answer this call, enabling the conversion of natural renewable resources, originating from the biological world, into energy sources [23,24]. The bioeconomy provides suitable answers to the requirements of sustainability in several areas [25,26].

Biomethane can be considered a reference model for the circular bioeconomy [27] through which sustainable best practices could be followed. Indeed, composed of different substrates (e.g., crop leftovers, organic components of municipal solid waste, etc.), it allows the displacement of non-renewable resources with biological ones, triggering the use of biomass and shrinking biowaste presence in landfills [28]. In addition, biomethane can be allocated for different uses (e.g., natural gas grids, energy source for vehicles) and/or converted into feed cogeneration units [29].

Even if different benefits deriving from biomethane adoption can be envisaged for different stakeholders of society levels [30–32], a quite heterogeneous and relevant set of issues can be registered that hinder its full diffusion (e.g., improvement in cleaning and upgrading technologies and processes [33], distrust towards related production plants, and regulation inefficiency [34]). To bridge all the issues related to the full adoption of biomethane towards a widespread solution in current society, pushing the criticalities analysis and leading policymakers to identify guidelines to be implemented to foster biomethane development, innovative frameworks are needed.

Indeed, there is a clear need for understanding how to distribute and deliver the latent value coming from biomethane among all the stakeholders in society through a sustainable hand [22]. The main issue is that to manage to achieve sustainability effectively, the huge quantity of different flows involved (of resources, energy, and wastes) needs a systematization of the related data and information [35]. The main means of bolstering this mechanism are not only digital technologies, which need to be implemented and exploited along the entire product lifecycle [36], but also innovation ecosystems, which are necessary to trigger cooperation and foster the birth of communities able to propose sustainable solutions addressing present and future necessities.

The sustainability bioeconomy aims to provide concrete answers to current problems but also to meet the needs of the literature [22]. For this reason, biomethane is proposed as a virtuous example of a circular bioeconomy, and its potential within an energy community is identified. A topic that needs attention and that finds little space is the use of biomass. Grounded on the concept of energy communities, the goal of this work is to build, propose, and assess a meso-level framework that is needed in CE models to analyze the large number of variables affecting rising biomethane communities. The framework is also engaged to explore the advantages deriving from cooperation dynamics characterizing the MMAP regions model.

The paper is structured as follows. Section 2 presents the research process, detailing the criteria chosen in the hybrid AHP and 10-point scale methodology conducted that are useful for building the biomethane framework development and for its application to the

Italian context of the central and southern Adriatic regions. Section 3 shows the results of the survey related to the framework proposed, detailed with the category, local, and global priority (in Sections 3.1–3.3, respectively). It is also proposed in Section 3.4 a discussion to drive a future direction towards the sustainability of the biomethane community composed by the MMAP regions in Italy. Finally, Section 4 concludes the paper, raising the gaps of the research and unveiling that there is room for further studies to address the challenges to be addressed in the near future to bolster local communities in approaching sustainability.

2. Materials and Methods

A framework gains greater visibility when composed of a substantial number of criteria. The number of criteria can better explain the differences among conflicting criteria or to better delineate a topic. In this paper, we follow the Triple Bottom Line but choose not to propose the three dimensions as reference categories. We aim to identify how other categories influence these dimensions. The need to individuate categories arises from the substantial number of criteria chosen. The AHP is characterized by a small number of criteria and the aggregation of criteria within categories allows for comparison. The local priority and global priority method has precisely this objective.

This section is divided into six parts. In Section 2.1, the hybrid AHP and 10-point scale methodology are proposed with an identification of the experts in Section 2.2. A framework for assessing the impact of biomethane development is proposed in Section 2.3. Furthermore, the environmental and economic values associated with biomethane plants are shown in Sections 2.4 and 2.5. Finally, the case study is proposed in Section 2.6.

2.1. Hybrid AHP and 10-Point Scale Methodology

Decision models allow for the comparison of alternative solutions and decision making through quantitative data. These methods include AHP, which is used to evaluate energy projects [37,38]. The method proposed by Saaty [39] allows obtaining a priority list to be identified through pairwise comparisons based on expert judgments. The AHP, also known as the analytical hierarchy process, is a structured technique for organizing and analyzing complex decisions and is based on mathematics and psychology [40].

The dimension of the AHP comparison matrix ranges from one to ten factors but is typically set to seven $\pm$ two. When the number of criteria is very large, some authors have suggested the use of an integrative method applied to the biogas based on local–global priority [41]. The local priority measures the relevance of a criterion within the same category. The category priority evaluates the relevance of each category and, finally, the global priority is obtained as a product of the local priority and the category priority.

For each matrix, a nine-level scale can be used [42]: 1 → equally preferred; 2 → equally to moderately; 3 → moderately preferred; 4 → moderately to strongly; 5 → strongly preferred; 6 → strongly to very strongly; 7 → very strongly preferred; 8 → very strongly to extremely; and 9 → extremely preferred.

All values are normalized to 1, and to verify the goodness of the results, the consistency ratio (CR) is calculated. The CR is calculated as the ratio between the consistency index and the random inconsistency. This value must be less than 0.10 [42]. This verification occurs automatically during the survey such that experts are not disturbed by an inconsistent assessment.

As a result, this work proposes a hybrid method in which AHP is used to assess local priority. Instead, the 10-point scale—which ranges from not at all (1) to extremely (10)—is chosen to evaluate category priority. This new approach emerged from the need to propose new methodological approaches that maintain a high quality of the result obtained but also from the time requirements related to experts (as highlighted in the pre-check phase of the survey).

2.2. Identification of Experts

The experts for this work were chosen, leading to 10 academic profiles. The invitation was made through an e-mail to those who have developed a Special Issue on scientific journals in the domains of biomethane, biogas, and, more generally, of bioenergy. The invitation was addressed to European colleagues and contained within it the objective of this work. Participation required a minimum number of years of experience equal to 10 and it was reported that the accessions would be chosen in chronological order. Once the adherence was received, the authors proceeded to organize an interview through Skype or Google Meet, lasting up to 1 h, in which they specifically reported what the study was aimed at, and feedback was also collected. It was important that the excel sheet includes only the necessary values and the self-check associated with the consistency ratio. Table A1 proposes the list of the experts with relative data on the role, country of work, and years of experience.

It should be noted that two of these ten experts were chosen in a phase called the pre-check phase of the survey in which the questionnaire was presented to receive feedback and the methodology was proposed in detail. In this phase, the necessity to replace the AHP methodology with the 10-point value emerged because it was considered more accessible. The motivation was not only due to the potential 9×9 matrix but also to the fact that, with a 10-point scale, the criteria were seen as non-conflicting. This information was re-proposed to the other experts as a modification of the first e-mail.

2.3. Biomethane Framework Development

The goal of this work was to assess a meso-level that is needed in CE models in order to provide a comprehensive framework for analysis [43]. To this end, the framework was built through an analysis of the literature [34,44–47] and the expertise acquired by the experts involved in this research, which led to considering a large number of variables and to ground the framework on the global/local priority approach [48].

Once the criteria were identified, it was decided to divide them into categories to allow for their comparison. To allow for a uniform comparison when evaluating global priority, an identical number of criteria per each category (in this case, chosen equal to three) was considered. As the criteria were chosen, the categories that could contain them were also identified. For the category priority, the number of nine was chosen.

Consequently, this survey consisted of nine 3×3 matrices related to the assessment of local priority conducted through the AHP method, while a 10-point scale was more preferred than an AHP-based 9×9 matrix to evaluate category priority.

It should be noted that the two pre-survey experts (see Section 2.2) validated the initial list of criteria but indicated changes to make the questions clearer. In addition, the two methodologies were compared (AHP and 10-point scale) to evaluate the category priority, from which emerged a preference for the second one. Furthermore, the values were then normalized to 1 to maintain consistency with what was obtained in the local priority evaluation phase and to make the calculation of global priority consistent.

The chosen categories addressed:

(i) Two phenomena that typically hinder plant deployment, e.g., Not in my back yard (Nimby) and Not in my term of office (Nimto);

(ii) Four variables that are characteristic of biomethane plants, e.g., size, substrate, final use, and technology;

(iii) Three variables that can help explain aggregation phenomena, e.g., energy community, regulation, and communication.

Among the categories, the three dimensions of sustainability were not considered to avoid triggering discussions about how the other categories might be viewed in contrast. However, all 27 criteria have an influence on the three dimensions of sustainability. In some categories, emphasis was placed on subsidies that are currently in place even in developed markets such as Italy. Table 1 proposes the list of all twenty-seven criteria, which is the first result of this work that was to be used as input for the survey.

Table 1. Biomethane framework.

Category	Acronym	Criteria
Regulation	R1	Self-sufficiency principle (regional level)
	R2	Self-sufficiency principle (inter-regional level, but not national)
	R3	Self-sufficiency principle (national level)
Substrates	S1	All substrates available
	S2	Only sustainable substrates (regional or inter-regional)
	S3	Only sustainable substrates (national)
Plant size	P1	More significant subsidies for small–medium size
	P2	More significant subsidies for large size
	P3	Subsidies not differentiated by plant size
Final use	F1	Electricity
	F2	Transport
	F3	Mix
Technology	T1	Enterprise–university relationships
	T2	Mature technology
	T3	Internally produced plant components
Nimby	N1	Stakeholders' engagement
	N2	Nimby with residues produced in your area
	N3	Nimby with residues not produced in your area
Energy community	E1	Bonus for installations in an energy community (tax deduction)
	E2	Bonus for installations in an energy community (subsidies)
	E3	No bonus for installations in an energy community
Nimto	I1	Nimto determined by local politicians
	I2	Nimto determined by national politicians
	I3	Nimto has not relevance
Communication	C1	Organization of webinars/public meetings
	C2	Transparent site in which to report the results
	C3	No additional actions required

2.4. Environmental Analysis

The environmental benefits associated with biomethane over the use of fossil sources are verified in several works: 23 $\frac{\text{g CO}_2\text{eq}}{\text{MJ}}$ [49], 40 $\frac{\text{g CO}_2\text{eq}}{\text{MJ}}$ [50], 53 $\frac{\text{g CO}_2\text{eq}}{\text{MJ}}$ [51], and 62 $\frac{\text{g CO}_2\text{eq}}{\text{MJ}}$ [52]. However, it is necessary to specify that several values are proposed for these analyses. The International Renewable Energy Agency (IRENA) estimates the greenhouse gas (GHG) emissions of biomethane according to the feedstock type: liquid manure 33 $\frac{\text{g CO}_2\text{eq}}{\text{km}}$, organic waste 48 $\frac{\text{g CO}_2\text{eq}}{\text{km}}$, and maize 66 $\frac{\text{g CO}_2\text{eq}}{\text{km}}$. These values are significantly lower than those produced by fossil sources: methane 124 $\frac{\text{g CO}_2\text{eq}}{\text{km}}$, diesel 156 $\frac{\text{g CO}_2\text{eq}}{\text{km}}$, and petrol 164 $\frac{\text{g CO}_2\text{eq}}{\text{km}}$ [53]. These data show that methane has a less negative impact on the environment than other sources and that biomethane has significant reductions. The highest value recorded for maize is caused by cultivation and harvesting processes.

In this research, three scenarios are considered, defined according to the literature [44]:

- The baseline green scenario, in which the unitary value of reduction in GHG emissions is assumed equal to 83.5 $\frac{\text{g CO}_2\text{eq}}{\text{km}}$.
- The alternative green scenario, in which the unitary value of reduction in GHG emissions is assumed equal to 76 $\frac{\text{g CO}_2\text{eq}}{\text{km}}$.
- The alternative strongly green scenario, in which the unitary value of reduction in GHG emissions is assumed equal to 91 $\frac{\text{g CO}_2\text{eq}}{\text{km}}$.

According to D'Adamo et al. [44], we defined a model to assess the potential reduction in terms of GHG emissions. It is necessary to make some assumptions to define the environmental savings associated with the use of a certain amount of biomethane. Considering that a natural gas vehicle (NGV) has an annual mileage of 20,000 km and multiplying this value with that of the three scenarios examined, it is possible to calculate how much less impact an NGV has on the environment if it is fueled by biomethane compared to natural gas. The next step is to estimate how many NGVs can be used by dividing the potential biomethane calculated in the previous subsection and the consumption of one NGV (1333 m^3). Finally, it is possible to estimate the overall emissions reduction by calculating the savings associated with a single NGV by their total number.

2.5. Economic Analysis

Economic analysis related to biomethane plants can be conducted on estimated production costs equal to 0.54–0.73 $\frac{EUR}{m^3}$ [29], 0.5–1.5 $\frac{\$US}{m^3}$ [53], and 90 $\frac{EUR}{MWh}$ [54], as well as on profitability values with a Net Present Value (NPV) that ranges from −585 thousand USD if subsidies were not provided to 5667 thousand USD [55], or from 0.49 million EUR to 132.7 million EUR based upon the mix of recovered waste [56]. However, in the analysis of the costs, the value associated with the externalities is not considered, and for this reason, it is suggested a minimal subsidy equal to 0.13 $\frac{EUR}{m^3}$ for biomethane production systems [57].

In March 2018, the Italian government adopted a policy decree (GU (Official Journal) no. 65 of 19 March 2018) to stimulate the development of biomethane [58]. This provides a value of incentive equal to 0.305 $\frac{EUR}{m^3}$ (single-counting) for the first ten years. Furthermore, this value is assumed equal to 0.61 $\frac{EUR}{m^3}$ (double-counting) if using some sustainable substrates (i.e., the Organic Fraction of Municipal Solid Waste (ofmsw) and by-products). The decree does not differentiate incentives by plant size.

In this research, two scenarios are considered, defined according to the literature [44]:

- The minimum scenario where the minimum size for which biomethane plants are profitable (200 $\frac{m^3}{h}$ and 350 $\frac{m^3}{h}$ for the ofmsw and by-products, respectively).
- The maximum scenario in which the size chosen for large plants is considered acceptable by citizens (500 $\frac{m^3}{h}$ for both substrates).

In the recent period, the conflict in Ukraine has led to a rise in costs, to which speculation has also been added, leading to the biomethane selling price (virtual trading point) being estimated at different values. The base value of 0.25 $\frac{EUR}{m^3}$ [44] is modified to 0.375 $\frac{EUR}{m^3}$ and 0.50 $\frac{EUR}{m^3}$ [22]. Table 2 proposes the economic profitability associated with the minimum and maximum scenarios for two distinct substrates as a function of biomethane selling price.

Table 2. NPV of biomethane plants [22,44]. Data are expressed in thousand EUR.

Biomethane Selling Price	0.25 $\frac{EUR}{m^3}$	0.375 $\frac{EUR}{m^3}$	0.50 $\frac{EUR}{m^3}$
ofmsw 200 $\frac{m^3}{h}$	421	2199	3779
ofmsw 500 $\frac{m^3}{h}$	8016	11,733	15,450
By-products 350 $\frac{m^3}{h}$	131	3028	5581
By-products 500 $\frac{m^3}{h}$	1656	5623	9131

Another useful indicator to monitor the delay in the realization of the projects is the Discounted Do Nothing Cost (DDNC), which, when considered for 1 year, presents the following values [44]:

- Six kEUR and 20 kEUR for the 350 $\frac{m^3}{h}$ by-products plant, and 200 $\frac{m^3}{h}$ for the ofmsw in the minimum scenario, respectively.
- Seventy-nine kEUR and 382 kEUR for the 500 $\frac{m^3}{h}$ by-products plant and 500 $\frac{m^3}{h}$ for the ofmsw in the maximum scenario, respectively.

It should be noted that the decree provides a bonus for an alternative "biomethane producer and distributor" business model. This is an important aspect as the number of points of sale is currently not suitable to meet the needs of consumers, especially in some areas of the country.

2.6. The Italian Context: Central and Southern Adriatic Regions

The 2019 Integrated National Plan for Energy and Climate assigns a priority role to renewable gas in order to achieve the biofuel release targets set by the directive on the promotion of the use of energy from renewable sources—RED II—in the European Union. The total theoretical potential, which is calculated with data updated to 2016, is estimated to be approximately 6.2 billion m^3 per year of advanced biomethane [59]. In particular, these authors have considered the following substrates: (i) straw; (ii) residues from the grape-wine chain; (iii) tomato peel residues; (iv) citrus juice residues; (v) residues from the olive oil industry; (vi) solid urban waste; (vii) zootechnical waste; (viii) sludge from wastewater purification; and (ix) milk whey. The aim of the Italian government is to inject 2.3 billion m^3 of biomethane into the gas network by 2026, plus 1.1 billion m^3 in transport within the Next Generation EU. The National Federation of Methane Distributors and Transporters (*Federmetano*) proposes the potential production of 8 billion m^3 by 2030.

In this context, the regions of central–southern Italy on the Adriatic side (MMAP) have started forms of collaboration to strengthen the Adriatic ridge and constitute a point of overall connection with all of Italy and Europe [60]. This is a political gesture among regions of different political orientation but that are united in building the future of their communities. Italy's population exceeds 60 million, and MMAP regions account for about 12% (Figure 1).

Figure 1. MMAP regions.

Table 3 proposes data on the potential of biomethane from substrates relative to the regions examined [59]. However, for the ofmsw, the values proposed by *Istituto superiore per la protezione e la ricerca ambientale* (ISPRA) were considered [61]—Table 4. The conversion factor ($\frac{tons\ of\ msw}{m^3}$ biomethane) proposed in Table 3 was considered, 75% of separate collection was assumed for all regions (a value currently achieved by the Veneto region), and the weight of organic waste on the total calculated in Table 3 was considered.

Table 3. Potential biomethane in MMAP regions (data in thousand m^3) [59].

	Marche	Abruzzo	Molise	Puglia	MMAP
Straw	119,106	51,554	31,044	328,206	529,910
Residues from the grape-wine chain	1071	3303	1766	14,512	20,652
Tomato peel residues	2	102	69	3660	3833
Citrus juice residues	0	4	0	12,559	12,563
Residues from the olive oil industry	358	1881	1993	14,252	18,484
Solid urban waste	21,573	14,098	1284	23,395	60,530
Zootechnical waste	30	21	18	197	266
Sludge from wastewater purification	141	22	104	109	376
Milk whey	3786	708	3904	9569	17,967
Total	146,247	71,693	40,182	406,459	664,581

Table 4. Data on waste in MMAP regions [61].

	Marche	Abruzzo	Molise	Puglia	Total
Total (kt)	797	600	111	1872	
Separate collection (%)	70.3	62.7	50.4	50.6	
Total separate collection (kt)	560	376	56	947	
Total organic fraction (kt)	248	162	23	383	
%organic/total	44.3	43.1	41.1	40.4	
Potential biomethane (thousand m^3)—estimated	23,558	15,432	2109	36,424	77,523

Thus, our estimates of an increase in separate collection to 75% for all regions predict an increase in the value of 60,530 thousand m^3 of biomethane. The final value is 77,523 thousand m^3 of biomethane (an increase of about 17 million m^3). Consequently, the potential biomethane in MMAP regions is assumed equal to 681,574 million m^3.

3. Results

This section presents results regarding the relevance of the framework criteria associated with biomethane. Sections 3.1 and 3.2 show the results for category and local priority, respectively, while Section 3.3 shows the results for global priority. Section 3.4 shows the sustainable benefits associated with MMAP regions.

3.1. The Assessment of Category Priority

The aggregation phase of the judgments provided by the experts consists of receiving the weighted data for each category (Table A2) and each criterion (Table A3). Their product determines the global priority, the result of which obviously depends on the incidence of the two specific components. In particular, the weight of each expert is the same during the aggregation phase. The experts evaluated the incidence of the categories and, in six cases, the maximum value of 10 was assigned to both regulation and the energy community. It should be pointed out that, in four cases, this value was also assigned to substrates and plant size. For the other five categories, the maximum value has never been assigned. It should be pointed out that normalization for this weight does not take place in such a way that the sum of all is equal to 1 but is conducted by comparing the absolute value obtained by the experts divided by the maximum value (equal to 10).

The results show that the four categories proposed above have the highest value that is also close to 10 (Figure 2): regulation (0.96), energy community (0.95), and substrates and plant size (both 0.93). However, all nine categories were considered to be of some relevance as the lowest value is given to communication with a weight of 0.74. Other categories have the following weight: technology (0.88), final use (0.84), Nimby (0.83), and Nimto (0.82).

Figure 2. Category priority.

Regulation is seen as a key element as it dominates the sector, defining the guidelines to be followed. Similarly, the theme of the energy community has recently been introduced by the European Union in the Clean Energy for all Europeans package of 2019. It is worth highlighting how the energy community theme has been characterized by the presence of subsidies that play a key role in the development of the sector [48,62]. The same subsidies are used to assess the theme of plant size and for substrates (although not implicitly for the latter category).

3.2. The Assessment of Local Priority

The analysis conducted within each category made it possible to highlight the role of the specific criteria. In this subsection, the specific result is assessed accordingly, while in the next subsection, the interconnections are identified.

As far as regulation is concerned, all the experts assigned the greatest importance to R2—self-sufficiency principle (inter-regional level and not national)—which assumed a weight of 0.54, followed by R1—self-sufficiency principle (regional level)—with 0.30 (Figure 3). Therefore, the meso-approach is preferred to a macro or micro model as it is considered that the theme of sustainability cannot be circumscribed to a restricted geographical area, which is, in any case, called upon to meet objectives with respect to national performance. In particular, the relationship between local and regional authorities is seen as a strategic lever for moving towards ambitious goals. Some authors have highlighted the relevance of quantifying the potential of biomethane for strategic choices by defining the relevance of the territorial context of reference [63].

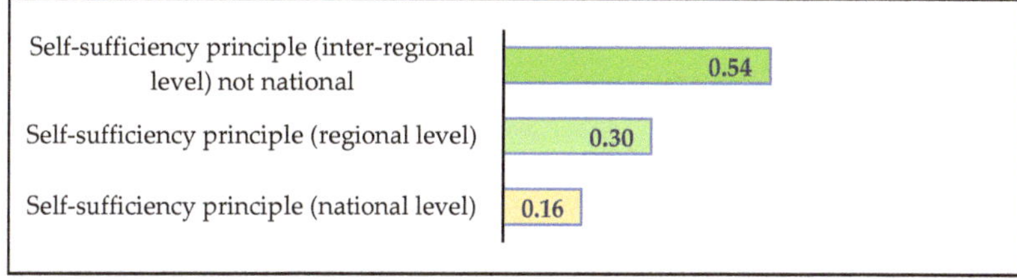

Figure 3. Local priority (regulation).

The analysis of substrates shows that unsustainable choices cannot be made and, therefore, each area must deal with residues, waste, and raw materials that do not, however, lead to a worsening of eco-systems. This weight of 0.93 is divided between S2—only sustainable substrates (regional or inter-regional)—with 0.51, and S3—only sustainable substrates (national)—with 0.42 (Figure 4). In particular, the sustainability of the resources used also suggests the use of resources coming from outside the area in question as the environmental balance is, in any case, compensated. In fact, the recovery of resources is linked to a longer transport than expected. The idea prevails that the inability of some areas to grasp sustainable opportunities can be a source of competitive advantage for others. Criterion S2 was considered most important to eight experts, while criterion S3 was considered most important to the other two. The relevance of substrates for energy and sustainable purposes is also evident in the different contributions that can be provided by resources that are classified as first, second, and third generation substrates [64].

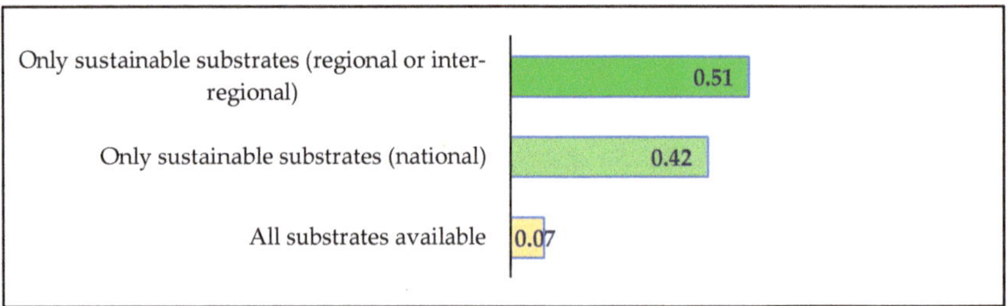

Figure 4. Local priority (substrates).

Regarding the size of the plants, it is clear that larger plants present significant economic results due to the economies of scale. In addition, this can also depend on incentive policies that do not distinguish between subsidies according to size. Experts did not consider this a particularly critical aspect (P3—subsidies not differentiated by plant size—with 0.30). However, they assigned greater importance to P1—more significant subsidies for small–medium size—with 0.62 (Figure 5). In particular, it emerges that small-scale plants can support sustainable development in areas that do not have a large quantity of raw materials. The risk run is that the non-profitable nature of such plants would rightly not allow their realization. All experts agreed on the greater relevance of criterion P1. The role of size should be properly identified in order to capture both the exploitation of available resources and public acceptance. This is an essential aspect that could be reflected in causing significant delays to the implementation of the work. However, such delays can also be determined by aspects related to bureaucracy [44].

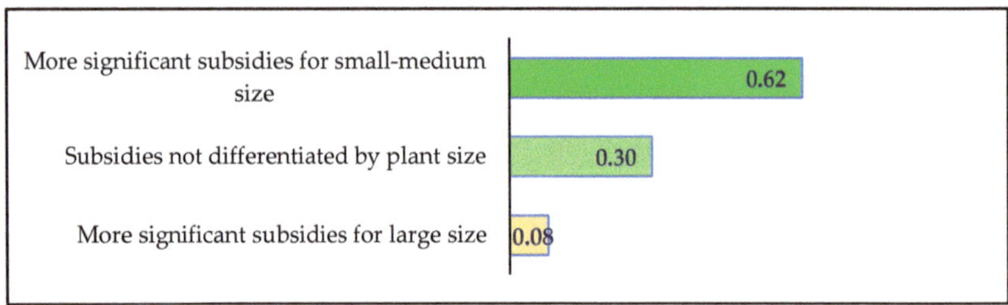

Figure 5. Local priority (plant size).

The issue of final use is one of the great advantages of biomethane that takes advantage of its flexibility. However, the perception is that electrical use can be reduced given the presence of other renewable sources and it is hoped that greater use will be made of transport—F2 with 0.60 (Figure 6). The contribution of renewables in the transport sector is still marginal in most European countries. However, the experts did not underestimate the flexible role of biomethane and, therefore, the mix is still given relevance, albeit less significant—F3 with 0.33. All experts agreed on the greater relevance of criterion F2. The purposes of biomethane use are all relevant as they contribute to the reduction in fossil fuels in different applications. However, a comparison of them should be verified in the function of the energy portfolio of the country under analysis [65].

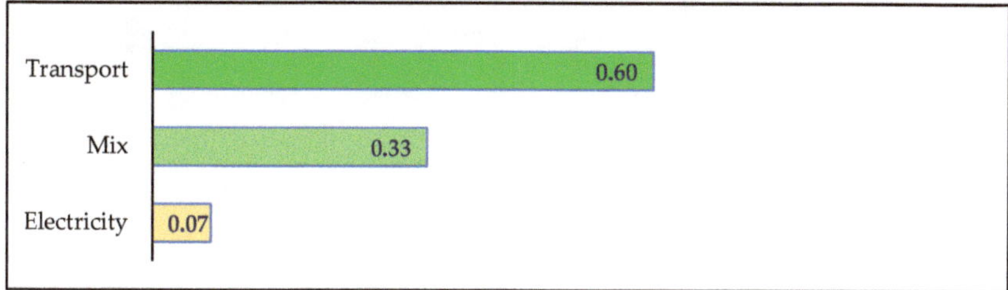

Figure 6. Local priority (final use).

The technology category shows that the biogas–biomethane chain is considered mature, but there is space for improvement. In particular, during the interviews with experts, it emerged that particular studies should be placed on how to increase the yield of raw materials. Therefore, the maturity of the technology takes a low weight not because it is considered a weak factor, but because other aspects must be valued. The academic world clearly highlights the importance of collaboration and, therefore, factor T1—enterprise–university relationships—takes on a weight of 0.42. However, the factor considered most important was T3—internally produced plant components—with 0.50 (Figure 7). This figure should also be understood in the period in which the survey was conducted, in which the importance for companies of internally producing the components they use most emerged. Experts were split almost down the middle on the most relevant criterion (six of them identified T3 while the others identified T1). Studies of biomethane in Europe provide experiences that can be examined in other countries as well. The relevance of the connection between technology, subsidies, and domestic production of components plays a favorable role in supporting the development of the sector [66].

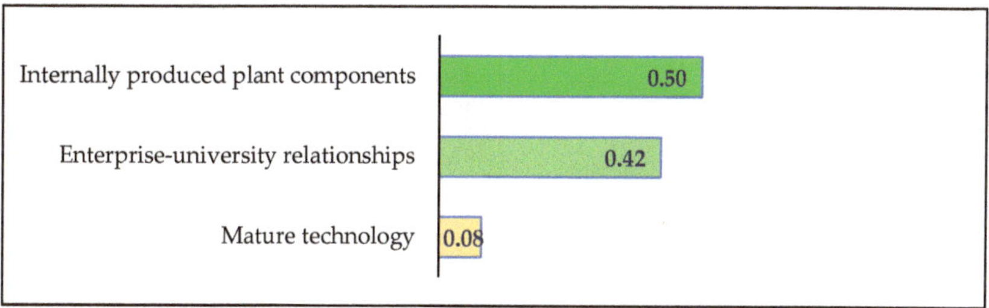

Figure 7. Local priority (technology).

As far as Nimby is concerned, the experts unanimously penalized this phenomenon if it develops for residues produced in their own area of competence, while they assigned a slight relevance to criterion N3—Nimby with residues not produced in your area—with 0.25. On the other hand, all agreed to consider criterion N1—stakeholders' engagement—the most relevant in this category and the weight assigned was decidedly very significant compared to the others, with an average weight of 0.70 (Figure 8). In particular, sustainable development is also associated with inclusive development and provides for decision-making models that involve all stakeholders. It was highlighted by experts as often being underestimated, but the achievement of which allows implementing good actions and to be a model of best practice. The correct balance between personal interest and the interest of an organization theorized by the school of human relations is the basis of this concept. The topic of stakeholder engagement is not well discussed in the literature and is one of the topics that will need to be explored [67]. In fact, this aspect turns out to be a decisive factor towards the implementation of energy community models [22].

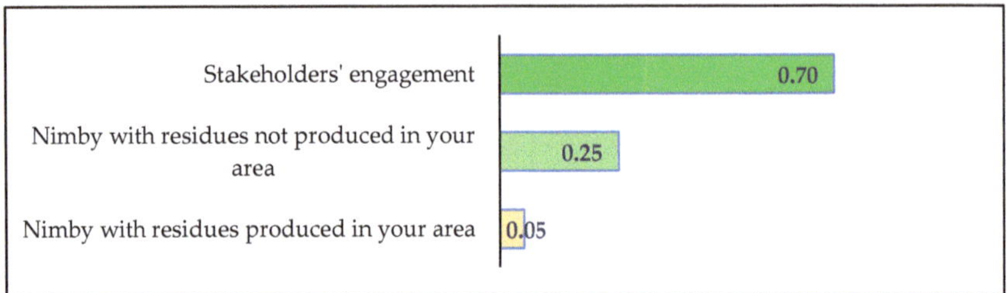

Figure 8. Local priority (Nimby).

The topic of the energy community is analyzed by considering whether its aggregation should be encouraged. The answer provided by the experts was clear. However, the way in which this happened sees seven of them favoring the subsidy while three favored the tax deduction. These aspects are reflected in the final weight, where, however, the difference is not so great: criterion E2—bonus for installations in an energy community (subsidies)—has a weight of 0.51, which is greater than the 0.44 associated with criterion E1—bonus for installations in an energy community (tax deduction)—Figure 9. The subsidy is seen with interest because it stimulates greater production of output. Instead, tax deduction typically affects investment costs. However, the need emerges to introduce a form of incentive to stimulate this ecological transition and to encourage such models of aggregation. A crucial aspect is therefore the contribution of local community energy initiatives to support a decentralized sustainable energy system. The key elements highlighted are the type of organization, the level of activities, and the development of a shared vision. All this translates into a very communicative slogan "power to the people" [68].

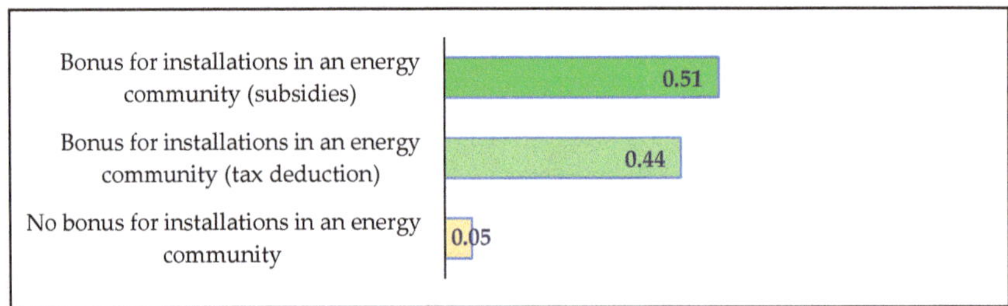

Figure 9. Local priority (energy community).

With regard to Nimto, it is highlighted how this aspect also influences the results. The judgment of the experts on this aspect was unanimous. However, the judgment changed as to who should be attributed greater responsibility in the non-realization of strategic work. In fact, criterion I1—Nimto determined by local politicians—has a weight of 0.51 and was considered the most relevant by seven experts. The others assigned it to criterion I2—Nimto determined by national politicians—with a weight of 0.44 (Figure 10). The rationale that emerges is that we do not always see a sharp distinction between these two categories based on the actual responsibilities that are laid out in the laws. What does emerge, however, is how political non-decision making can delay change towards ecosystem defenses. Support schemes and innovation are among the main forces that drive investment in renewable energy technologies, and both involve considerable uncertainty. These aspects identify the ability to pursue the fight against climate change. However, one element that can destabilize these choices is uncertainty [69].

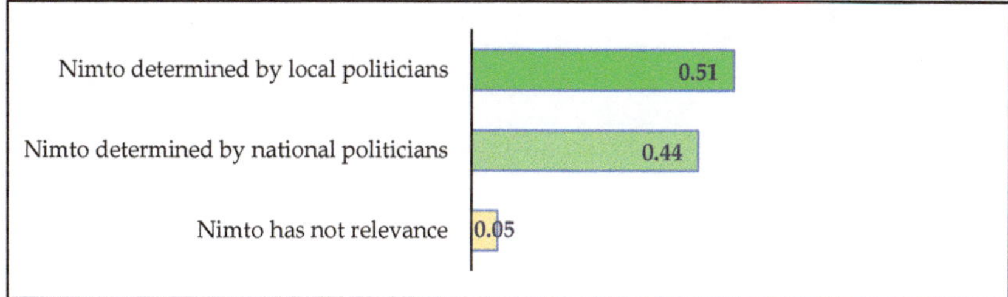

Figure 10. Local priority (Nimto).

Finally, the communication aspects show that, additionally in this case, the experts agreed on the need to implement changes compared to the current situation. Criterion C2—transparent site in which to report the results—achieved by the sorting and recovery of waste has a weight of 0.42, while criterion C1—organization of webinars/public meetings—is considered more relevant with 0.54 (Figure 11). These data highlight, on the one hand, that autonomy is given to people, to their digital skills. Similarly, interaction, both in human and digital form, is important as it is considered fundamental for dispelling doubts and perplexities. Almost all the experts (nine) identified C1 as the most relevant. However, key aspects of the development of biomethane plants are represented by appropriate communication models [22,70]. It should be emphasized that the literature shows that the topic of the circular economy needs to be more appropriately addressed [71].

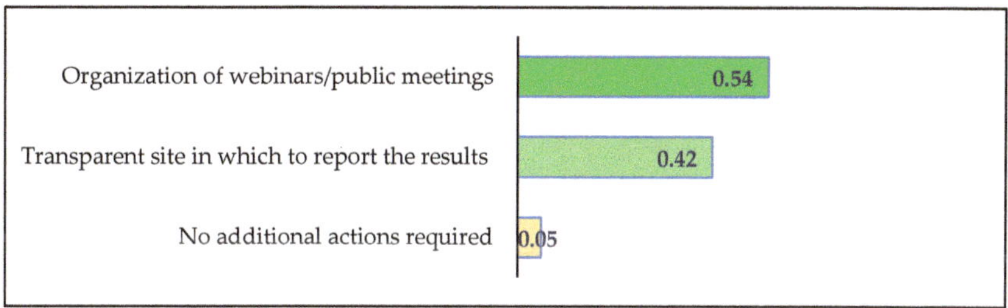

Figure 11. Local priority (communication).

3.3. The Assessment of Local Priority

The results show that global priority does not follow the order defined by category priority. In fact, in first place in the ranking, we find criterion N1 (with a value of 0.581), despite the fact that the Nimby category occupies only the seventh position in the relative ranking associated with the categories. This figure can be explained by what was determined within the local priority for this category: N1 has a weight of 70%, which is the most significant among all the cases analyzed.

In numerical terms, in second place in the global ranking is placed criterion P1 (with a value of 0.577). Additionally in this case, with 62%, this occupies the second most significant weight in terms of local priority when considering all the numerical values, even if the plant size category is the third most significant. In third place in terms of local priority, criterion F2 with 60% occupies the fourth place in the global ranking with 0.504, with the category to which it belongs positioned in sixth place. The most relevant category, regulation, sees its most significant criterion (R2) in third place in the global ranking with 0.518.

Among the top ten criteria in the global ranking (which have a value of at least 0.400), we find all the criteria positioned in first place in all categories. The only category that has two criteria is the energy community with criterion E2 positioned in fifth place in the global ranking with 0.485. This result can be explained by the minimal difference between the first two criteria that have a distance of seven percentage points (the same happens for the Nimto category).

In addition, the nine least relevant criteria for each category occupy the last positions in the overall ranking. It is worth noting that eight of the nine criteria have a weight as local priority less than 0.10, and all these criteria have a global priority value of less than 0.100. The exception is criterion R3 (regulation—self-sufficiency principle (national level)), which has a local weight of 0.16 and a global weight of 0.154—Table 5.

Table 5. Global priority.

Acronym	Criteria	Global Weight
N1	Stakeholders' engagement	0.581
P1	More significant subsidies for small–medium size	0.577
R2	Self-sufficiency principle (inter-regional level, but not national	0.518
F2	Transport	0.504
E2	Bonus for installations in an energy community (subsidies)	0.485
S2	Only sustainable substrates (regional or inter-regional)	0.474
T3	Internally produced plant components	0.440
I1	Nimto determined by local politicians	0.418
E1	Bonus for installations in an energy community (tax deduction)	0.418
C1	Organization of webinars/public meetings	0.400
S3	Only sustainable substrates (national)	0.391
T1	Enterprise–university relationships	0.370
I2	Nimto determined by national politicians	0.361
C2	Transparent site in which to report the results	0.311
R1	Self-sufficiency principle (regional level)	0.288
P3	Subsidies not differentiated by plant size	0.279
F3	Mix	0.277
N3	Nimby with residues not produced in your area	0.208
R3	Self-sufficiency principle (national level)	0.154
P2	More significant subsidies for large size	0.074
T2	Mature technology	0.070
S1	All substrates available	0.065
F1	Electricity	0.059
E3	No bonus for installations in an energy community	0.048
N2	Nimby with residues produced in your area	0.042
I3	Nimto has not relevance	0.041
C3	No additional actions required	0.037

These results highlight that the theme of the biomethane community can be defined as an enabling factor towards sustainability. The integration between the different criteria allows for highlighting how the aggregation between regions is a winning element because it allows for pursuing the spirit of a European community and can be suitable for intercepting public funds available. In addition, collaboration would allow the aggregation of skills and resources and could have greater weight as a leading player in a market.

The essential elements for an energy community foresee the collaboration among all the actors involved in which one could think of a bottom-up decision-making model that is not, however, confirmed by the relevance given to Nimto (the joint weight given both to the political and national class should be underlined), which could push to a top-down model. Clearly, the solution lies in stakeholder engagement where choices are made with everyone's input, but if a final choice is not reached (and it is necessary to make such a choice in order to not reduce future opportunities), a leader who can synthesize is needed. Incentives should privilege small plants to allow a greater spectrum of choices by providing some territories lacking raw materials with this option as well. The choice of biomethane towards transport appears to be the desired end use, and substrates from one's own geographical area can be preferred but others can be accepted if they follow sustainability criteria. The formation of an energy community should be subsidized and, in particular, there is a push for a choice that would increase the production of this energy carrier. Another aspect considered critical is to strengthen the industrial fabric to increase the competitive advantage of a territory.

3.4. Biomethane Community Composed of MMAP Regions: A Future Direction towards Sustainability

The results obtained from Section 3.3 have highlighted how the theme of the energy community, and specifically that represented by biomethane, is capable of providing various points for reflection. Sustainability requires the contribution of everyone, and during the green transition, it is clear that global scenarios based on competitiveness will change. The recent war in Ukraine, which has led to the death of many children, has highlighted the need for a change in mentality. A sustainable approach as reported in the editorial in which this new section within the *Sustainability* journal was introduced [22] has the ambition of being a meeting point among different stakeholders. A metaphor could indicate a port where ships from different parts of the world can dock, where ideas, projects, and ambitions of everyone can be gathered, but with a special eye on new generations. The Adriatic corridor and its four regions, MMAP, are fortunate to have uncontaminated territories where the relationship between man and nature is not disfigured. There are ideas for improvement that can be pursued and circular bioeconomy models can support the achievement of the Sustainable Development Goals [3].

The strategy can proceed in this direction by expanding the fields of collaboration and creating what could be called the sustainable innovation hub, the idea of an innovation ecosystem based on the involvement of different stakeholders, a result that emerges unequivocally from this research. Sustainable innovation hubs are grounded on the concept of Digital Innovation Hubs (DIHs), i.e., ecosystems that assist companies to improve their competitiveness through innovations and fostering the implementation of up-to-date digital technologies [72,73]. Involving different stakeholders belonging to a heterogeneous ecosystem, DIHs provide a set of supportive services that help companies to become more competitive by improving their business by means of digital technology [74].

The four regions have already signed important agreements in terms of infrastructure, transport, and communications, leaving aside political affiliation, which is certainly an element to be stressed. The need for teamwork is emphasized. This work, in a simple way, aims to unite this great strategic project with the results obtained previously, highlighting how the theme of biomethane development can be an element of sustainable success. This is confirmed not only by the environmental and economic results associated with them but also by a substantial number of social opportunities and related economic opportunities induced that could be connected. In addition, it would allow resources to be acquired at lower values and would allow all interested parties to present valid projects by choosing appropriate technologies, adopting models of public participation and reasoning on how this meso-approach reaches the macro-level of sustainability and is then architected at the micro-level on where to install these plants.

Economic estimates are that ofmsw is treated separately while the other substrates are all incorporated at the by-products level. Environmental estimates consider one cubic meter of biomethane regardless of its source of origin. Figures 12 and 13 provide environmental and economic results.

The obtained results are based on an estimate of potential biomethane production of 681.6 million m^3 for the four regions examined (an increase of 17 million compared to values proposed by Pierro et al. [59]). It is evident that this value also includes raw materials currently used for other purposes (e.g., for electricity production through biogas) and that all raw materials are recovered at the technical factors assumed in the reference study. It is also true that an increase in the percentage of separate waste collection has been considered. These data clearly call for the virtuous contribution of all stakeholders. It is clear that ofmsw plants are more profitable than by-products, but a reduction in the profitability of these plants should be envisaged with a simultaneous reduction in the bill paid by citizens, rewarded for good separate collection. In fact, nowadays, it is not only necessary to differentiate but also to do it correctly. This obviously translates into lower costs for the companies that treat such waste in a better energy yield from a technical point of view. As for by-products, it is necessary to inform all stakeholders of the advantages they could have from the recovery of these residues and their sending first to an anaerobic digestion plant and then to an upgrading plant.

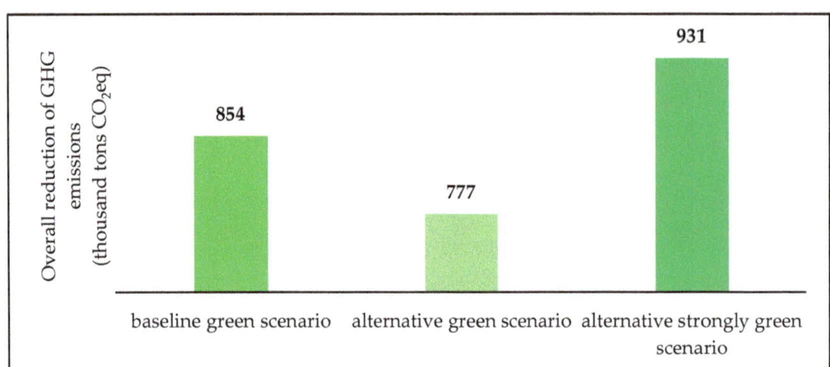

Figure 12. Potential environmental results (thousand tons CO_2eq).

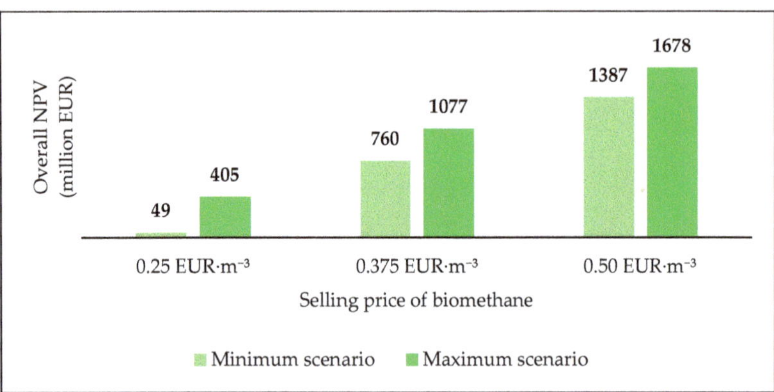

Figure 13. Potential economic results (million EUR).

The economic results also show that as the sale price of biomethane increases, the plants would become more profitable for investors. However, a higher sales price would have a negative impact on consumers. Thus, it is necessary to be vigilant of these aspects. Sustainability is the meeting point of all stakeholders, thus it cannot only support renewable plant owners but also consumers. Taxes that are paid on resources that are more environmentally damaging are considered appropriate. In addition, attention should be paid to the less affluent income groups who do not have the opportunity to take advantage of these sources. It could happen that these groups would have to pay higher prices (taxes on fossil fuels), which would worsen their difficult economic situation.

In this period, great attention has been given to the development of electric vehicles, which, if powered by green sources and with proper battery disposal, are able to provide important environmental results. However, no less important is the role that can be played by NGVs, which are known to pollute less than diesel and gasoline, are a widely known technology, and are typically chosen by consumers for their savings. Evidently, these aspects can be confirmed if natural gas is gradually replaced by green gas. However, we would like to point out that, currently, the energy policy of countries should be projected towards a strong expansion of renewables but the transition should not disregard the use of some fossil sources. Among these, gas has the lowest impact, and the European Commission has indicated this source as necessary for this transition.

The reduction in geopolitical risk and the production of domestically produced energy are necessary steps for the energy development of a country and for energy communities to move towards these directions. The previous analysis has underlined the role of subsidies, which are justified if it is considered at present that several fossil sources are agreed. However, their value cannot be fixed forever in the long term but should be reduced gradually while assuming, however, a stable regulatory framework.

The results of this work assume that at the usual selling price of biomethane and not subject to increases, there would be 49 million EUR in the minimum scenario and 405 million EUR in the maximum scenario (see Figure 13). The choice of plant size must be referred to individual territories. However, some considerations emerge:

- Where it is possible to build a larger plant, it is desirable for the territory to provide significant raw materials;
- Incentives for small plants should be provided to allow the recovery of these residues that in the absence of economic unprofitability could lead to the non-realization of the plants;
- Technical analysis should be conducted on how the different substrates can be mixed as clearly a shared plant allows for taking advantage of different opportunities.

The associated DDNC 1-year amounts to 2263 million EUR in the minimum scenario and 19,333 million EUR in the maximum scenario. The delay in the non-construction of the plants has led to greater energy dependence on foreign countries, greater dependence on speculative aspects related to the different actors involved in the chain of sale of the final product, and, above all, a non-contradiction of environmental aspects. The results of this work show a reduction of 854 thousand $\frac{\text{tons CO}_2\text{eq}}{\text{year}}$ in the baseline green scenario and 777 $\frac{\text{tons CO}_2\text{eq}}{\text{year}}$ and 931 thousand $\frac{\text{tons CO}_2\text{eq}}{\text{year}}$ in the alternative ones (see Figure 12).

4. Conclusions

This work contributes to clarifying how the theme of biomethane, a virtuous model of the circular bioeconomy, is relevant to the reduction in geopolitical risks. The transport sector is called to reduce its environmental impact. The biogas–biomethane supply chain moves towards this direction as it can be used as a vehicle fuel, but it can also be used for other purposes (e.g., it can be distributed into the natural gas grid or converted into cogeneration units).

At present, those who have economic interests in fossil fuels are speculating because they are aware that, today, many countries depend on this raw material. The future foresees the presence on the market of an alternative resource, and this will inevitably lead to a reduction in economic opportunities associated with those who use fossil fuels. It is necessary to identify the transition to advance renewable sources and, in the same way, allow businesses and citizens to have controlled prices.

From a methodological point of view, this work proposes a hybrid approach based on AHP and 10-point value to determine the incidence of the criteria, and this model is suitable when it comes to consider weights based on the local–global priority in which the category priority is calculated considering a consistent number of criteria.

From a conceptual point of view, the model of a united Europe is based on the breaking down of barriers among several countries. The idea of a collaboration of the four regions of the Adriatic corridor (Marche, Molise, Abruzzo, and Puglia) has proved to be a winning strategic idea. The strategy can proceed in this direction by expanding the fields of collaboration and creating what could be called the sustainable innovation hub.

From an operational point of view, the energy communities are able to create new forms of market and alternatives to centralized structures to combine the interests of multiple subjects. However, their realization is by no means simple as it requires an approach based on the concept of shared value. Biomethane is an example of a shared resource that can affect these territories.

The work has some limitations:

- Need to conduct an up-to-date analysis on the energy yields of substrates;
- Environmental performance as a function of specific substrates;
- Economic evaluations applied to specific substrates;
- Development of a network of plants distributed throughout the territory in order to maximize available resources;
- Communication and dissemination models to inform stakeholders and citizens about these changes;
- Research method hybridization has a major impact and changing the combined method may result in different results.

However, this work reports economic and environmental results that are verified with an implementation plan for the development of biomethane, whose values may vary as a result of changes in critical variables. Nevertheless, findings underline that the biomethane community is seen as an enabling factor towards sustainable development.

The ecological transition is a great challenge and the increase in energy costs cannot be described as being caused by the development of sustainable sources. Obviously, this change will affect the interests of some companies and investors whose portfolios were based on fossil sources. However, the transition cannot be to the disadvantage of companies and citizens. For this reason, it should be monitored, and cooperation between the private and public sectors is required. Renewables can take advantage of the funds that are made available, but these projects should be implemented quickly in order to avoid generating costs of doing nothing. In addition, the intervention of a third party is required where abnormal market phenomena are created. This outcome would allow for proper movement towards the goals of the Next Generation EU.

The MMAP project may be able to combine tourism and industry to attract young people from all over the world and it may represent a strategic crossroads within the corridors of integrated logistics. The element that can make the difference is the ability to be a team creating a strong, cohesive, resilient, inclusive, and sustainable community.

Author Contributions: Conceptualization, I.D. and C.S.; methodology, I.D.; formal analysis, C.S.; data curation, C.S.; writing—original draft preparation, I.D. and C.S.; writing—review and editing, I.D. and C.S.; supervision, I.D. All authors have read and agreed to the published version of the manuscript.

Funding: This research received no external funding.

Institutional Review Board Statement: Not applicable.

Informed Consent Statement: Not applicable because the survey is not involved with any data of human subjects (1) through intervention or interaction with the individual, or (2) identifiable private information.

Data Availability Statement: Not applicable.

Acknowledgments: We are grateful to experts used in the AHP methodology.

Conflicts of Interest: The authors declare no conflict of interest.

Appendix A

Table A1. Survey participants.

No.	Role	Country	No. Years
1	Full Professor	Sweden	18
2	Full Professor	Spain	19
3	Full Professor	Italy	20
4	Full Professor	Denmark	16
5	Associate Professor	Finland	12

Table A1. *Cont.*

No.	Role	Country	No. Years
6	Associate Professor	Germany	15
7	Associate Professor	Switzerland	16
8	Associate Professor	Greece	11
9	Associate Professor	Italy	12
10	Associate Professor	France	14

Table A2. Category weights provided by ten experts.

Category	1	2	3	4	5	6	7	8	9	10
Regulation	10	9	10	9	9	10	10	10	10	9
Substrates	8	10	9	9	10	9	9	10	9	10
Plant size	8	10	10	9	9	9	10	9	10	9
Final use	9	8	8	8	9	9	9	8	8	8
Technology	9	9	9	9	9	8	8	9	9	9
Nimby	8	8	8	9	9	9	8	8	8	8
Energy community	9	8	9	10	9	10	10	10	10	10
Nimto	8	9	9	8	8	8	8	9	7	8
Communication	8	7	8	8	7	7	7	7	8	7

Table A3. Local weights provided by ten experts.

Criteria	1	2	3	4	5	6	7	8	9	10
R1	30	25	30	20	35	35	30	30	30	35
R2	60	60	60	60	45	50	55	50	60	40
R3	10	15	10	20	20	15	15	20	10	25
S1	5	10	5	5	5	5	10	5	10	10
S2	55	50	55	55	60	40	50	55	40	50
S3	40	40	40	40	35	55	40	40	50	40
P1	62	70	60	55	60	55	70	50	65	60
P2	8	5	10	5	10	5	10	10	10	5
P3	30	25	30	40	30	40	20	40	25	35
F1	5	5	10	5	5	5	10	15	5	5
F2	75	50	55	65	65	65	50	60	50	65
F3	20	45	35	30	30	30	40	25	45	30
T1	55	35	40	30	50	40	40	50	30	50
T2	5	10	5	10	5	10	5	10	10	10
T3	40	55	55	60	45	50	55	40	60	40
N1	60	75	70	70	65	75	70	70	75	70
N2	5	5	5	5	5	5	5	5	5	5
N3	35	20	25	25	30	20	25	25	20	25
E1	50	45	60	40	55	40	40	40	35	35
E2	45	50	35	55	40	55	55	55	60	60
E3	5	5	5	5	5	5	5	5	5	5
I1	45	50	55	40	55	60	45	50	55	55
I2	50	45	40	55	40	35	50	45	40	40
I3	5	5	5	5	5	5	5	5	5	5
C1	55	50	45	55	55	60	55	50	60	55
C2	40	45	50	40	50	35	40	45	35	40
C3	5	5	5	5	5	5	5	5	5	5

References

1. Tang, M.; Liao, H.; Wan, Z.; Herrera-Viedma, E.; Rosen, M.A. Ten Years of Sustainability (2009 to 2018): A Bibliometric Overview. *Sustainability* **2018**, *10*, 1655. [CrossRef]
2. World Commission on Environment and Development. *Our Common Future*; Oxford University Press: Oxford, UK, 1987; ISBN 019282080X.
3. D'Adamo, I.; Gastaldi, M.; Imbriani, C.; Morone, P. Assessing regional performance for the Sustainable Development Goals in Italy. *Sci. Rep.* **2021**, *11*, 24117. [CrossRef] [PubMed]
4. United Nations. THE 17 GOALS | Sustainable Development. Available online: https://sdgs.un.org/goals (accessed on 18 February 2022).
5. The Ellen MacArthur Foundation. *Towards the Circular Economy Economic and Business Rationale for an Accelerated Transition*; Ellen MacArthur Foundation: Cowes, UK, 2013; pp. 21–34.
6. Bocken, N.M.P.; Short, S.W.; Rana, P.; Evans, S. A literature and practice review to develop sustainable business model archetypes. *J. Clean. Prod.* **2014**, *65*, 42–56. [CrossRef]
7. Velenturf, A.P.M.; Purnell, P. Principles for a sustainable circular economy. *Sustain. Prod. Consum.* **2021**, *27*, 1437–1457. [CrossRef]
8. Bocken, N.M.P.; de Pauw, I.; Bakker, C.A.; van der Grinten, B. Product design and business model strategies for a circular economy. *J. Ind. Prod. Eng.* **2016**, *33*, 308–320. [CrossRef]
9. Kirchherr, J.; Reike, D.; Hekkert, M. Conceptualizing the circular economy: An analysis of 114 definitions. *Resour. Conserv. Recycl.* **2017**, *127*, 221–232. [CrossRef]
10. Aldieri, L.; Ioppolo, G.; Vinci, C.P.; Yigitcanlar, T. Waste recycling patents and environmental innovations: An economic analysis of policy instruments in the USA, Japan and Europe. *Waste Manag.* **2019**, *95*, 612–619. [CrossRef]
11. Caruso, G.; Fortuna, F. Mediterranean diet Patterns in the Italian Population: A functional data analysis of Google Trends. In *Decisions and Trends in Social Systems, Innovative and Integrated Approaches of Care Services*; Springer: Cham, Switzerland, 2020; pp. 1–10.
12. Cambini, C.; Congiu, R.; Jamasb, T.; Llorca, M.; Soroush, G. Energy Systems Integration: Implications for public policy. *Energy Policy* **2020**, *143*, 111609. [CrossRef]
13. Van Bommel, N.; Höffken, J.I. Energy justice within, between and beyond European community energy initiatives: A review. *Energy Res. Soc. Sci.* **2021**, *79*, 102157. [CrossRef]
14. Leonhardt, R.; Noble, B.; Poelzer, G.; Fitzpatrick, P.; Belcher, K.; Holdmann, G. Advancing local energy transitions: A global review of government instruments supporting community energy. *Energy Res. Soc. Sci.* **2022**, *83*, 102350. [CrossRef]
15. Pena-Bello, A.; Parra, D.; Herberz, M.; Tiefenbeck, V.; Patel, M.K.; Hahnel, U.J.J. Integration of prosumer peer-to-peer trading decisions into energy community modelling. *Nat. Energy* **2022**, *7*, 74–82. [CrossRef]
16. Heuninckx, S.; te Boveldt, G.; Macharis, C.; Coosemans, T. Stakeholder objectives for joining an energy community: Flemish case studies. *Energy Policy* **2022**, *162*, 112808. [CrossRef]
17. Ceglia, F.; Marrasso, E.; Roselli, C.; Sasso, M. Small Renewable Energy Community: The Role of Energy and Environmental Indicators for Power Grid. *Sustainability* **2021**, *13*, 42137. [CrossRef]
18. Bianco, G.; Bonvini, B.; Bracco, S.; Delfino, F.; Laiolo, P.; Piazza, G. Key Performance Indicators for an Energy Community Based on Sustainable Technologies. *Sustainability* **2021**, *13*, 68789. [CrossRef]
19. Böttcher, C.; Müller, M. Insights on the impact of energy management systems on carbon and corporate performance. An empirical analysis with data from German automotive suppliers. *J. Clean. Prod.* **2016**, *137*, 1449–1457. [CrossRef]
20. Rauf, A.; Ozturk, I.; Ahmad, F.; Shehzad, K.; Chandiao, A.A.; Irfan, M.; Abid, S.; Jinkai, L. Do Tourism Development, Energy Consumption and Transportation Demolish Sustainable Environments? Evidence from Chinese Provinces. *Sustainability* **2021**, *13*, 12361. [CrossRef]
21. Villa, S.; Sassanelli, C. The Data-Driven Multi-Step Approach for Dynamic Estimation of Buildings' Interior Temperature. *Energies* **2020**, *13*, 6654. [CrossRef]
22. D'Adamo, I.; Gastaldi, M.; Morone, P.; Rosa, P.; Sassanelli, C.; Settembre-blundo, D.; Shen, Y. Bioeconomy of Sustainability: Drivers, Opportunities and Policy Implications. *Sustainability* **2022**, *14*, 200. [CrossRef]
23. Castaldi, M.; van Deventer, J.; Lavoie, J.M.; Legrand, J.; Nzihou, A.; Pontikes, Y.; Py, X.; Vandecasteele, C.; Vasudevan, P.T.; Verstraete, W.; et al. Progress and Prospects in the Field of Biomass and Waste to Energy and Added-Value Materials. *Waste Biomass Valorization* **2017**, *8*, 1875–1884. [CrossRef]
24. Atabani, A.E.; Pugazhendhi, A.; Al-Muhtaseb, A.H.; Kobayashi, T.; Lee, C. Editorial Preface to the Special Issue on "The 2nd International Conference on Alternative Fuels and Energy: Futures and Challenges (ICAFE 2017)" 23rd–25th October 2017, Daegu, Republic of Korea. *Waste Biomass Valorization* **2020**, *11*, 1017. [CrossRef]
25. Tsalidis, G.A. Human Health and Ecosystem Quality Benefits with Life Cycle Assessment Due to Fungicides Elimination in Agriculture. *Sustainability* **2022**, *14*, 846. [CrossRef]
26. Sili, M.; Dürr, J. Bioeconomic Entrepreneurship and Key Factors of Development: Lessons from Argentina. *Sustainability* **2022**, *14*, 2447. [CrossRef]
27. Kardung, M.; Cingiz, K.; Costenoble, O.; Delahaye, R.; Heijman, W.; Lovrić, M.; van Leeuwen, M.; M'Barek, R.; van Meijl, H.; Piotrowski, S.; et al. Development of the Circular Bioeconomy: Drivers and Indicators. *Sustainability* **2021**, *13*, 10413. [CrossRef]

28. Barragán-Escandón, A.; Olmedo Ruiz, J.M.; Curillo Tigre, J.D.; Zalamea-León, E.F. Assessment of Power Generation Using Biogas from Landfills in an Equatorial Tropical Context. *Sustainability* **2020**, *12*, 2669. [CrossRef]

29. Rotunno, P.; Lanzini, A.; Leone, P. Energy and economic analysis of a water scrubbing based biogas upgrading process for biomethane injection into the gas grid or use as transportation fuel. *Renew. Energy* **2017**, *102*, 417–432. [CrossRef]

30. Haider, J.; Qyyum, M.A.; Kazmi, B.; Ali, I.; Nizami, A.-S.; Lee, M. Simulation study of deep eutectic solvent-based biogas upgrading process integrated with single mixed refrigerant biomethane liquefaction. *Biofuel Res. J.* **2020**, *7*, 1245. [CrossRef]

31. Zhu, T.; Curtis, J.; Clancy, M. Promoting agricultural biogas and biomethane production: Lessons from cross-country studies. *Renew. Sustain. Energy Rev.* **2019**, *114*, 109332. [CrossRef]

32. Mohammed, M.N.; Atabani, A.E.; Uguz, G.; Lay, C.H.; Kumar, G.; Al-Samaraae, R.R. Characterization of Hemp (*Cannabis sativa* L.) Biodiesel Blends with Euro Diesel, Butanol and Diethyl Ether Using FT-IR, UV–Vis, TGA and DSC Techniques. *Waste Biomass Valorization* **2020**, *11*, 1097–1113. [CrossRef]

33. Awe, O.W.; Zhao, Y.; Nzihou, A.; Minh, D.P.; Lyczko, N. A Review of Biogas Utilisation, Purification and Upgrading Technologies. *Waste Biomass Valorization* **2017**, *8*, 267–283. [CrossRef]

34. Budzianowski, W.M.; Brodacka, M. Biomethane storage: Evaluation of technologies, end uses, business models, and sustainability. *Energy Convers. Manag.* **2017**, *141*, 254–273. [CrossRef]

35. Acerbi, F.; Sassanelli, C.; Terzi, S.; Taisch, M. A Systematic Literature Review on Data and Information Required for Circular Manufacturing Strategies Adoption. *Sustainability* **2021**, *13*, 42047. [CrossRef]

36. Sassanelli, C.; Rossi, M.; Terzi, S. Evaluating the smart maturity of manufacturing companies along the product development process to set a PLM project roadmap. *Int. J. Prod. Lifecycle Manag.* **2020**, *12*, 185–209. [CrossRef]

37. Ikram, M.; Sroufe, R.; Awan, U.; Abid, N. Enabling Progress in Developing Economies: A Novel Hybrid Decision-Making Model for Green Technology Planning. *Sustainability* **2022**, *14*, 258. [CrossRef]

38. Kyriakopoulos, G.L.; Kapsalis, V.C.; Aravossis, K.G.; Zamparas, M.; Mitsikas, A. Evaluating Circular Economy under a Multi-Parametric Approach: A Technological Review. *Sustainability* **2019**, *11*, 6139. [CrossRef]

39. Saaty, T.L. *The Analytic Process: Planning, Priority Setting, Resources Allocation*; McGraw: New York, NY, USA, 1980.

40. De Felice, F.; Petrillo, A.; Saaty, T. *Applications and Theory of Analytic Hierarchy Process: Decision Making for Strategic Decisions*; BoD–Books on Demand: Norderstedt, Germany, 2016; ISBN 9535125605.

41. Brudermann, T.; Mitterhuber, C.; Posch, A. Agricultural biogas plants—A systematic analysis of strengths, weaknesses, opportunities and threats. *Energy Policy* **2015**, *76*, 107–111. [CrossRef]

42. Saaty, T.L. Decision making with the analytic hierarchy process. *Int. J. Serv. Sci.* **2008**, *1*, 83–98. [CrossRef]

43. Nikolaou, I.E.; Tsagarakis, K.P. An introduction to circular economy and sustainability: Some existing lessons and future directions. *Sustain. Prod. Consum.* **2021**, *28*, 600–609. [CrossRef]

44. D'Adamo, I.; Falcone, P.M.; Huisingh, D.; Morone, P. A circular economy model based on biomethane: What are the opportunities for the municipality of Rome and beyond? *Renew. Energy* **2021**, *163*, 1660–1672. [CrossRef]

45. Baena-Moreno, F.M.; Malico, I.; Marques, I.P. Promoting Sustainability: Wastewater Treatment Plants as a Source of Biomethane in Regions Far from a High-Pressure Grid. A Real Portuguese Case Study. *Sustainability* **2021**, *13*, 68933. [CrossRef]

46. Wall, D.M.; McDonagh, S.; Murphy, J.D. Cascading biomethane energy systems for sustainable green gas production in a circular economy. *Bioresour. Technol.* **2017**, *243*, 1207–1215. [CrossRef]

47. Rasi, S.; Timonen, K.; Joensuu, K.; Regina, K.; Virkajärvi, P.; Heusala, H.; Tampio, E.; Luostarinen, S. Sustainability of Vehicle Fuel Biomethane Produced from Grass Silage in Finland. *Sustainability* **2020**, *12*, 3994. [CrossRef]

48. D'Adamo, I.; Falcone, P.M.; Gastaldi, M.; Morone, P. RES-T trajectories and an integrated SWOT-AHP analysis for biomethane. Policy implications to support a green revolution in European transport. *Energy Policy* **2020**, *138*, 111220. [CrossRef]

49. Ammenberg, J.; Feiz, R. Assessment of feedstocks for biogas production, part II—Results for strategic decision making. *Resour. Conserv. Recycl.* **2017**, *122*, 388–404. [CrossRef]

50. Collet, P.; Flottes, E.; Favre, A.; Raynal, L.; Pierre, H.; Capela, S.; Peregrina, C. Techno-economic and Life Cycle Assessment of methane production via biogas upgrading and power to gas technology. *Appl. Energy* **2017**, *192*, 282–295. [CrossRef]

51. Vo, T.T.Q.; Wall, D.M.; Ring, D.; Rajendran, K.; Murphy, J.D. Techno-economic analysis of biogas upgrading via amine scrubber, carbon capture and ex-situ methanation. *Appl. Energy* **2018**, *212*, 1191–1202. [CrossRef]

52. Valli, L.; Rossi, L.; Fabbri, C.; Sibilla, F.; Gattoni, P.; Dale, B.E.; Kim, S.; Ong, R.G.; Bozzetto, S. Greenhouse gas emissions of electricity and biomethane produced using the Biogasdoneright™ system: Four case studies from Italy. *Biofuels Bioprod. Biorefining* **2017**, *11*, 847–860. [CrossRef]

53. IRENA. *Renewable Capacity Statistics*; IRENA: Abu Dhabi, United Arab Emirates, 2021.

54. IEA World Energy Outlook. 2019. Available online: https://www.iea.org/ (accessed on 12 February 2019).

55. Gutiérrez, E.C.; Wall, D.M.; O'Shea, R.; Novelo, R.M.; Gómez, M.M.; Murphy, J.D. An economic and carbon analysis of biomethane production from food waste to be used as a transport fuel in Mexico. *J. Clean. Prod.* **2018**, *196*, 852–862. [CrossRef]

56. O'Shea, R.; Wall, D.; Kilgallon, I.; Murphy, J.D. Assessment of the impact of incentives and of scale on the build order and location of biomethane facilities and the feedstock they utilise. *Appl. Energy* **2016**, *182*, 394–408. [CrossRef]

57. Rajendran, K.; Browne, J.D.; Murphy, J.D. What is the level of incentivisation required for biomethane upgrading technologies with carbon capture and reuse? *Renew. Energy* **2019**, *133*, 951–963. [CrossRef]

58. MISE Interministerial Decree of 2 March 2018. Promotion of the Use of Biomethane and Other Advanced Biofuels in the Transportation Sector. Available online: https://www.mise.gov.it/index.php/it/ (accessed on 5 June 2019).
59. Pierro, N.; Giocoli, A.; De Bari, I.; Agostini, A.; Motola, V.; Dipinto, S. *Potenziale Teorico di Biometano Avanzato in Italia*; ENEA: Rome, Italy, 2021.
60. Cianciotta, S.; D'Adamo, I. The Evolution of Sustainability: The Automotive Supply Chain Opportunity in Southern Italy. *Sustainability* **2021**, *13*, 10930. [CrossRef]
61. ISPRA Urban Waste Report. Available online: https://www.isprambiente.gov.it/files2020/pubblicazioni/rapporti/rapportorifiutiurbani_ed-2020_n-331-1.pdf (accessed on 12 March 2022).
62. Baena-Moreno, F.M.; Malico, I.; Rodríguez-Galán, M.; Serrano, A.; Fermoso, F.G.; Navarrete, B. The importance of governmental incentives for small biomethane plants in South Spain. *Energy* **2020**, *206*, 118158. [CrossRef]
63. Smyth, B.M.; Smyth, H.; Murphy, J.D. Determining the regional potential for a grass biomethane industry. *Appl. Energy* **2011**, *88*, 2037–2049. [CrossRef]
64. Allen, E.; Wall, D.M.; Herrmann, C.; Murphy, J.D. A detailed assessment of resource of biomethane from first, second and third generation substrates. *Renew. Energy* **2016**, *87*, 656–665. [CrossRef]
65. Khan, M.U.; Lee, J.T.E.; Bashir, M.A.; Dissanayake, P.D.; Ok, Y.S.; Tong, Y.W.; Shariati, M.A.; Wu, S.; Ahring, B.K. Current status of biogas upgrading for direct biomethane use: A review. *Renew. Sustain. Energy Rev.* **2021**, *149*, 111343. [CrossRef]
66. Xue, S.; Zhang, S.; Wang, Y.; Wang, Y.; Song, J.; Lyu, X.; Wang, X.; Yang, G. What can we learn from the experience of European countries in biomethane industry: Taking China as an example? *Renew. Sustain. Energy Rev.* **2022**, *157*, 112049. [CrossRef]
67. Guerin, T.F. Business model scaling can be used to activate and grow the biogas-to-grid market in Australia to decarbonise hard-to-abate industries: An application of entrepreneurial management. *Renew. Sustain. Energy Rev.* **2022**, *158*, 112090. [CrossRef]
68. Van der Schoor, T.; Scholtens, B. Power to the people: Local community initiatives and the transition to sustainable energy. *Renew. Sustain. Energy Rev.* **2015**, *43*, 666–675. [CrossRef]
69. Sendstad, L.H.; Chronopoulos, M. Sequential investment in renewable energy technologies under policy uncertainty. *Energy Policy* **2020**, *137*, 111152. [CrossRef]
70. Shanmugam, K.; Baroth, A.; Nande, S.; Yacout, D.M.M.; Tysklind, M.; Upadhyayula, V.K.K. Social Cost Benefit Analysis of Operating Compressed Biomethane (CBM) Transit Buses in Cities of Developing Nations: A Case Study. *Sustainability* **2019**, *11*, 4190. [CrossRef]
71. Mies, A.; Gold, S. Mapping the social dimension of the circular economy. *J. Clean. Prod.* **2021**, *321*, 128960. [CrossRef]
72. Crupi, A.; Del Sarto, N.; Di Minin, A.; Gregori, G.L.; Lepore, D.; Marinelli, L.; Spigarelli, F. The digital transformation of SMEs—A new knowledge broker called the digital innovation hub. *J. Knowl. Manag.* **2020**, *24*, 1263–1288. [CrossRef]
73. Sassanelli, C.; Terzi, S.; Panetto, H.; Doumeingts, G. Digital Innovation Hubs supporting SMEs digital transformation. In Proceedings of the 27th ICE/IEEE International Technology Management Conference, Cardiff, UK, 21–23 June 2021; IEEE: Piscataway, NJ, USA, 2021; pp. 1–8.
74. Hervas-Oliver, J.L.; Gonzalez-Alcaide, G.; Rojas-Alvarado, R.; Monto-Mompo, S. Emerging regional innovation policies for industry 4.0: Analyzing the digital innovation hub program in European regions. *Compet. Rev.* **2021**, *31*, 106–129. [CrossRef]

Article

In-Line Monitoring of Carbon Dioxide Capture with Sodium Hydroxide in a Customized 3D-Printed Reactor without Forced Mixing

Emmanouela Leventaki [1], Francisco M. Baena-Moreno [1,*], Gaetano Sardina [2], Henrik Ström [2], Ebrahim Ghahramani [2], Shirin Naserifar [1,3], Phuoc Hoang Ho [1], Aleksandra M. Kozlowski [1] and Diana Bernin [1,*]

[1] Department of Chemistry and Chemical Engineering, Chalmers University of Technology, 412 96 Gothenburg, Sweden
[2] Department of Mechanical and Maritime Sciences, Chalmers University of Technology, 412 96 Gothenburg, Sweden
[3] Wallenberg Wood Science Center, Chalmers University of Technology, 412 96 Gothenburg, Sweden
* Correspondence: francisco.baena@chalmers.se (F.M.B.-M.); diana.bernin@chalmers.se (D.B.)

Citation: Leventaki, E.; Baena-Moreno, F.M.; Sardina, G.; Ström, H.; Ghahramani, E.; Naserifar, S.; Ho, P.H.; Kozlowski, A.M.; Bernin, D. In-Line Monitoring of Carbon Dioxide Capture with Sodium Hydroxide in a Customized 3D-Printed Reactor without Forced Mixing. *Sustainability* 2022, 14, 10795. https://doi.org/10.3390/su141710795

Academic Editors: Idiano D'Adamo and Massimo Gastaldi

Received: 9 August 2022
Accepted: 26 August 2022
Published: 30 August 2022

Abstract: Many industrial processes make use of sodium because sodium is the fifth most abundant metal and the seventh most abundant element on Earth. Consequently, there are many sodium-containing industrial wastes that could potentially be used for carbon capture, paving the way towards a circular and biobased economy. For example, a common industrial chemical is NaOH, which is found in black liquor, a by-product of the paper and pulp industry. Nonetheless, the literature available on CO_2 absorption capacity of aqueous NaOH is scarce for making a fair comparison with sodium-containing waste. Therefore, to fill this gap and set the foundation for future research on carbon capture, the CO_2 absorption capacity of NaOH solutions in a concentration range of 1–8 *w/w%* was evaluated, a wider range compared with currently available data. The data set presented here enables evaluating the performance of sodium-based wastes, which are complex mixtures and might contain other compounds that enhance or worsen their carbon capture capacity. We designed a customized reactor using a 3D-printer to facilitate in-line measurements and proper mixing between phases without the energy of stirring. The mixing performance was confirmed by computational fluid dynamics simulations. The CO_2 absorption capacity was measured via weight analysis and the progress of carbonation using a pH meter and an FTIR probe in-line. At 5 *w/w%* NaOH and higher, the reaction resulted in precipitation. The solids were analyzed with X-ray diffraction and scanning electron microscope, and nahcolite and natrite were identified. With our setup, we achieved absorption capacities in the range of 9.5 to 78.9 g CO_2/L for 1 *w/w%* and 8 *w/w%* of NaOH, respectively. The results are in fair agreement with previously reported literature, suggesting that non-forced mixing reactors function for carbon capture without the need of stirring equipment and a possible lower energy consumption.

Keywords: carbon capture and storage; aqueous bases; reactor design; chemical absorption; FTIR; absorption capacity; in-line measurements; 3D-printed reactors

1. Introduction

In recent decades, environmental changes due to increased greenhouse gas (GHG) emissions have been intensely studied [1]. Among these GHGs, it has become evident that the increased level of CO_2 in the atmosphere is one of the main contributors to these changes [2]. Consequently, interest has been growing towards developing technologies for carbon capture and storage (CCS) as a temporary but necessary solution to capture the emitted CO_2 [3,4].

Even though many different CCS technologies have been proposed to date, only a few have reached commercial status, highlighting CO_2 chemical absorption [5]. Currently, solvents such as ammonia, piperazine and amines have been proposed for their absorption of industrial CO_2 emissions, with commercial focus on the latter group [6,7]. Monoethanolamine (MEA) is one example of a commercial amine with a high capture capacity [8–10]. Despite its popularity, MEA has some significant drawbacks, such as high regeneration costs, emissions from the MEA production process, and the formation of nitrosamine—a carcinogenic compound—due to amine degradation [11,12].

Hydroxide-based solvents (i.e., NaOH or KOH) are another interesting group of solvents for CO_2 chemical absorption [11]. Indeed, NaOH has been proposed previously as an effective solvent for CO_2 capture [13–17]. Many industrial processes make use of sodium-based chemicals because sodium is the fifth most abundant metal and the seventh most abundant element on Earth. Consequently, there are many sodium-containing wastes that could potentially be used for carbon capture, paving the way towards a circular and biobased economy. The promotion of bioeconomy strategies plays a key role in the development of our society towards sustainability [18–20]. A common industrial chemical is NaOH, which is found in black liquor, a biobased residue of the paper and pulp industry [21,22]. Another example of high sodium content waste is saline industrial residue. On the other side, the regeneration of NaOH usually requires either a high energy penalty (thermal regeneration) or a chemical reaction with other compounds (chemical regeneration) [11]. Nonetheless, if NaOH comes from wastes, the carbonated solution due to CO_2 capture does not need to be regenerated. Hence, the formed carbonates could be directly stored or even used as added-value products. This is a clear example of strategies to promote bioeconomy-based and sustainable societies.

The use of an aqueous NaOH solution for CO_2 absorption has been reported in the literature [23–25]. For example, CO_2 capture from coal-fired flue gas was investigated in a continuous bubble column scrubber with efficiencies in the range of 30–98% to produce Na_2CO_3 for algae production [26]. In addition, parameters such as absorption efficiency and mass-transfer coefficient were evaluated [26]. A comparison of removal efficiencies of CO_2 between aqueous ammonia and an aqueous NaOH solution in a fine spray column was carried out [27]. These authors targeted a CO_2 removal efficiency of 90% and reported the corresponding mole ratios of solvent to CO_2 to be 4.43 and 9.68 for NaOH and aqueous ammonia, respectively. These findings further suggest the advantages of NaOH for CO_2 capture over other solvents. Another very interesting study proposed capturing CO_2 from air by spraying NaOH. The expenses of this CO_2 capturing alternative ranged from 53 to 127 \$/ton CO_2 [28]. Aqueous NaOH was also used to capture CO_2 in a two-stage combined process, in which $Ca(OH)_2$ was utilized to obtain $CaCO_3$ and regenerate the NaOH. The authors obtained CO_2 absorption efficiencies of 67.85% and regeneration rates for NaOH of 85.37% [29]. As summarized, the use of NaOH for CO_2 capture has been studied.

Nonetheless and despite the importance of aqueous NaOH solutions for CO_2 capture with sodium-based wastes and to the best of our knowledge, only one work has dealt with the estimation of absorption capacities of CO_2 in aqueous NaOH solutions. This is crucial to estimate the performance of other sodium-based wastes and to validate the performance of alternative reactor designs. Yoo Miran et al. evaluated the absorption capacity of a gas stream containing 30 *v/v*% CO_2 for different concentrations of aqueous NaOH solutions (1–5 *w/w*%) [17]. In this work, authors measured the absorption capacity through a combination of experiments i.e., the CO_2 concentration in the outcoming gas stream, the pH value and the electrical conductivity of the solution over time. Furthermore, they predicted the evolution of chemical species in the solutions with time. However, for a mechanistic optimization of future carbonation reactors, it is crucial to monitor the chemical reactions occurring during absorption. Furthermore, it is necessary to expand the studied concentration range to values typically found in waste solutions [21,22]. Hence, to fulfill these knowledge gaps, the present study measured the absorption of CO_2 in aqueous solutions of NaOH up to 8 *w/w*% at ambient pressure and temperature. Importantly, the absorption was monitored using an analytical balance and the progress of the chemical reactions inside the reactor was followed with a pH meter and an FTIR probe. We observed the formation of carbonate and bicarbonate ions and absorption of CO_2 depending on pH. This opens new avenues for optimization either for maximum absorption capacity or the formation of value-added products.

In addition, we customized the reactor design for non-forced mixing, which reduces the amount of equipment and possibly the amount of energy needed. The reactor enabled also in-line measurements and was 3D printed using a resin printer. Its mixing performance was corroborated by computational fluid dynamics (CFD) simulations. All experiments were conducted in this customized 3D-printed bubble batch reactor using bubble motion to create mixing.

At solutions of 5 *w/w*% NaOH and higher, solid carbonate and bicarbonate species were formed due to saturation. The solids were analyzed with X-ray diffraction and scanning electron microscopy and natrite (Na_2CO_3) and nahcolite ($NaHCO_3$) were identified. With our setup, we achieved absorption capacities in the range of 9.5 to 78.9 g CO_2/L for 1 *w/w*% and 8 *w/w*% of NaOH, respectively. The results are in fair agreement with previously reported literature findings that non-forced mixing reactors function for carbon capture and may entail economic advantages compared to stirring-based ones. This assessment is however out of the scope of this study. With this work, we aimed to lay the foundation for future works for sustainable CO_2 capture using wastes with a high sodium content. Furthermore, the results enable finding suitable conditions for reaching either a high absorption rate, a maximum absorption capacity or value-added species formed by absorption.

2. Materials and Methods

2.1. Materials

Aqueous NaOH solutions were prepared by dissolving high purity NaOH (VWR, 99% purity) in distilled water to give concentrations ranging from 1–8 *w/w*%. The gas mixture had a composition of 30 *v/v*% CO_2 and 70 *v/v*% N_2. This composition has been previously used by other authors [17]. A stereolithography (SLA) 3D-printer Form 3+ (Formlabs) was used to print a custom designed reactor of the UV curable resin Rigid 10K (Formlabs). The reactor was designed using the CAD software Autodesk Fusion 360.

2.2. Experimental Setup, Procedures, and Reactor Design

Figure 1 shows the experimental setup used for CO_2 carbonation experiments. The gas mixture of CO_2 and N_2 was flowing at a steady rate of 200 mL/min adjusted with a pressure regulator and monitored with an in-line flowmeter. The gas passed through a sparger into the 3D-printed reactor, which was filled with 60 mL solution. For each NaOH concentration, experiments were carried out using two different setups. The first setup included the pH-meter (HQ430D, HACH, Loveland, CO, USA) and the FTIR (ReactIR 702L, Mettler Toledo, Greifensee, Switzerland) probe immersed in the solution to record pH and FTIR spectra with time (Figure 1A). The experiments were stopped once the pH reached a value of 8 because below this threshold the amount of absorbed CO_2 is insignificant. To measure the absorption capacity, a second experiment was performed for the same duration as the first with the 3D-printed reactor mounted on a balance (QUINTIX2102-1S, Sartorius, Göttingen, Germany), which was connected to a computer for data logging (Figure 1B). The pH value was checked at the end of this second run to corroborate that an identical pH value was reached.

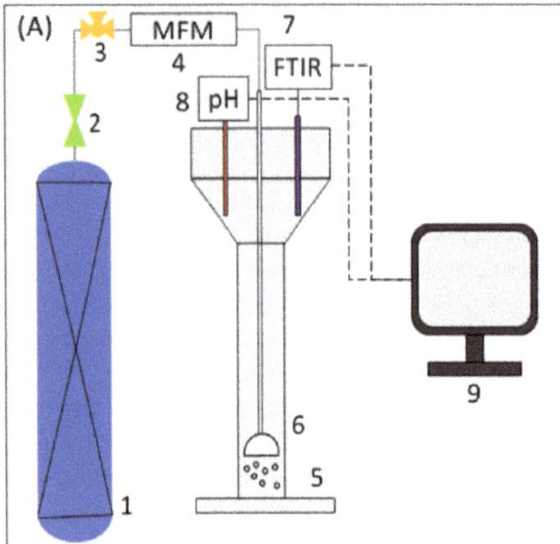
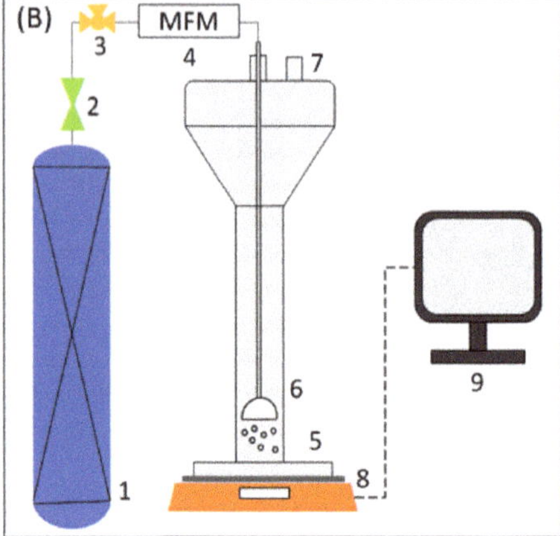

Figure 1. Experimental setups: (**A**) The first run includes 1. gas bottle (30 *v/v*% CO_2, 70 *v/v*% N_2), 2. pressure regulator, 3. three-way valve, 4. mass flow meter, 5. reactor, 6. sparger, 7. FTIR, 8. pH-meter, 9. computer for data logging; (**B**) The second run includes 1. gas bottle (30 *v/v*% CO_2, 70 *v/v*% N_2), 2. pressure regulator, 3. three-way valve, 4. mass flow meter, 5. reactor, 6. sparger, 7. gas outlet, 8. balance, 9. computer for data logging.

The FTIR spectrum of air was subtracted as background from the obtained FTIR spectra prior to analysis. For 5 *w/w*% NaOH and higher, precipitation occurred. The formed solid product was filtered and dried prior to X-ray Diffraction (XRD) and Scanning Electron Microscope (SEM) characterization. X-ray diffraction (XRD) measurements were performed on the D8 Discover Bruker instrument. The patterns were recorded for a diffraction angle range of 2θ from 10 to 70° with a scan step of 0.02° per second. The peaks of the obtained phases were matched with peaks present in the database (PDF-4-2021) from the International Centre for Diffraction Data (ICDD). SEM images were obtained using a Phenom ProX Desktop SEM (ThermoFisher Scientific, Waltham, MA, USA).

Concerning the reactor design, there are two important processes during optimized gas absorption, (1) enhancing the mass transfer of the gas into the liquid through the bubble surface and (2) avoiding saturation of the absorbed gas in the liquid phase in the reactor. Raising bubbles introduce a fluid motion upwards followed by a down-coming fluid motion elsewhere by continuity. However, these upwards and downwards flows depend on the size of the reactor cross-section. The reactor was designed to have a narrow and long body with the gas sparged from the bottom enabling homogenous mixing caused by the bubble motion without the need for forced mixing. This reduces the need of additional equipment and energy for stirring. The 3D-printed reactor design and its dimensions can be seen in Figure 2.

The top of the vessel is wider to allow the pH meter and FTIR probe to be immersed in the solution. The non-absorbed gas escaped to the atmosphere through the gas outlet and could have carried a small amount of solution resulting in a loss of weight. Evaporation of the solvent also occurred during absorption. Hence, blank experiments were carried out to evaluate the losses. These experiments were conducted by sparging the same gas mixture through distilled water, with the reactor mounted on the analytical balance and measuring the weight loss with time. The obtained weight loss trend from the blank runs was included in the data treatment to compensate for the solvent losses.

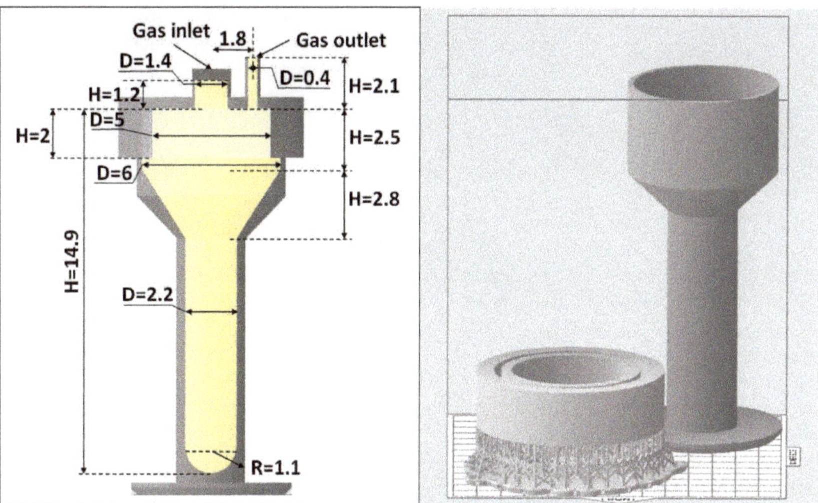

Figure 2. Dimensions (**left**) and design of reactor in Autodesk Fusion 360 (**right**).

2.3. Computational Fluid Dynamics (CFD) Methodology

To qualitatively address the flow field inside the reactor geometry and evaluate the degree of mixing of the solution, we employed CFD using an unstructured mesh of 11,428 cells to solve the non-reactive multiphase model equations in the reactor. We employed a two-fluid Eulerian-Eulerian model to describe the suspension of CO_2 and N_2 bubbles in water. In particular, the CO_2 and N_2 components are treated as a gas phase while the water is the liquid phase. Gas and liquid volume fractions measure the relative concentration of the two components in each part of the reactor. Following the experimental procedure, the gas phase is injected into the domain from the sparger after starting to rise along the liquid phase driven by buoyancy effects. Turbulence in the liquid phase was modeled using a modified version of the k-ε model, adapted for turbulent bubbly suspensions [30]. The momentum exchange term between the two-phases includes drag, lift and virtual mass forces, moreover, turbulent dispersion was also included using the Favre-averaged drag relationship developed in [31].

We used the software ANSYS Fluent R2021 V2 for our simulations. The temporal discretization is first-order implicit with a time step of 10^{-4} s, the pressure-velocity coupling is handled by the Phase Coupled scheme SIMPLE, and the pressure interpolation is body force weighted. Convective terms and volume fractions are discretized using the QUICK scheme while the turbulent equations employ a first-order upwind in the non-linear terms.

3. Theoretical Considerations of Carbonation Reactions of CO_2 in Aqueous Bases

For a comprehensive explanation of the theory involved in carbonation reactions, readers are referred to reference [17]. In this section, the essentials of carbonation reactions of CO_2 in aqueous bases are described. As the system has two phases, liquid and gas, the process is dictated by two phenomena, the physical transition of CO_2 from the gas bubble to the liquid, and the chemical reaction between the absorbed CO_2 and the OH^-. The following three different forms, the carbonate ion CO_3^{2-}, the bicarbonate ion HCO_3^- and carbonic acid H_2CO_3, exist in an aqueous solution depending on the pH [32].

At pH above 12.3, CO_2 reacts with OH^- ions to form CO_3^{2-} ions according to Equation (1).

$$CO_2 + 2OH^- \leftrightarrow CO_3^{2-} + H_2O \tag{1}$$

As the OH^- ions are consumed i.e., the pH is lowered, the CO_2 converts to both carbonate and bicarbonate ions. The two species exist in equilibrium dependent on pH and below a specific pH only bicarbonate ions are formed according to Equation (2).

$$CO_2 + OH^- \leftrightarrow HCO_3^- \tag{2}$$

Finally, at pH = 4.3 and lower, all of the CO_2 turns into carbonic acid (Equation (3)).

$$CO_2 + H_2O \leftrightarrow H_2CO_3 \tag{3}$$

The carbonation of CO_2 with NaOH occurs in two steps to form sodium carbonate (Equation (4)) and bicarbonate (Equation (5)) [15].

$$CO_2(g) + 2NaOH(aq) \leftrightarrow Na_2CO_3(aq) + H_2O(l) \tag{4}$$

$$Na_2CO_3(aq) + H_2O(l) + CO_2(g) \leftrightarrow 2NaHCO_3(aq) \tag{5}$$

Sodium carbonate and bicarbonate have a high solubility in water up to 30.7 g/100 g water and 10.3 g/100 g water respectively at room temperature [33]. In contrast, carbonic acid has very low solubility i.e., once the pH drops to favor the production of carbonic acid, the absorption of CO_2 is reduced significantly [34].

4. Results

4.1. Mixing in the Reactor

The first step is to corroborate that the reactor shape allows suitable mixing, which was confirmed in CFD simulations. Figure 3 shows the flow fields computed using the CFD numerical approach. Panel (A) represents the gas volume fraction inside the reactor along a vertical section. The gas bubbly phase is injected from the sparger, close to the bottom of the reactor. The highest relative concentration is found near the injection point, as indicated by the red-orange color. Afterwards, the gas rises along the vertical axis, decreasing the relative concentration compared to the liquid phase (represented by green and blue colors). The order of magnitude of the gas velocities inside the reactor can be seen in panel (B) of Figure 3, where the highest velocity of around 0.3 m/s is found close to the inner wall of the reactor immediately on top of the sparger. Finally, to evaluate the degree of mixing of the gas phase inside the reactor, we compare the ratio between the turbulent viscosity and molecular viscosity of the suspension (Panel C).

Since the reactor does not have rotating stirring parts, turbulence is needed to reach a good mixing because molecular diffusion is too slow. The regions inside the reactor with a high ratio between turbulent and molecular viscosity will give an indication of the preferential zones where mixing is achieved. From Figure 3C, it can be observed that the lower part of the reactor, on top of the sparger, is characterized by a turbulent viscosity ratio that is up to one order of magnitude the molecular viscosity of the water, suggesting that this is the region with the highest degree of mixing inside the reactor. The good level of mixing is confirmed by looking at the gas volume fraction distribution in Figure 3A where, after the mixing region, a uniform gas concentration in the horizontal plane is achieved at around 50% of the volume fraction (green contour). Overall, these results confirm that the reactor design is adequate: the bubble motion creates a suitable mixing environment without the need for any additional power input by stirring.

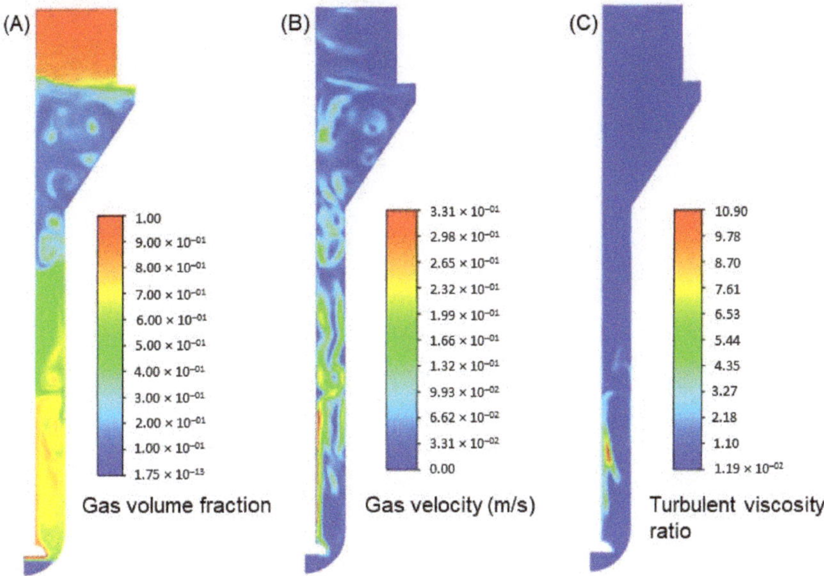

Figure 3. Contour plots of the gas volume fraction (**A**), gas velocity (**B**) and turbulent viscosity ratio (**C**) obtained from the CFD simulations.

4.2. Carbonation Reactions and Reactor Performance

The FTIR probe provided insights for understanding the occurring chemical reactions over time. Carbonate ions appear with a strong peak at around 1380 cm^{-1}, bicarbonate ions with a lower intensity at around 1360 cm^{-1}, 1300 cm^{-1} and 1008 cm^{-1} and water at around 1635 cm^{-1}, as highlighted in Figure 4E [35]. Unfortunately, the bicarbonate ion peaks at 1360 cm^{-1} and 1300 cm^{-1} overlap with the carbonate ion peak, which complicates quantification of each species. FTIR spectra as a function of wavenumber and time of all NaOH solutions are shown in Figure 4. The carbonate ion peak appears from 2 *w/w%* (B) while the bicarbonate ion peak starts to be visible from 3 *w/w%* (C). Furthermore, the carbonate ion peak grows first with time before a shift to lower wavenumbers is observed. This is true for all samples.

It is noteworthy that the intensities of the carbonate ion peaks in comparison with the water peak increased up to a NaOH concentration of 6 *w/w%* (F). For example, 2 *w/w%* NaOH (A), the carbonate ion peak at 1380 cm^{-1} is only slightly noticeable in contrast to 4 *w/w%* NaOH (D) revealing strong carbonate and bicarbonate ion peaks. This might indicate that the intensity of the observed carbonate ion peaks albeit overlapping signals may be correlated with its amount not only with time but also between samples. This observation might provide insights for CO$_2$ absorption in sodium-based wastes. At 5 *w/w%* NaOH and higher, precipitation occurred. This might be attributed to a stagnation of the intensity increase of the carbonate ion peak compared with the water peak. In the case of the 8 *w/w%* NaOH, the pattern of the FTIR spectra changed, which might be initiated by precipitation. Due to the gas bubbling, most of the solid particles were most likely dispersed in the reactor causing an interference of the FTIR measurement.

The formed solids that precipitated for the 6 *w/w%* were analyzed with XRD and SEM. The obtained SEM images presented in Figure 5A–D reveal irregular, rectangular and needle shapes. The obtained XRD pattern corresponds to a mix of natrite (main peaks at 26°, 27.5°, 33°, 35°, 38°, 40°, 41.5°, 46.5°, 47°, 48°, and all the other peaks from 53.5°), and to a minor extend of nahcolite (main peaks at 30°, 34.5°, and 44.5°) as shown in Figure 5E. Similar shapes and crystal structures were also found in previous works both from an after-process product [36] and commercial NaHCO$_3$ production before utilization [37].

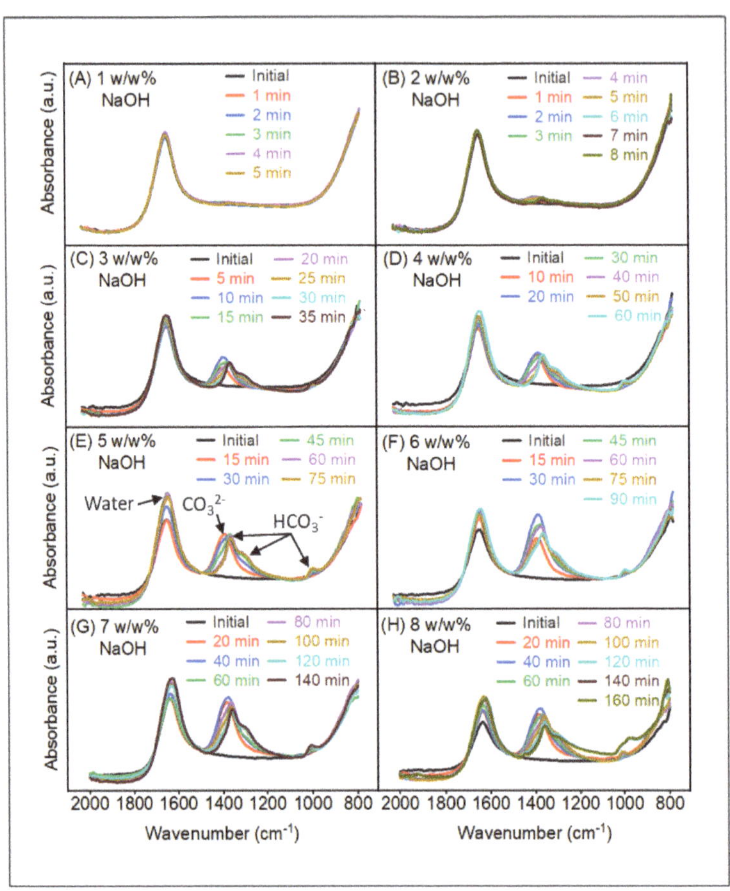

Figure 4. FTIR spectra of CO_2 absorption in NaOH solutions with time for (**A**) 1 *w/w%*; (**B**) 2 *w/w%*; (**C**) 3 *w/w%*; (**D**) 4 *w/w%*; (**E**) 5 *w/w%*; (**F**) 6 *w/w%*; (**G**) 7 *w/w%*; (**H**) 8 *w/w%*.

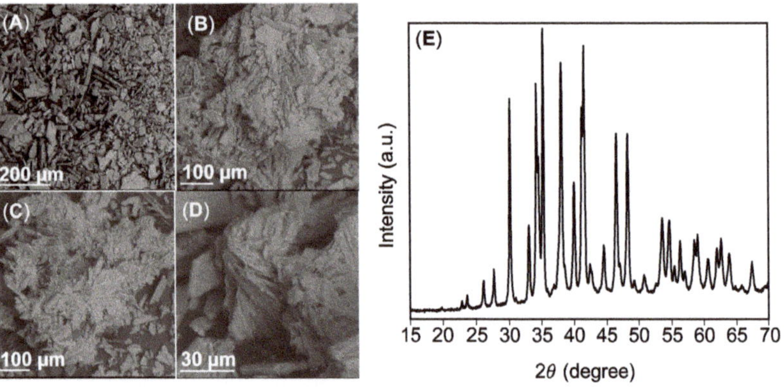

Figure 5. SEM images (**A–D**) and XRD diffractogram (**E**) of the powder obtained for 6 *w/w%* NaOH.

The CO_2 carbonation and absorption of the NaOH solutions were successfully monitored using an analytical balance, pH meter and FTIR probe. The increase in weight with time corresponded directly to the captured amount of CO_2. The pH and FTIR data agreed with the results of the balance and provided insight into the chemical reactions occurring during the absorption, as explained in Section 4.2. The pH value prior to absorption is 12.4 for the 1 *w/w%* NaOH solutions and 13.5 for the 8 *w/w%*. It decreased upon the consumption of OH$^-$ ions due to the formation of carbonate and bicarbonate ions. This formation could be followed by FTIR as explained in Section 4.2 (see typical FTIR spectrum in Figure 4E). Hence, the sum of absorbance values in the regions 1372–1388 cm^{-1} for the carbonate ion and 1000–1016 cm^{-1} for the bicarbonate ion were plotted with time to reflect their formation during absorption.

Figure 6 shows an example of all measurements obtained with time for 4 *w/w%* NaOH providing a clear picture of the progress of absorption. In the beginning the absorption rate of CO$_2$ was high and all absorbed CO$_2$ was converted into carbonate ions (A). The highest absorption rate occurred within the first 6 min until a pH value of 12.9. Up to that point, the CO$_2$ absorption was 12.4 g/L. After approximately 11 min, the pH dropped from 12.3 to 10.2 with the highest drop rate at pH 11.2 (B). At the same point in time, according to the sum of absorbance intensities (C), the formation of carbonate ions reached a maximum, and bicarbonate ions started to emerge (D). The absorption of CO$_2$ up to that point was 19.8 g/L. After 24.5 min, the pH continued dropping at a much slower rate and it took 35 additional minutes to decrease from 9.6 to 8.0 (B). During that time, 11.5 g/L of CO$_2$ was absorbed until the experiment was stopped at pH 8. These absorption trends were previously reported by [17], which confirms that the proposed reactor design produces the same results as forced-mixing reactors. Nevertheless, our results including FTIR verified that the formation of carbonate and bicarbonate ions depends on pH and absorption capacity. This wealth of data enables optimizing the duration of the reaction to maximum absorption rate or capacity or value-added products.

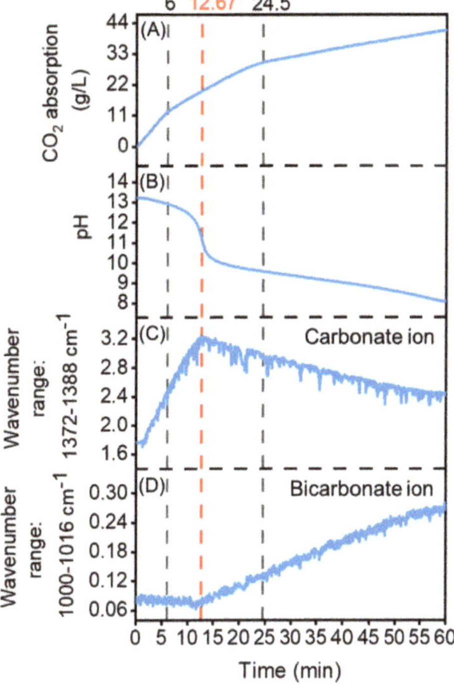

Figure 6. CO$_2$ absorption (**A**), pH (**B**), the sum of the absorbance intensities of the wavenumber range 1372–1388 cm^{-1} corresponding to the carbonate ion region (**C**) and the sum of the absorbance intensity of the wavenumber range 1000–1016 cm^{-1} corresponding to the bicarbonate ion region (**D**) with time for 4 *w/w%* NaOH.

Solutions of 1 to 8 *w/w*% NaOH were prepared to obtain a mechanistic understanding of the concentration-absorption rate dependence. The increase in weight with time due to CO_2 absorption of three solutions of 1, 4 and 8 *w/w*% NaOH is shown in Figure 7. The total duration of the experiments to reach pH 8 became longer with increasing NaOH concentration. For the 1 *w/w*% NaOH solution, it took 4.5 min (A) while for the 8 *w/w*% it lasted 2 h and 20 min (C). Furthermore, for the 1 *w/w*% NaOH, the rate of absorption was relatively constant throughout the experiment in contrast to the 4 and 8 *w/w*% NaOH solutions for which the rate levelled out with time. Interestingly, the same amount of CO_2 absorption for the 4 *w/w*%, 40 g/L, was reached after roughly 1/3 of the duration for 8 *w/w*% NaOH. This finding suggests that the amount of OH^- ions present in solution i.e., pH, is the driving force for the absorption rate. Below pH 8, CO_2 is still being absorbed at a very low rate until complete saturation and the weight stabilizes. These experiments were not run until rate depletion because such a low rate is not of interest from an industrial perspective.

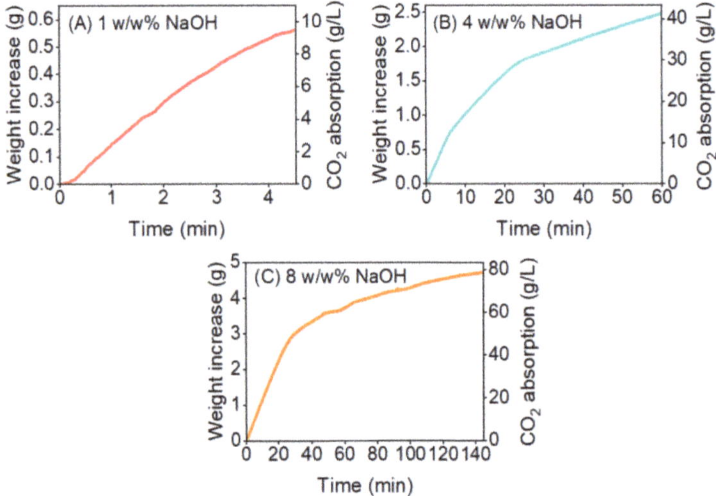

Figure 7. Weight increase of the reactor and the corresponding CO_2 absorption for 1 (**A**), 4 (**B**) and 8 (**C**) *w/w*% NaOH.

Figure 8A compares the experimental absorption capacities with the theoretical values if all CO_2 reacts forming carbonate and bicarbonate ions, respectively. These theoretical absorption capacities were calculated as the stoichiometric amount of CO_2 that reacts completely towards carbonates or bicarbonates according to Equations (1) and (2). This can only be used as a guideline because for most of the studied pH range here, carbonate and bicarbonate ions will form. The absorption capacities are between the two theoretical values, indicating that a mixture of carbonate and bicarbonate ions was produced at the end of the reaction. This agrees with the chemical reactions occurring at this pH (Equations (1) and (2)) and has also been verified by the FTIR spectra as we will explain in the next section. Additionally, it appeared that the absorption capacity has an almost linear correlation with the concentration of NaOH as shown in Figure 8B. This correlation and the absorption capacities for 1–5 *w/w*% NaOH agreed well with the trends and values reported in the literature [17]. This further proves that the 3D-printed reactor without forced mixing produces the same absorption capacities as those reactors employing forced mixing. Consequently, the presented solution has a clear advantage from an energy consumption perspective.

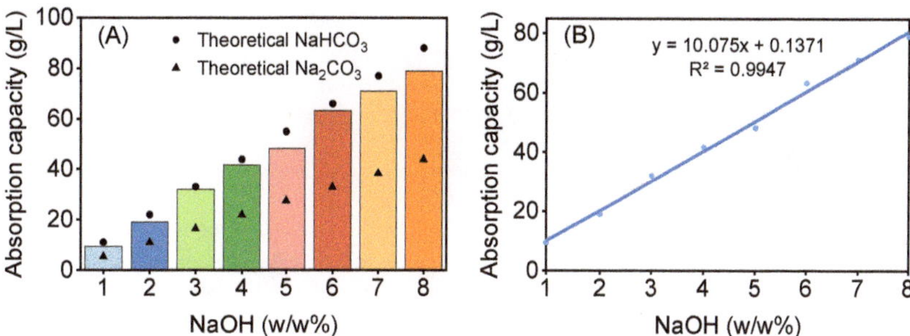

Figure 8. (**A**) Obtained and theoretical absorption capacities for NaOH solutions of 1 to 8 *w/w%* NaOH. A total conversion of CO_2 to carbonate and bicarbonate ions was assumed to estimate the theoretical capacities using Equations (1) and (2). (**B**) Linear regression based on the experimental data, indicating that the concentration of NaOH and the absorption capacity correlate linearly.

5. Conclusions and Future Remarks

Herein the absorption capacity of aqueous NaOH solutions for CO_2 capture was assessed for a wide range of concentrations (1–8 *w/w%*) with a 3D-printed reactor that avoids forced-mixing. The experimental setup included an in-line FTIR probe and pH meter to monitor the reaction over time and an analytical balance to estimate the absorption capacity through weight increase. Overall, the obtained CO_2 absorption capacities were in the range of 9.5 to 78.9 g/L for 1 *w/w%* and 8 *w/w%*, respectively. Reported capacities and crystal structures agreed with our results confirming that the non-forced mixing reactor design functions for CO_2 capture, which is an advantage from an energy and equipment savings perspective. Furthermore, monitoring the formation of carbonate and bicarbonate ions concurrently with pH and CO_2 absorption paves the way for optimizing the duration of the reaction for the maximum absorption rate or capacity or value-added products. The further optimization of absorption rate can be performed by adjusting the NaOH concentration. The 3D-printed design allows for easy implementation and hence applications for sodium-based wastes. The absorption capacity of such wastes can now be compared to aqueous NaOH solutions and optimized for different purposes and processes, such as the use of industrial waste to promote circularity in our societies. There are many sodium-containing industrial waste and its use for carbon capture undoubtedly boost the transition towards a biobased economy.

Author Contributions: Conceptualization, E.L., F.M.B.-M. and D.B.; methodology, E.L., F.M.B.-M., G.S. and H.S.; software, G.S., H.S. and E.G.; validation, F.M.B.-M. and D.B.; formal analysis, E.L., F.M.B.-M. and D.B.; investigation, E.L., F.M.B.-M., G.S., H.S. and D.B.; resources, G.S., H.S. and D.B.; data curation, E.L., F.M.B.-M., S.N., P.H.H. and A.M.K.; writing—original draft preparation, E.L., F.M.B.-M., G.S. and D.B.; writing—review and editing, E.L., F.M.B.-M. and D.B.; visualization, E.L., F.M.B.-M. and D.B.; supervision, G.S., H.S. and D.B.; project administration, G.S., H.S. and D.B.; funding acquisition, G.S., H.S. and D.B.. All authors have read and agreed to the published version of the manuscript.

Funding: We acknowledge the Area of Advance Energy, Chalmers University of Technology and Energimyndigheten (P2021-00009) for financial support.

Institutional Review Board Statement: Not applicable.

Informed Consent Statement: Not applicable.

Data Availability Statement: Not applicable.

Conflicts of Interest: The authors declare no conflict of interest.

References

1. Ritchie, H.; Roser, M.; Rosado, P. CO_2 and Greenhouse Gas Emissions. Available online: https://ourworldindata.org/co2-and-other-greenhouse-gas-emissions (accessed on 18 July 2022).
2. Witte, K. Social Acceptance of Carbon Capture and Storage (CCS) from Industrial Applications. *Sustainability* **2021**, *13*, 12278. [CrossRef]
3. Fichera, A.; Samanta, S.; Volpe, R. Exergetic Analysis of a Natural Gas Combined-Cycle Power Plant with a Molten Carbonate Fuel Cell for Carbon Capture. *Sustainability* **2022**, *14*, 533. [CrossRef]
4. Kheirinik, M.; Ahmed, S.; Rahmanian, N. Comparative Techno-Economic Analysis of Carbon Capture Processes: Pre-Combustion, Post-Combustion, and Oxy-Fuel Combustion Operations. *Sustainability* **2021**, *13*, 13567. [CrossRef]
5. Vega, F.; Baena-Moreno, F.M.; Gallego Fernández, L.M.; Portillo, E.; Navarrete, B.; Zhang, Z. Current Status of CO_2 Chemical Absorption Research Applied to CCS: Towards Full Deployment at Industrial Scale. *Appl. Energy* **2020**, *260*, 114313. [CrossRef]
6. Mohd Pauzi, M.M.; Azmi, N.; Lau, K.K. Emerging Solvent Regeneration Technologies for CO_2 Capture through Offshore Natural Gas Purification Processes. *Sustainability* **2022**, *14*, 4350. [CrossRef]
7. Sánchez-Bautista, A.; Palmero, E.M.; Moya, A.J.; Gómez-Díaz, D.; la Rubia, M.D. Characterization of Alkanolamine Blends for Carbon Dioxide Absorption. Corrosion and Regeneration Studies. *Sustainability* **2021**, *13*, 4011. [CrossRef]
8. Ji, L.; Yu, H.; Li, K.; Yu, B.; Grigore, M.; Yang, Q.; Wang, X.; Chen, Z.; Zeng, M.; Zhao, S. Integrated Absorption-Mineralisation for Low-Energy CO_2 Capture and Sequestration. *Appl. Energy* **2018**, *225*, 356–366. [CrossRef]
9. Hong, S.; Sim, G.; Moon, S.; Park, Y. Low-Temperature Regeneration of Amines Integrated with Production of Structure-Controlled Calcium Carbonates for Combined CO_2 Capture and Utilization. *Energy Fuels* **2020**, *34*, 3532–3539. [CrossRef]
10. Arti, M.; Youn, M.H.; Park, K.T.; Kim, H.J.; Kim, Y.E.; Jeong, S.K. Single Process for CO_2 Capture and Mineralization in Various Alkanolamines Using Calcium Chloride. *Energy Fuels* **2017**, *31*, 763–769. [CrossRef]
11. Baena-Moreno, F.M.; Rodríguez-Galán, M.; Vega, F.; Ramirez-Reina, T.; Vilches, L.; Navarrete, B. Understanding the Influence of the Alkaline Cation K^+ or Na^+ in the Regeneration Efficiency of a Biogas Upgrading Unit. *Int. J. Energy Res.* **2019**, *43*, 1578–1585. [CrossRef]
12. Luis, P. Use of Monoethanolamine (MEA) for CO_2 Capture in a Global Scenario: Consequences and Alternatives. *Desalination* **2016**, *380*, 93–99. [CrossRef]
13. Mahmoudkhani, M.; Keith, D.W. Low-Energy Sodium Hydroxide Recovery for CO_2 Capture from Atmospheric Air—Thermodynamic Analysis. *Int. J. Greenh. Gas Control.* **2009**, *3*, 376–384. [CrossRef]
14. Hikita, H.; Asai, S.; Takatsuka, T. Absorption of Carbon Dioxide into Aqueous Sodium Hydroxide and Sodium Carbonate-Bicarbonate Solutions. *Chem. Eng. J.* **1976**, *11*, 131–141. [CrossRef]
15. Shim, J.-G.; Lee, D.W.; Lee, J.H.; Kwak, N.-S. Experimental Study on Capture of Carbon Dioxide and Production of Sodium Bicarbonate from Sodium Hydroxide. *Environ. Eng. Res.* **2016**, *21*, 297–303. [CrossRef]
16. Lucile, F.; Cézac, P.; Contamine, F.; Serin, J.-P.; Houssin, D.; Arpentinier, P. Solubility of Carbon Dioxide in Water and Aqueous Solution Containing Sodium Hydroxide at Temperatures from (293.15 to 393.15) K and Pressure up to 5 MPa: Experimental Measurements. *J. Chem. Eng. Data* **2012**, *57*, 784–789. [CrossRef]
17. Yoo, M.; Han, S.-J.; Wee, J.-H. Carbon Dioxide Capture Capacity of Sodium Hydroxide Aqueous Solution. *J. Environ. Manag.* **2013**, *114*, 512–519. [CrossRef]
18. D'Adamo, I.; Gastaldi, M.; Morone, P.; Rosa, P.; Sassanelli, C.; Settembre-Blundo, D.; Shen, Y. Bioeconomy of Sustainability: Drivers, Opportunities and Policy Implications. *Sustainability* **2021**, *14*, 200. [CrossRef]
19. D'Adamo, I.; Sassanelli, C. Biomethane Community: A Research Agenda towards Sustainability. *Sustainability* **2022**, *14*, 4735. [CrossRef]
20. D'Adamo, I.; Gastaldi, M. Sustainable Development Goals: A Regional Overview Based on Multi-Criteria Decision Analysis. *Sustainability* **2022**, *14*, 9779. [CrossRef]
21. Cardoso, M.; de Oliveira, É.D.; Passos, M.L. Chemical Composition and Physical Properties of Black Liquors and Their Effects on Liquor Recovery Operation in Brazilian Pulp Mills. *Fuel* **2009**, *88*, 756–763. [CrossRef]
22. Reyes, L.; Nikitine, C.; Vilcocq, L.; Fongarland, P. Green Is the New Black—A Review of Technologies for Carboxylic Acid Recovery from Black Liquor. *Green Chem.* **2020**, *22*, 8097–8115. [CrossRef]
23. Kordylewski, W.; Sawicka, D.; Falkowski, T. Laboratory Tests on the Efficiency of Carbon Dioxide Capture from Gases in Naoh Solutions. *J. Ecol. Eng.* **2013**, *14*, 54–62. [CrossRef]
24. Salmón, I.; Cambier, N.; Luis, P. CO_2 Capture by Alkaline Solution for Carbonate Production: A Comparison between a Packed Column and a Membrane Contactor. *Appl. Sci.* **2018**, *8*, 996. [CrossRef]
25. Azizi, F.; Kaady, L.; Al-Hindi, M. Chemical Absorption of CO_2 in Alkaline Solutions Using an Intensified Reactor. *Can. J. Chem. Eng.* **2022**, *100*, 2172–2190. [CrossRef]
26. Chen, P.C.; Huang, C.F.; Chen, H.-W.; Yang, M.-W.; Tsao, C.-M. Capture of CO_2 from Coal-Fired Power Plant with NaOH Solution in a Continuous Pilot-Scale Bubble-Column Scrubber. *Energy Procedia* **2014**, *61*, 1660–1664. [CrossRef]
27. Yincheng, G.; Zhenqi, N.; Wenyi, L. Comparison of Removal Efficiencies of Carbon Dioxide between Aqueous Ammonia and NaOH Solution in a Fine Spray Column. *Energy Procedia* **2011**, *4*, 512–518. [CrossRef]
28. Stolaroff, J.K.; Keith, D.W.; Lowry, G.V. Carbon Dioxide Capture from Atmospheric Air Using Sodium Hydroxide Spray. *Environ. Sci. Technol.* **2008**, *42*, 2728–2735. [CrossRef]

29. Zhou, P.; Wang, H. Carbon Capture and Storage—Solidification and Storage of Carbon Dioxide Captured on Ships. *Ocean. Eng.* **2014**, *91*, 172–180. [CrossRef]
30. Pfleger, D.; Becker, S. Modelling and Simulation of the Dynamic Flow Behaviour in a Bubble Column. *Chem. Eng. Sci.* **2001**, *56*, 1737–1747. [CrossRef]
31. Burns, A.D.; Frank, T.; Hamill, I.; Shi, J.M. The Favre Averaged Drag Model for Turbulent Dispersion in Eulerian Multi-Phase Flows. In Proceedings of the Fifth International Conference on Multiphase Flow, Yokohama, Japan, 30 May–4 June 2004.
32. Wolf-Gladrow, D.A.; Zeebe, R.E.; Klaas, C.; Körtzinger, A.; Dickson, A.G. Total Alkalinity: The Explicit Conservative Expression and Its Application to Biogeochemical Processes. *Mar. Chem.* **2007**, *106*, 287–300. [CrossRef]
33. *CRC Handbook of Chemistry and Physics*; Haynes, W.M. (Ed.) CRC Press: Boca Raton, FL, USA, 2014; ISBN 9780429170195.
34. Wishart, D.S.; Guo, A.; Oler, E.; Wang, F.; Anjum, A.; Peters, H.; Dizon, R.; Sayeeda, Z.; Tian, S.; Lee, B.L.; et al. HMDB 5.0: The Human Metabolome Database for 2022. *Nucleic Acids Res.* **2022**, *50*, D622–D631. [CrossRef]
35. National Institute of Standards and Technology NIST Chemistry WebBook. Available online: http://webbook.nist.gov (accessed on 18 July 2022).
36. Salvador Cob, S. *Towards Zero Liquid Discharge in Drinking Water Production*; TU Delft: Delft, The Netherlands, 2014.
37. Graham, M.; Allington-Jones, L. The Air-Abrasive Technique: A Re-Evaluation of Its Use in Fossil Preparation. *Palaeontol. Electron.* **2018**, 1–15. [CrossRef]

Review

Sustainable Supply Chain Management, Performance Measurement, and Management: A Review

Anup Kumar [1], Santosh Kumar Shrivastav [2], Avinash K. Shrivastava [3,*], Rashmi Ranjan Panigrahi [4], Abbas Mardani [5] and Fausto Cavallaro [6]

1   IMT Nagpur, Nagpur 441502, Maharashtra, India
2   IMT Ghaziabad, Ghaziabad 201001, Uttar Pradesh, India
3   International Management Institute Kolkata, Kolkata 700027, West Bengal, India
4   GITAM School of Business, Visakhapatnam 530045, Andhra Pradesh, India
5   Muma Business School, University of South Florida (USF), Tampa, FL 33612, USA
6   Department of Economics, University of Molise, Via De Sanctis, 86100 Campobasso, Italy
*   Correspondence: kavinash1987@gmail.com; Tel.: +91-98-9191-5372

Abstract: The research highlights the importance of sustainable supply chain management (SSCM), technology adoption (TA), and performance measurement in promoting sustainability and improving supply chain performance. By incorporating sustainable practices and utilizing digital technologies, organizations can create a more sustainable future and improve their overall performances. This study conducted an in-depth review of the literature to investigate the presence of TA in SSCM with a focus on digital-based supply chains. The review used both bibliometric and content analysis methods to analyze relevant research articles, with the goal of providing a comprehensive understanding of the current state of research in the field, identifying any gaps in the literature, and providing direction for future research. The content analysis of the literature showed the absence of concrete frameworks for SSCM and the need for clearer and more applicable sustainability measurement indices. To address this gap, the study proposed a framework for achieving sustainable development goals through SSCM. In addition, a framework for deploying sustainability indicators was presented. The proposed framework can be used by practitioners to develop practical and comprehensive measures for their respective industries.

Keywords: sustainable supply chain management; sustainable development; sustainability performance; social responsibility; environmental issues; sustainability

Citation: Kumar, A.; Shrivastav, S.K.; Shrivastava, A.K.; Panigrahi, R.R.; Mardani, A.; Cavallaro, F. Sustainable Supply Chain Management, Performance Measurement, and Management: A Review. *Sustainability* **2023**, *15*, 5290. https://doi.org/10.3390/su15065290

Academic Editors: Idiano D'Adamo and Massimo Gastaldi

Received: 20 November 2022
Revised: 25 February 2023
Accepted: 9 March 2023
Published: 16 March 2023

1. Introduction

The Sustainable Development Goals (SDGs) of 2030 and digital technology are closely interconnected. Digital technology can play a significant role in achieving SDGs by enabling faster and more efficient solutions to global challenges such as poverty, inequality, and climate change. Sustainable business practices or otherwise, though not a new concept, have gained importance with time, looking at the adverse impacts on mother earth and the environment causing a threat to the existence of humankind.

Sustainable supply chain management (SSCM) can facilitate the adoption and implementation of supply chain integration using digital technologies. By incorporating sustainable practices and principles into their operations, organizations can improve their overall supply chain performance, reduce waste, and increase efficiency. Technology Adoption (TA) such as automation, data analytics, and the Internet of Things (IoT) can help organizations to achieve these goals by providing real-time data, facilitating communication and collaboration, and streamlining processes. As a result, suppliers can become more competitive in terms of their operational performance, as they can better meet customer demand, reduce costs, and improve their overall competitiveness [1]. Value creation is essential in sustainable model research and other management fields [1]. This realization and sustainability awareness are restricted to the manufacturers and across the value chain, both upstream and downstream [2]. Organizations are now more concerned with long-term survival than with competitive advantage. Concerns about supply chain sustainability have grown even in emerging economies. Digitalization can potentially restructure how procurement departments and vendors collaborate in supply chains [3], including sustainability. Sharing and collaboration should be encouraged as a result of digitalization. The role of Industry 4.0 in sustainable supply chain management has been investigated. Most businesses anticipate that digitalization will improve sustainability by utilizing big data analytics for energy management or facilitating chain knowledge transfer [4]. The traditional objective has been to make these elements and the chain efficient to garner higher profits; however, now, the sustainable supply chain is in focus, and Supply Chain Functions (SCFs) emphasize sustainability and profits [5,6] both. With the added dimension of sustainability, environmental, social, and economics in the business, the importance of measuring the performance of the supply chain has grown over recent periods in emerging countries such as India, China, and other Southeast Asian nations [2,7]. Measuring the performance of the supply chain helps firms analyze and improve Supply Chain Performance (SCP) and is an effective tool [8]. SCP is found to be helpful in measuring a business's entire value network within and outside its boundaries.

The Sustainable Supply Chain Management (SSCM) concept has evolved to integrate the supply chain's social, environmental, and financial concerns. A systematic review has been conducted to develop SSCM performance management and set a future research agenda. There are substantial reviews of the literature on SSCM available [9]; however, this study is focused on SSCM and measurement matrices and related management frameworks. A review of the literature is one of the principal methodologies to accumulate better knowledge and understanding of past work and related key issues.

Supply chain management systems use digital technologies to track and optimize the environmental impact of products. Integrating digital technologies with sustainability practices can involve the use of technology to optimize resource usage, reduce waste and emissions, and improve overall sustainability performance. Over time, research is evident that there is the potential for digital technologies to improve sustainability performance in supply chains, while also providing valuable insights and guidance for organizations looking to adopt AI-based technologies.

Digital technologies play a significant role in SSCM. In the area of supply chain management (SCM), digitization has allowed organizations to have a more comprehensive and real-time view of their supply chains, enabling them to make informed decisions and respond quickly to changes in demand and supply. For instance, digitization has enabled organizations to use data and analytics to optimize their inventory levels, improve delivery times, and reduce waste [10].

Moreover, digitization has made it possible for organizations to adopt new business models such as a circular economy, where the focus is on reducing waste, conserving resources, and creating sustainable value. In this context, digitization has enabled organizations to better track and manage the life cycle of products, from raw materials to end-of-life, thereby reducing their environmental impact and promoting sustainability [10]. The Internet of Things (IoT) and big data analytics (BDA) are indeed two of the most prevalent and impactful digital technologies in the current business landscape. The IoT refers to the network of connected devices that can collect and transmit data, enabling organizations to gain real-time insights into their operations and make informed decisions [3]. Blockchain technology has gained significant attention in recent years for its potential to revolutionize supply chain management. Its key features, such as transparency and data sharing, make it well-suited for promoting greater visibility and accountability in the supply chain. This has the potential to help companies identify and address sustainability challenges and improve the overall sustainability of their supply chains [11].

SCM theory suggests that effective management of the four critical elements of the supply chain is essential for ensuring that the supply chain operates efficiently and effectively and that the customer's needs are met in a timely and cost-effective manner. Theory suggests that planning, sourcing, manufacturing, and delivery are critical elements of any supply chain function [12].

Sustainable development is essential for addressing the environmental and social challenges facing the world, and it requires a holistic approach that balances economic, social, and environmental considerations. The goal of sustainable development is to create a more resilient and equitable world for current and future generations (MDG 2015) (http://www.un.org/millenniumgoals/2015_MDG_Report/pdf/MDG%202015%20rev%20(July%201).pdf (accessed on 20 September 2022)) report, people living in slum conditions are growing worldwide. The World Social Situation report by the U.N. reveals that more than 40% of the world population earns less than 1.25 dollars per day. Sustainability has been defined differently and needs to be conceptualized in totality. Measuring or quantifying sustainable development is the challenge for the day. Therefore, with the growing need for a sustainable supply chain, it is imperative to examine the role and importance of Sustainability Indicator Functions in driving a supply chain's sustainability performance and deploying sustainable indicators.

An extensive review of the literature has been undertaken to address the sustainability measurement issues from a different perspective of a supply chain and connect sustainable development goals through sustainable supply chain management by its various stockholders, such as employees, buyers, sellers, government, and society. The integration of digital technology and SSCM has the potential to revolutionize the way companies manage their supply chains, enabling them to create more sustainable, efficient, and responsible supply chains. As companies continue to adopt digital technologies, it is likely that the role of SSCM in supply chain management will become increasingly important, leading to greater sustainability and a more responsible and sustainable future.

There is less time to come up with answers to sustainability problems; no one has a clear understanding of how much the issue has actually impacted society and business [13]. In this paper, the authors made an attempt to identify the novelty of the study related to (1) the sustainability and digital supply chain (DSCM) prospectus of TA and (2) how the adoption of technology related to DCM business can achieve SSCM. There is little research that evidences digital technologies' potential impact on supply chain sustainability. These aspects highlight the need for further research to better understand the intersection of sustainability and digital technologies in supply chains and to develop practical guidelines for their effective integration.

Based on the above discussion, this study investigates the following research questions (RQ) to set future research direction:

RQ 1. How can organizations effectively integrate digital technologies and sustainability practices to create more sustainable and environmentally responsible supply chains?

RQ 2. What are the best practices for deploying the sustainability functions of a supply chain?

RQ 3. How can blockchain and other digital technologies help improve supply chain transparency and enhance sustainability performance?

2. Materials and Methods

The systemic review conducted by [14] aimed to establish a link between Sustainable Supply Chain Management (SSCM) and sustainable development, digital enablers, and supply chain management [10]. The review used the Scopus database and searched for articles using the keywords "Sustainability and sustainable supply chain management" and "Digital Supply chain" from 2011 to 2022. A total of 543 articles were reviewed out of 989. The review aimed to provide insights into the current state of research on SSCM and its association with sustainable development. The results of the review could be used to inform future research and guide organizations in their efforts to adopt sustainable supply chain practices.

A systemic review has been conducted to establish an association between Sustainable Supply Chain Management (SSCM) and sustainable development [14], sustainable supply chain management, and food company performance with linking to quality assurance [15]. The review on Sustainable Supply Chain Management and Digital Transformation [3] literature has been searched using the keywords sustainability and sustainable supply chain management and digital supply chain on the most widely accepted and accessed data source, i.e., Scopus database during 2011–2022. A total of 543 articles have been reviewed out of 989 articles.

A reliability test for content analysis has been carried out using the method proposed by [16] in SPSS. The alpha value of the test result was 0.83, which is acceptable. The frequencies of different sustainability themes in supply chain management are listed in Figure 1 (the x-axis represents the frequency of words).

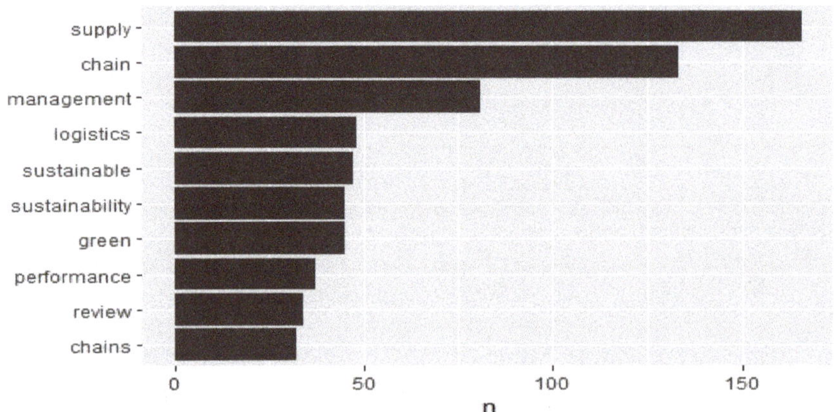

Figure 1. Frequency of keywords in reviewed papers.

2.1. Descriptive Statistics of the Reviewed Papers

For our supply chain management research, we rely on Scopus because of its comprehensive coverage of peer-reviewed studies [17]. Now that we have been exposed to newer research, we use the most commonly used keywords [18–20]. Table 1 shows our systematic approach to compiling the final data sets of five hundred forty-three research papers. In the realm of academic research in quantitative science studies, Scopus serves as a content-rich and comprehensive bibliometric data source [21]. The Scopus database has increased interest more than the WOS and PubMed data sources of its worldwide access and diverse research platform [22]. So, the bibliometric analysis of this study is conducted using the Scopus database. Due to the initial manuscript being retrieved from the Scopus database via software, bibliometric and bibliographic information have also been utilized [17]. When such documents are converted without first cleaning the data, there is the risk that an unsupported assertion will be made. Thus, we examine the data's references to clean it up. This course of action is made possible by the expansion, visualization, and comprehension of bibliographic data [23]. To obtain the above-filtered data, we used the search words "sustainability" and "sustainable supply chain management".

Table 1. The summary statistics of literature used for bibliometric analysis are listed below.

Filtering Criteria	Reject	Accept
Search parameters.		
Search engine: Scopus		
Date of data Extraction: 28 June 2022		
Author's keywords: ("sustainability" AND "sustainable supply chain management")		989
Subject area: "Business, Management and Accounting", "Economics", "Econometrics and Finance", "Social Sciences", "Decision Sciences", and "Environmental Science".	158	831
Search year: 2011–28 June 2022	30	801
Language screening: "English".	4	797
Document type: "Articles", "Book",	188	609
Publication state: "Final" and "Article in the press".	40	569
Content screening: Include articles if "Titles, abstracts, and keywords" indicate relevance to the scope of the study (i.e., sustainable supply chain management)	26	543
Authors of single-authored documents	-	1000
Minimum two documents as an author with one citation	-	222

Note(s): Table showing the last collection of (543) articles from the Scopus database.

The annual percentage growth rate of the studied literature was 56.7164. Sustainability Switzerland (76 publications), the Journal of Cleaner Production (66 publications), the International Journal of Production Economics (22 publications), Business Strategy and The Environment (19 publications), Resources Conservation and Recycling (18 publications), Supply Chain Management (15 publications), International Journal of Operations and Production Management (13 publications), and International Journal of Production Research and Transportation Research Part E Logistics And Transportation Review (11 publications) are among the top contributors (10 publications each) to the Forum on the Supply Chain (07), as shown in Figure 2.

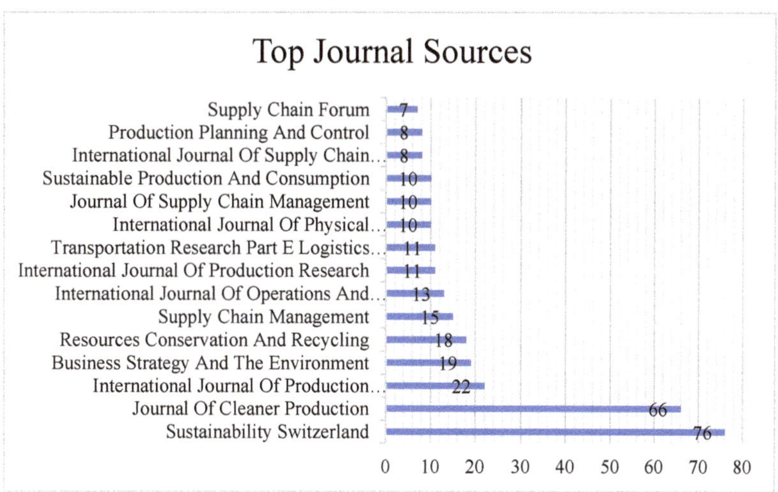

Figure 2. Statistics of reviewed papers from top journals.

The environmental dimension of supply chain management has a higher correlation (72.48%) with sustainability, followed by economic (70%) and sustainable supply chain (56%) dimensions, as shown in the correlation-based network diagram of keywords in

Figure 3. The year-wise distribution of papers reviewed is presented in Figure 4, which indicates a sudden (almost double) increase in the articles published in this area from 2011 to 2012 and demonstrates that the topic has gained paramount importance in the past few years. The tools and methods for reviewing papers are classified in Table 2.

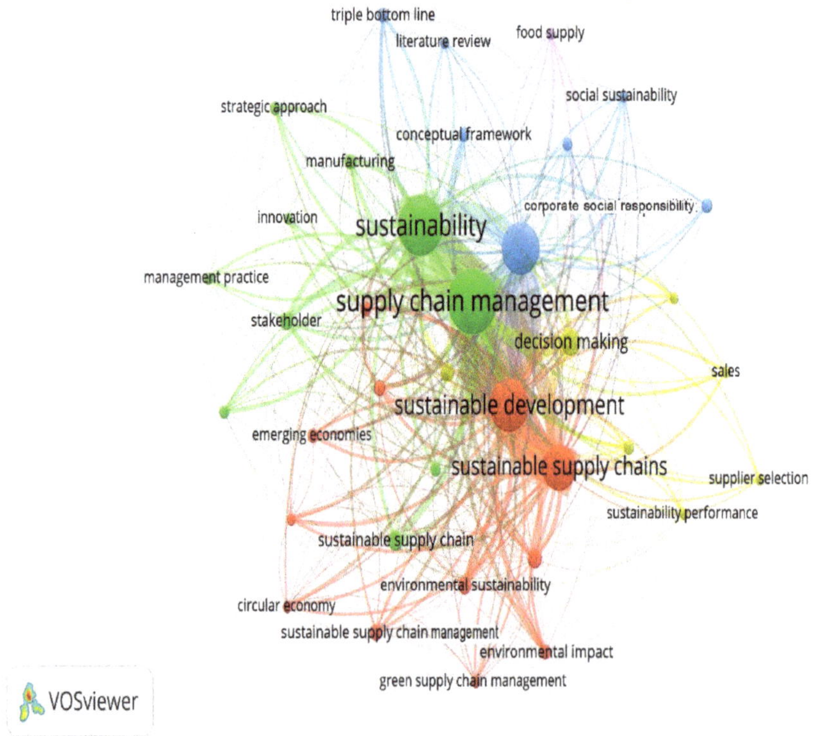

Figure 3. The directed network of supply chain management, sustainability, sustainable development, sustainable supply chain management, and sustainable supply chains dimensions in the reviewed literature.

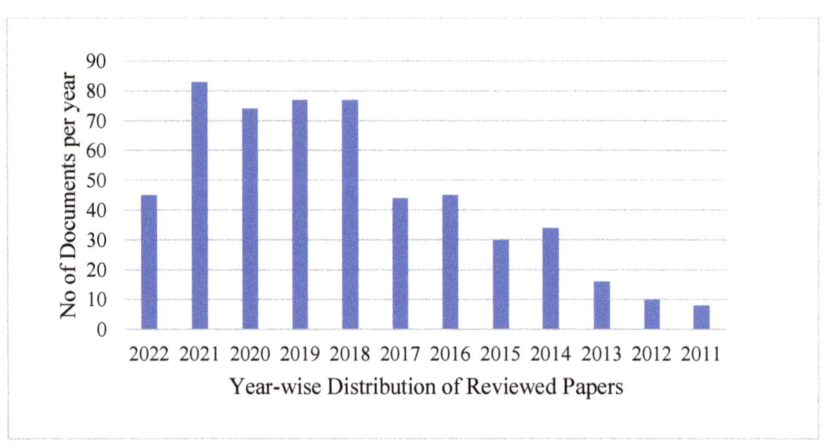

Figure 4. Year-wise distribution of reviewed papers.

Table 2. Methods used for performance measurement/management in the reviewed papers.

Tools/Methods	Papers	Counts
AHP, AHP-QFD, FAHP, QFD-SFD	[24–33]	16
Balance Score Card, Association Rule, Bibliometric Network Analysis	[34–40]	12
BSC, Fuzzy Delphi Method (FDM), Fuzzy MCDM, Fuzzy-Based DEMATEL, Grey-Based MCDM, Fuzzy ANP	[41–48]	23
Content Analysis, Econometric Method, Conceptual Framework, Empirical Study and Modelling, Data Analysis, Inferential Statistics	[3,8,49–93]	64
DEA, ISM, LCA, EIO, VCA, Matrices, Factor Analysis	[53,59,76,94–110]	18
Quantitative Modelling, TBL, Simulation, and Modelling, Social Network Analysis	[39,111–120]	22
Qualitative Framework, Grey-Based SCOR, MCDM, CFA, Complexity Analysis, MILP, Bibliometric Analysis	[35,121–129]	17

2.2. Inferential Statistics

The study undertakes content analysis for data synthesis and identification of emerging themes, which was considered the most reliable approach for this kind of study, as suggested by [130]. Content analysis has been performed among 543 reviewed articles to verify that Sustainable Supply Chain Management positively supports sustainability (Figure 5).

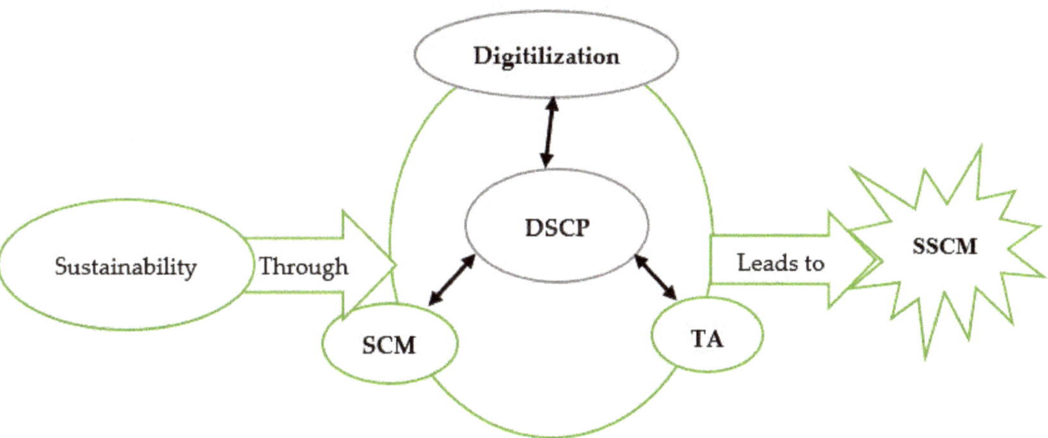

Figure 5. Directional relationship between Sustainability and SSCM. Model referred from [131].

From text analytics, we generate the directional network as shown in Figure 3; 48.96 percent of the reviewed literature has taken all three dimensions of sustainability; therefore, it is legitimate to test the stated hypothesis among 48.96 percent of the literature [132]. Out of 71 pieces of literature with all dimensions of sustainability, 43 noted a positive association between sustainable development (SD) and SSCM, while 10 stated negative and 18 said nothing about sustainability. Of 543 pieces of literature with at least one sustainability dimension, 73 endorsed a positive association between sustainability and SSCM. Most of the reviewed literature has supported the positive association between SD and SSCM. The reviewed literature has also indicated increased citations on sustainability science. Theorists of SSCM should seek out those institutional entrepreneurs who actively reshape the institutional conditions in which they are situated [133].

Establishing a relationship between SSCM practices and sustainability performance management has been challenging. The authors have tested the association between SSCM practices and sustainability performance with widely stated variables in Table 1. The principle SSCM methods, such as green procurement, green manufacturing, green distribution, green logistics, and recycling, were tested for association with sustainability performance (economic performance, environmental performance, and social performance). The Chi-square association test shows a relationship between SSCM practices and sustainability performance (Tables 3–5). The *p*-value of the test is more than 0.05, which shows the strong relationship between SSCM practices and sustainability performances. SSCM has become an increasingly complex and globally accepted phenomenon in the area of sustainability [31,134]. Global supply chains face difficulties assessing their performance if the multidimensional and transnational aspects are ignored (GSCs) [135].

Table 3. Cross tabulation results.

Sustainable Supply Chain Practices × Sustainability Performance Cross Tabulation				
Count				
		Sustainability Performance		Total
		Economic Performance	Environmental Performance	Social Performance	
	Green Distribution	40	30	33	103
Sustainable Supply Chain Practices	Green Logistics	30	35	34	99
	Green Manufacturing	35	40	30	105
	Green Procurement	29	42	40	111
	Recycling	45	45	35	125
Total		179	192	172	543

Table 4. Chi-Square test results.

	Value	Def	Asymp. Sig. (Two-Sided)
Pearson Chi-Square	7.342 [a]	8	0.500
Likelihood Ratio	7.469	8	0.487
No. of Valid Cases	466		

[a.] 0 cells (. 0%) have an expected count of less than five. The minimum expected count is 26.50.

Table 5. List of sustainability indices.

S.No.	Index	Acronym/Institution
1	EPI	Environmental Performance Index
2	NRMI	Natural Resource Management Index
3	H.F.	Human Footprint
4	GRI	Global Reporting Initiatives
5	GPI	Genuine Progress Index
6	DJSI	Dow Jones Sustainability Index
7	CSD	Commission on Sustainable Development
8	Dashboard	Developed by Consultee Group of Sustainable Development Indicators
9	Barometer	Developed by the World Conservation Institute
10	TBL	Triple Bottom Line Index
11	Ethos	Corporate Social Responsibility Indicators
12	IChemE	Institute of Chemical Engineers

2.2.1. Sustainability and Supply Chain Management

This study conducted a structured review to determine whether millennium development could be achieved through Sustainable Supply Chain Management (SSCM). The word sustainable development indicates the participation of every member of society in economic activity. [136] have advocated traditional farming systems and other activities such as craftsmanship (e.g., manufacturing and local food production), which essentially lead to human integration with nature. The Common Agricultural Policy (CAP), formulated in Europe, challenges participatory planning and attitudes toward traditional work. Participatory economic activity and collaboration are significant indicators of sustainable supply chain management; a few examples, like AMUL and GOPALJEE dairy supply chains from India, could be stated in support of the argument. The human race has endless needs, and to satisfy these needs, a supply chain management system has been developed and defined with linkages (networks) for different human activities. Therefore, researchers must address those endless needs legitimately for the sustainable development theory. A system view of supply chain management is shown in Figure 6, followed by a process view of the supply chain and SSCM process in Figures 7 and 8.

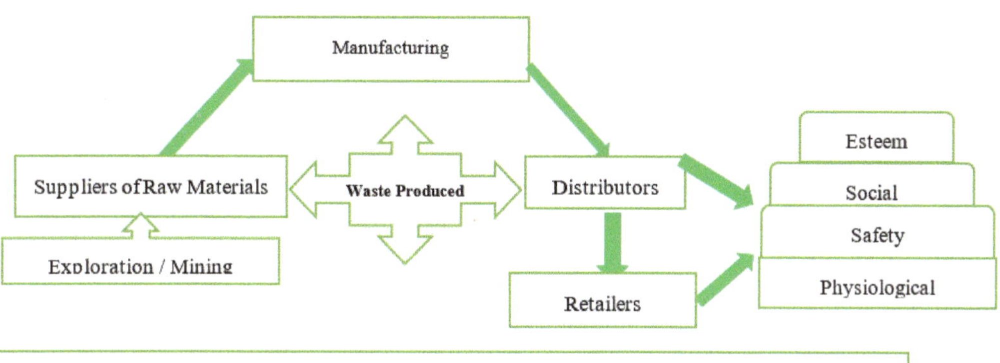

Figure 6. A simple supply chain.

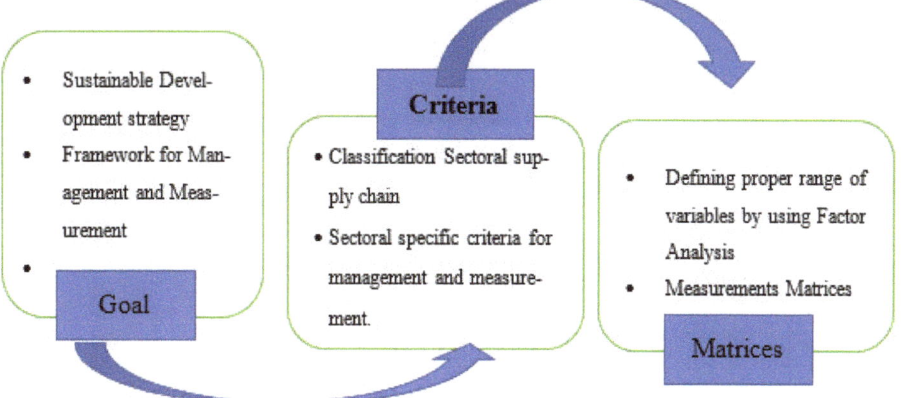

Figure 7. SSCM process.

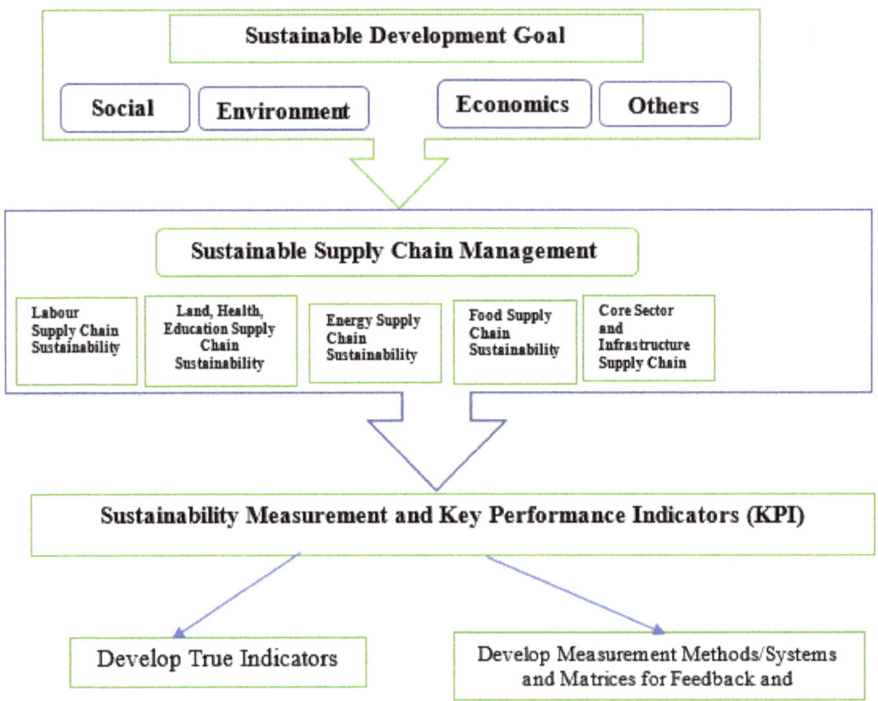

Figure 8. Process view of attaining sustainable development through SSCM.

The objective of a supply chain is to maximize supply chain surplus, which is defined by the following:

$$\text{Supply Chain Surplus} = \text{Customer value} - \text{Supply Chain Cost}$$

The value of the product or service fulfilled by a supply chain may vary according to the need. Therefore, for sustainable supply chain management, supply chain costs should be minimized; costs may function as environmental, social, and economic concerns, including waste produced, carbon emission, etc.

2.2.2. Sustainability and Digital Supply Chain Management (DSCM)

Sustainability and digital supply chain management (SCM) are closely related concepts that can be used to improve the overall efficiency and impact of businesses. Sustainability refers to the responsible use of resources [131], the reduction of negative environmental impact, and promotion of social well-being. Digital solutions can provide visibility and traceability in the supply chain, enabling companies to make more informed decisions and better assess the sustainability of their suppliers and operations [136]. Digitizing the global value chain has grown significantly [131]. Digital transformation leveraging sophisticated technologies like cloud computing, big data, and blockchain simplifies and accelerates supply chain integration [131] with Industry 4.0. By integrating materials and information and synchronizing inter-organizational activities, firms have tried to improve supply chain operations by collaborating with suppliers and customers [137]. It is important for companies to have a clear and comprehensive understanding of the potential benefits and risks of digitization and to develop effective change management strategies to ensure successful adoption.

Table 5 shows the different measurement indices used for measuring sustainability [138]; most companies use these indices for their sustainability initiatives. [139] argued that these indices have failed to provide a minimum acceptable level of indicators. Using these indices in the deployment of sustainability indicators is also unclear.

The authors have surveyed sustainability initiatives of the top ten Fortune 500 companies, as listed in Table 6. The sustainability reporting initiatives contain CSR as well. CSR initiatives are now reported and designed as sustainability programs for companies. The shift from CSR to sustainability reporting is evident in Table 6. Sustainability reporting practices are random; they should be structured and contain concrete evidence of sustainability initiatives.

Table 6. Sustainability reporting program initiatives from top 10 Fortune 500 companies.

Rank	Company	Website	Sustainability Initiatives		
			Social	**Environmental**	**Economic**
1	Walmart Stores	www.corporate.walmart.com	The Walmart Foundation meets the needs of the under-served by directing charitable giving toward our core areas of focus: Opportunity, Sustainability, and Community.		
2	Exxon Mobil	www.exxonmobil.com	Malaria, corporate citizenship, human rights, local Economic development, Indigenous peoples, cultural heritage and diversity, land use and resettlement, transparency and corruption, community relations	Air emissions reductions, ecosystem services, environmental drilling initiatives, environmental stewardship, freshwater management, managing arctic resources, site remediation, spill performance	Environment and safety, operations, policy, technology, flexibility, energy efficiency
3	Chevron	www.chevron.com	Health, education, volunteerism, freshwater	Freshwater, biodiversity	Energy efficiency, land management, IPIECA/API/OGP reporting standard
4	Berkshire Hathaway	www.berkshirehathaway.com	Investors health		
5	Apple	www.apple.com	Environment, supplier responsibility, accessibility, privacy, inclusion and diversity, education, reuse and recycling		
6	Phillips 66	www.phillips66.com	Operating excellence, people, transparency & accountability, community investments, supplier diversity, governance & ethics, environmental metrics.		
7	General Motors	www.gm.com	Sustainability considers environmental, social, and economic opportunities and supports the long-term success of our company. Value is created through top-line growth opportunities, bottom-line improvements, and risk mitigation. Our customers drive how much weight is designed—everything starts and ends with our customers. At G.M., we view sustainability as a business approach that makes long-term stakeholder value. It is an approach executed by every function at every level of our company.		
8	Ford Motor	www.ford.com	Investors health		
9	General Electric	www.ge.com	At G.E., Sustainability means aligning our business strategy to meet societal needs while minimizing environmental impact and advancing social development. This commitment is embedded at every level of our company.		
10	Valero Energy	www.valero.com	Social health and safety, community, valero benefit of children, volunteerism, employee benefit	Environmental matrices, environmental awards	Investors, financial reports, filings & statements, industry fundamentals, investor faqs

3. Measurement of Sustainability and Its Indicators

Today, sustainable development and equitable growth are extremely important. The authors discovered evidence and a link between Sustainable Supply Chain Management and Sustainable Development, as indicated in the research question. The United Nations General Assembly approved a set of global Sustainable Development Goals (SDGs) consisting of 17 goals and 169 targets. In addition, in March 2015, an initial set of 330 indicators was introduced. Some SDGs were based on previous Millennium Development Goals, while others were based on fresh ideas. A rigorous examination of these indicators was conducted by [140].

Many researchers have indicated the sustainability of the energy supply chain, infrastructure needs, road, and transports mechanism could enhance inclusive growth [141–145]. A few researchers have tried to capture sustainability matrices for measurement. Refs. [53,146] have personated a well-structured review of different sustainability measurement indices and frameworks such as GRI, IChemE, DJSI, CSD, Dashboard, Barometer, TBL, and Ethos. While measuring and managing the performance of sustainability, the following questions need to be answered, as discussed by [138]. Ref. [138] have discussed performance measurement and management tools in SSCM, and the same, with some additions, which are presented in Table 7.

- What type of performance is being measured in SSCM?
- Which indicators should be used?
- Who is measuring performance?
- What type of SSCM Performance Measurement indicators and methods will be used by which business sector?

Table 7. Overview of performance measurement and management tools in SSCM.

	Social	Environmental	Economic	Other	Integrative
Tools	Social LCA, Social Audit, Social Benchmarking, Stakeholder Dialog, Social Reporting	Life Cycle Assessment (LCA), Eco-audit, Environmental Benchmarking, Environmental Reporting	Cost-benefit Analysis, Economic Input–Output Analysis, Financial Reporting, Risk Analysis	Transparency Deployment, Quality Sustainability, Role of I.T. in Sustainability	Sustainability Audit, Sustainability Benchmarking, Sustainability Reporting
Concept	Corporate Citizenship	Design of the Environment	SCOR Framework, Financial Audit	Digitalization	Sustainability Balanced Scorecard (SBSC)
System	Social Management System (SMS) Occupational Health and Safety	Environmental Management System (EMS)	Quality Management System(QMS)	Yet to be Developed	Integrated Management System
Standard	SA 8000(SMS), OHSAS 18001 (OHS)	ISO 14001 (EMS), ISO 14040 (LCA), ISO 14064	ISO 9001 (QMS)		Global Reporting Initiatives (report), U.N. Global Compact

3.1. Social Indicators

A reference model for social sustainability measurement has been proposed by [146] that indicates the dimensions and level of sustainability measurement. The reference model proposed could be linked to the sustainable social supply chain management model, as suggested in Figure 9. A list of social sustainability indicators is presented in Table 8.

Table 8. Social Indicators of SSCM.

S.No.	Indicator	Description
1	Workforce participation	Describe joint management and workforce health and safety programs and processes to facilitate the crew's participation in health and safety dialogues at all levels.
2	Workforce health	Describe programs and processes for identifying and addressing significant workforce health issues, especially at the community and country levels.
3	Occupational injury and illness incidents	Report health and safety data on workforce injuries or illnesses resulting from occupational incidents.
4	Product stewardship	Describe the company's approach to assessing and communicating product health, safety, and environmental risks.
5	Process safety	Report the number of Tier 1 process safety events (an unplanned or uncontrolled loss of primary containment) with a narrative and report per business activity (refining, upstream, etc.).
6	Local community impacts and engagement	Describe policies, strategies, and procedures to understand and address local community impacts and engage with affected stakeholders.
7	Indigenous peoples	Describe policies, programs, and procedures to engage with indigenous peoples and address their concerns and expectations.
8	Involuntary resettlement	Describe policies, programs, and procedures related to involuntary resettlement.
9	Social investment	Describe strategies, programs, and procedures relating to social investment and its effectiveness.
10	Local content practices	Describe policies and practices related to local content.
11	Local hiring practices	Describe the company's approach and programs for providing employment opportunities to residents of host countries.
12	Local procurement and supplier development	Describe the company's programs and processes to improve the ability of local suppliers and contractors to support operations and carry out projects.
13	Human rights due diligence	Describe policies, programs, and procedures the company has in place to respect human rights—including workers' rights—in its operations.
14	Human rights and suppliers	Describe policies, programs, and procedures to promote respect for human rights by suppliers.
15	Security and human rights	Describe policies, programs, and procedures related to security and human rights.
16	Preventing corruption	Describe policies, programs, and procedures to prevent bribery and corruption and monitor compliance.
17	Preventing corruption involving business partners	Describe anti-corruption policies and procedures for business partners, including suppliers and contractors.
18	Transparency of payments to host governments	Describe policies, initiatives, or advocacy programs to promote revenue transparency.
19	Public advocacy and lobbying	Describe the company's approach to managing public advocacy, lobbying, and political contributions.
20	Workforce diversity and inclusion	Describe policy and procedures promoting diversity and inclusion.
21	Workforce engagement	Describe policies, programs, and procedures on engagement and workforce satisfaction.
22	Workforce training and development	Describe policies and procedures for providing workforce training and development opportunities.
23	Non-retaliation and grievance system	Describe the non-retaliation policy and confidential workforce grievance system.

Figure 9. Social framework of sustainable development using SSCM.

3.2. Environmental Indicators

The concept of 'Green Supply Chain Management (GSCM)' was introduced to include environmental concerns of the supply chain, and plenty of scholarly studies and formulations have been carved out in this area. The significant ideas from these studies are remanufacturing, recycling, green production, waste management, green energy, reverse logistics, etc. [99]. A sustainability reporting framework is proposed (Figure 10) based on the content analysis of the reviewed literature. A list of Contemporary Environmental Indicators is identified in Table 9.

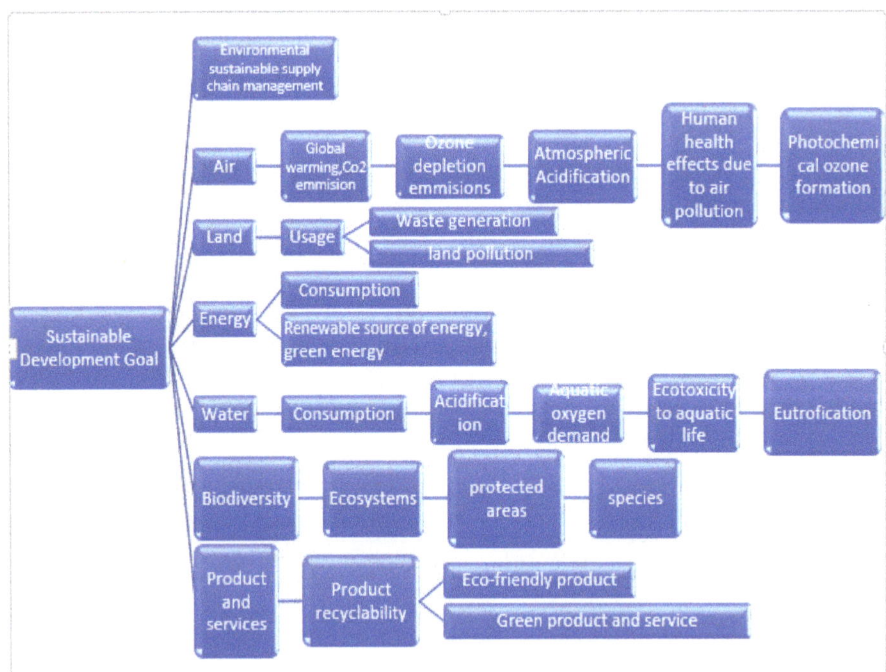

Figure 10. Environmental framework of sustainable development using SSCM.

Table 9. Environmental indicators of SSCM.

S.No.	Indicator	Description
1	Natural recourse management	Report the effective use of natural resources
2	Land management and greenery	Report the effective use of land
3	Environmental governance	Report environmental awareness
4	Energy management	Report the effort to use renewable energy
5	Greenhouse gas emissions	Report the quantity of greenhouse gas emissions from combustion and other processes, including carbon dioxide and methane
6	Eco-efficiency	Report total energy consumed in oil and gas operations or other business activities
7	Alternative energy sources	Report qualitatively on company research, plans, or current initiatives related to alternative or renewable energy sources
8	Flared gas	Report the quantity of hydrocarbon gas flared into the atmosphere from operations
9	Biodiversity and ecosystem services	Qualitatively describe how the company addresses risks and opportunities related to biodiversity and ecosystem services
10	Freshwater	Report the quantity of freshwater withdrawn or consumed by operations
11	Other air emissions	Report quantities of emissions to the atmosphere from operations
12	Spills to the environment	Quantify spills to the environment from operations and describe significant falls and response measures
13	Discharges to water	Quantify hydrocarbon discharges to a water environment from operations
14	Waste and recycling	Report quantities of waste disposed of, resulting from operations and methods of recycling
15	Environmental health	Report the health of the air, water, and land

3.3. Economic Indicators

Economic efficiency has been the primary performance criteria for a supply chain that includes cost, service, and operational efficiencies until the emergence of Supply Chain Management, Global Commodity Chains, Global Value Chains, and Global Production Networks [54]. After 2000, a sudden increase in the literature on sustainability and a worldwide debate on sustainability has forced supply chain planners and designers to design holistic performance criteria in sync with economic, social, and environmental performance. Economic indicators alone have been used for performance measurement by about 40% of the reviewed literature. The rest of the reviewed literature has suggested combining three dimensions of sustainability. After a rigorous review of the 145 pieces of the stated literature, a list of economic indicators has been compiled and shown in Table 10. A conceptual framework for economic sustainability management has also been proposed and demonstrated in Figure 11.

3.4. Other Indicators

A supply network is created to satisfy various customers' needs; therefore, the networks have different issues for different regions. Hence, it is imperative to devise a few need-based sustainability dimensions for specific sectors. The House of Sustainability concept is proposed by [147] for the coal sector, which advocates including moral sustainability factors in dimensions of sustainability. Ethical sustainability is essential for companies to grow ethically and socially. Transparency sustainability is necessary for Food Supply Chains such as meat, milk, wine, frozen food, and perishable vegetables, while some authors have suggested technical and political sustainability and information transparency for the future sustainable dimensions [28,37,60].

Table 10. Economic indicators of SSCM.

S.No.	Component	Indicator
1		Inventory
2	Cost	Transportation
3		Facilities and Handling
4		Information
5		Productivity
6		Response Time
7		Product Variety
8	Service	Product Availability
9		Customer Experience
10		Time to Market
11		Order Visibility
12		Return Ability
13		Return on Investment
14	Financial	Return on Equity
15		Return on Assets
16		Accounts Payable Turnover

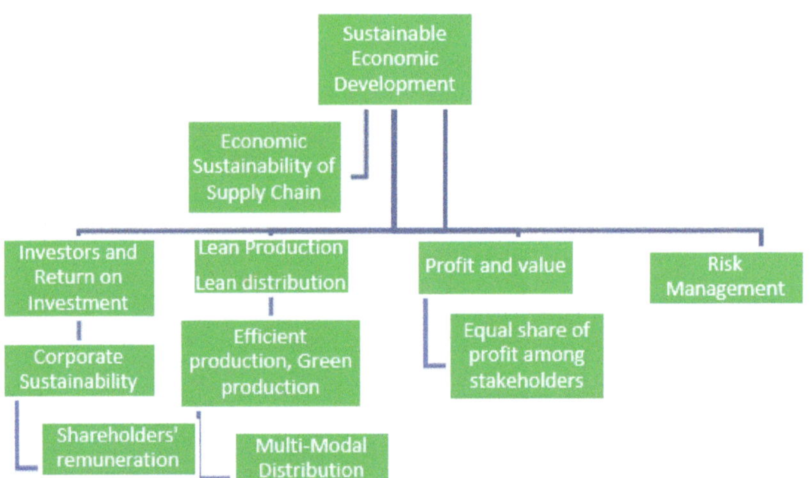

Figure 11. Economic Framework of Sustainable Development.

4. Discussions

From production to operations through end-of-life management, green supply chain management (GSCM) integrates the principle into conventional supply chains to create environmentally friendly processes (reduce, reuse, recycle, reclaim, and degrade). The goal of green supply chain management is to aid businesses in operating more sustainably and effectively. It is sometimes referred to as green logistics. The purpose of green SCM goes beyond only being environmentally friendly. Enhancing sustainability and increasing efficiency in operations are other important goals. The three pillars of sustainability—people, earth, and profits—are all considered by an all-encompassing green approach. It is evident from the literature that Green Supply Chain Management (GSCM) is a subset of Sustainable Supply Chain Management (SSCM), as SSCM covers all dimensions of sustainability [148].

Forty-eight percent of the reviewed literature has endorsed that SSCM could achieve sustainability goals proposed by the United Nations MDG report. A framework showing an association between sustainable development and SSCM is established in Figure 12.

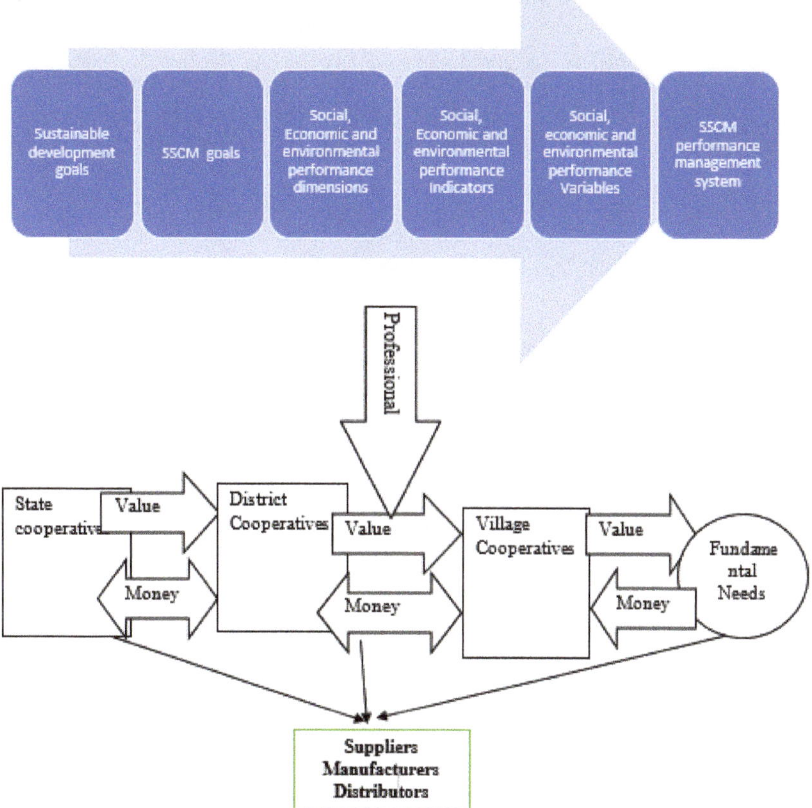

Figure 12. Sustainable development and SSCM.

4.1. Sustainable Development and SSCM

The research objective of this study is investigated through content analysis of the reviewed literature. Most of the literature that addresses the SSCM has mentioned sustainable development (80%). The impact of sustainability initiatives on other functions and capabilities of a supply chain has been studied and concluded to have a positive association with some lag, while some authors have shown a negative impact [52,58,67,106,149–152].

Social, environmental, and economic indicators and variables for SSCM are a subset of 300 indicators of sustainable development mentioned in the UN MDG Report 2015. Overall, the data and content analysis reveal that through SSCM and government and NGO initiatives, sustainable development could be achieved; a framework for achieving sustainable development is shown in Figure 12 and answers the RQ1. Moreover, a systemic process could be adopted to measure and manage SSCM, as shown in Figure 13. An SSCM Management System may contain performance measurement methods, matrices, and indices for different dimensions of social, economic, and environmental sustainability development.

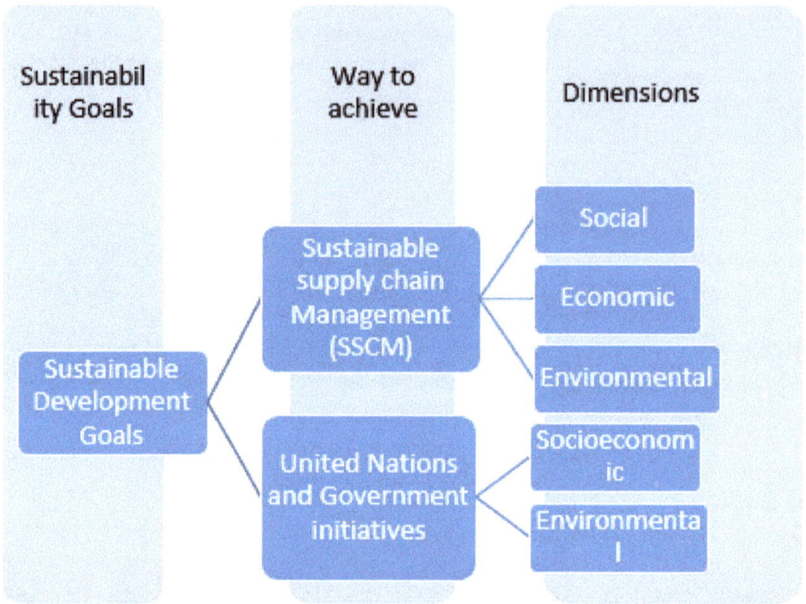

Figure 13. SSCM deployment process.

4.2. Deployment of SSCM

Ascertaining questions are still important, such as who is going to measure sustainability? What should be the appropriate framework for different sectors or entities of a supply chain? The solutions are unclear, and each stage of a supply network has another framework and sustainability indicators. Therefore, the SSCM deployment framework is proposed to address the above questions, as shown in Figure 12, and answers the RQ2.

Ref. [24] have proposed a four-stage QFD that is confusing and tedious to implement. Ref. [27] presented the Sustainability Deployment Function (SDF) based on customers' needs. However, while implementing this framework, there may be a possibility that some dimensions of sustainability may be ignored; therefore, it is essential to deploy the sustainability functions from the company's perspective, not from the customers' perspective.

Hence, the authors proposed a house of a sustainability framework; the proposed House of Sustainability framework may be easy to deploy for all stakeholders of a supply network. It allows the flexibility to choose specific sustainability indicators for their basket and requirement. The following steps could be adapted for the deployment of sustainability:

- Identify the sustainability indicators for the company using AHP, ISM, or some clustering tools;
- Each indicator has variables to measure or deploy, so identify the variables of each of the arrows;
- Construct the House of Sustainability similar to the House of Quality proposed by [147];
- Find out the suitable variables to deploy or report, or measure for the company using the following formula:

$$F_i = d_i \sum_{j=1}^{n} v_{ij}$$

Where
$F_i = total$ impact score on i^{th} *indicator*
d_i = Relative importance or weight of ith *sustainability indicator*
V_{ij} = impact of jth variable on i^{th} sustainability Indicator

Considering the total impact score and the company's sustainability reporting strategy, the company should deploy the required sustainability indicators.

5. Conclusions, Limitation and Scope of Future Research

Existing reviews have focused on the practices of SSCM. This study has shown the road for the future in the SSCM area. Content analysis of the literature reviewed showed a positive association between sustainable development and Sustainable Supply Chain Management. The suggested framework demonstrates how various supply chain management stakeholders, including employees, suppliers, buyers, the government, and society, may work together to accomplish the sustainable development aim. Additionally, it aids in the achievement of the sustainable development goal through the use of sustainable supply chain management. The author argues that sustaining a supply network might lead to sustainability and suggests a comprehensive methodology for integrating sustainability indicators into a supply chain. In addition, the study provided a framework for using sustainability indicators.

The reviewed literature has also indicated the sharp growth of research on sustainability associated with SSCM. The United Nations' sustainability goals could be achieved through SSCM; hence, SSCM will also guide future research. The study proposes the 'House of Sustainability' framework for deploying sustainability. The proposed frameworks to achieve 17 SDG goals are formulated in this literature concerning SSCM for all three dimensions of sustainability. It has also been shown statistically that through SSCM practices, sustainability performance could be affected. A report piloted by Elsevier in collaboration with SciDev.Net (https://www.elsevier.com/__data/assets/pdf_file/0018/119061/SustainabilityScienceReport-Web.pdf (accessed on 20 September 2022)) has indicated a baseline in both the definition and the understanding of sustainability science. This report helps to follow its progression and trajectory. The report examines six key themes encompassing the 17 United Nations Sustainability Development Goals: Dignity, People, Prosperity, Planet, Justice, and Partnership. The key findings of the report are (https://www.elsevier.com/research-intelligence/resource-library/sustainability-2015 (accessed on 20 September 2022)):

- Sustainability science has a high growth rate (7.6%) in research output;
- Research output in sustainability science attracts 30% more citations than an average research paper;
- Research in sustainability science is highly collaborative;
- Sustainability science is less interdisciplinary than the world average.

The report has also indicated that the high growth rate is contributed mainly by developed countries, showing the collaboration between the northern and southern hemispheres. The key findings indicate the importance of the topic for the research; it also points out that it is less interdisciplinary and can expand. The analysis of this paper also presented the same story as the above report. Practitioners may create thorough and realistic strategies for their specific sectors using the author's suggested framework for a sustainable supply chain. Sustainability indicator deployment and sustainability reporting have different formats in different countries, and business entities along the supply chain face the challenge of deploying sustainability indicators.

Since the actual data for the developed framework in this study is based on published literature indexed in the SCOPUS database, we did not include any research in our analysis that were not SCOPUS-indexed. Future studies, including publications indexed in other databases, should be looked into to gain a more comprehensive understanding of particular issues. In the future, by including the voice of supply chain managers of various firms and social media data, a more comprehensive framework and sustainability deployment mechanism can be developed to eliminate the grey areas of sustainability, DSCM, and sustainable supply chain management.

Funding: This research received no external funding.

Data Availability Statement: No new data were created or analyzed in this study. Data sharing is not applicable to this article.

Conflicts of Interest: The authors declare no conflict of interest.

References

1. Boruchowitch, F.; Fritz, M.M. Who in the firm can create sustainable value and for whom? A single case-study on sustainable procurement and supply chain stakeholders. *J. Clean. Prod.* **2022**, *363*, 132619. [CrossRef]
2. Geng, J.; Long, R.; Chen, H.; Li, W. Exploring the motivation-behavior gap in urban residents' green travel behavior: A theoretical and empirical study. *Resour. Conserv. Recycl.* **2017**, *125*, 282–292. [CrossRef]
3. Stroumpoulis, A.; Kopanaki, E. Theoretical Perspectives on Sustainable Supply Chain Management and Digital Transformation: A Literature Review and a Conceptual Framework. *Sustainability* **2022**, *14*, 4862. [CrossRef]
4. Kunkel, S.; Matthess, M.; Xue, B.; Beier, G. Industry 4.0 in sustainable supply chain collaboration: Insights from an interview study with international buying firms and Chinese suppliers in the electronics industry. *Resour. Conserv. Recycl.* **2022**, *182*, 106274. [CrossRef]
5. Hassini, E.; Surti, C.; Searcy, C. A literature review and a case study of sustainable supply chains with a focus on metrics. *Int. J. Prod. Econ.* **2012**, *140*, 69–82. [CrossRef]
6. Reefke, H.; Sundaram, D. Key themes and research opportunities in sustainable supply chain management–identification and evaluation. *Omega* **2017**, *66*, 195–211. [CrossRef]
7. Jayaram, J.; Avittathur, B. Green supply chains: A perspective from an emerging economy. *Int. J. Prod. Econ.* **2015**, *164*, 234–244. [CrossRef]
8. Shepherd, C.; Günter, H. Measuring supply chain performance: Current research and future directions. *Int. J. Product. Perform. Manag.* **2006**, *55*, 242–258. [CrossRef]
9. Martins, C.L.; Pato, M. Supply chain sustainability: A tertiary literature review. *J. Clean. Prod.* **2019**, *225*, 995–1016. [CrossRef]
10. Attaran, M. Digital technology enablers and their implications for supply chain management. *Supply Chain Forum Int. J.* **2020**, *21*, 158–172. [CrossRef]
11. Zhang, J.; Zhang, X.; Liu, W.; Ji, M.; Mishra, A.R. Critical success factors of blockchain technology to implement the sustainable supply chain using an extended decision-making approach. *Technol. Forecast. Soc. Chang.* **2022**, *182*, 121881. [CrossRef]
12. Difrancesco, R.M.; Luzzini, D.; Patrucco, A.S. Purchasing realized absorptive capacity as the gateway to sustainable supply chain management. *Int. J. Oper. Prod. Manag.* **2022**, *42*, 603–636. [CrossRef]
13. D'Adamo, I.; Gastaldi, M.; Morone, P.; Rosa, P.; Sassanelli, C.; Settembre-Blundo, D.; Shen, Y. Bioeconomy of Sustainability: Drivers, Opportunities and Policy Implications. *Sustainability* **2021**, *14*, 200. [CrossRef]
14. Koberg, E.; Longoni, A. A systematic review of sustainable supply chain management in global supply chains. *J. Clean. Prod.* **2019**, *207*, 1084–1098. [CrossRef]
15. Kuwornu, J.K.; Khaipetch, J.; Gunawan, E.; Bannor, R.K.; Ho, T.D. The adoption of sustainable supply chain management practices on performance and quality assurance of food companies. *Sustain. Futur.* **2023**, *5*, 100103. [CrossRef]
16. Hayes, A.F.; Krippendorff, K. Answering the Call for a Standard Reliability Measure for Coding Data. *Commun. Methods Meas.* **2007**, *1*, 77–89. [CrossRef]
17. Goodell, J.W.; Kumar, S.; Lim, W.M.; Pattnaik, D. Artificial intelligence and machine learning in finance: Identifying foundations, themes, and research clusters from bibliometric analysis. *J. Behav. Exp. Finance* **2021**, *32*, 100577. [CrossRef]
18. Asgari, A.; Abdul Hamid, A.B.; Ebrahim, N.A. Supply Chain Integration: A Review and Bibliometric Analysis. *Int. J. Econ. Manag. Sci.* **2017**, *6*, 1000447. [CrossRef]
19. El Baz, J.; Iddik, S. Green supply chain management and organizational culture: A bibliometric analysis based on Scopus data (2001–2020). *Int. J. Organ. Anal.* **2022**, *30*, 156–179. [CrossRef]
20. Hushko, S.; Botelho, J.M.; Maksymova, I.; Slusarenko, K.; Kulishov, V. Sustainable development of global mineral resources market in Industry 4.0 context. *IOP Conf. Ser. Earth Environ. Sci.* **2021**, *628*, 012025. [CrossRef]
21. Baas, J.; Schotten, M.; Plume, A.; Côté, G.; Karimi, R. Scopus as a curated, high-quality bibliometric data source for academic research in quantitative science studies. *Quant. Sci. Stud.* **2020**, *1*, 377–386. [CrossRef]
22. AlRyalat, S.A.S.; Malkawi, L.W.; Momani, S.M. Comparing Bibliometric Analysis Using PubMed, Scopus, and Web of Science Databases. *J. Vis. Exp.* **2019**, *152*, e58494. [CrossRef]

23. Donthu, N.; Kumar, S.; Mukherjee, D.; Pandey, N.; Lim, W.M. How to conduct a bibliometric analysis: An overview and guidelines. *J. Bus. Res.* **2021**, *133*, 285–296. [CrossRef]

24. Dai, J.; Blackhurst, J. A four-phase AHP–QFD approach for supplier assessment: A sustainability perspective. *Int. J. Prod. Res.* **2012**, *50*, 5474–5490. [CrossRef]

25. Dey, P.K.; Cheffi, W. Green supply chain performance measurement using the analytic hierarchy process: A comparative analysis of manufacturing organisations. *Prod. Plan. Control* **2012**, *24*, 702–720. [CrossRef]

26. Yu, V.F.; Tseng, L.-C. Measuring social compliance performance in the global sustainable supply chain: An AHP approach. *J. Inf. Optim. Sci.* **2014**, *35*, 47–72. [CrossRef]

27. Zhang, Z.; Awasthi, A. Modelling customer and technical requirements for sustainable supply chain planning. *Int. J. Prod. Res.* **2014**, *52*, 5131–5154. [CrossRef]

28. Gopal, P.R.C.; Thakkar, J. Development of composite sustainable supply chain performance index for the automobile industry. *Int. J. Sustain. Eng.* **2015**, *8*, 366–385. [CrossRef]

29. Colicchia, C.; Melacini, M.; Perotti, S. Benchmarking supply chain sustainability: Insights from a field study. *Bench Marking Int. J.* **2011**, *18*, 705–732. [CrossRef]

30. Rajesh, R. Sustainable supply chains in the Indian context: An integrative decision-making model. *Technol. Soc.* **2020**, *61*, 101230. [CrossRef]

31. Yontar, E.; Ersöz, S. Sustainability assessment with structural equation modeling in fresh food supply chain management. *Environ. Sci. Pollut. Res.* **2021**, *28*, 39558–39575. [CrossRef]

32. Kumar, A.; Srivastava, S.K.; Singh, S. How blockchain technology can be a sustainable infrastructure for the agrifood supply chain in developing countries. *J. Glob. Oper. Strat. Sourc.* **2022**, *15*, 380–405. [CrossRef]

33. Shrivastav, S.K. Exploring the application of analytics in supply chain during COVID-19 pandemic: A review and future research agenda. *J. Glob. Oper. Strat. Sourc.* 2022; *ahead-of-print*. [CrossRef]

34. Chia, A.; Goh, M.; Hum, S.H. Performance measurement in supply chain entities: Balanced scorecard perspective. *Benchmarking Int. J.* **2009**, *16*, 605–620. [CrossRef]

35. Fahimnia, B.; Sarkis, J.; Davarzani, H. Green supply chain management: A review and bibliometric analysis. *Int. J. Prod. Econ.* **2015**, *162*, 101–114. [CrossRef]

36. Touboulic, A.; Walker, H. Love me, love me not: A nuanced view on collaboration in sustainable supply chains. *J. Purch. Supply Manag.* **2015**, *21*, 178–191. [CrossRef]

37. Ting, S.; Tse, Y.; Ho, G.; Chung, S.; Pang, G. Mining logistics data to assure the quality in a sustainable food supply chain: A case in the red wine industry. *Int. J. Prod. Econ.* **2014**, *152*, 200–209. [CrossRef]

38. Hallinger, P. A Meta-Synthesis of Bibliometric Reviews of Research on Managing for Sustainability, 1982–2019. *Sustainability* **2021**, *13*, 3469. [CrossRef]

39. Shashi; Centobelli, P.; Cerchione, R.; Mittal, A. Managing sustainability in luxury industry to pursue circular economy strategies. *Bus. Strategy Environ.* **2021**, *30*, 432–462. [CrossRef]

40. Biswal, J.N.; Muduli, K.; Satapathy, S.; Yadav, D.K. A TISM based study of SSCM enablers: An Indian coal- fired thermal power plant perspective. *Int. J. Syst. Assur. Eng. Manag.* **2019**, *10*, 126–141. [CrossRef]

41. Kumar, A.; Shrivastav, S.K.; Oberoi, S.S. Application of Analytics in Supply Chain Management from Industry and Academic Perspective. *FIIB Bus. Rev.* **2021**, 23197145211028041. [CrossRef]

42. Sharma, M.; Kamble, S.; Mani, V.; Sehrawat, R.; Belhadi, A.; Sharma, V. Industry 4.0 adoption for sustainability in multi-tier manufacturing supply chain in emerging economies. *J. Clean. Prod.* **2021**, *281*, 125013. [CrossRef]

43. Tseng, M.; Lim, M.; Wong, W.P. Sustainable supply chain management: A closed-loop network hierarchical approach. *Ind. Manag. Data Syst.* **2015**, *115*, 436–461. [CrossRef]

44. Chithambaranathan, P.; Subramanian, N.; Gunasekaran, A.; Palaniappan, P.K. Service supply chain environmental performance evaluation using grey based hybrid MCDM approach. *Int. J. Prod. Econ.* **2015**, *166*, 163–176. [CrossRef]

45. Govindan, K.; Khodaverdi, R.; Vafadarnikjoo, A. Intuitionistic fuzzy based DEMATEL method for developing green practices and performances in a green supply chain. *Expert Syst. Appl.* **2015**, *42*, 7207–7220. [CrossRef]

46. Bhattacharya, A.; Mohapatra, P.; Kumar, V.; Dey, P.K.; Brady, M.; Tiwari, M.K.; Nudurupati, S.S. Green supply chain performance measurement using fuzzy ANP-based balanced scorecard: A collaborative decision-making approach. *Prod. Plan. Control* **2014**, *25*, 698–714. [CrossRef]

47. Uysal, F. An Integrated Model for Sustainable Performance Measurement in Supply Chain. *Procedia Soc. Behav. Sci.* **2012**, *62*, 689–694. [CrossRef]

48. Erol, I.; Sencer, S.; Sari, R. A new fuzzy multi-criteria framework for measuring sustainability performance of a supply chain. *Ecol. Econ.* **2011**, *70*, 1088–1100. [CrossRef]

49. Bush, S.R.; Oosterveer, P.; Bailey, M.; Mol, A.P. Sustainability governance of chains and networks: A review and future outlook. *J. Clean. Prod.* **2015**, *107*, 8–19. [CrossRef]

50. Khalid, R.U.; Seuring, S. Analyzing Base-of-the-Pyramid Research from a (Sustainable) Supply Chain Perspective. *J. Bus. Ethic* **2019**, *155*, 663–686. [CrossRef]

51. Singh, S.; Srivastava, S.K. Decision support framework for integrating triple bottom line (TBL) sustainability in agriculture supply chain. *Sustain. Account. Manag. Policy J.* **2022**, *13*, 387–413. [CrossRef]

52. Meixell, M.J.; Luoma, P. Stakeholder pressure in sustainable supply chain management: A systematic review. *Int. J. Phys. Distrib. Logist. Manag.* **2015**, *45*, 69–89. [CrossRef]

53. Tajbakhsh, A.; Hassini, E. Performance measurement of sustainable supply chains: A review and research questions. *Int. J. Product. Perform. Manag.* **2015**, *64*, 744–783. [CrossRef]

54. Shrivastav, S.K. How The TQM Journal has addressed "quality": A literature review using bibliometric analysis. *TQM J.* 2023; ahead-of-print. [CrossRef]

55. Chin, T.A.; Tat, H.H.; Sulaiman, Z. Green Supply Chain Management, Environmental Collaboration and Sustainability Performance. *Procedia CIRP* **2015**, *26*, 695–699. [CrossRef]

56. Günther, H.-O.; Kannegiesser, M.; Autenrieb, N. The role of electric vehicles for supply chain sustainability in the automotive industry. *J. Clean. Prod.* **2015**, *90*, 220–233. [CrossRef]

57. Leigh, M.; Li, X. Industrial ecology, industrial symbiosis and supply chain environmental sustainability: A case study of a large U.K. distributor. *J. Clean. Prod.* **2015**, *106*, 632–643. [CrossRef]

58. Luzzini, D.; Brandon-Jones, E.; Brandon-Jones, A.; Spina, G. From sustainability commitment to performance: The role of intra- and inter-firm collaborative capabilities in the upstream supply chain. *Int. J. Prod. Econ.* **2015**, *165*, 51–63. [CrossRef]

59. Mani, V.; Agrawal, R.; Sharma, V. Supply Chain Social Sustainability: A Comparative Case Analysis in Indian Manufacturing Industries. *Procedia Soc. Behav. Sci.* **2015**, *189*, 234–251. [CrossRef]

60. Mol, A. Transparency and value chain sustainability. *J. Clean. Prod.* **2015**, *107*, 154–161. [CrossRef]

61. Nappi, V.; Rozenfeld, H. The Incorporation of Sustainability Indicators into a Performance Measurement System. *Procedia CIRP* **2015**, *26*, 7–12. [CrossRef]

62. Shaharudin, M.R.; Govindan, K.; Zailani, S.; Tan, K.C. Managing product returns to achieve supply chain sustainability: An exploratory study and research propositions. *J. Clean. Prod.* **2015**, *101*, 1–15. [CrossRef]

63. Silvestre, B.S. Sustainable supply chain management in emerging economies: Environmental turbulence, institutional voids and sustainability trajectories. *Int. J. Prod. Econ.* **2015**, *167*, 156–169. [CrossRef]

64. Xu, X.; Gursoy, D. Influence of sustainable hospitality supply chain management on customers' attitudes and behaviors. *Int. J. Hosp. Manag.* **2015**, *49*, 105–116. [CrossRef]

65. Bloemhof, J.M.; van der Vorst, J.G.; Bastl, M.; Allaoui, H. Sustainability assessment of food chain logistics. *Int. J. Logist. Res. Appl.* **2015**, *18*, 101–117. [CrossRef]

66. Tachizawa, E.M.; Wong, C.Y. Towards a theory of multi-tier sustainable supply chains: A systematic literature review. Supply Chain Manag. Int. J. **2014**, *19*, 643–663. [CrossRef]

67. Ortas, E.; Moneva, J.M.; Álvarez, I. Sustainable supply chain and company performance: A global examination. *Supply Chain Manag. Int. J.* **2014**, *19*, 332–350. [CrossRef]

68. Varsei, M.; Soosay, C.; Fahimnia, B.; Sarkis, J. Framing sustainability performance of supply chains with multidimensional indicators. *Supply Chain. Manag. Int. J.* **2014**, *19*, 242–257. [CrossRef]

69. Azfar, K.R.W.; Khan, N.; Gabriel, H.F. Performance Measurement: A Conceptual Framework for Supply Chain Practices. *Procedia Soc. Behav. Sci.* **2014**, *150*, 803–812. [CrossRef]

70. Bourlakis, M.; Maglaras, G.; Aktas, E.; Gallear, D.; Fotopoulos, C. Firm size and sustainable performance in food supply chains: Insights from Greek SMEs. *Int. J. Prod. Econ.* **2014**, *152*, 112–130. [CrossRef]

71. Bourlakis, M.; Maglaras, G.; Gallear, D.; Fotopoulos, C. Examining sustainability performance in the supply chain: The case of the Greek dairy sector. *Ind. Mark. Manag.* **2014**, *43*, 56–66. [CrossRef]

72. Li, Y.; Zhao, X.; Shi, D.; Li, X. Governance of sustainable supply chains in the fast fashion industry. *Eur. Manag. J.* **2014**, *32*, 823–836. [CrossRef]

73. Turker, D.; Altuntas, C. Sustainable supply chain management in the fast fashion industry: An analysis of corporate reports. *Eur. Manag. J.* **2014**, *32*, 837–849. [CrossRef]

74. Marshall, D.; McCarthy, L.; Heavey, C.; McGrath, P. Environmental and social supply chain management sustainability practices: Construct development and measurement. *Prod. Plan. Control* **2015**, *26*, 673–690. [CrossRef]

75. Xu, X.; Gursoy, D. A Conceptual Framework of Sustainable Hospitality Supply Chain Management. *J. Hosp. Mark. Manag.* **2014**, *24*, 229–259. [CrossRef]

76. Taticchi, P.; Tonelli, F.; Pasqualino, R. Performance measurement of sustainable supply chains: A literature review and a research agenda. *Int. J. Product. Perform. Manag.* **2013**, *62*, 782–804. [CrossRef]

77. Winter, M.; Knemeyer, A.M. Exploring the integration of sustainability and supply chain management: Current state and opportunities for future inquiry. *Int. J. Phys. Distrib. Logist. Manag.* **2013**, *43*, 18–38. [CrossRef]

78. Cucchiella, F.; D'Adamo, I.; Gastaldi, M.; Koh, S.L. Implementation of a real option in a sustainable supply chain: An empirical study of alkaline battery recycling. *Int. J. Syst. Sci.* **2013**, *45*, 1268–1282. [CrossRef]

79. Bastian, J.; Zentes, J. Supply chain transparency as a key prerequisite for sustainable agri-food supply chain management. *Int. Rev. Retail Distrib. Consum. Res.* **2013**, *23*, 553–570. [CrossRef]

80. Wognum, P.; Bremmers, H.; Trienekens, J.H.; van der Vorst, J.G.; Bloemhof, J.M. Systems for sustainability and transparency of food supply chains–Current status and challenges. *Adv. Eng. Informatics* **2011**, *25*, 65–76. [CrossRef]

81. Genovese, A.; Lenny Koh, S.C.; Kumar, N.; Tripathi, P.K. Exploring the challenges in implementing supplier environmental performance measurement models: A case study. *Prod. Plan. Control* **2014**, *25*, 1198–1211. [CrossRef]

82. Miemczyk, J.; Johnsen, T.E.; Macquet, M. Sustainable purchasing and supply management: A structured literature review of definitions and measures at the dyad, chain and network levels. *Supply Chain Manag. Int. J.* **2012**, *17*, 478–496. [CrossRef]

83. Giunipero, L.C.; Hooker, R.E.; Denslow, D. Purchasing and supply management sustainability: Drivers and barriers. *J. Purch. Supply Manag.* **2012**, *18*, 258–269. [CrossRef]

84. Corbière-Nicollier, T.; Blanc, I.; Erkman, S. Towards a global criteria based framework for the sustainability assessment of bioethanol supply chains: Application to the Swiss dilemma: Is local produced bioethanol more sustainable than bioethanol imported from Brazil? *Ecol. Indic.* **2011**, *11*, 1447–1458. [CrossRef]

85. Eltayeb, T.K.; Zailani, S.; Ramayah, T. Green supply chain initiatives among certified companies in Malaysia and environmental sustainability: Investigating the outcomes. *Resour. Conserv. Recycl.* **2011**, *55*, 495–506. [CrossRef]

86. Wong, W.P.; Wong, K.Y. Supply chain performance measurement system using DEA modeling. *Ind. Manag. Data Syst.* **2007**, *107*, 361–381. [CrossRef]

87. Baddeley, J.; Font, X. Barriers to Tour Operator Sustainable Supply Chain Management. *Tour. Recreat. Res.* **2011**, *36*, 205–214. [CrossRef]

88. Kalenoja, H.; Kallionpää, E.; Rantala, J. Indicators of energy efficiency of supply chains. *Int. J. Logist. Res. Appl.* **2011**, *14*, 77–95. [CrossRef]

89. Shaw, S.; Grant, D.B.; Mangan, J. Developing environmental supply chain performance measures. *Benchmarking Int. J.* **2010**, *17*, 320–339. [CrossRef]

90. Bayo-Moriones, A.; Bello-Pintado, A.; Merino-Díaz-de-Cerio, J. Quality assurance practices in the global supply chain: The effect of supplier localisation. *Int. J. Prod. Res.* **2010**, *49*, 255–268. [CrossRef]

91. Pullman, M.E.; Maloni, M.J.; Dillard, J. Sustainability Practices in Food Supply Chains: How is Wine Different? *J. Wine Res.* **2010**, *21*, 35–56. [CrossRef]

92. Vachon, S.; Klassen, R.D. Supply chain management and environmental technologies: The role of integration. *Int. J. Prod. Res.* **2007**, *45*, 401–423. [CrossRef]

93. Yakovleva, N. Measuring the Sustainability of the food supply chain: A case study of the U.K. *J. Environ. Policy Plan.* **2007**, *9*, 75–100. [CrossRef]

94. Dubey, R.; Gunasekaran, A. The sustainable humanitarian supply chain design: Agility, adaptability and alignment. *Int. J. Logist. Res. Appl.* **2015**, *19*, 62–82. [CrossRef]

95. Khodakarami, M.; Shabani, A.; Farzipoor Saen, R.; Azadi, M. Developing distinctive two-stage data envelopment analysis models: An application in evaluating the sustainability of supply chain management. *Measurement* **2015**, *70*, 62–74. [CrossRef]

96. Eskandarpour, M.; Dejax, P.; Miemczyk, J.; Péton, O. Sustainable supply chain network design: An optimization-oriented review. *Omega* **2015**, *54*, 11–32. [CrossRef]

97. Acquaye, A.A.; Yamoah, F.A.; Feng, K. An integrated environmental and fairtrade labelling scheme for product supply chains. *Int. J. Prod. Econ.* **2015**, *164*, 472–483. [CrossRef]

98. Mani, V.; Agarwal, R.; Gunasekaran, A.; Papadopoulos, T.; Dubey, R.; Childe, S.J. Social sustainability in the supply chain: Construct development and measurement validation. *Ecol. Indic.* **2016**, *71*, 270–279. [CrossRef]

99. Arif-Uz-Zaman, K.; Nazmul Ahsan, A.M.M. Lean supply chain performance measurement. *Int. J. Product. Perform. Manag.* **2014**, *63*, 588–612. [CrossRef]

100. Bai, C.; Sarkis, J. Determining and applying sustainable supplier key performance indicators. *Supply Chain Manag. Int. J.* **2014**, *19*, 275–291. [CrossRef]

101. Egilmez, G.; Kucukvar, M.; Tatari, O.; Bhutta, M.K.S. Supply chain sustainability assessment of the U.S. food manufacturing sectors: A life cycle-based frontier approach. *Resour. Conserv. Recycl.* **2014**, *82*, 8–20. [CrossRef]

102. Sueyoshi, T.; Wang, D. Sustainability development for supply chain management in U.S. petroleum industry by DEA environmental assessment. *Energy Econ.* **2014**, *46*, 360–374. [CrossRef]

103. Luthra, S.; Garg, D.; Haleem, A. Critical success factors of green supply chain management for achieving Sustainability in Indian automobile industry. *Prod. Plan. Control* **2015**, *26*, 339–362.

104. Gallear, D.; Ghobadian, A.; Li, Y.; O'Regan, N.; Childerhouse, P.; Naim, M. An environmental uncertainty-based diagnostic reference tool for evaluating the performance of supply chain value streams. *Prod. Plan. Control* **2014**, *25*, 1182–1197. [CrossRef]

105. Freidberg, S. Calculating sustainability in supply chain capitalism. *Econ. Soc.* **2013**, *42*, 571–596. [CrossRef]

106. Qrunfleh, S.; Tarafdar, M. Lean and agile supply chain strategies and supply chain responsiveness: The role of strategic supplier partnership and postponement. *Supply Chain Manag. Int. J.* **2013**, *18*, 571–582. [CrossRef]

107. Fearne, A.; Martinez, M.G.; Dent, B. Dimensions of sustainable value chains: Implications for value chain analysis. *Supply Chain Manag. Int. J.* **2012**, *17*, 575–581. [CrossRef]

108. Ross, A.D.; Parker, H.; Benavides-Espinosa, M.D.M.; Droge, C. Sustainability and supply chain infrastructure development. *Manag. Decis.* **2012**, *50*, 1891–1910. [CrossRef]

109. De Coster, R.J.; Bateman, R.J.; Plant, A.V.C. Supply Chain Implications of Sustainable Design Strategies for Electronics Products. *Int. J. Adv. Logist.* **2012**, *1*, 1–20. [CrossRef]

110. Menon, R.R.; Ravi, V. Analysis of enablers of sustainable supply chain management in electronics industries: The Indian context. *Clean. Eng. Technol.* **2021**, *5*, 100302. [CrossRef]

111. Martín-Gómez, A.; Aguayo-González, F.; Luque, A. A holonic framework for managing the sustainable supply chain in emerging economies with smart connected metabolism. *Resour. Conserv. Recycl.* **2019**, *141*, 219–232. [CrossRef]

112. Boukherroub, T.; Ruiz, A.; Guinet, A.; Fondrevelle, J. An integrated approach for sustainable supply chain planning. *Comput. Oper. Res.* **2015**, *54*, 180–194. [CrossRef]

113. Bing, X.; Bloemhof-Ruwaard, J.; Chaabane, A.; van der Vorst, J. Global reverse supply chain redesign for household plastic waste under the emission trading scheme. *J. Clean. Prod.* **2015**, *103*, 28–39. [CrossRef]

114. Ahi, P.; Searcy, C. Measuring social issues in sustainable supply chains. *Meas. Bus. Excel.* **2015**, *19*, 33–45. [CrossRef]

115. Burritt, R.; Schaltegger, S. Accounting towards sustainability in production and supply chains. *Br. Account. Rev.* **2014**, *46*, 327–343. [CrossRef]

116. Lee, K.-H.; Wu, Y. Integrating sustainability performance measurement into logistics and supply networks: A multi-methodological approach. *Br. Account. Rev.* **2014**, *46*, 361–378. [CrossRef]

117. Taticchi, P.; Garengo, P.; Nudurupati, S.S.; Tonelli, F.; Pasqualino, R. A review of decision-support tools and performance measurement and sustainable supply chain management. *Int. J. Prod. Res.* **2015**, *53*, 6473–6494. [CrossRef]

118. Awudu, I.; Zhang, J. Uncertainties and sustainability concepts in biofuel supply chain management: A review. *Renew. Sustain. Energy Rev.* **2012**, *16*, 1359–1368. [CrossRef]

119. Hollos, D.; Blome, C.; Foerstl, K. Does sustainable supplier co-operation affect performance? Examining implications for the triple bottom line. *Int. J. Prod. Res.* **2012**, *50*, 2968–2986. [CrossRef]

120. van der Vorst, J.G.A.J.; Tromp, S.-O.; Zee, D.-J.V.d. Simulation modelling for food supply chain redesign; integrated decision making on product quality, Sustainability and logistics. *Int. J. Prod. Res.* **2009**, *47*, 6611–6631. [CrossRef]

121. Chalmeta, R.; Barqueros-Muñoz, J.-E. Using Big Data for Sustainability in Supply Chain Management. *Sustainability* **2021**, *13*, 7004. [CrossRef]

122. Creighton, R.; Jestratijevic, I.; Lee, D. Sustainability supplier scorecard assessment tools: A comparison between apparel retailers. *J. Glob. Fash. Mark.* **2021**, *13*, 61–74. [CrossRef]

123. Ebrahimi, F.; Saen, R.F.; Karimi, B. Assessing the sustainability of supply chains by dynamic network data envelopment analysis: A SCOR-based framework. *Environ. Sci. Pollut. Res.* **2021**, *28*, 64039–64067. [CrossRef] [PubMed]

124. Fahimnia, B.; Tang, C.S.; Davarzani, H.; Sarkis, J. Quantitative models for managing supply chain risks: A review. *Eur. J. Oper. Res.* **2015**, *247*, 1–15. [CrossRef]

125. Bai, C.; Sarkis, J.; Wei, X.; Koh, L. Evaluating ecological sustainable performance measures for supply chain management. *Supply Chain. Manag. Int. J.* **2012**, *17*, 78–92. [CrossRef]

126. Chaabane, A.; Ramudhin, A.; Paquet, M. Designing supply chains with sustainability considerations. *Prod. Plan. Control* **2011**, *22*, 727–741. [CrossRef]

127. Hall, J.; Matos, S.; Silvestre, B. Understanding why firms should invest in sustainable supply chains: A complexity approach. *Int. J. Prod. Res.* **2012**, *50*, 1332–1348. [CrossRef]

128. Ramudhin, A.; Chaabane, A.; Paquet, M. Carbon market sensitive sustainable supply chain network design. *Int. J. Manag. Sci. Eng. Manag.* **2010**, *5*, 30–38. [CrossRef]

129. Hutchins, M.J.; Sutherland, J.W. An exploration of measures of social sustainability and their application to supply chain decisions. *J. Clean. Prod.* **2008**, *16*, 1688–1698. [CrossRef]

130. Prashar, A.; Sunder, M.V. A bibliometric and content analysis of sustainable development in small and medium-sized enterprises. *J. Clean. Prod.* **2020**, *245*, 118665. [CrossRef]

131. Büyüközkan, G.; Göçer, F. Digital Supply Chain: Literature review and a proposed framework for future research. *Comput. Ind.* **2018**, *97*, 157–177. [CrossRef]

132. Isaksson, R.B.; Garvare, R.; Johnson, M. The crippled bottom line–measuring and managing Sustainability. *Int. J. Product. Perform. Manag.* **2015**, *64*, 334–355. [CrossRef]

133. Prajapati, D.; Chan, F.T.S.; Chelladurai, H.; Lakshay, L.; Pratap, S. An Internet of Things Embedded Sustainable Supply Chain Management of B2B E-Commerce. *Sustainability* **2022**, *14*, 5066. [CrossRef]

134. Qorri, A.; Gashi, S.; Kraslawski, A. Performance outcomes of supply chain practices for sustainable development: A meta-analysis of moderators. *Sustain. Dev.* **2021**, *29*, 194–216. [CrossRef]

135. Huang, Y.-C.; Huang, C.-H. Examining the antecedents and consequences of sustainable green supply chain management from the perspective of ecological modernization: Evidence from Taiwan's high-tech sector. *J. Environ. Plan. Manag.* **2021**, *65*, 1579–1610. [CrossRef]

136. Lee, S.-Y. Sustainable Supply Chain Management, Digital-Based Supply Chain Integration, and Firm Performance: A Cross-Country Empirical Comparison between South Korea and Vietnam. *Sustainability* **2021**, *13*, 7315. [CrossRef]

137. Queiroz, M.M.; Pereira, S.C.F.; Telles, R.; Machado, M.C. Industry 4.0 and digital supply chain capabilities. *Benchmarking Int. J.* **2021**, *28*, 1761–1782. [CrossRef]

138. Beske-Janssen, P.; Johnson, M.P.; Schaltegger, S. 20 years of performance measurement in sustainable supply chain management-what has been achieved? *Supply Chain Manag. Int. J.* **2015**, *20*, 664–680. [CrossRef]

139. Afful-Dadzie, A.; Afful-Dadzie, E.; Turkson, C. A TOPSIS extension framework for re-conceptualizing sustainability measurement. *Kybernetes* **2016**, *45*, 70–86. [CrossRef]

140. Hák, T.; Janoušková, S.; Moldan, B. Sustainable Development Goals: A need for relevant indicators. *Ecol. Indic.* **2016**, *60*, 565–573. [CrossRef]
141. Emodi, N.V.; Boo, K.-J. Sustainable energy development in Nigeria: Current status and policy options. *Renew. Sustain. Energy Rev.* **2015**, *51*, 356–381. [CrossRef]
142. Gaigalis, V.; Markevicius, A.; Skema, R.; Savickas, J. Sustainable energy strategy of Lithuanian Ignalina Nuclear Power Plant region for 2012–2035 as a chance for regional development. *Renew. Sustain. Energy Rev.* **2015**, *51*, 1680–1696. [CrossRef]
143. García-Álvarez, M.T.; Moreno, B.; Soares, I. Analyzing the sustainable energy development in the EU-15 by an aggregated synthetic index. *Ecol. Indic.* **2016**, *60*, 996–1007. [CrossRef]
144. Salvi, B.; Subramanian, K. Sustainable development of road transportation sector using hydrogen energy system. *Renew. Sustain. Energy Rev.* **2015**, *51*, 1132–1155. [CrossRef]
145. Zhang, X.; Yu, Y.; Zhang, N. Sustainable supply chain management under big data: A bibliometric analysis. *J. Enterp. Inf. Manag.* **2020**, *34*, 427–445. [CrossRef]
146. Delai, I.; Takahashi, S. Sustainability measurement system: A reference model proposal. *Soc. Responsib. J.* **2011**, *7*, 438–471. [CrossRef]
147. Mukherjee, K. House of Sustainability (HOS): An Innovative Approach to Achieve Sustainability in the Indian Coal Sector. In *Handbook of Corporate Sustainability*; Edward Elgar Publishing: Cheltenham, UK, 2011. [CrossRef]
148. Okongwu, U.; Morimoto, R.; Lauras, M. The maturity of supply chain sustainability disclosure from a continuous improvement perspective. *Int. J. Prod. Perform. Manag.* **2013**, *62*, 827–855. [CrossRef]
149. Ahi, P.; Searcy, C. Assessing sustainability in the supply chain: A triple bottom line approach. *Appl. Math. Model.* **2015**, *39*, 2882–2896. [CrossRef]
150. Artsiomchyk, Y.; Zhivitskaya, H. Designing Sustainable Supply Chain under Innovation Influence. *IFAC-PapersOnLine* **2015**, *48*, 1695–1699. [CrossRef]
151. Vasileiou, K.; Morris, J. The sustainability of the supply chain for fresh potatoes in Britain. *Supply Chain Manag. Int. J.* **2006**, *11*, 317–327. [CrossRef]
152. Wang, Z.; Sarkis, J. Investigating the relationship of sustainable supply chain management with corporate financial performance. *Int. J. Prod. Perform. Manag.* **2013**, *62*, 871–888. [CrossRef]

 *sustainability*

Article

Carving out a Niche in the Sustainability Confluence for Environmental Education Centers in Cyprus and Greece

Filippos Eliades [1], Maria K. Doula [1], Iliana Papamichael [1], Ioannis Vardopoulos [2], Irene Voukkali [1] and Antonis A. Zorpas [1,*]

[1] Laboratory of Chemical Engineering and Engineering Sustainability, Faculty of Pure and Applied Sciences, Open University of Cyprus (OUC), Giannou Kranidioti 33, Latsia, Nicosia 2220, Cyprus; filippos.eliades@gmail.com (F.E.); mdoula@otenet.gr (M.K.D.); iliana.papamichael@gmail.com (I.P.); voukkei@yahoo.gr (I.V.)

[2] Department of Economics and Sustainable Development, School of Environment, Geography and Applied Economics, Harokopio University (HUA), 70 Eleftheriou Venizelou Avenue, 17676 Kallithea, Attica, Greece; ivardopoulos@post.com

[*] Correspondence: antoniszorpas@yahoo.com or antonis.zorpas@ouc.ac.cy

Abstract: Given the environmental issues that today's societies confront, such as climate change, waste management, ecosystem deterioration, etc., environmental education is becoming increasingly important. Adoption of environmental education as an integral part of the educational system is required for the Environmental Education Center (EEC) to be able to provide knowledge, skills, and values so that society can become active and environmentally responsible through awareness-raising. According to the scholarly published research, EECs can positively affect local communities and create an environmentally friendly culture. In addition, given that EECs can even play a significant part in the development of lifelong learning activities at the education and sustainable development nexus, it is considered critical to establishing future potentials and dynamics. Thus, aiming to analyze EECs' strengths, weaknesses, opportunities, and threats (SWOT analysis) within our complex and ever-evolving world, educators, students, and other fellow citizens in Cyprus and Greece participated in a survey in which they were asked to fill in a questionnaire, specifically developed for each group category. The findings of this study provide a deeper understanding of the implications arising as a result of effective environmental education absence, as well as the importance of a holistic approach through EECs. Moreover, it offers the research community a solid framework for future innovation in citizen engagement and training.

Keywords: sustainability; environmental awareness; behavior change; SWOT analysis; qualitative analysis; education for sustainable development; green movement; environmentalism; environmental science; environmental education

Citation: Eliades, F.; Doula, M.K.; Papamichael, I.; Vardopoulos, I.; Voukkali, I.; Zorpas, A.A. Carving out a Niche in the Sustainability Confluence for Environmental Education Centers in Cyprus and Greece. *Sustainability* **2022**, *14*, 8368. https://doi.org/10.3390/su14148368

Academic Editors: Idiano D'Adamo and Massimo Gastaldi

Received: 3 June 2022
Accepted: 4 July 2022
Published: 8 July 2022

Publisher's Note: MDPI stays neutral with regard to jurisdictional claims in published maps and institutional affiliations.

1. Introduction

The personal relationship of individuals with the environment in the present time proves to be the beginning of all the problems that govern it [1–3]. The increasingly rapid growth rates and the lack of rational management of environmental issues have the effect of continually degrading and destroying natural resources, resulting in the degradation and deterioration of ecosystems [4,5].

According to Global Environment Outlook 5 (2012) [6], the five environmental issues that require immediate management in Europe are air quality, biodiversity, chemicals and waste, climate change and drinking water [7,8]. An important role in tackling these problems on a local scale, which is the beginning of the widening of initiatives and actions at a broader level (regional, national, transnational), is held by local communities [9]. The first and most important stage in the development of active and sensitized local societies is their training as well as their practical application to local, initial environmental issues, more generally or more broadly [10].

Sustainable development seeks to secure a new policy where economic development interacts in harmony with environmental protection and social well-being [11,12]. With "lighthouse" in the triptych Economy, Society and Environment, Sustainable Development is defined as "development that satisfies the present, without affecting the potential of future generations to meet its needs" [13,14]. Such an approach effectively limits the uncontrolled economic development that has the sole purpose of economic prosperity, without taking into account the environmental parameter and corresponding limits to environmental protection that only concentrates on the environment [15].

In this framework, it is considered that the education of citizens regarding sustainability constitutes one of the most powerful drivers to combat the areas of concern and a critical intervention from local authorities to establish the potential impact on climate change mitigation [16,17]. An important pillar of boosting awareness among citizens includes prevention activities. According to the Waste Framework Directive [18], waste prevention measures need to be taken before the categorization of material or discarded products (i.e., chemicals, materials, substances). On the contrary, prevention activities should aim at the reduction of the quantity of waste beforehand. Therefore, waste prevention activities should not be based on general observations and implementations that might confuse the learning process of the citizens but should focus on individual waste streams, behaviors, local culture and waste characteristics (i.e., composition, production, recycling, and recovery, etc.) [17,19]. Awareness activities relating to sustainable development and waste reduction (i.e., recycle, reuse, refurbish and recovery) are key components to the transition towards sustainable development [20–22].

According to Pappas et al. (2022) [23], both industrial and domestic alterations are able to reduce greenhouse gas emissions and aid the development of strategies toward sustainability. An important component of the implementation of such activities is the acceptance of them on behalf of the public through a willingness to contribute to such a concept and not only the enforcement of policies towards the final goal of climate mitigation. Understanding the problems and opportunities to the transition towards sustainability from all sides comes only through education of the citizens and the integration of individual responsibility and thoughtful actions before an item becomes waste [23]. Areas of investigation include the environment (i.e., limited resources, health implications due to climate change and waste accumulation), society (i.e., social behavior and acceptance towards sustainability, job openings due to new sustainable development strategies) and economy (i.e., waste disposal fees, increase in prices due to limited resources).

Because of the importance of social interaction and innovation to address sustainability challenges, environmental and sustainable education has been deemed of utmost importance in higher education institutions. The main objectives of environmental education include not merely increasing eco-awareness among citizens but also creating sensitivity around environmental topics. It aims at improving critical thinking and interactions between the public and the environment, but also, it assists individuals in more research regarding environmental issues [17,24]. Education regarding sustainable development empowers learners to make informed critical decisions while taking both individual and collective actions to change society regarding the environment. Environmental education is a lifelong process not bound by age which enhances the cognitive, socio-emotional, socio-economic and socio-environmental progression of learning. It responds to urgent and dramatic challenges faced by the globe due to continuous and collective human activity with catastrophic environmental, social and economic consequences which endanger the survival of our own species. It provided the learners with the necessary skills, values and knowledge to address global issues like climate change, resource usage and waste, loss of biodiversity, etc. It is necessary for the comprehension of consequences and opportunities of everyday actions on behalf of the citizens, while at the same time, can act as a pressure lever towards inert governmental bodies from the citizens and vice versa [25,26].

The integration of sustainable education in the curriculum aims to cultivate social innovators and contributors who are willingly participating in local and EU strategies and plans regarding sustainable development [27]. The development of effective environmental education programs is a tool for the enhancement of positive environmental attitudes and values, boosts knowledge and skills for individuals and communities and strengthens the collaboration of the public, authorities, stakeholders, decision-makers, scientists and academia for a positive environmental action [28]. At the same time, according to Ardoin and Bowers (2020) [29], environmental education in the early childhood years gives even better results than in higher years and has gained increasing momentum. This is due to the combination of persistent environmental challenges of the world along with existing interests exhibited by children. In their systematic review, Ardoin and Bowers (2020) [29] investigated empirical studies of environmental education programs focusing on a 25-year span. Results from 66 investigated studies indicated that participants of earlier ages (0–8 years old) revealed the development of environmental literacy, cognitive development as well as emotional and social development. However, according to Moustairas et al. (2022) [30], obstacles to public acceptance of the development of education centers are affected by many variables and mostly demographics, environmental consciousness, recycling behavior and economic incentives.

Emerging from this growing interest in environmental education, environmental education centers (EEC) were established. These institutions are highly connected to environmental international and local movements and are considered one of the most important factors of extra-curriculum education [31,32]. They are encouraged to cooperate with citizens and prolong the learning experience to replace traditional learning [33]. They have substantial power to influence local communities, students and, by extension, policymakers in the transition towards sustainability [34]. EECs of all countries take the largest share in the realization of the vision of Environmental Education/Education for Sustainable Development [35,36]. EECs are standard learning organizations and constitute a network of decentralized public educational structures for environmental education and sustainability and support at the local, national and international levels. EECs serve the following goals [34–36]:

1. The development and implementation of educational programs for students of all levels of education,
2. The development of training programs and the provision of training to educators, either by living or distance, or by combining them,
3. The production and distribution of educational material in schools and the community in paper and/or digital form,
4. The development of national and regional thematic and/or methodological networks,
5. Cooperation with schools and general structures and education staff with the community and public and private bodies,
6. Cooperation with Universities and Technological Educational Institutes for relevant research and educational activities.

This research aims to analyze and evaluate the contribution of EECs to environmental education in Cyprus and Greece, as well as new trends in education, the adoption of which by EECs will further strengthen their role.

2. Literature Review

This study explores EECs whose primary purpose is to deliver environmental education in order to promote environmental awareness. This criterion is satisfied by a group of institutions that differ in terms of organizational structure and form, teaching goals, objective(s) and means, as well as content and accessibility. However, the scope of this analysis exempts academic entities with a primary commitment to research.

Continuous and lifelong learning constitute the main characteristics of sustainable development education. Therefore, activities of EECs are to be developed in a holistic and expansive manner to include all three pillars of sustainability (environmental, social, and economic). To benefit from such establishments, social learning should be a key ingredient where people from different backgrounds and experiences, values and learning perspectives come together and commit to problem-solving of not yet found sonot-yet for sustainability. The purpose of EECs is also the motivation of civilians to not only participate but to innovate regarding pre-existing perspectives and designs and come up with new ideas derived from everyday life [37].

The United States, Canada, Australia, and several European Union member states have established both publicly as well as privately funded EECs. Despite the considerable differences, similar trends and patterns could well be identified around the planning and development of EECs. Several EECs are working to promote environmental education and awareness. The types of these EECs vary according to their origin, environment and funding as well as continued development over the years. They include school farms, eco-museums connected to natural spaces, interpretation and natural centers [38,39].

For instance, support for environmental education is state-dependent within the systems of government in the United States, Canada, and Australia, resulting in various degrees of EEC growth [40–43]. In Europe, and in particular Italy, Slovenia, Spain, Cyprus, and Greece, governing systems benefit from the EU's supporting policies and, consequently, monetary resources in order to develop EECs [44]. Conversely, when central government financing ran out in Great Britain, the entire EEC network established to that point was privatized [45]. Thus, it is difficult to avoid the view that the passive support of higher administrative political systems not only supports the EECs development but also reinforces those establishments' resilience and sustainability.

International research on EECs mainly adopts a qualitative perspective and has only marginally addressed the perceptions of EECs' functions and contributions. For the needs of the present research, a review of certain previous studies on the perceptions of the direct and indirect recipients of environmental education, as well as the direct and indirect outcomes makes much sense.

Simmons (1991) [46] surveyed EEC workers, finding substantial support for goals and objective(s) set by environmental education institutions related to behavioral change, owning to heightened environmental awareness, although with doubts about localized environmental issues with political overtones. Hart (1996) [47] observed a widespread absence of empirical studies monitoring teachers' perceptions attitudes, pedagogical tools and methods and levels of engagement in environmental education. Nevertheless, according to Mavrikaki et al. (2004) [48], environmental education is a valuable educational procedure that, however, engages just a minority of instructors due to its elective nature. Flogaitis, Daskolia and Liarakou (2005) [49] used a survey to assess the environmental education approaches, motives and counter-motives linked to environmental education. Based on the empirical findings, the researchers argue that praxis-focused training sessions might help environmental education diffusion. Following that, Erickson and Erickson (2006) [50] explored the aspects related to the EECs' performance from the management point of view, concluding that strong personnel (i.e., educators) is the most important aspect. Around the same period, Ballantyne and Packer (2006) [43] conducted in-depth interviews with EECs' senior management in order to understand and assess the various ways in which such centers partnered with or worked within compulsory education establishments. Their findings support that the cooperation mechanics developed establish fully-fledged relations in a way that environmental education consistency, advancement and continuance are ensured, even after the completion of the individual environmental education programs.

Ernst (2009) [51] rendered the fine nuances between environment-based education and general environmental education and is currently investigating the qualitative and quantitative aspects of environmental education activity in the United States, supporting that lack of funds was the primary constraint on environmental education. Conversely, in 2012, within the context of a larger research project on environmental and outdoor/nature-based education, a study attempted—to the best of the authors' knowledge—to collect and analyze educators' perspectives for the first time. Despite the fact that the majority of the survey respondents indicated that outdoor/nature-based training was not a part of their initial education syllabus, their replies suggest that outdoor/nature-based education plays a significant part in the overall learning outcomes [52].

A recent (2018) study, upon comparing EECs, supports that such centers primarily focus on providing a supporting mechanism for other organizations, the formation of networks, and developing initiatives to address local environmental demands, rather than contesting a strategic implementation role directly conducting environmental education [53]. In addition, it is stressed that EECs should invest in developing national, regional and local environmental education policies in close cooperation with the competent bodies, as well as seek stronger and more direct participation in achieving sustainable development goals.

Building on previous learnings, Moustairas et al. (2022) [30] attempted to determine the elements influencing public support for establishing EECs. According to their findings, the formation of an EEC is influenced by demographics (admittedly country/region and culture specific), environmental awareness, recycling behavior, and economic incentives. This conclusion reflects on the public's feeling of individual efficacy and sense of responsibility, supporting that environmentally friendly activities do not solely fall within a bourgeoisie of environmental gurus, thus arguing that those principles have a limited role in establishing a society's "eco-consciousness." Furthermore, the authors, through their statistical analysis, indicate that the acceptability rates of establishing an EEC will rise if further importance is placed on local sustainable development, effectively complementing the Agenda 2030 sustainable development goals as well as the circular economy model.

While EECs are encouraged to engage in lifelong and conventional training, they are also encouraged to join forces with regional and local authorities and the forces of production, as well as civil society in making decisions directly or indirectly relating to their quality of living, in order to become a channel in achieving sustainable development [54]. An empirical study was conducted in 2013 attempting to investigate the role of EECs in remote regions [37]. The research suggests that the activity of EECs in remote areas positively affects the local community, both in terms of direct benefits for consumers, businesses, and the economy as well as indirect benefits, such as creating desirable places to live with a strong identity and pride in the community, and—perhaps most importantly— that young learners are becoming environmentally aware and can develop into prospective enforcers of sustainability.

However, when the scope is narrowed from the importance of the development of lifelong learning environmental education activities to the benchmarking of the defining potential and future dynamics, pertinent research becomes scarce, thus leaving room for the current research to fill this gap, attempting to assess the various arguments in favor of and against EECs' course of action within a constantly changing world and under the pressure of environmental degradation.

3. Materials and Methods

To assess the role and contribution of EECs as well as their greater influence on society, this study focuses on the following research questions [55,56]; (a) What is the degree of influence of the EECs in shaping environmental consciousness? (b) What are EECs' contributions to the dissemination of sustainability principles? (c) What are the learners' views on environmental issues and to what extent are they influenced by EECs actions? (d) What are the limiting factors, the necessary structures and processes that will contribute to the modernization of the EECs in order to gain momentum and increase their impact?

In order to identify issues encountered by EECs in their operations and attainment of their goals, as well as contemporary developments in environmental education, this study attempts to comparatively determine those aspects between Greece and Cyprus that are considered an advantage and which are not. To do so, the current research adopts data from the internationally published scholarly literature on the role and the actions of EECs in Greece and Cyprus. In addition, the Strategic Planning of the European Union, which contains policies, directives, and regulations, as well as information from the Pedagogical Institutes of the two countries, were used.

A strategic research approach was developed, and the means to collect the necessary information were defined as follows:

1. Target Audience: For the scope of the current research, the audience that was targeted was EECs educators, students of the Lower and Upper Secondary School age (i.e., 12–18 years old), as well as their parents (including custodians).

2. Collect Responses: The target audience was invited to participate in a survey. The survey included closed-ended questions that will be addressed in the form of questionnaires in each target group. The survey was designed and conducted in accordance with the Greek and Cypriot legislation and with institutional requirements. All participants gave their informed consent to participate prior to the start of the survey. They were informed about the aim of the study and that they could terminate their participation at any time without negative consequences. The survey was anonymous, and the participants gave consent to use their answers for research. The participation of both countries is presented in Table 1. The survey was conducted during the spring and summer 2020 terms, namely from mid-January to late July.

3. Data Processing: Survey data processing, including sorting and cleaning, and converting data into a usable format, towards transforming the raw data into structured information that can then be analyzed for insights.

4. Exploratory Analysis: This part concerns the collection, organization, analysis, interpretation and presentation of the data in relation to the research issues on which the research focuses.

5. SWOT Analysis: A framework for identifying Strengths, Weaknesses, Opportunities and Threats (herein after; SWOT Analysis) was conducted to identify the strengths and weaknesses of the contribution of EECs to environmental education, as well as the threats and opportunities for overcoming any weaknesses and addressing threats towards a more effective, creative, and innovative operation of these educational institutions. SWOT analysis is a strategic planning tool used that constitutes a cognitive process that studies the relations between external and internal factors and surroundings of a territory, business, product, or organization based on mixed subjective and objective evaluation [57,58]. The SWOT acronym is derived from the English words: Strengths, Weaknesses, Opportunities and Threats [59,60] and is a key to identifying strengths, eliminating weaknesses, as well as seizing opportunities and responding to threats. Strengths are positive aspects of the analysis, while weaknesses are negatives related to the system's internal aspects. On the other hand, opportunities and threats refer to the external aspects and refer to positive or negative interactions with the system respectively [61]. The internal evaluation of the environment under consideration aims to identify the strengths and weaknesses mainly of internal resources such as personnel, products and services [62], while also examining threats and risks due to the external competitive environment [63].

6. TOWS Matrix: SWOT analysis has been adopted by many organizations [64] and businesses, mainly because of its simplicity and the advantages it offers, such as addressing weaknesses that affect the achievement of goals, understanding the business or organization's philosophy, the prospect of exploiting the strengths to achieve the desired goal [65,66]. Coupling the S-W-O-T parameters by two, after identifying them and ranking them, yields four coupled parameters, each corresponding to a strategy type. The characterization and type of strategies resulting from this coupling, namely the TOWS matrix, where the strategic planning framework is specifically at the decision-making stage, are [67–70]:

   - Development Strategy (maxi–maxi), "S-O": This strategy focuses on exploiting the strong points to seize opportunities.
   - Corrective Strategy (mini–maxi), "W-O": This strategy seeks to minimize and improve weaknesses that hinder opportunities.

- Maximizing Strategy (maxi–mini), "S-T": This strategy focuses on exploiting the strong points to address and minimize threats to the environment.
- Defense Strategy (mini–mini), "W-T": The goal of the strategy is to minimize weaknesses in order to avoid threats.

Table 1. Involvement of social groups in research.

Country	Students	Parents	Educators
Cyprus	72	72	9
Greece	77	-	12

The right strategy choice is derived from the image described by the SWOT matrix and results in proportion to the overriding parameter. If the "S-O" Development Strategy prevails, then the organization has many opportunities as well as advantages and is recommended to adapt. A "W-O" Correction Strategy is appropriate when the organization's internal weaknesses prevent it from taking advantage of existing opportunities. Therefore, in this case, this strategy focuses on addressing and correcting the weaknesses as much as possible. If the "S-T" Improvement Strategy prevails, it means that although the organization faces many external threats, it has strong points and advantages in mitigating them to achieve its goals. The "W-T," Defense Strategy is chosen if an organization receives many external threats and its internal forces are insufficient to deal with them. The overall research roadmap for the current research is graphically presented in Figure 1.

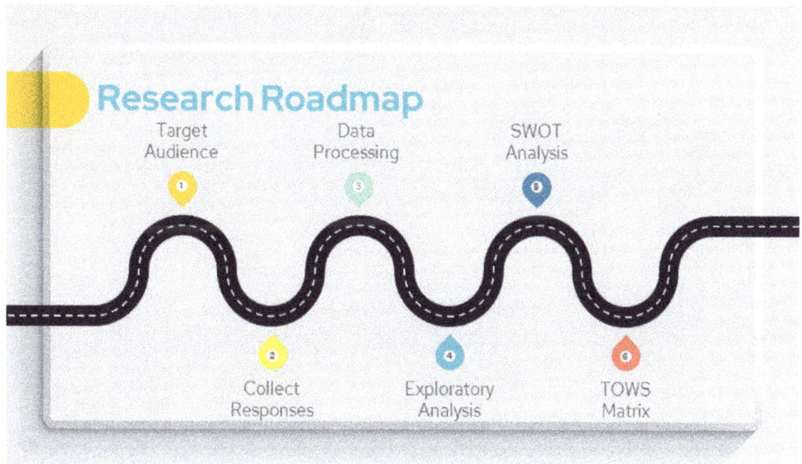

Figure 1. Graphical illustration of the research roadmap.

4. Results

The founding of the EECs emerged in the 1970s when the idea of Environmental Education was reinforced through the conferences of Tbilisi (1977) and Moscow (1987) [71]. The enlargement of Environmental Education was primarily aimed at implementing actions in direct contact with the environment, which would be ensured by the operation of the EECs. This enabled the educational community and students [43] to develop skills on the subject of environmental protection, supplying it with knowledge and techniques that theoretical education could not provide in the past. These newly acquired skills will include awareness and sensitivity, knowledge and understanding of key concepts (i.e., circular economy, sustainable development, etc.), cultivation of talented individuals regarding sectors of environmental protection (i.e., environmental engineering, governmental bodies, law, etc.), new attitudes towards sustainability, willingness to actively participate in national or EU legislation and action plans as well as the necessary critical thinking skills to identify challenges and opportunities for improving individual prevention activities of everyday life [72,73]. EECs evolved further in 1990, in the sense of "Sustainable Development" as a central term of Environmental Education, while the idea came to fruition in the mid-decade with the Rio Agenda 21 conference in 1992, stressing the urgent need to create EECs as facilitators in linking education and society [74]. Since then, EECs have become the connecting link between education and the social community, offering all social strata the opportunity to directly interact with environmental issues, educating and directing critical thinking, awareness and participation.

In Cyprus, the establishment of the EECs was launched in 2000, when the Ministry of Education of Cyprus developed the "National Strategic Plan for Environmental Education Focusing on Sustainable Development," which was completed in 2004. With the adoption of EU policies, the implementation of its planning integrates Environmental Education into the education system of Cyprus, aiming to inform students and the general public about environmental issues through the actions developed by Tosh and extra curriculum activities [75]. There are five EECs in Cyprus today, and two more centers are expected to open and operate. Each one of Cyprus' EECs supports specific programs that can be taught by students in all schools in the country but also functions as a network, thereby achieving constructive cooperation between them [76].

In Greece, EECs were established by law 1892/90. Following the implementation of this Law, the creation of EECs began, setting out the aims, functions and actions around the EU as well as the qualified staff for the Centers [77]. Sixty EECs operate in Greece today throughout its land and island area. The first EEC was established in 1993 in Achaea's Kleitoria [44] and others throughout the country. Moving on to the idea, the main directions of the Ministry of National Education and Religious Affairs were the rational distribution of the student population, the genuine interest in local communities and the utilization of existing infrastructure. Thus, eight more EECs were established in 1995, twelve in 1997, seventeen in 1998, thirty-one in 2003 and fourteen in 2006 [78]. The EECs are distributed so that each prefecture of Greece is supported by one EEC, and each district by nine. However, the student population is not generally evenly distributed, with the result that some EECs are underperforming.

4.1. Level of Training and Evaluation of EECs Educators

As regards the possession of postgraduate degrees by educators and as shown in Figures 2 and 3 for the question "What is the specialization of your postgraduate and doctoral studies, if any?" an advantage is the possession of postgraduate studies in the subject. Namely, 67% of Cypriots declared that they hold a postgraduate degree in Environmental Education and Sustainable Development and 11% in Ecology, Management and Protection of the Natural Environment, Natural Sciences Teaching and Environmental Education and Biotechnology, Ecology and Biodiversity. On the contrary, in the case of Greece, 76% of the respondents did not hold a postgraduate degree, which is a weakness of the EECs with regard to the existing training and specialization of their educators, while 8% of the educators hold the title of Educational Planning, Investment Finance in REN (Renewable Energy) and Models of Design and Development of Educational Units, Physical Education and Quality of Life. A big difference is presented in the two cases as Cyprus has a clear advantage over Greece, as shown by the results.

To the question "How do you assess the level of training of your educators?" in the case of Cyprus, 49% of students consider their level very high, 33% high, 17% satisfactory and 1% moderate. In the case of Greece, only 18% consider that the level is very high, 43% high, 34% satisfactory and then 3% moderate and low (Figure 4). Although objectively, students are not able to evaluate the training of their instructors, it is a very serious threat, especially for Greece, the non-acceptance and assessment of the level of training of the educators by the students. In the case of Cyprus, the training of educators is better, according to the viewpoint of the student learners; however, improvements are needed. This indicates significant internal weakness of structures to highlight the skills of educators or to select managers with knowledge transfer skills to gain the acceptance of learners, a parameter necessary to accept the knowledge being passed.

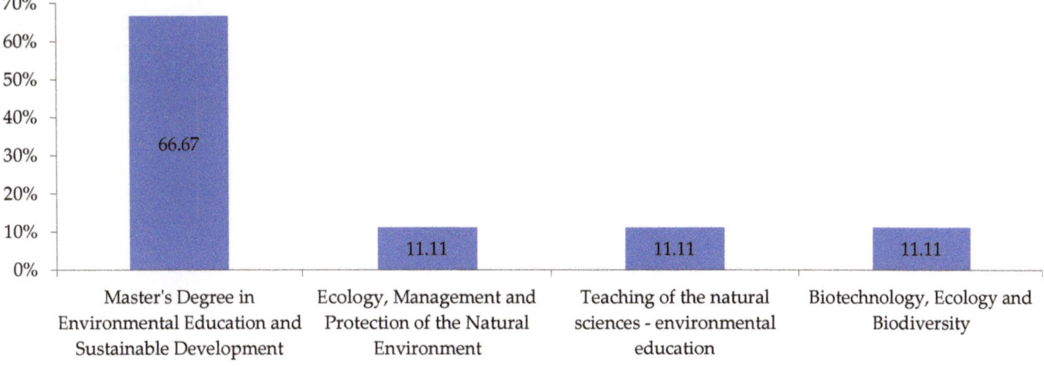

Figure 2. Level and type EECs educators in Cyprus.

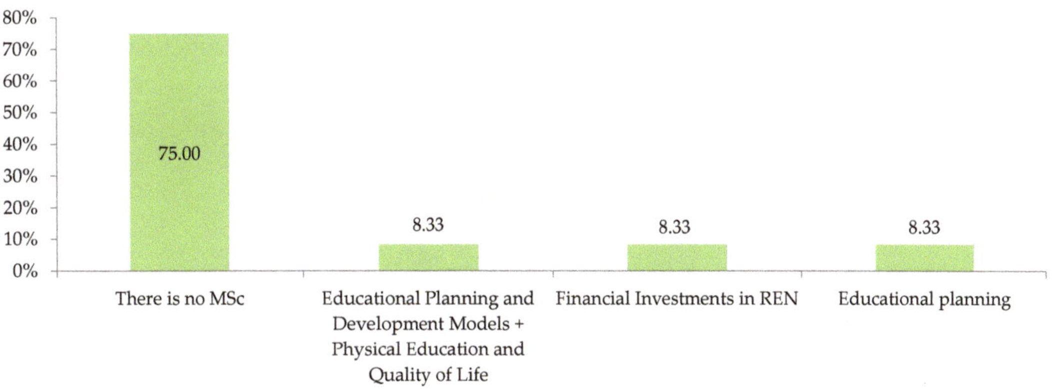

Figure 3. Level and type EECs educators in Greece.

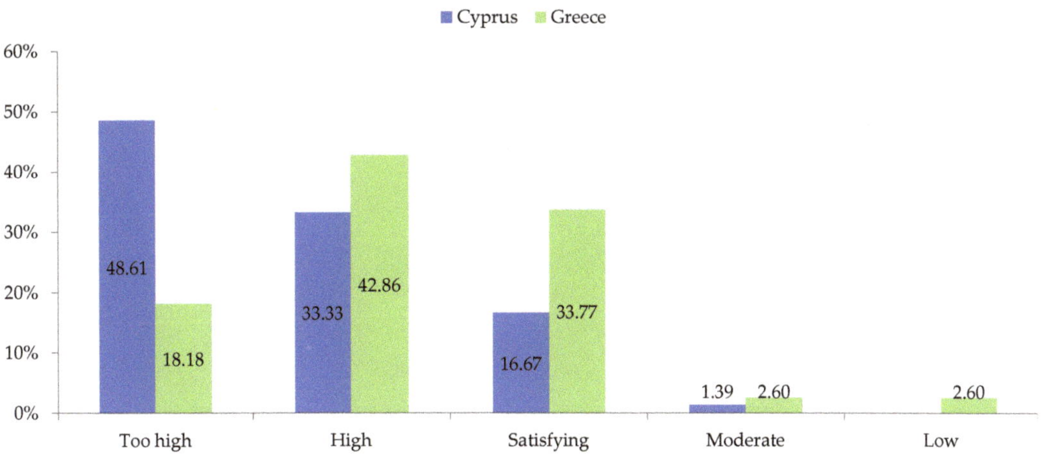

Figure 4. Evaluation of EECs educators by students in Cyprus and Greece.

4.2. Knowledge of the Differences between Different Forms of Environmental Education

An overwhelming majority of 89% of educators in Cyprus (Figure 5), answered the question "Do you know the characteristics that differentiate Environmental Education from Education for Sustainable Development?" stating that they know the differences between the two terms, shaping the answer to a free text in the questionnaire, indicative of the level of training. In the case of Greece, half of the responders (Figure 5) responded that they are not aware of the characteristics that differentiate environmental education from education to sustainable development, which demonstrates the need for training and is currently assessed as a weakness.

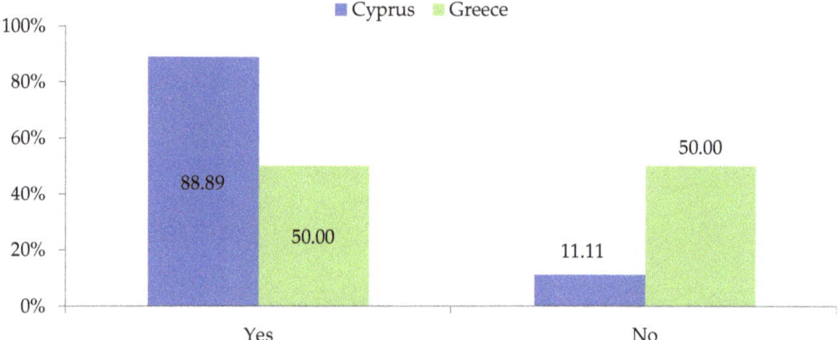

Figure 5. Differences between Environmental Education and Education for Sustainable Development in Cyprus and Greece.

4.3. Feedback in EECs by Students

For both countries, there is a lack of feedback from trainees or school units about their satisfaction after the end of their visit to the EEC. Relatively to Cyprus, 67% say it is sometimes, 22% is always done, while the remaining 11% is rare. A total of 33% (Figure 6) of educators say they receive feedback sometimes or rarely and, perhaps for this reason, a large percentage of them are satisfied with the degree of influence of EECs but feel that improvements are needed. In fact, the degree of influence may be partly exacerbated by the degree of learners' satisfaction, which many times educators may not know.

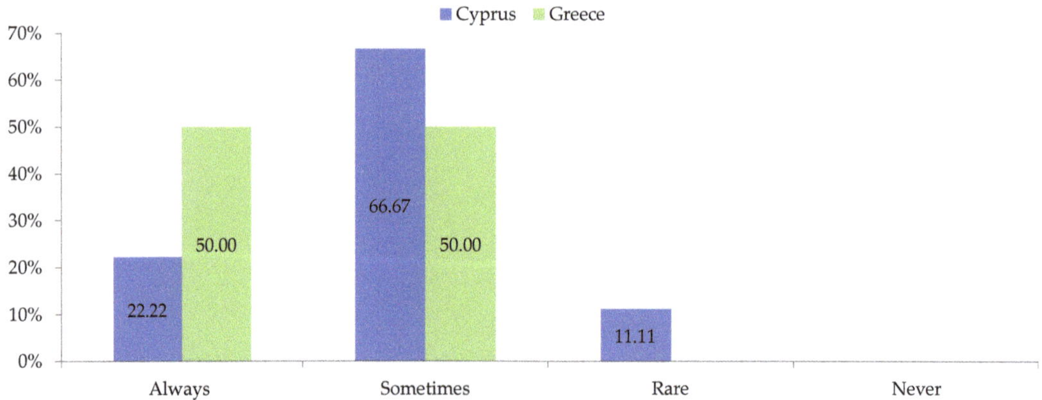

Figure 6. The feedback received in Cyprus and Greece.

For Greece, half of educators said they always receive feedback (Figure 6) on students' satisfaction while half stated that they do not receive any.

The lack of feedback deprives the EEC of both countries of knowing their outward image, which would help them to understand the degree of influence they exert. This may be the result of possible incomplete cooperation between EECs and school units or the absence of the implementation of an effective operational protocol.

4.4. Satisfaction of Trainees with the Practical Part of Education

The students who participated in activities with a practical part also replied to the question "If your training had a practical part, are you satisfied with your degree of participation in it?." According to Figure 7, 74% of students responded in Cyprus that they are very satisfied with their participation in the practical part, while 23% answered moderately and 3% little. In Greece, satisfaction for the practical part is expressed by the students at 86%, while 11% express moderate satisfaction. Greece's dissatisfaction corresponds to 3% of the respondents. Therefore, in the case of student satisfaction for practical actions, 26% of the non-positive answers should be recorded as a weakness in the case of Cyprus, it needs attention and especially redesign of the actions so that the EECs can gain all the students.

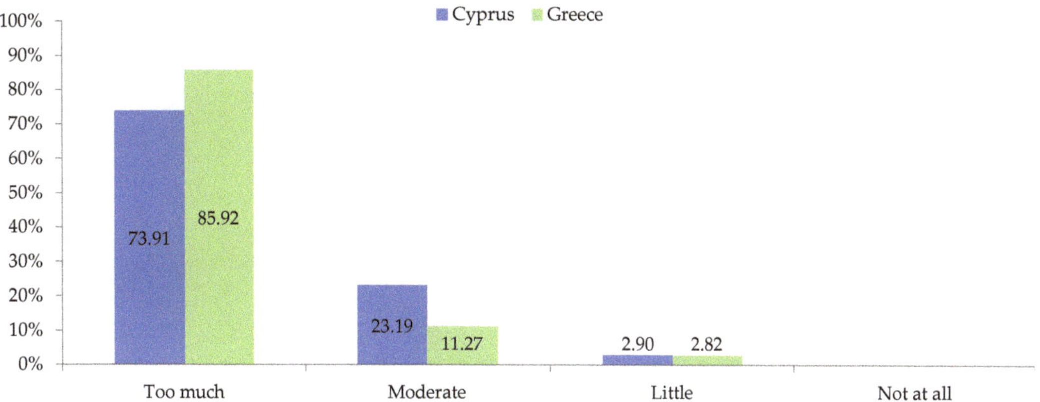

Figure 7. Satisfaction of trainees with the practical part of education in Cyprus and Greece.

There is a contradiction in Cyprus with the question "From the image you have acquired to this day, to what extent learners are satisfied in the case of practical training?" as 100% of educators (Figure 8) replied that the students are very satisfied, confirming the conclusion that led to the following questions that Cypriot educators seem to have no real picture of the degree of EEC influence on students, which is a serious threat in this case. On the contrary, in the case of Greece, educators seem to have become more aware and understand their influence on students and therefore also an opportunity that, if successful, will be able to redefine new strategies to approach the issues they are developing and approaching the target group.

Cypriot educators appear not to have a real situation of the degree of influence of EECs on students, which is a major threat in this case. On the contrary, in the case of Greece, educators seem to have become more aware and understanding of their influence on students, and therefore this is also an opportunity that they will be able to redefine if they use it, new strategies for approaching the issues they are developing and targeting target groups.

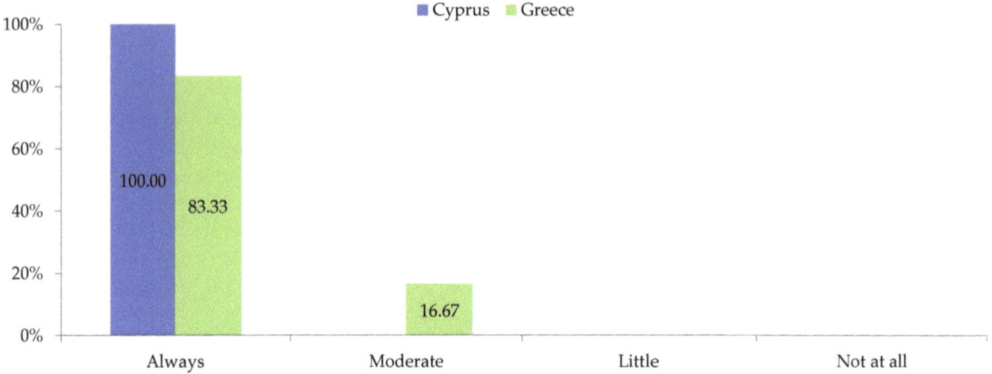

Figure 8. Satisfaction of trainees with the practical part of education according to educators in Cyprus and Greece.

4.5. General SWOT Results

Proceeding to the SWOT matrix, the results are derived, by conjugating Opportunities with Strengths (S–O), Opportunities with Weaknesses (O–W), Threats with Strengths (T–S) and Threats with Weaknesses (T–W). The results in Cyprus are 30 cases/opportunities, 17 threats/strengths, 21 opportunities/weaknesses and 12 threats/weaknesses. The case of Greece, there are 13 cases/opportunities, 11 threats/strengths, 10 opportunities/weaknesses and 32 threats/weaknesses.

4.6. SWOT Matrix Findings Analysis

At the end of the analysis, consequently, with the extraction of results from the SWOT matrix, the division of factors into quadrants resulted in the case of Cyprus having the "O–S" Development Strategy while in the case of Greece, the dominant factor distribution is that of "T–W", Shrink and Defense.

It is clear that in the case of Cyprus, the combination of high teacher education with the interest shown by pupils and parents is the key to a sound Development Strategy, where by adopting innovative teaching methods, they will be able to raise the awareness of the public [50,79] and reverse any negative situation. The willingness of Cypriot teachers to collaborate with both research and academic institutions and similar infrastructures can offer a variety of contemporary environmental topics, providing new knowledge to the public and exerting a positive influence on them. Equally important in raising the awareness of trainees are the appropriate staff and infrastructure available in the Cypriot EECs, enhancing their role and strength. Finally, funding through European programs and/or by the state will add additional momentum and weight to the work that Cypriot EECs have to do with the adoption of Strategic Development (Figure 9).

In the case of Greece, the adoption of Strategic Defense seems to be the safest solution, as due to the economic crisis, the EECs could not remain unaffected, resulting in a lack of infrastructure and basic tools to complete their environmental education actions. Despite the public awareness and cooperation with academic institutions, non-adoption of the Strategic Planning proposed by the European Union has had a negative impact on the proper functioning of EECs, which has weakened their contribution to the public and thus it also diminished public interest, jeopardizing its viability. It is easily understood that the adoption of a Strategic Defense will "freeze" the adverse situation [80] in which the EECs are located and will have the time needed to adopt radical changes that could reverse the negative condition (Figure 9).

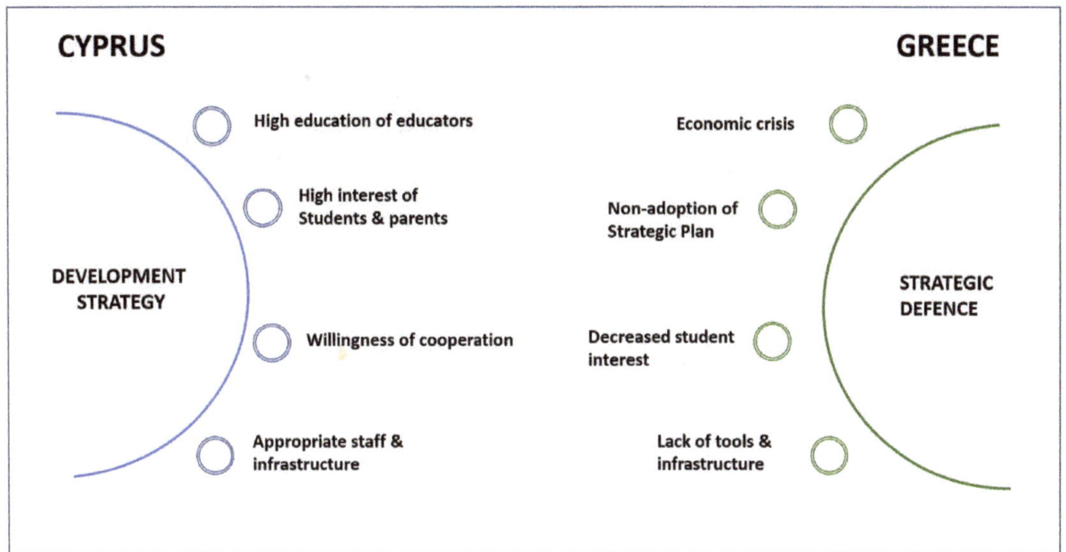

Figure 9. Outline of recommendations regarding the strategy to be followed in Cyprus (Development Strategy) and Greece (Strategic defense) and key aspects of them.

5. Conclusions

In the case of Cyprus, the recommended strategy is a Development Strategy because of the combination of high education of educators and the interest shown by students and parents. At the same time, the willingness of Cypriot educators to cooperate with both research and university institutions and similar infrastructures can offer a variety of modern environmental issues, providing new knowledge to the public and exerting a positive influence on them. An equally important parameter for educating learners is the appropriate staff and infrastructure in Cypriot EECs, enhancing their role and strength.

Regarding Greece, the adoption of Strategic Defense seems to be the safest solution because of the economic crisis as EECs could not remain unaffected, resulting in the lack of infrastructure and key tools for completing their educational environmental actions. Furthermore, despite public awareness and cooperation with university institutions, non-adoption of the Strategic Plan proposed by the European Union has had a negative impact on the proper functioning of the EECs and therefore the impact of EECs is decreasing and the students' interest also.

It is easy to see that the adoption of a Strategic Defense will "freeze" the situation in which the EECs are located and will have the required time to adopt radical changes that could overturn the negative situation. In any case, the establishment of EECs in Greece and Cyprus as well as any part of the world requires further action on behalf of different districts which wish to start up such centers. Environmental awareness and education in the form of EECs can provide the setting stone for the transition to a more sustainable society, if the EECs are established and modified wisely, according to the area's profile. It is strongly suggested that further research regarding specific areas for establishing EECs in Greece and Cyprus are investigated, for a prosperous and worthwhile goal toward climate mitigation and sustainability.

Author Contributions: Conceptualization, M.K.D. and A.A.Z.; methodology, F.E., M.K.D., I.V. (Ioannis Vardopoulos), I.V. (Irene Voukkali), I.P. and A.A.Z.; validation, F.E., M.K.D., I.P. and A.A.Z.; resources, M.K.D. and A.A.Z.; investigation, F.E., M.K.D., I.V. (Ioannis Vardopoulos), I.V. (Irene Voukkali), I.P. and A.A.Z.; data curation, F.E., M.K.D., I.V. (Ioannis Vardopoulos), I.V. (Irene Voukkali), I.P. and A.A.Z.; writing—original draft preparation, F.E., M.K.D. and A.A.Z.; writing—review and editing, I.V. (Ioannis Vardopoulos), I.V. (Irene Voukkali), I.P. and A.A.Z.; supervision, M.K.D. and A.A.Z.; project administration, M.K.D. and A.A.Z.; funding acquisition, A.A.Z. All authors have read and agreed to the published version of the manuscript.

Funding: This research received no external funding.

Institutional Review Board Statement: Ethical review and approval were waived for this study. The survey was anonymous, gathering no sensitive data, and the participants gave consent to use their answers for research.

Informed Consent Statement: Informed consent was obtained from all subjects involved in the study.

Data Availability Statement: The data presented in this study are available upon substantiated request from the corresponding author.

Acknowledgments: The authors would like to acknowledge all, who have directly or indirectly helped in carrying out this study.

Conflicts of Interest: The authors declare no conflict of interest.

References

1. Loughland, T.; Reid, A.; Walker, K.; Petocz, P. Factors influencing young people's conceptions of environment. *Environ. Educ. Res.* **2003**, *9*, 3–19. [CrossRef]
2. D'Adamo, I.; Gastaldi, M.; Morone, P. Economic sustainable development goals: Assessments and perspectives in Europe. *J. Clean. Prod.* **2022**, *354*, 131730. [CrossRef]
3. Cotler, H.; Cuevas, M.L.; Landa, R.; Frausto, J.M. Environmental Governance in Urban Watersheds: The Role of Civil Society Organizations in Mexico. *Sustainability* **2022**, *14*, 988. [CrossRef]
4. Weart, S.R. The idea of anthropogenic global climate change in the 20th century. *WIREs Clim. Chang.* **2010**, *1*, 67–81. [CrossRef]
5. Colasante, A.; D'Adamo, I.; Morone, P.; Rosa, P. Assessing the circularity performance in a European cross-country comparison. *Environ. Impact Assess. Rev.* **2022**, *93*, 106730. [CrossRef]
6. UNEP. *Global Environment Outlook 5. Environment for the Future We Want*; Progress Press: Valletta, Malta, 2012.
7. Beg, N. Linkages between climate change and sustainable development. *Clim. Policy* **2002**, *2*, 129–144. [CrossRef]
8. D'Adamo, I.; Gastaldi, M.; Morone, P.; Rosa, P.; Sassanelli, C.; Settembre-Blundo, D.; Shen, Y. Bioeconomy of Sustainability: Drivers, Opportunities and Policy Implications. *Sustainability* **2022**, *14*, 200. [CrossRef]
9. Cifrić, I. Environmental literacy between cultural tradition and environmental routine. *Soc. Ekol.* **1996**, *5*, 403–421.
10. Sathiendrakumar, R. Greenhouse emission reduction and sustainable development. *Int. J. Soc. Econ.* **2003**, *30*, 1233–1248. [CrossRef]
11. Lélé, S.M. Sustainable development: A critical review. *World Dev.* **1991**, *19*, 607–621. [CrossRef]
12. D'Adamo, I.; Gastaldi, M.; Imbriani, C.; Morone, P. Assessing regional performance for the Sustainable Development Goals in Italy. *Sci. Rep.* **2021**, *11*, 24117. [CrossRef] [PubMed]
13. Lior, N. Sustainable energy development: The present (2009) situation and possible paths to the future. *Energy* **2010**, *35*, 3976–3994. [CrossRef]
14. Vardopoulos, I.; Stamopoulos, C.; Chatzithanasis, G.; Michalakelis, C.; Giannouli, P.; Pastrapa, E. Considering urban development paths and processes on account of adaptive reuse projects. *Buildings* **2020**, *10*, 73. [CrossRef]
15. Giddings, B.; Hopwood, B.; O'Brien, G. Environment, economy and society: Fitting them together into sustainable development. *Sustain. Dev.* **2002**, *10*, 187–196. [CrossRef]
16. Álvarez-Nieto, C.; Richardson, J.; Navarro-Perán, M.Á.; Tutticci, N.; Huss, N.; Elf, M.; Anåker, A.; Aronsson, J.; Baid, H.; López-Medina, I.M. Nursing students' attitudes towards climate change and sustainability: A cross-sectional multisite study. *Nurse Educ. Today* **2022**, *108*, 105185. [CrossRef] [PubMed]
17. Zorpas, A.A.; Voukkali, I.; Loizia, P. Effectiveness of waste prevention program in primary students' schools. *Environ. Sci. Pollut. Res.* **2017**, *24*, 14304–14311. [CrossRef] [PubMed]
18. European Commission. European Commission Directive 2008/98/EC of the European Parliament and of the Council of 19 November 2008 on waste and repealing certain directives. *Off. J. EU* **2008**, *34*, 99–126.
19. Zorpas, A.A.; Navarro-Pedreño, J.; Jeguirim, M.; Dimitriou, G.; Almendro Candel, M.B.; Argirusis, C.; Vardopoulos, I.; Loizia, P.; Chatziparaskeva, G.; Papamichael, I. Crisis in leadership vs. waste management. *Euro-Mediterranean J. Environ. Integr.* **2021**, *6*, 80. [CrossRef]
20. Abeliotis, K.; Chroni, C.; Lasaridi, K. Consumers' Behavior Regarding Food Waste Prevention. In *Encyclopedia of Food Security and Sustainability*; Elsevier: Amsterdam, The Netherlands, 2019; pp. 510–514. [CrossRef]
21. Vardopoulos, I.; Konstantopoulos, I.; Zorpas, A.; Limousy, L.; Bennici, S.; Inglezakis, V.; Voukkali, I. Sustainable metropolitan areas perspectives through assessment of the existing waste management strategies. *Environ. Sci. Pollut. Res.* **2021**, *28*, 24305–24320. [CrossRef]

22. Xu, L.; Zhong, Y.; He, X.; Shi, X.; Song, Q. Perception and Behavioural Changes of Residents and Enterprises under the Plastic Bag Restricting Law. *Sustainability* **2022**, *14*, 7792. [CrossRef]

23. Pappas, G.; Papamichael, I.; Zorpas, A.; Siegel, J.E.; Rutkowski, J.; Politopoulos, K. Modelling Key Performance Indicators in a Gamified Waste Management Tool. *Modelling* **2021**, *3*, 27–53. [CrossRef]

24. Kougias, K.; Sardianou, E.; Saiti, A. Attitudes and Perceptions on Education for Sustainable Development. *Circ. Econ. Sustain.* **2022**, 1–21. [CrossRef]

25. UNESCO. What Is Education for Sustainable Development? Available online: https://www.unesco.org/en/education/sustainable-development/need-know (accessed on 26 June 2022).

26. Sardianou, E. Sustainable Development Integration in Higher Education. In *Oxford Research Encyclopedia of Education*; Oxford University Press: Oxford, UK, 2020; ISBN 9780190264093.

27. Wang, H.; Jiang, X.; Wu, W.; Tang, Y. The effect of social innovation education on sustainability learning outcomes: The roles of intrinsic learning motivation and prosocial motivation. *Int. J. Sustain. High. Educ.* 2022; *ahead of print*. [CrossRef]

28. Ardoin, N.M.; Bowers, A.W.; Gaillard, E. Environmental education outcomes for conservation: A systematic review. *Biol. Conserv.* **2020**, *241*, 108224. [CrossRef]

29. Ardoin, N.M.; Bowers, A.W. Early childhood environmental education: A systematic review of the research literature. *Educ. Res. Rev.* **2020**, *31*, 100353. [CrossRef]

30. Moustairas, I.; Vardopoulos, I.; Kavouras, S.; Salvati, L.; Zorpas, A.A. Exploring factors that affect public acceptance of establishing an urban environmental education and recycling center. *Sustain. Chem. Pharm.* **2022**, *25*, 100605. [CrossRef]

31. Gomes, L.A.; Brasileiro, T.S.A.; Caeiro, S.S.F.S. Sustainability in Higher Education Institutions in the Amazon Region: A Case Study in a Federal Public University in Western Pará, Brazil. *Sustainability* **2022**, *14*, 3155. [CrossRef]

32. Theodoropoulou, E.; Vardopoulos, I.; Sardianou, E.; Mitoula, R.; Kavouras, S. Sustainable development: Setting the curricula for environmental and occupational safety and health skills and education. In Proceedings of the 2019 International Institute of Engineers and Researchers International Conference on Natural Science and Environment, Florence, Italy, 19–20 December 2019; pp. 22–26.

33. Grigoriou, M.; Efstathiadou, E. Environmental Education Centers and Experiential Learning: Views and Opinions of an EEC Director. Available online: https://ejournals.lib.uoc.gr/index.php/edusci/article/view/1218 (accessed on 1 July 2022).

34. Jorgenson, S.N.; Stephens, J.C.; White, B. Environmental education in transition: A critical review of recent research on climate change and energy education. *J. Environ. Educ.* **2019**, *50*, 160–171. [CrossRef]

35. Caeiro, S.; Azeiteiro, U.M. Sustainability Assessment in Higher Education Institutions. *Sustainability* **2020**, *12*, 3433. [CrossRef]

36. Kavouras, S.; Vardopoulos, I.; Mitoula, R.; Zorpas, A.A.; Kaldis, P. Occupational health and safety scope significance in achieving sustainability. *Sustainability* **2022**, *14*, 2424. [CrossRef]

37. Pitoska, E.; Lazarides, T. Environmental Education Centers and Local Communities: A Case Study. *Procedia Technol.* **2013**, *8*, 215–221. [CrossRef]

38. Medir, R.M.; Heras, R.; Geli, A.M. Guiding documents for environmental education centres: An analysis in the Spanish context. *Environ. Educ. Res.* **2014**, *20*, 680–694. [CrossRef]

39. Chao, Y.-L. A Performance Evaluation of Environmental Education Regional Centers: Positioning of Roles and Reflections on Expertise Development. *Sustainability* **2020**, *12*, 2501. [CrossRef]

40. Clair, R.S. Words for the world: Creating critical environmental literacy for adults. *New Dir. Adult Contin. Educ.* **2003**, *2003*, 69–78. [CrossRef]

41. Marcinkowski, T.J. Contemporary Challenges and Opportunities in Environmental Education: Where Are We Headed and What Deserves Our Attention? *J. Environ. Educ.* **2009**, *41*, 34–54. [CrossRef]

42. Eagles, P.F.J.; Richardson, M. The Status of Environmental Education at Field Centers of Ontario Schools. *J. Environ. Educ.* **1992**, *23*, 9–14. [CrossRef]

43. Ballantyne, R.; Packer, J. Promoting Learning for Sustainability: Principals' Perceptions of the Role of Outdoor and Environmental Education Centres. *Aust. J. Environ. Educ.* **2006**, *22*, 15–29. [CrossRef]

44. Yanniris, C. 20+ Years of Environmental Education Centers in Greece: Teachers' Perceptions and Future Challenges. *Appl. Environ. Educ. Commun.* **2015**, *14*, 149–166. [CrossRef]

45. Higgins, P.; Kirk, G. Sustainability Education in Scotland: The Impact of National and International Initiatives on Teacher Education and Outdoor Education. *J. Geogr. High. Educ.* **2006**, *30*, 313–326. [CrossRef]

46. Simmons, D.A. Are We Meeting the Goal of Responsible Environmental Behavior? An Examination of Nature and Environmental Education Center Goals. *J. Environ. Educ.* **1991**, *22*, 16–21. [CrossRef]

47. Hart, P. Problematizing Enquiry in Environmental Education: Issues of Method in a Study of Teacher Thinking and Practice. *J. Environ. Educ.* **1996**, *1*, 56–88.

48. Mavrikaki, E.; Kyridis, A.; Tsakiridou, E.; Golia, P. Greek educators' attitudes and beliefs about the application of environmental education in elementary school. In *International Perspectives in Environmental Education*; Leal Filho, W., Littledyke, M., Eds.; Peter Lang: Frankfurt, Germany, 2004; pp. 29–36.

49. Flogaitis, E.; Daskolia, M.; Liarakou, G. Greek kindergarten teachers' practice in environmental education. *J. Early Child. Res.* **2005**, *3*, 299–320. [CrossRef]

50. Erickson, E.; Erickson, J. Lessons Learned from Environmental Education Center Directors. *Appl. Environ. Educ. Commun.* **2006**, *5*, 1–8. [CrossRef]
51. Ernst, J. Influences on US middle school teachers' use of environment-based education. *Environ. Educ. Res.* **2009**, *15*, 71–92. [CrossRef]
52. Pedretti, E.; Nazir, J.; Tan, M.; Bellomo, K.; Ayyavoo, G. A baseline study of ontario teachers' views of environmental and outdoor education. *Pathw. Ont. J. Outdoor Educ.* **2012**, *24*, 4–12.
53. Yu-Long, C.; Heekyung, K.; Chankook, K. A comparison of the regional environmental education centers in Korea and Taiwan: Systems, roles, and practices. *J. Environ. Educ. Res.* **2018**, *14*, 121–152. [CrossRef]
54. Wals, A.E.J.; Hoeven, N.; Blanken, H. *The Acoustics of Social Learning*; Wageningen Academic Publishers: Wageningen, The Netherlands, 2009.
55. Eliadis, F.; Doula, M.; Zorpas, A. The Role of Environmental Education Centers in Climate Change Education and Awareness Raising of the Society. The Cases of Cyprus and Greece. 2019. Available online: https://www.researchgate.net/publication/332259354_The_role_of_Environmental_Education_Centers_in_climate_change_education_and_awareness_raising_of_the_society_The_cases_of_Cyprus_and_Greece/citations (accessed on 3 July 2022).
56. Eshun, F.; Wotorchie, R.K.; Buahing, A.A.; Harrison-Afful, A.A.; Atiatorme, W.K.; Amedzake, G.; Adofo-Yeboah, Y.; Mante, V. A Survey of the Role of Environmental Education in Biodiversity Conservation in the Greater Accra Region of Ghana. *Conservation* **2022**, *2*, 297–304. [CrossRef]
57. Amin, S.H.; Razmi, J.; Zhang, G. Supplier selection and order allocation based on fuzzy SWOT analysis and fuzzy linear programming. *Expert Syst. Appl.* **2011**, *38*, 334–342. [CrossRef]
58. Voukkali, I.; Zorpas, A. Evaluation of urban metabolism assessment methods through SWOT analysis and analytical hierocracy process. *Sci. Total Environ.* **2021**, *807*, 150700. [CrossRef]
59. Posthuma-Trumpie, G.A.; Korf, J.; van Amerongen, A. Lateral flow (immuno) assay: Its strengths, weaknesses, opportunities and threats. A literature survey. *Anal. Bioanal. Chem.* **2009**, *393*, 569–582. [CrossRef]
60. Vardopoulos, I.; Tsilika, E.; Sarantakou, E.; Zorpas, A.; Salvati, L.; Tsartas, P. An integrated SWOT-PESTLE-AHP model assessing sustainability in adaptive reuse projects. *Appl. Sci.* **2021**, *11*, 7134. [CrossRef]
61. Symeonides, D.; Loizia, P.; Zorpas, A.A. Tire waste management system in Cyprus in the framework of circular economy strategy. *Environ. Sci. Pollut. Res.* **2019**, *26*, 35445–35460. [CrossRef] [PubMed]
62. Dyson, R.G. Strategic development and SWOT analysis at the University of Warwick. *Eur. J. Oper. Res.* **2004**, *152*, 631–640. [CrossRef]
63. Phadermrod, B.; Crowder, R.M.; Wills, G.B. Importance-Performance Analysis based SWOT analysis. *Int. J. Inf. Manag.* **2019**, *44*, 194–203. [CrossRef]
64. Helms, M.M.; Nixon, J. Exploring SWOT analysis—Where are we now? *J. Strateg. Manag.* **2010**, *3*, 215–251. [CrossRef]
65. Ławińska, O.; Korombel, A.; Zajemska, M. Pyrolysis-Based Municipal Solid Waste Management in Poland—SWOT Analysis. *Energies* **2022**, *15*, 510. [CrossRef]
66. Tsangas, M.; Jeguirim, M.; Limousy, L.; Zorpas, A. The Application of Analytical Hierarchy Process in Combination with PESTEL-SWOT Analysis to Assess the Hydrocarbons Sector in Cyprus. *Energies* **2019**, *12*, 791. [CrossRef]
67. Ghazinoory, S.; Esmail Zadeh, A.; Memariani, A. Fuzzy SWOT analysis. *J. Intell. Fuzzy Syst.* **2007**, *18*, 99–108.
68. Seebohm, L. SWOT/TOWS. Collaborative Tools for Stragegic Line Planning. 2014. Available online: https://dokumen.tips/documents/swot-tows-concurrent-collaborative-tools-for-strategic-line-planning-presented.html?page=2 (accessed on 3 July 2022).
69. Oxford College of Marketing. TOWS Analysis: A Step By Step Guide. Available online: https://blog.oxfordcollegeofmarketing.com/2016/06/07/tows-analysis-guide/ (accessed on 3 July 2022).
70. Weihrich, H. The TOWS matrix—A tool for situational analysis. *Long Range Plann.* **1982**, *15*, 54–66. [CrossRef]
71. Loubser, C.P.; Ferreira, J.G. Environmental Education in South Africa in Light of the Tbilisi and Moscow Conferences. *J. Environ. Educ.* **1992**, *23*, 31–34. [CrossRef]
72. United States Environmental Protection Agency (EPA). What Is Environmental Education. Available online: https://www.epa.gov/education/what-environmental-education (accessed on 26 June 2022).
73. Sun, C.; Liu, J.; Razmerita, L.; Xu, Y.; Qi, J. Higher Education to Support Sustainable Development: The Influence of Information Literacy and Online Learning Process on Chinese Postgraduates' Innovation Performance. *Sustainability* **2022**, *14*, 7789. [CrossRef]
74. Hollweg, K.S. One Environmental Education Center's Industry Initiative: Collaborating to Create More Environmentally and Economically Sustainable Businesses. *Appl. Environ. Educ. Commun.* **2009**, *8*, 67–77. [CrossRef]
75. Zachariou, A.; Symeou, L. The Local Community as a Means for Promoting Education for Sustainable Development. *Appl. Environ. Educ. Commun.* **2009**, *7*, 129–143. [CrossRef]
76. Viloria, L.A. A network of regional centres of environmental education and training: A strategy for developing countries. *Mar. Pollut. Bull.* **1991**, *23*, 633–635. [CrossRef]
77. Zikas, V. Environmental Education and Physical Education. A Work Plan for Lyceum. Master's Thesis, University of the Aegean, Mitilini, Greece, 2008. (In Greek).
78. Antonopoulos, G.; Skanavis, C. Promoting an environmental awareness centre to enhance educational activities in Linaria port, Skyros. *Int. J. Green Econ.* **2020**, *14*, 95. [CrossRef]

79. Abeliotis, K.; Goussia-Rizou, M.; Sdrali, D.; Vassiloudis, I. How parents report their environmental attitudes: A case study from Greece. *Environ. Dev. Sustain.* **2010**, *12*, 329–339. [CrossRef]
80. Xingang, Z.; Jiaoli, K.; Bei, L. Focus on the development of shale gas in China—Based on SWOT analysis. *Renew. Sustain. Energy Rev.* **2013**, *21*, 603–613. [CrossRef]

Article

Building the Value Proposition of a Digital Innovation Hub Network to Support Ecosystem Sustainability

Claudio Sassanelli [1,*] and Sergio Terzi [2,*]

[1] Department of Mechanics, Mathematics and Management, Politecnico di Bari, Via Orabona 4, 70125 Bari, Italy
[2] Department of Management, Economics and Industrial Engineering, Politecnico di Milano, Piazza Leonardo da Vinci 32, 20133 Milan, Italy
* Correspondence: claudio.sassanelli@poliba.it (C.S.); sergio.terzi@polimi.it (S.T.)

Abstract: Digital Innovation Hubs (DIHs) play a key role in bolstering European companies to overwhelm innovation barriers and drive Europe as the world's primary leader in the Industry 4.0 digital revolution; they are one-stop-shop ecosystems able to provide four main functionalities (test before investing, support to find investments, innovation ecosystems, and networking, skills and training). Even if a surge in their diffusion has been registered, their sustainability is still far from being well defined in a structured way. Several approaches and methods are available from literature to ground the sustainability plan of companies' business. Among them, the first activity to be addressed is the value proposition (VP) analysis, and the most diffused approach is the Value Proposition Canvas (VPC); this paper proposes the application of the VPC (jointly used with other methods from the VP literature) to build the VP of the HUBCAP network (supporting European small and medium-sized enterprises in the adoption of model-based design methods and tools to support cyber-physical system technologies) per each of its four main customer segments (DIHs, academic partners and research and technology organizations, technology/tool providers and technology/tool users). Results highlight the need to characterize the analysis per each of these customers, open up new opportunities to build a structured business model of the network, and constitute a basis for assessing the potential synergies with similar DIH networks. The method proposed can be applied to any other DIH or network of DIH to define their specific VP, ground the strategy to reach their sustainability, and trigger collaborations with each of the four customer segments considered in the analysis.

Keywords: digital innovation hub; value proposition canvas; ecosystem sustainability; digital transformation; model-based design; cyber-physical system; collaboration platform

Citation: Sassanelli, C.; Terzi, S. Building the Value Proposition of a Digital Innovation Hub Network to Support Ecosystem Sustainability. *Sustainability* **2022**, *14*, 11159. https://doi.org/10.3390/su141811159

Academic Editor: Antonis A. Zorpas

Received: 27 July 2022
Accepted: 31 August 2022
Published: 6 September 2022

Publisher's Note: MDPI stays neutral with regard to jurisdictional claims in published maps and institutional affiliations.

1. Introduction

Digital Innovation Hubs (DIHs) play a key role in bolstering European companies to overwhelm innovation barriers and drive Europe as the world's primary leader in the Industry 4.0 digital revolution [1]; they are one-stop-shop ecosystems able to provide four main functionalities (test before invest, support to find investments, innovation ecosystems, and networking, skills and training) [2] through a certain set of assets (competences and skills, technologies, services, etc.) [3]. To be recognized, DIHs need to be part of a regional, national or European policy initiative to digitize industry, be non-for-profit organizations, have a physical presence in the region, present an updated website and have at least three examples of how they have helped a company in the digital transformation referring to publicly available information [2,4–7]. Recently, several projects have been financed by the European Commission (EC) to push them from a regional towards a wider pan-European impact [8,9], and several results have been obtained in terms of supporting models and methods to both configure their portfolios [10] and decode and build customer journeys (CJs), bridging their provision [11]; however, being not-for-profit organizations and being, so far, always financed by EC projects, the sustainability of DIHs (and of their networks) is still far from being well defined in a structured way [12]. On the other side, from a wider perspective, several approaches and methods are available in the literature to ground the sustainability plan and the business model of organizations and companies. Among them, the first activity to be addressed is the value proposition (VP) analysis, and the most diffused approach is the Value Proposition Canvas (VPC) [13].

Among the several projects funded by the EC, the recent HUBCAP project [14] has constituted a network composed of DIHs, technology/tool providers and users, and academic partners/research and technology organizations (RTOs); this network still represents a niche because it deals with the adoption of strong technical and specialized "test before invest" model-based design (MBD) assets (models and tools) to support small and medium-sized enterprises (SMEs) in the development and adoption of cyber-physical systems (CPSs); however, even if the HUBCAP network managed to configure its service portfolio and the related typical CJs useful for the provision of these services [11], the need to enhance its capability to attract new potential stakeholders and customers with the final aim of establishing the foundation for its sustainability plan has been unveiled. Therefore, to ease such dynamics, to be able to attract more effectively new stakeholders in the innovative HUBCAP ecosystem, and to push the impact of its digital collaboration platform [15], the HUBCAP VP should be clearly detailed.

Nevertheless, in the extant literature, there is a lack of information regarding how DIH sustainability could be reached and how VP analysis in the DIH domain could be performed. To verify its effectiveness for these ecosystems, and based on the positive experience coming from the DIH4CPS project [16], this paper proposes the application of the VPC template (jointly used with other methods from the VP literature) to build the VP of the HUBCAP network. The results highlight the need to characterize the analysis per each of the four customer profiles detected, opening up new opportunities to build a structured business model of the network and constituting a basis for assessing the potential synergies with other DIH networks. Four main slogans have been developed to represent the VP of the HUBCAP DIH network, highlighting the different perspectives to be taken when assets are offered to DIHs (a trustworthy platform-driven collaboration on CPS innovation), academic partners and RTOs (a recognized network of experts to spread MBD assets adoption), tool/technology providers (a sandbox providing a tool repository able to address end-user requests), and tool/technology users (a secure and intuitive environment capable of offering a multi-user assets catalogue and related training and knowledge to test facilities).

The paper is structured as follows. Section 2 presents the literature review to ground the research (i.e., DIHs in European digitization, MBD models and tools in the CPS domain and the VPC template) and the research method adopted. Section 3 explains the research process and Section 4 the results obtained, detailing the VP per each of the DIH network's customers. Finally, Section 5 discusses the results and concludes the paper, unveiling limitations and opening room for further developments.

2. Literature Review and Research Method

2.1. Digital Innovation Hubs and European SMEs Digitization

Society's daily lives, the way people work and do business, how they comprehend and utilise the environment and natural resources, and how people interact, communicate and educate themselves are all being dramatically altered by digital technologies. DIHs assist European companies, and in particular SMEs, in deploying and employing digital technologies to enhance business/production processes, products, or services by giving access to technical knowledge, experimentation, and the opportunity to "test before investing". A successful digital transformation requires innovation services like finance guidance, training, and talent development, which they also offer [17]. Environmental considerations are also made, particularly in relation to energy usage [18,19] and reduced carbon emissions [20,21].

The characteristics of DIHs and their interactions with stakeholders have a significant impact on how SMEs approach the digital transformation process [3]. The diversity of these ecosystems aligns with the main objectives of the EC, which include promoting their growth, expanding the network of DIHs that already exists, and establishing an integrated, flexible, and interoperable platform for DIHs from various, primarily digitally underdeveloped industries and regions. A vast pan-European ecosystem of DIHs is the end result that the EC aims to achieve. Indeed, each DIH is unique, located in specific areas, and concentrates on a variety of sectors and digital technologies. By developing, providing, and matching services jointly with other DIHs, the forthcoming European DIH (EDIH) can spark dynamics of innovation-driven cooperation. If this goal is accomplished, DIHs will not try to simultaneously fulfil all four roles; instead, they will concentrate more on the function that is most representative of them, relying on the connections and collaborations with other ecosystems [22].

In this instance, the definition of a clear VP, able to fully represent the characteristics and the assets provided by the single DIHs, can have a key role in supporting the success of such ecosystems and triggering the creation of collaborating and integrated communities specialized in different topics, industries, technologies, and approaches.

2.2. CPSs Supported by MBD Methods and Tools through the HUBCAP Network and Collaboration Platform

MBD is a visual method to address the design of complex control, signal and communication systems based on mathematical models [23–25]. With MBD, it is possible to prescribe the use of models through the whole development process, representing the system structure and behaviour, providing a basis for machine-assisted analysis of system properties, and supporting design decisions for technology refinement.

HUBCAP creates a collaborative environment between DIHs and SMEs, inspired by enterprise social software [26]; this environment can be accessed through a web portal [15,26], where "Access to" and "Collaborate with" services can be found. SMEs can hence access MBD assets that can offer great support in their digital transformation process, triggering a more aware test before invest phase of the technologies they want to employ in their organizations and processes. Indeed, the chance for collaboration, through the use of tools and models provided by the platform, can support companies through the whole transformation path (i.e., piloting, testing before investing, and experimenting). HUBCAP offers two catalogues of MBD assets, one composed of tools (software packages and their dependencies that enable the development, analysis and simulation of models) and the other of models (mathematical or formal abstractions of system elements [components or subsystems]). The MBD techniques offered by the DIH network have four main types (i.e., simulation, model checking, contract-based analysis and model-based safety assessment) and can be specified in specific techniques (e.g., model checking can be split into invariant model checking, linear temporal logic model checking and deadlock checking) and models (e.g., deadlock checking techniques can be divided into CHESS, mXmv, HyComp and COMPASS [detailed in several models such as GPS, Engine, etc.]). The SMEs accessing the platform can browse both of them to perform tests and experiments using a sandbox without the need to incur high investments. The sandbox of the HUBCAP platform was inspired by the results of the INTO-CPS project [27]. In it, MBD tools were integrated into a single application [28] aggregating the different models using the Functional Mock-up Interface (FMI)-based co-simulation orchestration engine [29]; this type of MBD setup offers, together with MBD tools and models, the third class of assets: operating systems (OSs), which refer to a software environment providing libraries and dependencies needed to run the tools. The main objective of the sandbox is to enable users to add MBD assets to a cart and launch them in a cloud environment, which caters to the different technicalities required and allows several users to co-develop a model by sharing the user interface of machines hosted in a cloud environment and accessible via a web browser. HUBCAP allows SMEs to have access to MBD tools and models they can utilize to foster their digitalization process; it is intended to support trial experiments that can design and develop innovative

CPS solutions; however, a solid VP must still be defined to both ground a sustainability plan and attract new customers/stakeholders.

2.3. VP of Innovation Ecosystems and VPC

Innovation ecosystems, often intended as boundary organizations [30], are increasingly regarded as important vehicles to create and capture value from complex VPs [31]. Dedehayir et al. [32,33] defined four main roles of innovation ecosystems (leadership roles, direct value creation roles, value creation support roles, and entrepreneurial ecosystem roles) and examined the impact of disruption on the innovation ecosystem in its entirety (i.e., the group of organizations that collaborate in creating a holistic VP for the end user). Talmar et al. [34] developed a strategy tool (called the Ecosystem Pie Model [EPM]) to map, analyse and design (i.e., model) innovation ecosystems, focusing on the constructs and relationships that capture how actors in an ecosystem interact in creating and capturing value. Instead, the most used and valuable tool able to design the characterizing VP of an organization in relation to its main stakeholders is the VPC [13] (Figure 1); it has two sides: the customer profile, clarifying the customer understanding, and the value map, describing how it is intended to create value for that customer. The customer (segment) profile describes a specific customer segment in the business model; it breaks customers down into their jobs, pains and gains:

- Gains describe the outcomes customers want to achieve or the concrete benefits they are seeking,
- Pains describe bad outcomes, risks and obstacles related to customer jobs.
- Customer jobs describe what customers are trying to get done in their work, expressed in their own words.
- The value (proposition) map describes the features of a specific VP, breaking it down into products and services, pain relievers, and gain creators:
- Gain creators describe how products and services create customer gains,
- Pain relievers describe how products and services alleviate customer pains.
- Products and services refer to a list of all the products and services around which a VP is built.

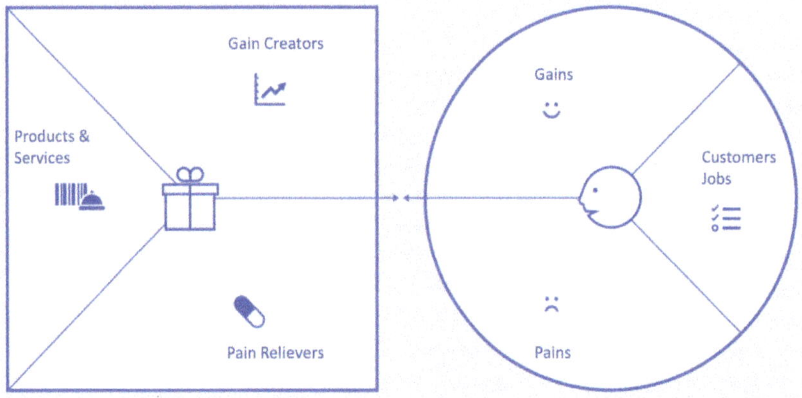

Figure 1. The VPC template (adapted from [14]).

The fit is achieved when the value map aligns with the customer profile (i.e., when the products and services produce pain relievers and gain creators that match one or more of the jobs, pains and gains that are important to the customer).

2.4. Research Method

This sub-section is aimed at describing how the VP has been built for the DIH network of HUBCAP, also based on the positive experience coming from the DIH4CPS project [16]. Four main customer segments (DIHs, academic partners and RTOs, technology/tool providers, and technology/tool users) were detected for this network. Therefore, using the Mural online collaboration platform, a workshop for each of them was conducted to explain the VPC and apply it. Each workshop lasted about 3 h, and each involved, on average, 10 people belonging to different organizations of each customer category of the HUBCAP network. In addition, more time (about 1 week) was left to them to allow each of the organizations involved to brainstorm internally to fill out the VPC template and conclude the activity. Furthermore, back-office work (about 10 h for each of the two researchers involved per VPC customer) was needed to group all the input received during the workshops. The main categories for each field of the VPC of each customer segment were defined. Then, choosing their recurrency as a driver of prioritization, groups were ranked. Afterwards, for each customer profile, the main items characterizing the VPC dimensions were defined, and fitting was done between the items referring to the dimensions composing the customer profiles (jobs, pains and gains) and those of the VP maps (products and services, pain relievers and gain creators). In particular, the resulting VP items were arranged in a pyramid-shaped template (shown in Figure 2) composed of three levels, as follows:

- Details: items contributing to a detailed description of the VP. From a customer perspective, they report the benefits coming from HUBCAP outcomes, the problems solved by them, and the arguments that prove that the organization delivering the VP is doing better than its competitors.
- Summary: two sentences synthesizing the HUBCAP offer content (answering the questions "For whom?" and "Why is it useful?").
- Title/Slogan: a sentence summarizing the VP in which the ultimate benefit of the HUBCAP offer is expressed to attract customers' attention and curiosity.

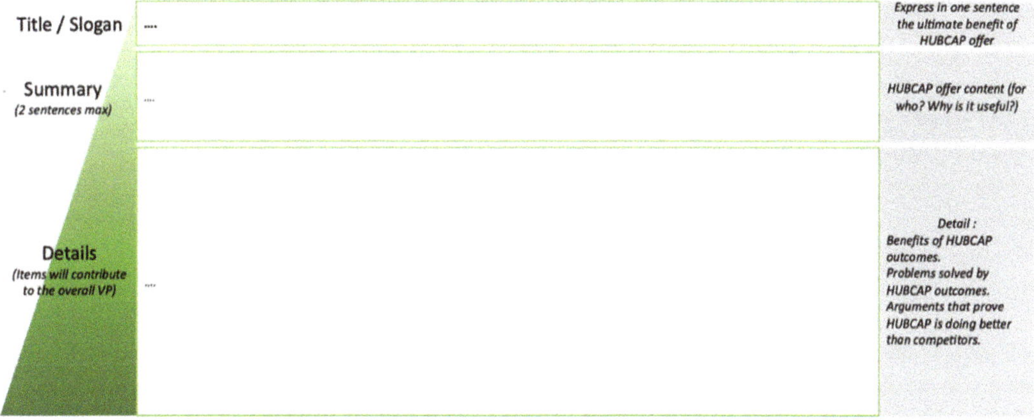

Figure 2. The framework used to structure the VP of each customer segment.

To translate the VP items obtained through the VPC into direct and effective statements that everyone could understand, even if they are not part of the MBD niche in the CPS domain, different theories were adopted:

- The "Value Positioning Statement" [35]: "For (target customer) who (statement of the need or opportunity), our (product/service name) is (product category) that (statement of benefit)"; it can also be simplified in the "XYZ": "We help X do Y doing Z".

- the "Jobs-to-be-done" (JTBD) theory [36]: "Action verb-Object of action-Contextual identifier".

Finally, a validation workshop was organized to check the results with the participants of the previous workshops.

3. Results

As mentioned in Section 1, four main customer segments were considered in the analysis of the VP of the network of DIHs providing MBD assets. Results are shown as follows. As an example, the building of the HUBCAP network VP related to the DIHs customer profile is shown step by step to better explain the research method adopted. In particular, Figure 3 illustrates the results obtained through the Mural platform, showing the detected VP items in the grey squares. After that, the two researchers gathered, grouped and ranked the input received during the workshop. The grey squares represent the VP items reported in the polished VPC in Figure 4.

Finally, Figure 5 shows the fit between the items belonging to the dimensions of the customer profile (jobs, pains and gains) with those related to the VP map (products and services, pain relievers and gain creators) per the DIH customer profile. In this figure, the items are flanked by a weight which expresses how many times the workshop participants provided input for that specific VP item. In addition, the products and services, pain creators, and gain relievers that fit the customer profile items better have been detected. The items with higher importance contribute more to the construction of the top two levels ("Summary" and "Title/Slogan") of the pyramid in Figure 2. Figure 5 shows the single case of the specific customer segment of DIHs, also demonstrating that most customer needs were covered by the HUBCAP DIH network offer. For example, the fit at the top of the figure demonstrates that the DIH ecosystem analysed is able to offer network, brokering and matchmaking to both enable other DIHs to easily find training, skills and competencies in the CPS technology/MBD domain and to improve the visibility to attract new customers. In addition, the MBD assets (tools, training, success stories, etc.) provided by the DIH network address the networking and matchmaking need to find synergies with both academic and industrial partners. Instead, the fit at the centre shows that a gain creator as an improved catalogue of services, MBD tools and knowledge triggers multiple gains in terms of an increase of knowledge and opportunity for training, an increase in the reach of MBD and specialized services, and an improvement of sales and monetization of DIH customers (SMEs). Finally, the fit at the bottom shows that a pain reliever in the form of MBD experts able to provide missing knowledge could address multiple pains for new DIHs, such as lack of skills, knowledge and technical complexity; MBD costs; and lack of confidence in MBD tools.

For the sake of completeness, the results of the VPC application for the other three customer profiles (academic partners and RTOs, technology/tool providers, and technology/tool users) are shown respectively in Figures 6–8; it can be noted that, per each type of stakeholder of the DIH network, different results have been obtained. Indeed, for academic partners and RTOs (Figure 6), the main customer jobs to be addressed by the DIH network are knowledge creation and transferring for teaching MBD, networking with MBD users, exploitation of existing MBD tools and models in applied research, and so on. To enable their satisfaction, the DIH network provides training and demonstration, community and networking services, different and customizable models and tools, and events and dissemination channels. For this kind of customer, the main pains are the effort to learn and adopt MBD, and potential adopters' lack of interest and understanding of MBD. The DIH network is capable of defusing them through effective collaboration through the sandbox, a low bureaucracy and high safety, indirect advertising, and existing teaching and training material. At the same time, this offers increased visibility, enhances the network, develops more skilled personnel and improves research quality.

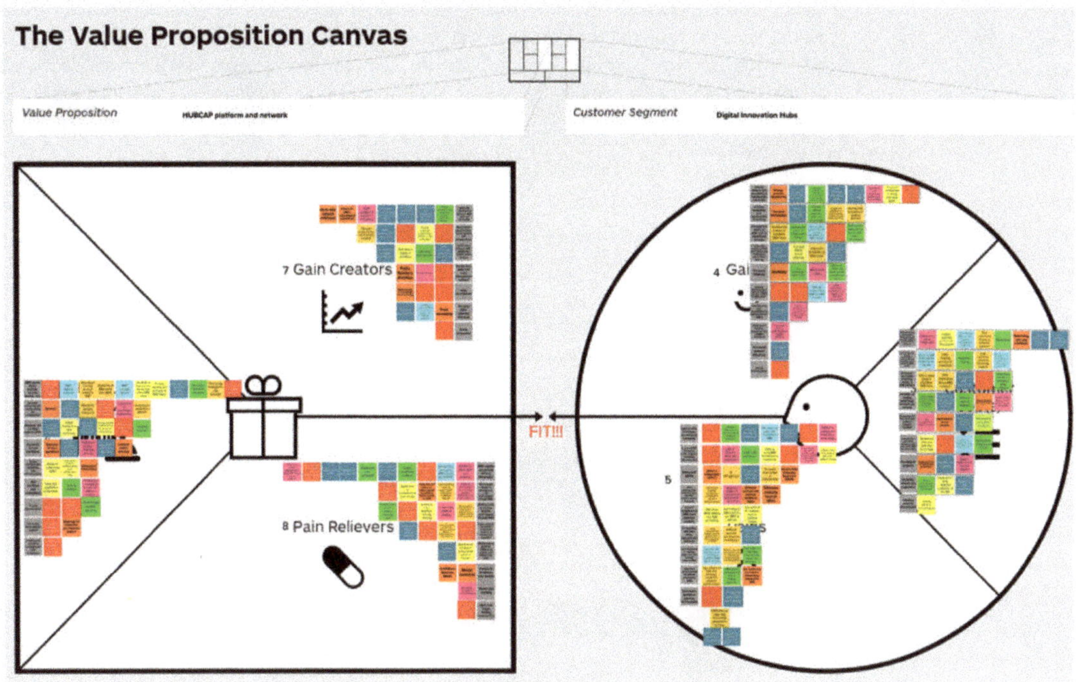

Figure 3. Workshop on the Mural online collaboration platform: results for VPC related to the DIH customer segments.

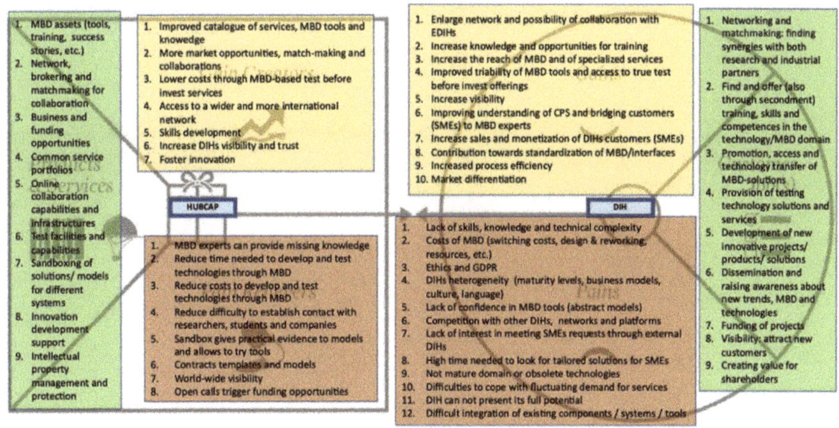

Figure 4. Main categories of the HUBCAP VP for the DIH customer segments.

For technology/tool providers (Figure 7), it is necessary to install, sell and provide tools and toolchains in a homogeneous environment, to attract new partners and establish network cooperation, or to provide access to a new platform with new tools, infrastructures, data and capabilities. In this case, the DIH network provides a collaboration platform as a tool repository/catalogue, a network collaboration and ecosystem service, and the dissemination of a set of best practices and success stories. Pains such as poor performance of the environment and embedded tools, a bad user experience, and a poor platform performance can be defused by low infrastructure cost and no dependency on expensive proprietary platforms, the possibility to support customers with remote assistance and manuals, and resource use monitoring, user statistics and bug detection. Multiple gains are brought to the customer as, for example, a set of remote tools that are easily accessible, lower barriers for new users, and more visibility of new application cases and markets.

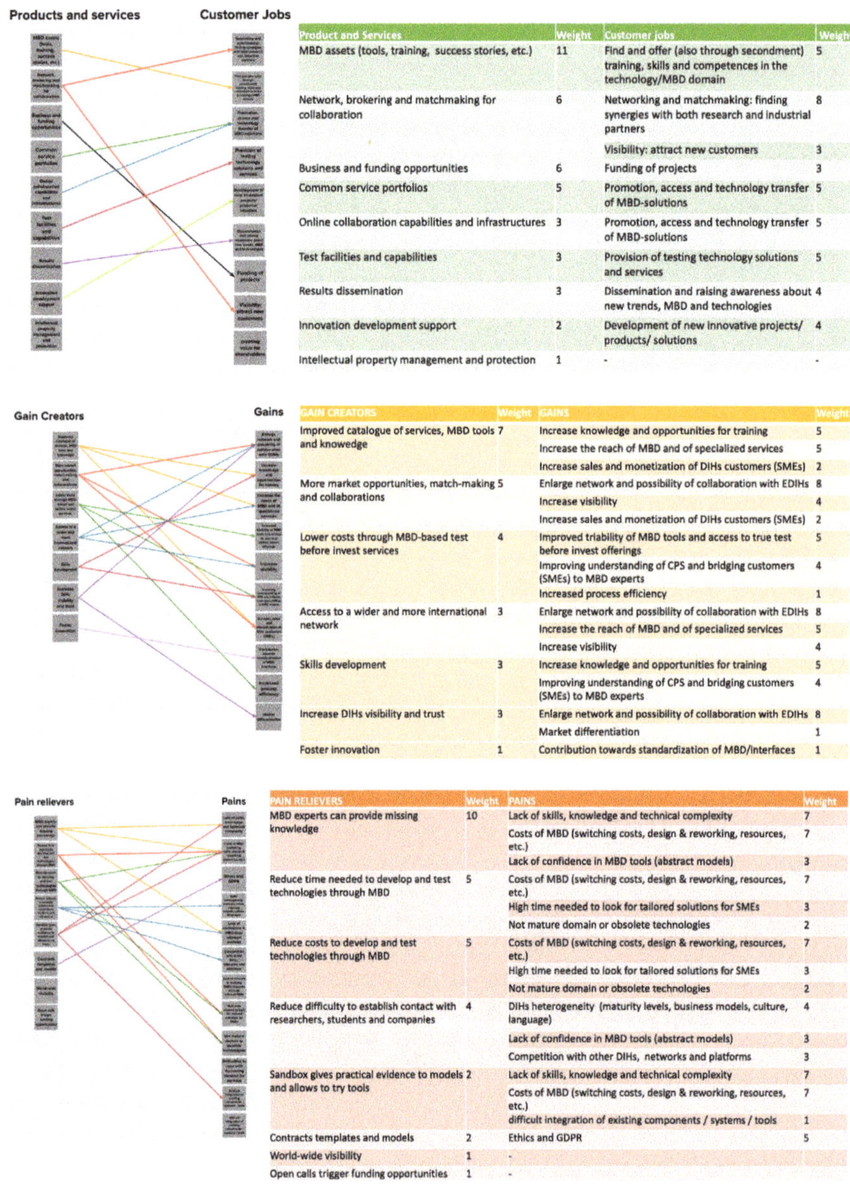

Product and Services	Weight	Customer jobs	Weight
MBD assets (tools, training, success stories, etc.)	11	Find and offer (also through secondment) training, skills and competences in the technology/MBD domain	5
Network, brokering and matchmaking for collaboration	6	Networking and matchmaking: finding synergies with both research and industrial partners	8
		Visibility: attract new customers	3
Business and funding opportunities	6	Funding of projects	3
Common service portfolios	5	Promotion, access and technology transfer of MBD-solutions	5
Online collaboration capabilities and infrastructures	3	Promotion, access and technology transfer of MBD-solutions	5
Test facilities and capabilities	3	Provision of testing technology solutions and services	5
Results dissemination	3	Dissemination and raising awareness about new trends, MBD and technologies	4
Innovation development support	2	Development of new innovative projects/ products/ solutions	4
Intellectual property management and protection	1	-	-

GAIN CREATORS	Weight	GAINS	Weight
Improved catalogue of services, MBD tools and knowedge	7	Increase knowledge and opportunities for training	5
		Increase the reach of MBD and of specialized services	5
		Increase sales and monetization of DIHs customers (SMEs)	2
More market opportunities, match-making and collaborations	5	Enlarge network and possibility of collaboration with EDIHs	8
		Increase visibility	4
		Increase sales and monetization of DIHs customers (SMEs)	2
Lower costs through MBD-based test before invest services	4	Improved triability of MBD tools and access to true test before invest offerings	5
		Improving understanding of CPS and bridging customers (SMEs) to MBD experts	4
		Increased process efficiency	1
Access to a wider and more international network	3	Enlarge network and possibility of collaboration with EDIHs	8
		Increase the reach of MBD and of specialized services	5
		Increase visibility	4
Skills development	3	Increase knowledge and opportunities for training	5
		Improving understanding of CPS and bridging customers (SMEs) to MBD experts	4
Increase DIHs visibility and trust	3	Enlarge network and possibility of collaboration with EDIHs	8
		Market differentiation	1
Foster innovation	1	Contribution towards standardization of MBD/interfaces	1

PAIN RELIEVERS	Weight	PAINS	Weight
MBD experts can provide missing knowledge	10	Lack of skills, knowledge and technical complexity	7
		Costs of MBD (switching costs, design & reworking, resources, etc.)	7
		Lack of confidence in MBD tools (abstract models)	3
Reduce time needed to develop and test technologies through MBD	5	Costs of MBD (switching costs, design & reworking, resources, etc.)	7
		High time needed to look for tailored solutions for SMEs	3
		Not mature domain or obsolete technologies	2
Reduce costs to develop and test technologies through MBD	5	Costs of MBD (switching costs, design & reworking, resources, etc.)	7
		High time needed to look for tailored solutions for SMEs	3
		Not mature domain or obsolete technologies	2
Reduce difficulty to establish contact with researchers, students and companies	4	DIHs heterogeneity (maturity levels, business models, culture, language)	4
		Lack of confidence in MBD tools (abstract models)	3
		Competition with other DIHs, networks and platforms	3
Sandbox gives practical evidence to models and allows to try tools	2	Lack of skills, knowledge and technical complexity	7
		Costs of MBD (switching costs, design & reworking, resources, etc.)	7
		difficult integration of existing components / systems / tools	1
Contracts templates and models	2	Ethics and GDPR	5
World-wide visibility	1	-	
Open calls trigger funding opportunities	1	-	

Figure 5. FIT for DIH customer segments among the customer profile dimensions (jobs, pains and gains) and of the VP map (products and services, pain relievers and gain creators).

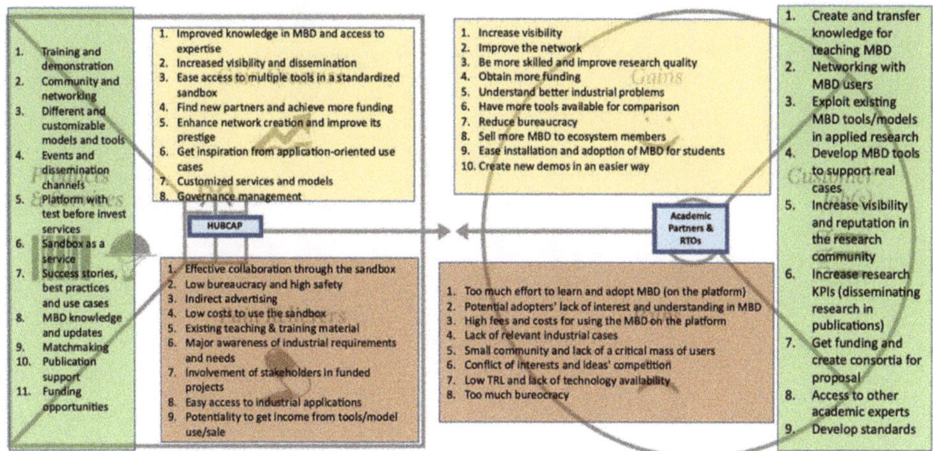

Figure 6. Main categories of the HUBCAP VP for the academic partner and RTO customer segments.

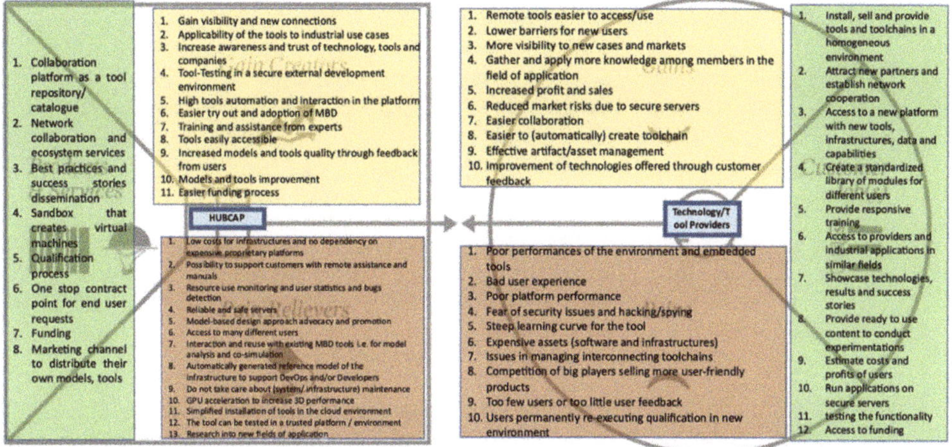

Figure 7. Main categories of the HUBCAP VP for the technology/tool provider customer segment.

In the case of the technology/tool users customer segment (Figure 8), the main jobs are to solve a specific problem regarding accessing new tools, prototype testing and reaching new domains, partners and networks to be more innovative and competitive. The DIH network achieves these activities by providing training knowledge and expert support, a multi-user assets catalogue to test facilities in several operating systems, and more. The main pains of this segment are IP protection, overly complex technologies/models/tools/platform/sandbox, and lack of competencies/skills/knowledge/support. The DIH network can defuse them through specialized and real-time knowledge and support, application examples for many different domains, and easy access to test technologies.

The HUBCAP DIH network, thanks to this analysis, can understand how to satisfy each of its customers, defusing pains and creating gains for each of them.

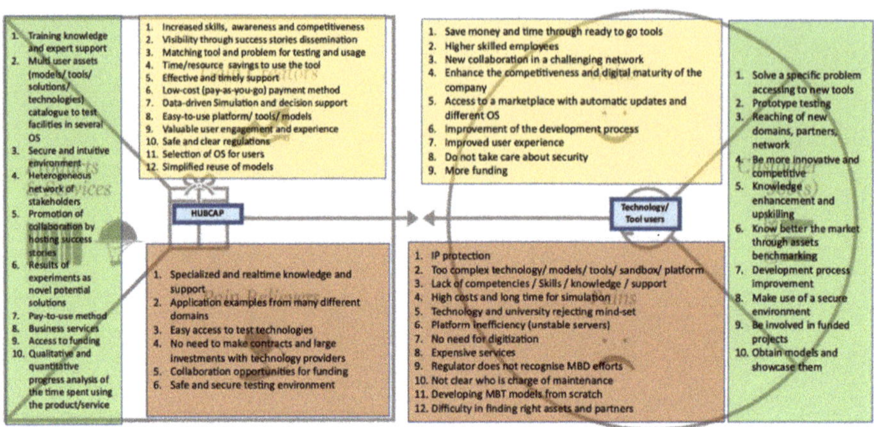

Figure 8. Main categories of the HUBCAP VP for the technology/tool user customer segment.

4. Discussion

Figure 9 shows the VPs related to the four customer segments considered in this research: the DIHs (top left), academic partners and RTOs (top right), technology/tool providers (bottom left), and technology/tool users (bottom right), useful to the HUBCAP DIH network for attracting stakeholders belonging to each of these categories; it can be noted that depending on the customers considered, the VP of the HUBCAP network changes. In particular, for the DIH customer profile, the slogan is "Leverage MBD assets to trigger a trustworthy platform-driven collaboration on CPS innovation among DIHs". In particular, in the summary, there are two main elements that HUBCAP could offer to attract this customer profile: the provision of MBD assets (tools, training, success stories, etc.) in a common service portfolio to easily test facilities and capabilities in the CPS innovation development through online collaboration infrastructures and DIHs' actions (networking, brokering and matchmaking) to provide business and funding opportunities and to manage and protect intellectual property. For academic partners and RTOs, the main slogan is "Create a recognized network of experts to spread MBD assets adoption and dissemination". Indeed, the HUBCAP offer for this customer profile in the summary is characterized by two main needs: the provision of MBD-based assets (tools and methods, training material, success stories, best practices and use cases) to foster publications through the exploitation of a sandbox as a service on a platform and the provision of matchmaking, events and dissemination channels, and funding opportunities services to support community development and networking. Concerning the technology/tools providers, the HUBCAP slogan is "Collaboration platform to address end-user requests through a sandbox providing a tool repository/catalogue". The HUBCAP network in this case offers two main elements to engage new tool providers to participate in its digital platform: a collaboration platform as a one-stop contact point to address end-user requests through a sandbox providing a tool repository/catalogue and a marketing channel to disseminate models and tools, best practices and success stories, training and funding opportunities. Finally, dealing with technology/tool users, the slogan is "Multi-user assets catalogue and related training and knowledge to test facilities in several OSs in a secure and intuitive environment through a pay-to-use method"; indeed, the strengths of the HUBCAP offer to attract new users belonging to this customer profile are, on the one hand, success stories, training knowledge and expert support provided by a collaborative and heterogeneous network of stakeholders, and on the other hand, multi-user assets (models/tools/solutions/technologies) catalogue to test facilities in several OSs in a secure and intuitive environment through a pay-to-use method.

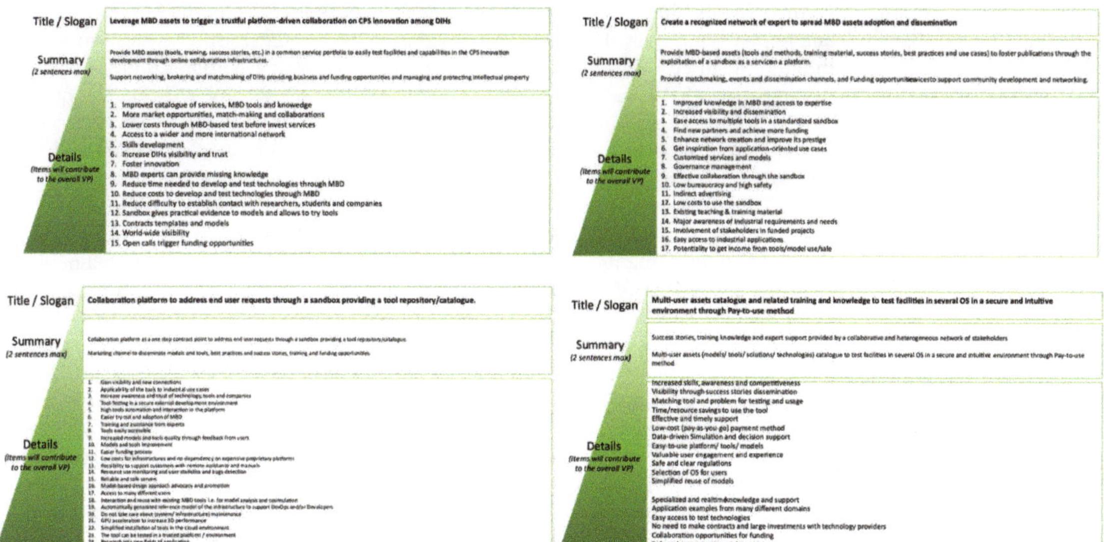

Figure 9. VP of the HUBCAP network for DIHs (**top left**), academic partners and RTOs (**top right**), technology/tool providers (**bottom left**), and technology/tool users (**bottom right**).

This research contributes to knowledge in multiple ways; it explores the VP items characterizing each of the four main customer profiles of a network of DIHs operating in the CPS domain and specialized in MBD and tools; it proposes a structured method to build the VP of a network of DIHs combining methods and approaches previously implemented in the business domain; it lays the groundwork to develop a sustainability plan of a network of DIHs and to develop the related business model. In the specific case of MBD models and tools provided by the HUBCAP network, this research contributes in raising the importance of MBD assets to innovate in digital technologies and impels companies to collaborate in multiple domains, putting together different and complementary competences.

From a practical perspective, this research contributes by providing a well-defined and structured VP, tailored to each customer profile, to a network of DIHs operating in the CPS domain and offering a set of MBD assets, making the network more appealing and enhancing its attractiveness to customers as well as to other potential collaborating DIH networks (because it highlights the strengths and weaknesses of its offer), as for example Circular Economy and bioeconomy [37,38]. Indeed, this research contributes to pushing MBD assets adoption to test digital technologies before actually investing in them (preventing companies from useless or bad investments) and triggers cross-sectorial and domain collaboration among multiple users through the use of the sandbox on the HUBCAP digital platform.

Finally, from a managerial perspective, the governance and managers of the DIH network offering the VP defined in this research are guided by these results to push their network in the market, develop a solid business model, make agreements with other networks, and finally achieve solid sustainability.

5. Conclusions

This paper, applying the VPC template, developed the VP (detailed in a three-level pyramid) of the HUBCAP DIH network aiming to support European SMEs in the adoption of MBD methods and tools to develop CPSs. In the analysis, four different customer segments were considered (i.e., DIHs, academic partners and RTOs, technology/tool providers, and technology/tool users). The research proposes the VPC approach to obtain and prioritize the VP items per customer segment; then, it develops a VP pyramid, structured on three levels (title, summary, details), which is representative of the DIH network offer and appeals to attract stakeholders belonging to its different customer segments. Going through the three levels with a bottom-up approach (design vision) helps the organization by offering the VP to build it in a more structured way. Instead, going through a top-down approach (customer vision) allows the organization to catch more effectively the customers to make them adhere to the VP. Indeed, the mission of the VP statements gathered in this template is to create visibility for the HUBCAP network (constituted by DIHs and services and MBD assets providers in the CPS technology domain across Europe) to allow SMEs access to this set of assets. Therefore, this research sought not only to test the VPC in the DIH ecosystem domain but also to propose a more complete approach to systematize its results in a VP pyramid. The results show that depending on the customers considered, the VP of the DIH network changes. Notwithstanding the novelty of this research in defining a structured VP of a network of DIHs, limitations can also be described. Indeed, in this paper, only the results related to the HUBCAP network (which represents a niche in the CPS domain) are presented. The same analysis could be performed for different networks operating either in the same domain, the CPS, or others (e.g., artificial intelligence). A comparison among the value propositions of these different networks could be performed to unveil the peculiar characteristics of each of them and detect the touch points among them to encourage future collaborations. Once the collaborations among different networks are materialized, the overall VP can be defined, providing room to bolster the actions of future European DIHs. Finally, this analysis represents only the first step towards the sustainability project of the HUBCAP network of DIHs. For each customer considered, a well-defined business model will be developed. Together, an analysis of the network assets (MBD and tools, competences and services) will be performed. The service offering will be analysed, detecting the services that contribute more to creating revenues among those constituting the HUBCAP service portfolio (decoded through the Data-based Business-Ecosystem-Skills-Technology [D-BEST] reference model); it must be highlighted that in the HUBCAP VP, a key role is played by the MBD assets (models and tools) because they are an effective means for testing before investing and thus are able to trigger along the CJ a high quantity of connected services to address the digital transformation.

Author Contributions: Conceptualization, C.S. and S.T.; methodology, C.S.; validation, C.S.; formal analysis, C.S.; investigation, C.S.; data curation, C.S.; writing—original draft preparation, C.S.; writing—review and editing, C.S.; visualization, C.S.; supervision, S.T. All authors have read and agreed to the published version of the manuscript.

Funding: This research was funded by the European Union's Horizon 2020 research and innovation programme under grant agreement No 872698 (HUBCAP Innovation Action).

Informed Consent Statement: Informed consent was obtained from all subjects involved in the study.

Conflicts of Interest: The authors declare no conflict of interest. The funders had no role in the design of the study; in the collection, analyses, or interpretation of data; in the writing of the manuscript; or in the decision to publish the results.

References

1. Sassanelli, C.; Rossi, M.; Terzi, S. Evaluating the Smart Maturity of Manufacturing Companies along the Product Development Process to Set a PLM Project Roadmap. *Int. J. Prod. Lifecycle Manag.* **2020**, *12*, 185–209. [CrossRef]
2. European Commission. *Smart Anything Everywhere—Digital Innovation Hubs—Accelerators for the Broad Digital Transformation of the European Industry*; European Commission: Brussels, Belgium, 2018.
3. Crupi, A.; Del Sarto, N.; Di Minin, A.; Gregori, G.L.; Lepore, D.; Marinelli, L.; Spigarelli, F. The Digital Transformation of SMEs—A New Knowledge Broker Called the Digital Innovation Hub. *J. Knowl. Manag.* **2020**, *24*, 1263–1288. [CrossRef]
4. European Commission. Digital Innovation Hubs—Smart Specialisation Platform. Available online: https://s3platform.jrc.ec.europa.eu/digital-innovation-hubs-tool (accessed on 25 November 2020).
5. European Commission. *Digital Innovation Hubs in Smart Specialisation Strategies. Early Lessons from European Regions*; European Commission: Brussels, Belgium, 2018.
6. European Commission. *Digitising European Industry. Reaping the Full Benefits of a Digital Single Market*; European Commission: Brussels, Belgium, 2016.

7. European Commission. *European Digital Innovation Hubs in Digital Europe Programme—Draft Working Document*; European Commission: Brussels, Belgium, 2020.

8. Doyle, F.; Cosgrove, J. Steps towards Digitization of Manufacturing in an SME Environment. *Procedia Manuf.* **2019**, *38*, 540–547. [CrossRef]

9. Hervas-Oliver, J.L.; Gonzalez-Alcaide, G.; Rojas-Alvarado, R.; Monto-Mompo, S. Emerging Regional Innovation Policies for Industry 4.0: Analyzing the Digital Innovation Hub Program in European Regions. *Compet. Rev.* **2021**, *31*, 106–129. [CrossRef]

10. Sassanelli, C.; Terzi, S. The D-BEST Reference Model: A Flexible and Sustainable Support for the Digital Transformation of Small and Medium Enterprises. *Glob. J. Flex. Syst. Manag.* **2022**, *40171*, 1–26. [CrossRef]

11. Sassanelli, C.; Terzi, S. The D-BEST Based Digital Innovation Hub Customer Journey Analysis Method: Configuring DIHs Unique Value Proposition. *Int. J. Eng. Bus. Manag.* **2022**, forthcoming. [CrossRef]

12. Zamiri, M.; Ferreira, J.; Sarraipa, J.; Sassanelli, C.; Gusmeroli, S.; Jardim-Goncalves, R. Towards A Conceptual Framework for Developing Sustainable Digital Innovation Hubs. In Proceedings of the 27th ICE/IEEE International Technology Management Conference, Cardiff, UK, 23 June 2021; IEEE: New York, NY, USA, 2021; pp. 1–7.

13. Osterwalder, A.; Pigneur, Y.; Bernarda, G.; Smith, A. *Value Proposition Design*; Wiley: Hoboken, NJ, USA, 2014; ISBN 9781118973103.

14. HUBCAP Project. Available online: https://www.hubcap.eu/ (accessed on 14 January 2022).

15. Larsen, P.G.; Macedo, D.H.; Fitzgerald, J.; Pfeifer, H.; Benedict, M.; Tonetta, S.; Marguglio, A.; Gusmeroli, S.; Suciu, G. A Cloud-Based Collaboration Platform for Model-Based Design of Cyber-Physical Systems. In Proceedings of the 10th International Conference on Simulation and Modeling Methodologies, Technologies and Applications, Lieusaint, France, 8–10 July 2020.

16. DIH4CPS Project. Available online: http://dih4cps.eu/ (accessed on 9 May 2020).

17. European Commission. European Digital Innovation Hubs—Shaping Europe's Digital Future. Available online: https://digital-strategy.ec.europa.eu/en/activities/edihs (accessed on 19 August 2022).

18. Egiluz, Z.; Cuadrado, J.; Kortazar, A.; Marcos, I.; Blanco Álvarez, A.; Bosch, P.; Pujadas Álvarez, P. Multi-Criteria Decision-Making Method for Sustainable Energy-Saving Retrofit Façade Solutions. *Sustainability* **2021**, *13*, 13168. [CrossRef]

19. Sassanelli, C.; Arriga, T.; Zanin, S.; D'Adamo, I.; Terzi, S. Industry 4.0 Driven Result-Oriented PSS: An Assessment in the Energy Management. *Int. J. Energy Econ. Policy* **2022**, *12*, 186–203. [CrossRef]

20. Aragonés, M.M.; Nieto, G.D.L.V.; Fajardo, M.N.; Rodríguez, D.P.; Gaffey, J.; Attard, J.; McMahon, H.; Doody, P.; Ugarte, J.A.; Pérez-Camacho, M.N.; et al. Digital Innovation Hubs as a Tool for Boosting Biomass Valorisation in Regional Bioeconomies: Andalusian and South-East Irish Case Studies. *J. Open Innov. Technol. Mark. Complex.* **2020**, *6*, 115. [CrossRef]

21. Tsangas, M.; Jeguirim, M.; Limousy, L.; Zorpas, A. The Application of Analytical Hierarchy Process in Combination with PESTEL-SWOT Analysis to Assess the Hydrocarbons Sector in Cyprus. *Energies* **2019**, *12*, 791. [CrossRef]

22. Quesado, P.; Silva, R. Activity-Based Costing (ABC) and Its Implication for Open Innovation. *J. Open Innov. Technol. Mark. Complex.* **2021**, *7*, 41. [CrossRef]

23. Reedy, J.; Lunzman, S.; Mekari, B. *Model Based Design Accelerates the Development of Mechanical Locomotive Controls*; SAE Technical Paper; SAE: Warrendale, PA, USA, 2010. [CrossRef]

24. Ahmadian, M. Model Based Design and SDR. In Proceedings of the 2nd IEE/EURASIP Conference on DSPenabledRadio; Institution of Engineering and Technology (IET), Southampton, UK, 19–20 September 2005.

25. MATLAB & Simulink. Why Adopt Model-Based Design? Available online: https://www.mathworks.com/content/dam/mathworks/white-paper/why-adopt-model-based-design-white-paper.pdf (accessed on 10 March 2022).

26. Barker, P. Enterprise 2.0: How Social Software Will Change the Future of Work. *Electron. Libr.* **2010**, *28*, 350–351. [CrossRef]

27. Larsen, P.G.; Fitzgerald, J.; Woodcock, J.; Fritzson, P.; Brauer, J.; Kleijn, C.; Lecomte, T.; Pfeil, M.; Green, O.; Basagiannis, S.; et al. Integrated Tool Chain for Model-Based Design of Cyber-Physical Systems: The INTO-CPS Project. In Proceedings of the 2016 2nd International Workshop on Modelling, Analysis, and Control of Complex CPS, CPS Data 2016, Vienna, Austria, 11 April 2016; Institute of Electrical and Electronics Engineers Inc.: New York, NY, USA, 2016.

28. Macedo, H.D.; Rasmussen, M.B.; Thule, C.; Larsen, P.G. Migrating the Into-Cps Application to the Cloud. In Proceedings of the Formal Methods 2019 International Workshops, Porto, Portugal, 7–11 October 2019; Springer: Berlin/Heidelberg, Germany, 2020; Volume 12233, pp. 254–271.

29. Thule, C.; Palmieri, M.; Gomes, C.; Lausdahl, K.; Macedo, H.D.; Battle, N.; Larsen, P.G. Towards Reuse of Synchronization Algorithms in Co-Simulation Frameworks. In *Software Engineering and Formal Methods*; Springer Science and Business Media: Berlin/Heidelberg, Germany, 2020; Volume 12226, pp. 50–66. ISBN 9783030575052.

30. O'Mahony, S.; Bechky, B.A. Boundary Organizations: Enabling Collaboration among Unexpected Allies. *Adm. Sci. Q.* **2008**, *53*, 422–459. [CrossRef]

31. Dattee, B.; Alexy, O.; Autio, E. Maneuvering in Poor Visibility: How Firms Play the Ecosystem Game When Uncertainty Is High. *Acad. Manag. J.* **2018**, *61*, 466–498. [CrossRef]

32. Dedehayir, O.; Ortt, J.R.; Seppänen, M. Disruptive Change and the Reconfiguration of Innovation Ecosystems. *J. Technol. Manag. Innov.* **2017**, *2017*, 12. [CrossRef]

33. Dedehayir, O.; Mäkinen, S.J.; Roland Ortt, J. Roles during Innovation Ecosystem Genesis: A Literature Review. *Technol. Forecast. Soc. Change* **2018**, *136*, 18–29. [CrossRef]

34. Talmar, M.; Walrave, B.; Podoynitsyna, K.S.; Holmström, J.; Romme, A.G.L. Mapping, Analyzing and Designing Innovation Ecosystems: The Ecosystem Pie Model. *Long Range Plann.* **2020**, *53*, 1–9. [CrossRef]

35. Moore, G.A. *Crossing the Chasm: Marketing and Selling Disruptive Products to Mainstream Customers*, 3rd ed.; Harperbusiness, Ed.; HarperCollins: Warsaw, Poland, 2014; ISBN 0062292986.
36. Christensen, C.M.; Hall, T.; Dillon, K.; Duncan, D.S. (Eds.) *Competing against Luck: The Story of Innovation and Customer Choice*, 1st ed.; Harper Business: New York, NY, USA, 2016; ISBN 0062435612.
37. D'Adamo, I.; Gastaldi, M.; Morone, P.; Rosa, P.; Sassanelli, C.; Settembre-blundo, D.; Shen, Y. Bioeconomy of Sustainability: Drivers, Opportunities and Policy Implications. *Sustainability* **2022**, *14*, 200. [CrossRef]
38. Appolloni, A.; Chiappetta Jabbour, C.J.; D'Adamo, I.; Gastaldi, M.; Settembre-Blundo, D. Green recovery in the mature manufacturing industry: The role of the green-circular premium and sustainability certification in innovative efforts. *Ecol. Econ.* **2022**, *193*, 1–9. [CrossRef]

 *sustainability*

Article

An IVIF-Distance Measure and Relative Closeness Coefficient-Based Model for Assessing the Sustainable Development Barriers to Biofuel Enterprises in India

Arunodaya Raj Mishra [1], **Pratibha Rani** [2], **Fausto Cavallaro** [3,*] **and Ibrahim M. Hezam** [4]

[1] Department of Mathematics, Government College Raigaon, Satna 485441, Madhya Pradesh, India
[2] Department of Engineering Mathematics, Koneru Lakshmaiah Education Foundation, Vaddeswaram 522302, Andhra Pradesh, India
[3] Department of Economics, University of Molise, Via De Sanctis, 86100 Campobasso, Italy
[4] Department of Statistics & Operations Research, College of Sciences, King Saud University, Riyadh 11495, Saudi Arabia
* Correspondence: cavallaro@unimol.it

Abstract: Biofuel can become a favorable sustainable energy resource in India by relieving conventional fossil fuels. However, biofuel enterprises (BEs) are still in the preliminary phase because of sustainable development barriers (SDBs) in environmental, technological, economic, social, and regulatory aspects. In the paper, nineteen SDBs to biofuels are identified by studying the literature and decision experts' (DEs') views. Considering the involvement of multiple tangible and non-tangible barriers, the assessment of SDBs to BEs can be taken as a multi-attribute decision-analysis (MADA) problem. Since ambiguity and imprecision generally ensue in the assessment of SDBs to BEs, the doctrine of interval-valued intuitionistic fuzzy sets (IVIFSs) has been recognized as a more sensible and proficient way to tackle uncertain MADA problems. Then, an integrated approach with IVIF-distance measure and IVIF-relative closeness coefficient models is presented to form associations between the SDBs to recognize the most important SDBs. The outcomes of this study show that four SDBs, i.e., "lack of effective storage facilities (EC-2), lack of investors (EC-3), technical issues associated with conversion technologies (T-2), and lack of trust between local societies, agencies, and developers (S-4)" are the leading obstacles. The paper also discusses some policies that can be utilized as a managing stage by the DEs to articulate guidelines for the operational exclusion of SDBs to biofuel enterprises.

Keywords: interval-valued intuitionistic fuzzy sets; distance measure; closeness coefficient; sustainability; barriers; MADA; biofuel sector

Citation: Mishra, A.R.; Rani, P.; Cavallaro, F.; Hezam, I.M. An IVIF-Distance Measure and Relative Closeness Coefficient-Based Model for Assessing the Sustainable Development Barriers to Biofuel Enterprises in India. *Sustainability* **2023**, *15*, 4354. https://doi.org/10.3390/su15054354

Academic Editors: Idiano D'Adamo and Massimo Gastaldi

Received: 20 January 2023
Revised: 22 February 2023
Accepted: 23 February 2023
Published: 28 February 2023

1. Introduction

Nowadays, people are relentlessly concerned about global warming because of the increase in the world population, the diminution of customary "fossil fuels (FFs)", and pollution produced by automobiles [1]. Henceforth, conventional fuels need to be replaced by "renewable energy resources (RESs)". As an RES, biomass can represent a solution that contributes to supplying the electricity demands and producing high-density fuels. However, biomass is an inadequate RES and can only complement variable RESs for electricity generation and electrification of transportation sectors [2,3]. Biomass-based fuel blends as an alternative can play an important part in achieving "sustainable development (SD)" and improving energy safety. Biofuels and electric vehicles have started to be used to reach the targets of "sustainable development goals (SDGs)" [4]. At present, biodiesel is the major biofuel utilized in the European Union for transport [5]. Moreover, ethanol is usually utilized as a blend with a low proportion of fossil fuels. Consequently, scientists and governments have started to pay attention to hydrocarbon fuels as the target product of biomass refinement technology.

Since India is a huge agricultural nation, the growth of the biofuel economy would cause savings in traditional fuels and support the alleviation of pollution concerns to a certain level. The utilization of biomass considerably supports the reduction of carbon emissions and allows the growth of rural regions because biofuels can be utilized for manufacturing transportation fuel, electricity, and heat [6–8]. The agriculture of biomass delivers rural progress prospects and agriculture variation, and the energy achieved from biomass will have further social recognition, as it is associated with the diverse practices of RESs [9]. Biofuels are a promising energy resource produced from different plant oils, waste oils, microbial lipids, food crops, agricultural residues, and animal fats, and have a massive possibility to meet more than a quarter of the world demand for transportation fuels by 2050, especially in developing countries such as India. India has around 500 MTs of biomass produced per year, out of which 120 to 150 MTs are additional. Additionally, 12.83% of the whole RE production is funded by biofuels only [10]. Furthermore, advanced adaptation efficiencies and minor costs are the noteworthy drivers of bioenergy abstraction [11].

India is the fourth-place net importer and user of crude oil and petroleum goods behind the USA, China, and Japan [12]. Moreover, India is the fourth-place emitter of "greenhouse gases (GHGs)", and the nation's transportation field produces 13% of CO_2 emissions [8,13]. These emissions owing to transportation can be minimized using sustainable methods, for example, the practice of public transit, more and more use of biofuels, and refining vehicle proficiencies. Since oil is the second-biggest energy source next to coal with a share of 30.5% of prime energy consumption in India [12], the growth of RESs or other alternatives needs to be produced successfully and resourcefully to replace or enhance petrol family oils. Biofuels are evolving as the most favorable energy choices to conventional fuels. Biofuels have momentous benefits for national energy safety, alleviation of GHGs, and rural growth [14]. Biomass-based energy can address several concerns associated with energy safety and organization [15]. This produces an important barrier for biofuel energy abstraction policies. It is noticed that energy abstraction from biomass should not be taken as an economic or technological issue. Additionally, it comprises public views about the risks elaborated in the procedure of growth. Additionally, it is claimed that the utilization of comestible biomass for making fuel may have an adversative impact on food safety and inflate food costs [7,12]. Furthermore, agriculture imitative biofuels are related to ecological damage and they are less efficient as compared to traditional fuel oils, which raises biofuel consumption [16].

Sustainability mobility raises a certain number of questions. Over the last decades, most industrialized countries have introduced strict regulations limiting the environmental impact produced by combustion engines. Governments pressing the car industry to electrify their products in order to assure the ecological transition of mobility. This has forced car manufacturers to invest heavily in research and development into alternative fuels and new propulsion systems. To encourage this transition, governments offer incentives and benefits, such as tax exemption (ownership tax) or free access to restricted traffic zones (ZTLs). As argued by D'Adamo et al. [17], the real weight of this transition is charged to the customers that are stimulated to change their traditional vehicles but at the same time, they have to buy hybrid/electric cars with high costs. Ecological benefits could, of course, be obtained by considering alternative fuels, but also by considering the "end of life" strategies based on the circular economy approach [18]. The sustainable transition also needs an original social approach involving citizens in the decision-making activity. Moreover, incentives and economic measures should be provided to stimulate the dissemination of small-scale plants in the territory and the creation of energy communities [17].

It may be seen that the "biofuel enterprise (BE)" is in its initial phase in India because of numerous "sustainable development barriers (SDBs)" in different aspects of sustainability. The SDBs are inter-reliant, and there exists a cause-and-influence association with the SDBs. Assessment of SDB pillars considers several barriers in the biofuel industry. Since the assessment of SDBs involves diverse aspects and uncertainty, it can be considered a "multi-attribute decision analysis (MADA)" issue with uncertainty. The "interval-valued intuitionistic fuzzy sets (IVIFSs)" [19] can treat imprecise and uncertain data in numerous realistic settings. As the generalization of the fuzzy set, the theory of IVIFS is characterized by the membership grade (MG) and non-membership grade (NG), expressed in the form of intervals rather than exact numbers. As IVIFSs can effectively deal with the MADA concern with ambiguity and fuzziness, some extant studies have applied IVIFS-based models from the perspective of SD [20–22].

In this research, the precarious SDBs to BEs in India are recognized and demonstrated using an integrated MADA procedure for identifying the limitations and their cause-influence association, which is missing in earlier studies. This paper also plans to offer a few policies, which may be utilized by "decision experts (DEs)" and executives to articulate appropriate strategies for the operative eradication of recognized SDBs. We present the notable research contributions of the paper as follows:

- This study classifies the crucial SDBs to the biofuel industry in India and assesses the association between the recognized barriers. However, existing methods given by [1,23–25] are not able to identify and assess the SDBs in the biofuel industry.
- Distance measure, as one of the important information measures, plays a vital role in real-life problems such as decision-making, pattern recognition, texture recognition, and so forth. In this study, we propose a new IVIF-distance measure with enviable properties to measure the degree of discrimination between IVIFSs.
- Direct assumption of decision experts' (DEs') weights results in loss of information while making decisions. Thus, it is very important to determine the weights of DEs during the process of decision-making. In this paper, we propose a new IVIF-score value and rank sum (RS) model-based weighting approach to derive the DEs' weights within the IVIFS context.
- In order to consider the relative closeness coefficient of barriers, this paper presents a new IVIF-distance-based model and uses it to find the objective and subjective weight of barriers to prioritize the SDBs in the biofuel industry.

2. Literature Review

2.1. Studies on the Biofuel Sector

This section entails a review of the biofuel sector. With the use of the gray DEMATEL approach, Liang et al. [1] identified the critical success factors for enhancing the sustainability of China's biofuel sector. That study has effectively recognized the success factors but does not consider the barriers to SD in China's biofuel sector. Jernstrom et al. [23] identified the opportunities and analyzed the barriers to entry for small/medium enterprises in biofuel-based sectors. However, their study is unable to express the uncertainty and vagueness of real-life problems, while practical decision-making problems usually involve uncertainty due to time limitations and the subjectivity of the human mind. Saravanan et al. [24] studied strategy barriers to biofuel marketing from an Indian perspective and they underlined the efforts of the public and government to overcome these barriers. That study only considers an empirical study but does not provide any tool to identify the strategy barriers. In addition, their study is not able to handle the biofuel marketing decision-making problems from an uncertainty perspective. Malode et al. [25] provided the theoretical aspects of recent advances and the possibility of biofuel production in the biofuel sector. Unfortunately, there is no study that identifies the critical SDBs to the biofuel industry in India from an uncertainty perspective.

2.2. Review on IVIFSs and MADA

The theory of IVIFS has been given by Atanassov and Gargov [19] for treating uncertain information in realistic MADA problems. Numerous scholars have employed IVIFSs to develop MADA models for handling realistic issues with uncertain settings [26]. Firstly, Xu [27] discussed diverse basic "aggregation operators (AOs)" to aggregate the information and score and accuracy functions to rank the IVIFNs. Wang and Mendel [28] proposed a decision-making method based on the Lukasiewicz triangular norm. Moreover, their study presented the drawbacks of existing studies on IVIFSs. In a study, Hu et al. [29] developed a novel entropy-weighted TOPSIS methodology with interval-valued intuitionistic fuzzy information. Their application presented the assessment of technology portfolios of clean energy-driven desalination-irrigation systems. Mishra et al. [30] introduced a divergence and entropy measure-based decision support system for assessing the service quality problem. For this purpose, they proposed some divergence measures to quantify the degree of discrimination between IVIFSs and entropy measures to quantify the uncertainty of IVIFSs. Oraki et al. [31] defined some frank t-norm and t-conorm operations on IVIFNs. Further, they proposed a list of frank AOs by analyzing the limitations of existing AOs under the IVIFS context. Bharati [32] studied a new ranking method by means of the law of trichotomy. In addition, their applicability has been tested on a transportation problem under an interval-valued intuitionistic fuzzy environment. As per our investigation, no study has used the theory of IVIFS for assessing SDBs in the biofuel industry.

3. IVIF-Distance Measure

3.1. Preliminaries

Here, some essential concepts of IVIFSs are discussed.

Atanassov and Gargov [19] extended IVIFSs based on IFSs to handle the uncertainty, which is exemplified by the "membership grade (MG)" and "non-membership grade (NG)" in interval form.

Definition 1 [19]. *Let $\Omega = \{x_1, x_2, \ldots, x_n\}$ be a fixed set. An IVIFS P on Ω is described as $P = \{\langle x_i, \mu_P(x_i), v_P(x_i)\rangle : x_i \in \Omega\}$, where $\mu_P(x_i) = [\mu_P^-(x_i), \mu_P^+(x_i)] : \Omega \to [0, 1]$ and $v_P = [v_P^-(x_i), v_P^+(x_i)] : \Omega \to [0, 1]$ hold $\sup(\mu_P(x_i)) + \sup(v_P(x_i)) \leq 1$. The intervals $\mu_P(x_i)$ and $v_P(x_i)$ indicate the MG and NG of the variable x_i in Ω, respectively.*

The interval $\pi_P(x_i) = [\pi_P^-(x_i), \pi_P^+(x_i)] = [1 - \mu_P^+(x_i) - v_P^+(x_i), 1 - \mu_P^-(x_i) - v_P^-(x_i)]$ signifies the "hesitancy grade (HG)" of x_i to P. The pair $\left([\mu_P^-(x_i), \mu_P^+(x_i)], [v_P^-(x_i), v_P^+(x_i)]\right)$ is termed an IVIFN [27] and is commonly denoted by $\theta = ([p, q], [0, 1])$, where $[p, q] \subset [0, 1]$, $[r, s] \subset [0, 1]$, and $q + s \leq 1$.

Definition 2 [17]. *Let $P, Q \in IVIFSs(\Omega)$. Some basic operations on IVIFSs are defined as*

(a) $P \subseteq Q$ if and only if $\mu_P^-(x_i) \leq \mu_Q^-(x_i)$, $\mu_P^+(x_i) \leq \mu_Q^+(x_i)$, $v_P^-(x_i) \geq v_Q^-(x_i)$ and $v_P^+(x_i) \geq v_Q^+(x_i), \forall x_i \in \Omega,$

(b) $P = Q$ if and only if $P \subseteq Q$ and $P \supseteq Q,$

(c) $P^c = \{\langle \alpha_i, [v_P^-(\alpha_i), v_P^+(\alpha_i)], [\mu_P^-(\alpha_i), \mu_P^+(\alpha_i)]\rangle \mid \alpha_i \in \alpha\},$

(d) $P \cup Q = \left\{\left\langle \alpha_i, \begin{array}{l}[\mu_P^-(\alpha_i) \vee \mu_Q^-(\alpha_i), \mu_P^+(\alpha_i) \vee \mu_Q^+(\alpha_i)], \\ [v_P^-(\alpha_i) \wedge v_Q^-(\alpha_i), v_P^+(\alpha_i) \wedge v_Q^+(\alpha_i)]\end{array}\right\rangle \middle| \alpha_i \in \alpha\right\},$

(e) $P \cap Q = \left\{\left\langle \alpha_i, \begin{array}{l}[\mu_P^-(\alpha_i) \wedge \mu_Q^-(\alpha_i), \mu_P^+(\alpha_i) \wedge \mu_Q^+(\alpha_i)], \\ [v_P^-(\alpha_i) \vee v_Q^-(\alpha_i), v_P^+(\alpha_i) \vee v_Q^+(\alpha_i)]\end{array}\right\rangle \middle| \alpha_i \in \alpha\right\}.$

Definition 3 [27]. *Consider $\theta = ([p, q], [r, s])$ be an IVIFN, then $\mathbb{S}(\theta) = \frac{1}{2}(p + q - r - s)$ and $\hbar(\theta) = \frac{1}{2}(p + q + r + s)$ are said to be IVIF-score and IVIF-accuracy values of θ, respectively.*

Bai [33] pioneered the improved score value using the HD between the BD and ND of IVIFNs.

Definition 4 [33]. *Let $\theta = ([p, q], [r, s])$ be an IVIFN. Then,*

$$\mathbb{S}^*(\theta) = \frac{p + p(1 - p - r) + q + q(1 - q - s)}{2} \tag{1}$$

is known as an improved score function, where $\mathbb{S}^*(\theta) \in [0, 1]$.

Definition 5 [27]. *For a set of IVIFNs* $P = \{P_1, P_2, \ldots, P_\ell\}$, *where* $P_k = ([p_k, q_k], [r_k, s_k])$, $k = 1, 2, \ldots, \ell$, *the IVIFWA operator is given*

$$\overset{\ell}{\underset{k=1}{\oplus}} \xi_k P_k = \left(\left[1 - \prod_{k=1}^{\ell}(1 - p_k)^{\xi_k}, 1 - \prod_{k=1}^{\ell}(1 - q_k)^{\xi_k} \right], \left[\prod_{k=1}^{\ell}(r_k)^{\xi_k}, \prod_{k=1}^{\ell}(s_k)^{\xi_k} \right] \right). \quad (2)$$

Along a similar line, the IVIFWG operator is given by

$$\overset{\ell}{\underset{k=1}{\otimes}} \xi_k P_k = \left(\left[\prod_{k=1}^{\ell}(p_k)^{\xi_k}, \prod_{k=1}^{\ell}(q_k)^{\xi_k} \right], \left[1 - \prod_{k=1}^{\ell}(1 - r_k)^{\xi_k}, 1 - \prod_{k=1}^{\ell}(1 - s_k)^{\xi_k} \right] \right). \quad (3)$$

Definition 6 [34]. *An IVIF-distance measure* $d : IVIFSs(\Omega) \times IVIFSs(\Omega) \to [0, 1]$ *is a real-valued mapping that holds*

(C_1). $0 \leq d(P, Q) \leq 1$,
(C_2). $d(P, Q) = 0 \Leftrightarrow P = Q$,
(C_3). $d(P, Q) = 1 \Leftrightarrow Q = P^c$,
(C_4). $d(P, Q) = d(Q, P)$,
(C_5). If $P \subseteq Q \subseteq H$, then $d(P, H) \geq d(P, Q)$ and $d(P, H) \geq d(Q, H)$, for all $F, G, H \in IVIFSs(\Omega)$.

3.2. Proposed IVIF-Distance Measure

The objective of the section is to develop new IVIF-distance measures and then employ them to originate the attribute weight in the next section. Based on Tripathi et al. [35], a distance measure is developed for IVIFSs.

For $P, Q \in IVIFSs(\Omega)$, we develop a new IVIF-distance measure for estimating the discrimination between two IVIFSs, given as

$$d_1(P, Q) = \frac{1 - \exp\left[-\frac{1}{2} \left(\sum_{i=1}^{t} \left(\begin{array}{c} \left| \mu_P^-(x_i) - \mu_Q^-(x_i) \right|^\gamma + \left| \mu_P^+(x_i) - \mu_Q^+(x_i) \right|^\gamma + \left| \nu_P^-(x_i) - \nu_Q^-(x_i) \right|^\gamma \\ + \left| \nu_P^+(x_i) - \nu_Q^+(x_i) \right|^\gamma + \left| \pi_P^-(x_i) - \pi_Q^-(x_i) \right|^\gamma + \left| \pi_P^+(x_i) - \pi_Q^+(x_i) \right|^\gamma \end{array} \right) \right)^{1/\gamma} \right]}{1 - \exp\left(-(t)^{1/\gamma} \right)}, \quad (4)$$

where $\gamma > 0$, $\gamma \neq 1$.

Lemma 1. *If* $h(\lambda) = 1 - \dfrac{1 - \exp(-\lambda)}{1 - \exp\left(-(t)^{1/\gamma} \right)}$, *then*

$$\min_{\lambda \in [0, t]} h(\lambda) = h(0) = 0 \text{ and } \max_{\lambda \in [0, t]} h(\lambda) = h(t) = 1.$$

Proof. Since $h'(\lambda) = \dfrac{\exp(-\lambda)}{1 - \exp\left(-(t)^{1/\gamma} \right)} < 0$, $\forall \lambda \in [0, t]$, therefore, $h(\lambda)$ is increasing in $[0, t]$. $\square$

Theorem 1. *The measure* $d_1(P, Q)$ *in Equation (4) is a valid IVIF-distance measure.*

Proof. In this regard, $d_1(P, Q)$ must fulfill the axioms (C_1)–(C_5) of Definition 6. (C_1). Let $P, Q \in IVIFSs(\Omega)$, and

$$\lambda = \frac{1}{2}\left(\sum_{i=1}^{t}\left(\begin{array}{l} \left|\mu_P^-(x_i) - \mu_Q^-(x_i)\right|^{\gamma} + \left|\mu_P^+(x_i) - \mu_Q^+(x_i)\right|^{\gamma} + \left|v_P^-(x_i) - v_Q^-(x_i)\right|^{\gamma} \\ + \left|v_P^+(x_i) - v_Q^+(x_i)\right|^{\gamma} + \left|\pi_P^-(x_i) - \pi_Q^-(x_i)\right|^{\gamma} + \left|\pi_P^+(x_i) - \pi_Q^+(x_i)\right|^{\gamma} \end{array} \right) \right)^{1/\gamma}.$$

Since $\lambda \in [0, t]$, therefore, $d_1(P, Q) = h(\lambda)$. Thus, utilizing Lemma 1, we have $0 \leq d_1(P, Q) \leq 1$.

(C_2). Let $P = Q$. Then $\mu_P^-(x_i) = \mu_Q^-(x_i)$, $\mu_P^+(x_i) = \mu_Q^+(x_i)$, $v_P^-(x_i) = v_Q^-(x_i)$ and $v_P^+(x_i) = v_Q^+(x_i)$, $\forall_{x_i \in \Omega}$. Then, it is obvious from Equation (4) that $d_1(P, Q) = 0$.

Let $d_1(P, Q) = 0$. From Equation (4), we have

$$\frac{1 - \exp\left[-\frac{1}{2}\left(\sum_{i=1}^{t}\left(\begin{array}{l} \left|\mu_P^-(x_i) - \mu_Q^-(x_i)\right|^{\gamma} + \left|\mu_P^+(x_i) - \mu_Q^+(x_i)\right|^{\gamma} + \left|v_P^-(x_i) - v_Q^-(x_i)\right|^{\gamma} \\ + \left|v_P^+(x_i) - v_Q^+(x_i)\right|^{\gamma} + \left|\pi_P^-(x_i) - \pi_Q^-(x_i)\right|^{\gamma} + \left|\pi_P^+(x_i) - \pi_Q^+(x_i)\right|^{\gamma} \end{array} \right) \right)^{1/\gamma} \right]}{1 - \exp\left(-(t)^{1/\gamma} \right)} = 0.$$

It implies that

$$\sum_{i=1}^{t}\left(\begin{array}{l} \left|\mu_P^-(x_i) - \mu_Q^-(x_i)\right|^{\gamma} + \left|\mu_P^+(x_i) - \mu_Q^+(x_i)\right|^{\gamma} + \left|v_P^-(x_i) - v_Q^-(x_i)\right|^{\gamma} \\ + \left|v_P^+(x_i) - v_Q^+(x_i)\right|^{\gamma} + \left|\pi_P^-(x_i) - \pi_Q^-(x_i)\right|^{\gamma} + \left|\pi_P^+(x_i) - \pi_Q^+(x_i)\right|^{\gamma} \end{array} \right) = 0, \forall_{x_i \in \Omega}.$$

Hence $P = Q$.

(C_3). It is obvious from the definition that $d_1(P, Q) = 1 \Leftrightarrow Q = P^c$.

(C_4). Clearly, $d_1(P, Q) = d_1(Q, P)$.

(C_5). Let $P \subseteq Q \subseteq H$, then $\mu_P^-(x_i) \leq \mu_Q^-(x_i) \leq \mu_H^-(x_i)$, $\mu_P^+(x_i) \leq \mu_Q^+(x_i) \leq \mu_H^+(x_i)$, $v_P^-(x_i) \geq v_Q^-(x_i) \geq v_H^-(x_i)$ and $v_P^+(\alpha_i) \geq v_Q^+(\alpha_i) \geq v_H^+(\alpha_i)$, $\forall_{\alpha_i \in \alpha}$.

Now,

$$\lambda_1 = \frac{1}{2}\sum_{i=1}^{t}\left(\begin{array}{l} \left|\mu_P^-(x_i) - \mu_Q^-(x_i)\right|^{\gamma} + \left|\mu_P^+(x_i) - \mu_Q^+(x_i)\right|^{\gamma} + \left|v_P^-(x_i) - v_Q^-(x_i)\right|^{\gamma} \\ + \left|v_P^+(x_i) - v_Q^+(x_i)\right|^{\gamma} + \left|\pi_P^-(x_i) - \pi_Q^-(x_i)\right|^{\gamma} + \left|\pi_P^+(x_i) - \pi_Q^+(x_i)\right|^{\gamma} \end{array} \right)$$

$$\leq \lambda_2 = \frac{1}{2}\sum_{i=1}^{t}\left(\begin{array}{l} \left|\mu_P^-(x_i) - \mu_H^-(x_i)\right|^{\gamma} + \left|\mu_P^+(x_i) - \mu_H^+(x_i)\right|^{\gamma} + \left|v_P^-(x_i) - v_H^-(x_i)\right|^{\gamma} \\ + \left|v_P^+(x_i) - v_H^+(x_i)\right|^{\gamma} + \left|\pi_P^-(x_i) - \pi_H^-(x_i)\right|^{\gamma} + \left|\pi_P^+(x_i) - \pi_H^+(x_i)\right|^{\gamma} \end{array} \right), \forall_{x_i \in \Omega}.$$

From Lemma 1, we find $d_1(P, Q) = h(\lambda_1) \leq h(\lambda_2) = d_1(P, H)$. In the same way, we can prove that $d_1(Q, H) \leq d_1(P, H)$. Hence, $d_1(P, Q)$ is a suitable IVIF-distance measure. $\square$

Next, an IVIF-distance measure between two matrices is discussed as follows:

Let $P = \left(p_{ij} \right)$ and $Q = \left(q_{ij} \right)$, $i = 1(1)s, j = 1(1)t$ be two IVIF matrices, where $p_{ij} = \left(\left[\mu_{ij}^{-p}, \mu_{ij}^{+p} \right], \left[v_{ij}^{-p}, v_{ij}^{+p} \right] \right)$ and $q_{ij} = \left(\left[\mu_{ij}^{-q}, \mu_{ij}^{+q} \right], \left[v_{ij}^{-q}, v_{ij}^{+q} \right] \right)$ are IVIFNs. Thus, the distance measure between P and Q is proposed as

$$d_2(P, Q) = \frac{1 - \exp\left[-\frac{1}{2st}\left(\sum_{i=1}^{s}\sum_{j=1}^{t}\left(\begin{array}{l} \left|\mu_{ij}^{-p} - \mu_{ij}^{-q}\right|^{\gamma} + \left|\mu_{ij}^{+p} - \mu_{ij}^{+q}\right|^{\gamma} + \left|v_{ij}^{-p} - v_{ij}^{-q}\right|^{\gamma} \\ + \left|v_{ij}^{+p} - v_{ij}^{+q}\right|^{\gamma} + \left|\pi_{ij}^{-p} - \pi_{ij}^{-q}\right|^{\gamma} + \left|\pi_{ij}^{+p} - \pi_{ij}^{+q}\right|^{\gamma} \end{array} \right) \right)^{1/\gamma} \right]}{1 - \exp(-1)}, \tag{5}$$

where $\gamma > 0, \gamma \neq 1$.

Theorem 2. *The measure $d_2(P, Q)$ in Equation (5) is a valid IVIF-distance measure.*

Proof. The proof is omitted. $\square$

4. Proposed IVIF-DM-Relative Closeness Coefficient Model

This section suggests an integrated decision-analysis model known as the IVIF-DM-relative closeness coefficient model. The development of the IVIF-DM-relative closeness coefficient model is presented and depicted in Figure 1.

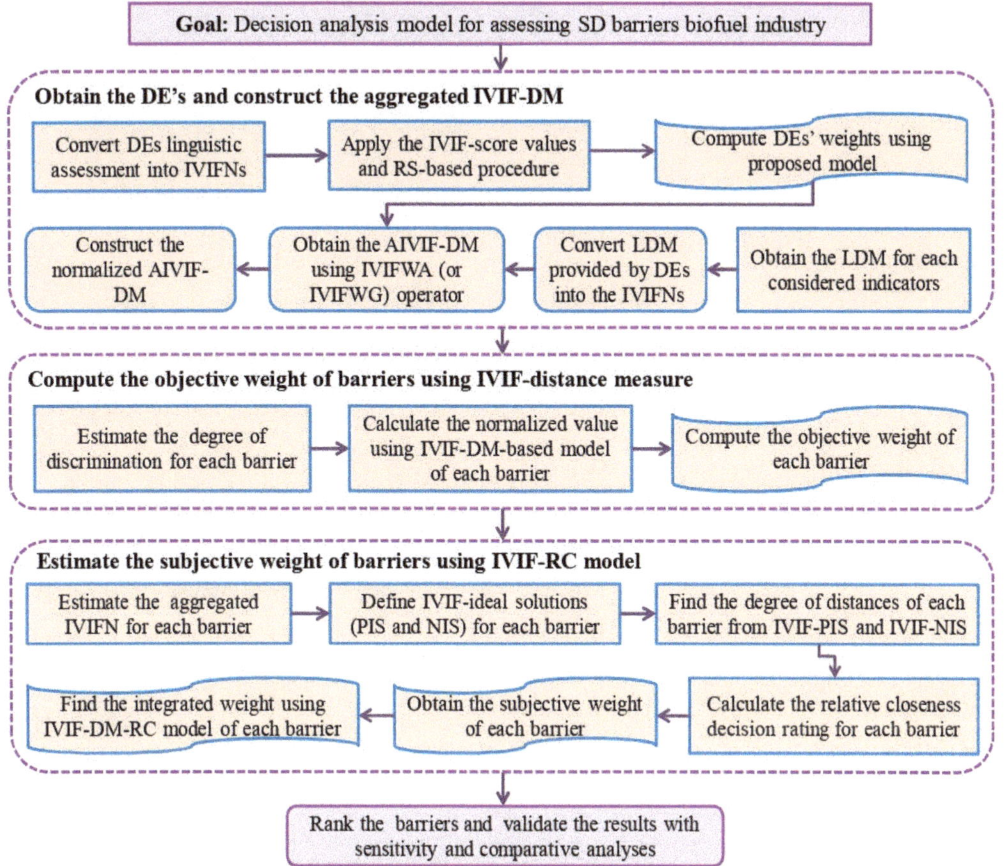

Figure 1. Flowchart of the developed IVIF-DM-relative closeness coefficient model.

Step 1: Create a "linguistic decision matrix (LDM)".

Consider a set of n criteria/SDB $Q = \{q_1, q_2, \ldots, q_n\}$. We create a set of DEs $E = \{e_1, e_2, \ldots, e_l\}$ to evaluate the SDBs to the biofuel industry in India. An LDM is created based on DEs' opinions in which each DE presents a "linguistic rating (LR)" for each criterion q_j with respect to different alternatives/firms of the biofuel industry.

Step 2: Obtain the DE's weight (λ_k).

Initially, the evaluation ratings of DEs are defined as the LRs and then changed into IVIFNs. Let $\alpha_k = \left([\mu_k^-, \mu_k^+], [v_k^-, v_k^+]\right)$, $k = 1, 2, \ldots, l$ be the corresponding IVIFN and then the expression for finding DE's weight is given by

Step 2a: Find the IVIF-score matrix.

The normalized IVIF-score value $(\bar{\alpha}_k)$ of each IVIFN α_k is calculated as follows:

$$\bar{\alpha}_k = \frac{\mu_k^- + \mu_k^-\left(1 - \mu_k^- - v_k^-\right) + \mu_k^+ + \mu_k^+\left(1 - \mu_k^+ - v_k^+\right),}{\sum\limits_{k=1}^{l}\left(\mu_k^- + \mu_k^-\left(1 - \mu_k^- - v_k^-\right) + \mu_k^+ + \mu_k^+\left(1 - \mu_k^+ - v_k^+\right)\right)}, \quad k = 1, 2, \ldots, l. \tag{6}$$

Step 2b: Estimate the ranking of relevant assessment DE and find the DE's weight $l - r_k + 1$, where r_k is the priority of *kth* criterion. Each weight is normalized as follows:

$$(\overline{\alpha}_k^r) = \frac{l - r_k + 1}{\sum\limits_{k=1}^{l}(l - r_k + 1)}, k = 1, 2, \ldots, l. \tag{7}$$

Step 2c: Calculation of the expert's weight.

To find the DE's weight, we combine Equations (6) and (7) as follows:

$$\lambda_k = \frac{1}{2}((\overline{\alpha}_k) + (\overline{\alpha}_k^r)), k = 1, 2, \ldots, l, \text{ where } \lambda_k \geq 0 \text{ and } \sum\limits_{k=1}^{l}\lambda_k = 1. \tag{8}$$

Step 3: Create an "aggregated IVIF-DM (AIVIF-DM)".

All the IVIF-DMs are operated into AIVIF-DM. The IVIFWA (or IVIFWG) operator is utilized to generate the AIVIF-DM $Z = \left(\xi_{ij}\right)_{m \times n}$, where

$$\xi_{ij} = \left(\left[\mu_{ij}^-, \mu_{ij}^+\right], \left[v_{ij}^-, v_{ij}^+\right]\right) = IVIFWA_{\lambda_k}\left(\xi_{ij}^{(1)}, \xi_{ij}^{(2)}, \ldots, \xi_{ij}^{(l)}\right) \text{ or } IVIFWG_{\lambda_k}\left(\xi_{ij}^{(1)}, \xi_{ij}^{(2)}, \ldots, \xi_{ij}^{(l)}\right). \tag{9}$$

Step 4: Obtain the objective weight using the IVIF-distance measure weighting model.

The formula of the IVIF-distance-based weight-determining model for SDBs is presented as

$$w_j^o = \frac{\frac{1}{m-1}\sum\limits_{i=1}^{m}\sum\limits_{k=1}^{m} d_1\left(\xi_{ij}, \xi_{kj}\right)}{\sum\limits_{j=1}^{n}\left(\frac{1}{m-1}\sum\limits_{i=1}^{m}\sum\limits_{k=1}^{m} d_1\left(\xi_{ij}, \xi_{kj}\right)\right)}, j = 1(1)n. \tag{10}$$

where $\sum\limits_{j=1}^{n} w_j^o = 1$ and $w_j^o \in [0, 1]$.

Step 5: Estimate the A-IVIFNs by combining the LDM assessment degrees provided by DEs using the IVIFWA operator and obtained $G = \left(z_j\right)_{1 \times n}$.

Step 6: Describe the IVIF-ideal ratings.

An IVIFN has a positive ideal rating (IVIF-PIR) and a negative ideal rating (IVIF-NIR), which define grades $\phi^+ = (1, 0, 0)$ and $\phi^- = (0, 1, 0)$, respectively, while IVIF-PIR and IVIF-NIR are considered by maximum and minimum operators and it is found that there is no substantial gap in their results.

Step 7: Derive the degrees of discrimination of each SDB from IVIF-PIR and IVIF-NIR.

To compute the discrimination value, the proposed IVIF-distance measure is applied. Here, p_j^+ and p_j^- denote the positive and negative distance measures from $G = \left(\xi_j\right)_{1 \times n}$, therefore, the IVIF-PIR and IVIF-NIR, respectively.

$$p_j^+ = \frac{1 - \exp\left[-\frac{1}{2}\left(\sum\limits_{j=1}^{n}\left(\left|\mu_{\xi_j}^- - \mu_{\phi^+}^-\right|^\gamma + \left|\mu_{\xi_j}^+ - \mu_{\phi^+}^+\right|^\gamma + \left|v_{\xi_j}^- - v_{\phi^+}^-\right|^\gamma + \left|v_{\xi_j}^+ - v_{\phi^+}^+\right|^\gamma + \left|\pi_{\xi_j}^- - \pi_{\phi^+}^-\right|^\gamma + \left|\pi_{\xi_j}^+ - \pi_{\phi^+}^+\right|^\gamma\right)\right)^{1/\gamma}\right]}{1 - \exp\left(-(n)^{1/\gamma}\right)}, \tag{11}$$

$$p_j^- = \frac{1 - \exp\left[-\frac{1}{2}\left(\sum\limits_{i=1}^{n}\left(\left|\mu_{\xi_j}^- - \mu_{\phi^-}^-\right|^\gamma + \left|\mu_{\xi_j}^+ - \mu_{\phi^-}^+\right|^\gamma + \left|v_{\xi_j}^- - v_{\phi^-}^-\right|^\gamma + \left|v_{\xi_j}^+ - v_{\phi^-}^+\right|^\gamma + \left|\pi_{\xi_j}^- - \pi_{\phi^-}^-\right|^\gamma + \left|\pi_{\xi_j}^+ - \pi_{\phi^-}^+\right|^\gamma\right)\right)^{1/\gamma}\right]}{1 - \exp\left(-(n)^{1/\gamma}\right)}. \tag{12}$$

Step 8: Compute the relative closeness-decision rating (RC-DR).

$$rc_j = \frac{p_j^-}{p_j^- + p_j^+}, j = 1, 2, \ldots, n. \tag{13}$$

The RC-DR also states the optimization type (beneficial or non-beneficial) of each SDB to BEs.

Step 9: Obtain the subjective weight $\left(w_j^s\right)$ of each SDB as follows:

$$w_j^s = \frac{rc_j}{\sum_{j=1}^{n} rc_j}. \tag{14}$$

Step 10: Calculate the criteria weights by the IVIF-DM-relative closeness coefficient-based model.

To find the SDBs' weights, the IVIF-DM-relative closeness coefficient-based model is applied. Let $w = (w_1, w_2, \ldots, w_n)^T$ be the weight value of SDBs with $\sum_{j=1}^{n} w_j = 1$ and $w_j \in [0, 1]$. Then, the process for determining the attribute weight by the IVIF-DM-relative closeness coefficient-based model is discussed. With the use Equations (10)–(14), the integrated weight of SDB is defined as

$$w_j = \gamma w_j^s + (1 - \gamma)w_j^o, \, j = 1, 2, \ldots, n, \tag{15}$$

where $\gamma \in [0, 1]$ is the decision precision factor.

Step 11: Rank the SDBs.

Once all assessment degrees are calculated, finally, SDBs are ranked in descending order with their assessment scores. It should be stated that the SDBs with the largest degrees are the biggest obstacles among the other SDBs in the biofuel industry.

5. Case Study: Assessment of SDBs to BEs in India

In this article, twenty-five critical SDBs to BEs were recognized through the survey and DEs' opinions. Then, a questionnaire was created by inviting DEs from the enterprise and academia with at least fifteen years of experience. A DEs team (e_1, e_2, e_3, e_4, and e_5) is comprised of four sets: the first set contains two "supply chain (SC)" and logistics experts from case enterprise, the second set contains a professor from the agricultural science sector, the third set takes three farmers and environmental NGOs, and the fourth set contains a professor from the industrial engineering sector. The questionnaire then abridged the crucial SDBs to nineteen. The considered SDBs with five aspects, economic (*Ec*), environmental (*En*), social (*S*), technological (*T*), and regulatory (*R*), are revealed in Table 1. The respondent of each SDB is assessed using a 9-stage scale, where EL means extremely low and EH means extremely high, as presented in Tables 2 and 3.

Table 1. The assessment of SDBs to biofuel enterprises.

Dimensions	Barriers	Meaning	References
Economic (Ec)	Financial concerns during the whole lifespan of the plant (EC-1)	Financial problems that impact the SC performance and ambiguity related to return on investment are continuously an issue for stakeholders.	[9,36–38]
	Lack of effective storage services (EC-2)	Storage services need to be require enhanced, especially in the biomass zone.	[12]
	Lack of investors (EC-3)	The biofuel region has good prospects and investors must be fascinated to fund.	[12]
	High logistics costs (EC-4)	Logistic charge rises because of a lack of significantly sized resources, namely biomass.	[39,40]
Environmental (En)	By-products disposal with their chemical properties (EN-1)	Disposing of by-products is a key issue because of environmental pollution and chemical impacts.	[15,36,41]
	Emission of light at night (EN-2)	People continuously complain related to the emission of light at night from the biofuel plant.	[12,15]
	The minimum energy density of bioenergy (EN-3)	Fossil fuels ease effective transport; however, biomass has a minimum energy density problem.	[39–42]
	Emissions (water vapor and GHG) (EN-4)	Emission lessening must be taken into consideration for a "green image" of the enterprise.	[15,36–38,41]
Social (S)	Lack of entrepreneurship assistance (S-1)	Developing nations such as India can utilize social entrepreneurship.	[12,42]
	Unfriendly odor, noise, and vibration from the power plant (S-2)	Noise and vibration at power plants may cause accidents. The issue of odor must be addressed for a healthier working situation.	[15,34,41]
	Fear of public health and safety hazards (S-3)	Safety assessments must be conducted periodically to deal with the concern of public health and hazards.	[15,38,41]
	Lack of trust between local societies, enterprises, and inventors (S-4)	Owing to the lack of trust of diverse stakeholders, there is a suspension in plant expansion.	[38,41,43]
	Lack of public awareness of bioenergy technologies (S-5)	Government organizations and NGOs must be conducted awareness programs about bioenergy technologies.	[36,38]
Technological (T)	Seasonality of biomass (T-1)	Seasonality is an appropriate (weekly, monthly, or quarterly) occurrence of variation that ensued in a year. There are important technical and technological concerns.	[12,36,44]
	Technical issues about the conversion technologies (T-2)	Technical concerns in biofuel comprise fuel chain assessment, prolonged problems, and life cycle. Modern technological developments can be supportive.	[36,38,43]
	Lack of professional training institutions (T-3)	Training organizations must assist specialists, scholars, and DEs in training and education.	[12]
Regulatory (R)	Lack of administrative standards on SC coordination (R-1)	SC about the conversion, transport, records, and farming provide their own standards.	[36,38]
	Lack of biomass SC standards (R-2)	SC benchmarks must be defined predominantly for SC functioning in rural regions. SCM doctrines must be used by the inventors.	[36,38,41]
	Lack of governmental support for SSC solutions (R-3)	The Indian government must assist in solutions for SSC of effective employment in bioenergy.	[36,38,44]

Table 2. LRs for assessment of DEs.

LRs	IVIFNs
Extremely significant (ES)	([0.90, 0.95], [0.00, 0.05])
Very very significant (VVS)	([0.80, 0.85], [0.05, 0.10])
Very significant (VS)	([0.75, 0.85], [0.10, 0.15])
Significant (S)	([0.60, 0.70], [0.15, 0.30])
Moderate (M)	([0.50, 0.60], [0.30, 0.40])
Insignificant (I)	([0.30, 0.45], [0.45, 0.50])
Very insignificant (VI)	([0.20, 0.30], [0.50, 0.60])
Very very insignificant (VVI)	([0.10, 0.20], [0.60, 0.75])
Extremely insignificant (EI)	([0.00, 0.05], [0.80, 0.95])

Table 3. LRs for SDBs to the biofuel sector.

LRs	IVIFNs
Extremely good (EG)	([0.90, 0.95], [0.0, 0.05])
Very good (VG)	([0.80, 0.90], [0.05, 0.10])
Good (G)	([0.70, 0.80], [0.10, 0.15])
Slightly good (SG)	([0.65, 0.70], [0.15, 0.25])
Average (A)	([0.55, 0.65], [0.20, 0.35])
Slightly Low (SL)	([0.40, 0.50], [0.40, 0.45])
Low (L)	([0.25, 0.40], [0.45, 0.50])
Very Low (VL)	([0.15, 0.20], [0.60, 0.75])
Extremely Low (EL)	([0.05, 0.10], [0.80, 0.90])

Step 1: Each DE executes his/her views about the grading of SDBs to BEs. Here, Table 2 signifies the LRs in terms of IVIFNs to determine the weight value of DEs [30]. Table 3 articulates the LRs for evaluating the SDBs in the biofuel industry. Table 4 expresses the LDM of each DE's opinion for each SDB related to the different biofuel industries.

Table 4. The LDM for SDBs to the biofuel sector by DEs.

Barriers	T_1	T_2	T_3	T_4
q_1	(A,VG,SG,G,G)	(G,A,G,VG,SG)	(G, SG,A,G,A)	(SG,G,G,VG,SL)
q_2	(SL,G,A,VG,A)	(G,G,VL,SG,A)	(A,G,SG,SL,SL)	(SG,G,SG,VG,L)
q_3	(L,VG,SL,SG,G)	(SG,SL,G,VG,G)	(SL, G,VG,L,SG)	(VL,SL,VG,G,VG)
q_4	(VL,SL,A,G,VG)	(VL,G,VG,SL,G)	(VG,A,SL,SL,G)	(A,VG,SG,SL,SG)
q_5	(G,SG,A,SL,VG)	(VG,SG,A,A,G)	(A,SG,G,SG,SG)	(VG,G,G,SG,A)
q_6	(VG, G,VG,A,SG)	(SL,G,A,VG,SG)	(VG,SG,A,G,A)	(G,G,A,VG,SG)
q_7	(VG,SG,SL,L,VG)	(VG,SG,A,SL,SL)	(VG,VG,SG,SL,L)	(SL,G,VG,G,A)
q_8	(VL,SL,SG,VG,G)	(SL,L,SL,G,SG)	(L,SL,A,VG,VG)	(L,VG,A,SL,VG)
q_9	(L,SL,A,G,VG)	(A,SL,SL,VG,G)	(L,SG,G,SG,A)	(L,SL,SG,A,VG)
q_{10}	(A,SG,G,VG,L)	(VG,G,G,SL,VL)	(SG,SL,VG,G,A)	(SG,G,SL,G,A)
q_{11}	(VG,G,SG,G,A)	(SG,G,VG,SG,A)	(L,G,SG,G,SL)	(G,SG,G,VG,VL)
q_{12}	(L,SL,A,SG,VG)	(SL,SG,G,SL,G)	(G,A,SG,SL,G)	(A,SL,SG,A,VG)
q_{13}	(SG,G,A,L,VG)	(VG,G,A,A,SG)	(A,G,G,SL,SG)	(SG,G,SG,SG,A)
q_{14}	(G, SG,VG,A,SL)	(L,VG,A,G,SG)	(VG,SL,A,VG,A)	(SG,G,A,VG,G)
q_{15}	(SG,G,SL,VL,VG)	(G,SG,A,SG,SL)	(VG,G,SL,SL,L)	(SL,SG,VG,G,A)
q_{16}	(L,SL,G,VG,A)	(L,SL,SG,G,G)	(SL,SL,A,G,VG)	(SL,VG,A,SL,G)
q_{17}	(VL,SL,A,SG,VG)	(A,SG,SL,G,G)	(SL,SG,G,G,A)	(L,A,SG,A,VG)
q_{18}	(A,G,SG,VG,SL)	(SG,G,G,A,VL)	(G,SL,G,SG,A)	(SG,A,SG,G,A)
q_{19}	(G,G,VG,SG,A)	(G,SG,VG,G,A)	(SL,SG,VG,G,L)	(A,G,SG,VG,L)

Step 2: Based on the IVIFN scale given in Table 2 and Equations (6)–(8), the weights of DEs are computed and presented in Table 5 for the performance of SDBs selection of the biofuel industry.

Table 5. DEs' weights for SDBs to the biofuel sector.

	e_1	e_2	e_3	e_4	e_5
LRs	Significant	Moderate	Very Significant	Extremely significant	Very very significant
IVIFNs	([0.60, 0.70], [0.15, 0.30])	([0.50, 0.60], [0.30, 0.40])	([0.75, 0.85], [0.10, 0.15])	([0.90,0.95], [0.0,0.05])	([0.80,0.85], [0.05, 0.10])
$\bar{\alpha}_k$	0.7250	0.6000	0.8562	0.9700	0.9063
r_k	4	5	3	1	2
Weights	0.1560	0.1073	0.2055	0.2862	0.2450

Step 3: Applying Equation (9) and Tables 3 and 4, the aggregated IVIF-DM is constructed and shown in Table 6.

Table 6. The AIVIF-DM for SDBs to the biofuel sector.

Barriers	T_1	T_2	T_3	T_4
q_1	([0.684, 0.780], [0.112, 0.174])	([0.710, 0.808], [0.098, 0.157])	([0.634, 0.731], [0.143, 0.227])	([0.676, 0.781], [0.123, 0.183])
q_2	([0.654, 0.757], [0.151, 0.232])	([0.571, 0.657], [0.192, 0.279])	([0.523, 0.614], [0.253, 0.326])	([0.646, 0.751], [0.137, 0.199])
q_3	([0.601, 0.701], [0.175, 0.236])	([0.705, 0.807], [0.101, 0.157])	([0.585, 0.697], [0.183, 0.248])	([0.683, 0.800], [0.112, 0.181])
q_4	([0.626, 0.741], [0.149, 0.234])	([0.604, 0.720], [0.171, 0.243])	([0.586, 0.701], [0.191, 0.265])	([0.600, 0.684], [0.185, 0.256])
q_5	([0.634, 0.743], [0.151, 0.228])	([0.651, 0.753], [0.132, 0.220])	([0.647, 0.717], [0.144, 0.235])	([0.675, 0.769], [0.119, 0.188])
q_6	([0.698, 0.798], [0.105, 0.177])	([0.664, 0.766], [0.130, 0.202])	([0.656, 0.759], [0.128, 0.213])	([0.698, 0.797], [0.104, 0.171])
q_7	([0.611, 0.738], [0.162, 0.233])	([0.550, 0.658], [0.226, 0.310])	([0.575, 0.692], [0.195, 0.263])	([0.660, 0.770], [0.132, 0.202])
q_8	([0.651, 0.756], [0.137, 0.205])	([0.564, 0.654], [0.214, 0.272])	([0.673, 0.797], [0.117, 0.195])	([0.602, 0.729], [0.170, 0.256])
q_9	([0.633, 0.752], [0.143, 0.219])	([0.647, 0.762], [0.141, 0.215])	([0.594, 0.681], [0.176, 0.249])	([0.609, 0.718], [0.164, 0.249])
q_{10}	([0.638, 0.755], [0.138, 0.211])	([0.557, 0.672], [0.207, 0.286])	([0.664, 0.766], [0.127, 0.200])	([0.609, 0.691], [0.168, 0.242])
q_{11}	([0.679, 0.776], [0.116, 0.184])	([0.674, 0.762], [0.123, 0.193])	([0.577, 0.677], [0.193, 0.251])	([0.649, 0.759], [0.133, 0.204])
q_{12}	([0.617, 0.722], [0.160, 0.238])	([0.586, 0.687], [0.263, 0.368])	([0.606, 0.700], [0.174, 0.239])	([0.639, 0.741], [0.145, 0.236])
q_{13}	([0.607, 0.724], [0.159, 0.239])	([0.643, 0.739], [0.139, 0.229])	([0.595, 0.687], [0.183, 0.252])	([0.634, 0.702], [0.154, 0.222])
q_{14}	([0.627, 0.734], [0.155, 0.237])	([0.626, 0.727], [0.149, 0.221])	([0.676, 0.791], [0.117, 0.207])	([0.703, 0.804], [0.105, 0.181])
q_{15}	([0.568, 0.677], [0.200, 0.282])	([0.589, 0.671], [0.190, 0.262])	([0.504, 0.631], [0.257, 0.325])	([0.655, 0.760], [0.137, 0.221])
q_{16}	([0.633, 0.754], [0.143, 0.223])	([0.615, 0.715], [0.159, 0.216])	([0.646, 0.759], [0.140, 0.216])	([0.576, 0.688], [0.198, 0.278])
q_{17}	([0.609, 0.709], [0.182, 0.298])	([0.625, 0.725], [0.192, 0.258])	([0.625, 0.724], [0.154, 0.226])	([0.621, 0.729], [0.152, 0.243])
q_{18}	([0.652, 0.757], [0.139, 0.212])	([0.555, 0.649], [0.201, 0.297])	([0.627, 0.716], [0.154, 0.226])	([0.634, 0.718], [0.148, 0.224])
q_{19}	([0.681, 0.777], [0.115, 0.184])	([0.690, 0.792], [0.107, 0.175])	([0.609, 0.726], [0.163, 0.227])	([0.632, 0.745], [0.144, 0.217])

Step 4: Applying Equation (10), we compute the objective weight of each SDB using the developed IVIF-distance measure (4) (or (5)) as follows (see Figure 2):

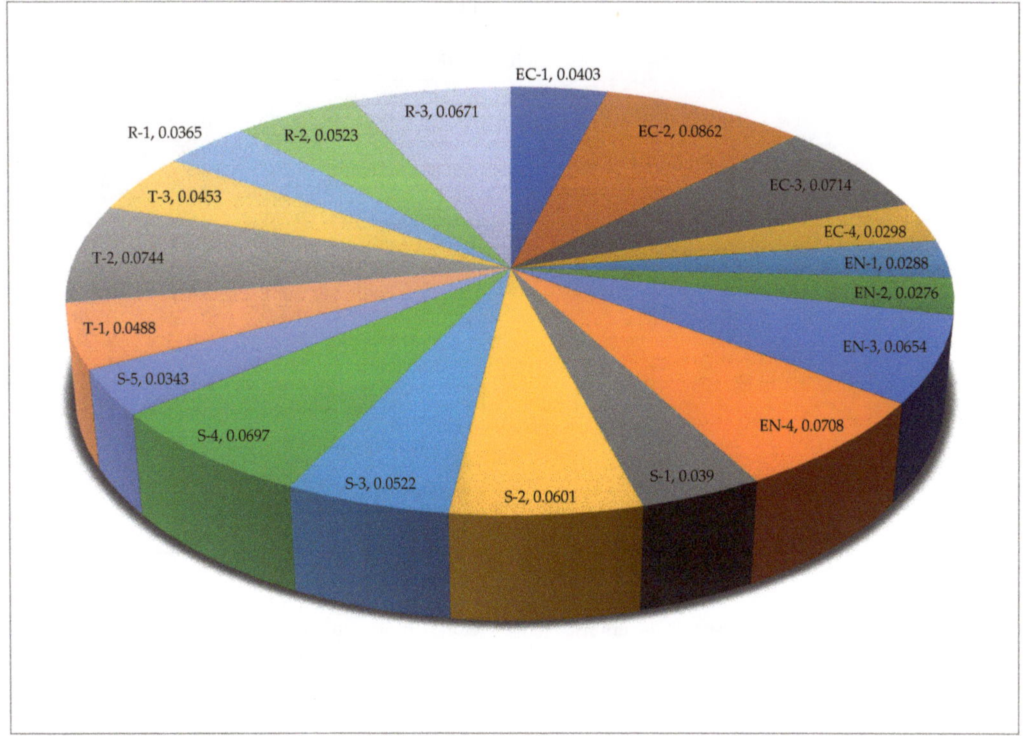

Figure 2. Objective weight of SDBs to the biofuel sector.

$$w_j^o = (0.0403, 0.0863, 0.0714, 0.0298, 0.0288, 0.0276, 0.0654, 0.0708, 0.0390, 0.0601, 0.0522, 0.0697, 0.0343, 0.0488, 0.0744, 0.0453, 0.0365, 0.0523, 0.0671).$$

Here, Figure 2 shows the SDBs' criteria weights with respect to the outcome. The lack of effective storage facilities (EC-2), with a weight of value 0.0863, has come out to be the most important SDB to the biofuel industry. Lack of investors (EC-3), with a weight of 0.0714, is the second most significant SDB in the biofuel industry. Technical problems related to conversion technologies (T-2) is third, with a weight value of 0.0744. Emissions (water vapor and greenhouse gases) (EN-4) is fourth, with a weight of 0.0708; lack of trust between local societies, enterprises, and inventors (S-4) with a weight of 0.0697 is the fifth most significant SDB to the biofuel industry and others are considered crucial SDBs to the biofuel industry.

Step 5: Estimate the AIVIF-DM $G = \left(z_j\right)_{1 \times n}$ by combining the LDM assessment degrees for SDBs provided by DEs using the operator in Equation (9) and presented in Table 7.

Step 6: Define the IVIF-ideal ratings.

We define the IVIF-PIR $\phi^+ = (1, 0, 0)$ and IVIF-NIR $\phi^- = (0, 1, 0)$ for SDBs in the biofuel industry.

Step 7: Derive the degrees of discrimination of each SDB from IVIF-PIR and IVIF-NIR.

From Table 7 and Equations (11) and (12), the discrimination of AIVIF-DM from IVIF-PIR and IVIF-NIS is calculated.

Step 8: The IVIF-relative closeness coefficient rc_j is estimated using Equation (13) and mentioned in Table 7.

Step 9: The subjective weight of the criteria is computed using Equation (14) and is presented as follows:

$$w_j^s = (0.0534, 0.0530, 0.0560, 0.0559, 0.0509, 0.0523, 0.0532, 0.0518, 0.0520, 0.0537, 0.0545, 0.0478, 0.0544, 0.0507, 0.0550, 0.0511, 0.0477, 0.0540, 0.0525).$$

Table 7. Weight of SDBs in the form of LRs.

Barriers	d_1	d_2	d_3	d_4	d_5	AIVIF-DM	p_{ij}^+	p_{ij}^-	rc_j	w_j^s
q_1	G	VG	G	A	G	([0.677, 0.782], [0.113, 0.183])	0.346	0.872	0.716	0.0534
q_2	SG	A	VG	SG	SG	([0.679, 0.757], [0.123, 0.184])	0.352	0.866	0.711	0.0530
q_3	VG	SL	VG	G	G	([0.721, 0.828], [0.090, 0.146])	0.297	0.897	0.751	0.0560
q_4	A	A	VG	G	VG	([0.722, 0.830], [0.088, 0.156])	0.298	0.896	0.750	0.0559
q_5	SG	SL	SG	SG	G	([0.643, 0.713], [0.151, 0.203])	0.392	0.844	0.683	0.0509
q_6	G	SL	SG	A	VG	([0.661, 0.765], [0.130, 0.207])	0.365	0.859	0.702	0.0523
q_7	VG	SG	SL	VG	A	([0.675, 0.787], [0.121, 0.199])	0.348	0.868	0.714	0.0532
q_8	SL	VG	SG	VG	SL	([0.651, 0.761], [0.144, 0.221])	0.374	0.851	0.695	0.0518
q_9	VG	SG	G	SL	G	([0.651, 0.756], [0.139, 0.199])	0.371	0.856	0.697	0.0520
q_{10}	A	VG	VG	A	G	([0.684, 0.794], [0.109, 0.192])	0.340	0.874	0.720	0.0537
q_{11}	VG	SG	G	SG	G	([0.701, 0.789], [0.105, 0.158])	0.324	0.883	0.731	0.0545
q_{12}	A	G	SG	L	G	([0.571, 0.675], [0.186, 0.256])	0.452	0.810	0.642	0.0478
q_{13}	G	L	G	VG	SG	([0.694, 0.796], [0.106, 0.163])	0.327	0.882	0.729	0.0544
q_{14}	SG	SG	G	SG	A	([0.639, 0.713], [0.148, 0.216])	0.395	0.842	0.680	0.0507
q_{15}	G	A	SG	G	VG	([0.707, 0.805], [0.099, 0.158])	0.315	0.887	0.738	0.0550
q_{16}	L	VG	G	SG	SG	([0.640, 0.727], [0.146, 0.202])	0.388	0.848	0.686	0.0511
q_{17}	G	SG	SG	L	SG	([0.575, 0.657], [0.193, 0.249])	0.455	0.808	0.640	0.0477
q_{18}	A	VG	VG	G	A	([0.689, 0.798], [0.106, 0.186])	0.334	0.877	0.724	0.0540
q_{19}	SG	G	SG	G	SG	([0.671, 0.744], [0.128, 0.179])	0.361	0.863	0.705	0.0525

The value of the subjective weight of SDBs to the biofuel industry is depicted in Figure 3. Here, Figure 3 shows the criteria weights with respect to the outcome. Lack of investors (EC-3) with a weight of value 0.0560 has come out to be the most important SDB to the biofuel industry. High logistics costs (EC-4), with a weight of 0.0559, is the second most significant SDB. Technical problems related to conversion technologies (T-2) is third, with a weight value of 0.0550. Fear of public health and safety hazards (S-3) is fourth, with a weight of 0.0545; lack of public awareness of bioenergy technologies (S-5), with a weight of 0.0544, is the fifth most significant SDB to the biofuel industry and others are considered crucial SDBs to the biofuel industry.

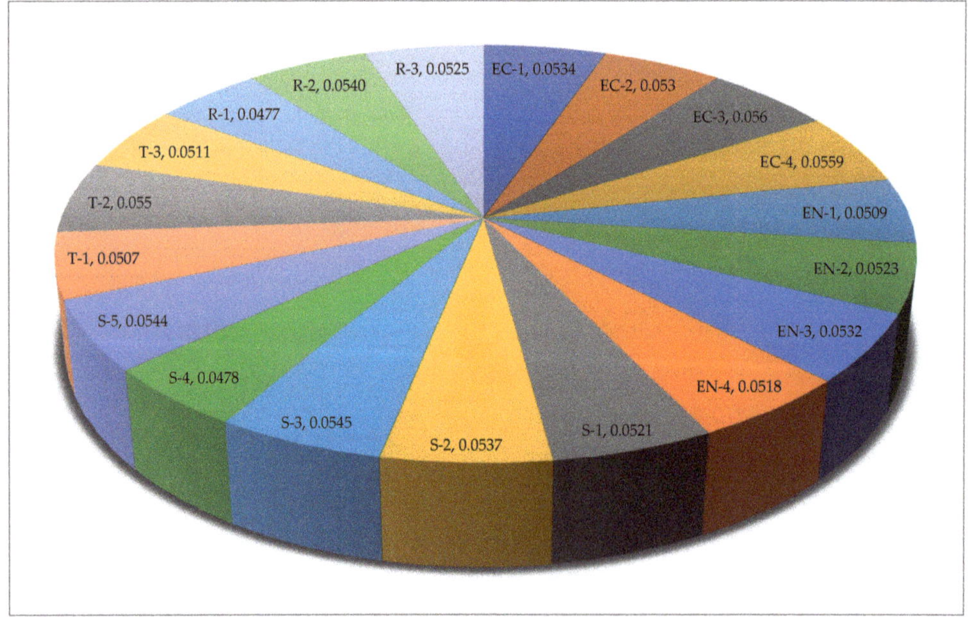

Figure 3. Subjective weight of SDBs in the biofuel sector.

Step 10: We combine the IVIF-DM-based weighting model and the IVIF-relative closeness coefficient-based model with the use of Equation (15). Hence, the combined weight of SDBs for $\tau = 0.5$ is depicted in Figure 4 and presented as:

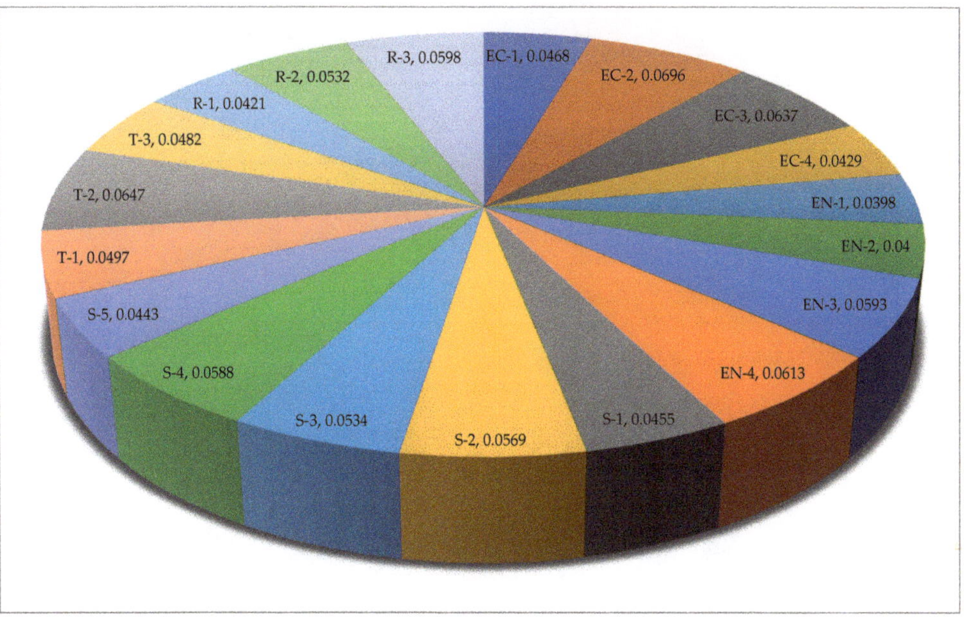

Figure 4. Combined weight of SDBs in the biofuel sector.

w_j = (0.0468, 0.0696, 0.0637, 0.0429, 0.0398, 0.0400, 0.0593, 0.0613, 0.0455, 0.0569, 0.0534, 0.0588, 0.0443, 0.0497, 0.0647, 0.0482, 0.0421, 0.0532, 0.0598).

Here, Figure 4 shows the SDBs' criteria weights with respect to the outcome. The lack of effective storage facilities (EC-2), with a weight of value 0.0696, has come out to be the most important SDB in the biofuel industry. Technical problems related to conversion technologies (T-2), with a weight of 0.0647, is the second most significant SDB in the biofuel industry. Lack of investors (EC-3) is third with a weight value of 0.0637. Emissions (water vapor and greenhouse gases) (EN-4) is fourth, with a weight of 0.0613; lack of governmental support for SSC solutions (R-3), with a weight of 0.0598, is the fifth most significant SDB in the biofuel industry and others are considered crucial SDBs in the biofuel industry.

Step 11: From Figure 5, we find that the lack of effective storage facilities (EC-2) is the most important SDB im the biofuel industry of the proposed model and IVIF-distance measure model, while the lack of investors (EC-3) has come out to be the most important SDB om biofuel industry based on the proposed IVIF-relative closeness coefficient model.

5.1. Sensitivity Analysis

In the current section, we discuss the variation of objective and subjective weighting models for considered SDBs in the developed weight-determining model. The analyses are performed by considering the following cases. In these two cases, we examine the usage of DEs' views in the subjective weighting tool while giving the assessment rating of each SDB and also modeling the data of the objective weighting tool of SDBs with changing $\gamma \in [0, 1]$ values.

Case-1. This case considers the SDBs' weight in the biofuel industry with the objective weighting model (when $\gamma = 0.0$) in place of an integrated weighting tool. Thus, the assessment ratings and priority of SDBs are estimated and given in Table 8. The lack of investors (EC-3), with a weight of value 0.0560, has come out to be the most important SDB in the biofuel industry.

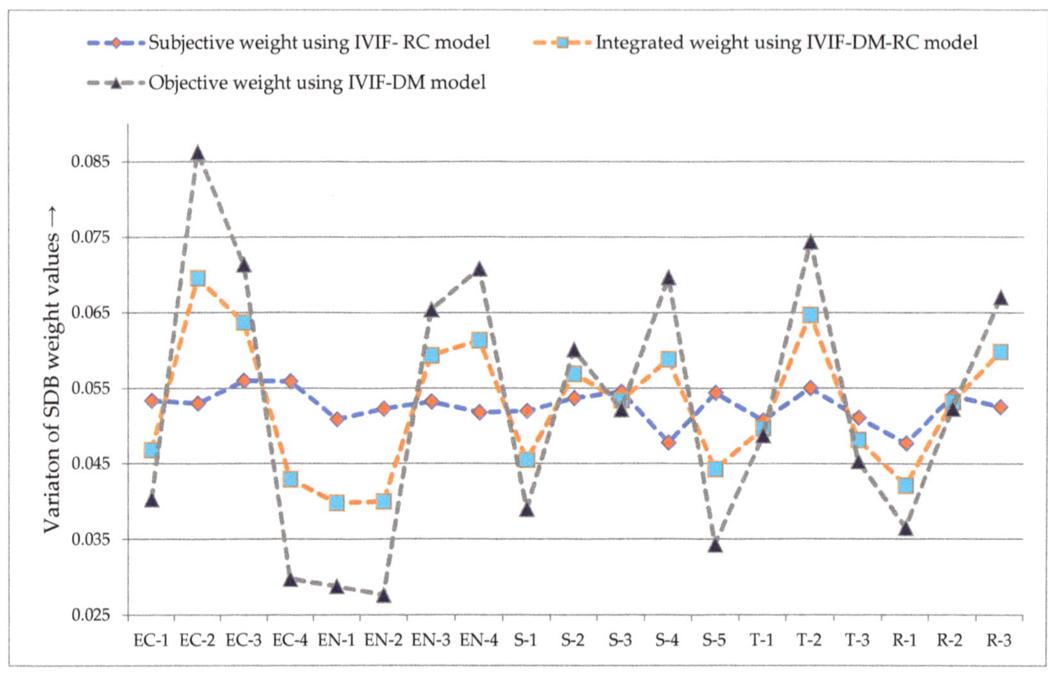

Figure 5. Variation of SDB values in the biofuel sector with the proposed model.

Table 8. Variation of SDB weights of the proposed IVIF-DM-RC method.

	$\gamma = 0.0$	0.1	0.2	0.3	0.4	0.5	0.6	0.7	0.8	0.9	1.0
q_1	0.0534	0.0521	0.0508	0.0495	0.0482	0.0468	0.0455	0.0442	0.0429	0.0416	0.0403
q_2	0.0530	0.0563	0.0596	0.0630	0.0663	0.0696	0.0730	0.0763	0.0796	0.0829	0.0863
q_3	0.0560	0.0575	0.0591	0.0606	0.0621	0.0637	0.0652	0.0668	0.0683	0.0698	0.0714
q_4	0.0559	0.0533	0.0507	0.0481	0.0455	0.0429	0.0403	0.0376	0.0350	0.0324	0.0298
q_5	0.0509	0.0487	0.0465	0.0443	0.0421	0.0398	0.0376	0.0354	0.0332	0.0310	0.0288
q_6	0.0523	0.0499	0.0474	0.0449	0.0425	0.0400	0.0375	0.0350	0.0326	0.0301	0.0276
q_7	0.0532	0.0544	0.0556	0.0569	0.0581	0.0593	0.0605	0.0617	0.0629	0.0642	0.0654
q_8	0.0518	0.0537	0.0556	0.0575	0.0594	0.0613	0.0632	0.0651	0.0670	0.0689	0.0708
q_9	0.0520	0.0507	0.0494	0.0481	0.0468	0.0455	0.0442	0.0429	0.0416	0.0403	0.0390
q_{10}	0.0537	0.0543	0.0550	0.0556	0.0562	0.0569	0.0575	0.0582	0.0588	0.0595	0.0601
q_{11}	0.0545	0.0543	0.0541	0.0538	0.0536	0.0534	0.0531	0.0529	0.0527	0.0524	0.0522
q_{12}	0.0478	0.0500	0.0522	0.0544	0.0566	0.0588	0.0609	0.0631	0.0653	0.0675	0.0697
q_{13}	0.0544	0.0524	0.0504	0.0483	0.0463	0.0443	0.0423	0.0403	0.0383	0.0363	0.0343
q_{14}	0.0507	0.0505	0.0503	0.0501	0.0499	0.0497	0.0495	0.0494	0.0492	0.0490	0.0488
q_{15}	0.0550	0.0569	0.0589	0.0608	0.0627	0.0647	0.0666	0.0686	0.0705	0.0724	0.0744
q_{16}	0.0511	0.0505	0.0500	0.0494	0.0488	0.0482	0.0476	0.0471	0.0465	0.0459	0.0453
q_{17}	0.0477	0.0466	0.0454	0.0443	0.0432	0.0421	0.0410	0.0399	0.0387	0.0376	0.0365
q_{18}	0.0540	0.0538	0.0537	0.0535	0.0533	0.0532	0.0530	0.0528	0.0527	0.0525	0.0523
q_{19}	0.0525	0.0540	0.0555	0.0569	0.0584	0.0598	0.0613	0.0627	0.0642	0.0656	0.0671

Case-2. This case shows the SDB weight using the subjective weighting model (when $\gamma = 1.0$) rather than the integrated weighting tool. Hence, the assessment ratings and priority of SDBs are presented in Table 8. The lack of effective storage facilities (EC-2) with a weight of value 0.0696 has come out to be the most important SDB in the biofuel industry.

Based on the aforementioned discussion, we find the following outcomes: (i) prioritization of SDBs in case-1 demonstrates the performance of the considered SDBs from the objectivity perspective of DEs; (ii) prioritization of SDBs in case-2 illustrates the importance of DEs for SDBs from a subjectivity perspective. The following two cases elucidate that when changing the SDB weights, we obtain a diverse preference order of SDBs. Thus, due to this reason, we believe that we can select the most suitable SDBs in the biofuel industry by considering the combined IVIF-DM-relative closeness coefficient model. The outcomes of the analysis with anticipated weights are presented in Figure 6. According to the aforesaid discussion, it is concluded that considering the diverse ratings of parameters will enhance the strength of the proposed IVIF-DM-relative closeness coefficient model.

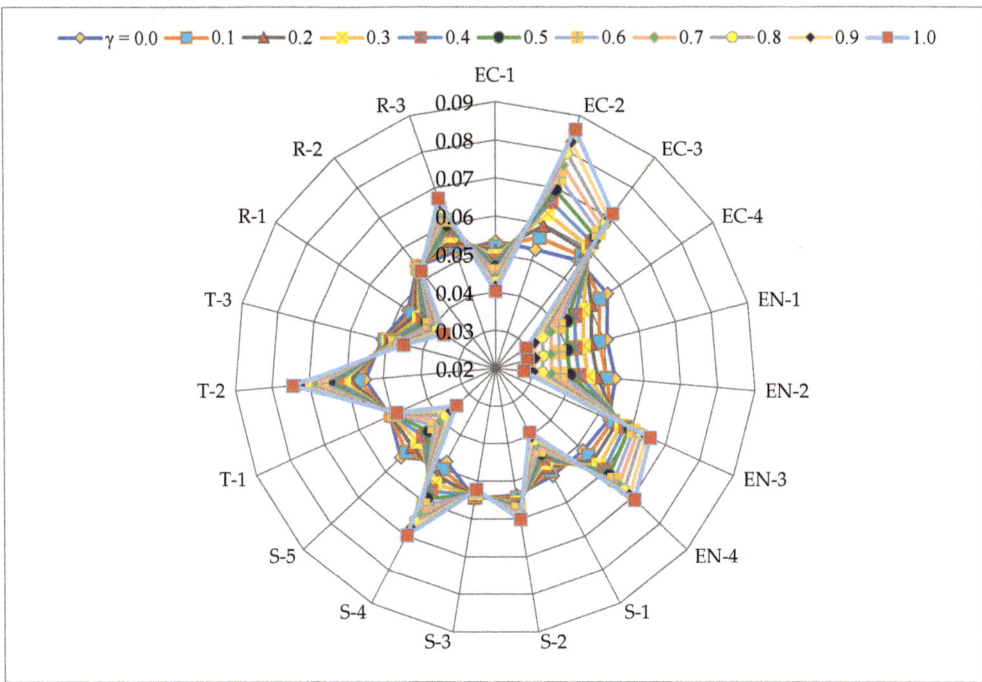

Figure 6. Sensitivity analysis of SDB degrees over various parameter (γ) values.

5.2. Discussion and Implications

The weighting outcomes revealed that the preservation of the social (S) factor had become the most significant pillar for the present SDB assessment (see Figure 7) in the biofuel industry. As a result, this aspect of SDB assessment should be taken sincerely, while economic (Ec), environmental (En), technological (T), and regulatory (R) should be also emphasized. Based on the aforesaid discussion, it can be concluded that the lack of effective storage facilities (EC-2), technical issues related to conversion technologies (T-2), lack of investors (EC-3), emissions (water vapor and greenhouse gases) (EN-4), and lack of governmental support for SSC solutions (R-3) are the most significant influencing SDBs in the biofuel industry for the given case study. From Figure 4, we can find the other important SDBs from a sustainability perspective. By means of the concept of IVIF-DM-relative closeness coefficient framework, we have combined the weight-determining models based on distance measure and the relative closeness coefficient model, which reduces information loss during the process of making a decision.

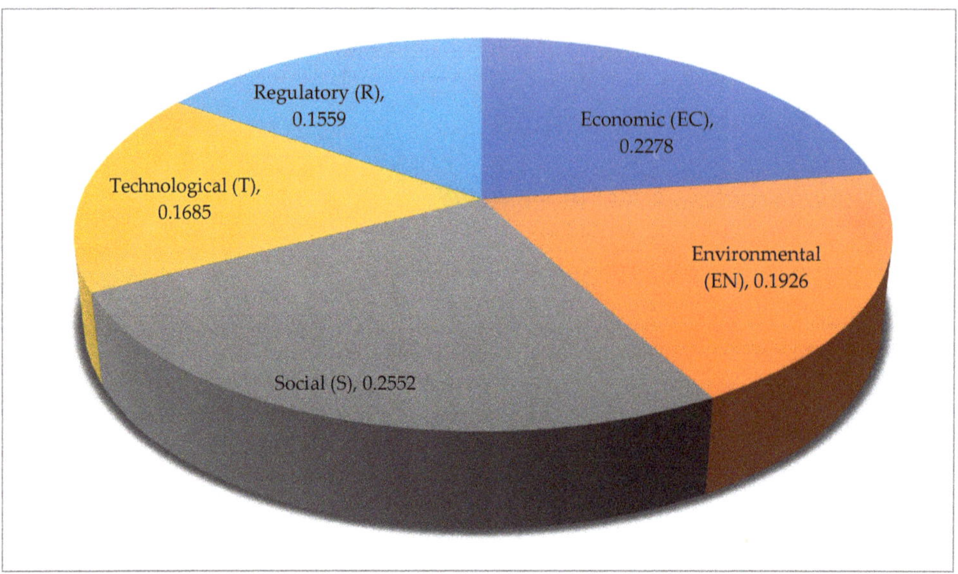

Figure 7. Depiction of the significance degrees of different sustainability aspects.

The development and implementation of the hybrid framework is the key contributions in this study, which can discuss the dual nature (qualitative and quantitative, precise and fuzzy) of semantic judgment in more practical situations. This procedure was documented based on the discussion with four DEs, in which they supposed the assessment of mixed information could better describe the decision. Here, we consider three kinds of criteria degrees, exact, interval, and fuzzy numbers. Quantitative assessments are described by precise and interval numbers, while qualitative assessments are discussed by IVIFNs. The utilization of IVIFNs makes it easier and faster for experts to make decisions and avoid errors caused by indeterminacy and non-intuition. In the integration of IVIF-DM and IVIF-relative closeness coefficient, qualitative and quantitative SDBs are categorized by different data types that can be estimated and compared in a similar dimension, which increases the efficiency and comprehensive assessment of SDB selection.

To show the effectiveness of the IVIF-DM-relative closeness coefficient framework, we relate the outcomes of the developed model with some of the extant models such as the "IVPF-SWARA [45]" and "IVIF-distance measure-entropy [21]" models. The comparative outcomes are presented in Figure 8 and Table 9. The purpose of choosing the IVPF-SWARA model is that the approach employs the subjective assessment of SDBs. In comparison with the IVPF-SWARA and IVIF-distance measure-entropy models, the proposed approach has the following advantages:

— In the present work, we determine a systematic assessment of the DEs' weights using the IVIF-score value and IVIF-rank sum model, which reduces the imprecision and biases in the MADA procedure, while existing studies do not provide this information.
— The developed method determines the integrated weights (combination of objective and subjective weighting) of SDBs using the IVIF-DM-relative closeness coefficient-based tool. In contrast, in IVPF-SWARA, the subjective weighting of SDB is estimated with the SWARA model, and in the IVIF-distance measure-entropy model, the objective weights of the SDBs are obtained using distance measure and entropy-based approach.
— Liang et al. [1] used the gray DEMATEL approach for assessing the success factors of the biofuel industry in China. In comparison with [1], the proposed approach has simpler computational steps and is easy to understand for decision experts during the assessment of SDBs.

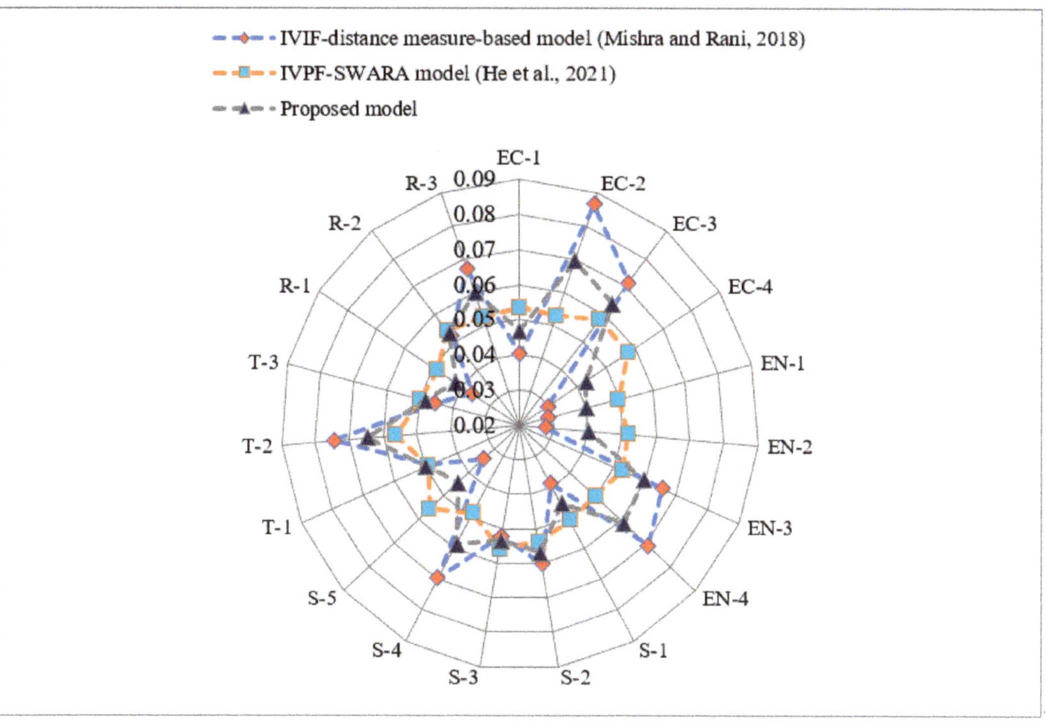

Figure 8. Comparison of SDB weights obtained by proposed and extant models [21,45].

Table 9. Comparison with existing methodologies.

Parameters	He et al. [45]	Mishra and Rani [21]	Proposed Model
Benchmark	IVPF-SWARA model	IVIF-distance measure-entropy model	IVIF-DM-relative closeness model
Alternatives/criteria assessments	IVPFSs	IVIFSs	IVIFSs
Criteria weight	Subjective weight	Objective weight	Integrated weight using objective and subjective weights
DMEs' weights	Assumed	Considered	IVIF-rank sum andScore degree-based model
Decision-making process	Group	Group	Group

As India is an energy-lacking nation, RESs are strategically important. It is noteworthy that the biofuel region can manufacture a sufficient amount of energy for considering numerous requirements of energy, namely industrial heat services, fuel for vehicles, electricity production, and others. The expansion of the biofuel zone will result in job creation, alleviation of climate change, enhanced industrial competitiveness, better exports, the establishment of infrastructure in the area, and better living standards. However, bio-waste assessment is a crucial issue that requires substantial attention from DEs.

Electricity alone will not be able to ensure the complete decarbonization process of energy systems, due to the presence of final energy uses such as maritime and air transport, which require synergy with other commodities. For a fully decarbonized transport sector, biofuels are expected to support the energy transition, particularly for some sectors. For a concrete and rational "green transition", a crucial role could be played by alternative biofuels which fit into a circular economy approach coherently with the European Green Deal and strategic plans of the European Commission. The large-scale production of biofuels must be designed in a sustainable way including the preservation of biodiversity, optimal water utilization, air quality, soil conservation, social issues, and fair labor practices [46]. Moreover, it is important to support sustainable agriculture and forestry to stimulate growth and employment, particularly in rural areas. Regardless, we must pay attention to the fact that since biodiesel production has risen globally, the prices for food and vegetable oils have also grown. So the principle that connects the ideas of circular economy, green economy, and bioeconomy is to find the right equilibrium between economic, environmental, and social objectives [46].

The current study has addressed three purposes, as mentioned in the introduction of this study. The first aim, i.e., recognizing the crucial SDBs to BEs, is solved in Section 5, where nineteen SDBs are recognized. Additionally, the second aim, i.e., assessing the significant degrees of identified SDBs using the proposed IVIF-DM-relative closeness coefficient tool, was carried out, and not only forms associations but also offers the integrated weights of the identified SDBs. Further, the sensitivity assessment and comparison of SDBs are also performed to show the utility of the developed model. Third, few policies are presented for disabling the substantial SDBs, which are given as the government can propose operative biomass SD benchmarks and execute stern procedures and guidelines to avoid and combat fraudulent actions. The constitutional authorities can articulate strategies/schemes/rules for the growth of the biofuel industry by harmonizing all the sustainability aspects.

6. Conclusions

The study presents the procedure for recognizing critical SDBs in the biofuel industry in India. Overall, nineteen SDBs are assessed using the proposed IVIF-DM-relative closeness coefficient approach. The top five significant SDBs are lack of effective storage facilities (EC-2), technical issues related to conversion technologies (T-2), lack of investors (EC-3), emissions (water vapor and greenhouse gases) (EN-4), and lack of governmental support for SSC solutions (R-3). Further, a new distance measure is developed with some elegant properties for evaluating the objective weighting of different SDBs in the biofuel sector. The present paper aims to assist the executives of the biofuel sector in understanding the impact of SDBs on the biofuel industry. The novel plans and schemes may be framed, or extant policies may be reformed to deal with concerns of the biofuel sector by reducing or dropping the impact of the considered SDBs. To lift or improve the biofuel industry in India, there is a requirement to promote biofuel practices by evolving compulsory strategies of integrating biofuels with traditional fuels. Furthermore, there is a requirement to promote the usage of biofuels amongst automobile operators with education and promotional activities. The management must sponsor the prospects of biomass energy, i.e., that bioenergy is environmentally beneficial, reasonable, entirely practical, and real. It may be distinguished that the sustainability of bioenergy is crucial and appropriate care should be considered to confirm the reduction GHGs in bioenergy tools. Additionally, comparisons with extant tools and sensitivity assessment have been studied to expose the rationality and solidity of the obtained outcomes. The findings of this study prove that the developed method has great significance and solidity, and is more dependable than extant tools.

This work has some limitations:

(i)   The considered evaluation criteria are not inter-dependent;
(ii) Risk aspects of sustainability are missing during the assessment of SDBs;
(iii) The proposed work is not able to express uncertain, indeterminate, and inconsistent information simultaneously.

In the future, it would be exciting to use the introduced model for other decision-making scenarios. In addition, we extend the proposed model under different disciplines, namely, interval-valued Pythagorean fuzzy sets, q-ROFSs, FFSs, complex q-ROFSs, and others. Furthermore, we will try to use new technologies that encourage the energy conversion of different types of biomass such as woody energy crops, agricultural residues, and forestry residues.

Author Contributions: Conceptualization, A.R.M. and P.R.; methodology, A.R.M.; software, P.R.; validation, F.C., I.M.H. and A.R.M.; formal analysis, A.R.M.; investigation, F.C.; resources, I.M.H.; data curation, P.R.; writing—original draft preparation, A.R.M. and F.C.; writing—review and editing, P.R. and I.M.H.; visualization, A.R.M.; supervision, F.C. and I.M.H.; project administration, P.R. and F.C.; funding acquisition, I.M.H. All authors have read and agreed to the published version of the manuscript.

Funding: This paper is funded by "Researchers Supporting Project number (RSP2023R389), King Saud University, Riyadh, Saudi Arabia".

Institutional Review Board Statement: Not applicable.

Informed Consent Statement: Not applicable.

Data Availability Statement: Not applicable.

Conflicts of Interest: The authors declare no conflict of interest.

References

1. Liang, H.; Ren, J.; Gao, Z.; Gao, S.; Luo, X.; Dong, L.; Scipioni, A. Identification of critical success factors for sustainable development of biofuel industry in China based on grey decision-making trial and evaluation laboratory (DEMATEL). *J. Clean. Prod.* **2016**, *131*, 500–508. [CrossRef]
2. Mortensen, A.W.; Mathiesen, B.V.; Hansen, A.B.; Pedersen, S.L.; Grandal, R.D.; Wenzel, H. The role of electrification and hydrogen in breaking the biomass bottleneck of the renewable energy system—A study on the Danish energy system. *Applied Energy* **2020**, *275*, 115331. [CrossRef]
3. Zhang, X.; Li, H.; Harvey, J.T.; Butt, A.A.; Jia, M.; Liu, J. A review of converting woody biomass waste into useful and eco-friendly road materials. *Transp. Saf. Environ.* **2021**, *4*, tdab031. [CrossRef]
4. Khan, M.A.H.; Bonifacio, S.; Clowes, J.; Foulds, A.; Holland, R.; Matthews, J.C.; Percival, C.J.; Shallcross, D.E. Investigation of biofuel as a potential renewable energy source. *Atmosphere* **2021**, *12*, 1289. [CrossRef]
5. Cadillo-Benalcazar, J.J.; Bukkens, S.G.F.; Ripa, M.; Giampietro, M. Why does the European Union produce biofuels? Examining consistency and plausibility in prevailing narratives with quantitative storytelling. *Energy Res. Soc. Sci.* **2021**, *71*, 101810. [CrossRef]
6. Bhutto, A.W.; Bazmi, A.A.; Zahedi, G. Greener energy: Issues and challenges for Pakistan—Biomass energy prospective. *Renew. Sustain. Energy Rev.* **2011**, *15*, 3207–3219. [CrossRef]
7. Bravo, M.D.L.; Naim, M.M.; Potter, A. Key issues of the upstream segment of biofuels supply chain: A qualitative analysis. *Logist. Res.* **2012**, *5*, 21–31. [CrossRef]
8. Altan, A.; Karasu, S.; Bekiros, S. Digital currency forecasting with chaotic metaheuristic bio-inspired signal processing techniques. *Chaos, Solit. Fractals* **2019**, *126*, 325–336. [CrossRef]
9. Piterou, A.; Shackley, S.; Upham, P. Project ARBRE: Lessons for bio-energy developers and policy-makers. *Energy Policy* **2008**, *36*, 2044–2050. [CrossRef]
10. Kumar, A.; Kumar, N.; Baredar, P.; Shukla, A. A review on biomass energy resources, potential, conversion and policy in India, Renew. *Sustain. Energy Rev.* **2015**, *45*, 530–539. [CrossRef]
11. Singh, J.; Gu, S. Biomass conversion to energy in India-A critique. *Renew. Sustain. Energy Rev.* **2010**, *14*, 1367–1378. [CrossRef]
12. Narwane, V.S.; Yadav, V.S.; Raut, R.D.; Narkhede, B.E.; Gardas, B.B. Sustainable development challenges of the biofuel industry in India based on integrated MCDM approach. *Renew. Energy* **2021**, *164*, 298–309. [CrossRef]
13. IEA. *World Energy Outlook 2014*; Int. Energy Agency (IEA): Paris, France, 2014. Available online: http://www.iea.org/publications/freepublications/publication/WEO2014.pdf (accessed on 5 January 2023).
14. Purohit, P.; Dhar, S. *Promoting Low Carbon Transport in India, Biofuel Roadmap for India*; UNEP: Nairobi, Kenya, 2015.
15. Upreti, B.R. Conflict over biomass energy development in the United Kingdom: Some observations and lessons from England and Wales. *Energy Policy* **2004**, *32*, 785–800. [CrossRef]
16. Charles, M.B.; Ryan, R.; Ryan, N.; Oloruntoba, R. Public policy and biofuels: The way forward? *Energy Policy* **2007**, *35*, 5737–5746. [CrossRef]
17. D'Adamo, I.; Gastaldi, M.; Morone, P.; Rosa, P.; Sassanelli, C.; Settembre-Blundo, D.; Shen, Y. Bioeconomy of Sustainability: Drivers, Opportunities and Policy Implications. *Sustainability* **2022**, *14*, 200. [CrossRef]
18. D'Adamo, I.; Rosa, P. A structured literature review on obsolete electric vehicles management practices. *Sustainability* **2019**, *11*, 6876. [CrossRef]
19. Atanassov, K.; Gargov, G. Interval valued intuitionistic fuzzy sets. *Fuzzy Sets Syst.* **1989**, *31*, 343–349. [CrossRef]
20. Kumar, K.; Chen, S.-M. Multiattribute decision making based on interval-valued intuitionistic fuzzy values, score function of connection numbers, and the set pair analysis theory. *Inf. Sci.* **2021**, *551*, 100–112. [CrossRef]
21. Mishra, A.R.; Rani, P. Interval-valued intuitionistic fuzzy WASPAS method: Application in reservoir flood control management policy. *Group Decis. Negot.* **2018**, *27*, 1047–1078. [CrossRef]
22. Alrasheedi, M.; Mardani, A.; Mishra, A.R.; Streimikiene, D.; Liao, H.; Al-nefaie, A.H. Evaluating the green growth indicators to achieve sustainable development: A novel extended interval-valued intuitionistic fuzzy-combined compromise solution approach. *Sustain. Dev.* **2021**, *29*, 120–142. [CrossRef]
23. Jernstrom, E.; Karvonen, V.; Kassi, T.; Kraslawski, A.; Hallikas, J. The main factors affecting the entry of SMEs into bio-based industry. *J. Clean. Prod.* **2017**, *141*, 1–10. [CrossRef]
24. Saravanan, A.P.; Mathimani, T.; Deviram, G.; Rajendran, K.; Pugazhendhi, A. Biofuel policy in India: A review of policy barriers in sustainable marketing of biofuel. *J. Clean. Prod.* **2018**, *193*, 734–747. [CrossRef]

25. Malode, S.J.; Prabhu, K.K.; Mascarenhas, R.J.; Shetti, N.P.; Aminabhavi, T.M. Recent advances and viability in biofuel production. *Energy Convers. Manag. X* **2021**, *10*, 100070. [CrossRef]
26. Hajek, P.; Froelich, W. Integrating TOPSIS with interval-valued intuitionistic fuzzy cognitive maps for effective group decision making. *Inf. Sci.* **2019**, *485*, 394–412. [CrossRef]
27. Xu, Z. Methods for aggregating interval-valued intuitionistic fuzzy information and their application to decision making. *Control. Decis.* **2007**, *22*, 215–219.
28. Wang, W.; Mendel, J.M. Interval-valued intuitionistic fuzzy aggregation methodology for decision making with a prioritization of criteria. *Iran. J. Fuzzy Syst.* **2019**, *16*, 115–127.
29. Hu, K.; Tan, Q.; Zhang, T.; Wang, S. Assessing technology portfolios of clean energy-driven desalination-irrigation systems with interval-valued intuitionistic fuzzy sets. *Renew. Sustain. Energy Rev.* **2020**, *132*, 109950. [CrossRef]
30. Mishra, A.R.; Rani, P.; Pardasani, K.R.; Mardani, A.; Stević, Ž.; Pamučar, D. A novel entropy and divergence measures with multi-criteria service quality assessment using interval-valued intuitionistic fuzzy TODIM method. *Soft Comput.* **2020**, *24*, 11641–11661. [CrossRef]
31. Oraki, M.; Gordji, M.E.; Ardakani, H. Some frank aggregation operators based on the interval-valued intuitionistic fuzzy numbers. *Int. J. Nonlinear Anal. Appl.* **2021**, *12*, 325–342.
32. Bharati, S.K. Transportation problem with interval-valued intuitionistic fuzzy sets: Impact of a new ranking. *Prog. Artif. Intell.* **2021**, *10*, 129–145. [CrossRef]
33. Bai, Z.-Y. An interval-valued intuitionistic fuzzy TOPSIS method based on an improved score function. *Sci. World J.* **2013**, *2013*, 879089. [CrossRef] [PubMed]
34. Xu, Z. A method based on distance measure for interval-valued intuitionistic fuzzy group decision making. *Inf. Sci.* **2010**, *180*, 181–190. [CrossRef]
35. Tripathi, D.; Nigam, S.K.; Mishra, A.R.; Shah, A.R. A novel intuitionistic fuzzy distance measure-SWARA-COPRAS method for multi-criteria food waste treatment technology selection. *Oper. Res. Eng. Sci. Theory Appl.* **2022**, *in press*. [CrossRef]
36. McCormick, K.; Kåberger, T. Key barriers for bioenergy in Europe: Economic conditions, know-how and institutional capacity, and supply chain co-ordination. *Biomass Bioenergy* **2007**, *31*, 443–452. [CrossRef]
37. Popp, J.; Lakner, Z.; Harangi-Rakos, M.; Fari, M. The effect of bioenergy expansion: Food, energy, and environment. *Renew. Sustain. Energy Rev.* **2014**, *32*, 559–578. [CrossRef]
38. Mafakheri, F.; Nasiri, F. Modeling of biomass-to-energy supply chain operations: Applications, challenges and research directions. *Energy Policy* **2014**, *67*, 116–126. [CrossRef]
39. Saidur, R.; Abdelaziz, E.A.; Demirbas, A.; Hossain, M.S.; Mekhilef, S. A review on biomass as a fuel for boilers. *Renew. Sustain. Energy Rev.* **2011**, *15*, 2262–2289. [CrossRef]
40. Thornley, P. Increasing biomass based power generation in the UK. *Energy Policy* **2006**, *34*, 2087–2099. [CrossRef]
41. Faaij, A.P.C.; Domac, J. Emerging international bio-energy markets and opportunities for socio-economic development. *Energy Sustain. Dev.* **2006**, *10*, 7–19. [CrossRef]
42. Engelken, M.; Romer, B.; Drescher, M.; Welpe, I.M.; Picot, A. Comparing drivers, barriers, and opportunities of business models for renewable energies: A review. *Renew. Sustain. Energy Rev.* **2016**, *60*, 795–809. [CrossRef]
43. Adams, P.W.; Hammond, G.P.; McManus, M.C.; Mezzullo, W.G. Barriers to and drivers for UK bioenergy development. *Renew. Sustain. Energy Rev.* **2011**, *15*, 1217–1227. [CrossRef]
44. Gegg, P.; Budd, L.; Ison, S. The market development of aviation biofuel: Drivers and constraints. *J. Air Transport. Manag.* **2014**, *39*, 34–40. [CrossRef]
45. He, J.; Huang, Z.; Mishra, A.R.; Alrasheedi, M. Developing a new framework for conceptualizing the emerging sustainable community-based tourism using an extended interval-valued Pythagorean fuzzy SWARA-MULTIMOORA. *Technol. Forecast. Soc. Change* **2021**, *171*, 120955. [CrossRef]
46. Khan, N.; Sudhakar, K.; Mamat, R. Role of biofuels in energy transition, green economy and carbon neutrality. *Sustainability* **2021**, *13*, 12374. [CrossRef]

Review

Water Quality Assessment and Monitoring in Pakistan: A Comprehensive Review

Love Kumar [1,*], Ramna Kumari [2], Avinash Kumar [3], Imran Aziz Tunio [4] and Claudio Sassanelli [5]

[1] Soil, Water and Ecosystem Science, University of Florida, Gainesville, FL 32603, USA
[2] Computer System Engineering, Quaid-e-Awam University of Engineering, Science and Technology, Nawabshah 67450, Pakistan
[3] Chemistry Department, Florida Atlantic University, Boca Raton, FL 33431, USA
[4] Institute of Environmental Engineering & Management, Mehran University of Engineering and Technology, Jamshoro 76062, Pakistan
[5] Department of Mechanics, Mathematics and Management, Politecnico di Bari, 70126 Bari, Italy
*   Correspondence: lovekumar@ufl.edu

Abstract: Water quality has been a major problem in Pakistan owing to a mix of factors such as population expansion, industrial units in urban areas, and agricultural activities. The purpose of this research is to conduct a comprehensive evaluation of water quality monitoring and assessment in Pakistan. The article begins by examining the water sources of Pakistan (i.e., surface water, groundwater, and rainwater). The paper then discusses the methods used by researchers in Pakistan for water quality monitoring and assessment, including chemical, physical, and biological methods. It has been determined that in certain regions in Pakistan, the concentration of arsenic present in the groundwater exceeds the national and international prescribed maximum limits. The range of arsenic concentrations in the Punjab province can vary from 10 to 200 µg/L, while higher concentrations of up to 1400 µg/L have been recorded in Sindh. In the Punjab province, fluoride concentrations vary from 0.5 to 30 mg/L, while in Sindh, the levels can reach up to 18 mg/L. In addition, some of the research has talked about bacteria. A 2017 study found that the fecal coliform concentrations in certain water in different cities of Pakistan surpassed limits and were as high as 1100 CFU/100 mL. Additionally, natural factors such as geological formations and high salinity in some areas contribute to the contamination of water. The effect of water pollution on public health has the potential to cause harm. It is critical to investigate creative strategies for improving water quality, and it is necessary to make investments in research and development, which could include the implementation of sophisticated technologies and the conception of new treatment processes. The review performed in this paper facilitates an understanding of the current water quality in Pakistan, including the types and magnitudes of contaminants present in the water sources. Subsequently, the assessment emphasizes deficiencies and challenges in the existing water quality monitoring frameworks and provides suggestions for improving them. This review is also of significant benefit to all the stakeholders involved in ensuring clean and safe water for human consumption and other purposes in Pakistan, such as policymakers, water managers, researchers, and other stakeholders.

Keywords: water quality; pollutant; surface water; monitoring; Pakistan; drinking water

Citation: Kumar, L.; Kumari, R.; Kumar, A.; Tunio, I.A.; Sassanelli, C. Water Quality Assessment and Monitoring in Pakistan: A Comprehensive Review. *Sustainability* **2023**, *15*, 6246. https://doi.org/10.3390/su15076246

Academic Editor: Nallapaneni Manoj Kumar

Received: 17 February 2023
Revised: 2 April 2023
Accepted: 3 April 2023
Published: 5 April 2023

1. Introduction

Pakistan experiences severe shortages of water and water contamination, just like other developing nations throughout the world. The country is regarded as water-stressed and is expected to experience a water shortage in the coming years since its existing water resources have practically depleted [1,2]. The government and officials are in charge of providing inhabitants with access to potable drinking water, but unfortunately, in Pakistan, water shortages and pollution brought on by ineffective water management by the nation's government agencies are posing a threat to human existence [3]. According to community health surveys and statistics, drinking water of inadequate quality causes 40% of deaths and 50% of waterborne infections in Pakistan [4]. The presence of arsenic in high concentrations in Pakistan's drinking water may affect up to 60 million people [5]. In the last 4 years, roughly 1832 children lost their lives to waterborne illnesses and drought [6]. Currently, only a fifth of the population has access to safe drinking water, while the rest are forced to rely on water that is contaminated by various pollutants such as chemicals, fertilizer, industrial waste, and sewage [7]. The availability of insufficient and substandard drinking water is a serious human health issue. The release of hazardous industrial chemicals and contaminated trash into water bodies lowers the water quality and has detrimental effects on individual health. Lack of knowledge, outdated treatment methods and apparatuses, unskilled laborers, and insufficient inspection all play a role. It is likely that a war on water will begin both domestically and globally if the current scenario worsens [8]. Issues of sanitation, water, and hygiene have also caused a number of dangers to population health since the probability of contracting waterborne diseases is increasing fast. In Pakistan, 2.5 million cases of diarrheal disease-related fatalities were recorded in 2017; 40% of these deaths and 50% of the country's disorders are brought on by drinking polluted water [4].

Numerous contaminants, including pathogens and microorganisms, metal poisons, household and industrial waste, pharmaceuticals, and other dangerous medications, are present in the water [9] and need to be monitored [10] and managed [11–13]. Changes in climate that affect the annual precipitation, the inadequate construction of water storage structures, and governmental pressures are some of the causes of current water issues in Pakistan [4] Other factors include a rise in demand brought on by a significant expansion of industry and population [14]. Due to insufficient water sanitization, Pakistan faced major economic difficulties in 2019 [15], costing around PKR 343.7 billion (USD 1.5 billion). Additionally, from 2016 to 2017, albeit with UNICEF's assistance, the cost of distributing facilities for cleaner water grew from PKR 48 billion to PKR 72 billion [15]. It might be claimed that the provision of clean water in Pakistan will need financing because poor facilities are offered to the entire country. Many of the existing problems, including unemployment, the incidence of sickness, and economic instability, are expected to become significantly worse in the near future if things do not change [16].

This paper intends to provide a comprehensive review of the published scientific studies on monitoring, determining, and methodologies related to water quality. Surface water, drinking water, and groundwater sources are the focus of much of this study. Study results on water quality assessment and monitoring in Pakistan during the period of 2012 to 2022 are reviewed in-depth and methodically in this study. Surface water, groundwater, and rainwater are the three categories of water sources in both urban and rural Pakistan, and these categories were used to separate the subject in this paper.

Geography and Water Resources of Study Area

This section provides information about the topography and water resources of the study area. These are the critical characteristics of surface water, groundwater, and rainfall supplies and their present state. The weather data set the tone for readers to better understand the debates on surface and groundwater in Pakistan. The hydrological and geological knowledge is particularly significant since it will assist readers in understanding how hydrogeochemical processes impact the country's water quality.

Pakistan is a country in southern Asia that has borders with Afghanistan, India, and China (Figure 1). The Himalayas and the Karakoram are located in eastern Pakistan. There are mountain ranges in the north (Hindukush), hills in the northwest, and a highland plateau in Baluchistan. There is a wide range in the average rainfall, from arid to semiarid, across Pakistan [17]. The Indus River is Pakistan's most important waterway; it originates in the northern Karakorum Mountains, meanders its way south, and empties into the Arabian Sea. Pakistan's agricultural sector is vital to the nation's economy. Wheat, maize, rice, cotton, and sugarcane are the most common agricultural products, and they are grown in 27% of the total area. Increasing the agricultural yield is a common practice to meet the needs of a growing population. Industrial powerhouses such as the textile, pesticide, and fertilizer sectors may be found in the urban hubs of Pakistan [4]. Pakistan has two major river systems: those that flow into the Arabian Sea and those that flow into the Endorheic River basin. The former includes the Indus River basin, the Lyari river, and the Hingol and Hub rivers. The latter includes Mashkal, the Siastan Basin, and the Indus Plain [17]. The Indus Basin, the Kharan Desert system, and Makran coastal drainage make up Pakistan's three hydrological groups. The Indus Basin is where the majority of surface and groundwater resources are found. Few inter-basin diversions are technically or economically viable due to Pakistan's topography [18]. The overall water resource quantity is relatively unclear due to a lack of water data (particularly in Balochistan, where the hydrology is very unpredictable) and a lack of solid water accountancy, resulting in only estimated resources. Furthermore, surface groundwater fluxes are poorly measured [18].

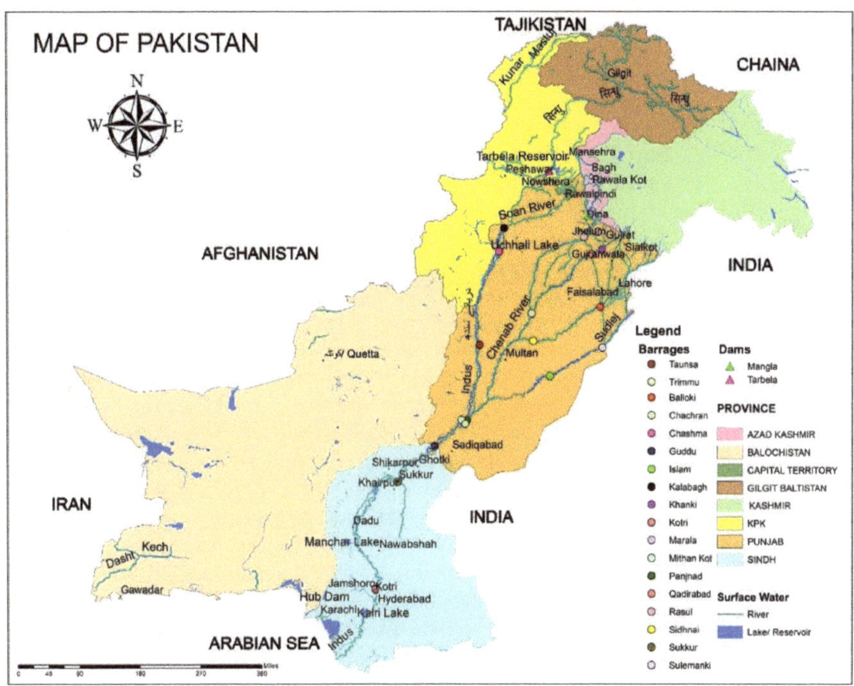

Figure 1. Geography and water resource map of the study area.

Approximately 70% of the yearly precipitation falls between June and September. This results in the majority of rainwater from the lower Indus valleys being drained into the sea. The pattern of mean annual precipitation in Pakistan exhibits substantial geographical variation. It varies from 125 mm in Balochistan (southeast) to 750 mm in the northwest [19,20]. In the lowlands of Sindh, heavy precipitation falls in July and August, and its strength decreases from the coast to the center of the province. The yearly precipitation in southern Punjab and northern Sindh is less than 152 mm. The regions above the Salt Range, such as some districts of Punjab and Sindh, receive a mean of more than 635 mm of precipitation annually [21]. Winter precipitation is often widespread. During the wintertime, the northern and northwestern regions of NWFP and the northern regions of Balochistan receive a relatively large amount of precipitation. Yearly rainfall over almost 21 million hectares (Mha) of the Indus Plains and Peshawar Basin averages 26 million acre-feet (MAF) (Figure 2). The rainfall yield to crops in irrigated areas is expected to be around 6 MAF [22].

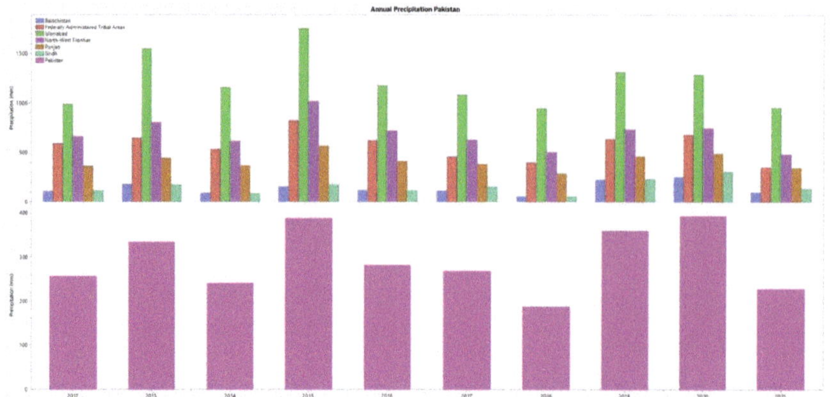

Figure 2. Annual precipitation in Pakistan and its provinces (source: World Bank Climate change knowledge portal, downloaded on 20 March 2023).

Surface water resources are mostly conditional on the tributaries of the Indus River. The total distance of the Indus River is 2900 km, and its basin is around 966,00 square km. Five main streams join its eastern side: the Jhelum, Chenab, Ravi, Beas, and Sutlej [23]. Additionally, three smaller rivers flow from hilly regions: the Soan, Harow, and Siran. In addition, a number of little streams enter the Indus on its west side. The largest of these is the Kabul River. Rivers in Pakistan have distinct charge characteristics, but all of them begin to increase in the spring and early summer due to the rainy season and snowmelt from the mountains, and they reach their stream flow in July and August. During the months of November through February, the average monthly flows are around one-eighth of those that occur during summertime. In addition to the larger rivers, there are countless tiny rivers and streams whose flow is only seasonal and dependent on precipitation; they carry little flow throughout the winter [24]. Furthermore, the majority of Pakistani groundwater resources are located on the Indus Plain, which stretches from the foothills of the Himalayas to the Arabian Sea and are kept in alluvial layers [25]. The Indus Plain is blessed with a huge aquifer system, which is quickly replacing it as a secondary supply of water for agriculture. It is around 1600 km wide and spans a region of 21 Mha. The aquifer was built over the course of the previous 90 years as a result of direct recharge from rainfall, river flow, and ongoing seepage from a conveyance network of canals, irrigation canals, waterways, and application losses in the irrigated regions [26].

Pakistan is usually regarded as both water-scarce (limited per capita water availability) and water-stressed. Despite this, crucial facets of Pakistan's water issues are frequently disregarded. Most evaluations of the water shortage exclude 24 percent of the total supply that is created locally (including the recharging of groundwater by precipitation) [27]. Decades-long population growth has led to a decline in mean availability, which also changes yearly with climatic changes [18]. The present water supply is around 79%. Public health is very at risk when drinking water is supplied improperly or poorly. The release of hazardous chemicals from urban areas and industries into bodies of water without proper treatment causes water pollution and has detrimental effects on human health. Due to rising demand, Pakistan's water and sanitation administration has been emphasizing the quantity of water over water quality [4]. This is all the result of a lack of knowledge, treatment technology, tools, trained staff, and quality control. One of the main factors in community health in Pakistan is highly polluted water. In this review, we have discussed water quality and the methods used to measure it from 2012 to 2022.

2. Water Quality Parameters and Limits in Pakistan

According to the World Health Organization, "safe drinking water" is any water that does not pose a significant threat to health during use, taking into account any variations in sensitivity between different life stages. Since water is the most frequently consumed liquid, it is thought to be the main means of disease transmission [28]. Bacterial pollution in water causes 80% of all human illnesses in developing countries. Access to potable water is one of Pakistan's key public health challenges, since the country faces water quality and quantity problems, as recorded in much research. The majority of the water supply, over 70%, is derived from groundwater aquifers throughout the nation [29]. Pakistan is a rich country with many policies. Several policies are directly and indirectly linked to maintaining the standards of water quality, but concerns about drinking water quality in the nation receive less attention since water system organizations prioritize volume over the quality of drinking water. Additionally, the lack of a legislative framework for drinking water quality concerns, poor institutional arrangements, a lack of well-equipped facilities, a lack of routine water quality monitoring, and inadequate institutional arrangements have made the situation worse [30]. In order to solve the major challenges and difficulties in the supply of clean drinking water, the Federal Cabinet of Pakistan passed the National Drinking Water Policy (NDWP) in September 2009. The overarching objective of the policy is to provide cheap, accessible, sustainable, and adequate access to potable water for the entire population and decrease the rate of waterborne disease-related morbidity and mortality. According to the policy, the federal government is in charge of creating special plans of action for underserved and poorly served regions; saline water regions; places vulnerable to natural disasters such as earthquakes, floods, and drought; and areas where women must travel more than 0.5 km to access clean drinking water. A study on Pakistani water quality was published in 2004 by the Pakistan Council of Research in Water Resources (PCRWR) and included suggestions for creating national drinking water quality standards. The World Health Organization (WHO), the Pakistani government, and the Ministry of Health examined the drinking water standards in place for quality assurance, revised them to reflect WHO drinking water quality requirements, and then finalized them (Pak-EPA 2008). The first safe drinking water legislation in Pakistan would be drafted and passed in accordance with the 2009 National Drinking Water Strategy. The policy also proclaimed access to clean water as a basic human right. In order to make the most of the resources available to the local government, cost-effective solutions will be used. The communication of information to all stakeholders regarding the Pakistan National Standards for Drinking Water was held to be the responsibility of the Ministry of Environment (currently the Ministry of Climate Change), the PCRWR, and the Pakistan Standards and Quality Control Authority (PSQCA) [31].

After Pakistan's 18th amendment to the constitution [32], province EPAs/EPDs were tasked with creating and executing environmental policies within their particular jurisdictions under the Provincial Environmental Protection Act and Policies. Additionally, the law divides up governing duties between the federation and the provinces. Following the passage of the Constitution (Eighteenth Amendment) Act of 2010 (the "18th Amendment"), Parliament is now able to pass laws, exercise executive power over the matters included in the Federal Legislative List in the Fourth Schedule to the Constitution and create extraterritorial legislation. On the other hand, regional assemblies have the authority to enact laws and exercise executive authority over any matter not included in the Federal Legislative List. Furthermore, the Constitution also gives the provinces the freedom to ask Parliament to pass legislation on any issue falling under their purview. The water quality standards of all four provinces and the Pakistan EPA can be found in the following list of websites (Table 1).

Table 1. Water quality standards.

Agency/Department Name	Website
Pakistan Environmental Protection Agency (PEPA)	https://www.elaw.org/system/files/Law-PEPA-1997.pdf, accessed on 20 March 2023
Sindh Environmental Protection Agency (SEPA)	https://hseforum.pk/forums/discussion/sindh-environmental-quality-standards/, accessed on 20 March 2023
Punjab Environment Protection Department (PEPA)	https://epd.punjab.gov.pk/system/files/Punjab%20Environmental%20Quality%20Standards%20for%20%20Drinking%20Water.pdf, accessed on 20 March 2023
KPK Environmental Protection Agency	https://epakp.gov.pk/, accessed on 20 March 2023
Baluchistan Environmental Protection Agency	https://bepa.gob.pk/, accessed on 20 March 2023

Samples are typically obtained, categorized for various contaminants and characteristics, compared to all these standards, and highlighted when the findings depart from the benchmarks. In Pakistan, this strategy is used for all sources of drinking water (surface water, groundwater, and rainwater) [33]. Various indicators (such as pH, TDS, heavy metals, and fecal coliforms) have also been used in other research to track water contamination. Studying the macroinvertebrate community, as well as other groups of microorganisms, is part of this. The number and variety of macroinvertebrates and microorganisms are utilized to inductively assess the water quality source even though they are not specifically employed as an indicator [34].

The establishment of the above water quality standards is regarded as a positive move that might serve as a model in addressing Pakistan's water security issue [35]. Regardless of having extensive national laws and regulations for groundwater resources and control of pollution and a policy structure for the protection of the environment, Pakistan's organizational and implementation capacity is severely deficient [36]. Local authorities in Pakistan are accountable for water supply and sanitation (WSS); solid waste disposal; and the disposal, treatment, and management of wastewater. The municipal administrations of Tehsil, as well as water and sanitation organizations, are in charge of implementation. Their effectiveness, unfortunately, is generally unregulated. Various national regulations control the water, sanitation, and hygiene (WASH) system; however, because water is a provincial issue, the influence of national organizations is limited. The NSP 2006 was created to encourage public participation, ownership, and management while also freeing up resources to spend on closing the clean water deficit [37]. The Pakistani government established the National Drinking Water Policy (NDWP) in September 2009 with the goal of supplying safe drinking water to all by 2025, with a focus on the poor and vulnerable [31]. The major goal is to distinguish between service provision and control. The right to drink water has priority over all other activities, such as agricultural or industrial water usage. In order to address the issue of water security, Pakistan established the new National Water Policy of Pakistan (NWPP) in 2018. The strategy highlights the need for Pakistan to address present and future challenges to its energy, water, and food security. It addresses a wide range of water-related issues in Pakistan, including the distribution and usage of water based on economic priorities, the sustainability of water basins, agriculture, changing climate impacts, drinking water and sanitation, groundwater, water rights and obligations, hydropower, conservation, sustainable water infrastructure, water-related risks, and regulatory frameworks [38]. Still, there is a lack of a clear framework and a sufficient implementation plan to address the problem, especially in organizational governance systems. However, due to insufficient law enforcement, none of the government's various projects to address water shortages and pollution in Pakistan have been successfully carried out, and the problems continue to exist. Along with legislation, public awareness is necessary to combat community indifference toward water contamination. Furthermore, official support is presently weak regarding laws, management efforts, and law enforcement, all of which are necessary for citizen education.

3. Methodology

In order to enable future research, recognize gaps in knowledge, and effectively specify study focuses, the goal of this review is to consolidate and categorize research results. This study consistently highlights the future prognosis for water quality monitoring and evaluation, and it goes on to concentrate on the importance of water quality in the environmental sustainability of Pakistan. The literature published on the topic over the previous 10 years was thoroughly searched on Google Scholar and Scopus to find data for this study. The relevant methods have been used in other research, as per Mosley, 2015 and Razali et al., 2018 [39,40]. The search terms used were "water," "quality," and "Pakistan"; the country name (Pakistan) was included for this purpose. After that, we narrowed it down to domestic articles in English, along with some other criteria such as the location, origin, document type, and publication year (2012–2022). This was successful in removing any important papers without omitting the pertinent ones. In the context of Pakistan, there are no such contemporary evaluations of this topic, nor are there any with the present breadth of coverage. This supports the originality and importance of such a complex strategy and emphasizes its applicability to a broad variety of Pakistani demographics, including academics, students, authorities, entrepreneurs, and common citizens.

4. Water Quality Assessment in Pakistan

4.1. Surface Water Quality Assessment

The examination and monitoring of the quality of surface waters in Pakistan have been discussed in light of the reported source of contamination, sample location, pollutant parameters, and relevant health issues. Table 2 includes, in reverse chronological order, complete information on the various water sources analyzed in the nation over the last decade (2022–2012), as well as the discovered contaminants. According to an older study by Sial et al. (2006), 1228 of Pakistan's 6634 recognized enterprises are classified as severely hazardous [2,41]. Industries became a major cause of water pollution in Pakistan because of the high load of organic and poisonous compounds in their wastewater effluents [42]. Textiles, leather, paper, pharmaceuticals, ceramics, petrochemicals, food, steel, oil mills, the sugar industry, and fertilizer plants are the main sectors causing water pollution [43]. These industries generate hundreds of thousands of tons of wastewater loaded with heavy substances such as mercury, magnesium, lead, cadmium, nitrites, chromium, and many more [44]. The majority of companies in Pakistan are concentrated in or near large cities. They directly discharge their waste effluents into surrounding sewers, streams, creeks, lakes, canals, and open or agricultural land. According to Fida et al. (2022) [44], waterborne diseases are caused by bacterial pollutants, including coliforms; toxic components such as Fe, Ni As, Cl, Hg, and F; and pesticides, which have been identified in surface water and groundwater. Forty-one locations all along the Indus River and its tributaries were sampled by Ahmad et al. (2021) [45] to determine the surface quality of the water in Sindh. Only four of the ten physiochemical properties tested (average values of 857 S/cm, 378 mg/L, 8.5 mg/L, and 220 mg/L for EC, TDS, pH, and COD, respectively) fulfilled the standards set by the WHO and the NSDWQ. Contrarily, the levels of TA (1400 mg/L), TH (584 mg/L), DO (6.40 mg/L), and NTU (219) were all above the recommended limits set by the WHO.

Table 2. Research on surface water quality in Pakistan.

Year	Sample Location	Number of Samples	Pollution Source	Flagged Pollutants and Parameters	Method Used	Health Issues Assessment	References
2022	Basho Valley, Gilgit Baltistan	23	Incoming pollution load from upstream channels	Cu, Zn, Mn, Mo, TCC, TFC, and TFS	Lab analysis, WQI, and statical analysis	x	[46]
	Kallar Kahar wetland, Punjab	5	Tourism, fishing, urban discharge, and open littering	Microplastics	Lab analysis and comparison	-	[47]
	Indus Drainage System	26	Anthropogenic	Mn, Co, Cu, Zn, Cr Ni, Cd, Hg, and Pb	Lab analysis, risk assessment	x	[48]
	Punjnad Headworks	27	Anthropogenic	Pb, As, Al, and Ba	Lab and statistical analyses, and risk assessment	x	[49]
	Tehsil Swabi	15	Flood runoff, biological contamination, and anthropogenic activities	Mg and HMs	Lab, statistical analysis, and risk assessment	x	[50]
	Chu Tran Valley	24	Agricultural activities, erosion, and domestic discharge	Heavy metals	WQI, PCA, and IDW	x	[51]
	Malakand	27	Anthropogenic	Cu, Fe, K, Mg, Al, Ca, Cr, Mn, Ni, P, and Zn	Lab and statistical analyses, and risk assessment	x	[52]
	Sutlej River	400	Livestock and human sources	Fecal coliform bacteria	Lab analysis, SWAT, and SSPs	x	[53]
	Eastern Peshawar basin	34	Industrial outlets	Heavy metals	Lab and statistical analyses	-	[54]
	Khyber Pakhtunkhwa	120	Not specified	Cu, Zn, Pb, Zn, Pb, Cu, Cr, Co, Ni, Hg, Cr, Co, Ni, Hg, Zn, Cu, Cr, Pb, Co, Ni, and Hg	Lab experiments and risk assessment	x	[55]
	Phuleli Canal	8	Industrial outlets	pH, EC, TDS, COD, CA, Mg, Na, K, BOD, COD, and fecal coliforms	Lab experiments	x	[56]
	Mirpurkhas	8	-	Salinity	Lab experiments	-	[57]
	Faisalabad	2	Industrial outlets	Sulfate ions, TDS, and barium ions	Lab analysis	x	[58]
2021	Kot Addu, Punjab	90	Not specified	As, Cd, and Pb	Lab analysis and multivariate statistical analysis	x	[59]
	Bajaur	75	Geogenic and anthropogenic activities	Cd, Pb, and Mn	Lab analysis, risk assessment, and questionnaire Survey	x	[34]
	Swat	9	Sewer waters	Escherichia coli	Lab analysis, statistical analysis, and questionnaire Survey	x	[60]
	Skardu	45	Not specified	Cd	Lab analysis, spatial analysis, and statistical analysis	-	[61]

Table 2. *Cont.*

Year	Sample Location	Number of Samples	Pollution Source	Flagged Pollutants and Parameters	Method Used	Health Issues Assessment	References
	Rawal Dam	148	Tourist	pH, EC, TDS, alkalinity, hardness, Cl, DO, SO_4, Ca, Mg, Fe, Cd, Ni, Cu, As, and F	Lab analysis, machine learning techniques, and WQI	-	[62]
	Ravi River	15	Industrial, domestic, and natural	TDS, COD, BOD, DO, TN, Cu, Pb, As, Cr, Mn, Zn, and Cd	Lab analysis, GIS, and statistical analysis	x	[63]
	Upper Indus River	642	Minerals, natural and anthropogenic	Bicarbonate	Gene expression programming, artificial neural networks, and linear and non-linear regression models	-	[64]
	RightBank Outfall Drain and Manchar Lake	14	Irrigation	EC, TDS, TH, SO_4, Cl, BOD, COD, Na, K, Ca, Mg, Cr, Ni, and Al, Pb, Cd, and fecal coliform	Lab analysis, WQI, and statistical analysis	-	[65]
	Sutlej River	97	Anthropogenic	Ca^-, Mg^-, HCO_3^-, Ca^-, Mg^-, Cl^-, and SO_4	Lab analysis, WQI, and statistical analysis	x	[66]
	Nawabshah	18	Agro-industrial and sewage	pH, TDS, EC, total hardness, and calcium	Lab analysis	-	[67]
2020	Harnai, Balochistan	24	Not specified	Nitrate	Lab analysis	x	[68]
	Abbottabad	100	Not specified	Turbidity, EC, and pH	Lab analysis	x	[69]
	Hattar Industrial Estate	60	Industries	Cd, Cr, Ni, Pb, and Zn	Lab analysis, risk assessment, and statistical analysis	x	[70]
	Lower Jhelum Canal	20	Industrial pollution	*E. coli*	Lab analysis, WQI, and GIS	x	[71]
	Swat River	30	Natural and anthropogenic	Cu, Ni, and Pb	Lab analysis, risk assessment, and statistical analysis	x	[72]
	Central Indus Basin	12	Mining activities	Fe, Mn, Ni, Cd, and Se	Lab analysis	x	[73]
2019	KhyberPakhtunkhwa and Gilgit-Baltistan	181	Anthropogenic	EC, turbidity, and As	Lab analysis risk assessment, and statistical analysis	x	[74]
	Faisalabad	37	Textiles, ice, pharmaceuticals, flour, cotton, sugar, and food	Cu, Fe, Pb, Al, As, Ba, Cd, Cr, Ni, and Zn	Lab analysis and health risk assessment	x	[75]
	Lower Sindh	8	Industry and agriculture	Cu, Mn, Ni, and Zn	Lab analysis and health risk assessment	x	[76]
	Lower Indus Plain	360	Irrigation	As	Lab analysis	x	[77]
	Industrial Hub of Pakistan	13	Industries	Cr	Lab analysis	x	[78]

Table 2. *Cont.*

Year	Sample Location	Number of Samples	Pollution Source	Flagged Pollutants and Parameters	Method Used	Health Issues Assessment	References
	Mangla Lake, Rawal Lake, and Simly Lake	6	Anthropogenic	Cr, Cd, Co, Pb, and Ni	Lab analysis, risk assessment, and statistical analysis	x	[79]
	Swat	58	Human and animal fecal material	Coliform bacteria	Lab analysis, questionnaire survey, and statistical analysis	x	[80]
	Jaffarabad	25	Natural disasters, residues from pesticides, fertilizers, and other domestic and industrial wastes	Turbidity, hardness, TDS, Cl, SO_4^{-2}, and Fe	Lab analysis and risk assessment	x	[81]
2018	Indus Delta	50	Industrial sewage and agricultural and industrial wastes	As	Lab analysis, statistical analysis, WQI, and SPI	-	[82]
	Indus River Basin	84	Geogenic and anthropogenic	TDS, pH	Lab analysis	-	[83]
	Lahore	15	Domestic and industrial	BOD_5, COD, Cu, Zn, Fe, Pb, Co, Ni, and Cd	Lab and statistical analyses	-	[84]
	Lower Indus Plain	48	Domestic and industrial	EC, TDS, hardness, cations, and anions	Lab analysis	x	[85]
	Rasul Barrage, River Jhelum	18	Not specified	TDS, turbidity, Fe, Cr, and Ni	Lab and statistical analyses	x	[86]
2017	Mansehra	24	Human	NO_3^-, PO_4^-, Fe, Pb, and Cd	Lab and statistical analyses	x	[87]
	Jhelum	292	Natural and anthropogenic	TDS	Lab and Statistical analysis	-	[88]
	Faisalabad	92	Textile industries	Pb, Cd, COD, and COD	GIS, lab analysis, and statistical analysis	x	[89]
	Coastal Areas of Sindh	34	Industrial waste	As, Mg, Mn, Cr, Cu, Fe, Pb, and Ni	Lab and statistical analyses	x	[90]
	Mangla Dam	120	Anthropogenic	Cd, Cu, Zn, Co, Pb, Ni, As, Fe, Cd, Co, Pb, Ni, As, Fe, Zn, Cu, and Mn	Lab and statistical analyses	-	[91]
2016	Peshawar	50	Natural	TDS, Cu, Pb, Cr, Cl, and Nitrites	Lab analysis	x	[92]
	Faisalabad	225	Not specified	TDS	Lab analysis	-	[93]
	Jamshoro	67	Irrigation	TDS, EC, salinity, turbidity, and pH	Lab and statistical analyses	x	[94]
	Gulshan-e-Iqbal Karachi	6	Industrial and domestic	pH, EC, and TDS	Lab analysis	x	[95]
	Punjab	1600	Domestic	TDS, F, and Fe	Lab and statistical analyses	-	[96]
2015	Pakistan	753	Natural and anthropogenic	As and c	Lab and statistical analyses and risk prediction	x	[97]
	Northern Pakistan	33	Natural and anthropogenic	Mn, Fe, Ni, Cr, and Co	Lab and statistical analyses and risk prediction	x	[98]

Table 2. *Cont.*

Year	Sample Location	Number of Samples	Pollution Source	Flagged Pollutants and Parameters	Method Used	Health Issues Assessment	References
	Rawalpindi and Islamabad	30	Not specified	TTHMs and chloroform	Lab analysis and risk assessment	x	[99]
	Lahore	9	Not specified	Cr, Fe, and Cu	Lab analysis	-	[100]
	Nawabshah	60	Point and nonpoint sources	TDS, pH, bicarbonates, Hardness, Cl, and SO$_4$.	Lab analysis	x	[101]
	Nowshera	39	Domestic	Cr, Ni, Pb, Cd, and As	Lab analysis and risk assessment	x	[102]
	Quetta City	16	Agriculture and domestic	TDS and EC	Lab analysis	x	[103]
	Panjkora River, Lower Dir	6	Not specified	ZN, Ir, and Mg	Lab analysis	x	[104]
	Haripur Basin	32	Natural and anthropogenic	Pb, Co, Cr, Cu, Cd, As, Hg Fe, Mn, Zn, Ni, and Zn	Lab and statistical analyses and risk assessment	x	[105]
2014	Tando Muhammad Khan	14	Not specified	TDS, Cl, SO4, Ca, Na, hardness, and As.	Lab analysis	-	[106]
	Manchar Lake	10	Anthropogenic	pH, EC, salinity, TDS, Alkalinity, nitrates, Cl, and total hardness	Lab analysis	-	[107]
	Hingol River, Balochistan,	22	Natural and anthropogenic	Pb, Ni, Zn, Mn, Fe, As, Cr, and Cu	Lab analysis	-	[108]
	Mianwali	28	Anthropogenic	BOD, COD, TDS, EC, pH, and heavy metals	Lab analysis	-	[109]
2013	Pakistan	747	Natural and anthropogenic	Fluoride	Lab analysis	x	[110]
	Bahawalpur	20	Natural and anthropogenic	EC, TDS, hardness, pH, Ni, Na, and K	Lab analysis	x	[111]
	Badin	18	Anthropogenic	pH, EC, and TDS	Lab and statistical analyses	x	[112]
2012	Jamshoro	57	Natural and anthropogenic	As	Lab analysis	-	[113]
	Indus River	27	Natural and anthropogenic	Cd, Pb, Hg, and Cu	Lab and statistical analyses	-	[114]
Total	73	7452				48	

x—included health issues in paper. —did not include health issues.

Major Pollutants and Their Sources in the Surface Water of Pakistan

Effective water contaminants may be responsible for the pollution and transmission of infection [115]. Bacteria (microorganisms and infections), chemical contaminants (toxic and heavy metals, salts, and acids), and cations and anions are the most prevalent contaminants [44]. These compounds are toxic if they exceed limits and represent significant physical concerns to individuals and other species in the environment if they exceed certain values [116]. Microbiological contamination; heavy metals, including Pb, As, Hg, and Cd; pesticides; and other anions such as NO$_3$ and F impact the environment and human health throughout Pakistan [8]. Table 2 depicts the primary pollutants discovered in several cities and provinces, as well as the ratios of samples that exceeded the requirements.

The most likely source of drinking water contamination in Pakistan has been identified as coliform pollution [117]. According to numerous studies, the country's drinking water is heavily contaminated by bacteria [71]; many of the documented bacterial species can have serious health consequences (Table 2). Most parts of the country's sources of water, including streams, waterways, rivers, and lakes, are heavily contaminated with bacteria. Out of 73 reviewed papers in this study, the authors found 12 studies that found bacterial contamination in their samples (Table 2). These studies also talked about health issues due to this [118]. Ref [2] studied bacteria contamination in the Sutlej River. In their research, they collected 400 samples from 13 different locations around the river. Their model found that *E. coli* levels in a more polluted environment will show 108% and 173% increases in the near- and mid-future, respectively, and a 251% rise in the distant future compared with the reference time (baseline) levels. In Phuleli Canal, Sindh Province, out of eight water samples, most were found to contain fecal coliforms [56]. Another study in Khairpur found fecal coliform and coliform bacteria in 100% of water samples taken from the water channel, the distribution system, and household taps [119]. The scenario is similar to all other major cities across the country, including Islamabad, Quetta, and Hyderabad. All of these cities' drinking water was determined to be polluted with bacteria. The major sources of these contaminants were domestic anthropogenic activities and old, open water supply lines [119]. Furthermore, several studies have found health issues due to using bacteria-contaminated water, such as asthma and infections related to the urinary tract, circulation, and other organs. An invasive pathogen creates a range of bloodstream pathogens in patients' urinary tracts, respiratory tracts, blood, and other ordinarily sterile areas [33].

The physical and chemical characteristics of water, such as pH, EC, DO, and BOD, and the presence of things such as TDS and TSS are used to establish standards and evaluate the physical and chemical water quality. Pesticides and fertilizers are examples of toxic elements that come from factories, ground sediments, and fertilizer waste and end up in water supplies nearby [4]. These water sources have trace-metal contaminants because, as water moves downhill in a hydrological cycle, it dissolves these things [120]. In addition, these metals enter both surface water and groundwater due to human actions such as dumping untreated municipal and industrial waste and using a lot of chemicals in farming. Many of these elements are important for individual health, but too many of them can pollute water and make people and other living things sick [121]. Table 2 is a summary of the different studies that have been conducted from 2012 to 2022 in Pakistan on water pollution caused by physicochemical pollution, such as EC, pH, TDS, TH, and TA. A study by Jabeen et al. (2014) [105] of the Haripur Basin on physicochemical parameters found that a stream sample from the Hattar Industrial Park region had the lowest pH (5.4), whereas a river sample from next to a marble industry area had the highest pH (9.20). Moreover, the surface water had an electrical conductivity range of 0.180 to 1.182 mS/cm. Surface samples from areas near industrial areas often have higher EC values. TDS concentrations in surface water can vary from 116 to 956 mg/L (mean, 322). Mean pH, EC, and TDS values in surface water have been reported to be below the WHO (2008) recommendations. The surface water quality of the Indus River in Sindh was studied by sampling 41 locations all along the river and its tributaries. Only four physicochemical characteristics, namely, EC, TDS, pH, and COD, with average values of 857 S/cm, 378 mg/L, 8.5 mg/L, and 220 mg/L, respectively, were within the WHO-permitted standards. TA (1400 mg/L), TH (584 mg/L), DO (6.40 mg/L), and NTU (219) values, on the other hand, were above the WHO-permitted criteria [45]. Arshad and Imran (2017) [122] found significant As concentrations in two villages in the central Punjab region. Out of 73, most of the studies found low-to-high concentrations of As, COD, BOD, TDS, and other chemicals (Table 2). Furthermore, the sources of these contaminations were industrial and anthropogenic activities and, in some areas, natural inputs [123].

The accumulation of harmful compounds in surface water, particularly harmful heavy metals, is a major environmental and social problem across the world [44]. Heavy metal traces, such as Pb, Mn, Cu, Ni, Fe, and Cr, are substantial environmental contaminants, especially in regions subjected to intense anthropogenic pressure [124]. These pollutants are capable of causing several negative health consequences in humans through the use of heavy-metal-contaminated water [125]. Although certain toxic substances are required for life in low quantities, others, such as Cd, Ni, Hg, and Pb, are physiologically nonessential and extremely dangerous to living creatures [126]. Recent research has found heavy metal pollution in the drinking water of Pakistan, as detailed in Table 2. In the majority of Barandu River samples, heavy metal concentrations of Fe, Pb, Ni, Cr, Mn, and Cd were found to be higher than the WHO-permitted levels for drinking water. People and other living things may have issues as a result of this high concentration of polluted water. Fe is one of the required elements, but if it is present in excess, it can have a negative impact on aquatic life and human health. High levels of Fe harm gastrointestinal tract cells and prevent them from controlling Fe absorption, which results in an unwarranted increase in blood's Fe concentration [127]. Another study on the Indus Drainage System found that the concentrations of most metals in riverine surface water were well below (Table 2) the guidelines established by the WHO and national agencies, but they showed wide interannual variation when compared with the other river systems. Along with the Indus Drainage System, the general riverine water contamination was higher than that observed in the surface water of the Shatt Al-Arab River [128], the South China Pearl River [129], and Argentine rivers [130] but smaller than those of the stream water of the Ganga River, India [131], and Gadea, Spain [132]. Since the Indus River primarily flows through the populated industrial zones of Faisalabad, Jhang, Sialkot, and Kabul, the heavy metal contamination level of the Chenab River was found to be lower than that of the Indus River [118]. Heavy metals in drinking water can pose significant health risks, including damage to the nervous system, kidney and liver damage, and cancer. In Pakistan, studies conducted between 2012 and 2022 found that high levels of heavy metals, such as lead, chromium, and arsenic, have been detected in drinking water sources (Table 2). It is crucial for Pakistan to tackle heavy metal contamination in drinking water in order to safeguard public health.

4.2. Groundwater Quality Assessment

Pakistan is one of the world's driest regions, and the majority of the people rely on groundwater to meet their water needs for drinking and other household reasons [133]. Groundwater is a vital resource in Pakistan, providing drinking water and irrigation for agriculture; however, the overuse and overexploitation of this resource have led to a decline in both water quantity and quality in the country [134]. Research on groundwater has largely been studied by authors in Pakistan. We found 97 studies conducted from 2012 to 2022 (using the above method), and 18,220 water samples were collected for water quality analysis (Table 3). Furthermore, these studies were important because groundwater levels in Pakistan have been declining at an average rate of 0.5 m per year over the past decade [135]. This decline is due to the overuse of groundwater resources for irrigation, as well a lack of effective management and regulation in the groundwater sector [66]. Several studies have also warned that the depletion of groundwater resources could have severe consequences for food security and economic development [136,137]. Groundwater in Pakistan is also being contaminated by a variety of pollutants, including agricultural chemicals, industrial effluents, and sewage (Table 3). Our comprehensive review found that these pollutants are affecting both the quality and the quantity of the groundwater, making it unsuitable for use. Furthermore, studies over the last decade have reported high levels of nitrates, arsenic, and fluoride in groundwater, making it unsafe for drinking and irrigation [138]. To address these issues, researchers have studied the implementation of sustainable groundwater management strategies in Pakistan. This includes the development of water quality assessment methods, policies, and regulations to better manage and conserve groundwater resources, as well as the use of new technologies and practices to reduce water loss and improve water quality. All the information on the many places examined throughout the country over the last decade, as well as the discovered contaminants, can be found in Table 3, listed in reverse chronological order.

Table 3. Research on groundwater quality in Pakistan.

Year	Sample Location	Number of Samples	Pollution Source	Flagged Pollutants and Parameters	Method Used	Health Issues Assessment	References
	Tehsil Swabi	26	Flood runoff, biological contamination, and anthropogenic activities	Mg and HMs	Lab and statistical analyses	x	[50]
	Eastern Peshawar Basin	36	Industrial outlets	Heavy metals	Lab and statistical analyses	-	[54]
	Industrial Zone of Faisalabad	60	Madhuana drain's recharge	pH, EC, Fe, Mn, Cu, and Cr	Lab experiments and GIS	x	[139]
	Khyber Pakhtunkhwa	120	Not specified	Co, Ni, Cu, Cr, Pb, Hg, Cu, Zn, Pb, Cu, Cr, Zn, Pb, C, Co, Ni, Hg, Zn, Co, Ni, and Hg	Lab experiments and risk assessment	x	[55]
	Taluka Ratodero	25	Agricultural fertilizers, industrial waste, and drainage	EC, TDS, Cl, Fu, Pb, Ni, and Cd	Lab experiments and WQI	x	[140]
	Gujranwala	200	Not specified	pH, EC, TH, Ca, Mg, Cl, and heavy metals	Lab experiments, risk assessment, and statistical and spatial data analyses	x	[141]
2022	Tharparkar	25	Natural and anthropogenic	EC, TDS, F, and As	Lab experiments, risk assessment, and statistical and spatial data analyses	x	[2]
	Rajanpur	200	Not specified	Ca, Na, HCO_3, Cl, Mn, and SO_4	Lab experiments and statistical and spatial data analyses	x	[142]
	Badin	1	Not specified	-	VES	-	[143]
	Larkana	43	Anthropogenic	EC, Cl, Ca, Mg, and hardness	Lab experiments, WQI, and SPI	-	[144]
	Nowshera KPK	48	Not specified	TDS and Cl	Lab experiments and map	x	[145]
	Kamber-Shahdadkot	46	Industrialization and urbanization	pH, TDS	Lab experiments, WQI, and GIS	x	[146]
	Batkhela	60	Anthropogenic and natural	$CaHCO_3$, $NaHCO_3$, and NaCl	Geostatistical analysis and risk assessment	x	[147]
	Bahawalnagar	40	Natural	As	Lab and statistical analyses	x	[148]
	Pakistan	2160	Natural and anthropogenic	F^-	Lab analysis and modeling	x	[135]
	Jhelum	82	Natural	F^-, As, nitrate, and bacteriological	Lab experiments, WQI, and risk assessment	x	[118]
	Mardan	13	Landfilling	(TDS), pH, EC, TH, NO_3^-, TC, Cr, Ni, Zn, Cd, TC SO_4^{-2}, NO^{2-}, Ca^{+2}, and Na^+,	Questionnaire survey and lab and statistical analyses	-	[149]

Table 3. *Cont.*

Year	Sample Location	Number of Samples	Pollution Source	Flagged Pollutants and Parameters	Method Used	Health Issues Assessment	References
2021	Kot Addu, Punjab	90	Not specified	As, Cd, and Pb	Lab analysis and multivariate statistical analysis	x	[59]
	Peshawar	217	Agricultural activities, sewerage lines, toilets, seepage, and percolation of polluted water	Nitrate and Ca	Lab analysis, spatial analysis, and risk assessment	x	[150]
	Bajaur	88	Geogenic and anthropogenic activities	Cd, Pb, and Mn	Lab analysis, risk assessment, and questionnaire Survey	x	[34]
	Central Punjab	20	Leaching	Zn, Fe, Cu, Cr, Ni, Cd, Co, and Pb	Lab analysis	x	[151]
	Swat	32	Sewer water	*Escherichia coli*	Lab analysis, statistical analysis, and questionnaire survey	x	[60]
	Rawal Dam	30	Tourist	pH, EC, TDS, Tur, alkalinity, hardness, Cl, DO, Fe, F, SO_4, Ca, Mg, Cd, Ni, Cu, Mn, and As	Lab analysis, machine learning techniques, and WQI	-	[62]
	Dokri	40	Precipitation	Cl and As	Lab analysis, GIS, and WQI	-	[152]
	Isa Khel, Mianwali	43	Not specified	EC and fluoride	Lab analysis	-	[153]
	Jhelum Basin	59	Air pollution through factories	Na^+, Ca^{2+}, Mg^{2+}, K^+, HCO_3, SO_4, Cl, NO_3, F, and As	Lab analysis, GIS, and WQI	x	[21]
	Lahore	39	Anthropogenic activities	Pb, Cr, Ni, and Cd	Lab analysis	-	[154]
	River Sutlej	111	Anthropogenic activities	EC, HCO_3, Cl, and SO_4	Lab analysis	x	[155]
	Southern Punjab	68	Silicate minerals and anthropogenic	HCO_3^-, SO_4^{2-}, Cl^-, F^-, Na^+, Ca^{2+}, Mg^{2+}, and K^+	Lab and Statistical analyses	-	[156]
	Lahore	1305	Textile mills and paper, electronic, plastic, paint, and pharmaceutical industries	Fe, NO_3^-, K, F, SO_4^{2-}, and As	Lab analysis, WQI, and entropy water quality index	x	[157]
	Sanghar	61	Anthropogenic and geogenic cause	As	Lab and statistical analyses and hadrochemical facies	-	[158]
2020	Sindh	425	Not specified	TDS, EC, Cl, turbidity, and hardness	Lab analysis and WQI	x	[159]
	Sindh Industrial Trading Estate, Karachi	24	Industrial	pH, EC, TDS, TH, Na, K, Ca, M, Cl, SO_4, HCO_3, NO_3, Fe, and Zn	Lab analysis and GIS	-	[160]
	Vehari	129	Not specified	Pb, Cd, and Fe	Lab analysis and risk assessment	x	[161]
	Vehari	75	Anthropogenic	As	Lab analysis and risk assessment	x	[162]

Table 3. *Cont.*

Year	Sample Location	Number of Samples	Pollution Source	Flagged Pollutants and Parameters	Method Used	Health Issues Assessment	References
	Malir Karachi	8	Domestic and industrial	*E. coli*	Lab analysis and WQI	x	[163]
	Lower Jhelum Canal	20	Industrial pollution	*E. coli*	Lab analysis, WQI, and GIS	x	[71]
	Sujawal	94	Anthropogenic and geogenic causes	Ca	Lab analysis, WQI, and SPI	x	[164]
	Central Indus Basin	50	Mining activities	Fe, Mn, Ni, Cd, and Se	Lab analysis	x	[73]
	Islamkot, Tharparkar	40	Natural	pH, EC, TDS, salinity, Cl, total alkalinity, Fl, and As	Lab analysis, WQI, and GIS	x	[165]
	Punjab	242	Agriculture	EC and residual sodium carbonate (SAR)	Lab and spatial variability analyses	-	[166]
	Karachi	42	Plumbing	As, TDS, hardness, and chloride	Lab analysis	x	[167]
	Indus Delta	180	Anthropogenic and geogenic causes	TDS, Cl, Ca, and Mg	Lab analysis, WQI, and SPI	x	[164]
	Thatta	100	Not specified	TDS, Cl, and Ca	Lab analysis, WQI, GIS, and SPI	-	[168]
	Punjab and Sindh	6	Anthropogenic	TSS, *E. coli*, $HCO_3{}^-$, $SO_4{}^{2-}$	Lab analysis	-	[169]
	Balochistan	30	Anthropogenic and geogenic causes	F^-, As, Hg, Ni, Cd, Cr, Fe, and Pb	Lab and statistical analyses	-	[170]
	Faisalabad	48	Textiles, ice, pharmaceuticals, flour, cotton, sugar, and food	Al, As, Ba, Cd, Cr, Cu, Fe, Pb, Ni, and Zn	Lab analysis, WQI, and health risk assessment	x	[75]
2019	Bajaur Agency	44	Anthropogenic	Na^+	Lab analysis and HCA	x	[171]
	Gujranwala	08	Anthropogenic	Bacterial, Cr, Cu, Zn, As, Co, Ni, and Cd.	Lab analysis and GIS	x	[172]
	Lower Indus Plain	360	Irrigation	As	Lab analysis	x	[77]
	Central Sindh	59	Anthropogenic and natural sources	Ca, Mg, Cl, and Na-Cl	Lab analysis and WQI	x	[173]
	Industrial Hub of Pakistan	31	Industries	Cr	Lab analysis	x	[78]
	Tharparkar	2170	Natural	TDS and F	Lab and statistical analyses	-	[174]
	Gujrat	10	Industries	Heavy metals	Lab and statistical analyses and PCA	x	[175]
	Peshawar	52	Anthropogenic and natural sources	Fe, Cu, pH, TSS, Cl, Cu Zn, Ni, and Pb	Lab and statistical analysis	x	[176]
2018	Swat	139	Human and animal fecal material	Coliform bacteria	Lab analysis, questionnaire survey, and statistical analysis	x	[80]
	Jaffarabad	50	Natural disasters, residues from pesticides, fertilizers, and other domestic and industrial waste	Turbidity, hardness, TDS, Cl, $SO_4{}^{-2}$, and Fe	Lab analysis and risk assessment	x	[81]

Table 3. *Cont.*

Year	Sample Location	Number of Samples	Pollution Source	Flagged Pollutants and Parameters	Method Used	Health Issues Assessment	References
	Eastern Punjab, Pakistan	66	Not specified	As	Lab analysis, saturation indices, and statistical analysis	-	[177]
	New Karachi Town	25	Domestic	TDS, Mg, K, Ca, Na, Cl, SO$_4$, and HCO$_3$	Lab analysis and WQI	-	[178]
	Nagarparkar	29	Natural	EC, TDS, Ca, lead, Ni, and Zn	Lab analysis and WQI	-	[179]
	Sindh	13	Domestic and pit	Alkalinity, HCO$_3$, Ca, CO$_3$, Turb, Cl, Mg, pH, K, Na, TDS, SO$_4$, NO$_3$, and microbials	Lab analysis	x	[119]
	Sukkur	20	Not specified	TDS, sodium, fluoride, and magnesium	Lab analysis	-	[180]
	Shaheed Benazir Abad	40	Agriculture and industry	TDS, Cl, sulfate, Na, and hardness	Lab analysis	x	[25]
	Lahore	380	Natural and anthropogenic	As	Lab analysis and GIS	x	[181]
	Indus Valley	1200	Natural and anthropogenic	As and pH	Lab analysis and GIS	x	[182]
	Mansehra	40	Human	NO$_3$$^-$, PO$_4$$^{3-}$, Fe, Pb, and Cd	Lab and statistical analysis	x	[87]
2017	Hakra Command Area	134	Not specified	EC	Lab analysis	-	[133]
	Lahore	73	Natural and anthropogenic	TDS and turbidity	Lab analysis, WQI, and GIS	x	[88]
	Khipro	39	Geological processes and source rock	pH, EC, TDS, TH, Na$^+$, K$^+$, Ca^{2+}, Mg^{2+}, Cl$^-$, SO$_4$$^{2-}$, and HCO$_3$$^-$	Lab analysis	-	[183]
	Peshawar	74	Industries	Pb, Cr, Cd, and Ni	Lab, statistical analysis, pollution index, and risk assessment	x	[126]
	Lahore	983	Anthropogenic	pH, TDS, and Mg	Lab analysis, WQI, and GIS	x	[184]
	Southern Lahore	50	Natural	Fluoride	Lab analysis	x	[185]
	Lakki Marwat	17	Not specified	Zn, lead, and Cd	Lab analysis	x	[186]
2016	Khyber Pakhtunkhwa	54	Solid waste and sewage	Giardia, Crypto, T. Gondi, Fasciola, B. coli, and entamoeba	Lab and statistical analysis	x	[187]
	Lower Indus Plain	218	Sewage, urban runoff, and industrial wastewater	EC, TDS, Na, Cl, SO$_4$, HCO$_3$, turbidity, and hardness	Lab analysis, WQI, and GIS	x	[188]
	Mailsi, Punjab	44	Anthropogenic and natural sources	As	Lab and statistical analyses	x	[189]
	Islamabad	42	Domestic	Fecal coliform bacteria	Lab analysis	-	[190]

Table 3. *Cont.*

Year	Sample Location	Number of Samples	Pollution Source	Flagged Pollutants and Parameters	Method Used	Health Issues Assessment	References
	Tharparkar	200	Natural	As	Lab analysis	x	[191]
	Sindh	200	Natural	As	Lab analysis	x	[192]
2015	Pakistan	1903	Natural and anthropogenic	Ca, Cr, Fe, Ni, and Pb	Lab and statistical analyses and risk prediction	x	[97]
	Northern Pakistan	82	Natural and anthropogenic	Mn, Fe, Ni, Cr, and Co	Lab and statistical analyses and risk prediction	x	[98]
	Nawabshah	65	Point and nonpoint sources	TDS, pH, bicarbonates, hardness, chloride, and sulfate	Lab analysis	x	[101]
	Badin	170	Not specified	TDS, turbidity, and pH	Lab analysis	x	[193]
2014	Peshawar	105	Not specified	TDS, EC, hardness, Ca, Mg, and Cl	GIS and lab and statistical analyses	-	[138]
	Haripur Basin	98	Natural and anthropogenic	Fe, Mn, Zn, Ni, Pb, Co, Cr, Cu, Cd, As, Hg, and Zn	Lab and statistical analyses and risk assessment	x	[105]
	Rawalpindi	262	Natural and anthropogenic	pH, TDS, and EC	GIS, lab analysis, and WQI	-	[194]
	Lahore	340	Poor drainage systems	TDS, pH, alkalinity, and turbidity	Lab analysis and GIS	-	[195]
	Punjab	36	Human sewage, agricultural	Fe, Ir, Ba, Al, and Cr	Lab analysis	-	[196]
	Dir Lower	11	Not specified	EC	Lab and Statistical analyses	-	[197]
	Bannu	197	Refuse dump and domestic sewage	pH, EC, TDS, hardness, salinity, alkalinity, Na, K, Li, Ca, Mg, Ba, Cu, Fe, Mn, Ni, and Zn	Lab and statistical analyses	-	[198]
2013	Charsadda	951	Anthropogenic	Pb, Cd, Fe, Ni, and Zn.	GIS, lab and statistical analyses, and risk assessment	x	[199]
	Tharparkar	30	Natural	As and F	Lab and statistical analyses	-	[200]
2012	Tharparkar	99	Natural	Fe, Ca, Cu, and Zn	Lab analysis	-	[201]
	Rawalpindi	96	Anthropogenic	TDS, turbidity, TOC, and E-Coli	Lab analysis	x	[202]
	Hangu	35	Natural and anthropogenic	Fecal coliforms, pH, and turbidity	Lab analysis	x	[203]
Total	97	18220				63	

x—included health issues in paper. —did not include health issues.

Major Pollutants and Their Sources in the Groundwater of Pakistan

Chemical (organic and inorganic) and biological pollutants are the two categories of natural pollutants found in groundwater. While inorganic pollutants come from natural origins, the majority of organic pollutants in groundwater are formed by anthropogenic activities. Numerous substances have been identified as inorganic pollutants and claimed to be sensitive aquatic pollutants. Organic pollutants such as oil and pesticides are the most frequent pollutants. Pathogens (bacteria, protozoa, and viruses), water-soluble radioactive compounds, and anions and cations are examples of toxic elements [204]. These compounds are dangerous and may seriously harm people and other ecosystem inhabitants if any amount of them exceeds the allowable limit. Heavy metals, including As, Cd, Pb, and Ni, as well as anions such as NO_3 and F, are prominent pollutants and are seen as a danger to groundwater quality in Pakistan [205,206]. Solid trash has increased as a result of population expansion and radical efforts to raise life quality [207]. Municipal waste is distinguished as household, commercial, or institutional garbage, whereas solid waste is categorized as dangerous, clinical, urban, or radioactive [208]. As is clear from the preceding paragraph, solid waste dumps primarily exist in urban areas of developing countries and pose serious risks to groundwater sources, which are an important source of residential water supplies in these places [209]. Groundwater supplies may become contaminated by bacteria and other diseases from municipal waste dumps [149].

Regarding human health, ensuring the bacteriological safety and purity of drinkable water is an ongoing challenge. The transfer of harmful germs, such as microorganisms, infections, and protozoa, occurs through water. In the entire world, 80 percent of illnesses are caused by contaminated water [113]. Fecal and total coliforms in drinking water are indications of the existence of disease-causing microorganisms and pathogens; microbiological safety and groundwater quality are established and monitored by analyzing their occurrence [119]. In both urban and rural parts of Pakistan, bacterial pollution in drinkable water is a major public health concern since it carries germs that can cause infectious illnesses [44]. Multiple studies have found bacterial pollution in groundwater; the country ranks 80th out of 122 nations with low-quality drinking water [117]. In Islamabad, nearly half the samples were polluted with *E. coli* and fecal coliform bacteria, rendering the water unfit for consumption [190]. A systematic review in 2011 was carried out in four provinces of Pakistan. The study found that coliform bacteria were present in 64% of the samples from Punjab Province, 67% of samples from Khyber Pakhtunkhwa Province, 83% of samples from Sindh Province, and 80% of the samples from Balochistan [210]. Empirical research was conducted recently in Kohistan (MIS1), Shangla (MIS2), and two organizations, namely, Malakand (MIS3) and Mohmand (MIS4). Fecal coliform bacteria were present in the study region at concentrations of BDL-60.0, BDL-32.0, BDL-97.0, and BDL-89.0 colonization units per 100 mL (CFU/100 mL), respectively. The water samples from MIS2 were found to have the lowest mean concentration of fecal bacteria (8.45 CFU/100 mL), whereas the water samples from MIS3 had the highest mean concentration (18.25 CFU/100 mL). In the research region, 78% of groundwater sources had coliform bacteria contamination when contrasted with the permissible threshold (0 CFU/100 mL) of water established by the WHO and Pak-EPA. This high fecal contamination bacteria pollution may be caused by poor sanitation, poor-quality sewage tanks, and animal and human waste [211]. In the Sindh province of Pakistan, this is due to the application of fertilizers in agriculture and the discharge of untreated wastewater into nearby water bodies [125]. A study also found that the presence of nitrates was higher in groundwater samples collected from areas with a higher population density and intensive agricultural activities [4]. In conclusion, the biological contamination of groundwater in Pakistan is a serious problem that is caused by a variety of factors, including the discharge of untreated wastewater, the presence of septic tanks and latrines, the disposal of industrial waste, the use of pesticides in agriculture, and the application of synthetic fertilizers [7]. This problem is particularly severe in areas with a higher population density and intensive agricultural activities, and it poses a significant threat to the health and well-being of the population (Table 3).

The physical and chemical characteristics of groundwater that can determine its quality and appropriateness for different purposes are referred to as physicochemical parameters. Temperature, pH, conductivity, turbidity, dissolved oxygen, and concentrations of different dissolved ions and compounds, such as nitrates, chlorides, and heavy metals, are some of the characteristics that make up this list [87]. These variables can be used to evaluate the general condition of a groundwater system and spot any possible deterioration or pollution. Monitoring these variables over time can also assist in observing changes in groundwater quality and spotting any possible problems. Chemical contaminants such as fertilizers and pesticides come through industry, soil sediments, and fertilizer runoff and then reach surrounding sources of water [44]. These sources of water include harmful metal contaminants as a result of the dissolution of these chemicals during a hydrological cycle [4,44]. Research was conducted to assess the total dissolved solids, hardness, pH, alkalinity, and turbidity of groundwater. Groundwater chemistry analysis data from Lahore were used to assess the quality of drinking water regions using a GIS. The study found that 61% of the zone was excellent, 27% was good, 9% was moderate, and 3% had poor quality. However, when using PSQCA guidelines, only 5% of the region is perfect, 29% is satisfactory, 34% is medium, and 32% is not acceptable for drinking [195]. In a study by Shahab et al. (2016) [188], water samples collected in Sindh were found to be unsafe to drink. In addition, 62.84 percent of EC samples, 34.86 percent of TDS samples, 43 percent of Na^+ samples, 17.88 percent of Cl^- samples, 26.60 percent of SO_4^{2-} samples, 39.44 percent of HCO_4^- samples, 41.7% of turbidity samples, and 35.32 percent of hardness samples were above the WHO standard limit for drinking water. The highest concentrations were recorded in lower Sindh (Thatta, Badin), Tharparker, and central Sindh, where seawater intrusion occurs. Principal component analysis and correlation studies confirm the positive link between As and Fe, which may be the reason for As mobilization in Sindh's water table. Moreover, in Malakand city, water samples were taken from 75 groundwater sources to determine the quality of drinking water by analyzing the physiochemical characteristics. Among the basic measures, only TH (1500 mg/L) surpassed the WHO and NSDWQ guidelines. These sources of drinking water represent a grave risk to the community's health [212]. By collecting 10 samples of drinking water from different sources, the physicochemical quality of drinking water in new urban areas of Peshawar was analyzed. The pH fell within the WHO-permitted range; however, TSS levels in five samples and NTU levels in six samples exceeded WHO and NSDWQ recommendations. Pathogens and bacterial illnesses, including diarrhea, nausea, and abdominal pains, can be caused by elevated levels of NTU and TSS [213].

Heavy metal contamination in groundwater is a major concern in Pakistan, and studies have reported the presence of heavy metals such as lead, cadmium, chromium, and nickel in many samples above the permissible limits set by the WHO [161,175]. The sources of heavy metal contamination are varied, including industrial activities, agricultural practices, and municipal waste disposal. It is important to continue research on the extent and sources of the contaminations and implement measures to protect groundwater resources [142]. Trace metals are elements such as iron (Fe), cobalt (Co), and zinc (Zn) that are necessary for the proper growth and functioning of the human body in extremely minute quantities. Heavy metal often relates to a metallic element with a very high density or that is hazardous in trace levels, although excessive amounts of these substances might have negative consequences. Heavy metal examples contain lead (Pb), mercury (Hg), and arsenic (As) [214]. Numerous study studies have been published between 2012 and 2022 stressing the adverse consequences of heavy metals found in drinking water, particularly groundwater. As a result of a lack of economic and technological means, protecting the human population from the harmful consequences of metal poisoning is far more difficult in developing countries than in industrialized nations. Table 3 shows a brief description of the major and heavy metals present in different areas of Pakistan groundwater. Research has examined the levels of various pollutants present in the drinking water and the potential health hazards in Charsadda District, Khyber Pakhtunkhwa, Pakistan. The levels of nitrates in 13 locations exceeded the limit of 10 mg/L set by the US EPA, with concentrations ranging from 10.3 to 14.84 mg/L. Similarly, sulfate levels in nine sites exceeded the limit of 500 mg/L established by the WHO, with concentrations ranging from 505 to 555 mg/L. Additionally, concentrations of Pb, Cd, Ni, and Fe exceeded the permissible limits set by various organizations in certain areas [215]. Nickle is present in groundwater from all of Pakistan's main cities, with Lahore and Karachi being the worst affected (Table 3). The concentrations range from 0.001 to 3.66 mg/L. The WHO-recommended level is 0.07 mg/L, but PCSQA's allowed limit is 0.02 mg/L [204]. When it comes to the poisoning of groundwater with arsenic and fluoride,

Pakistan is ranked fourth internationally [216]. Regions of Sindh and Punjab have the highest levels of arsenic and fluoride contamination. In Sindh, arsenic and fluoride groundwater contamination has been recorded in the districts of Sukkur, Karachi, Tharparkar, and Hyderabad [204,210]. The groundwater in the Pakistani province of Punjab has significant arsenic concentrations in the districts of Kasur, Lahore, Multan, Sheikhupura, GI Khan, Mianwali, and Vehari. A little under 30,000 wells have been examined as part of this general investigation. As was found in 79% of the wells; 11% had almost 0.01 and 0.05 mg/L, and 10% of the samples had As concentrations of more than 0.05 mg/L [217].

4.3. Research on Filtration Plant Water Quality

The development of water treatment facilities and the encouragement of secure water storage and hygienic practices are only a few of the actions taken by the government of Pakistan to improve the quality of the water supply. However, a lack of financing and infrastructure has hampered many of these initiatives. In Pakistan, there are also a lot of nongovernmental groups that are striving to raise water quality through community-based projects and education. Few studies have been conducted to assess filtration plant water quality (Table 4). One study focused on evaluating the water quality of filtration plants in two populated cities of Pakistan, Rawalpindi and Islamabad. Samples were collected from plants run by the Capital Development Authority in Islamabad and the Water and Sanitation Agency in Rawalpindi. The results showed that several physiochemical parameters and metals were above the acceptable limit set by the World Health Organization. Many samples in both cities were found to be of poor quality with a water quality index greater than 100. The study also found that children are more vulnerable to health hazards from contaminated water. The study suggests that the proper management of limited underground water resources is necessary to ensure the sustainability of safe drinking water in these cities [123]. Another study in the same city assessed the quality of drinking water in the distribution network by collecting 80 samples of water in triplicate from treatment plants and residential taps. Samples were analyzed for various parameters such as free total coliforms, chlorine residue, chloramines, trihalomethanes, total chlorine, total organic carbon, fecal coliforms, and turbidity. The results revealed that some areas of Islamabad City, such as sectors F-6, F-10, and F-11, were free of fecal contamination, while other areas, such as E-7 and F-7, had contamination at a few stations among all the collected samples. The most contaminated areas were found to be sectors F-8, G-9, G-8, I-9, and I-8 of Islamabad, with high levels of *E. coli* at most sampling stations. Additionally, high concentrations of trihalomethanes and chloroform were detected in sector F-8 [218] (Table 4). It is crucial to evaluate the water quality in filtration facilities since it has a direct influence on people's health and well-being. The sources of water for these filtration plants were groundwater and surface water [123]. It is essential to systematically research the water quality in filtration facilities in Pakistan, where access to clean drinking water is a key challenge. The scarce water resources are under stress due to a growing population and human activities, making it even more crucial to monitor and preserve them. In developing nations, where water quality monitoring and maintenance are frequently insufficient, the situation is more serious. Given the importance of this problem, it is essential that further research be conducted to thoroughly examine the water quality of Pakistani filtration facilities. The microbiological and health concerns related to water should also be studied, in addition to the physiochemical characteristics. Researchers can then pinpoint issues and provide workable remedies to raise the nation's water quality. In order to guarantee the population's access to potable drinking water for the foreseeable future, the research should also take sustainable management of subsurface water resources into account. Furthermore, evaluating the water quality in filtration facilities in Pakistan is essential to safeguarding the population's health and welfare. To thoroughly examine the water quality and provide practical strategies to enhance it, further research is required. In order to guarantee the population's access to clean drinking water over the long term, this research should also take sustainable water resource management into account.

Table 4. Research on filtration plant water quality in Pakistan.

Year	Sample Location	Number of Samples	Pollution Source	Flagged Pollutants and Parameters	Method Used	Health Issues Assessment	References
2020	Rawalpindi and Islamabad	85	Not specified	Electrical conductivity, alkalinity, and arsenic	Lab analysis	x	[123]
	Pakistan	638	Not specified	Microbiological, Fe, and Mn	Lab analysis	x	[219]
2014	Islamabad	80	Not specified	*E. coli*	Lab analysis	x	[218]
Total	03	803				03	

x—included health issues in paper.

4.4. Research on Rainwater Quality

Rainwater quality assessment is an important field of research in Pakistan since it is a key supply of water for many people. The country's semiarid and desert regions rely largely on precipitation as a supply of water for drinking, irrigation, and other purposes. However, numerous variables, such as industrial and agricultural operations, urbanization, and climate change, can have an impact on the quality of rainfall [220]. There have been several studies on the rainwater quality in Pakistan in recent years. These studies have focused on various aspects of rainwater quality, including chemical and physical parameters, as well as microbiological and health risks. In order to solve the water scarcity problem in the chosen research region, a study was undertaken to analyze the quality of roof-harvested rainwater. The research team also attempted to discover any potential health risks linked with rainwater drinking in the study location. With the exception of pH, turbidity, and trace amounts of certain elements such as Fe and Pb, the examination of the samples found that all of the physicochemical parameters were within the allowed range stated in the World Health Organization's drinking water standards. The samples' mean pH values varied from 5.18 to 6.26, representing mild acidity. The greatest mean turbidity level measured was 5.77 NTU. Furthermore, the mean amounts of Fe and Pb were 0.95 mg/L and 0.056 mg/L, respectively, exceeding the World Health Organization's drinking water recommendations. The study's findings show that, while roof-harvested rainwater can be a possible supply of drinking water, it may require further treatment to fulfill the WHO's drinking water quality requirements [220].

Several contaminants and characteristics have been identified in rainwater in Pakistan that may constitute a health risk if swallowed. Due to industrial pollution and pesticide usage in agriculture, rainwater in Pakistan has been discovered to have high amounts of heavy metals, such as Pb, Cd, and Cr [221]. Rainwater in Pakistan has also been shown to be rich in bacteria, including *E. coli* and fecal coliforms, which can cause waterborne illnesses. Furthermore, organic contaminants such as pesticides and herbicides are found in significant concentrations in rainfall samples. These contaminants have the potential to harm aquatic life and taint drinking water supplies [222]. Several studies evaluating rainwater quality in Pakistan have been conducted recently. The studies in Table 5 concentrated on different elements of rainwater quality and demonstrated that a number of variables, such as industrial and agricultural activity, urbanization, and climate change, can have an impact on rainwater quality. According to the research, the good management of rainwater resources is necessary to guarantee that there is enough clean drinking water for the entire population [220].

Table 5. Research on rainwater quality in Pakistan.

Year	Sample Location	Number of Samples	Pollution Source	Flagged Pollutants and Parameters	Method Used	Health Issues Assessment	References
2022	Lower Dir	35	Vehicular and industrial pollution and spray drift	Fe and Pb	Lab analysis and risk assessment	x	[220]
2019	Punjab	20	Industries, fuel burning, and vehicles	Cd and Pb	Lab analysis	-	[221]
2018	Jamshoro	4	Natural	Alkalinity, nitrogen, sodium, Cl, silicates, and phosphate	Lab analysis	x	[222]
2017	Toba Tek Singh	72	Not specified	pH, EC, and TDS	Lab analysis	-	[223]
	Tharparkar	9	Natural	As and F	Lab analysis and statistical analysis	-	[224]
2014	Karachi	54	Industries, fuel burning, and vehicles	NO_3^-	Lab analysis	x	[225]
	Karachi	35	Industries, fuel burning, and vehicles	TDS and F	Lab analysis	x	[226]
Total	7	229				4	

x—included health issues in paper. —did not include health issues.

5. Pollution Sources, Assessment Techniques, and Sustainable Development Importance

5.1. Water Pollution Sources in Pakistan

Rising water scarcity and pollution in Pakistan pose a serious danger to the economy and human life, unveiling the importance of the bioeconomy to sustainability [227,228]. Water contamination has been a major concern in the country as the population and economy have grown [44]. The situation is exacerbated by a lack of water treatment technology and knowledge [229]. The overuse of chemical fertilizers and pesticides in agriculture [178], the discharge of industrial waste [175] and untreated sewage into nearby bodies of water [209], and contamination from leaks and malfunctioning pipes used for water delivery are the primary water pollution sources [87]. Industrial wastewater is directly discharged into near water bodies. Pollutants can enter surface water and then groundwater from industrial areas such as textile mills, tanneries, and chemical plants due to leaks or inappropriate waste disposal. Toxic materials such as heavy metals, chemicals, and other harmful substances are examples of pollutants [2,230]. Sewage discharge is another type of groundwater pollution in Pakistan, where improper sewage disposal can contaminate groundwater with bacteria and other contaminants. This can happen when sewage is not adequately treated before disposal or is not disposed of in a way that prevents it from leaking into groundwater [231]. In Pakistan, landfills and the open dumping of solid waste are also sources of surface contamination. Pakistan produces over 48.5 million tons of solid waste yearly [149,156]. Pakistan faces many of the same environmental concerns as other developing countries due to a lack of garbage disposal facilities. The majority of municipal garbage is often burnt, discarded, or dumped in vacant lots in many areas, which has an impact on the well-being of the populace. Major cities in Pakistan are estimated to produce 87,000 tons of solid trash daily by the government (GOP) [51]. Solid waste placed in landfills can produce leachate and harm groundwater [232]. This can happen if the landfill is not adequately planned or managed or if garbage is not confined effectively.

Water quality can be changed by a large number of factors, including agricultural practices, industrial activity, and poor waste disposal. Agricultural activities are a major source of groundwater and surface water pollution in Pakistan [156]. Farmers frequently rely on agrochemicals to keep their crops healthy and protect them from bacterial and insect attacks. However, these pollutants can leach into the soil and eventually pollute groundwater [233]. Furthermore, these chemicals combine with precipitation and run into water bodies, eventually leaking into the sea and contributing to water contamination. Furthermore, pesticides can combine with irrigation and rainwater and penetrate aquifers, polluting these crucial water sources even further. Pesticides and fertilizers include a wide range of hazardous substances that can affect human health and aquatic life [234]. Improper sewage disposal can contaminate groundwater with bacteria and other contaminants. This can occur when sewage is not adequately treated before disposal or is not disposed of in a way that prevents it from leaking into groundwater. This form of pollution may have a considerable influence on the population's health and well-being, as well as the environment [235].

Mining is another source of groundwater contamination. Water contamination can result from mining minerals in mines [73]. This may happen if the mining activity is not sufficiently monitored or if the mining site is not properly cleaned up afterward. Furthermore, Pakistan is facing many climate change issues. Climate change is also altering the quality and availability of water resources in Pakistan [44,134]. Surface water supplies are being reduced as snow and ice cover in the Himalayas and Hindu Kush Mountains, which feed the Indus River system, decreases. The melting of glaciers has a significant influence on the supply of water for agriculture and residential usage, as well as the region's hydropower potential [236]. The contamination of both groundwater and surface water is a major concern in Pakistan, putting people's health and well-being at risk. To address this challenge, important steps must be taken, such as strengthening environmental laws and policies, initiating public education programs, and undertaking more research to identify effective solutions.

5.2. Method Used for the Water Quality Assessment

Water quality assessment is a crucial aspect of ensuring the safety and health of the population. Pakistan faces a number of challenges in terms of providing safe drinking water, including a lack of infrastructure and limited resources [97,237]. However, there have been several methods used in different studies in the period from 2012 to 2022 to assess the drinking water quality in Pakistan. One of the most commonly used methods is lab analysis [92,118]. This involves collecting samples of water from different study areas (such as lakes and rivers) and analyzing them for various contaminants (such as physicochemical ones, bacteria, and heavy metals). Water testing can be performed using various techniques, such as PCR, ELISA, and microscopy. Another method for assessing drinking water quality is through the use of water quality indicators [82,238]. While all WQIs have a similar framework, each was created with one of two goals. These goals may be broader, such as measuring the quality of life in a specific area, or more particular, such as implementing a new water treatment system. These methods have been used for both surface and groundwater. Furthermore, Tables 2–5 also list, for each of the WQIs under evaluation, the context in which it was created or is being used.

A comprehensive assessment is meant to provide individuals with an idea of the overall water quality, while specialized evaluations are meant to ensure that water is "suitable" for those who want to do certain things with it [65,71]. Some authors also used some recently developed modeling-based methods (SWAT and AI) to assess rain surfaces and groundwater quality [50]. Different studies have used geographical information systems (GISs) to create maps and visualize data, such as water quality test results, and can help identify patterns and trends in water quality. This is useful for identifying areas where water quality is poor and developing plans to improve it [54,239]. Furthermore, community-based monitoring is another important method for drinking water quality assessment in Pakistan. This approach involves engaging local communities in the assessment and monitoring of surface and groundwater quality. Different studies have used the questionnaire survey method to understand water quality and health issues [153,240]. Furthermore, to evaluate data and detect patterns and trends in water quality, statistical analysis methods are extensively employed in water quality evaluation in Pakistan. Principal component analysis (PCA) [168]; cluster analysis (CA) [87]; multivariate statistical techniques (MST), such as factor analysis (FA) and discriminant analysis (DA) [84,198]; and statistical process control are among these methodologies (SPC). In addition, along with these, various other methods are now being used to characterize and evaluate the quality of groundwater. Some of these methods include the blind number approach, the fuzzy mathematical method, cluster analysis, and a collection of pair analysis approaches [241]. These approaches can be used to detect probable sources of pollution, assess the efficacy of water treatment processes, classify water samples, and track water quality over time. Remote sensing technology can be used by future researchers in surface water quality analysis. This involves using satellite imagery and other remote sensing techniques to map and monitor water resources, such as surface water and groundwater bodies. This can provide information on water availability and quality, as well as help identify potential sources of pollution [242,243]. It is important to note that the evaluation of water quality in Pakistan also takes into account other variables, including TDS, BOD, and DO. To obtain a thorough evaluation of water quality in Pakistan, these factors are employed in addition to the ones described above.

Moreover, there are various methods that can be used in the future to assess drinking water quality in Pakistan. These methods can be used in combination to provide comprehensive knowledge of water quality and to identify potential sources of contamination. Recent years have observed an expanding necessity for a real-time sensing system capable of appositely recognizing occurrences of contamination in water distribution systems. A study by Piazza et al. (2020) [9] suggests a numerical optimization methodology that involves the utilization of the NSGA-II genetic algorithm combined with a diffusive–dispersive hydraulic simulator to optimize the efficiency of sensor configurations [244–246]. Due consideration of diffusion in the placement of water quality sensors is of the greatest importance if reliable monitoring networks are to be established, according to the findings of the research. Furthermore, the identification of sources of contamination is essential for the successful management of water pollution, which is an important issue in the water sector. The Bayesian method has also been used by different researchers to identify the most contaminating effluents in pressurized distribution systems and urban drainage networks. This can be achieved by the installation of real-time water-quality-measuring sensors [247]. Researchers in Pakistan can potentially utilize these methods in the future to assess water quality.

5.3. Importance of Water Quality in Combating Climate Change and Achieving Sustainable Development

Water quality is an important component of both sustainable development and climate change mitigation [248,249]. Water is a fundamental human right that is required for the existence and well-being of all living creatures. Poor water quality can cause health issues, financial losses, and environmental deterioration. The United Nations Sustainable Development Goals (SDGs) comprise explicit aims to achieve universal access to and the sustainable management of water and sanitation [250]. This includes Objective 6.1, which seeks to provide universal and equal access to clean and cheap drinking water for all, and Target 6.2, which aims to provide universal and equitable access to sanitation and hygiene for all [251,252]. The lack of access to water, whether resulting from human activity or natural phenomena, can have catastrophic consequences for the health, dignity, and economic prosperity of billions of people around the globe. SDG 6, commonly referred to as the "water goal", serves as a guideline for achieving water security and is indispensable for the achievement of the goals of the remaining SDGs [253]. Climate change also has an impact on water quality in a variety of ways. Changes in precipitation patterns and rising temperatures can cause more frequent and severe flooding, droughts, and other extreme weather events, contaminating water supplies and disrupting water infrastructure. These occurrences can also cause increased sedimentation and erosion, which can damage water infrastructure and make it more difficult to provide populations with clean and safe water [251,254]. Furthermore, climate change can cause changes in water temperature and chemistry, which can harm aquatic habitats and the animals that rely on them. This can eventually disrupt the entire water cycle, resulting in water shortages and compromising sustainable development goals. Improving water quality is thus critical for attaining a variety of long-term development goals and mitigating the effects of climate change. Investing in water infrastructure and improving water management methods, as well as decreasing pollution and safeguarding water sources, are all part of this [255]. This also entails striving to reduce the effects of climate change by lowering greenhouse gas emissions and adapting to existing changes [256,257]. Water quality is critical not just for human consumption but also for agricultural and industrial applications. Poor water quality can result in lower agricultural yields and higher expenses for farmers and companies. This has an impact on food security and may result in economic losses. Water shortages and poor water quality can also boost rivalry and conflict over water resources, undermining long-term development [258]. Water quality is an important component of both sustainable development and climate change mitigation. Addressing water quality challenges necessitates a multifaceted approach that includes establishing integrated water resource management, investing in water infrastructure, and employing cutting-edge technology. This can eventually lead to better access to clean and safe water, higher food security, and long-term economic prosperity. Furthermore, it is critical to reducing the consequences of climate change, which is increasing water quality problems.

6. Conclusions

This comprehensive review of water quality assessments and monitoring in Pakistan has offered a complete picture of the country's present water quality situation. In order to properly manage and safeguard Pakistan's water resources, improved monitoring and evaluation methodologies are required, as evidenced by the data in this research. This review indicates that Pakistan has substantial water quality challenges, including high levels of industrial and agricultural pollution. In addition, restricted access to reliable data and a lack of cooperation between many stakeholders are significant obstacles to successfully resolving these concerns. The assessment found that the quality of water resources in Pakistan, both surface and groundwater, is highly inconsistent and often falls short of the standards set by the WHO and national standards. The presence of pollutants in water sources, including heavy metals, pesticides, and microorganisms, poses a significant danger to human health and the environment. Industrial and agricultural activities have been found to have a serious impact on the water quality of Pakistan's major rivers, including the Indus and the Chenab. Furthermore, the water quality of groundwater resources is at risk due to excessive extraction and inadequate management. The low availability of reliable data and information on water quality is another key problem. Despite the presence of several monitoring programs, the data collected is frequently insufficient, inconsistent, and difficult for the public to access. This makes it challenging for policymakers and decision-makers to make informed water management and protection decisions. Moreover, a lack of coordination between many stakeholders, such as government agencies, industry, and local populations, limits the effective management of water resources.

This examination also highlighted the necessity of enhanced monitoring and evaluation procedures. Actual monitoring and evaluation methods in Pakistan are frequently out of date and do not correctly reflect the current water quality situation. In addition, our analysis uncovered a lack of ability and resources for monitoring and evaluation at the local level. This makes it challenging for local communities to monitor and manage their water supplies properly. This review recommends a variety of initiatives for governments, businesses, and local communities to adopt to address these challenges. Industry must be held accountable for its effect on water quality. The government should impose more stringent rules on industrial and agricultural activities and hold enterprises accountable for any water contamination caused by their operations. Finally, local communities should be given the authority to actively monitor and safeguard their water supplies. The government should assist local communities with the skills and resources necessary to successfully monitor and manage their water supplies. Additionally, local communities should be included in water management and protection decision-making processes.

Limitations of This Study

Several intriguing discoveries have been made due to this thorough evaluation of water quality monitoring and assessment in Pakistan that might be pursued in future research. The Indus River is the most polluted river in terms of microplastic particles in the world. This river is the principal source of potable water for agriculture and human consumption in Pakistan. Our water sources have not been studied for these contaminants in the past. Future research can examine the presence of pharmaceuticals, soaps, detergents, body lotion, and other personal care items in sources of drinking water. GISs have been used in water quality monitoring and evaluation, although this is still uncommon. This method may be used to collect data on water contamination across a larger area, as opposed to still-popular city-by-city study investigations. In addition, this study has mostly focused on surface and groundwater resources and may have overlooked other significant water sources, such as bottled water, filtered water, rainfall harvesting, and desalination. These water sources are gaining significance, particularly in countries where water shortages are a significant concern. Consequently, future studies may expand into these areas. High levels of pollution from industrial and agricultural sources have been highlighted as the most significant water quality challenges in Pakistan. However, no concrete advice for correcting these difficulties is provided. Therefore, further studies and research are required to comprehend the complexities of water quality challenges and find viable strategies to address them. Furthermore, the economic component of water quality monitoring and evaluation has not been explored. The cost of adopting the proposed solutions has not been analyzed, nor have the potential benefits and costs of these solutions. It is crucial to assess the economic viability of the suggested alternatives. Lastly, there is a lack of studies on the relationship between water quality and its health effects. Therefore, future research should focus on finding the link between water and health.

Author Contributions: L.K., conceptualization, methodology, data curation, and writing—original draft preparation; R.K., writing, resources, and visualization; A.K., literature review, writing, and editing original draft; I.A.T., literature review and editing; C.S., review and editing. All authors have read and agreed to the published version of the manuscript.

Funding: This research received no external funding.

Institutional Review Board Statement: Not applicable.

Informed Consent Statement: Not applicable.

Data Availability Statement: All data used have been added to the manuscript.

Conflicts of Interest: The authors declare no conflict of interest.

Abbreviations

Ministry of Environment, Pakistan (MOE-PAK), biochemical oxygen demand (BOD), copper (Cu), lead (Pb), zinc (Zn), dissolved oxygen (DO), manganese (Mn), molybdenum (Mo), water quality index (WQI). multivariate analysis comprising principal component analysis (PCA), National Water Quality Monitoring (NWQM), spatial distribution using inverse distance weight (IDW), Pakistan Council for Research in Water Resources (PCRWR), electrical conductivity (EC), soil and water assessment tool (SWAT), shared socioeconomic pathways (SSPs), vertical electrical soundings (VES), total suspended solids (TSS), geographical information system (GIS), hierarchical cluster analysis (HCA), pollution source apportionment methodology (PCA/MLR), chemical oxygen demand (COD), Pakistan Environmental Protection Agency (PAK-EPA), World Health Organization (WHO).

References

1. Hashmi, I.; Farooq, S.; Qaiser, S. Chlorination and water quality monitoring within a public drinking water supply in Rawalpindi Cantt (Westridge and Tench) area, Pakistan. *Environ. Monit. Assess.* **2009**, *158*, 393–403. [CrossRef] [PubMed]
2. Kumar, L.; Deitch, M.J.; Tunio, I.A.; Kumar, A.; Memon, S.A.; Williams, L.; Tagar, U.; Kumari, R.; Basheer, S. Assessment of physicochemical parameters in groundwater quality of desert area (Tharparkar) of Pakistan. *Case Stud. Chem. Environ. Eng.* **2022**, *6*, 100232. [CrossRef]
3. Ebrahim, Z.T. Is Pakistan running dry? In *Water Issues in Himalayan South Asia*; Palgrave Macmillan: Singapore, 2019; pp. 153–181. [CrossRef]
4. Daud, M.K.; Nafees, M.; Ali, S.; Rizwan, M.; Bajwa, R.A.; Shakoor, M.B.; Arshad, M.U.; Chatha, S.A.S.; Deeba, F.; Murad, W.; et al. Drinking Water Quality Status and Contamination in Pakistan. *BioMed Res. Int.* **2017**, *2017*, 7908183. [CrossRef] [PubMed]
5. Guglielmi, G. Arsenic in drinking water threatens up to 60 million in Pakistan. *Science* **2017**, *24*, 15–16. [CrossRef]
6. Nabi, G.; Ali, M.; Khan, S.; Kumar, S. The crisis of water shortage and pollution in Pakistan: Risk to public health, biodiversity, and ecosystem. *Environ. Sci. Pollut. Res.* **2019**, *26*, 10443–10445. [CrossRef]
7. Ilyas, M.; Ahmad, W.; Khan, H.; Yousaf, S.; Yasir, M.; Khan, A. Environmental and health impacts of industrial wastewater effluents in Pakistan: A review. *Rev. Environ. Health* **2019**, *34*, 171–186. [CrossRef]
8. Kalair, A.R.; Abas, N.; Ul Hasan, Q.; Kalair, E.; Kalair, A.; Khan, N. Water, energy and food nexus of Indus Water Treaty: Water governance. *Water Energy Nexus* **2019**, *2*, 10–24. [CrossRef]
9. Piazza, S.; Blokker, E.J.M.; Freni, G.; Puleo, V.; Sambito, M. Impact of diffusion and dispersion of contaminants in water distribution networks modelling and monitoring. *Water Sci. Technol. Water Supply* **2020**, *20*, 46–58. [CrossRef]
10. Hazbavi, Z.; Sadeghi, S.H.; Gholamalifard, M.; Davudirad, A.A. Watershed health assessment using the pressure–state–response (PSR) framework. *Land Degrad. Dev.* **2020**, *31*, 3–19. [CrossRef]
11. Abdul, N.A.; Talib, S.A.; Amir, A. Removal Kinetics of Chromium by Nano-Magnetite in Different Environments of Groundwater. *J. Environ. Eng.* **2020**, *146*, 04019111. [CrossRef]
12. Kolli, M.K.; Opp, C.; Karthe, D.; Kumar, N.M. Web-Based Decision Support System for Managing the Food–Water–Soil–Ecosystem Nexus in the Kolleru Freshwater Lake of Andhra Pradesh in South India. *Sustainability* **2022**, *14*, 2044. [CrossRef]
13. Kumar, N.M.; Kanchikere, P.M.J. Floatovoltaics: Towards improved energy efficiency, land and water management. *Int. J. Civ. Eng. Technol.* **2018**, *9*, 1089–1096.
14. Zhang, D.; Sial, M.S.; Ahmad, N.; Filipe, J.A.; Thu, P.A.; Zia-Ud-din, M.; Caleiro, A.B. Water scarcity and sustainability in an emerging economy: A management perspective for future. *Sustainability* **2021**, *13*, 144. [CrossRef]
15. Qamar, K.; Nchasi, G.; Mirha, H.T.; Siddiqui, J.A.; Jahangir, K.; Shaeen, S.K.; Islam, Z.; Essar, M.Y. Water sanitation problem in Pakistan: A review on disease prevalence, strategies for treatment and prevention. *Ann. Med. Surg.* **2022**, *82*, 104709. [CrossRef]
16. Howard, G. The future of water and sanitation: Global challenges and the need for greater ambition. *J. Water Supply Res. Technol.* **2021**, *70*, 438–448. [CrossRef]
17. Ranjan, A. Inter-Provincial water sharing conflicts in Pakistan. *Pak. A J. Pak. Stud.* **2012**, *4*, 102–122.
18. Young, W.J.; Anwar, A.; Bhatti, T.; Borgomeo, E.; Davies, S.; Garthwaite, W.R., III; Gilmont, E.M.; Leb, C.; Lytton, L.; Makin, I.; et al. *Pakistan Getting More from Water*; World Bank: Washington, DC, USA, 2019. [CrossRef]
19. Akhtar, I.u.H.; Athar, H. Water supply and effective rainfall impacts on major crops across irrigated areas of Punjab, Pakistan. *Theor. Appl. Climatol.* **2020**, *142*, 1097–1116. [CrossRef]
20. Salma, S.; Rehman, S.; Shah, M.A. Rainfall Trends in Different Climate Zones of Pakistan. *Pak. J. Meteorol.* **2012**, *9*, 37–47.
21. Abbas, M.; Shen, S.L.; Lyu, H.M.; Zhou, A.; Rashid, S. Evaluation of the hydrochemistry of groundwater at Jhelum Basin, Punjab, Pakistan. *Environ. Earth Sci.* **2021**, *80*, 300. [CrossRef]
22. Ali, G.; Sajjad, M.; Kanwal, S.; Xiao, T.; Khalid, S.; Shoaib, F.; Gul, H.N. Spatial–temporal characterization of rainfall in Pakistan during the past half-century (1961–2020). *Sci. Rep.* **2021**, *11*, 6935. [CrossRef]
23. Ashfaq, M.; Li, Y.; Rehman, M.S.U.; Zubair, M.; Mustafa, G.; Nazar, M.F.; Yu, C.P.; Sun, Q. Occurrence, spatial variation and risk assessment of pharmaceuticals and personal care products in urban wastewater, canal surface water, and their sediments: A case study of Lahore, Pakistan. *Sci. Total Environ.* **2019**, *688*, 653–663. [CrossRef] [PubMed]
24. Irfan, M.; Qadir, A.; Mumtaz, M.; Ahmad, S.R. An unintended challenge of microplastic pollution in the urban surface water system of Lahore, Pakistan. *Environ. Sci. Pollut. Res.* **2020**, *27*, 16718–16730. [CrossRef] [PubMed]
25. Bhatti, N.K.; Saand, A.; Keerio, M.A.; Ali, A.; Bhatti, N.-K.; Samo, S.R.; Bhuriro, A.A. Ground Water Quality Assessment of Daur Taluka, Shaheed Benazir Abad. *Eng. Technol. Appl. Sci. Res.* **2018**, *8*, 2785–2789. [CrossRef]
26. Yu, C.-H.; Wu, X.; Zhang, D.; Chen, S.; Zhao, J. Demand for green finance: Resolving financing constraints on green innovation in China. *Energy Policy* **2021**, *153*, 112255. [CrossRef]

27. Chilton, P.J.; Jamieson, D.; Abid, M.S.; Milne, C.J.; Ince, M.E.; Aziz, J.A. *Pakistan Water Quality Mapping and Management Project*; Water Engineering Development Centre, Loughborough University & London School of Hygiene and Tropical Medicines: Loughborough, UK, 2001.

28. Ullah, S.; Javed, M.W.; Shafique, M.; Khan, S.F. An integrated approach for quality assessment of drinking water using GIS: A case study of Lower Dir. *J. Himal. Earth Sci.* **2014**, *47*, 163–174.

29. Butt, M.; Khair, S.M. Cost of Illness of Water-borne Diseases: A Case Study of Quetta. *J. Appl. Emerg. Sci.* **2016**, *5*, 133–143. [CrossRef]

30. Haydar, S.; Arshad, M.; Aziz, J.A. Evaluation of Drinking Water Quality in Urban Areas of Pakistan: A Case Study of Southern Lahore. *Pak. J. Eng. Appl. Sci.* **2009**, *5*, 16–23.

31. Khwaja, M.A.; Aslam, A. *Comparative Assessment of Pakistan National Drinking Water Quality Standards with Selected Asian Countries and World Health Organization*; Sustainable Development Policy Institute (SDPI): Islamabad, Pakistan, 2018.

32. Khan, M.; Chaudhry, M.N. Role of and challenges to environmental impact assessment proponents in Pakistan. *Environ. Impact Assess. Rev.* **2021**, *90*, 106606. [CrossRef]

33. Khalid, S.; Murtaza, B.; Shaheen, I.; Ahmad, I.; Ullah, M.I.; Abbas, T.; Rehman, F.; Ashraf, M.R.; Khalid, S.; Abbas, S.; et al. Assessment and public perception of drinking water quality and safety in district Vehari, Punjab, Pakistan. *J. Clean. Prod.* **2018**, *181*, 224–234. [CrossRef]

34. Khan, M.H.; Nafees, M.; Muhammad, N.; Ullah, U.; Hussain, R.; Bilal, M. Assessment of Drinking Water Sources for Water Quality, Human Health Risks, and Pollution Sources: A Case Study of the District Bajaur, Pakistan. *Arch. Environ. Contam. Toxicol.* **2021**, *80*, 41–54. [CrossRef]

35. Weitzberg, E.; Lundberg, J.O. Novel Aspects of Dietary Nitrate and Human Health. *Annu. Rev. Nutr.* **2013**, *33*, 129–159. [CrossRef]

36. Nadeem, O.; Hameed, R. Evaluation of environmental impact assessment system in Pakistan. *Environ. Impact Assess. Rev.* **2008**, *28*, 562–571. [CrossRef]

37. Haider, H.; Ali, W. Sustainability of Sanitation Systems in Pakistan Solid Waste Management View Project. 2009. Available online: https://www.researchgate.net/publication/256191560_Sustainability_of_Sanitation_Systems_in_Pakistan (accessed on 24 February 2023).

38. Arfan, M.; Ansari, K.; Ullah, A.; Hassan, D.; Siyal, A.; Water, S.J. Agenda setting in water and IWRM: Discourse analysis of water policy debate in Pakistan. *Water* **2020**, *12*, 1656. [CrossRef]

39. Mosley, L.M. Drought impacts on the water quality of freshwater systems; review and integration. *Earth Sci. Rev.* **2015**, *140*, 203–214. [CrossRef]

40. Razali, A.; Ismail, S.N.S.; Awang, S.; Praveena, S.M.; Abidin, E.Z. Land use change in highland area and its impact on river water quality: A review of case studies in Malaysia. *Ecol. Process.* **2018**, *7*, 19. [CrossRef]

41. Sial, R.A.; Chaudhary, M.F.; Abbas, S.T.; Latif, M.I.; Khan, A.G. Quality of effluents from Hattar Industrial Estate. *J. Zhejiang Univ. Sci. B* **2006**, *7*, 974–980. [CrossRef]

42. Nasrullah, R.; Bibi, H.; Iqbal, M.; Durrani, M.I. Pollution load in industrial effluent and ground water of Gadoon Amazai Industrial Estate (GAIE) Swabi, NWFP. *J. Agric. Biol. Sci.* **2006**, *1*, 18–24.

43. Kumar, L.; Kamil, I.; Ahmad, M.; Naqvi, S.A.; Deitch, M.J.; Amjad, A.Q.; Kumar, A.; Basheer, S.; Arshad, M.; Sassanelli, C. In-house resource efficiency improvements supplementing the end of pipe treatments in textile SMEs under a circular economy fashion. *Front. Environ. Sci.* **2022**, *10*, 1002319. [CrossRef]

44. Fida, M.; Li, P.; Wang, Y.; Alam, S.M.K.; Nsabimana, A. Water Contamination and Human Health Risks in Pakistan: A Review. *Expo. Health* **2022**, *2022*, 1–21. [CrossRef]

45. Ahmad, W.; Zubair, A.; Abbasi, H.N.; Nasir, M.I. Water Study of Physical, Chemical and Heavy Metals Parameters in River Indus and its Tributaries, Sindh, Pakistan. *Pak. J. Sci. Ind. Res. Ser. A Phys. Sci.* **2021**, *64*, 103–109. [CrossRef]

46. Fatima, S.U.; Khan, M.A.; Siddiqui, F.; Mahmood, N.; Salman, N.; Alamgir, A.; Shaukat, S.S. Geospatial assessment of water quality using principal components analysis (PCA) and water quality index (WQI) in Basho Valley, Gilgit Baltistan (Northern Areas of Pakistan). *Environ. Monit. Assess.* **2022**, *194*, 151. [CrossRef] [PubMed]

47. Dilshad, A.; Taneez, M.; Younas, F.; Jabeen, A.; Rafiq, M.T.; Fatimah, H. Microplastic pollution in the surface water and sediments from Kallar Kahar wetland, Pakistan: Occurrence, distribution, and characterization by ATR-FTIR. *Environ. Monit. Assess.* **2022**, *194*, 511. [CrossRef] [PubMed]

48. Khan, K.; Younas, M.; Sharif, H.M.A.; Wang, C.; Yaseen, M.; Cao, X.; Zhou, Y.; Ibrahim, S.M.; Yvette, B.; Lu, Y. Heavy metals contamination, potential pathways and risks along the Indus Drainage System of Pakistan. *Sci. Total Environ.* **2022**, *809*, 151994. [CrossRef] [PubMed]

49. Naz, S.; Mansouri, B.; Chatha, A.M.M.; Ullah, Q.; Abadeen, Z.U.; Khan, M.Z.; Khan, A.; Saeed, S.; Bhat, R.A. Water quality and health risk assessment of trace elements in surface water at Punjnad Headworks, Punjab, Pakistan. *Environ. Sci. Pollut. Res.* **2022**, *29*, 61457–61469. [CrossRef]

50. Ahmad, L.; Waheed, H.; Gul, N.; Sheikh, L.; Khan, A.; Iqbal, H. Geochemistry of subsurface water of Swabi district and associated health risk with heavy metal contamination. *Environ. Monit. Assess.* **2022**, *194*, 480. [CrossRef]

51. Fatima, S.U.; Khan, M.A.; Alamgir, A.; Mahmood, N.; Sulman, N. Multivariate and spatial methods-based water quality assessment of Chu Tran Valley, Gilgit Baltistan. *Appl. Water Sci.* **2022**, *12*, 129. [CrossRef]

52. Khan, A.; Khan, M.S.; Egozcue, J.J.; Shafique, M.A.; Nadeem, S.; Saddiq, G. Irrigation suitability, health risk assessment and source apportionment of heavy metals in surface water used for irrigation near marble industry in Malakand, Pakistan. *PLoS ONE* **2022**, *17*, e0279083. [CrossRef]

53. Iqbal, M.S.; Islam, M.; Hassan, M.; Bilal, H.; Shah, I.A.; Ourania, T. Modeling the fecal contamination (fecal coliform bacteria) in transboundary waters using the scenario matrix approach: A case study of Sutlej River, Pakistan. *Environ. Sci. Pollut. Res.* **2022**, *29*, 79555–79566. [CrossRef]

54. Daud, S.; MonaLisa; Nisar, U. Bin Integrated geophysical, geochemical, and geospatial tools to characterize water resources in GAIE, Eastern Peshawar basin, Pakistan. *Environ. Earth Sci.* **2022**, *81*, 390. [CrossRef]

55. Nawab, J.; Rahman, A.; Khan, S.; Ghani, J.; Ullah, Z.; Khan, H.; Waqas, M. Drinking Water Quality Assessment of Government, Non-Government and Self-Based Schemes in the Disaster Affected Areas of Khyber Pakhtunkhwa, Pakistan. *Expo. Health* **2022**, 1–17. [CrossRef]

56. Panhwar, M.Y.; Panhwar, S.; Keerio, H.A.; Khokhar, N.H.; Shah, S.A.; Pathan, N. Water quality analysis of old and new Phuleli Canal for irrigation purpose in the vicinity of Hyderabad, Pakistan. *Water Pract. Technol.* **2022**, *17*, 529–536. [CrossRef]

57. Mastoi, S.T.; Channa, A.S.; Qureshi, K.M.; Khokhar, W.A. Assessment of Water Quality and Quantity of Surface and Subsurface Drainage System in the Command Area of Bareji Distributary Mirpurkhas, Sindh, Pakistan. *QUEST Res. J.* **2022**, *20*, 127–137. [CrossRef]

58. Amrane, A.; Khellaf, N.; Khan, R.U.; Hamayun, M.; Altaf, A.A.; Kausar, S.; Razzaq, Z.; Javaid, T. Assessment and Removal of Heavy Metals and Other Ions from the Industrial Wastewater of Faisalabad, Pakistan. *Processes* **2022**, *10*, 2165. [CrossRef]

59. Abbas, Z.; Imran, M.; Natasha, N.; Murtaza, B.; Amjad, M.; Shah, N.S.; Khan, Z.U.H.; Ahmad, I.; Ahmad, S. Distribution and health risk assessment of trace elements in ground/surface water of Kot Addu, Punjab, Pakistan: A multivariate analysis. *Environ. Monit. Assess.* **2021**, *193*, 351. [CrossRef]

60. Salam, M.; Alam, F.; Hossain, M.N.; Saeed, M.A.; Khan, T.; Zarin, K.; Rwan, B.; Ullah, W.; Khan, W.; Khan, O. Assessing the drinking water quality of educational institutions at selected locations of district Swat, Pakistan. *Environ. Earth Sci.* **2021**, *80*, 322. [CrossRef]

61. Ahsan, W.A.; Ahmad, H.R.; Farooqi, Z.U.R.; Sabir, M.; Ayub, M.A.; Rizwan, M.; Ilic, P. Surface water quality assessment of Skardu springs using Water Quality Index. *Environ. Sci. Pollut. Res.* **2021**, *28*, 20537–20548. [CrossRef]

62. Ahmed, M.; Mumtaz, R.; Zaidi, S.M.H. Analysis of water quality indices and machine learning techniques for rating water pollution: A case study of Rawal Dam, Pakistan. *Water Supply* **2021**, *21*, 3225–3250. [CrossRef]

63. Khan, A.; Khan, A.; Khan, F.A.; Shah, L.A.; Rauf, A.U.; Badrashi, Y.I.; Khan, W.; Khan, J. Assessment of the Impacts of Terrestrial Determinants on Surface Water Quality at Multiple Spatial Scales. *Pol. J. Environ. Stud.* **2021**, *30*, 2137–2147. [CrossRef]

64. Shah, M.I.; Javed, M.F.; Abunama, T. Proposed formulation of surface water quality and modelling using gene expression, machine learning, and regression techniques. *Environ. Sci. Pollut. Res.* **2021**, *28*, 13202–13220. [CrossRef]

65. Khuhawar, M.Y.; Lanjwani, M.F.; Khuhawar, T.M.J. Assessment of variation in water quality at Right Bank Outfall Drain, including Manchar lake, Sindh, Pakistan. *Int. J. Environ. Anal. Chem.* **2021**, *00*, 1–23. [CrossRef]

66. Setia, R.; Lamba, S.; Chander, S.; Kumar, V.; Dhir, N.; Sharma, M.; Singh, R.P.; Pateriya, B. Hydrochemical evaluation of surface water quality of Sutlej river using multi-indices, multivariate statistics and GIS. *Environ. Earth Sci.* **2021**, *80*, 565. [CrossRef]

67. Akhtar, F.; Ahmed, M.; Akhtar, M.N. Drinking, Tap and Canal Water Quality Analysis for Human Consumption: A Case Study of Nawabshah City, Pakistan. *Mehran Univ. Res. J. Eng. Technol.* **2021**, *40*, 392–398. [CrossRef]

68. Rehman, J.U.; Ahmad, N.; Ullah, N.; Alam, I.; Ullah, H. Health Risks in Different Age Group of Nitrate in Spring Water Used for Drinking in Harnai, Balochistan, Pakistan. *Ecol. Food Nutr.* **2020**, *59*, 462–471. [CrossRef] [PubMed]

69. Jadoon, S.; Wang, J.; Mahmood, Q.; Li, X.D.; Zeb, B.S.; Naseem, I.; Hayat, M.T.; Nawazish, S.; Ditta, A. Association of Nephrolithiasis with Drinking Water Quality and Diet in Pakistan. *Environ. Eng. Manag. J.* **2020**, *19*, 1289–1297. [CrossRef]

70. Jehan, S.; Khattak, S.A.; Muhammad, S.; Ali, L.; Rashid, A.; Hussain, M.L. Human health risks by potentially toxic metals in drinking water along the Hattar Industrial Estate, Pakistan. *Environ. Sci. Pollut. Res.* **2020**, *27*, 2677–2690. [CrossRef]

71. Bashir, N.; Saeed, R.; Afzaal, M.; Ahmad, A.; Muhammad, N.; Iqbal, J.; Khan, A.; Maqbool, Y.; Hameed, S. Water quality assessment of lower Jhelum canal in Pakistan by using geographic information system (GIS). *Groundw. Sustain. Dev.* **2020**, *10*, 100357. [CrossRef]

72. Jehan, S.; Ullah, I.; Khan, S.; Muhammad, S.; Khattak, S.A.; Khan, T. Evaluation of the Swat River, Northern Pakistan, water quality using multivariate statistical techniques and water quality index (WQI) model. *Environ. Sci. Pollut. Res.* **2020**, *27*, 38545–38558. [CrossRef]

73. Khan, A.J.; Akhter, G.; Gabriel, H.F.; Shahid, M. Anthropogenic effects of coal mining on ecological resources of the central indus basin, Pakistan. *Int. J. Environ. Res. Public Health* **2020**, *17*, 1255. [CrossRef]

74. Sohail, M.T.; Aftab, R.; Mahfooz, Y.; Yasar, A.; Yen, Y.; Shaikh, S.A.; Irshad, S. Estimation of water quality, management and risk assessment in Khyber Pakhtunkhwa and Gilgit-Baltistan, Pakistan. *Desalination Water Treat.* **2019**, *171*, 105–114. [CrossRef]

75. Mahfooz, Y.; Yasar, A.; Sohail, M.T.; Tabinda, A.B.; Rasheed, R.; Irshad, S.; Yousaf, B. Investigating the drinking and surface water quality and associated health risks in a semi-arid multi-industrial metropolis (Faisalabad), Pakistan. *Environ. Sci. Pollut. Res.* **2019**, *26*, 20853–20865. [CrossRef]

76. Imran, U.; Ullah, A.; Shaikh, K.; Mehmood, R.; Saeed, M. Health risk assessment of the exposure of heavy metal contamination in surface water of lower Sindh, Pakistan. *SN Appl. Sci.* **2019**, *1*, 589. [CrossRef]

77. Shahab, A.; Qi, S.; Zaheer, M. Arsenic contamination, subsequent water toxicity, and associated public health risks in the lower Indus plain, Sindh province, Pakistan. *Environ. Sci. Pollut. Res.* **2019**, *26*, 30642–30662. [CrossRef]

78. Mehmood, K.; Ahmad, H.R. Saifullah Quantitative assessment of human health risk posed with chromium in waste, ground, and surface water in an industrial hub of Pakistan. *Arab. J. Geosci.* **2019**, *12*, 283. [CrossRef]

79. Saleem, M.; Iqbal, J.; Shah, M.H. Seasonal variations, risk assessment and multivariate analysis of trace metals in the freshwater reservoirs of Pakistan. *Chemosphere* **2019**, *216*, 715–724. [CrossRef]

80. Khan, K.; Lu, Y.; Saeed, M.A.; Bilal, H.; Sher, H.; Khan, H.; Ali, J.; Wang, P.; Uwizeyimana, H.; Baninla, Y.; et al. Prevalent fecal contamination in drinking water resources and potential health risks in Swat, Pakistan. *J. Environ. Sci.* **2018**, *72*, 1–12. [CrossRef]

81. Sarfraz, M.; Sultana, N.; Jamil, M. Groundwater Quality and Health Risk Assessment in Rural Areas of District Jaffarabad, Baluchistan (Pakistan). *Pak. J. Anal. Environ. Chem.* **2018**, *19*, 79–85. [CrossRef]

82. Solangi, G.S.; Siyal, A.A.; Babar, M.M.; Siyal, P. Evaluation of surface water quality using the water quality index (Wqi) and the synthetic pollution index (spi): A case study of indus delta region of pakistan. *Desalination Water Treat.* **2018**, *118*, 39–48. [CrossRef]

83. Qaisar, F.U.R.; Zhang, F.; Pant, R.R.; Wang, G.; Khan, S.; Zeng, C. Spatial variation, source identification, and quality assessment of surface water geochemical composition in the Indus River Basin, Pakistan. *Environ. Sci. Pollut. Res.* **2018**, *25*, 12749–12763. [CrossRef]

84. Majeed, S.; Rashid, S.; Qadir, A.; Mackay, C.; Hayat, F. Spatial patterns of pollutants in water of metropolitan drain in Lahore, Pakistan, using multivariate statistical techniques. *Environ. Monit. Assess.* **2018**, *190*, 128. [CrossRef]

85. Shahab, A.; Qi, S.; Zaheer, M.; Rashid, A.; Talib, M.A.; Ashraf, U. Hydrochemical characteristics and water quality assessment for drinking and agricultural purposes in District Jacobabad, lower Indus Plain, Pakistan. *Int. J. Agric. Biol. Eng.* **2018**, *11*, 115–121. [CrossRef]

86. Iqbal, H.H.; Shahid, N.; Qadir, A.; Ahmad, S.R.; Sarwar, S.; Ashraf, M.R.; Arshad, H.M.; Masood, N. Hydrological and ichthyological impact assessment of rasul barrage, river jhelum, Pakistan. *Pol. J. Environ. Stud.* **2017**, *26*, 107–114. [CrossRef] [PubMed]

87. Raza, A.; Farooqi, A.; Javed, A.; Ali, W.; Zafar, M.I. Effect of human settlements on surface and groundwater quality: Statistical source identification of heavy and trace metals of Siran River and its catchment area Mansehra, Pakistan. *J. Chem. Soc. Pak.* **2017**, *39*, 296–308.

88. Javed, S.; Ali, A.; Ullah, S. Spatial assessment of water quality parameters in Jhelum city (Pakistan). *Environ. Monit. Assess.* **2017**, *189*, 119. [CrossRef] [PubMed]

89. Noreen, M.; Shahid, M.; Iqbal, M.; Nisar, J.; Abbas, M. Measurement of cytotoxicity and heavy metal load in drains water receiving textile effluents and drinking water in vicinity of drains. *Measurement* **2017**, *109*, 88–99. [CrossRef]

90. Alamgir, A.; Khan, M.A.; Manino, I.; Shaukat, S.S.; Shahab, S. Vulnerability to climate change of surface water resources of coastal areas of Sindh, Pakistan. *Desalination Water Treat.* **2016**, *57*, 18668–18678. [CrossRef]

91. Saleem, M.; Iqbal, J.; Shah, M.H. Assessment of water quality for drinking/irrigation purpose from Mangla dam, Pakistan. *Geochem. Explor. Environ. Anal.* **2016**, *16*, 137–145. [CrossRef]

92. Iftikhar, B.; Bashirullah, N.; Ishtiaq, M.; Khan, S.A.; Siddique, S.; Ayaz, T. Chemical quality assessment of drinking water in district peshawar, pakistan. *Khyber Med. Univ. J.* **2016**, *8*, 1–6.

93. Zulfiqar, H.; Abbas, Q.; Raza, A.; Ali, A. Determinants of Safe Drinking Water in Pakistan: A Case Study of Faisalabad. *J. Glob. Innov. Agric. Soc. Sci.* **2016**, *04*, 40–45. [CrossRef]

94. Ghanghro, A.B.; Jahangir, T.M.; Memon, A.H.; Jahangir, T.M.; Lund, G.M.; Memon, A.H. Arsenic Contamination in Drinking Water of District Jamshoro Arsenic Contamination in Drinking Water of District Jamshoro, Sindh, Pakistan. *Biomed. Lett.* **2016**, *2*, 31–37.

95. Hussain, S.A.; Hussain, S.A.; Hussain, A.; Fatima, U.; Ali, W.; Hussain, A.; Hussain, N. Evaluation of drinking water quality in urban areas of Pakistan: A case study of Gulshan-e-Iqbal Karachi, Pakistan. *J. Biodivers. Environ. Sci.* **2016**, *8*, 64–76.

96. Nazir, H.M.; Hussain, I.; Zafar, M.I.; Ali, Z.; AbdEl-Salam, N.M. Classification of Drinking Water Quality Index and Identification of Significant Factors. *Water Resour. Manag.* **2016**, *30*, 4233–4246. [CrossRef]

97. Bhowmik, A.K.; Alamdar, A.; Katsoyiannis, I.; Shen, H.; Ali, N.; Ali, S.M.; Bokhari, H.; Schäfer, R.B.; Eqani, S.A.M.A.S. Mapping human health risks from exposure to trace metal contamination of drinking water sources in Pakistan. *Sci. Total Environ.* **2015**, *538*, 306–316. [CrossRef]

98. Begum, S.; Shah, M.T.; Khan, S. Role of mafic and ultramafic rocks in drinking water quality and its potential health risk assessment, Northern Pakistan. *J. Water Health* **2015**, *13*, 1130–1142. [CrossRef]

99. Abbas, S.; Hashmi, I.; Saif, M.; Rehman, U.; Qazi, I.A.; Awan, M.A.; Nasir, H. Monitoring of chlorination disinfection by-products and their associated health risks in drinking water of Pakistan. *J. Water Health* **2015**, *13*, 270–284. [CrossRef]

100. Mumtaz, M.W.; Adnan, A.; Mukhtar, H.; Danish, M.; Raza, M.A. Determination of toxic metals in water of Lahore canal by atomic absorption spectroscopy. *J. Water Chem. Technol.* **2015**, *37*, 73–77. [CrossRef]

101. Kandhro, A.J.; Rind, A.M.; Mastoi, A.A.; Almani, K.F.; Meghwar, S.; Laghari, M.A.; Rajpout, M.S. Physico-Chemical Assessment of Surface and Ground Water for Drinking Purpose in Nawabshah City, Sindh, Pakistan. *Am. J. Environ. Prot.* **2015**, *4*, 62–69. [CrossRef]

102. Khan, S.; Shah, I.A.; Muhammad, S.; Malik, R.N.; Shah, M.T. Arsenic and Heavy Metal Concentrations in Drinking Water in Pakistan and Risk Assessment: A Case Study. *Hum. Ecol. Risk Assess.* **2015**, *21*, 1020–1031. [CrossRef]
103. Khan, S.S.; Tareen, H.; Jabeen, U.; Mengal, F.; Masood, Z.; Ahmed, S.; Bibi, S.; Riaz, M.; Rizwan, S.; Mandokhail, F.; et al. Quality Assessment of Drinking Water from the Different Colonies of Quetta City, Pakistan according to WHO Standards. *Biol. Forum Int. J.* **2015**, *7*, 699–702.
104. Yousafzai, A.M.; Bari, F.; Ullah, T.; Pakistan, H.B. Assessment of heavy metals in surface water of River Panjkora Dir Lower, KPK Pakistannet. *J. Biodivers. Environ. Sci.* **2014**, *5*, 144–152.
105. Jabeen, S.; Shah, M.T.; Ahmed, I.; Khan, S.; Hayat, M.Q. Physico-chemical parameters of surface and ground water and their environmental impact assessment in the Haripur Basin, Pakistan. *J. Geochem. Explor.* **2014**, *138*, 1–7. [CrossRef]
106. Sheikh, S.A.; Panhwar, A.A.; Channa, M.J.; Merani, B.N.; Nizamani, S.M. Determination of Ground Water Quality for Agriculture and Drinking Purpose in Sindh, Pakistan. *J. Pharm. Nutr. Sci.* **2014**, *4*, 81–87. [CrossRef]
107. Channar, A.G.; Rind, A.M.; Mastoi, G.M.; Almani, K.F.; Lashari, K.H.; Qurishi, M.A.; Mahar, N. Comparative Study of Water Quality of Manchhar Lake with Drinking Water Quality Standard of World Health Organization. *Am. J. Environ. Prot.* **2014**, *3*, 68–72. [CrossRef]
108. Khan, M.A.; Lang, M.; Shaukat, S.S.; Alamgir, A.; Baloch, T. Water quality assessment of hingol river, balochistan, Pakistan. *Middle East J. Sci. Res.* **2014**, *19*, 306–313. [CrossRef]
109. Wattoo, F.H.; Wattoo, M.H.S.; Tirmizi, S.A.; Qadir, M.A. Monitoring of anthropogenic influences on underground and surface water quality of Indus River at district Mianwali-Pakistan. *Turk. J. Biochem. J Biochem.* **2013**, *38*, 25–30. [CrossRef]
110. Tahir, M.A.; Rasheed, H. Fluoride in the drinking water of Pakistan and the possible risk of crippling fluorosis. *Drink. Water Eng. Sci.* **2013**, *6*, 17–23. [CrossRef]
111. Mumtaz, A.; Mirjat, M.S.; Mangio, H.U.R.; Soomro, A. Assessment of Drinking Water Quality Status and its Impact on Health in Tandojam City. *Int. J. Humanit. Soc. Sci.* **2013**, *13*, 363–369. [CrossRef]
112. Ahmed, A.; Noonari, T.M.; Magsi, H.; Mahar, A. Risk assessment of total and faecal coliform bacteria from drinking water supply of Badin city, Pakistan. *J. Environ. Prof. Sri Lanka* **2013**, *2*, 52. [CrossRef]
113. Baig, J.A.; Kazi, T.G.; Shah, A.Q.; Kandhro, G.A.; Afridi, H.I.; Khan, S.; Kolachi, N.F.; Wadhwa, S.K. Arsenic speciation and other parameters of surface and ground water samples of jamshoro, Pakistan. *Int. J. Environ. Anal. Chem.* **2012**, *92*, 28–42. [CrossRef]
114. Farooq, M.A.; Shaukat, S.S.; Zafar, M.U.; Abbas, Q. Variation Pattern of Heavy Metal Concentrations During Pre- and Post-Monsoon Seasons in the Surface Water of River Indus (Sindh Province). *World Appl. Sci. J.* **2012**, *19*, 582–587. [CrossRef]
115. Li, P.; Wu, J. Medical Geology and Medical Geochemistry: An Editorial Introduction. *Expo. Health* **2022**, *14*, 217–218. [CrossRef]
116. Pal, A.; He, Y.; Jekel, M.; Reinhard, M.; Gin, K.Y.H. Emerging contaminants of public health significance as water quality indicator compounds in the urban water cycle. *Environ. Int.* **2014**, *71*, 46–62. [CrossRef]
117. Azizullah, A.; Khattak, M.N.K.; Richter, P.; Häder, D.P. Water pollution in Pakistan and its impact on public health—A review. *Environ. Int.* **2011**, *37*, 479–497. [CrossRef]
118. Rasheed, H.; Iqbal, N.; Ashraf, M.; ul Hasan, F. Groundwater quality and availability assessment: A case study of District Jhelum in the Upper Indus, Pakistan. *Environ. Adv.* **2022**, *7*, 100148. [CrossRef]
119. Khan, S.; Aziz, T.; Noor-Ul-Ain Ahmed, K.; Ahmed, I.; Nida; Akbar, S.S. Drinking Water Quality in 13 Different Districts of Sindh, Pakistan. *Health Care Curr. Rev.* **2018**, *6*, 4. [CrossRef]
120. Yang, X.; Liu, Q.; He, Y.; Luo, X.; Zhang, X. Comparison of daily and sub-daily SWAT models for daily streamflow simulation in the upper Huai River Basin of China. *Stoch. Environ. Res. Risk Assess.* **2016**, *30*, 959–972. [CrossRef]
121. Vardhan, K.H.; Kumar, P.S.; Panda, R.C. A review on heavy metal pollution, toxicity and remedial measures: Current trends and future perspectives. *J. Mol. Liq.* **2019**, *290*, 111197. [CrossRef]
122. Arshad, N.; Imran, S. Assessment of arsenic, fluoride, bacteria, and other contaminants in drinking water sources for rural communities of Kasur and other districts in Punjab, Pakistan. *Environ. Sci. Pollut. Res.* **2017**, *24*, 2449–2463. [CrossRef]
123. Sohail, M.T.; Mahfooz, Y.; Aftab, R.; Yen, Y.; Talib, M.A.; Rasool, A. Water quality and health risk of public drinking water sources: A study of fltration plants installed in Rawalpindi and Islamabad, Pakistan. *Desalination Water Treat.* **2020**, *181*, 239–250. [CrossRef]
124. He, X.; Li, P. Surface Water Pollution in the Middle Chinese Loess Plateau with Special Focus on Hexavalent Chromium (Cr^{6+}): Occurrence, Sources and Health Risks. *Expo. Health* **2020**, *12*, 385–401. [CrossRef]
125. Natasha; Shahid, M.; Khalid, S.; Murtaza, B.; Anwar, H.; Shah, A.H.; Sardar, A.; Shabbir, Z.; Niazi, N.K. A critical analysis of wastewater use in agriculture and associated health risks in Pakistan. *Environ. Geochem. Health* **2020**, 1–20. [CrossRef]
126. Khan, S.; Rauf, R.; Muhammad, S.; Qasim, M.; Din, I. Arsenic and heavy metals health risk assessment through drinking water consumption in the Peshawar District, Pakistan. *Hum. Ecol. Risk Assess.* **2016**, *22*, 581–596. [CrossRef]
127. Mulk, S.; Azizullah, A.; Korai, A.L.; Khattak, M.N.K. Impact of marble industry effluents on water and sediment quality of Barandu River in Buner District, Pakistan. *Environ. Monit. Assess.* **2015**, *187*, 8. [CrossRef] [PubMed]
128. Abdullah, E.J. Quality Assessment for Shatt Al-Arab River Using Heavy Metal Pollution Index and Metal Index. *J. Environ. Earth Sci.* **2013**, *3*, 114–120.
129. Geng, J.; Wang, Y.; Luo, H. Distribution, sources, and fluxes of heavy metals in the Pearl River Delta, South China. *Mar. Pollut. Bull.* **2015**, *101*, 914–921. [CrossRef]

130. Avigliano, E.; Schenone, N.F. Human health risk assessment and environmental distribution of trace elements, glyphosate, fecal coliform and total coliform in Atlantic Rainforest mountain rivers (South America). *Microchem. J.* **2015**, *122*, 149–158. [CrossRef]
131. Singh, H.; Pandey, R.; Singh, S.K.; Shukla, D.N. Assessment of heavy metal contamination in the sediment of the River Ghaghara, a major tributary of the River Ganga in Northern India. *Appl. Water Sci.* **2017**, *7*, 4133–4149. [CrossRef]
132. Cánovas, C.R.; Olías, M.; Nieto, J.M.; Galván, L. Wash-out processes of evaporitic sulfate salts in the Tinto river: Hydrogeochemical evolution and environmental impact. *Appl. Geochem.* **2010**, *25*, 288–301. [CrossRef]
133. Bhatti, M.T.; Anwar, A.A.; Aslam, M. Groundwater monitoring and management: Status and options in Pakistan. *Comput. Electron. Agric.* **2017**, *135*, 143–153. [CrossRef]
134. Alamgir, A.; Khan, M.A.; Schilling, J.; Shaukat, S.S.; Shahab, S. Assessment of groundwater quality in the coastal area of Sindh province, Pakistan. *Environ. Monit. Assess.* **2016**, *188*, 78. [CrossRef]
135. Ling, Y.; Podgorski, J.; Sadiq, M.; Rasheed, H.; Eqani, S.A.M.A.S.; Berg, M. Monitoring and prediction of high fluoride concentrations in groundwater in Pakistan. *Sci. Total Environ.* **2022**, *839*, 156058. [CrossRef]
136. Schewe, J.; Heinke, J.; Gerten, D.; Haddeland, I.; Arnell, N.W.; Clark, D.B.; Dankers, R.; Eisner, S.; Fekete, B.M.; Colón-González, F.J.; et al. Multimodel assessment of water scarcity under climate change. *Proc. Natl. Acad. Sci. USA* **2014**, *111*, 3245–3250. [CrossRef]
137. Khoso, S.; Wagan, F.H.; Tunio, A.H.; Ansari, A.A. An overview on emerging water scarcity in pakistan, its causes, impacts and remedial measures. *J. Appl. Eng. Sci.* **2015**, *13*, 35–44. [CrossRef]
138. Adnan, S.; Iqbal, J. Spatial analysis of the groundwater quality in the Peshawar district, Pakistan. *Procedia Eng.* **2014**, *70*, 14–22. [CrossRef]
139. Ullah, A.S.; Rashid, H.; Khan, S.N.; Akbar, M.U.; Arshad, A.; Rahman, M.M.; Mustafa, S. A Localized Assessment of Groundwater Quality Status Using GIS-Based Water Quality Index in Industrial Zone of Faisalabad, Pakistan. *Water* **2022**, *14*, 3342. [CrossRef]
140. Lanjwani, M.F.; Khuhawar, M.Y.; Jahangir Khuhawar, T.M. Assessment of groundwater quality for drinking and irrigation uses in taluka Ratodero, district Larkana, Sindh, Pakistan. *Int. J. Environ. Anal. Chem.* **2020**, *102*, 4134–4157. [CrossRef]
141. ur Rehman, H.; Munir, M.; Ashraf, K.; Fatima, K.; Shahab, S.; Ali, B.; Al-Saeed, F.A.; Abbas, A.M.; uz Zaman, Q. Heavy Metals, Pesticide, Plasticizers Contamination and Risk Analysis of Drinking Water Quality in the Newly Developed Housing Societies of Gujranwala, Pakistan. *Water* **2022**, *14*, 3787. [CrossRef]
142. Mughal, A.; Sultan, K.; Ashraf, K.; Hassan, A.; uz Zaman, Q.; Haider, F.U.; Shahzad, B. Risk Analysis of Heavy Metals and Groundwater Quality Indices in Residential Areas: A Case Study in the Rajanpur District, Pakistan. *Water* **2022**, *14*, 3551. [CrossRef]
143. Jamali, M.A.; Markhand, A.H.; Agheem, M.H.; Zardari, S.H.; Arain, A.Y.W. Spatial variation in groundwater quality with respect to surface water seepages in Kadhan area District Badin (Indus Delta), Sindh, Pakistan. *Int. J. Energy Water Resour.* **2022**, *7*, 105–117. [CrossRef]
144. Jamali, M.Z.; Solangi, G.S.; Keerio, M.A.; Keerio, J.A.; Bheel, N. Assessing and mapping the groundwater quality of Taluka Larkana, Sindh, Pakistan, using water quality indices and geospatial tools. *Int. J. Environ. Sci. Technol.* **2022**, 1–14. [CrossRef]
145. Nasir, M.J.; Tufail, M.; Ayaz, T.; Khan, S.; Khan, A.Z.; Lei, M. Groundwater quality assessment and its vulnerability to pollution: A study of district Nowshera, Khyber Pakhtunkhwa, Pakistan. *Environ. Monit. Assess.* **2022**, *194*, 692. [CrossRef]
146. Lanjwani, M.F.; Khuhawar, M.Y.; Lanjwani, A.H.; Khuahwar, T.M.J.; Samtio, M.S.; Rind, I.K.; Soomro, W.A.; Khokhar, L.A.; Channa, F.A. Spatial variability and risk assessment of metals in groundwater of district Kamber-Shahdadkot, Sindh, Pakistan. *Groundw. Sustain. Dev.* **2022**, *18*, 100784. [CrossRef]
147. Noor, S.; Rashid, A.; Javed, A.; Khattak, J.A.; Farooqi, A. Hydrogeological properties, sources provenance, and health risk exposure of fluoride in the groundwater of Batkhela, Pakistan. *Environ. Technol. Innov.* **2022**, *25*, 102239. [CrossRef]
148. Iqbal, Z.; Imran, M.; Natasha; Rahman, G.; Miandad, M.; Shahid, M.; Murtaza, B. Spatial distribution, health risk assessment, and public perception of groundwater in Bahawalnagar, Punjab, Pakistan: A multivariate analysis. *Environ. Geochem. Health* **2022**, *45*, 381–391. [CrossRef] [PubMed]
149. Israr, M.; Nazneen, S.; Raza, A.; Ali, N.; Khan, S.A.; Khan, H.; Khan, S.; Ali, J. Assessment of municipal solid waste landfilling practices on the groundwater quality and associated health risks: A case study of Mardan-Pakistan. *Arab. J. Geosci.* **2022**, *15*, 1445. [CrossRef]
150. Ahmad, W.; Iqbal, J.; Nasir, M.J.; Ahmad, B.; Khan, M.T.; Khan, S.N.; Adnan, S. Impact of land use/land cover changes on water quality and human health in district Peshawar Pakistan. *Sci. Rep.* **2021**, *11*, 16526. [CrossRef]
151. Hassan, M.; Khan, M.J.; Ali, S.S. Environmental security View project Energy and Environmental Security in Developing Countries View project. *Pak. J. Sci.* **2021**, *73*.
152. Lanjwani, M.F.; Khuhawar, M.Y.; Khuhawar, T.M.J.; Lanjwani, A.H.; Soomro, W.A. Evaluation of hydrochemistry of the Dokri groundwater, including historical site Mohenjo-Daro, Sindh, Pakistan. *Int. J. Environ. Anal. Chem.* **2021**, *103*, 1892–1916. [CrossRef]
153. Khan, H.; Khan, M.N.; Sirajuddin, M.; Salman, S.M.; Bilal, M. Assessment of Drinking Water Quality of Different Areas in Tehsil Isa Khel, Mianwali, Punjab, Pakistan. *Pak. J. Anal. Environ. Chem.* **2021**, *22*, 376–387. [CrossRef]
154. Jalees, M.I.; Farooq, M.U.; Anis, M.; Hussain, G.; Iqbal, A.; Saleem, S. Hydrochemistry modelling: Evaluation of groundwater quality deterioration due to anthropogenic activities in Lahore, Pakistan. *Environ. Dev. Sustain.* **2021**, *23*, 3062–3076. [CrossRef]

155. Ahmad, S.; Imran, M.; Murtaza, B.; Natasha; Arshad, M.; Nawaz, R.; Waheed, A.; Hammad, H.M.; Naeem, M.A.; Shahid, M.; et al. Hydrogeochemical and health risk investigation of potentially toxic elements in groundwater along River Sutlej floodplain in Punjab, Pakistan. *Environ. Geochem. Health* **2021**, *43*, 5195–5209. [CrossRef]

156. Iqbal, J.; Su, C.; Rashid, A.; Yang, N.; Baloch, M.Y.J.; Talpur, S.A.; Ullah, Z.; Rahman, G.; Rahman, N.U.; Earjh; et al. Hydrogeochemical Assessment of Groundwater and Suitability Analysis for Domestic and Agricultural Utility in Southern Punjab, Pakistan. *Water* **2021**, *13*, 3589. [CrossRef]

157. Ismail, S.; Ahmed, M.F. GIS-based spatio-temporal and geostatistical analysis of groundwater parameters of Lahore region Pakistan and their source characterization. *Environ. Earth Sci.* **2021**, *80*, 719. [CrossRef]

158. Ullah, Z.; Talib, M.A.; Rashid, A.; Ghani, J.; Shahab, A.; Irfan, M.; Rauf, A.; Bawazeer, S.; Almarhoon, Z.M.; Mabkhot, Y.N. Hydrogeochemical Investigation of Elevated Arsenic Based on Entropy Modeling, in the Aquifers of District Sanghar, Sindh, Pakistan. *Water* **2021**, *13*, 3477. [CrossRef]

159. Ahmed, J.; Ping Wong, L.; Piaw Chua, Y.; Channa, N. Drinking Water Quality Mapping Using Water Quality Index and Geospatial Analysis in Primary Schools of Pakistan. *Water* **2020**, *12*, 3382. [CrossRef]

160. Khan, M.J.; Shah, B.A.; Nasir, B. Groundwater quality assessment for drinking purpose: A case study from Sindh Industrial Trading Estate, Karachi, Pakistan. *Model. Earth Syst. Environ.* **2020**, *6*, 263–272. [CrossRef]

161. Khalid, S.; Shahid, M.; Natasha; Shah, A.H.; Saeed, F.; Ali, M.; Qaisrani, S.A.; Dumat, C. Heavy metal contamination and exposure risk assessment via drinking groundwater in Vehari, Pakistan. *Environ. Sci. Pollut. Res.* **2020**, *27*, 39852–39864. [CrossRef]

162. Murtaza, B.; Natasha; Amjad, M.; Shahid, M.; Imran, M.; Shah, N.S.; Abbas, G.; Naeem, M.A.; Amjad, M. Compositional and health risk assessment of drinking water from health facilities of District Vehari, Pakistan. *Environ. Geochem. Health* **2020**, *42*, 2425–2437. [CrossRef]

163. Seelro, M.A.; Ansari, M.U.; Manzoor, S.A.; Abodif, A.M.; Sadaf, A. Comparative Study of Ground and Surface Water Quality Assessment Using Water Quality Index (WQI) in Model Colony Malir, Karachi, Pakistan. *Environ. Contam. Rev.* **2020**, *3*, 4–12. [CrossRef]

164. Solangi, G.S.; Siyal, A.A.; Babar, M.M.; Siyal, P. Groundwater quality evaluation using the water quality index (WQI), the synthetic pollution index (SPI), and geospatial tools: A case study of Sujawal district, Pakistan. *Hum. Ecol. Risk Assess. Int. J.* **2019**, *26*, 1529–1549. [CrossRef]

165. Kumar, N.; Memon, S.A.; Mahessar, A.A.; Ansari, K.; Qureshi, A. Impact Assessment of Groundwater Quality using WQI and Geospatial tools: A Case Study of Islamkot, Tharparkar, Pakistan. *Technol. Appl. Sci. Res.* **2020**, *10*, 5288–5294. [CrossRef]

166. Shahzad, H.; Farid, H.U.; Khan, Z.M.; Anjum, M.N.; Ahmad, I.; Chen, X.; Sakindar, P.; Mubeen, M.; Ahmad, M.; Gulakhmadov, A. An integrated use of gis, geostatistical and map overlay techniques for spatio-temporal variability analysis of groundwater quality and level in the punjab province of pakistan, south asia. *Water* **2020**, *12*, 3555. [CrossRef]

167. Razzaq, S.S.; Naz, S.A.; Zubair, A.; Yasmeen, K.; Shafique, M.; Jabeen, N.; Magsi, A. Detection of Hazardous Contaminants in Ground Water Resources: An Alarming Situation for Public Health in Karachi, Pakistan. *Pak. J. Anal. Environ. Chem.* **2020**, *21*, 322–331. [CrossRef]

168. Solangi, G.S.; Munir, B.M.; Siyal, P.; Siyal, A.A.; Babar, M.M. Evaluation of drinking water quality using the water quality index (WQI), the synthetic pollution index (SPI) and geospatial tools in Thatta district, Pakistan. *Desalination Water Treat.* **2019**, *160*, 202–213. [CrossRef]

169. Deeba, F.; Abbas, N.; Butt, M.; Irfan, M. Ground Water Quality of Selected Areas of Punjab and Sind Provinces, Pakistan: Chemical and Microbiological Aspects. *SSRN Electron. J.* **2019**, *5*, 241–246. [CrossRef]

170. Khanoranga; Khalid, S. An assessment of groundwater quality for irrigation and drinking purposes around brick kilns in three districts of Balochistan province, Pakistan, through water quality index and multivariate statistical approaches. *J. Geochem. Explor.* **2019**, *197*, 14–26. [CrossRef]

171. Jehan, S.; Khan, S.; Khattak, S.A.; Muhammad, S.; Rashid, A.; Muhammad, N. Hydrochemical properties of drinking water and their sources apportionment of pollution in Bajaur agency, Pakistan. *Measurement* **2019**, *139*, 249–257. [CrossRef]

172. Mazhar, I.; Hamid, A.; Afzal, S. Groundwater quality assessment and human health risks in Gujranwala District, Pakistan. *Environ. Earth Sci.* **2019**, *78*, 634. [CrossRef]

173. Talib, M.A.; Tang, Z.; Shahab, A.; Siddique, J.; Faheem, M.; Fatima, M. Hydrogeochemical Characterization and Suitability Assessment of Groundwater: A Case Study in Central Sindh, Pakistan. *Int. J. Environ. Res. Public Health* **2019**, *16*, 886. [CrossRef]

174. Khuhawar, M.Y.; Khuhawar, M.J.; Ursani, H.-R.; Farooque, M.; Lanjwani; Mahessar, A.A.; Tunio, I.A.; Soomro, A.G.; Rind, I.K.; Brohi, R.-Z.; et al. Assessment of Water Quality of Groundwater of Thar Desert, Sindh, Pakistan. *J. Hydrogeol. Hydrol. Eng. Res.* **2019**, *7*, 1000171. [CrossRef]

175. Masood, N.; Farooqi, A.; Zafar, M.I. Health risk assessment of arsenic and other potentially toxic elements in drinking water from an industrial zone of Gujrat, Pakistan: A case study. *Environ. Monit. Assess.* **2019**, *191*, 95. [CrossRef]

176. Yousaf, S.; Ilyas, M.; Khan, S.; Khan Khattak, A.; Anjum, S. Measurement of physicochemical and heavy metals concentration in drinking water from sources to consumption sites in. *J. Himal. Earth Sci.* **2019**, *52*, 36–45.

177. Mushtaq, N.; Younas, A.; Mashiatullah, A.; Javed, T.; Ahmad, A.; Farooqi, A. Hydrogeochemical and isotopic evaluation of groundwater with elevated arsenic in alkaline aquifers in Eastern Punjab, Pakistan. *Chemosphere* **2018**, *200*, 576–586. [CrossRef]

178. Khan, A.; Qureshi, F.R. Groundwater Quality Assessment through Water Quality Index (WQI) in New Karachi Town, Karachi, Pakistan. *Asian J. Water Environ. Pollut.* **2018**, *15*, 41–46. [CrossRef]

179. Bhatti, N.B.; Siyal, A.A.; Qureshi, A.L. Groundwater Quality Assessment Using Water Quality Index: A Case Study of Nagarparkar, Sindh, Pakistan. *Sindh Univ. Res. J. SURJ Sci. Ser.* **2018**, *50*, 227–234.
180. Laghari, A.N.; Siyal, Z.A.; Bangwar, D.K.; Soomro, M.A.; Walasai, G.; Shaikh, F.A. Groundwater Quality Analysis for Human Consumption A Case Study of Sukkur City, Pakistan. *Technol. Appl. Sci. Res.* **2018**, *8*, 2616–2620. [CrossRef]
181. Imran, M.; Jahanzaib, S.; Ashraf, A. Using geographical information systems to assess groundwater contamination from arsenic and related diseases based on survey data in Lahore, Pakistan. *Arab. J. Geosci.* **2017**, *10*, 450. [CrossRef]
182. Podgorski, J.E.; Eqani, S.A.M.A.S.; Khanam, T.; Ullah, R.; Shen, H.; Berg, M. Extensive arsenic contamination in high-pH unconfined aquifers in the Indus Valley. *Sci. Adv.* **2017**, *3*, 845–850. [CrossRef]
183. Bashir, E.; Huda, S.N.; Naseem, S.; Hamza, S.; Kaleem, M. Geochemistry and quality parameters of dug and tube well water of Khipro, District Sanghar, Sindh, Pakistan. *Appl. Water Sci.* **2017**, *7*, 1645–1655. [CrossRef]
184. Mahmood, K.; Ul-Haq, Z.; Batool, S.A.; Rana, A.D.; Tariq, S. Application of temporal GIS to track areas of significant concern regarding groundwater contamination. *Environ. Earth Sci.* **2016**, *75*, 33. [CrossRef]
185. Farooq, A.; Zahid, F.; Asif, S.; Ali, H.Q. Estimation of Fluoride in Drinking Water in Selected Areas of Southern Lahore, Pakistan. *Sci. Int.* **2016**, *28*, 391–395.
186. Khan, R.U.; Khan, P.; Waheed, M.W.; Jan, R.; Author, C.; Ur Rehman, H.; Bibi, S.; Nazir, R.; Shakir, S.K.; Naz, S.; et al. Heavy Metals Analysis in Drinking Water of Lakki Marwat District, KPK, Pakistan. *World Appl. Sci. J.* **2016**, *34*, 15–19. [CrossRef]
187. Shah, A.A.; Khan, M.A.; Kanwal, N.; Bernstein, R. Assessment of safety of drinking water in tank district: An empirical study of water-borne diseases in rural Khyber Pakhtunkhwa, Pakistan. *Int. J. Environ. Sci.* **2016**, *6*, 418–428. [CrossRef]
188. Shahab, A.; Shihua, Q.; Rashid, A.; Ul Hasan, F.; Sohail, M.T.; Pakistan, R. Evaluation of Water Quality for Drinking and Agricultural Suitability in the Lower Indus Plain in Sindh Province, Pakistan. *Pol. J. Environ. Stud.* **2016**, *25*, 2563–2574. [CrossRef] [PubMed]
189. Lewis, K.A.; Tzilivakis, J.; Warner, D.J.; Green, A. An international database for pesticide risk assessments and management. *Hum. Ecol. Risk Assess. Int. J.* **2016**, *22*, 1050–1064. [CrossRef]
190. Mustafa, M.F.; Afreen, A.; Abbas, Y.; Afridi, Z.-U.-R.; Athar, W.; Ur Rehman, Z.; Khan, Q.; ur Rehaman, S.W.; Ali, A. A Study on physio-chemical and biological analysis of drinking water quality from the residential areas of Islamabad, Pakistan. *J. Biodivers. Environ. Sci.* **2016**, *109*, 109–125.
191. Brahman, K.D.; Kazi, T.G.; Afridi, H.I.; Baig, J.A.; Arain, S.S.; Talpur, F.N.; Kazi, A.G.; Ali, J.; Panhwar, A.H.; Arain, M.B. Exposure of children to arsenic in drinking water in the Tharparkar region of Sindh, Pakistan. *Sci. Total Environ.* **2016**, *544*, 653–660. [CrossRef]
192. Baig, J.A.; Kazi, T.G.; Mustafa, M.A.; Solangi, I.B.; Mughal, M.J.; Afridi, H.I. Arsenic Exposure in Children through Drinking Water in Different Districts of Sindh, Pakistan. *Biol. Trace Elem. Res.* **2016**, *173*, 35–46. [CrossRef]
193. Mahessar, A.A.; Memon, N.A.; Leghari, M.E.H.; Qureshi, A.L.; Arain, G.M. Assessment of Source and Quality of Drinking Water in Coastal Area of. *IOSR J. Environ. Sci. Toxicol. Food Technol.* **2015**, *9*, 9–15. [CrossRef]
194. Shahid, S.U.; Iqbal, J.; Hasnain, G. Groundwater quality assessment and its correlation with gastroenteritis using GIS: A case study of Rawal Town, Rawalpindi, Pakistan. *Environ. Monit. Assess.* **2014**, *186*, 7525–7537. [CrossRef]
195. Malik, M.A.; Tang, Z.; Mohamadi, B. Contamination Potential Assessment of Potable Groundwater in Lahore, Pakistan. *Pol. J. Environ. Stud.* **2014**, *23*, 1905–1916.
196. Hassan, A.; Nawaz, M. African Journal of Microbiology Research Microbiological and physicochemical assessments of groundwater quality at Punjab, Pakistan. *Afr. J. Microbiol. Res.* **2014**, *8*, 2672–2681. [CrossRef]
197. Ullah, S.; Akmal, M.; Aziz, F.; Ullah, S.; Khan, K.J. Hand Pumps' Water Quality Analysis for Drinking and Irrigation Purposes at District Dir Lower, Khyber Pakhtunkhwa Pakistan. *Eur. Acad. Res.* **2014**, *2*, 1560–1572.
198. Arain, M.B.; Ullah, I.; Niaz, A.; Shah, N.; Shah, A.; Hussain, Z.; Tariq, M.; Afridi, H.I.; Baig, J.A.; Kazi, T.G. Evaluation of water quality parameters in drinking water of district Bannu, Pakistan: Multivariate study. *Sustain. Water Qual. Ecol.* **2014**, *3–4*, 114–123. [CrossRef]
199. Khan, S.; Shahnaz, M.; Jehan, N.; Rehman, S.; Shah, M.T.; Din, I. Drinking water quality and human health risk in Charsadda district, Pakistan. *J. Clean. Prod.* **2013**, *60*, 93–101. [CrossRef]
200. Brahman, K.D.; Kazi, T.G.; Afridi, H.I.; Naseem, S.; Arain, S.S.; Ullah, N. Evaluation of high levels of fluoride, arsenic species and other physicochemical parameters in underground water of two sub districts of Tharparkar, Pakistan: A multivariate study. *Water Res.* **2013**, *47*, 1005–1020. [CrossRef]
201. Rashid, U.; Alvi, S.K.; Perveen, F.; Khan, F.A.; Bhutto, S.; Siddiqui, I.; Bano, A.; Usmani, T.H. Metal contents in the ground waters of Tharparkar district, Sindh, Pakistan, with special focus on arsenic. *Pak. J. Sci. Ind. Res.* **2012**, *55*, 49–56. [CrossRef]
202. Hashmi, I.; Qaiser, S.; Farooq, S. Microbiological quality of drinking water in urban communities, Rawalpindi, Pakistan. *Desalin. Water Treat.* **2012**, *41*, 240–248. [CrossRef]
203. Ahmad, M.; Ahmad, N. Potable Water Quality Characteristics of the Rural Areas of District Hangu, Khyber Pakhtunkhwa-Pakistan. *Int. J. Multidiscip. Sci. Eng.* **2012**, *3*, 7–9.
204. Raza, M.; Hussain, F.; Lee, J.Y.; Shakoor, M.B.; Kwon, K.D. Groundwater status in Pakistan: A review of contamination, health risks, and potential needs. *Crit. Rev. Environ. Sci. Technol.* **2017**, *47*, 1713–1762. [CrossRef]

205. Ayaz, T.; Khan, S.; Khan, A.Z.; Lei, M.; Alam, M. Remediation of industrial wastewater using four hydrophyte species: A comparison of individual (pot experiments) and mix plants (constructed wetland). *J. Environ. Manag.* **2020**, *255*, 109833. [CrossRef] [PubMed]

206. Memon, Y.I.; Qureshi, S.S.; Kandhar, I.A.; Qureshi, N.A.; Saeed, S.; Mubarak, N.; Khan, S.U.; Saleh, T.A. Statistical analysis and physicochemical characteristics of groundwater quality parameters: A case study. *Int. J. Environ. Anal. Chem.* **2021**, 1–22. [CrossRef]

207. Usmani, Z.; Kumar, V.; Varjani, S.; Gupta, P.; Rani, R.; Chandra, A. Municipal solid waste to clean energy system: A contribution toward sustainable development. *Curr. Dev. Biotechnol. Bioeng. Resour. Recovery Wastes* **2020**, *2020*, 217–231. [CrossRef]

208. Tang, Z.; Li, W.; Tam, V.W.Y.; Xue, C. Advanced progress in recycling municipal and construction solid wastes for manufacturing sustainable construction materials. *Resour. Conserv. Recycl. X* **2020**, *6*, 100036. [CrossRef]

209. Farid, S.; Baloch, M.K.; Ahmad, S.A.; Khan, I. Water pollution: Major issue in urban areas. *Int. J. Water Resour. Environ. Eng.* **2012**, *4*, 55–65. [CrossRef]

210. Soomro, M.; Khokhar, M.; Hussain, W.; Hussain, M. Drinking water Quality challenges in Pakistan. *Pak. Counc. Res. Water Resour. Lahore* **2011**, 17–28.

211. Nawab, J.; Khan, S.; Khan, M.A.; Sher, H.; Rehamn, U.U.; Ali, S.; Shah, S.M. Potentially Toxic Metals and Biological Contamination in Drinking Water Sources in Chromite Mining-Impacted Areas of Pakistan: A Comparative Study. *Expo. Health* **2017**, *9*, 275–287. [CrossRef]

212. Nawab, J.; Khan, S.; Ali, S.; Sher, H.; Rahman, Z.; Khan, K.; Tang, J.; Ahmad, A. Health risk assessment of heavy metals and bacterial contamination in drinking water sources: A case study of Malakand Agency, Pakistan. *Environ. Monit. Assess.* **2016**, *188*, 286. [CrossRef]

213. Roohul-Amin; Ali, S.S.; Anwar, Z.; Khattak, J.Z.K. Microbial Analysis of Drinking Water and Water Distribution System in New Urban Peshawar. *Curr. Res. J. Biol. Sci.* **2012**, *4*, 731–737. [CrossRef]

214. Noshin, M.; Batool, S.; Farooqi, A. Groundwater pollution in Pakistan. *Groundw. Pollut. Pak.* **2021**, *13*, 309–322. [CrossRef]

215. Khan, K.; Lu, Y.; Khan, H.; Zakir, S.; Ihsanullah; Khan, S.; Khan, A.A.; Wei, L.; Wang, T. Health risks associated with heavy metals in the drinking water of Swat, northern Pakistan. *J. Environ. Sci.* **2013**, *25*, 2003–2013. [CrossRef]

216. Jadhav, S.V.; Bringas, E.; Yadav, G.D.; Rathod, V.K.; Ortiz, I.; Marathe, K.V. Arsenic and fluoride contaminated groundwaters: A review of current technologies for contaminants removal. *J. Environ. Manag.* **2015**, *162*, 306–325. [CrossRef] [PubMed]

217. van Geen, A.; Farooqi, A.; Kumar, A.; Khattak, J.A.; Mushtaq, N.; Hussain, I.; Ellis, T.; Singh, C.K. Field testing of over 30,000 wells for arsenic across 400 villages of the Punjab plains of Pakistan and India: Implications for prioritizing mitigation. *Sci. Total Environ.* **2019**, *654*, 1358–1363. [CrossRef]

218. Qaiser, S.; Hashmi, I.; Nasir, H. Chlorination at Treatment Plant and Drinking Water Quality: A Case Study of Different Sectors of Islamabad, Pakistan. *Arab. J. Sci. Eng.* **2014**, *39*, 5665–5675. [CrossRef]

219. Waqar, A.; Ali, M. Different Perspectives on Water Quality of Local Filtration Plants in Pakistan. In Proceedings of the 2nd Conference on Sustainability in Civil Engineering, Islamabad, Pakistan, 12 August 2020; pp. 1–8.

220. Rawan, B.; Ullah, W.; Ullah, R.; Akbar, T.A.; Ayaz, Z.; Javed, M.F.; Din, I.; Ullah, S.; Aziz, M.; Mohamed, A.; et al. Assessments of Roof-Harvested Rainwater in Disctrict Dir Lower, Khyber Pakhtunkhwa Pakistan. *Water* **2022**, *14*, 3270. [CrossRef]

221. Yaqub, G.; Hamid, A.; Asghar, S. Rain water quality assessment as air quality indicator in Pakistan. *Bangladesh J. Sci. Ind. Res.* **2019**, *54*, 161–168. [CrossRef]

222. Pathan, M.A.; Lashari, R.A.; Maira, M. Physical and Chemical Contamination Studies of Drinking Water in the Vicinity of Jamshoro Area (Jetharo Village) Sindh, Pakistan. *Int. J. Eng. Sci. Comput.* **2018**, *8*, 17985.

223. Hasan, M.; Shang, Y.; Akhter, G.; Jin, W. Evaluation of Groundwater Suitability for Drinking and Irrigation Purposes in Toba Tek Singh District, Pakistan. *Irrig. Drain. Syst. Eng.* **2017**, *6*, 185. [CrossRef]

224. Brahman, K.D.; Kazi, T.G.; Afridi, H.I.; Rafique, T.; Baig, J.A.; Arain, S.S.; Ullah, N.; Panhwar, A.H.; Arain, S. Evaluation of fresh and stored rainwater quality in fluoride and arsenic endemic area of Thar Desert, Pakistan. *Environ. Monit. Assess.* **2014**, *186*, 8611–8628. [CrossRef]

225. Chughtai, M.; Mustafa, S.; Mumtaz, M. Study of Physicochemical Parameters of Rainwater: A Case Study of Karachi, Pakistan. *Am. J. Anal. Chem.* **2014**, *2014*, 235–242. [CrossRef]

226. Chughtai, M.; Mustafa, S.; Mahmood, R.; Mumtaz, M. Physicochemical Assessment of Rainwater of Karachi, Pakistan. *Eur. Acad. Res.* **2014**, *1*, 4099–4108.

227. D'Adamo, I.; Gastaldi, M.; Morone, P.; Rosa, P.; Sassanelli, C.; Settembre-blundo, D.; Shen, Y. Bioeconomy of Sustainability: Drivers, Opportunities and Policy Implications. *Sustainability* **2022**, *14*, 200. [CrossRef]

228. D'Adamo, I.; Gastaldi, M. Perspectives and Challenges on Sustainability: Drivers, Opportunities and Policy Implications in Universities. *Sustainability* **2023**, *15*, 3564. [CrossRef]

229. Ali, S.; Azam, F.; Naveed, H.M.; Abid, W. Impact of Prestigious Indicators on Sustainable Growth of Small and Medium-Sized Enterprises in Pakistan. *Asian J. Econ. Empir. Res.* **2020**, *7*, 251–257. [CrossRef]

230. Shehzadi, M.; Afzal, M.; Khan, M.U.; Islam, E.; Mobin, A.; Anwar, S.; Khan, Q.M. Enhanced degradation of textile effluent in constructed wetland system using Typha domingensis and textile effluent-degrading endophytic bacteria. *Water Res.* **2014**, *58*, 152–159. [CrossRef]

231. Shahid, M.; Niazi, N.K.; Dumat, C.; Naidu, R.; Khalid, S.; Rahman, M.M.; Bibi, I. A meta-analysis of the distribution, sources and health risks of arsenic-contaminated groundwater in Pakistan. *Environ. Pollut.* **2018**, *242*, 307–319. [CrossRef]

232. Abiriga, D.; Vestgarden, L.S.; Klempe, H. Groundwater contamination from a municipal landfill: Effect of age, landfill closure, and season on groundwater chemistry. *Sci. Total Environ.* **2020**, *737*, 140307. [CrossRef]

233. Sharma, A.; Kumar, V.; Shahzad, B.; Tanveer, M.; Sidhu, G.P.S.; Handa, N.; Kohli, S.K.; Yadav, P.; Bali, A.S.; Parihar, R.D.; et al. Worldwide pesticide usage and its impacts on ecosystem. *SN Appl. Sci.* **2019**, *1*, 1446. [CrossRef]

234. Khan, M.I.; Shoukat, M.A.; Alam, S. Use, Contamination and Exposure of Pesticides in Pakistan: A Review. *Pak. J. Agri. Sci* **2020**, *57*, 131–149. [CrossRef]

235. Qureshi, A.S.; McCornick, P.G.; Sarwar, A.; Sharma, B.R. Challenges and Prospects of Sustainable Groundwater Management in the Indus Basin, Pakistan. *Water Resour. Manag.* **2010**, *24*, 1551–1569. [CrossRef]

236. Mukherji, A.; Scott, C.; Molden, D.; Maharjan, A. Megatrends in Hindu Kush Himalaya: Climate Change, Urbanisation and Migration and Their Implications for Water, Energy and Food. In *Water Resources Development and Management*; Springer: Berlin/Heidelberg, Germany, 2018; pp. 125–146. [CrossRef]

237. Sleet, P. *Water Resources in Pakistan: Scarce, Polluted and Poorly Governed*; Future Directions International: Nedlands, Australia, 2019.

238. Haidary, A.; Amiri, B.J.; Adamowski, J.; Fohrer, N.; Nakane, K. Assessing the Impacts of Four Land Use Types on the Water Quality of Wetlands in Japan. *Water Resour. Manag.* **2013**, *27*, 2217–2229. [CrossRef]

239. Shakoor, M.B.; Niazi, N.K.; Bibi, I.; Shahid, M.; Sharif, F.; Bashir, S.; Shaheen, S.M.; Wang, H.; Tsang, D.C.W.; Ok, Y.S.; et al. Arsenic removal by natural and chemically modified water melon rind in aqueous solutions and groundwater. *Sci. Total Environ.* **2018**, *645*, 1444–1455. [CrossRef] [PubMed]

240. Shah, A.H.; Shahid, M.; Khalid, S.; Natasha; Shabbir, Z.; Bakhat, H.F.; Murtaza, B.; Farooq, A.; Akram, M.; Shah, G.M.; et al. Assessment of arsenic exposure by drinking well water and associated carcinogenic risk in peri-urban areas of Vehari, Pakistan. *Environ. Geochem. Health* **2020**, *42*, 121–133. [CrossRef] [PubMed]

241. Kamrani, S.; Rezaei, M.; Amiri, V.; Saberinasr, A. Investigating the efficiency of information entropy and fuzzy theories to classification of groundwater samples for drinking purposes: Lenjanat Plain, Central Iran. *Environ. Earth Sci.* **2016**, *75*, 1370. [CrossRef]

242. Elhag, M.; Gitas, I.; Othman, A.; Bahrawi, J.; Gikas, P. Assessment of Water Quality Parameters Using Temporal Remote Sensing Spectral Reflectance in Arid Environments, Saudi Arabia. *Water* **2019**, *11*, 556. [CrossRef]

243. Avdan, Z.Y.; Kaplan, G.; Goncu, S.; Avdan, U. Monitoring the Water Quality of Small Water Bodies Using High-Resolution Remote Sensing Data. *ISPRS Int. J. Geo Inf.* **2019**, *8*, 553. [CrossRef]

244. Villa, S.; Sassanelli, C. The Data-Driven Multi-Step Approach for Dynamic Estimation of Buildings' Interior Temperature. *Energies* **2020**, *13*, 6654. [CrossRef]

245. Sassanelli, C.; Arriga, T.; Zanin, S.; D'Adamo, I.; Terzi, S. Industry 4.0 Driven Result-oriented PSS: An Assessment in the Energy Management. *Int. J. Energy Econ. Policy* **2022**, *12*, 186–203. [CrossRef]

246. D'Adamo, I. The Profitability of Residential Photovoltaic Systems. A New Scheme of Subsidies Based on the Price of CO_2 in a Developed PV Market. *Soc. Sci.* **2018**, *7*, 148. [CrossRef]

247. De Paola, F.; Sambito, M.; Piazza, S.; Freni, G. Optimal Deployment of the Water Quality Sensors in Urban Drainage Systems. *Environ. Sci. Proc.* **2022**, *21*, 42. [CrossRef]

248. Le Blanc, D. Towards Integration at Last? The Sustainable Development Goals as a Network of Targets. *Sustain. Dev.* **2015**, *23*, 176–187. [CrossRef]

249. Osborn, D.; Cutter, A.; Ullah, F. Understanding the Transformational Challenge for Developed Countries. 2015. Available online: https://sustainabledevelopment.un.org/content/documents/1684SF_-_SDG_Universality_Report_-_May_2015.pdf (accessed on 1 March 2023).

250. Khan, M.A.; Khan, J.A.; Ali, Z.; Ahmad, I.; Ahmad, M.N. The challenge of climate change and policy response in Pakistan. *Environ. Earth Sci.* **2016**, *75*, 412. [CrossRef]

251. Alcamo, J. Water quality and its interlinkages with the Sustainable Development Goals. *Curr. Opin. Environ. Sustain.* **2019**, *36*, 126–140. [CrossRef]

252. Hoekstra, A.Y.; Chapagain, A.K.; van Oel, P.R. Advancing Water Footprint Assessment Research: Challenges in Monitoring Progress towards Sustainable Development Goal 6. *Water* **2017**, *9*, 438. [CrossRef]

253. Arora, N.K.; Mishra, I. Sustainable development goal 6: Global Water Security. *Environ. Sustain.* **2022**, *5*, 271–275. [CrossRef]

254. Herrera, V. Reconciling global aspirations and local realities: Challenges facing the Sustainable Development Goals for water and sanitation. *World Dev.* **2019**, *118*, 106–117. [CrossRef]

255. Azadi, F.; Ashofteh, P.S.; Loáiciga, H.A. Reservoir Water-Quality Projections under Climate-Change Conditions. *Water Resour. Manag.* **2019**, *33*, 401–421. [CrossRef]

256. Kumar, L.; Nadeem, F.; Sloan, M.; Restle-Steinert, J.; Deitch, M.J.; Naqvi, S.A.; Kumar, A.; Sassanelli, C. Fostering Green Finance for Sustainable Development: A Focus on Textile and Leather Small Medium Enterprises in Pakistan. *Sustainability* **2022**, *14*, 11908. [CrossRef]

257. Michels-Brito, A.; Rodriguez, D.A.; Cruz Junior, W.L.; de Souza Vianna, J.N. The climate change potential effects on the run-of-river plant and the environmental and economic dimensions of sustainability. *Renew. Sustain. Energy Rev.* **2021**, *147*, 111238. [CrossRef]
258. Mukate, S.; Wagh, V.; Panaskar, D.; Jacobs, J.A.; Sawant, A. Development of new integrated water quality index (IWQI) model to evaluate the drinking suitability of water. *Ecol. Indic.* **2019**, *101*, 348–354. [CrossRef]

Article

Synergetic Benefits for a Pig Farm and Local Bioeconomy Development from Extended Green Biorefinery Value Chains

James Gaffey [1,2,3,*], Cathal O'Donovan [4], Declan Murphy [5], Tracey O'Connor [1,2], David Walsh [6], Luis Alejandro Vergara [2], Kwame Donkor [7], Lalitha Gottumukkala [7], Sybrandus Koopmans [8], Enda Buckley [9], Kevin O'Connor [2] and Johan P. M. Sanders [8]

[1] Circular Bioeconomy Research Group, Shannon Application Biotechnology Centre, Munster Technology, V92 CX88 Tralee, Ireland
[2] BiOrbic Bioeconomy SFI Research Centre, O'Brien Centre for Science, University College Dublin, D04 V1W8 Dublin, Ireland
[3] Department of Environmental Engineering, University of Limerick, Plassey, V94 T9PX Limerick, Ireland
[4] Carhue Piggeries, Cooligboy, Timoleague, Co., P72 HD61 Cork, Ireland
[5] Makeway Nutrition, Unit 6, Riverstown Business Park, Tramore, Co., X91 TRF9 Waterford, Ireland
[6] Barryroe Co-Operative, Lislevane House, Tirnanean, Bandon, Co., P47 YW77 Cork, Ireland
[7] Celignis Analytical, Unit 11, Holland Rd., Castletroy, Plassey, Co., V94 7Y42 Limerick, Ireland
[8] Grassa BV, Villaflloraweg 1, 5928 SZ Venlo, The Netherlands
[9] Carbery Group, Phale Lower, Ballineen, Co., P47 YW77 Cork, Ireland
*   Correspondence: james.gaffey@mtu.ie; Tel.: +353-66-714-4254

Abstract: As the global population rises, agriculture and industry are under increasing pressure to become more sustainable in meeting this growing demand, while minimizing impacts on global emissions, land use change, and biodiversity. The development of efficient and symbiotic local bioeconomies can help to respond to this challenge by using land, resources, and side streams in efficient ways tailored to the needs of different regions. Green biorefineries offer a unique opportunity for regions with abundant grasslands to use this primary resource more sustainably, providing feed for cows, while also generating feed for monogastric animals, along with the co-production of biomaterials and energy. The current study investigates the impact of a green biorefinery co-product, leaf protein concentrate (LPC), for input to a pig farm, assessing its impact on pig diets, and the extended impact on the bioenergy performance of the pig farm. The study found that LPC replaced soya bean meal at a 50% displacement rate, with pigs showing positive performance in intake and weight gain. Based on laboratory analysis, the resulting pig slurry demonstrated a higher biogas content and 26% higher biomethane potential compared with the control slurry. The findings demonstrate some of the local synergies between agricultural sectors that can be achieved through extended green biorefinery development, and the benefits for local bioeconomy actors.

Keywords: bioeconomy; biorefinery; pigs; soya bean; grass; protein; biogas; biomethane

Citation: Gaffey, J.; O'Donovan, C.; Murphy, D.; O'Connor, T.; Walsh, D.; Vergara, L.A.; Donkor, K.; Gottumukkala, L.; Koopmans, S.; Buckley, E.; et al. Synergetic Benefits for a Pig Farm and Local Bioeconomy Development from Extended Green Biorefinery Value Chains. *Sustainability* **2023**, *15*, 8692. https://doi.org/10.3390/su15118692

Academic Editors: Idiano D'Adamo and Massimo Gastaldi

Received: 28 February 2023
Revised: 21 May 2023
Accepted: 22 May 2023
Published: 27 May 2023

1. Introduction

The world currently faces a climate and biodiversity emergency brought about, in part, by an unsustainable food system, which has an immense environmental footprint [1,2]. Globally, livestock farming is the single greatest source of environmental impacts in agriculture, accounting for 77% of all land used for food production, with 14.5% of global greenhouse gases (GHGs) linked to livestock production (rising to 25% if land use change is included) [3]. In Europe, agriculture is a significant contributor to GHGs with about 10% of all Europe's GHGs generated by agriculture [4]. This varies by country, with countries at the lower end, such as Netherlands, U.K., and Greece, below 10%, and other countries such as Lithuania and Latvia recording over 20% of their emissions from agriculture [5]. Ireland has the highest proportion of national agricultural emissions of all EU member states, with over 30% of total national emissions coming from agriculture [5,6], the vast majority of which arises from livestock sectors [6].

The European Green Deal sets out Europe's pathway towards a climate-neutral future, which involves reducing net emissions by at least 55% by 2030 and ultimately achieving climate neutrality by 2050 in line with its commitments under the Paris Climate Agreement [7]. These ambitious European targets place pressure on the agriculture sector to reduce its emissions at the member-state level, with several member states already taking sector-specific action in response.

The development of a circular bioeconomy underpinned by sustainable renewable biological resources from primary production and its associated technological, product, and systemic solutions has been proposed as offering a solution to support a reduction in emissions within the broader agricultural and livestock sectors, while at the same time offering new business and innovation opportunities in traditional primary sectors [8]. Solutions vary from synthetic and seaweed-based additives, which can potentially inhibit rumen methane emissions [9,10], to the displacement of fossil-based materials and fuels with new bio-based materials and biofuels produced from agricultural by-products [11–13], to nature-based solutions [14,15]. Many synergies for bioeconomy development exist within primary sectors, and some of these are explored in Figure 1. Using this approach, many local agricultural and societal needs may be met using available local primary resources, thereby increasing the self-sufficiency of the region, while reducing emissions by using local and shorter supply chains. This approach may be seen as an expansion of the concept of bio-districts or eco-regions which, as described by Dias et al. [16], aims to stimulate a collective approach to sustainable resource management by adopting organic farming practices at a territorial level to ensure benefits are distributed across the region, stimulating rural stakeholder networks and strengthening existing links between rural stakeholders. This can contribute to sustainable local development that prioritizes resource conservation and ecological integrity as inherent characteristics of economic logic, and can contribute to greater quality of life for direct stakeholders and local communities [16]. The potential of adapting this bio-district approach to support local bioeconomy development is exemplified in Figure 1 below, which shows potential synergies between (1.) grassland, (2.) monogastrics, and (3.) marine sectors, which can all provide goods and services for the overall bio-district.

One opportunity for potential local symbiosis between primary sectors exists in the development of green biorefinery models. Many protein sources are not being used optimally and may be more efficiently used by deploying biorefinery approaches [17]. In green biorefinery systems, green biomasses, such as grasses and legumes, can be processed into multiple co-products, including leaf protein concentrate (LPC), which is suitable for both ruminant and monogastric animals and for use in aquaculture [18–20]. This approach can help to improve the overall resource efficiency of the green biomass by providing an ensiled fiber press cake for cows, which can offer a reduction in nitrogen and phosphorous emissions while maintaining milk productivity [18,20,21], and by providing an additional LPC co-product that can potentially replace soya bean meal in pigs, poultry, and other monogastrics [22,23]. The opportunity to use available grasslands in Europe more efficiently to produce a local, homegrown protein could resolve a key challenge for European agriculture, which is heavily reliant on imported sources of unsustainable feed. The European Union (EU) is reliant on animal feed imports from North and South America for livestock production, due to domestic deficits [24]. In addition, such a model can provide further bio-materials including fiber for insulation materials, bio-composites and packaging, high-value prebiotic materials, and brown juice or grass whey [25,26]. This whey can subsequently be used in different applications, for example, in biogas production or as fertilizer [25,26]. The interest in biogas production and its upgrading to biomethane, which can substitute natural gas and vehicle fuel, has become a trend in Europe, but its potential is still underexploited [27–29]. This interest has been further heightened by the Russia–Ukraine war. The REPower EU Plan, introduced by the EU Commission in 2022, focuses on reducing energy dependence on Russia by setting a target of 35 billion cubic meters of biomethane production by 2030, thus replacing the need for the import of natural gas [30].

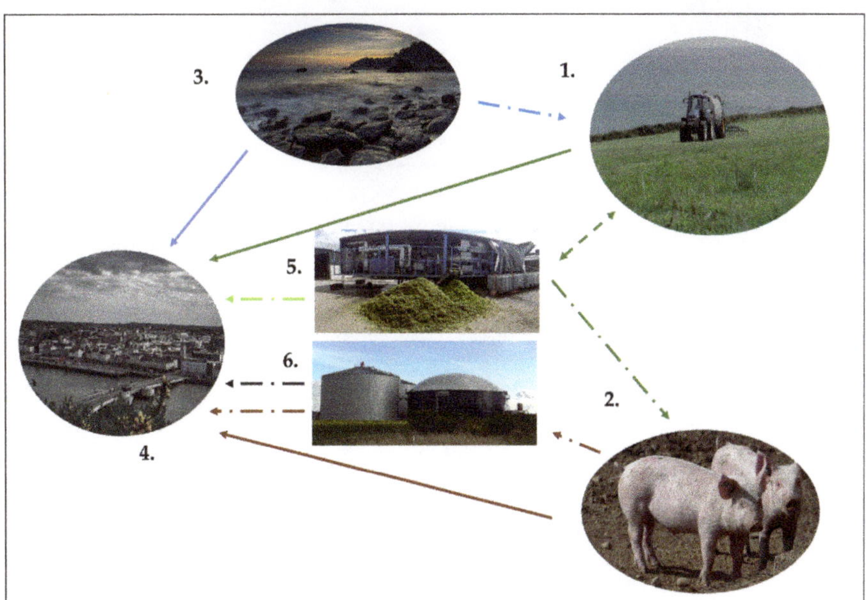

Figure 1. A bio-district approach for bioeconomy development. The figure shows potential synergies between (1.) grassland (ruminants), (2.) monogastrics, and (3.) marine sectors, which can collectively provide goods and services for the overall local bio-district (4.). Conventional products (traditional food products, such as milk, beef, fish, and pork) are highlighted by continuous lines while dotted lines indicate new value chains and products. A green biorefinery (5.) and anaerobic digestion unit (6.) are included as examples of enabling technologies that can support new local value chain development. In sector (1.), grass is supplied into a green biorefinery which can create low-emission feeds for ruminants and monogastrics (and potentially aquaculture feed), replacing imported feed sources. Other bio-material products may also be produced such as fiber insulation materials. Conventional dairy and meat products are also produced from the grassland chain. In sector (2.), pig slurry along with other animal slurries and food waste from the municipality can be supplied to an anaerobic digester to create heat, electricity, and/or biofuels for the bio-district. Pork is also produced. In sector (3.), coastal/marine resources can supply food products such as fish, while processed marine biomass, such as microalgae and Asparagopsis, can be used as fertilizer, feed, or anti-methanogenic additives for ruminants within the grassland chain.

This article explores the potential benefits of green biorefinery co-product LPC for a pig farm, partially displacing imported soya bean meal within a conventional pig diet. The impact of this diet change on pig performance is considered, as is the resulting pig slurry including its impact on the biomethane potential of slurry, a key factor for the pig farm which supplies its slurry to a local community anaerobic digestion plant. While a previous study has investigated the potential of green biorefinery LPC as an alternative pig feed [31], within this current study the inclusion rate of LPC was greatly increased to understand the impact of a higher inclusion rate. Furthermore, while other studies have investigated the biogas and biomethane potential of green biorefinery by-products such as grass whey, brown juice, de-FOS whey [25,26], and press cake [25], and various studies have looked at pig slurry [32–34], no similar studies investigating and analyzing the impact of the resulting biorefinery LPC pig slurry from this value chain were found within the literature This represents a novel aspect of the current research. In this way, this paper explores, in more detail, the potential benefits that green biorefineries implemented within the cattle sector can deliver for the pig sector.

2. Materials and Methods

Fresh grass was harvested from farms located within a 10 km radius of Afferden, Limburg, in the Netherlands during July and August of 2021. The feedstock, a 75–25% perennial ryegrass–clover mix, was processed within 12 h. The processing was carried out with the innovative "green biorefinery" process developed by Grassa BV. The biorefinery is a fixed unit demonstration facility with 4 tonnes per hour of fresh input processing capacity. A schematic of the process is included as part of Figure 2 below. Briefly, fresh grass was washed with water upon entry to the biorefinery to remove sand, with the water being recycled within the process. The grass was then subject to wet fractionation using an extrusion process. This created two primary products, a protein-rich "press cake" containing 50–60% of the initial grass protein, and a green, grass-derived juice. The press cake was further preserved through ensiling and baling, to be later used as ruminant feed, which exhibits a high nitrogen use efficiency and reduced nitrogen and phosphorous emissions [20]. The remaining protein is contained within the green juice. The protein fraction of the green juice was solidified using a heat coagulation process. This solid protein portion was extracted using a vacuum filter and then dried to in excess of 90% dry matter (DM) creating a storable LPC. The LPC was incorporated into appropriate feed for the pigs involved in this study. The remaining grass whey retains valuable sugars, e.g., fructans, and minerals after the protein fraction has been removed, and these can be extracted through further processing. The Grassa green biorefinery process is presented in Figure 3a below. The LPC was later tested as a feed ingredient at Carhue Piggeries in Timoleague, Co., Cork, Ireland. The piggery is also linked with Timoleague AgriGen Biodigester, where it currently supplies its generated pig slurry for the purposes of biogas production (Figure 3b below). The green biorefinery process is highlighted in Figure 2 below, with the main focus points of the current study, testing of LPC input to the pig farm and impacts on pig and biogas production, highlighted by a dotted line.

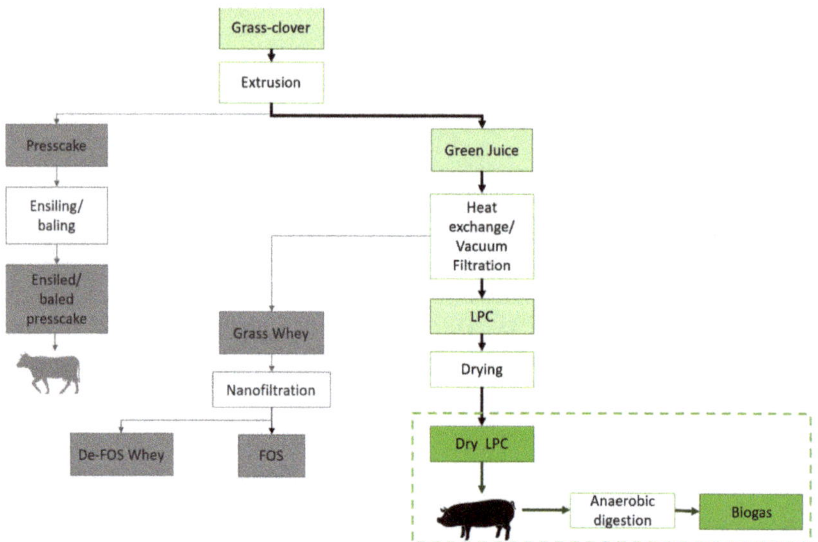

Figure 2. Schematic of green biorefinery value chain with main focus areas of the current study highlighted with a dotted line.

(**a**)   (**b**)

Figure 3. (**a**) Grassa 4-tonne per hour pilot green biorefinery plant production unit for LPC; (**b**) Timoleauge AgriGen Biodigester for supply of slurry to produce biogas.

2.1. Analysis of Green Biorefinery LPC as an Alternative Feed for Pigs

2.1.1. Characterization of LPC

A proximate analysis of LPC was performed by Dairygold's analytical laboratory at Mallow, Co., Cork, Ireland. Crude fiber was analyzed using the Fibretec method based on ISO 6865:2000 [35]. Protein was determined using the Semi-Micro-Kjeldahl method based on ISO 5983:2000 [36]. Calcium, magnesium, sodium, and potassium were determined using atomic absorption based on ISO 6869:2000 [37]. Moisture content was determined using an oven at 102°C based on ISO 6496:1999 [38]. Phosphorous was determined using colorimetric UV spectrophotometry following ISO 6491:1998 [39]. Starch was determined using the polarimetry method based on ISO 15914:2004 [40]. Oil content was determined using the Soxhlet method following ISO 6492:1999 [41]. Crude ash was determined using a muffle furnace based on ISO 5984:2002 [42].

The LPC was analyzed by Sciantec, North Yorkshire, U.K., to assess the amino acid profile. Cation exchange chromatography was used to measure the AA content, following 24 h of exposure of the LPC to acid hydrolysis.

2.1.2. LPC and Control Diet Pig Feed Trial

To explore the potential of the LPC biorefinery co-product as a partial substitute for soya bean meal in pig diets, pig feeding trials were conducted at a privately owned piggery, Carhue Piggeries at Timoleague, Co., Cork, Ireland. Pigs were fed from two separate silo's via Schauer's Spotmix feeding system, in which feed is dispensed dry and is combined with water at the valve above the feeding trough and is presented to the pigs as wet feed. This system facilitates the feeding of multiple diets across pen groups without cross contamination or feed freshness issues. Pigs were fed ad libitum with feed troughs probed by the computer multiple times per hour with the same probe preventing over-feeding by regulation by the same feeding probe. Based on the characterization of the LPC, treatment and control feed rations were designed by Makeway Nutrition Ltd. with input from contributors with a view to reducing the soya bean usage of a conventional pig weaner ration by 50%. The nutritional requirements of the weaner pigs were considered in the preparation of the treatment diet in order to make it comparable to the control diet and sufficient for the pigs' needs. The treatment diet is provided in Table 1 alongside the control diet.

Table 1. Treatment and Control diet formulations for weaner pigs.

Raw Material	Control (%)	Treatment (%)
Barley	30.00	29.20
Maize	10.00	10.00
Wheat	25.00	25.00
Molasses	2.00	2.00
Hipro soya	22.00	11.00
Soya hulls	1.00	1.00
Whey permeate powder	2.50	2.50
Soya oil	3.70	3.10
Grass protein (42.8% CP)	0.00	12.40
Limestone flour	0.80	0.68
Salt	0.40	0.40
Lysine hydrochloride	0.00	0.08
Methionine	0.00	0.04
104 Weaner + Vita GP (2.6%)	2.60	2.60
Total	100.0	100.0

The focus of the trial was on weaner pigs aged approximately 9 weeks old and weighing 17 kg on average at the start of the testing phase. The pigs were randomly split into a treatment group of 110 pigs to be fed the treatment feed and a control group with 110 pigs to be fed the control diet over a 31-day period. The control diet was comprised of wheat, maize, barley, molasses, soya bean meal (in the form of hipro soya), soy oil, soy hull, and minerals in recommended amounts (Table 2). The treatment feed included LPC replacing 50% of the hipro soya level contained in the control diet, which represented the most significant change between the two diets (Table 2). Small amounts of synthetic lysine and methionine (0.08% and 0.04% of diet, respectively) were added to the LPC diet to prevent the crude protein of the diet from exceeding 18% crude protein and differing greatly from the control diet. All other amino acids are non-limiting. Additionally, a small amount of limestone flour was removed from the grass protein diet due to the high calcium value within the LPC. A comparison of constituents from the control and treatment diet is displayed in Table 2. The two diets were dispensed from separate feed silos using a computerized feed system which dispensed weighed amounts of feed to each group of pigs.

Table 2. Main constituents of treatment and control diets.

Constituent	Value	Control (%)	Treatment (%)
Protein	%	17.87	18.00
Oil	%	5.34	6.07
Fiber	%	3.34	3.35
Ash	%	5.51	5.86
DE	MJ/kg	14.32	14.24
NE	MJ/kg	10.25	10.24
Lysine	%	1.26	1.25
ILD lysine	%	1.15	1.15
Calcium	%	0.70	0.71
Dig phos	%	0.36	0.36
Sodium	%	0.23	0.23

The weight of pigs and feed intake were recorded on a weekly basis. These data were necessary in order to calculate the feed intake and weight gain on a daily basis (daily feed intake and average daily gain), and to evaluate the feed conversion ratio per treatment.

The daily feed intake (DFI) [43] for each treatment was evaluated by dividing the total feed intake by the size of the group, i.e., the number of pigs. The total feed intake was calculated at the end of the trial by deducting the feed remaining from the amount of feed delivered.

$$\text{Daily Feed Intake} = \frac{\text{Total Feed Intake}}{\text{Number of pigs on treatment}}$$

The pigs were weighed throughout the trial, at the start and end of every week, to calculate the average daily gain (ADG).

To calculate the feed conversion ratio (FCR) [44], the DFI was divided by the ADG for each treatment group.

$$\text{Feed Conversion Ratio} = \frac{\text{Daily Feed Intake}}{\text{Average Daily Weight Gain}}$$

2.2. LPC and Control Slurry Biogas and Biomethane Analysis

To assess the potential effect of the diet change on the slurry feedstock from the perspective of anaerobic digestion, a biogas and biomethane analysis of the feedstock was undertaken. Carhue Piggeries currently sends its produced pig slurry to Timoleague Agri Gen, a community-based anaerobic digestion facility located at Timoleague Co., Cork, Ireland approximately 2 km from the piggery site. At this facility, the slurry is co-digested along with food waste to produce biogas which is converted to heat and electricity. The site produces 500 KW of renewable electricity. The digestate produced from the digestion process is rich in nitrogen (N), phosphate (P), and potassium (K), and this biofertilizer can be land spread as a substitute for chemical fertilizer. To complete the analysis, approximately 2 kg of mixed slurry resulting from the treatment and control pig feeding trial batches from Carhue piggeries were collected and sent for analysis by Celignis Analytical, located at Castletroy, Co., Limerick, Ireland.

2.2.1. Determination of Total and Volatile Solids for Slurries

Slurries from weaner pigs feeding on LPC and the traditional soya bean meal diets were analyzed for biomethane potential. The samples were extracted simultaneously from the manure cellars when the feeding period ended. Prior to the BMP analysis, both slurries underwent a proximate analysis for determination of the total solids, volatile solids, and ash content. The proximate analysis followed European standard protocols as described in reference methods EN 14774-1:2009 [45] and EN 14775:2009 [46].

2.2.2. Biomethane Potential (BMP) of Slurries

An Anaero BMP unit consisting of 15 slots for 1 L digesters (700 mL working volume) was utilized for determination of methane potential from the slurries. The digesters were constantly stirred with stainless steel paddle systems and were placed in a 37 °C water bath. The BMP analysis followed the German standard method protocols (VDI 4630). Celignis propriety inoculum, which treats different sources of substrates including grass, whey, sewage sludge, manure, and other lignocellulosic biomass, was utilized for the BMP analysis. The inoculum was sieved to remove residual organic matter and degassed for at least seven days to remove any remaining organic matter. An inoculum to substrate ratio of 4:1 was prescribed in the VDI 4630 reference method. The BMP analysis of each slurry was performed in triplicate with no pH adjustments. pH adjustments with special reagents were not required because the pH of the substrate–inoculum mix was within the optimum range for anaerobic digestion. The BMP analysis of LPC and soya bean meal pig weaner slurries was monitored within 28 days of digestion.

2.2.3. Biogas Analysis and Biomethane Calculations

In the Anaero BMP unit, a tipping bucket flowmeter system coupled with a recording computer was used to measure biogas production from the various digesters. The flow meter had 15 chambers (buoyancy bucket design in every chamber), and each chamber contained a salt solution that prevents the dissolving of carbon dioxide, hydrogen sulfide, and ammonia from the biogas generated during this study. The biogas from the flowmeter was collected in 2 L Tedlar bags and methane, carbon dioxide, hydrogen sulfide, ammonia, and oxygen were analyzed using a Biogas 5000 gas analyzer. On the 3rd, 7th, 14th, 21st, and 28th days of digestion, the gas composition was analyzed and utilized to determine the biomethane potential of the slurries, applying methods used by Ravindran et al. [25] to evaluate the biomethane potential of green biorefinery products (Equations (1)–(5) [47–51]).

$$\text{Cumulative biogas produced} = \sum_{\text{Day } 0}^{\text{Day } 28} (\text{biogas produced per day}\,(\text{mL})) \tag{1}$$

$$\text{Biogas yield} = \frac{\text{Cumulative biogas produced }(\text{mL})}{\text{FM or TS or VS fed }(\text{g})} \tag{2}$$

$$\text{Average CH}_4 \text{ percentage} = \text{Determined from Biogas analyzer 5000} \tag{3}$$

$$\text{Cumulative CH}_4 \text{ produced} = \sum_{\text{Day } 0}^{\text{Day } 28} (\%\text{ Average CH}_4 \times \text{biogas produced per day}\,(\text{mL})) \tag{4}$$

$$\text{Methane yield} = \frac{\text{Cumulative methane produced }(\text{L})}{\text{FM or TS or VS fed }(\text{kg})} \tag{5}$$

3. Results

3.1. LPC and Control Diet Pig Feed Trial Results

3.1.1. Characterization of Grass-Derived LPC

A number of important LPC properties including crude fiber, ash, protein, starch, and total solids were determined through proximate analysis. Based on the analysis, the LPC had a high DM content, containing only 5.5% moisture. The sample contained 42.8% crude protein on a DM basis, 3.9% crude fiber, 9.4% ash, 2.8% potassium, 0.37% phosphorous, and oil content of 12.1%. Neutral detergent fiber was 43.5%, and acid detergent fiber was 7.63%.

3.1.2. Amino Acid Profile of LPC

The amino acid profile of the LPC is provided in Table 3. The analysis shows that the LPC was rich in glutamic acid (3.7%), aspartic acid (3.53%), leucine (3.10%), and alanine (2.24%). Table 4 provides a comparison between the constituents of LPC compared with soya bean meal and other common feedstuffs. Overall, the LPC compares quite well with soya bean meal, providing comparable levels of protein, lysine, methionine, and threonine, although with lower cystine. The LPC also compares favorably with rapeseed meal, with LPC containing higher crude protein and crude fiber contents than rapeseed meal (Table 4). Comparing the composition of the LPC with the results of LPC produced by Ravindran et al. [31] from perennial rye grass, the overall CP is significantly higher (43% versus 34%) and more consistent with the CP content of soya bean meal, as indicated in Table 4. The sample also has a higher content of production-limiting amino acids lysine, methionine, and threonine. The increase in protein content may be partly resulting from the removal of small fibers from the protein products and soluble salts such as potassium, which are washed using a vacuum filter in the current biorefinery process. The DM content of the LPC from this study is higher when compared with Ravindran et al. [31] (94.5% versus 87%) due to a new drying process that was implemented in the current biorefinery process. The protein was dried at 70 °C in a belt dryer with a residence time of 24 h; Maillard reactions occurred, as well as polymerization reactions in unsaturated fats. The crude fiber of the current LPC is lower than that found by Ravindran et al. [31], but it is still in the comparable range with soya bean meal (Table 4) [33]. The ash content was reduced from 11.9% to 9%. This reduction may be attributed to lower sand contained on the feedstock or improved washing of feedstock during pre-processing.

Table 3. Amino acid profile of LPC.

Constituent	%
Cystine	0.21
Aspartic	3.53
Methionine	0.72
Threonine	1.71
Serine	1.54
Glutamic	3.77
Glycine	1.91
Alanine	2.24
Valine	2.09
Iso-leucine	1.71
Leucine	3.10
Tyrosine	1.00
Phenylalanine	2.07
Histidine	0.78
Lysine	2.03
Arginine	2.07
Proline	1.65

Table 4. Crude fiber, crude protein, and selected amino acid profile of various feedstuffs in comparison with LPC from the current study [52] (g/100 g).

Animal Feed Protein Sources	Unit	Crude Protein	Lysine	Methionine	Cysteine	Threonine	Crude Fiber
Soya bean meal	g/100 g	44–48	2.81–3.20	0.60–0.75	0.69–0.74	0.71–2.00	3.0–7.0
Sunflower meal	g/100 g	24–44	1.18–1.49	0.74–0.79	0.55–0.59	1.21–1.48	12.0–32.0
Rapeseed meal	g/100 g	34–36	2.00–2.12	0.67–0.75	0.54–0.91	1.53–2.21	10.0–15.0
Cottonseed meal	g/100 g	24–41	1.05–1.71	0.41–0.72	0.64–0.70	1.32–1.36	25.0–30.0
Grass protein concentrate	g/100 g	42.8	2.03	0.72	0.21	1.71	3.9

Once mixed into rations at the Barryroe feed mill, finished feed appeared dark green in color compared with the control sample (Figure 4 below).

Figure 4. Control feed (**left**) compared with treatment feed (**right**)—note the green color of the treatment feed.

3.1.3. Feeding Trial

Pig weights were recorded at the beginning of the trial. The initial average weight of the control group pigs was 17.388 kg, while the initial average weight for the treatment group pigs was 17.246 kg. As expected, feed intake increased in both groups during the course of the trial, and by week 5 the average trial end weight was 35.092 for the control diet and 35.450 for the treatment diet. The total weight gain per pig since weaning is presented in kg in Table 5and shows a comparable weight gain for pigs on the control diet versus those on the treatment diet, with pigs gaining 19.494 kg and 19.554 kg, respectively, by week 5.

Table 5. Total weight gain per pig since weaning on treatment and control diet.

| | Total Weight Gain Per Pig Since Weaning (kg) | |
Date of Weighing	Treatment	Control
Week 1	4.656	4.413
Week 2	8.557	9.178
Week 3	13.104	14.178
Week 4	18.204	17.704
Week 5	19.554	19.494

The dung consistency appeared similar in both batches,; however the treatment dung had a notable green color. Overall, the pigs on the treatment diet appeared healthy.

Daily Feed Intake

The DFI was measured weekly during the trial. The results indicate that the pigs readily accepted the treatment feed. During the first week, week 1, the control group DFI was 1.060 kg/d and the treatment group DFI was similar: 1.079 kg/d. During week 2, the DFI dipped lower (1.155 kg/d) for the treatment diet than the control diet (1.167 kg/d). Subsequently, the DFI of the treatment diet overtook that of the control diet from week 3 onwards and was 3% higher at the end of the trial (1.427 kg/d versus 1.391 kg/d, respectively, for treatment and control).

Overall, the difference in DFI between the treatment and control group was not significant. A slightly larger amount of the treatment feed was eaten by the pigs in the treatment group than the amount of control feed eaten by their counterparts in the control group, which suggests that the weaner pigs liked the treatment diet. These results are described in Table 6. A similar trend was also seen in the study by Ravindran et al. [31], with pigs also consuming more treatment feed [31]. These findings indicate that the incorporation of green protein in the diet enhances the attractiveness of the feed, e.g., due to changes in taste, smell, or other sensory characteristics. Stødkilde et al. [23] found that the addition of green protein to pig feed rations did not negatively change the taste, i.e., pigs were not discouraged from consuming it [23]. Moreover, inclusion of green protein from specific feedstocks, e.g., clover grass, has also been found to improve the meat yield and omega-3 fatty acid content of pig meat, which may have positive health benefits [24].

Table 6. Comparison of treatment and control groups describing daily feed intake, feed conversion efficiency, and average daily gain.

| | Daily Feed Intake (kg/day) | | Feed Conversion Efficiency | | Average Daily Gain (kg/day) | |
Date of Weighing	Treatment	Control	Treatment	Control	Treatment	Control
Week 1	1.079	1.060	1.62	1.68	0.665	0.630
Week 2	1.155	1.167	1.89	1.78	0.611	0.656
Week 3	1.285	1.265	2.06	1.87	0.624	0.675
Week 4	1.384	1.349	2.13	2.13	0.650	0.632
Week 5	1.427	1.391	2.19	2.14	0.652	0.650

Average Daily Weight Gain

Average daily gain (ADG) is an evaluation of the average daily increase in the live weight of an animal and is recorded across the growth period of the animal. This can provide insight into the animal's growth rate, and evaluate the time at which it should reach market weight [53]. Ravindran et al. (2021) noted a number of factors that can influence pig weight, including genetic differences, sex effects, weight at birth, age at weaning, feeding level, and specific amino acid content (e.g., arginine) in feed [31].

During the feed trials, pig weight was measured individually, first at the beginning of the trial and at the end of each following week. In week 1, pigs in the control group gained 0.630 kg/day, increasing to 0.656 kg/day after week 2, 0.675 kg/day after week 3, 0.632 kg/day after week 4, and 0.650 after the final week. For pigs on the treatment diet, the ADG started well at 0.665 kg/day at the end of week 1, but reduced to 0.611 kg/day by week 2. From that point, the ADG increased to 0.624 kg/day by end of week 3, 0.650 kg/day by end of week 4, and 0.652 kg/day by the end of the trial, slightly higher than the control diet. Overall, the variances in average weight gain between the weeks is considered to be negligible and are generally consistent across both diets. The results are presented in Table 6. Over the course of the trial, the analysis showed comparable performance in weight gain for pigs on the control and treatment diets, with pigs gaining 19.49 kg in the case of the control diet and 19.55 kg in the treatment diet by the end of the trial. The ADG of the treatment group pigs was marginally higher than the control group pigs during week 1, but dipped in the following week as the control ADG increased. This may have been a result of acclimatization of the pig's microbiome to the new protein diet. Over the course of the remaining weeks, the ADG increased in the treatment batch week on week, culminating at week 5, at which point the treatment pigs had a slightly higher (0.652 kg/day) ADG compared with the control batch (0.650 kg/day). Over the course of the trial, the ADG between both trials were quite evenly matched and without significant differences. While there were some variances in ADG week on week, these were relatively minor and did not indicate a major trend. Certain factors such as time of day with weighing or gut fill can have a big impact on the weekly weighings during trials; however, over an extended period these factors do not contribute the same variance.

Feed Conversion Ratio

The feed conversion ratio (FCR) describes the quantity of feed required for pig weight to increase by one pound. Lower FCR values indicate that pigs are efficiently converting feed into body weight, while a high FCR can be a sign of the pigs being unable to convert the feed into body weight effectively [31]. A number of factors influence the feed conversion ratio in pigs. Pierozan et al. (2020) found the number of pigs per pen, feeder type, origin, and sex of the animals to be determinant factors [54]. According to Hancock and Behnke (2000), feed conversion efficiency, and thus FCR, can be enhanced by pelleting feed as pigs will not sort and waste feed pellets [55]. Smaller pellets are also more digestible [55]. Lastly, dietary components such as lysine and phosphorous levels are important. Lysine is essential for pigs in order to effectively utilize other amino acids for growth [56]. Additionally, phosphorus at optimum levels is required for the proper development of muscle and optimal energy use, with excess phosphorus being excreted as feces [57].

The FCR increased weekly for both groups over the course of the trial. Initially, the FCR was recorded in week 1 as 1.68 for the control group, and 1.62 for the treatment group, which was slightly lower. In subsequent weeks, the FCR increase for both diets was quite comparable, with the FCR by the end of the trial being slightly lower for the control diet (2.14) compared with the treatment diet (2.19). The comparable FCR during the trial indicates that the treatment diet compares well to the conventional diet fed to the pigs. The comparative FCR by the end of the trial indicates a slight improvement in FCR compared with previous research on LPC [31]. The difference that exists may indicate that the regression equations used for energy estimation of the grass protein may need to be refined slightly. However, once again, the difference between the groups is minor and the performance is largely in line with the control diet. This is a field of primary concern to pig farmers as optimizing FCR can have a significant positive effect on overall profitability.

3.2. LPC and Control Slurry Biogas and Biomethane Analysis Results

The biomethane potential is a measure of the biomethane that can extracted from an organic substrate. Pig slurry is a well-recognized source of animal manure for biogas production when co-fed with carbohydrate-rich feedstock. Pig slurries have high biomethane potential (275–450 L/Kg VSfed) when compared with other animal manures except for poultry litter (460 L/Kg VSfed) [58].

The biogas and biomethane production profiles from both LPC and soya bean meal pig weaner slurries are shown in Figure 5. The maximum biogas and biomethane yields for both slurries was achieved after 25 days of digestion. The biogas production from both LPC and soya bean meal pig weaner slurries were determined to be 495 ± 12.52 L/Kg VSfed and 478 ± 7.93 L/Kg VSfed, respectively (Table 7). An ANOVA analysis conducted on the biogas yield showed no statistical significance (p-value (0.1206) > 0.05). This indicated no significant difference in biogas production from both LPC and soya bean meal pig weaner slurries. From Figure 5, it can be observed that 90% of biogas production was achieved after 10 to 11 days. The daily biogas production was similar for both LPC and soya bean slurries which recorded a high of 66 L/Kg VSfed. This aligns with the findings of Santos et al. [59] and Miroshnichenko et al. [60], which indicated high daily biogas production in anaerobic digestion of some pig slurries [59,60]. However, there was a significant difference in the daily methane production, which was between 10% to 50% higher in LPC slurry than soya bean meal slurry. This was attributed to the high methane content of produced biogas from the LPC, which was about 22% to 35% higher than soya bean meal slurry (statistical significance at p-value 0.0335 < 0.05) (Table 7). Biogas from anaerobic digestion of LPC slurry had methane contents ranging from 70% to 73% compared with the soya bean meal slurry which had biogas methane contents from 52% to 59%. The high methane content obtained for LPC could be attributed to the higher volatile solid content of the LPC pig slurry, especially with regard to the high-protein diet fed to the pigs. The significant difference in biogas methane content for both slurries led to a higher biomethane potential of 355 ± 9.45 L/Kg VSfed for LPC pig weaner slurry compared with 281 ± 7.11 L/Kg VSfed for soya bean meal pig weaner slurry (statistical significance at p-value 0.0004 < 0.05). The BMP results indicated that slurry from weaners feeding on the LPC treatment diet performed significantly better (26% increase in methane yield) than the conventional slurry from weaners on soya bean meal. This is potentially a major benefit in addition to the successful replacement of the soya bean mealdiet with the LPC treatment diet and suggests that co-digestion with carbohydrate-rich substrates at the Timoleague, Co., Cork, community anaerobic digester has the potential to yield improved performance with LPC pig slurry. Another key positive point was the high methane yield obtained from LPC slurry from pig weaners compared with the reported literature on methane production from pig slurry. Biomethane production from pig slurry/manure, especially from weaners, tends to have a lower methane yield compared with that obtained from this study. The studies by Browne et al. [61] and Miroshnichenko et al. [60] for pig slurries from weaners yielded considerably lower biomethane potentials of 38.0 L/Kg VSfed and 75.5 L/Kg VSfed, respectively [1,60]. On the other hand, studies from Santos et al. [59] and Rodríguez et al. [33] indicated high biomethane yield for pig slurries [33,59]. These studies mostly digested pig slurries from pig fatteners and pregnant sows which tend to yield high biomethane production. The differing biomethane yields for pig slurry/manure seem to be highly dependent on the type of pig (i.e., weaners, fatteners, pregnant sows, and suckling sows) excreting the slurry with another key factor being the type of meal fed to the varying range of pigs. Irrespective, the biomethane yield from the LPC pig weaner slurry performed considerably better than the reported literature on pig weaner slurries and was about 20% less than the highest reported study of various pig slurries/manures.

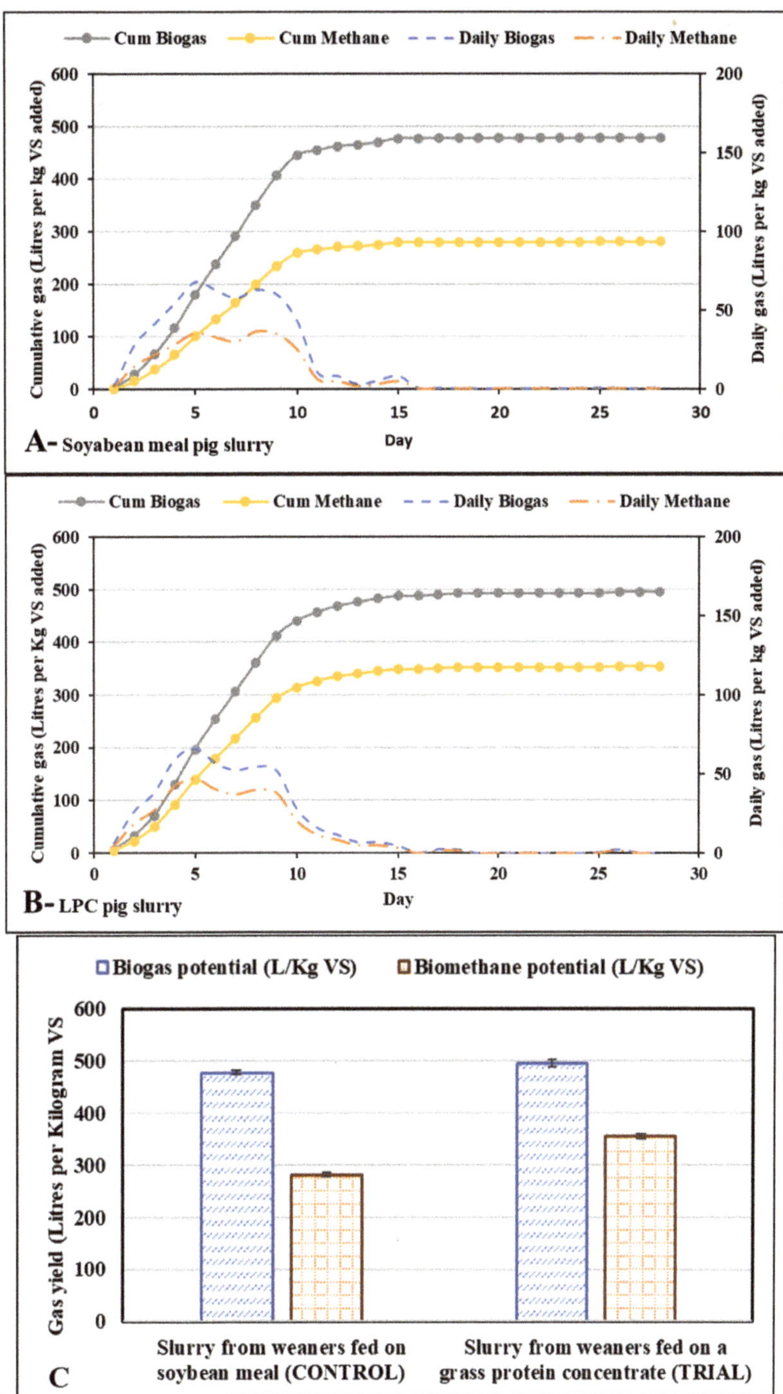

Figure 5. Daily and cumulative biogas and biomethane production profiles for (**A**) soya bean meal pig slurry; (**B**) LPC treatment meal pig slurry; and (**C**) biogas and biomethane potential of treatment and control pig slurry.

Table 7. Summary data for biogas and biomethane production from soya bean meal pig slurry and LPC treatment meal pig slurry.

Measurement	Control Diet Slurry	Treatment Diet Slurry
Biogas potential—replicate A (L/kg VS_{fed})	470.33	499.79
Biogas potential—replicate B (L/kg VS_{fed})	486.20	504.45
Biogas potential—replicate C (L/kg VS_{fed})	478.03	480.81
Average biogas potential (L/kg VS_{fed})	478.19 ± 7.93	495.02 ± 12.52
Average biomethane potential (L/kg VS_{fed})	280.85 ± 7.11	354.87 ± 9.45
Biogas potential (L/kg DM)	316.09 ± 5.25	349.65 ± 8.85
Biomethane potential (L/kg DM)	185.65 ± 4.70	250.66 ± 6.67
Biogas potential (L/kg FM)	8.52 ± 0.15	11.37 ± 0.29
Biomethane potential (L/kg FM)	5.01 ± 0.13	8.15 ± 0.21
Biogas composition (CH_4 %)	52.5–58.7	70.9–71.7
Biogas composition (CO_2 %)	41.3–47.5	28.3–29.1

3.3. Bioenergy Assessment of LPC Pig Slurry and Soya Bean Meal Pig Slurry Feed AD Systems: A Case Study

To demonstrate the potential impact of the above findings, in this case study, two AD scenarios are considered to ascertain bioenergy production from the usage of LPC pig slurry and soya bean meal pig slurry as feedstock to a community AD plant close to the pig farm where the LPC slurry was produced. This bioenergy assessment consists of mass and energy balance with the major assumptions listed in Table 8 [62]. The results of the bioenergy assessment are displayed in Figure 6 and discussed in Section 4.

Table 8. Assumptions made for bioenergy assessment of LPC pig slurry AD and soya bean meal pig slurry AD systems.

	Components	Conditions/Assumptions
Anaerobic digester	Operating temperature	36 °C
	Lower heating value (LHV) of methane	10 kWh/Nm^3 CH_4
Energy requirement of anaerobic digester	Electrical energy required for mixing slurry	10 $kWh_{electric}$/t slurry
	Thermal energy for heating digester	$E_{thermal} = Cp \times m \times \Delta T$
	Boiler efficiency	0.9
Energy requirement for digestate centrifugation	Electrical energy demand	3.5 $kWh_{electric}$/t digestate
	Moisture content in solid digestate	0.7
	Methane content in the upgraded biogas	96%
Energy requirement for amine scrubber biogas upgrading		Methane losses neglected
	Electrical energy demand	0.09 $kWh_{electric}$/m^3 biogas input
	Thermal energy demand	0.45 $kWh_{thermal}$/m^3 biogas input

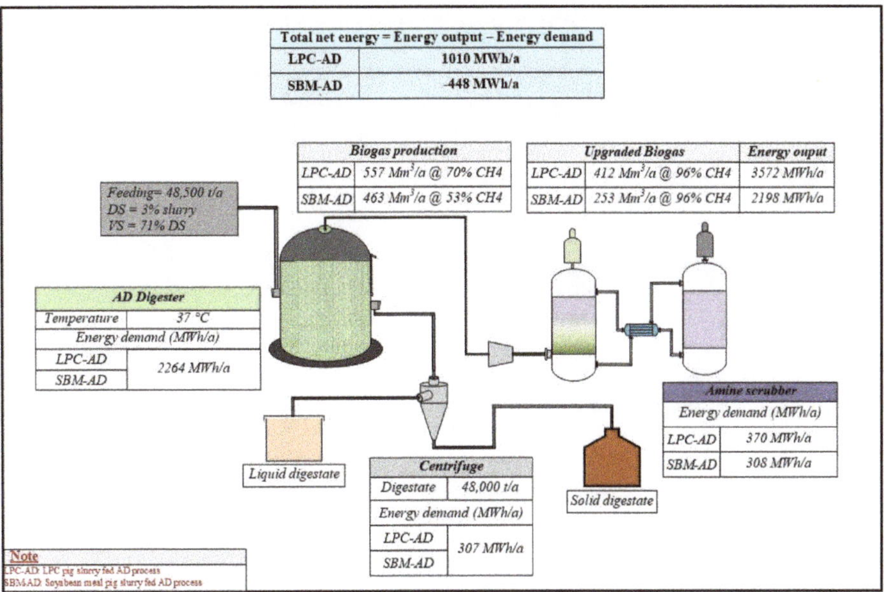

Figure 6. A bioenergy case study for an AD plant fed with LPC pig slurry and soya bean meal pig slurry.

4. Discussion

The results of these experiments demonstrate some potential for extended green biorefinery value chains to have positive benefits for the pig sector by supplying LPC as a sustainable alternative to soya bean meal and enhancing the potential for biogas production from pig slurry residues. Pig farmers, like most sectors of agriculture, are under pressure to become more sustainable. It is estimated that 68% of greenhouse gas emissions in the pork-production chain occur at the pre-farm gate phase [63]. A recent study from McAuliffe et al. [64] found that the primary aims of environmental performance improvements in the pig sector are reducing the crude protein content of pig feed and producing bioenergy through anaerobic digestion of pig slurry [64].

From a sustainability perspective, the potential to displace soya bean meal with an indigenous source of protein, such as LPC, could bring some key benefits. Soya bean meal is a major global commodity crop and is widely used in animal feed production, being one of the primary crops cultivated by farming communities across the world. Its production is mainly in the USA, Argentina, Brazil, China, and India, with only a small amount of cultivation taking place in Europe [17]. The soya bean market is closely associated with land use change for agricultural expansion and forest loss, particularly in South America [65]. It is estimated that approximately 2.31 million hectares of forest disappear annually to make way for soya bean production [66]. Despite a moratorium introduced in 2008 to avoid the purchase of soya beans from deforested land, by 2020 a further 133,000 ha of soy in the Amazon, planted on land deforested after this date, has been produced, which is linked to 69 million tonnes of CO_2 emissions [67]. A recent study from Franchi et al. [68], applying consequential life cycle assessment (LCA) to compare LPC and SBM for use in poultry diets, found a significantly lower environmental footprint in the case of LPC. A separate consequential LCA study from Parajuli et al. [69] found that a livestock production system that partially displaced Brazilian soya bean imports through integrated green biorefineries coupled with biogas facilities to produce feed and biomethane, and combined crop and livestock production, generated a lower environmental impact compared with the pre-existing "conventional" system (livestock production without biogas production or green biorefinery) [69].

Despite the negative impacts, South America still accounts for more than 50% of global soya bean production, of which 70% is exported, with around 21% of these exports coming to the EU for use in animal feed [65]. A recent study from Escobar et al. (2020), mapping carbon emissions embodied in Brazil's soy exports, found that the EU has a significantly higher footprint associated with soya bean imports compared with China; this mainly due to the source of soya beans being linked to deforested areas [70]. To highlight the impact that source can have on potential sustainability impact, referencing the life cycle inventory database Agri-Footprint 6, an economic allocation, and a point of substitution on the system, the footprints for Argentinean-, Brazilian-, and U.S.-sourced soya bean meal are 4.13 kg CO_2/kg, 4.28 kg CO_2/kg, and 0.53 kg CO_2/kg, respectively.

Despite these sustainability issues, the EU is still very dependent on imported soya beans. Soya bean meal is one of the most widely used individual protein sources in Europe, accounting for 29% of crude protein for animal feed in the EU (including the United Kingdom) during the period 2019–2020 [71]. This results in the average European person consuming 61 kg of soy indirectly every year. However, Europe has very low domestic soya bean production, accounting for just 3% of the total demand in the period 2019–2020, demonstrating a low self-sufficiency rate [71]. Overall, the EU's lack of indigenous protein results in approximately 17 million tonnes of proteins being imported on an annual basis, with the majority (13 million tonnes) deriving from soya bean [72]. This has prompted the European Commission to support the development of native protein ingredients such as peas, lupins, and faba beans [72]. On the other hand, grassland is a very available resource across the continent of Europe, with permanent grasslands making up 35% of total arable land use in EU-28 [73]. A study from Mandl [74] investigated the potential of grasslands to deliver additional protein for Europe. Focusing only on 15% surplus grassland, and assuming a yield of 8 t DM/ha, it is estimated that there is a grass surplus in Europe of approximately 20 million t DM/ha, creating an additional crude protein equivalent of approximately 3 million t DM [74]. This potential protein availability can be further increased to unlock more protein from all EU grasslands providing for ruminant and monogastric needs. The integration of legumes alongside grass for biorefining can further increase the sustainability of green biorefinery systems. If we consider, based on the more recent work of this paper, that by biorefining we can give almost 45% of the protein originally present in grass to pigs without reducing the impact on milk production from feeding the press-cake co-product to cows; then, assuming a grass yield of 10 t DM/ha/yr with 20% protein content, we can estimate an LPC biorefinery co-product yield from dairy farming of about 0.5 tonne protein/ha. This would mean that the 3 M tonnes can be obtained from only 6 Mha, or that the 17 M tonnes of protein that we import in Europe can be obtained from 34 Mha, being about half of all grassland in the EU.

Further environmental and self-sufficiency gains may be achieved using this extended green biorefinery model by increasing the potential of renewable energy from pig slurry, with the findings from this study indicating a 26% increase in biomethane yield from slurry produced from pigs on the treatment diet.

In Figure 6, a case study assesses two AD plant scenarios which considers the conventional soya bean meal pig slurry and the current study LPC pig slurry as AD feedstock to a community digester. The bioenergy case study includes the anaerobic digester, a centrifuge for digestate separation and an amine scrubber for biogas upgrading. The community digester can process a high throughput of 48,500 tonnes per annum of slurry, hence this feedstock capacity was assumed for both the LPC and soya bean meal pig slurries. Employing the biomethane yield from (Table 7), 357 and 220 thousand cubic meters of methane is produced per annum from LPC and soya bean meal slurries, respectively. This translates to 3572 MWh and 2198 MWh per annum of renewable bioenergy for the LPC and soya bean meal slurries, respectively. The gross bioenergy from the LPC pig slurry feed AD scenario was about 63% higher than that of the conventional soya bean meal pig slurry feed AD scenario. Furthermore, considering the parasitic energy demand of both scenarios, only the LPC pig slurry feed AD plant generated a positive net bioenergy of 1010 MWh per annum. This indicated the superiority and efficacy of the LPC pig slurry in utilization to produce renewable energy as compared with the conventional soymeal pig slurry.

According to the European Biogas Association [75], up to 41 billion cubic meters (bcm) of biomethane could be produced from sustainable feedstocks in Europe by 2030, rising to 151 bcm in 2050 [75]. These targets would enable the fulfillment of the European Commission target of 35 bcm biomethane production per annum by 2030, as outlined in the REPowerEU plan. Out of this total, the majority, 38 bcm by 2030 and 91 bcm by 2050, are estimated for anaerobic digestion, with 33% of the contribution anticipated to come from animal manure feedstocks [75].

While the overall CH_4 potential of pig slurry as a biogas feedstock can be comparatively low by comparison with other substrates, the utilization of pig slurry for on-farm energy production can have positive impacts on pig farm operations while contributing to a circular economy, also enabling the recirculation of nutrients to local farms [76]. Using LCA to investigate strategies for addressing greenhouse gas (GHG) emissions and fossil energy consumption in European pig production, Nguyen et al. [63] found that using pig manure as a feedstock source for anaerobic digestion had the ability to displace 53% of fossil energy usage and reduce GHG emissions of the farm by 27% over the baseline pig farm scenario [63]. Various other studies have investigated the feasibility and benefits of pig slurry anaerobic digestion for on-farm usage. Freitas et al. [77] investigated the potential of co-digestion with elephant grass silage, or corn silage, as well as the use of a biochar additive in comparison with monodigestion of pig slurry [77]. Using co-substrates and additives enabled greater production of electricity compared with monodigestion, but also resulted in significant environmental impacts associated with co-substrate use, primarily as a result of fossil fuel use during the silage production chain. When comparing digestion with the direct spreading of slurry, Zhang et al. [32] found that digestion of pig slurry alone resulted in a 48% decrease in direct emissions of GHGs (190 tonne CO_2e) compared with direct land application, due to the recovery of methane [34]. Using LCA, Jiang et al. [78] compared the co-digestion of pig manure and food waste with alternative management strategies including pig manure land use, food waste mono-digestion, and composting, and found that the co-digestion approach performed better in most environmental impact categories [78]. Using an average size piggery for comparison, comprising 762 sows with 16,000 t/yr pig manure, the same study found that the global warming potential (GWP) only became negative when the inclusion of food waste in feedstock was greater than 2000 tonnes per annum [78]. A further recent benefit of interlinking the green biorefinery approach with pig systems was reported by Regueiro et al. [79] who demonstrated that both unfermented and fermented whey or brown juice from the grass biorefinery can be used to stabilize pig slurry in storage by contributing to reducing the pH. A reduction below pH6 reduces ammonia emissions as well as methane emissions from the stored slurry.

The development of such anaerobic digestion models may also help to meet the needs of the local communities or districts in which they are based. Community-based anaerobic digestion models, which can help to meet the heating and electricity requirements of the community, for example, heating community buildings, are well established. In addition, such models may help to meet the mobility needs of the community, through the development of biomethane-based transport systems in pressurized or liquefied forms. In Europe, biomethane-based transport is at the furthest stage of development in Sweden, where half of the biogas production is used for transport [80], including public transport buses where it supplies the fuel for over 20% of the distance travelled [81]. Biomethane may also help to reduce the emissions and dependency of local manufacturing industries which are heavily dependent on natural gas, including pulp and paper, and some food processing industries.

In addition to the environmental sustainability and self-sufficiency benefits, the development of such local bioeconomy value chains may also help to relieve cost pressures for local pig farmers. According to the European Commission [82], EU pigmeat production is due to decrease by 5% in 2022 in large part due to rising input costs [82]. This situation has been compounded by the Russia–Ukraine war, as both Russia and Ukraine were key exporters of fertilizers and grain and oil crops, including wheat, maize, and sunflower, and Europe has also been dependent on natural gas from Russia. Between Q2 of 2021 and Q2 of 2022, following the Russian invasion of Ukraine, the average price of goods and services used within agriculture in the EU jumped by 36% for the same agricultural inputs [83]. An economic analysis of small-scale green biorefineries from Cong and Termansen [84] found that using LPC can be economically feasible for both pig farmers and the green biorefinery [84]. In order for these models to be successful and implementable, economic feasibility is an important outcome. The study found that LPC will decrease the average feed cost by 5%. Coupled with this, rising natural gas prices, and a need to reduce dependency on imported Russian gas, has changed the landscape for biomethane in Europe, making biogas more cost-competitive and even cheaper than natural gas in the current environment [85]. This combined with a significant increase in biomethane potential, demonstrated by this study, may offer a greater opportunity for pig farmers to produce, use, and supply renewable energy. While this paper primarily underlines the synergies and benefits that may be found by connecting ruminant and swine sectors in extended local green biorefinery models, as noted earlier, several additional synergies may be found by connecting additional value chains in local bioeconomy models, further improving the resilience of local regions.

5. Conclusions

Overall, this work demonstrates significant potential for the development of local bioeconomy value chains based on a green biorefinery model, increasing the resource efficiency of grasses and legumes, building synergies, and meeting input requirements to increase the sustainability and resilience of the pig sector. The LPC co-product from the green biorefinery was produced with a high-crude-protein content of 43.9% and has been integrated within the pig treatment diet replacing 50% of the soya bean meal present in the control diet, achieving a slightly higher daily feed intake and average weight gain compared with the control batch on conventional weaner diets. In addition, the slurry produced by pigs on the treatment diet achieved higher biogas and biomethane production rates compared with the slurry from pigs on the control diet. Since sustainability improvements in the pig diet and anaerobic digestion of slurries are identified as key components to increasing the sustainability of pig production systems, a local bioeconomy model may help to meet this objective, while providing additional co-products for use in the ruminant and industrial sectors. Given the abundance of green biomass that exists in Europe, the potential to unlock additional protein from these through local green biorefineries may offer significant potential to increase feed resilience for the pig and broader livestock sector.

Author Contributions: Conceptualization, J.G., C.O., D.M., K.D., L.G., S.K., E.B., K.O. and J.P.M.S.; methodology, J.G., C.O., D.M., D.W., K.D., L.G., S.K., E.B., K.O. and J.P.M.S.; formal analysis, J.G., C.O., D.M., D.W., K.D., L.G., S.K., E.B., K.O. and J.P.M.S.; investigation, J.G., C.O., D.M., D.W., K.D., L.G., S.K., E.B., K.O. and J.P.M.S.; resources, J.G., C.O., D.M., D.W., K.D., L.G., S.K., E.B., K.O. and J.P.M.S.; data curation, J.G., C.O., D.M., D.W., K.D., L.G., S.K., E.B., K.O. and J.P.M.S.; writing—original draft preparation, J.G.; writing—review and editing, J.G., T.O., C.O., D.M., D.W., L.A.V., K.D., L.G., S.K., E.B., K.O. and J.P.M.S.; supervision, J.G; project administration, J.G., E.B. and K.O.; funding acquisition, J.G., E.B. and K.O. All authors have read and agreed to the published version of the manuscript.

Funding: This research was funded through the Farm Zero C project by Science Foundation Ireland's Zero Emissions Challenge, grant number 19/FIP/ZE/7558. The authors are grateful to acknowledge this funding.

Institutional Review Board Statement: Not applicable.

Informed Consent Statement: Not applicable.

Data Availability Statement: Not applicable.

Conflicts of Interest: The authors declare no conflict of interest. The funders had no role in the design of the study; in the collection, analyses, or interpretation of data; in the writing of the manuscript; or in the decision to publish the results.

References

1. Rockström, J.; Steffen, W.; Noone, K.; Persson, Å.; Chapin, F.S.; Lambin, E.F.; Lenton, T.M.; Scheffer, M.; Folke, C.; Schellnhuber, H.J. A safe operating space for humanity. *Nature* **2009**, *461*, 472–475. [CrossRef]
2. Steffen, W.; Richardson, K.; Rockström, J.; Cornell, S.E.; Fetzer, I.; Bennett, E.M.; Biggs, R.; Carpenter, S.R.; De Vries, W.; De Wit, C.A. Planetary boundaries: Guiding human development on a changing planet. *Science* **2015**, *347*, 1259855. [CrossRef]
3. Clark, W.; Lenaghan, M. *The Future of Food: Sustainable Protein Strategies around the World*; Zero Waste Scotland: Glasgow, UK, 2020.
4. European Parliament. EU Agricultural Policy and Climate Change. 2020. Available online: https://www.europarl.europa.eu/RegData/etudes/BRIE/2020/651922/EPRS_BRI(2020)651922_EN.pdf (accessed on 13 January 2023).
5. Eurostat. Greenhouse Gas Emissions from Agriculture, by Country. 2012. Available online: https://ec.europa.eu/eurostat/statistics-explained/index.php?title=File:Greenhouse_gas_emissions_from_agriculture,_by_country,_2012.png (accessed on 15 January 2023).
6. Department of Communications, Climate Action and Environment. National Mitigation Plan. 2017. Available online: https://www.gov.ie/en/publication/48d4e-national-mitigation-plan/ (accessed on 18 January 2023).
7. European Commission. European Green Deal—Delivering on Our Targets. 2021. Available online: https://ec.europa.eu/commission/presscorner/detail/en/fs_21_3688 (accessed on 14 January 2023).
8. European Commission. A Sustainable Bioeconomy for Europe: Strengthening the Connection between Economy, Society and the Environment. 2018. Available online: https://op.europa.eu/en/publication-detail/-/publication/edace3e3-e189-11e8-b690-01aa75ed71a1/ (accessed on 18 January 2023).
9. Ahmed, E.; Batbekh, B.; Fukuma, N.; Hanada, M.; Nishida, T. Evaluation of Different Brown Seaweeds as Feed and Feed Additives Regarding Rumen Fermentation and Methane Mitigation. *Fermentation* **2022**, *8*, 504. [CrossRef]
10. Roque, B.M.; Venegas, M.; Kinley, R.D.; de Nys, R.; Duarte, T.L.; Yang, X.; Kebreab, E. Red seaweed (*Asparagopsis taxiformis*) supplementation reduces enteric methane by over 80 percent in beef steers. *PLoS ONE* **2021**, *16*, e0247820. [CrossRef]
11. Antoniêto, A.C.C.; Nogueira, K.M.V.; Mendes, V.; Maués, D.B.; Oshiquiri, L.H.; Zenaide-Neto, H.; de Paula, R.G.; Gaffey, J.; Tabatabaei, M.; Gupta, V.K. Use of carbohydrate-directed enzymes for the potential exploitation of sugarcane bagasse to obtain value-added biotechnological products. *Int. J. Biol. Macromol.* **2022**, *221*, 456–471. [CrossRef]
12. Usmani, Z.; Sharma, M.; Gaffey, J.; Sharma, M.; Dewhurst, R.J.; Moreau, B.; Newbold, J.; Clark, W.; Thakur, V.K.; Gupta, V.K. Valorization of dairy waste and by-products through microbial bioprocesses. *Bioresour. Technol.* **2021**, *346*, 126444. [CrossRef]
13. Wang, L.; Littlewood, J.; Murphy, R.J. Environmental sustainability of bioethanol production from wheat straw in the UK. *Renew. Sustain. Energy Rev.* **2013**, *28*, 715–725. [CrossRef]
14. Meza, L.E.; Rodríguez, A.G. Nature-Based Solutions and the Bioeconomy: Contributing to a Sustainable and Inclusive Transformation of Agriculture and to the Post-COVID-19 Recovery. 2022. Available online: https://www.cepal.org/en/publications/48101-nature-based-solutions-and-bioeconomy-contributing-sustainable-and-inclusive (accessed on 18 January 2023).
15. Neill, A.M.; O'Donoghue, C.; Stout, J.C. A Natural capital lens for a sustainable Bioeconomy: Determining the unrealised and unrecognised services from nature. *Sustainability* **2020**, *12*, 8033. [CrossRef]
16. Dias, R.S.; Costa, D.V.; Correia, H.E.; Costa, C.A. Building bio-districts or eco-regions: Participative processes supported by focal groups. *Agriculture* **2021**, *11*, 511. [CrossRef]
17. Mulder, W.; van der Peet-Schwering, C.; Hua, N.-P.; van Ree, R. *Proteins for Food, Feed and Biobased Applications: Biorefining of Protein Containing Biomass*; IEA Bioenergy Task 42; IEA Bioenergy: Enschede, The Netherlands, 2016.
18. Damborg, V.K.; Jensen, S.K.; Johansen, M.; Ambye-Jensen, M.; Weisbjerg, M.R. Ensiled pulp from biorefining increased milk production in dairy cows compared with grass–clover silage. *J. Dairy Sci.* **2019**, *102*, 8883–8897. [CrossRef]
19. Santamaría-Fernández, M.; Lübeck, M. Production of leaf protein concentrates in green biorefineries as alternative feed for monogastric animals. *Anim. Feed. Sci. Technol.* **2020**, *268*, 114605. [CrossRef]
20. Serra, E.; Lynch, M.; Gaffey, J.; Sanders, J.; Koopmans, S.; Markiewicz-Keszycka, M.; Bock, M.; McKay, Z.; Pierce, K. Biorefined press cake silage as feed source for dairy cows: Effect on milk production and composition, rumen fermentation, nitrogen and phosphorus excretion and in vitro methane production. *Livest. Sci.* **2022**, *267*, 105135. [CrossRef]
21. Pijlman, J.; Koopmans, S.; De Haan, G.; Lenssinck, F.; Van Houwelingen, K.; Sanders, J.; Deru, J.; Erisman, J. Effect of the grass fibrous fraction obtained from biorefinery on n and P utilization of dairy cows. In Proceedings of the 20th Nitrogen Workshop: "Coupling C-N-P-S cycles", Rennes, France, 25 June 2018; pp. 25–27.
22. Stødkilde, L.; Ambye-Jensen, M.; Krogh Jensen, S. Biorefined grass-clover protein composition and effect on organic broiler performance and meat fatty acid profile. *J. Anim. Physiol. Anim. Nutr.* **2020**, *104*, 1757–1767. [CrossRef]
23. Stødkilde, L.; Ambye-Jensen, M.; Jensen, S.K. Biorefined organic grass-clover protein concentrate for growing pigs: Effect on growth performance and meat fatty acid profile. *Anim. Feed Sci. Technol.* **2021**, *276*, 114943. [CrossRef]

24. Jensen, H.G.; Elleby, C.; Domínguez, I.P. Reducing the European Union's plant protein deficit: Options and impacts. *Agric. Econ.* **2021**, *67*, 391–398. [CrossRef]

25. Ravindran, R.; Donkor, K.; Gottumukkala, L.; Menon, A.; Guneratnam, A.J.; McMahon, H.; Koopmans, S.; Sanders, J.P.; Gaffey, J. Biogas, Biomethane and Digestate Potential of By-Products from Green Biorefinery Systems. *Clean Technol.* **2022**, *4*, 35–50. [CrossRef]

26. Santamaría-Fernández, M.; Molinuevo-Salces, B.; Lübeck, M.; Uellendahl, H. Biogas potential of green biomass after protein extraction in an organic biorefinery concept for feed, fuel and fertilizer production. *Renew. Energy* **2018**, *129*, 769–775. [CrossRef]

27. Igliński, B.; Pietrzak, M.B.; Kiełkowska, U.; Skrzatek, M.; Kumar, G.; Piechota, G. The assessment of renewable energy in Poland on the background of the world renewable energy sector. *Energy* **2022**, *261*, 125319. [CrossRef]

28. Igliński, B.; Kiełkowska, U.; Piechota, G.; Skrzatek, M.; Cichosz, M.; Iwański, P. Can energy self-sufficiency be achieved? Case study of Warmińsko-Mazurskie Voivodeship (Poland). *Clean Technol. Environ. Policy* **2021**, *23*, 2061–2081. [CrossRef]

29. D'Adamo, I.; Gastaldi, M.; Morone, P.; Rosa, P.; Sassanelli, C.; Settembre-Blundo, D.; Shen, Y. Bioeconomy of sustainability: Drivers, opportunities and policy implications. *Sustainability* **2021**, *14*, 200. [CrossRef]

30. Gaffey, J.; Rajauria, G.; McMahon, H.; Ravindran, R.; Dominguez, C.; Jensen, M.A.; Souza, M.F.; Meers, E.; Aragonés, M.M.; Skunca, D. Green Biorefinery systems for the production of climate-smart sustainable products from grasses, legumes and green crop residues. *Biotechnol. Adv.* **2023**, *66*, 108168. [CrossRef] [PubMed]

31. Ravindran, R.; Koopmans, S.; Sanders, J.P.; McMahon, H.; Gaffey, J. Production of Green Biorefinery Protein Concentrate Derived from Perennial Ryegrass as an Alternative Feed for Pigs. *Clean Technol.* **2021**, *3*, 656–669. [CrossRef]

32. Zhang, Y.; Jiang, Y.; Wang, S.; Wang, Z.; Liu, Y.; Hu, Z.; Zhan, X. Environmental sustainability assessment of pig manure mono-and co-digestion and dynamic land application of the digestate. *Renew. Sustain. Energy Rev.* **2021**, *137*, 110476. [CrossRef]

33. Rodríguez, A.; Ángel, J.; Rivero, E.; Acevedo, P.; Santis, A.; Rojas, I.C.; Acosta, M.; Hernández, M. Evaluation of the biochemical methane potential of pig manure, organic fraction of municipal solid waste and cocoa industry residues in Colombia. *Chem. Eng. Trans.* **2017**, *57*, 55–60.

34. Carrère, H.; Sialve, B.; Bernet, N. Improving pig manure conversion into biogas by thermal and thermo-chemical pretreatments. *Bioresour. Technol.* **2009**, *100*, 3690–3694. [CrossRef] [PubMed]

35. *ISO 6865:2000*; Animal Feeding Stuffs—Determination of Crude Fibre Content—Method with Intermediate Filtration. ISO: Geneva, Switzerland, 2000.

36. *ISO 5983-2:2009*; Animal Feeding Stuffs—Determination of Nitrogen Content and Calculation of Crude Protein Content—Part 2: Block Digestion and Steam Distillation Method. Animal Feeding Stuffs—Determination of Nitrogen Content and Calculation of Crude Protein Content—Part 2: Block Digestion and Steam Distillation Method. ISO: Geneva, Switzerland, 2009.

37. *ISO 6869:2000*; Animal Feeding Stuffs—Determination of the Contents of Calcium, Copper, Iron, Magnesium, Manganese, Potassium, Sodium and Zinc—Method Using Atomic Absorption Spectrometry. ISO: Geneva, Switzerland, 2000.

38. *ISO 6496:1999*; Animal Feeding Stuffs—Determination of Moisture and Other Volatile Matter Content. ISO: Geneva, Switzerland, 1999.

39. *ISO 6491:1998*; Animal Feeding Stuffs—Determination of Phosphorus Content—Spectrometric Method. ISO: Geneva, Switzerland, 1998.

40. *ISO 15914:2004*; Animal Feeding Stuffs—Enzymatic Determination of Total Starch Content. ISO: Geneva, Switzerland, 2004.

41. *ISO 6492:1999*; Animal Feeding Stuffs—Determination of fat Content. ISO: Geneva, Switzerland, 1999.

42. *ISO 5984:2002*; Animal Feeding Stuffs—Determination of Crude Ash. ISO: Geneva, Switzerland, 2002.

43. De Haer, L.; Merks, J. Patterns of daily food intake in growing pigs. *Anim. Sci.* **1992**, *54*, 95–104. [CrossRef]

44. Horodyska, J.; Hamill, R.M.; Varley, P.F.; Reyer, H.; Wimmers, K. Genome-wide association analysis and functional annotation of positional candidate genes for feed conversion efficiency and growth rate in pigs. *PLoS ONE* **2017**, *12*, e0173482. [CrossRef]

45. *EN 14774-1:2009*; Solid Biofuels-Determination of Moisture Content-Oven Dry Method-Part 1: Total Moisture-Reference Method. EN: Brussels, Belgium, 2010.

46. *EN 14775:2009*; Solid Biofuels–Determination of Ash Content. EU: Brussels, Belgium, 2009.

47. *ISO/TC 238*; Solid Biofuels–Determination of Total Content of Sulfur and Chlorine. ISO: Geneva, Switzerland, 2015.

48. *EN 15290:2011*; Solid Biofuels. Determination of Major Elements-Al, Ca, Fe, Mg, P, K, Si, Na and Ti. British Standards Institution: Chiswick, UK, 2011.

49. O'Dell, J. *The Determination of Chemical Oxygen Demand by Semi-Automated Colorimetry-Method 410.4*; Environmental Monitoring Systems Laboratory, Office of Research and Development, US Environmental Protection Agency: Cincinnati, OH, USA, 1993.

50. *ISO/TC 238*; Solid Biofuels—Determination of Minor Elements. ISO: Geneva, Switzerland, 2015.

51. Prochnow, A.; Heiermann, M.; Drenckhan, A.; Schelle, H. *Seasonal Pattern of Biomethanisation of Grass from Landscape Management*; International Commission of Agricultural Engineering: Liège, Belgium, 2005.

52. Florou-Paneri, P.; Christaki, E.; Giannenas, I.; Bonos, E.; Skoufos, I.; Tsinas, A.; Tzora, A.; Peng, J. Alternative Protein Sources to Soybean Meal in Pig Diets. *J. Food Agric. Environ.* **2014**, *12*, 655–660.

53. Pork Information Gateway. Timing the Purchase of Your Project Pig. Available online: https://porkgateway.org/resource/timing-the-purchae-of-your-project-pig/#:~:text=Average%20Daily%20Gain%20(ADG)&text=Understanding%20ADG%20is%20key%20to,approach%20their%20final%20market%20weight (accessed on 18 January 2023).

54. Pierozan, C.; Dias, C.; Temple, D.; Manteca, X.; da Silva, C. Welfare indicators associated with feed conversion ratio and daily feed intake of growing-finishing pigs. *Anim. Prod. Sci.* **2020**, *61*, 412–422. [CrossRef]

55. Hancock, J.D.; Behnke, K.C. Use of ingredient and diet processing technologies (grinding, mixing, pelleting, and extruding) to produce quality feeds for pigs. In *Swine Nutrition*; CRC Press: Boca Raton, FL, USA, 2000; pp. 489–518.

56. Hall, R.E.; Biehl, L.; Meyer, K. *Slaughter Checks—An Aid to Better Herd Health*; Pork Industry Handbook/Purdue University, Cooperative Extension Service: West Lafayette, IN, USA, 1978.

57. Patience, J. The influence of dietary energy on feed efficiency in grow-finish swine. In *Feed Efficiency in Swine*; Springer: Berlin/Heidelberg, Germany, 2012; pp. 101–129.

58. Vertes, A.A.; Qureshi, N.; Yukawa, H.; Blaschek, H.P. *Biomass to Biofuels: Strategies for Global Industries*; John Wiley & Sons: Hoboken, NJ, USA, 2011.

59. Santos, A.D.; Silva, J.R.; Castro, L.M.; Quinta-Ferreira, R.M. A biochemical methane potential of pig slurry. *Energy Rep.* **2022**, *8*, 153–158. [CrossRef]

60. Miroshnichenko, I.; Oskina, A.; Eremenko, E. Biogas potential of swine manure of different animal classes. *Energy Sources Part A Recovery Util. Environ. Eff.* **2020**, 1–12. [CrossRef]

61. Browne, J.D.; Allen, E.; Murphy, J.D. Evaluation of the biomethane potential from multiple waste streams for a proposed community scale anaerobic digester. *Environ. Technol.* **2013**, *34*, 2027–2038. [CrossRef]

62. Wu, B.; Lin, R.; O'Shea, R.; Deng, C.; Rajendran, K.; Murphy, J.D. Production of advanced fuels through integration of biological, thermo-chemical and power to gas technologies in a circular cascading bio-based system. *Renew. Sustain. Energy Rev.* **2021**, *135*, 110371. [CrossRef]

63. Nguyen, T.L.T.; Hermansen, J.E.; Mogensen, L. Fossil energy and GHG saving potentials of pig farming in the EU. *Energy Policy* **2010**, *38*, 2561–2571. [CrossRef]

64. McAuliffe, G.A.; Chapman, D.V.; Sage, C.L. A thematic review of life cycle assessment (LCA) applied to pig production. *Environ. Impact Assess. Rev.* **2016**, *56*, 12–22. [CrossRef]

65. Karlsson, J.O.; Parodi, A.; Van Zanten, H.H.; Hansson, P.-A.; Röös, E. Halting European Union soybean feed imports favours ruminants over pigs and poultry. *Nat. Food* **2021**, *2*, 38–46. [CrossRef] [PubMed]

66. Taelman, S.E.; De Meester, S.; Van Dijk, W.; Da Silva, V.; Dewulf, J. Environmental sustainability analysis of a protein-rich livestock feed ingredient in The Netherlands: Microalgae production versus soybean import. *Resour. Conserv. Recycl.* **2015**, *101*, 61–72. [CrossRef]

67. Stockholm Environment Institute. Connecting Exports of Brazilian Soy to Deforestation. 2022. Available online: https://www.sei.org/featured/connecting-exports-of-brazilian-soy-to-deforestation/ (accessed on 28 January 2023).

68. Franchi, C.; Brouwer, F.; Compeer, A. LCA summary report Grass protein versus Soy protein. Mechelen, Belgium. 2020. Available online: https://www.grasgoed.eu/wp-content/uploads/2020/06/GrasGoed-LCA-summary-report-chicken-feed-protein.pdf (accessed on 28 January 2023).

69. Parajuli, R.; Dalgaard, T.; Birkved, M. Can farmers mitigate environmental impacts through combined production of food, fuel and feed? A consequential life cycle assessment of integrated mixed crop-livestock system with a green biorefinery. *Sci. Total Environ.* **2018**, *619*, 127–143. [CrossRef]

70. Escobar, N.; Tizado, E.J.; zu Ermgassen, E.K.; Löfgren, P.; Börner, J.; Godar, J. Spatially-explicit footprints of agricultural commodities: Mapping carbon emissions embodied in Brazil's soy exports. *Glob. Environ. Chang.* **2020**, *62*, 102067. [CrossRef]

71. Kuepper, B.; Stravens, M. *Mapping the European Soy Supply Chain*; Profundo, Commissioned by WWF European Policy Office: Amsterdam, The Netherlands, 2022.

72. European Commission. Report from the Commission to the Council and the European Parliament on the Development of Plant Proteins in the European Union. *Eur. Comm.* **2018**, *757*, 1–15.

73. Schils, R.L.; Newell Price, P.; Klaus, V.; Tonn, B.; Hejduk, S.; Stypinski, P.; Hiron, M.; Fernández, P.; Ravetto Enri, S.; Lellei-Kovács, E. European Permanent Grasslands Mainly Threatened by Abandonment, Heat and Drought, and Conversion to Temporary Grassland. In Proceedings of the 28th General Meeting of the European Grassland Federation, Online, 19–21 October 2020; pp. 553–555.

74. Mandl, M.G. Status of green biorefining in Europe. *Biofuels Bioprod. Biorefin. Innov. A Sustain. Econ.* **2010**, *4*, 268–274. [CrossRef]

75. European Biogas Association. *Biomethane Production Potentials in the EU—Feasibility of REPowerEU 2030 Targets, Production Potentials in the Member States and Outlook to 2050*; European Biogas Association: Utrecht, The Netherlands, 2022.

76. Soares, C.A.; Viancelli, A.; Michelon, W.; Sbardelloto, M.; Camargo, A.F.; Vargas, G.D.L.; Fongaro, G.; Treichel, H. Biogas yield prospection from swine manure and placenta in real-scale systems on circular economy approach. *Biocatal. Agric. Biotechnol.* **2020**, *25*, 101598. [CrossRef]

77. Freitas, F.; Furtado, A.; Piñas, J.; Venturini, O.; Barros, R.; Lora, E. Holistic Life Cycle Assessment of a biogas-based electricity generation plant in a pig farm considering co-digestion and an additive. *Energy* **2022**, *261*, 125340. [CrossRef]

78. Jiang, Y.; Zhang, Y.; Wang, S.; Wang, Z.; Liu, Y.; Hu, Z.; Zhan, X. Improved environmental sustainability and bioenergy recovery through pig manure and food waste on-farm co-digestion in Ireland. *J. Clean. Prod.* **2021**, *280*, 125034. [CrossRef]

79. Regueiro, I.; Gómez-Muñoz, B.; Lübeck, M.; Hjorth, M.; Jensen, L.S. Bio-acidification of animal slurry: Efficiency, stability and the mechanisms involved. *Bioresour. Technol. Rep.* **2022**, *19*, 101135. [CrossRef]

80. Pääkkönen, A.; Aro, K.; Aalto, P.; Konttinen, J.; Kojo, M. The Potential of biomethane in replacing fossil fuels in heavy transport—A case study on Finland. *Sustainability* **2019**, *11*, 4750. [CrossRef]
81. Hagstroem, A. Prospects for Continued Use and Production of Swedish Biogas in Relation to Current Market Transformations in Public Transport. 2019. Available online: http://kth.diva-portal.org/smash/record.jsf?pid=diva2%3A1352422&dswid=7334 (accessed on 28 January 2023).
82. European Commission. Short-Term Outlook for EU Agricultural Markets in 2022. 2022. Available online: https://agriculture.ec.europa.eu/system/files/2022-10/short-term-outlook-autumn-2022_en_1.pdf (accessed on 29 January 2023).
83. Eurostat. EU Agricultural Prices Continued to Rise in Q2 2022. 2022. Available online: https://ec.europa.eu/eurostat/web/products-eurostat-news/-/ddn-20220930-3 (accessed on 29 January 2023).
84. Cong, R.-G.; Termansen, M. A bio-economic analysis of a sustainable agricultural transition using green biorefinery. *Sci. Total Environ.* **2016**, *571*, 153–163. [CrossRef]
85. European Biogas Association. A Way out of the EU Gas Price Crisis with Biomethane. 2022. Available online: https://www.europeanbiogas.eu/a-way-out-of-the-eu-gas-price-crisis-with-biomethane/ (accessed on 30 January 2023).

MDPI

Review

Industry 4.0 and Beyond: A Review of the Literature on the Challenges and Barriers Facing the Agri-Food Supply Chain

Arman Derakhti [1], Ernesto D. R. Santibanez Gonzalez [2,*] and Abbas Mardani [3]

[1] Industrial Engineering Department, Universidad de Talca, Curico 3340000, Chile
[2] Industrial Engineering Department, CES4.0, Faculty of Engineering, Universidad de Talca, Los Niches Km. 1, Edificio I+D, Curico 3340000, Chile
[3] Muma Business School, University of South Florida (USF), Tampa, FL 33612, USA
* Correspondence: santibanez.ernesto@gmail.com

Abstract: In recent years, the Industry 4.0 concept has gained considerable attention from professionals, researchers and decision makers. For its part, the COVID-19 pandemic has highlighted the importance of managing the agri-food supply chain to ensure the food that the population needs. Industry 4.0 and its extensions can address the needs of the agri-food supply chain by bringing new features such as security, transparency and traceability in line with sustainable development goals. This study aims to systematically analyze the literature to address the challenges and barriers against the application of industry 4.0 and its related technologies in the management of an agri-food supply chain. Currently, despite the large number of publications, there is no clear agreement on what Industry 4.0 is, and even less its extensions. The next revolution that includes new technologies and improves several existing technologies brings additional conceptual and practical complexity. Consequently, in this work we first determine the main components of I 4.0 and their extensions by studying the literature, and then, in the second step, define the agri-food supply chain on which I 4.0 technologies are applied. Two well-known databases—Web of Science and Scopus—were chosen to extract data for the systematic review of the literature. For the final evaluation, we identified 24 of 100 reviewed publications. The results provide an exhaustive analysis of the different I 4.0 technologies and their extensions that are applied in regards to the agri-food supply chain. In addition, we find 15 challenges that are classified into five major themes in the agri-food supply chain: technical, operational, financial, social and infrastructure. The four most important challenges identified are technological architecture, security and privacy, big data management and IoT (internet)-based infrastructure. Only a few articles addressed sustainability, which reaffirms and demonstrates a considerable gap in terms of the sustainable agri-food supply chain, with waste management being the one that has attracted the most attention. This review provides a roadmap for academics and practitioners alike, showing the gaps and facilitating the identification of I 4.0 technologies that can help address the challenges facing the efficient management of an agri-food supply chain.

Keywords: industry 4.0; agri-food supply chain; sustainability; agri-food 4.0 supply chain; agri-food 4.0; supply chain 4.0; food waste management; water management; agriculture 4.0

Citation: Derakhti, A.; Santibanez Gonzalez, E.D.R.; Mardani, A. Industry 4.0 and Beyond: A Review of the Literature on the Challenges and Barriers Facing the Agri-Food Supply Chain. *Sustainability* **2023**, *15*, 5078. https://doi.org/10.3390/su15065078

Academic Editor: Claudio Sassanelli

Received: 2 January 2023
Revised: 3 February 2023
Accepted: 20 February 2023
Published: 13 March 2023

1. Introduction

In recent years, the concept of Industry 4.0 (I 4.0) has attracted the attention of practitioners, researchers and decision makers, and its applications have been studied in multiple industrial sectors. Despite the large number of academic and non-academic publications, there is still no clear definition of I 4.0. Several technologies, such as Radio Frequency Identification, Internet of Things, Cloud Computing are considered I 4.0 components, and in some cases, by themselves define I 4.0. In order to standardize language and set the context for this article, we first tackle the problem of defining I 4.0.

In 2011 the German government representative used industry 4.0 term as a steadily growing industry that considerably affects our lives. This speech was about how digitalization and new technologies revolutionize the organization of global value chains [1]. As a pioneer country in manufacturing, Germany introduced the idea of integrated industry by launching I 4.0 initiatives in 2011 for its high-tech strategies [2]. I 4.0 is known with different terms in scientific publications such as "Fourth Industrial Revolution", "smart manufacturing", "Industrial internet" or "integrated industry" [3]. Moreover, the other terms suggested by [4] are "smart factories", smart industry", "digital manufacturing", and "smart production". I 4.0 itself is a concept that is applicable through different technologies. The number of technologies is growing and with emerging new technologies it would become more and more. We first determine its main component by studying the literature [3,4] and is more relevant to the supply chain research area.

1.1. Industry 4.0 Key Technologies

Four items were found as principal components, including cyber-physical system (CPS), internet of things (IoT), smart factory, and internet of services (IoS). The study [4] proposed the four mentioned technologies to introduce a coherent definition of I 4.0. Furthermore, [3] study suggested these four technologies as the fundamental technology components of I 4.0 in logistics which are explained as follows:

1.1.1. Cyber-Physical Systems (CP), and Their Application in the Agri-Food Industry

CPS has been invented to respond to the necessity of developing a connection between the physical and virtual worlds [5]. Thus, CPS is the transformative technology that manages the interaction between computational capabilities and physical assets [6]. CPS accomplishes its goal by using different sensors, communication devices, and actuators. The application of CPS in agri-food has recently been studied in two topics of smart agriculture and smart farming [7,8]. Precision agriculture is one of the achievements of CPS application in the agriculture industry, with more efficient performance and resource-saving outcomes [9,10]. Precision agriculture is possibly defined as Wireless Underground Sensor Networks, by implementing communication between computers and physical assets with sensors under soil [11] or underground sensor networks to control the quality of soils [11–13].

Furthermore, some papers discuss how CPS application can bring traceability to agri-food systems [14,15]. The application of such systems, in reality, is a challenging task [16–18].

1.1.2. Internet of Things (IoT), and Its Application in the Agri-Food Industry

IoT is considered the main initiator of I 4.0, which became popular in the early 21st century [19]. Physical devices such as equipment, machines, products, etc., are connected virtually at different and remote locations. These items that perform as physical access points are controlled and monitored by cyber systems [20,21]. "Things" are the entities with physical features in a physical property. These "Things" are incorporated flawlessly in a virtual network system that makes IoT, an information system [22].

IoT in the agri-food supply chain helps suppliers and consumers locate products quickly and display product details, leading to choosing fresher products with the help of sensors. IoT helps retailers monitor the food quality and let them waste management of the products that their expire date is close and reduce energy consumption by managing the temperature at the store, freezers, etc. IoT increases traceability, a prerequisite feature for accomplishing previous acts [23].

1.1.3. Internet of Services (IoS), and Its Application in the Agri-Food Industry

IoS might play a key role in the future of industry. Concepts such as software as a service (SaaS), service-oriented architecture (SOA), or business process outsourcing (BPO) are closely associated with the IoS. Barros and Oberle [24] (p. 6) propose a broader definition of the term service, namely "a commercial transaction where one party grants temporary access to the resources of another party in order to perform a prescribed function and a related benefit. Resources may be human workforce and skills, technical systems, information, consumables, land and others". IoS has not been discussed in the agri-food supply chain so far which shows a potential gap for this technology in this area.

1.1.4. Smart Factory and Its Application in the Agri-Food Industry

We have presented CPS, IoT, and IoS so far, which are the main components of I 4.0. The interaction of COS over the IoT and IoS enables a smart factory [3]. Smart factory works in decentralized manufacturing in which "human beings, machines, and resources communicate with each other as naturally as in a social network" [19] (p. 19). Smart factory in the agri-food 4.0 supply chain could be defined as "smart farming" or "smart agriculture".

Smart farming empowers farmers to apply more dependable control. Real-time, on-site processing data reduces time-consuming. Data is transmitted by cloud system for further analysis. IoT devices such as multiple sensors help cover more areas in the remote and expansive coverage areas [25,26]. Data stored within the cloud is also used by processing plants to resolve operational management problems [27].

This study aims to study the I 4.0 challenges in the agri-food supply chain. Therefore, it is necessary to define the agri-food supply chain precisely. Scholars in agricultural economics and management disciplines first proposed agri-food supply chain [28,29]. Agri-food supply chain management, first proposed by a group of Dutch researchers, manages the supply of raw materials for agricultural production, production processing, and product distribution and logistics [30,31]. This term has been mostly used in two research fields agricultural-related disciplines (e.g., agricultural science and agricultural economics), and business management disciplines (e.g., supply chain management and operational research). Based on [32], we consider agri-food supply chain as one of the four terms "agricultural supply chain", "agricultural value chain", "food supply chain", and "food value chain".

The application of I 4.0 in the agri-food supply chain is also known as the agri-food 4.0 supply chain. Furthermore, agri-food sector could be considered in bioeconomy definition [33] which defines an economy based on renewable biological resources.

2. Materials and Methods

In this study, we applied a systematic literature review (SLR) which was proposed by [34] and developed by [35]. This SLR mainly comprises five steps: research questions definition, search strategy design, study selection, quality assessment, and data extraction.

In the first step, we devise some research questions that should be addressed through this SLR. The questions are regarding the objective of this research. Afterward, in the second step, based on research questions, we come up with a search strategy to find the most relevant publications to the research questions. This step contains both sub-categories: finding the search keywords and determining the literature databases. In the third step, study selection criteria are formed to determine the narrow the most relevant study with respect to addressing the research questions. In the next step, we apply a quality assessment in which we set up some quality checklists to speed up the assessment process. The final data is gathered to answer the research questions in the last step, data extraction.

2.1. Research Questions

This SLR aims to recognize challenges ahead of industry 4.0 application in the agri-food supply chain. Towards this aim, six following research questions have been formed by the authors:

- RQ1: What classifications of agri-food products have been discussed with the emergence of industry 4.0? RQ1 aims to identify the agri-food products that have used the industry 4.0 context. By answering this question, scholars have a better understanding of potential research in the agri-food industry, and it demonstrates which products have adapted industry 4.0 technologies compared to others.

- RQ2: Among industry 4.0 technologies, which one has gained more attention in the agri-food supply chain considering product classification? (Which technology in which agri-food products). We defined four key technologies above: IoT, CPS, IOS, and smart factory for agri-food supply chain. It is essential to realize if there are only mentioned technologies in the agri-food supply chain or other technologies contribute to the supply chain.
- RQ3: What percentage of the literature addressed sustainability in the agri-food 4.0 supply chain? (Based on three aspects of sustainability). The all-new types of supply chains try to address sustainability in a specific way, and the new technologies facilitate this process with their unique features. This research question aims to find out how many of the selected articles addressed sustainability.
- RQ4: How does Industry 4.0 contribute to a sustainable agri-food supply chain? This research question aims to explore how industry 4.0 addresses sustainability in the supply chain. It focuses on the sub-classification of sustainability in the agri-food supply chain.
- RQ5: What challenges are ahead of applying industry 4.0 (I 4.0 adoption) in the agri-food supply chain? This research question aims to find the challenges of applying industry 4.0 in the agri-food supply chain. Practitioners need to know the challenges in advance to contemplate solutions.
- RQ 6: What are the main discussed themes in the agri-food 4.0 supply chain? Based on the answer to the previous question, this research question focuses on classifying challenges. We display a better perspective of challenges in an organized category with the answer.

I 4.0, with disruptive technologies and interconnected machinery, aims to improve production efficiency, which helps suppliers serve better to their customers. The proposed questions attempt to find out the obstacle against I 4.0 as well as how these technologies address sustainability within agri-food supply chain. In this regard, the first question classifies the agri-food products to find which agri-food products have taken advantage of I 4.0 technologies so far. The second question attempts to find all technologies used in agri-food supply chain and shows which technology has been used in what types of agri-food products to demonstrate the research in this context. The third and fourth questions address the sustainability and aim to find in which manner I 4.0 influences on sustainability of agri-food supply chain. Finally, the last two questions address the challenges and barriers of I 4.0 in the agri-food supply chain and picture different themes which will help practitioners and scholars to have a general picture of challenges from different perspectives.

2.2. Search Strategy

The search strategy includes three sub-classifications: search keywords and literature databases, which are explained in detail as follows:

2.2.1. Search Keywords

The following steps were done to find the search keywords [35]:

- Derive major keywords from the research questions;
- Identify alternative spellings and synonyms for principal keywords;
- Check the keywords in the relevant articles or publications;
- Use the Boolean OR to incorporate alternative spellings and synonyms;
- Use the Boolean ASTERISK to replace multiple characters to find the terms that have different appearances with the same meaning;
- Use the Boolean AND to connect the significant keywords;
- Find relevant references for defining the main scope of the research.

We added the last step as a new important step to address the scope of the research. In the scope of this SLR, industry 4.0 is a widely used concept that has several distinct technologies that contribute to I 4.0. As mentioned in the Introduction Section, it is necessary to define I 4.0 precisely; otherwise, the research cannot address the research questions properly because there would be many out-of-scope references in the result. To tackle this problem based on [3,4], I 4.0 has four main components, including CPS, IOT, IOS, and smart factory. However, after reading the articles by the authors, we concluded that the terms "smart farming" and "smart agriculture" are used interchangeably in addition to smart factory. Consequently, we added these two terms to the search keywords step. In the search keywords, in order to search for any group of characters before or after a term we add an asterisk (*). This function is available in both WOS and Scopus.

The search keywords result are shown as follows:

(*agri* OR "*food*" OR "*agro") AND ("internet of things" OR iot OR cps OR "cyber physical system*" OR ios OR "internet of services" OR "smart factory" OR "smart farm*" OR "smart agriculture ") AND ("supply chain*" OR " logistic*") AND ("industry 4*" OR "I 4.0" OR "agri-food 4" OR "agriculture 4" OR "the fourth industrial revolution" OR "smart manufacturing" OR "industrial internet" OR "integrated industry").

2.2.2. Literature Databases

Two well-known literature databases comprising the Web of Science (WOS) and SCOPUS were used for this SLR. Clearly, the two mentioned search engines are the most famous and completed search engines. This research aims to answer research questions considering the application of the fourth industrial revolution in the agri-food supply chain. Thus, the search keywords were used to search for peer-reviewed publications. The search keywords covered article title, abstract, and keywords provided by the author and publication for both WOS and SCOPUS. As the agri-food 4.0 supply chain concept and its applications are novel, we have not applied any time restrictions in our research.

2.3. Study Selection

Search phase 1 resulted in 101 peer-reviewed publications (see Figure 1). In both WOS and SCOPUS there is an option that lets the authors filter articles based on their document type, which is a part of the exclusion criteria. This SLR only considers journal articles, conference papers, and book chapters. After applying exclusion criteria and based on the document type criterion, 67 publications remained. We derived all references through the next step and eliminated duplicated publications by Mendeley software. Twelve duplicated articles were deleted through this step. These 55 remaining peer-reviewed publications went through the inclusion criteria, and only 24 articles remained in the pre-final step. In the last step, we applied a quality assessment. Although all publications received different scores, any publications have not been removed through this step and passed this assessment. The authors found out that the number the publications were few and this could be an explanation that is why all of them remained. The other description is that we define the scope of the research accurately, which leads us to a few numbers of articles. In other words, both preciseness of the scope of this research and the novelty of the topic concluded in a few publications but entirely related ones.

The defined inclusion and exclusion criteria are as follows:

Inclusion criteria:

- Only the studies that addressed the I 4.0 technologies based on the scope of this research will be included;
- Only the studies that addressed the agri-food supply chain will remain in this SLR;
- For the research that has both journal version and conference version, only the journal version will be included;
- For duplicated publications of the same study, only the newest and the complete one will be included.

Exclusion criteria:

- Duplicates are eliminated by Mendeley and a final revision by the authors;

- In the "Document type", we applied a filter by excluding article reviews, conference reviews, editorials, and short surveys in this research;
- Check the keywords in the relevant articles or publications; The authors omitted other languages such as Germany, Chinese, and Russian.

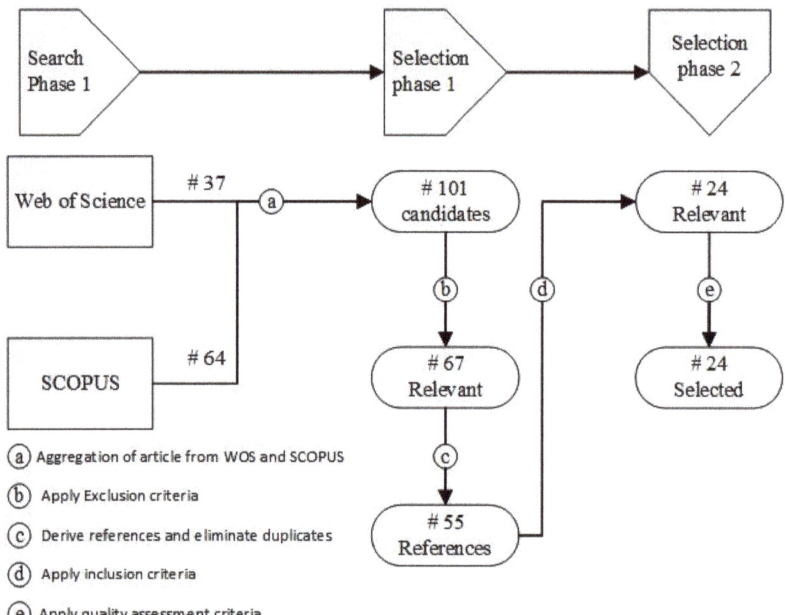

Figure 1. Search and selection process.

As we mentioned earlier, this SLR has two main scopes to find the industry 4.0 challenges in the agri-food supply chain. The first scope is about the I 4.0 that should discuss at least one of the four main components, IOT, CPS, IOS, and smart factory [3]. The second scope of this SLR is the agri-food supply chain, which we defined in the introduction. The included publications in this research should discuss at least one of the four supply chain as follow: "agricultural supply chain", "agricultural value chain", "food supply chain", and "food value chain". Finally, there is no time restriction regarding the novelty of the scope of this research. All publications have been published since 2015 based on our search keywords.

2.4. Study Quality Assessment

The quality assessment is a process in which we weigh the retrieved quantitative data in meta-analysis [35]. Since the results are too few, a meta-analysis is unsuitable for this SLR. Instead, we only use the result of quality assessment. After applying quality assessment, there would not be any changes in the number of outcomes. There is a reason to justify why there is no reduction in the selected articles after using quality assessment. The number of publications is too few, which shows how accurate and in detail the scope of this SLR is defined.

We devised five main questions and 17 criteria to assess the quality of selected articles which are shown in Table 1. Some of the questions were derived from [36]. Question 1 to question 4 are quantitative that has two answers: "Yes" or "No" which answers are scored as follow: "Yes = 1", and "No = 0". The other 17 criteria are scored in the same manner with answers "Yes = 1", and "No = 0". Question 5 is qualitative which is scored as follow: Excellent quality = 1, good quality = 0.67, Fair quality = 0.33, Poor quality = 0. To calculate the final score for a publication, each question has a weight, which is multiplied by the determined score for that question or criteria, and finally, we sum the scores. The weights are defined by authors based on their relevance to the scope of this research which are classified into two sub-categories; five main questions and 17 criteria. As the questions have higher importance, they have bigger weights than the criteria. These five questions aim to assess how close the selected articles are to the agri-food 4.0 supply chain area based on the above definition of the agri-food 4.0 supply chain.

Table 1. Quality assessment questions and criteria.

No.	Criteria or Question	Weight
QA1	Are the aims of the research clearly defined?	7
QA2	Is industry 4.0 adequately described?	10
QA3	Is the agri-food supply chain sufficiently defined?	10
QA4	Does the research address sustainability?	10
QA5	How well does the evaluation address its original aims and purpose?	8
C1	Less than 15 words	1
C2	Keyword in title	1
C3	Present a logical structure in the Abstract	2
C4	The introduction has a high-quality context	2
C5	The introduction mentions the Hypothesis	5
C6	The problem is defined in the Introduction	5
C7	State of the Art is in a logical order	5
C8	Has an appropriate Content of theoretical framework	4
C9	The methodology is explained in detail	5
C10	Data in the Results is available	3
C11	Results are consistent with the objectives	3
C12	Present complementary graphs for the text information	2
C13	Findings are discussed in relation to objectives	5
C14	Results are compared with the state of the art	3
C15	The conclusions correspond to the stated objective(s)	4
C16	Present future research	3
C17	References match	2

Moreover, 17 criteria assess how much the selected publications' general quality is close to scientific research, from article title to references. If a paper obtains 50% of the total score or more, it will be included in the SLR. In this study, all 24 articles received the minimum requirement. The criteria and the final score for each article are shown in Appendix A Table A1.

2.5. Data Extraction

After obtaining the target articles for answering the research questions, the articles were gathered into a separate excel sheet to find the challenges against the agri-food 4.0 supply chain.

3. Results and Discussion

This section includes four subsections that present our findings through this SLR. First, we explain the 24 selected articles in terms of publication year, and document type. Afterward, in the following parts, we answer the research questions. The last sub-section contains the answer to RQ5 and RQ6.

3.1. Overview of Selected Articles

We found 24 peer-reviewed publications that meet our SLR requirements. The first study was published in 2016. The distribution of the publications based on publication year is shown in Figure 2. There is enormous progress in the number of Publications between 2020 and 2021, which shows this area has gained more popularity since 2019. Thus, this Table affirms the novelty and importance of this new research area.

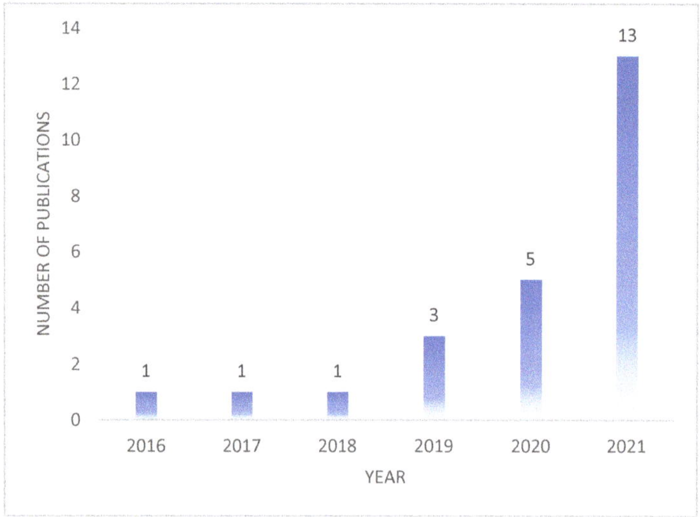

Figure 2. Distribution of publications based on year.

Regarding document types, Figure 3. displays the number of Publications according to their document type. The blue color shows the number of Articles which is equal to 18, and conference papers are displayed by orange color, which is equal to 5, and there is only one book chapter among this SLR.

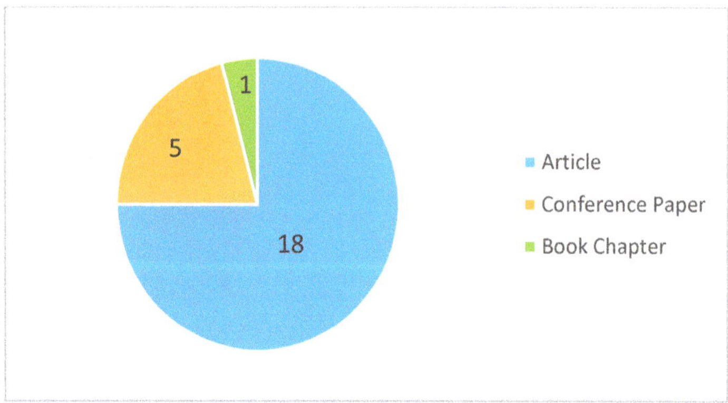

Figure 3. The number of articles based on the document type.

The quality assessment result is shown in Table 2. The quality score was calculated in scale 1 and all studies obtained more than 50% and were selected for final SLR. This Table demonstrates that more than 66% of studies have a high or very-high quality level.

Table 2. Quality assessment score result.

Quality Level	Of Studies	Percent
Very high (0.85 ≤ score ≤ 1)	3	0.125
high (0.7 ≤ score < 0.85)	13	0.54
Medium (0.5 ≤ score < 0.7)	8	0.33
Low (0 ≤ score < 0.5)	0	0
Total	24	100

The distribution of articles in journals is broad, and only Computers in Industry journal has two published articles among all selected studies.

3.2. Types of Agri-Food Products (RQ1)

This SLR aims to classify agri-food products. In the research [37], a general agri-food product classification was proposed by the authors, including (1) bulk cereals, (2) root vegetables and tubers, (3) sugar and sweeteners, (4) meat, (5) dairy products, (6) fruit, (7) vegetable oils, (8) other. We classified the 24 selected articles into five different categories in terms of agri-food products considering the classification in [37] which are shown in Figure 4. These five classifications include (1) fresh fruits or vegetables, (2) cold chain, (3) packaging, (4) food, and (5) agriculture. In [37], fruits and vegetables are two distinct classifications, however, we consider them in one category since the literature addresses their freshness as well as considering both of them in a single article. Thus, we consider fresh fruits and vegetables one class. Seven Publications refer to fresh fruits or vegetables that seem the most important or the most discussed subject in this area, and it is due to the perishability of these products. The second classification, cold chain, includes fisheries industry meats, or any other material that ship by cold supply chain and refrigerators, in which there are five articles. Packaging as a third classification is a new class we add to [37] because the only two articles that refer to packaging and discuss the importance of packaging in the agri-food supply chain in a general perspective. Finally, we proposed two general categories. Ten remained articles discuss food and agriculture products, respectively. These two terms, "food" and "agriculture" are general terms that do not refer to a specific product. We selected these titles because these two categories contain research papers more associated with general frameworks and review papers. Although these publications discuss the agri-food supply chain they do not concentrate on any specific products. This sub-section and Figure 4 answer the first research question.

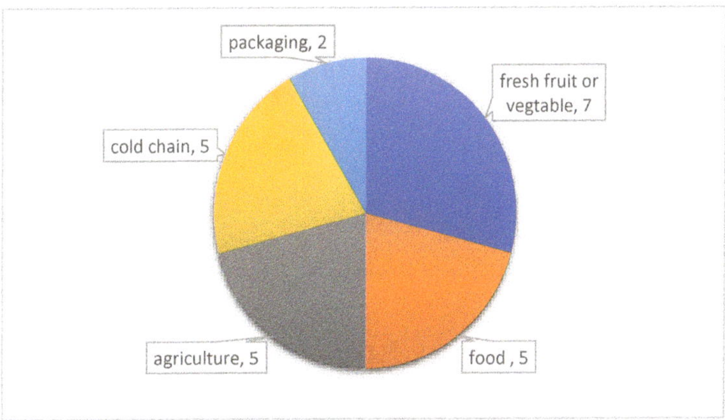

Figure 4. Classification of publications based on agri-food products based on the number of studies.

3.3. Types of Technologies (RQ2)

The answer to research question 2 is shown in Figure 5, which demonstrates the distribution of industrial 4.0 technologies based on agri-food products. As we defined before, industry 4.0 has four main Keys, including IoT CPS, IOS, and smart factory.

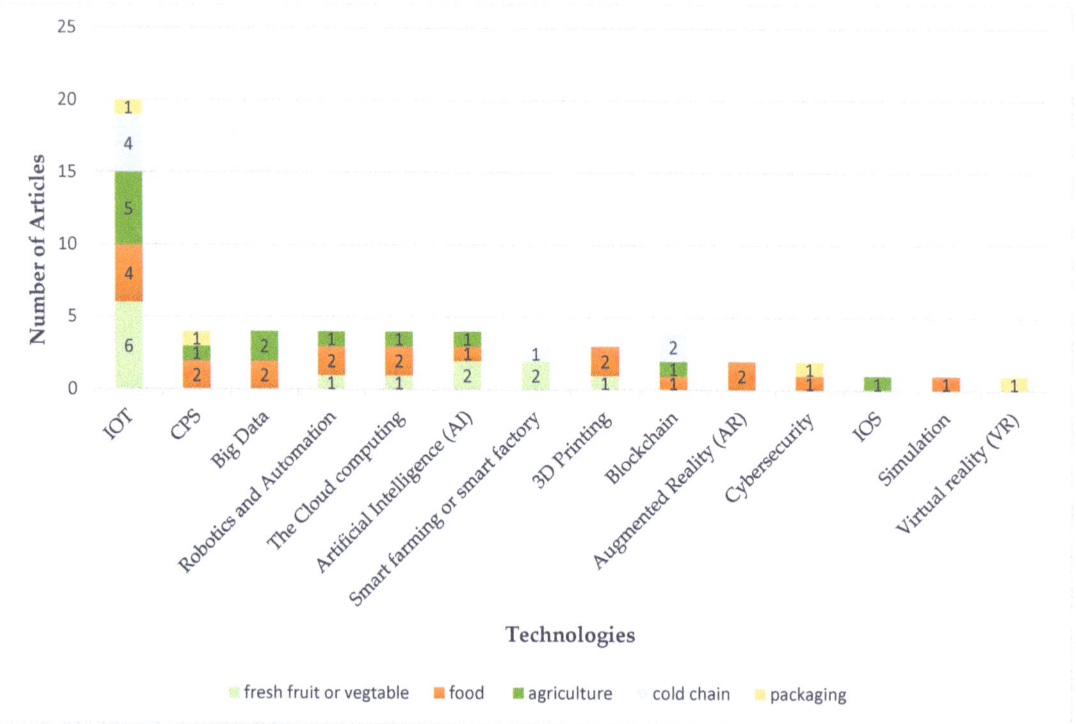

Figure 5. Industry 4.0 technology distribution based on agri-food products.

The authors read all articles through the selected publications and found out that some other technologies play roles in the agri-food 4.0 supply chain. IoT has attracted more attention compared to other technologies. Other technologies that the authors found are big data, robotics, and automation, cloud computing, artificial intelligence (AI), 3D printing, blockchain, augmented reality (AR), cybersecurity, simulation, and virtual reality (VR). Some articles addressed more than one technology in their research; for this reason, the number of technologies is more than the total number of selected papers. Our findings in this section cover nine technologies that were used to systemically review circular supply chains in and application of I 4.0 in the circular economy [38].

3.4. Sustainability Area (RQ3)

This section presents our findings in terms of sustainability. Only nine articles from 24 selected articles addressed sustainability which shows a gap regarding sustainability in this field. The distribution of sustainability is illustrated in Figure 6. Of these nine articles, only one addressed the social aspect, and all of them addressed the environmental aspect of sustainability.

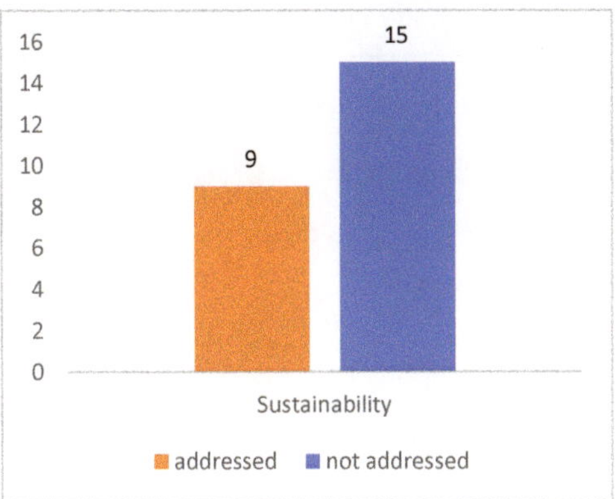

Figure 6. Distribution of sustainability in the literature.

3.5. The Contribution of I 4.0 in a Sustainable Agri-Food Supply Chain (RQ4)

This section aims to address research question 4; how I 4.0 contributes to the sustainable agri-food supply chain. When it comes to the environmental aspect of the agri-food supply chain, waste management attracts more attention due to the perishability of products in this value chain. Then, water management has the second step. Finally, only one paper addressed energy consumption management. The information is shown in Figure 7.

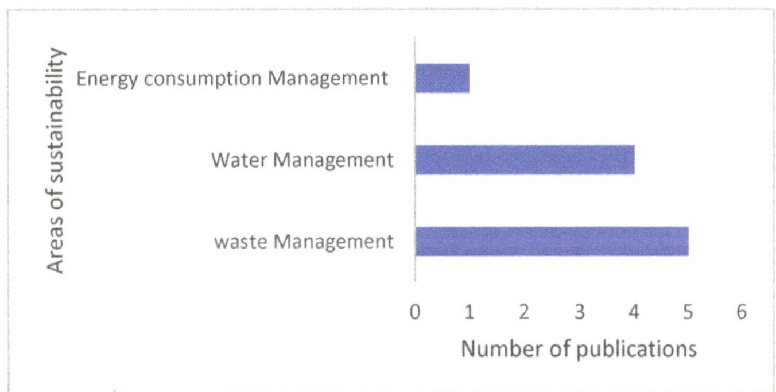

Figure 7. Areas of sustainability.

3.6. Challenges and Themes in the Agri-Food 4.0 Supply Chain (RQ5 and RQ6)

This study aims to find challenges in the agri-food 4.0 supply chain. To find the challenges, first, we read all the articles and found four papers that addressed challenges directly for different technologies but not specifically the industry 4.0 concept [13,16]. Then we reread the publications to see which article addresses which challenge. Among the four papers that addressed challenges directly, three [13–16] addressed challenges regarding IoT adoption in the agri-food supply chain. In [39], the authors studied challenges regarding different technologies, including IoT, Robotics and autonomous systems, AI, big data analytics, and blockchain. There are both common challenges and unique challenges mentioned in these four articles. We reread the challenges and tried to integrate the challenges that were mentioned with different names while they had the same characteristics. Finally, we reached 15 challenges and five themes. The themes and challenges are illustrated in Table 3.

Table 3. Challenges and themes in the agri-food 4.0 supply chain.

Themes	Challenges	References	Number of Observations
technical	Security and privacy	[20,23,39–46]	10
	Wireless power transfer and ambient energy harvesting	[39]	1
	Big data management	[20,47–50]	5
	Reliability, availability, and robustness	[20,46]	2
	Developing IoT-based cloud system	[40,43]	3
	Technological architecture	[20,27,42,45,49–57]	13
infrastructural	IoT-based infrastructure (Internet)	[39,40,44,45,48]	5
	lack of governmental regulations	[23,48]	2
	Standardization	[20,23,44]	3
operational	High energy consumption	[39,45,47]	3
	Scalability	[23,39,40,44]	4
	Interoperability	[20,39]	2
	Congestion and overload issues of IoT	[40]	1
financial	High implementation and operating costs	[23,41,58]	3
social	Lack of human skills and educational issues	[39,45,48,58]	4

As the number of challenges in this area is extensive, we aim to classify challenges to make them more sensible for scholars and future research. In [55], the article's authors classified IoT adoption challenges into five main themes, including technical, financial, operational, social, educational, and governmental, and fourteen challenges. We kept the four technical, financial, operational, and social themes, and we added a new classification called infrastructural themes. This study considers the educational theme as a subsection of the social theme because the terms education and society have been widely used close to each other. Moreover, the related governmental challenges are a sub-category of an infrastructural problem. We explain each of them separately below.

3.6.1. Technical Theme

After reading, analyzing, and interpreting the selected article, we concluded seven challenges that we consider technical challenges. Each challenge is described separately as below:

- Security and Privacy

Security and privacy have been widely proposed as a significant ongoing issue regarding I 4.0 technologies [43,44], more specifically, security issues associated with IoT technologies [20,23,40]. In industry 4.0, a huge amount of data is generated and distributed through a wireless sensor. Four extensive areas, including data authentication, resistance to attacks, client privacy, and access control regarding security and privacy, must be addressed [20,23]. Network and some parts of this data are about users and their private information, interaction, etc.

In other papers, security is considered cybersecurity [41,45,46] which refers to the security related to cyberspace and the link among physical assets and objects. According to the assertion in [41], insufficient cybersecurity awareness and a lack of educated and skilled people in the cybersecurity area can diminish security and raise risks. Due to an enormous number of stakeholders in the agri-food supply chain, such as farmers, manufacturers, wholesalers, retailers, and consumers, the supply chain is vulnerable to cyberattacks [46].

Blockchain, a fast-growing technology, is proposed as an infrastructure to share data in the agri-food 4.0 supply chain [39,42]. By combining IoT and blockchain, Grecuccio et al. [42] proposed a framework to reduce security in the food supply chain and increase traceability. Although blockchain, like other I 4.0 technologies, is vulnerable to cyber-attack, three key characteristics, including data storage, decentralized network, and peer-to-peer communication, intensify security and privacy considerably [39].

- Wireless power transfer and ambient energy harvesting

This challenge is only presented in [39], which seems a significant issue for applying I 4.0 appliances in the agri-food supply chain. Liu et al. in [39] propose a challenge for recharging small sensors necessary in farms such as underground, underwater, livestock, and trees. Replacing the batteries in these appliances seems unachievable. Wireless power transfer is a good solution by recharging the batteries through electromagnetic waves. However, underground, underwater, and long-distance wireless networks are issues ahead of this solution. Likewise, ambient energy harvesting is another potential solution for this challenge. Some research papers studied and showed that harvesting energy from rivers, movement of vehicles and fluid flow, and the ground surface is possible [39]. Nevertheless, as the converted electrical energy is limited, power conversion efficiency should be further improved.

- Big data management:

Big data has been at the center of attention in the SCM and many studies have addressed it recently. For example, in [59], the authors asserted that big data-driven SCM could facilitate the barriers. Moreover, big data management extensively used with I 4.0 in several contexts. In agri-food supply chain 4.0, it has been mentioned as big data, data management, and big data analytics [47–49]. The devices, appliances, and sensors generate numerous big data that need to be collected, stored, processed, and distributed through different networks. This abundance of data should be managed [20]. Some of the challenges in big data management are data storage, searching, sharing, analyzing, and data visualization. Since the amount of data rises daily, the mentioned challenges need high-end hardware and software and continuous upgrade of the system. The other challenge with respect to big data management is data insecurity [46].

- Reliability, availability, and robustness:

Reliability and availability of IoT devices and services are potential issues in the agri-food supply chain. There might be various failures in the system, such as hardware failure, software issues, malicious attacks, and limited energy. Consequently, the robustness and reliability of I 4.0 services are a significant challenge [20]. Both device failure and link failure, which lead to communication failure, can cause financial risk. In research by [20], authors propose a three-layer architecture for IoT-driven agriculture in which reliability is one of its main features.

- Developing IoT-based cloud system:

A cloud system is used in the I 4.0 context as an intermediary space to link devices and appliances through its network. Yadav and Garg, in their article [40], consider cloud-based development as a challenge ahead of IoT technologies. However, another research [40] considers cloud-based systems pricey and proposes a blockchain system and direct communication of IoT devices through blockchain.

- Technological architecture:

Khan and Altayar [20] propose technological architecture as an obstacle ahead of IoT adoption. They assert that it is necessary to develop, maintain and integrate a robust technological architecture that comprises all IoT-related technologies, such as cloud computing, artificial intelligence, blockchain, wireless technologies, machine learning, big data analytics, and data center and server technologies [20]. Thus, an open technological architecture ensures the integration of different technologies, scalability, mobility, interoperability, modularity, and openness in a heterogeneous environment. This study proves that technological architecture is the most important challenge based on our findings, because it has attained more attention in the literature than the other challenges in which 50 percent of the selected publications addressed this challenge.

A sustainable food supply chain was designed for IoT-enabled e-commerce food enterprises by applying a lateral inventory share policy [56]. Another study by Almadani and Mostafa [51] presents a systematic integration model of a multi-vendor agricultural production system that uses data distribution service middleware to facilitate communication among production systems.

The cocoa bean traditional assessment method is carried out by humans and takes a lot of time. Adhitya et al. [27] used artificial intelligence classifier methods for textural feature extraction to decrease the assessment time process and cocoa bean waste. Reducing food waste is a contribution to a sustainable supply chain. Jagtap et al. [52] designed a framework to diminish food waste as well as energy and water based on IoT-based devices. Other research [49], studied how cyber-physical systems can contribute to sustainable food systems by using machine learning techniques. Likewise, Mondragon et al. [50], proposed a two-layer conceptual approach in the fishery industry by using IoT devices and digitalization. Furthermore, a software framework that lets IoT devices communicate through blockchain was proposed by [42]. In [48], the author designed a module that uses AI-based IoT devices to ripen the fruits when they are shipped by in the container.

Moreover, Sharma et al. [54] proposed a framework with CPS concentration to improve productivity in the agricultural supply chain. In addition, a three-layer architecture proposed by [42] that is energy efficient, cost effective, heterogeneous, secure, and reliable to use IoT in agriculture based on cloud computing. Finally, the paper by [55] designed a model based on I 4.0 technologies to collect food traceability data through the supply chain and transform it to the consumer to improve the quality of products.

3.6.2. Infrastructural Theme

This theme presents three different infrastructural challenges. These challenges are IoT-based infrastructure, lack of governmental regulations, and standardization. Each challenge is discussed as follows:

- IoT-based infrastructure:

Infrastructure for IoT-based devices and services is another challenge in the agri-food 4.0 supply chain. Accessibility to the internet, the primary service of Industry 4.0, is low in the agri-food sector. In [40], the authors studied the lack of internet accessibility in the Indian agriculture supply chain. The availability and collaboration of a diverse range of services such as AI, CPS, IoT, IoS, blockchain, cloud systems, and so on need a high-quality infrastructure in the supply chain [48]. When it comes to infrastructure, implementation issues arise related to hardware installation challenges and changes in the processes [44]. Likewise, installing these infrastructures needs a high cost [45].

A robust wireless network is necessary to apply I 4.0 technologies in the agri-food supply chain. However, it is an issue because of several causes that negatively affect the wireless network. For example, temperature variations, humidity, human presence, and movements of animals lead to signal fluctuations [39]. Hence, a robust wireless network is vital to coping with weather conditions and the agricultural environment.

- Lack of governmental regulations:

The government plays a key role in the agri-food 4.0 supply chain. Poor regulations in IoT applications may create food safety issues and decrease traceability [60]. In [48], Yadav et al. demonstrated that the collaboration of government organizations, NGOS, and food processing organizations could enable I 4.0 in the agri-food supply chain. Laws and regulations should support the development and extension of I 4.0 in supply chain management, especially in the agri-food sector, which has important consequences on social health and poverty [23].

- Standardization:

Smart agriculture and global communication need standard protocols to prevent ambiguity and facilitate efficient and smooth integration among different vendors and data safety through cloud networks [23]. Standardization helps devices and digital appliances interact efficiently. This challenge becomes more important when it comes to the global supply chain between various continents with different regulations, responsibilities, and standards [44]. The complexity of I 4.0 technologies even makes it more difficult to define a unified standard for the interaction of sensors, software, devices, actuators, and networks with their own predefined protocols [20]. Lack of standardization has always been a challenge for most new technologies [23].

3.6.3. Operational Theme

This theme attempts to address the barriers regarding the operational aspect of I 4.0 technologies in the agri-food supply chain. The main challenges are high energy consumption, scalability, interoperability, proper connection of ASC entities and IoT technology, and IoT congestion and overload issues. The issues are explained as follows:

- High energy consumption:

Blockchain has gained more attention for this challenge. Due to the high energy consumption of blockchain technology, especially by coal and fossil-based energy, it is a thought-provoking barrier [39]. High energy consumption leads to a high cost of energy that makes it a barrier in the agri-food supply chain. To cope with this issue, scholars in [58] proposed adopting renewable energy to reduce costs in the long-term and meet sustainable development goals for reaching a green planet and mitigating climate change.

- Scalability:

Agri-food 4.0, with growing technologies such as IoT, blockchain, etc., includes an incredible size of devices and nodes that need to connect in the future. Thus, scalability has been mentioned as an ongoing challenge that should be addressed in agri-food 4.0. Some papers presented the scalability issue in IoT [23,40]. The middleware approach is proposed by [40] to provide a flexible service with a huge number of devices that can communicate with each other at one position,. The other research by [44] discusses the importance of servers' scalability in IoT. Scalability in blockchain is presented as an issue in [39] since the transaction speed in blockchain networks such as Bitcoin and Ethereum are so low compared to visa transactions.

- Interoperability:

The nature of I 4.0 works with devices that produce data, and they need to communicate with each other through diverse networks. When it comes to the agri-food supply chain, it contains several layers, including thousands of devices and communication. Therefore, I 4.0 technologies should have the capability to communicate and exchange data between devices. This capability is called interoperability, an issue in the agri-food 4.0 supply chain. The information exchange needs an interoperable environment. Interoperability generally has four types; technical, semantic, syntactic, and organizational [20]. This challenge has been studied for two technologies: IoT and blockchain [20,39]. Different sorts of blockchain networks can hardly communicate with each other, and there is a necessity for interoperable communication protocols [39].

- Congestion and overload issues of IoT:

This issue is similar to the operational challenge proposed by [61] to study collaboration in an industrial symbiosis network in the circular economy. The complexity of the supply chain and collaboration among its players was previously introduced as an issue. Given the complexity of the SCM network in which each layer in SCM uses several devices as well as fast-growing technologies and the existence of big data in the network system, facing congestion and overload of IoT devices is an inevitable challenge in the agri-food 4.0

supply chain. The congestion occurs when multiple devices produce information and want to load their data through the network [40]. It is necessary to find a solution for this issue in the future as an adoption barrier.

3.6.4. Financial Theme

The financial theme has only one challenge regarding general financial concerns in terms of novel technologies as follow:

- High implementation and operating costs:

Implementing I 4.0 requires new technologies and always emerging technologies have been pricy. Furthermore, the maintenance cost will be a challenge for organizations [23]. In a study by [41], active packaging is suggested as a new solution for sustainability in the post-harvest food supply chain. Although active packaging increases the quality, safety, and shelf life of packaged food, the high-cost implementation is a barrier to adopting this beneficial tech. On the other hand, the other paper proposes that although the implementation of I 4.0 technologies and digitalization is expensive, in the long term, these technologies are beneficial [58].

3.6.5. Social Theme

We name this theme social because it is about social challenges pertaining to humans. The issues show the lack of human skills and educational issues, which is explained as follow:

- Lack of human skills and educational issues:

It is essential to address the issues related to farmers and I 4.0. Farmers should be aware of the benefits of industry 4.0 context to participate in the supply chain. However, there is a lack of human skills in the agri-food supply chain as most data analysts and data scientists are not at agricultural universities or agri-food-related companies [39]. Likewise, universities have not yet prepared industry 4.0 courses or programs to provide sufficient human skills in this field. Furthermore, the lack of skilled labor is apparent, and it is essential to speed up preparing highly skilled experts and laborers in the agri-food supply chain [58]. In research by [48], education and training were found as the fifth most important enabler of the agri-food 4.0 supply chain among 14 enablers, proving the significance of this challenge.

4. Discussion and Conclusions

This systematic literature review aimed to address challenges regarding industry 4.0 application in the agri-food supply chain. In the first step, we defined industry 4.0 and its main technology components, including IoT, IOS, CPS, and smart factory. Then we described the agri-food supply chain containing four types of supply chain; "agriculture supply chain", "agriculture value chain", "food supply chain", and "food value chain". Afterward, we devised six research questions. We defined the search terms based on our definition of industry 4.0 and agri-food supply chain. In the final step of the systematic review, we applied a quality assessment. The result of SLR was 24 selected Publications. Although a few articles were selected in the final process (24), the quality assessment showed the selected publications are quite well matched to the scope of this study, which could be due to the novelty of the field and the preciseness of the scope of this SLR. We contribute to the literate by applying a systematic literature review into industry 4.0 application in the agri-food supply chain. A systematic review that concentrates on I 4.0 is scarce. Similar publications mainly discuss specific technologies such as IoT, CPS, or blockchain, and not the I 4.0 as a general concept. Hence, the findings in this study are worth reading, and it helps both scholars and practitioners in the agri-food supply chain.

We found 15 challenges regarding the agri-food 4.0 supply chain that each one is defined separately. Then, we classified them into five main themes. The principal findings of this review are summarized as follows:

- (RQ1) There are five categories in terms of agri-food products in the agri-food 4.0 supply chain. Three specific products and two general ones. The particular products contain fresh fruits and vegetables, cold chain, and packaging. On the other hand, in some studies, almost 40% of this SLR generally addresses food and agricultural products due to their concentration on frameworks or reviews. Since there have not been similar articles that classify products, our findings contribute to the agri-food supply chain by demonstrating the importance of the I 4.0 application for perishable products such as fresh fruits, vegetables, and cold chains.
- (RQ2) This research question aimed to find the most applicable technology in this field: IoT with its wide application. IoT has gotten considerable attention compared to other technologies. In this section, we found some other technologies in addition to the four leading mentioned ones; IoT, CPS, IoS, and smart agriculture. Other technologies are big data, robotics and automation, cloud computing, AI, 3D printing, blockchain, augmented reality (AR), cybersecurity, simulation, and VR. These technologies also play a role in the agri-food 4.0 supply chain. The results of this question are shown in Figure 5, another contribution of this study. We displayed which technology has been used on which type of products. For example, IoT and its application in fresh fruits and vegetables, agriculture, and food products have the highest number of articles, respectively.
- (RQ3) The answer to this research question identified a vast gap in sustainability in the agri-food 4.0 supply chain. When new technologies come up, sustainability should be considered. Nevertheless, some new technologies, such as a single AI, not only do not reduce carbon emissions but also can emit carbon as much as five cars through its lifetime [62]. The selected articles that addressed sustainability were 37.5% of selected studies. In this era, the importance of sustainability is inevitable in various contexts from climate change to the social terms. This gap is an opportunity for future research.
- (RQ4) The answer to this research question determined that waste management has gotten more attention due to the perishability of products in the agri-food supply chain. Then water resource management is another area that industry 4.0 contributes to the sustainable agri-food supply chain. Climate change has caused new challenges in the agri-food supply chain, such as water shortages and droughts. Scholars in this area should address how new technologies in I 4.0 would affect sustainability from different aspects. There is a necessity for the hard work of scholars and researchers to analyze I 4.0 impacts on sustainable development goals.
- (RQ5) In answer to this question, we found 15 challenges, including security and privacy, wireless power transfer and ambient energy harvesting, big data management, reliability, availability, and robustness, developing IoT-based cloud system, technological architecture, IoT-based infrastructure (Internet), lack of governmental regulations, standardization, high energy consumption, scalability, interoperability, congestion and overload issues of IoT, high implementing and operating costs, and lack of human skills and educational issues. The challenges are shown in Table A1 in which a column represents the number of articles discussing each challenge. With these numbers, we found the most notable barriers ahead of I 4.0 in the agri-food supply chains. Here we name the four most discussed challenges: technological architecture, security and privacy, big data management, and IoT-based infrastructure, respectively. Designing a professional framework or model in which technologies perform properly is the biggest challenge in the selected articles. Before doing any action, it is essential to have a map or an architecture that demonstrates how I 4.0 technologies should function. The second most discussed issue is security and privacy, which is a barrier for enterprises, farmers, and all layers of the agri-food supply chain to trust the new technologies. Security and privacy have always been a striking challenge for new technologies. Lastly, big data management and IoT-based infrastructure have the third important position in the list of challenges. It is vital to address this issue because data is the main component of this supply chain that facilitates procedures and speeds up functions.

Therefore, managing an immense of data is recognized as an issue. Furthermore, all efforts towards I 4.0 are useless without proper infrastructure like the internet.

- (RQ6) Finally, after reading the challenges, we classified them into five main classifications: technical, operational, social, infrastructural, and financial.

The findings of this study showed a considerable gap in the sustainability of agri-food 4.0 supply chain. Meanwhile, most of articles addressed I 4.0 application in the agri-food supply chain without addressing sustainability which draws attention and shows a considerable gap for future research in the agri-food 4.0 supply chain. On the other hand, among articles that addressed sustainability, waste management, and water management were found two sub-sections of sustainability that are more discussed and need to be further addressed. The two issues of water and waste management are the most notable challenges in the sustainable agri-food supply chain. More research is needed to study how I 4.0 is able to address water and waste management. Furthermore, I 4.0 is still emerging with the development of disruptive technologies and is in its early decade.

Moreover, based on agri-food product classification in this study we suggest further research on challenges regarding each product's category because first, this study considered challenges overall on agri-food supply chain and second, the character of each category could make define other challenges or more specific sub-challenges.

Furthermore, this study pictures a comprehensive classification of challenge themes which help practitioners and scholars to narrow the concentration on addressing the problem and to have a better understanding of the whole challenges in one frame. The presented themes outline the main barriers that need to be addressed by the agri-food industry in adopting industry 4.0 into their supply chain.

Funding: This research received no external funding.

Institutional Review Board Statement: Not applicable.

Data Availability Statement: No new data were created and Table A1 shows the selected references for this SLR.

Conflicts of Interest: The authors declare no conflict of interest.

Appendix A

Table A1. Quality assessment complete result of the selected studies.

ID	QA1	QA2	QA3	QA4	QA5	C1	C2	C3	C4	C5	C6	C7	C8	C9	C10	C11	C12	C13	C14	C15	C16	C17	SCORE	Ref
1	1	0	1	1	0.67	1	1	1	1	1	1	1	1	1	1	1	1	0	0	1	1	1	79.36	[56]
2	1	1	1	1	1	1	1	1	1	1	1	1	1	1	0	1	1	1	1	1	1	1	97	[51]
3	1	0	0	0	0.67	1	1	1	1	1	1	0	1	0	1	1	1	1	1	1	0	1	54.36	[27]
4	1	1	1	0	0.67	1	1	0	1	1	1	1	1	1	1	1	1	1	0	1	0	1	79.36	[47]
5	1	0	0	1	0.67	1	1	1	1	1	1	0	1	1	1	1	1	1	0	1	0	1	66.36	[52]
6	1	0	1	0	0.67	1	1	1	1	0	1	1	1	1	1	1	1	1	0	1	0	1	66.36	[23]
7	1	1	1	0	0.67	1	1	1	1	1	1	1	1	0	0	1	0	1	1	1	0	1	74.36	[46]
8	1	1	1	1	0.67	1	1	1	1	1	1	0	1	1	1	1	1	1	0	1	0	1	86.36	[63]
9	1	1	0	1	0.33	1	1	1	1	1	1	1	0	1	1	1	1	1	1	0	0	1	63.64	[48]
10	1	1	1	0	0.67	1	1	1	1	1	1	1	1	1	1	1	1	1	0	0	0	1	75.36	[49]
11	1	1	1	1	0.67	1	1	1	1	1	1	1	1	1	1	1	1	1	0	1	0	1	84	[39]
12	1	1	1	0	1	1	1	1	1	1	1	1	1	1	1	1	1	1	0	1	1	1	87	[50]
13	1	1	1	1	1	1	1	1	1	0	1	1	1	1	0	1	0	1	0	1	0	1	81.36	[41]
14	1	1	1	1	0.67	1	1	1	1	1	1	1	1	1	0	1	1	1	0	1	1	1	84	[42]
15	1	1	1	1	1	1	1	1	1	1	1	1	1	1	1	1	0	1	0	1	0	1	69.36	[40]
16	1	1	1	0	0.67	1	1	1	1	1	1	0	1	1	0	1	1	0	0	1	1	1	71.36	[43]
17	1	1	1	0	0.67	1	0	1	1	1	1	0	1	1	1	1	1	0	0	0	0	1	56.36	[53]
18	1	0	0	1	0.67	1	1	1	1	1	1	0	1	0	0	1	1	0	0	1	0	1	53.36	[54]
19	1	1	1	0	0.67	1	1	1	1	1	1	1	1	0	1	1	1	1	0	1	1	1	72	[20]
20	1	1	1	0	1	1	1	1	1	1	1	0	1	1	1	1	1	0	0	1	0	1	61.36	[44]
21	1	1	1	1	0.67	1	1	1	1	1	1	1	1	1	1	1	1	1	0	1	1	1	84	[45]
22	1	1	1	1	1	1	1	1	0	1	1	1	1	1	1	1	1	1	0	1	0	1	92	[58]
23	1	1	1	0	1	1	1	1	1	0	1	1	0	1	1	1	1	1	1	1	1	1	68.36	[58]
24	1	1	1	1	0.67	1	1	1	1	1	1	1	1	0	1	1	1	0	0	1	0	1	78.36	[55]

References

1. Kang, H.S.; Lee, J.Y.; Choi, S.; Kim, H.; Park, J.H.; Son, J.Y.; Kim, B.H.; Noh, S.D. Smart manufacturing: Past research, present findings, and future directions. *Int. J. Precis. Eng. Manuf. Technol.* **2016**, *3*, 111–128. [CrossRef]
2. Brettel, M.; Friederichsen, N.; Keller, M.; Rosenberg, M. How Virtualization, Decentralization and Network Building Change the Manufacturing Landscape: An Industry 4.0 Perspective. *World Acad. Sci. Eng. Technol. Int. J. Mech. Aerosp. Ind. Mechatron. Manuf. Eng.* **2014**, *8*, 37–44.
3. Hofmann, E.; Rüsch, M. Industry 4.0 and the current status as well as future prospects on logistics. *Comput. Ind.* **2017**, *89*, 23–34. [CrossRef]
4. Nosalska, K.; Piątek, Z.M.; Mazurek, G.; Rządca, R. Industry 4.0: Coherent definition framework with technological and organizational interdependencies. *J. Manuf. Technol. Manag.* **2020**, *31*, 837–862. [CrossRef]
5. Akanmu, A.; Anumba, C.J. Cyber-physical systems integration of building information models and the physical construction. *Eng. Constr. Archit. Manag.* **2015**, *22*, 516–535. [CrossRef]
6. Baheti, R.; Gill, H. Cyber-Physical Systems. *Impact Control Technol.* **2011**, *12*, 161–166.
7. An, W.; Wu, D.; Ci, S.; Luo, H.; Adamchuk, V.; Xu, Z. Agriculture Cyber-Physical Systems. In *Cyber-Physical Systems: Foundations, Principles and Applications*; Academic Press: Cambridge, MA, USA, 2017; pp. 399–417. [CrossRef]
8. Chowhan, R.S.; Dayya, P. Sustainable Smart Farming for Masses Using Modern Ways of Internet of Things (IoT) Into Agriculture. In *Research Anthology on Strategies for Achieving Agricultural Sustainability*; IGI Global: Hershey, PA, USA, 2019; pp. 189–219.
9. Kumar, S.A.; Ilango, P. The Impact of Wireless Sensor Network in the Field of Precision Agriculture: A Review. *Wirel. Pers. Commun.* **2018**, *98*, 685–698. [CrossRef]
10. Rad, C.-R.; Hancu, O.; Takacs, I.-A.; Olteanu, G. Smart Monitoring of Potato Crop: A Cyber-Physical System Architecture Model in the Field of Precision Agriculture. *Agric. Agric. Sci. Procedia* **2015**, *6*, 73–79. [CrossRef]
11. Vuran, M.C.; Salam, A.; Wong, R.; Irmak, S. Internet of underground things in precision agriculture: Architecture and technology aspects. *Ad Hoc Netw.* **2018**, *81*, 160–173. [CrossRef]
12. Hardie, M.; Hoyle, D. Underground Wireless Data Transmission Using 433-MHz LoRa for Agriculture. *Sensors* **2019**, *19*, 4232. [CrossRef]
13. Pandey, G.; Weber, R.J.; Kumar, R. Agricultural Cyber-Physical System: In-Situ Soil Moisture and Salinity Estimation by Dielectric Mixing. *IEEE Access* **2018**, *6*, 43179–43191. [CrossRef]
14. Chen, R.Y. Intelligent Predictive Food Traceability Cyber Physical System in Agriculture Food Supply Chain. *J. Phys. Conf. Ser.* **2018**, *1026*, 012017. [CrossRef]
15. Verdouw, C.N.; Wolfert, J.; Beulens, A.J.M.; Rialland, A. Virtualization of food supply chains with the internet of things. *J. Food Eng.* **2016**, *176*, 128–136. [CrossRef]
16. Verboven, P.; Defraeye, T.; Datta, A.K.; Nicolai, B. Digital twins of food process operations: The next step for food process models? *Curr. Opin. Food Sci.* **2020**, *35*, 79–87. [CrossRef]
17. Sundmaeker, H.; Verdouw, C.; Wolfert, S.; Freire, L.P. Internet of food and farm 2020. In *Digitising the Industry Internet of Things Connecting the Physical, Digital and Virtual Worlds*; River Publishers: Aalborg, Denmark, 2016.
18. Accorsi, R.; Tufano, A.; Gallo, A.; Galizia, F.G.; Cocchi, G.; Ronzoni, M.; Abbate, A.; Manzini, R. An application of collaborative robots in a food production facility. *Procedia Manuf.* **2019**, *38*, 341–348. [CrossRef]
19. Kagermann, H.; Wahlster, W.; Helbig, J. Recommendations for Implementing the Strategic Initiative INDUSTRIE 4.0. Final Report of the Industrie 4.0 Working Group. 2013. Available online: https://www.bibsonomy.org/bibtex/25c352acf1857c1c1839c1a11fe9b7e6c/flint63 (accessed on 8 April 2013).
20. Khan, S.; Altayar, M. Industrial internet of things: Investigation of the applications, issues, and challenges. *Int. J. Adv. Appl. Sci.* **2021**, *8*, 104–113. [CrossRef]
21. Lee, I.; Lee, K. The Internet of Things (IoT): Applications, investments, and challenges for enterprises. *Bus. Horiz.* **2015**, *58*, 431–440. [CrossRef]
22. Da Xu, L.; Xu, E.L.; Li, L. Industry 4.0: State of the art and future trends. *Int. J. Prod. Res.* **2018**, *56*, 2941–2962. [CrossRef]
23. Kamble, S.S.; Gunasekaran, A.; Parekh, H.; Joshi, S. Modeling the internet of things adoption barriers in food retail supply chains. *J. Retail. Consum. Serv.* **2019**, *48*, 154–168. [CrossRef]
24. Barros, A.; Oberle, D. (Eds.) *Handbook of Service Description*; Springer: New York, NY, USA, 2012.
25. Wang, P.; Valerdi, R.; Zhou, S.; Li, L. Introduction: Advances in IoT research and applications. *Inf. Syst. Front.* **2015**, *17*, 239–241. [CrossRef]
26. Codeluppi, G.; Cilfone, A.; Davoli, L.; Ferrari, G. LoRaFarM: A LoRaWAN-Based Smart Farming Modular IoT Architecture. *Sensors* **2020**, *20*, 2028. [CrossRef] [PubMed]
27. Adhitya, Y.; Prakosa, S.W.; Köppen, M.; Leu, J.-S. Feature extraction for cocoa bean digital image classification prediction for smart farming application. *Agronomy* **2020**, *10*, 1642. [CrossRef]
28. Marsden, T.; Banks, J.; Bristow, G. Food Supply Chain Approaches: Exploring their Role in Rural Development. *Sociol. Rural.* **2000**, *40*, 424–438. [CrossRef]
29. Salin, V. Information technology in agri-food supply chains. *Int. Food Agribus. Manag. Rev.* **1998**, *1*, 329–334. [CrossRef]
30. Apaiah, R.K.; Hendrix, E.M.T. Design of a supply chain network for pea-based novel protein foods. *J. Food Eng.* **2005**, *70*, 383–391. [CrossRef]

31. Van der Vorst, J.G.A.J.; Da Silva, C.A.; Trienekens, J.H. *Agro-Industrial Supply Chain Management: Concepts and Applications*; Agricultural management, marketing and finance occasional paper; FAO: Rome, Italy, 2007; p. 71. Available online: https://www.fao.org/publications/card/en/c/e35e5c44-8bc7-5abc-90a3-77cd1d6cd8bf/ (accessed on 1 January 2023).

32. Luo, J.; Ji, C.; Qiu, C.; Jia, F. Agri-Food Supply Chain Management: Bibliometric and Content Analyses. *Sustainability* **2018**, *10*, 1573. [CrossRef]

33. D'Adamo, I.; Gastaldi, M.; Morone, P.; Rosa, P.; Sassanelli, C.; Settembre-Blundo, D.; Shen, Y. Bioeconomy of Sustainability: Drivers, Opportunities and Policy Implications. *Sustainability* **2022**, *14*, 200. [CrossRef]

34. Kitchenham, B.; Charters, S. *Guidelines for Performing Systematic Literature Reviews in Software Engineering*; Version 2.3; EBSE Technical Report; Keele University: Keele, UK; University of Durham: Durham, UK, 2007.

35. Wen, J.; Li, S.; Lin, Z.; Hu, Y.; Huang, C. Systematic literature review of machine learning based software development effort estimation models. *Inf. Softw. Technol.* **2012**, *54*, 41–59. [CrossRef]

36. Dybå, T.; Dingsøyr, T. Empirical studies of agile software development: A systematic review. *Inf. Softw. Technol.* **2008**, *50*, 833–859. [CrossRef]

37. Serrano, R.; Pinilla, V. Changes in the structure of world trade in the agri-food industry: The impact of the home market effect and regional liberalization from a long-term perspective, 1963–2010. *Agribusiness* **2014**, *30*, 165–183. [CrossRef]

38. Taddei, E.; Sassanelli, C.; Rosa, P.; Terzi, S. Circular supply chains in the era of industry 4.0: A systematic literature review. *Comput. Ind. Eng.* **2022**, *170*, 108268. [CrossRef]

39. Liu, Y.; Ma, X.; Shu, L.; Hancke, G.P.; Abu-Mahfouz, A.M. From Industry 4.0 to Agriculture 4.0: Current Status, Enabling Technologies, and Research Challenges. *IEEE Trans. Ind. Inform.* **2021**, *17*, 4322–4334. [CrossRef]

40. Yadav, S.; Garg, D.; Luthra, S. Analysing challenges for internet of things adoption in agriculture supply chain management. *Int. J. Ind. Syst. Eng.* **2020**, *36*, 73–97. [CrossRef]

41. Fernandez, C.M.; Alves, J.; Gaspar, P.D.; Lima, T.M. Fostering awareness on environmentally sustainable technological solutions for the post-harvest food supply chain. *Processes* **2021**, *9*, 1611. [CrossRef]

42. Grecuccio, J.; Giusto, E.; Fiori, F.; Rebaudengo, M. Combining Blockchain and IoT: Food-Chain Traceability and Beyond. *Energies* **2020**, *13*, 3820. [CrossRef]

43. Mantravadi, S.; Moller, C.; Christensen, F.M.M. Perspectives on Real-Time Information Sharing through Smart Factories: Visibility via Enterprise Integration. In Proceedings of the 3rd International Conference on Smart Systems and Technologies (SST 2018), Osijek, Croatia, 10–12 October 2018; pp. 133–137. [CrossRef]

44. Capello, F.; Toja, M.; Trapani, N. A Real-Time Monitoring Service Based on Industrial Internet of Things to Manage Agrifood Logistics. 2016. Available online: https://www.scopus.com/inward/record.uri?eid=2-s2.0-84985961212&partnerID=40&md5=69c0be5721ad524f102f986f1dab7b85 (accessed on 1 January 2023).

45. Kuaban, G.S.; Czekalski, P.; Molua, E.L.; Grochla, K. An Architectural Framework Proposal for IoT Driven Agriculture. In Proceedings of the 26th International Conference on Computer Networks, CN 2019, Gliwice, Poland, 25–27 June 2019; Springer: Berlin/Heidelberg, Germany; Institute of Theoretical and Applied Informatics: Gliwice, Poland; Polish Academy of Sciences: Gdańsk, Poland, 2019; Volume 1039, pp. 18–33. [CrossRef]

46. Jagtap, S.; Bader, F.; Garcia-Garcia, G.; Trollman, H.; Fadiji, T.; Salonitis, K. Food Logistics 4.0: Opportunities and Challenges. *Logistics* **2021**, *5*, 2. [CrossRef]

47. Khan, P.W.; Byun, Y.-C.; Park, N. IoT-blockchain enabled optimized provenance system for food industry 4.0 using advanced deep learning. *Sensors* **2020**, *20*, 2990. [CrossRef]

48. Yadav, S.; Luthra, S.; Garg, D. Modelling Internet of things (IoT)-driven global sustainability in multi-tier agri-food supply chain under natural epidemic outbreaks. *Environ. Sci. Pollut. Res.* **2021**, *28*, 16633–16654. [CrossRef]

49. Smetana, S.; Aganovic, K.; Heinz, V. Food Supply Chains as Cyber-Physical Systems: A Path for More Sustainable Personalized Nutrition. *FOOD Eng. Rev.* **2021**, *13*, 92–103. [CrossRef]

50. Mondragon, A.E.; Mondragon, C.E.; Coronado, E.S. Managing the food supply chain in the age of digitalisation: A conceptual approach in the fisheries sector. *Prod. Plan. Control* **2021**, *32*, 242–255. [CrossRef]

51. Almadani, B.; Mostafa, S.M. IIoT Based Multimodal Communication Model for Agriculture and Agro-Industries. *IEEE Access* **2021**, *9*, 10070–10088. [CrossRef]

52. Jagtap, S.; Garcia-Garcia, G.; Rahimifard, S. Optimisation of the resource efficiency of food manufacturing via theInternet of Things. *Comput. Ind.* **2021**, *127*, 103397. [CrossRef]

53. Kanagachidambaresan, G.R. IoT Projects in Smart City Infrastructure. In *Internet of Things*; Springer: Berlin/Heidelberg, Germany; Department of CSE, Vel Tech Rangarajan Dr Sagunthala R&D Institute of Science and Technology: Avadi, India; Chennai, India; Tamil Nadu, India, 2021; pp. 199–215. [CrossRef]

54. Sharma, R.; Parhi, S.; Shishodia, A. Industry 4.0 Applications in Agriculture: Cyber-Physical Agricultural Systems (CPASs). In *Lecture Notes in Mechanical Engineering*; Springer: Singapore, 2021; pp. 807–813. [CrossRef]

55. Corallo, A.; Latino, M.E.; Menegoli, M. Agriculture 4.0: How Use Traceability Data to Tell Food Product to the Consumers. In Proceedings of the ICITM 2020—2020 9th International Conference on Industrial Technology and Management, Oxford, UK, 11–13 February 2020; pp. 197–201. [CrossRef]

56. Ekren, B.Y.; Mangla, S.K.; Turhanlar, E.E.; Kazancoglu, Y.; Li, G. Lateral inventory share-based models for IoT-enabled E-commerce sustainable food supply networks. *Comput. Oper. Res.* **2021**, *130*, 105237. [CrossRef]

57. Gao, Z.; Wang, D.; Zhou, H. Intelligent circulation system modeling using bilateral matching theory under Internet of Things technology. *J. Supercomput.* **2021**, *77*, 13514–13531. [CrossRef]
58. Boccia, F.; Covino, D.; Di Pietro, B. Industry 4.0: Food supply chain, sustainability and servitization. *Riv. Di Stud. Sulla Sostenibilita* **2019**, *2019*, 77–92. [CrossRef]
59. Jabbour, C.J.C.; Fiorini, P.D.C.; Ndubisi, N.O.; Queiroz, M.M.; Piato, E.L. Digitally-enabled sustainable supply chains in the 21st century: A review and a research agenda. *Sci. Total Environ.* **2020**, *725*, 138177. [CrossRef] [PubMed]
60. Aamer, A.M.; Al-Awlaqi, M.A.; Affia, I.; Arumsari, S.; Mandahawi, N. The internet of things in the food supply chain: Adoption challenges. *Benchmarking* **2021**, *28*, 2521–2541. [CrossRef]
61. Herczeg, G.; Akkerman, R.; Hauschild, M.Z. Supply chain collaboration in industrial symbiosis networks. *J. Clean. Prod.* **2018**, *171*, 1058–1067. [CrossRef]
62. Hao, K. Training a Single AI Model Can Emit as Much Carbon as Five Cars in Their Lifetimes. 2019. Available online: https://www.technologyreview.com/2019/06/06/239031/training-a-single-ai-model-can-emit-as-much-carbon-as-five-cars-in-their-lifetimes/ (accessed on 1 January 2023).
63. Vanderroost, M.; Ragaert, P.; Verwaeren, J.; De Meulenaer, B.; De Baets, B.; Devlieghere, F. The digitization of a food package's life cycle: Existing and emerging computer systems in the pre-logistics phase. *Comput. Ind.* **2017**, *87*, 15–30. [CrossRef]

 *sustainability*

Review

A Review of the Current Practices of Bioeconomy Education and Training in the EU

Bas Paris [1], Dimitris Michas [1], Athanasios T. Balafoutis [1], Leonardo Nibbi [2], Jan Skvaril [3], Hailong Li [3], Duarte Pimentel [4,5], Carlota da Silva [4], Elena Athanasopoulou [6], Dimitrios Petropoulos [7] and Nikolaos Apostolopoulos [8,*]

1   Institute of Bio-Economy & Agro-Technology, Centre of Research & Technology Hellas, Dimarchou Georgiadou 118, 38333 Volos, Greece
2   Department of Industrial Engineering, University of Florence, 50139 Florence, Italy
3   School of Business, Society and Engineering, Mälardalen University, 723 21 Västerås, Sweden
4   TERINOV—Parque de Ciência e Tecnologia da Ilha Terceira, 9700-702 Terra Chã, Portugal
5   Centro de Estudos de Economia Aplicada do Atlântico (CEEAplA), University of the Azores, 9500-321 Ponta Delgada, Portugal
6   Department of Business and Organizations Administration, University of Peloponnese, Antikalamos, 24100 Kalamata, Greece
7   Department of Agriculture, University of Peloponnese, Antikalamos, 24100 Kalamata, Greece
8   Department of Management Science and Technology, University of Peloponnese, 22100 Tripoli, Greece
*   Correspondence: anikos@uop.gr

Abstract: This study conducts a review of the current practices of bioeconomy education and training in the EU; as well as the associated methodologies; techniques and approaches. In recent years; considerable efforts have been made towards developing appropriate bioeconomy education and training programs in order to support a transition towards a circular bioeconomy. This review separates bioeconomy education approaches along: higher education and academic approaches, vocational education and training (VET) and practical approaches, short-term training and education approaches, and other approaches. A range of training methodologies and techniques and pedagogical approaches are identified. The main commonalities found amongst these approaches are that they are generally problem based and interdisciplinary, and combine academic and experiential. Higher education approaches are generally based on traditional lecture/campus-based formats with some experiential approaches integrated. In contrast, VET approaches often combine academic and practical learning methods while focusing on developing practical skills. A range of short-term courses and other approaches to bioeconomy education are also reviewed.

Keywords: bioeconomy; bioeconomy education; bioeconomy learning; higher education; vocational education and training

Citation: Paris, B.; Michas, D.; Balafoutis, A.T.; Nibbi, L.; Skvaril, J.; Li, H.; Pimentel, D.; da Silva, C.; Athanasopoulou, E.; Petropoulos, D.; et al. A Review of the Current Practices of Bioeconomy Education and Training in the EU. *Sustainability* **2023**, *15*, 954. https://doi.org/10.3390/su15020954

Academic Editors: Idiano D'Adamo and Massimo Gastaldi

Received: 1 December 2022
Revised: 22 December 2022
Accepted: 26 December 2022
Published: 4 January 2023

1. Introduction

In recent years the Bioeconomy has become a central topic in EU policymaking, especially in the context of the Green Deal which was approved by the European Commission in 2020 and includes a set of policy-related initiatives supporting the transition to carbon neutrality by 2050, and is expected to continue to receive increased national and EU-wide policy support in the coming years [1]. In this context, the bioeconomy is expected to underpin the transition from a linear economic model that is based on non-renewable resources, to a circular, low-carbon economy that relies heavily on the production and consumption of renewable, organic-based resources. Such a transition requires both a mentality change of the population and a practical transformation of its skillset; as the standards and the needs change, new job types arise across entire value chains. In order to effectively support such a transition, a comprehensive bioeconomy education and training system needs to be developed, one that will take into account the special needs and the interdisciplinary approach that the bioeconomy entails. Such a system requires a transformation of existing education methods and training approaches, as well as the development of new ones [2–5].

Considerable gaps exist in our understanding of the existing practices around bioeconomy-related training and education. This can partly be explained due to the relevant newness of the bioeconomy as a concept [6], but also because multiple educational practices and approaches, that are clearly located within the bioeconomy conceptually, are not explicitly labelled as such. A range of previous projects and studies have investigated bioeconomy education programs and approaches [7–10], however a comprehensive review of training methods and approaches across higher education (HE) and vocational education and training (VET) is lacking. This is especially relevant as previous studies have highlighted that education, training and knowledge development in bioeconomy sectors vary widely across Europe. According to the European Commission, there is a particular lack of training on bioeconomy for small enterprises and at low levels [2,6,11,12]. In addition, there is a concern that the labour force is not trained adequately for this transition to a bioeconomy, especially as it requires expertise that originates in different disciplines and combines a variety of skills across and along value chains.

This paper aims to fill this gap by conducting a review of the current practices of bioeconomy education and training in the EU, as well as the associated methodologies, techniques and approaches. This will contribute to our understanding of how bioeconomy education and training is conducted, evaluate the different approaches, and provide the background for the development of additional and necessary bioeconomy curricula. Furthermore, this study aims to conduct a review of current practices of bioeconomy education and training in the EU as well as the associated methodologies, techniques and approaches. By doing so this study provides an overview on and relevant categorisations of existing bioeconomy approaches and also highlights where developments in bioeconomy education and curriculum have been occurring and where they have been lacking in recent years. In addition, this study facilitates a discussion on the main commonalities and differences between the various educational approaches currently used in the bioeconomy and identifies various areas for future areas of research.

The paper is structured as follows: the rest of this section provides an overview of previous studies on bioeconomy training and education; Section 2 discusses the methodology used in this review; Section 3 presents the results of this study; Section 4 provides a discussion and macro analysis of trends in bioeconomy training and education in the context of previous studies; and Section 5 provides concluding remarks.

2. Literature Review

There isa range of bioeconomy policies and strategies developed at the EU level and country level that support related bioeconomy training and education processes. At EU level, the 2012 European Bioeconomy Strategy, which was updated in 2018 with the addition of a Bioeconomy Action Plan, which calls for the new education processes and the testing of new HE and vocational training curricula [13]. Beyond this, the bioeconomy is supported by other major EU policies and strategies, including the Green Deal, the Circular Economy Action Plan and the Farm to Fork Strategy [14,15]. There are also a few EU Member States with a national bioeconomy strategy, including Finland, Germany, Latvia, France, Spain, Italy, Ireland, Austria and the Netherlands, while Norway and the UK also feature their own strategies. Estonia, Hungary and Lithuania are in the process of developing national strategies [16]. Similarly, all these policies tend to include and promote relevant bioeconomy education and training in their strategies. In September 2022 the EU released the report on 'Promoting education, training and skills across the bioeconomy,' which assesses expert insights on current bioeconomy education practices and the future needs for bioeconomy related education until 2050 [17].

Various recent publications exist that investigate bioeconomy-related education and programmes. Pubule et al. (2020) analyse 10 Master programmes in bioeconomy and highlight the interdisciplinary nature of the approaches used [6], while Kalnbalkite et al. (2021)

look more widely at education for the bioeconomy [1]. Ruxandra et al. (2018) investigate the role of the University in developing human capital for a sustainable bioeconomy [18], and Masiero et al. (2020) look into bioeconomy-related perceptions among 1400 students in 29 Universities [19]. Less recently, Watkinson et al. (2012) conducted a review of advanced bioenergy education and training in Europe [11], and Ray et al. (2016) highlighted the importance of linking bioeconomy development with education in developing countries, emphasising the potential of e-learning courses [20].

A common theme among relevant journal publications is the importance of interdisciplinarity in any method or approach with the bioeconomy education [21]. Sacchi et al. (2021), investigate the educational processes required for a bio-based economy, and highlight that the main challenge for education related to the bio-based economy is the development of an effective framework that bridges the life sciences and the social sciences [22]. Onpraphai et al. (2021) argue for the importance of creating new education approaches that are able to support the shift to a sustainable bioeconomy [23]. It is documented that education for the bioeconomy is conceptually related to the attainment of the Sustainable Development Goals (SDGs) and education for sustainable development (ESD). Urmetzer et al. (2020) highlight that pedagogical approaches based on ESD that are interdisciplinary can support transformational bioeconomy education processes [3].

Lask et al. (2017), in looking at academic approaches to education for the bioeconomy, find that most HE programs, in particular at university level, are designed with I-shaped profiles whereby students specialise in one discipline and research field. However, bioeconomy programs are inherently interdisciplinary and aim to create a T-shaped profile, i.e., where integrative professionals are ideally also disciplinary experts, educated to incorporate and connect different disciplinary knowledge domains and methods. Therefore, the paper makes the case for a bioeconomy education that is multi- and transdisciplinary as well as practice-oriented in order to create bioeconomy professionals who, although specialised in one specific field, have an understanding of other related disciplines and are able to manipulate scientific jargon [24].

Moreover, a range of European and national projects are focused on supporting education and training in the bioeconomy. Examples of such projects include the European Bioeconomy Library [10], BIOEAST, The BRANCHES PROJECT [7], BE-Rural [8], MPowerBio [9]. Many of these projects include knowledge exchange and capacity building programmes for the bioeconomy. The ERASMUS+ project VET4BioECONOMY, which was completed in 2021, focused on VET programmes on forest bioeconomy, while the ERASMUS+ project FIELDS has a database on VET related programmes in the bioeconomy [25]. Extending the horizon beyond the EU landscape, the ERASMUS+ Capacity Building project BBChina established a 120 ECTS equivalent Master Program on Bioeconomy in three Chinese Universities through a cross-collaboration between European and Chinese Higher Education Institutions [26].

In recent years, a host of platforms have already proliferated that support education processes often by hosting and sharing relevant bioeconomy knowledge, including the Rural Bioeconomy Portal, the European Bioeconomy Network, the European Bioeconomy Library, the Bio-based Industry Consortium's (BIC) bioeconomy platform, the European Bioeconomy University, the European Bioeconomy Alliance, and the Circular Bioeconomy Alliance. These platforms, at times overlapping, generally support bioeconomy knowledge and information sharing and development, stakeholder collaboration, policy development and provide tools and support for relevant stakeholders.

These publications provide relevant insight into bioeconomy education but none of them conducts a comprehensive review of the entire bioeconomy education theme. Additionally, as the bioeconomy is such a wide concept, there are many educational programmes that are not labelled as 'bioeconomy' but clearly belong in this theme. For instance, a range of studies investigate training and education within the bioeconomy themes–the farming and agricultural sectors, water-based bioeconomy, forestry, bioindustry, bioenergy. However, these studies do not include courses in the bioeconomy per se–there are few curricula that are focused on the bioeconomy but there are many more that are related to the bioeconomy. This is an important finding, especially given the established significance for interdisciplinarity within bioeconomy-related courses, potentially entailing that bioeconomy education is inherently inter- and trans-disciplinary.

A common point between the relevant policies and strategies, the papers and the projects is that existing education and training approaches need to be transformed and new ones need to be developed to effectively train individuals for a sustainable bioeconomy–the bioeconomists. Relevantly, a recent report by the Global Bioeconomy Summit 2020 investigates how to shape education for a sustainable bioeconomy and identifies the relevant needs; "The expertise required in the [sustainable circular bioeconomy] SCB workforce in the industry by 2030+ was identified as: Knowledge transfer from lab to industry; critical thinking and problem-solving abilities, knowledge in business models and project management; knowledge about the principles of sustainable development and the circular economy; knowledge in bio-based markets and techno-economic expertise." This conclusion convincingly demonstrates the need for bridging theory with practice, life with social sciences, and cognitive with technical skills [27].

3. Methodology: Materials and Methods

3.1. Conceptual Approach

3.1.1. The Bioeconomy

There are a variety of definitions and approaches to 'bioeconomy' ranging from a broader, inclusive definition that focuses on biological resources and processes in general to narrower and more specific definitions that focus on particular sectors. The European Commission defines the bioeconomy as "using renewable biological resources from land and sea, like crops, forests, fish, animals and micro-organisms to produce food, materials and energy [28]." A more in-depth definition provided in the EU Bioeconomy Strategy is that the bioeconomy includes all systems that are dependent on biological resources (biomass, animals, plants, organisms), their functions and principles. This includes land and marine ecosystems; primary production sectors (agriculture, forestry, fisheries and aquaculture) and; all economic and industrial sectors using biological resources to produce food, feed, biobased products, energy and services [13].

Considering this definition, the bioeconomy can be conceptually categorised along 5 themes, and in line with the EU bioeconomy strategy. These are:

- Food/agriculture systems, which encompass the value chains related to farming, the production of food and biological feedstocks within the 10.5 million farms in the EU [29].
- Forestry/natural habitats systems, which are a source of environmental public goods, raw materials and services, and refer to processes of managing and utilising the EU's 182 million ha of forests [30].
- Water systems, which refer to all aquatic systems and respective economic activities (such as fisheries, aquaculture, aquatic biomass production, etc.).
- Bioenergy, which refers to all energy derived from organic sources, being the largest RES in the EU (12% of total energy demand).
- Biomaterials/bio-based products, which are derived from renewable biological feedstocks and are processed creating conventional products (e.g., timber and textiles) and advanced products (bio-plastics and pharmaceuticals).

It is well documented that the bioeconomy is interdisciplinary [12] and both transcends and interlinks between the themes (food/agriculture, forestry/natural habitats and aquatic/water system bioenergy and biomaterials), sectors (primary, secondary, tertiary) and the wider economy. The bioeconomy is inherently an inclusive concept as it is a response to the unsustainability of current economic processes that are often based on unsustainable production models and contribute to environmental degradation and societal fragmentation [3].

3.1.2. Educational Approaches and Methodologies

Bioeconomy education is generally distinguished by a number of characteristics. The previously cited studies indicate that the bioeconomy needs to adopt interdisciplinary learning and teaching approaches that are often focused on bringing together and integrating the STEM disciplines with the SSH (social sciences and humanities) disciplines.

This paper investigates training methodologies, techniques and approaches that are prevalent within bioeconomy education within the EU. A range of labelling of methodologies exists by different projects and studies. For this study, we prefer a relatively simple conceptual categorisation of educational approaches and methodologies, combining the insights and expertise from the RELIEF consortium as well as the research outputs of the Erasmus+ FIELDS project [31].

Training and education approaches are wide concepts that aim to broadly describe the pedagogical approaches to teaching. Based on a variety of sources, and in the framework of the RELIEF project, we categorise our research and findings along four main approaches: (i) the higher education and academic approaches, (ii) the VET and practical approaches, (iii) short-term approaches, and (iv) other approaches. Academic approaches are mainly based on cognitive approaches to education and learning knowledge and theory. These are knowledge-based approaches where the goal is to learn, integrate and apply knowledge in academic disciplines. This approach is often central to higher education institutions, such as universities, and teaching and learning generally occur over a number of years. Practical approaches are mainly centred around 'learning by doing' methods and are focused on supporting the development of an individual's skills, abilities and knowledge around practical, real-life (as opposed to theoretical, hypothetical) issues. These are competence-based approaches that are focused on supporting students in mastering specific skills and competencies. This approach is often used in VET institutions and learning duration ranges from a number of months to multiple years. This generally includes training in skills and knowledge for a trade or occupation and may be part of secondary or tertiary education programs or training for specific employment. Short-term approaches are time-bound, ranging from less than a day to a couple of months. These approaches are generally intensive, compressing large amounts of knowledge and learning in short time periods and may adopt a academic, practical or hybrid approach. Other approaches refer to other learning methodologies encountered in the literature, including do-it-yourself (DIY) learning, lifelong learning and informal education.

These are conceptual categorisations and considerable overlap between approaches exists, while each individual approach often adopts learning practices that are favoured by another approach. The categorisation has been determined based on the 'focus' of the given approach. For instance, in practice an academic bioeconomy University program may have a focus on theoretical concepts but generally also includes parts that are centred around problem or project or challenge based learning and also include practical learning and training. In addition, education and training for a sustainable and circular bioeconomy is closely related to education for sustainable development (ESD). ESD refers to incorporating key sustainable development factors into education programs including topics focused on climatic change, sustainable production and consumption and biodiversity. Similarly, it is recognized that ESD is interdisciplinary and requires participatory teaching and learning approaches [3].

Each approach adopts and favours a range of training methodologies and techniques. Training methodologies and techniques refer to those used in designing and implementing education and training, i.e., the pedagogical tools used to impart knowledge and skills in the context of each educational setting. These methodologies and techniques are explored as a sub-category of the four main training and education approaches identified. The training methodologies and techniques that were identified and referred to in the existing literature include: lecture and classroom-based formats, e-courses, virtual learning tutorials, on-the-job training, work placements/traineeships, practical workshops, in-person class, focus groups, panel discussions, mentoring, peer learning, participatory learning, on-site demonstration, gamification [31]. (See Figure 1). These methodologies and techniques are not mutually exclusive, meaning that more than one can be used in one education setting and overlap within one approach, while they are also not assigned specifically to any of the main approaches. For instance, a lecture/classroom-based University course may make use of participatory learning techniques, as can a VET course.

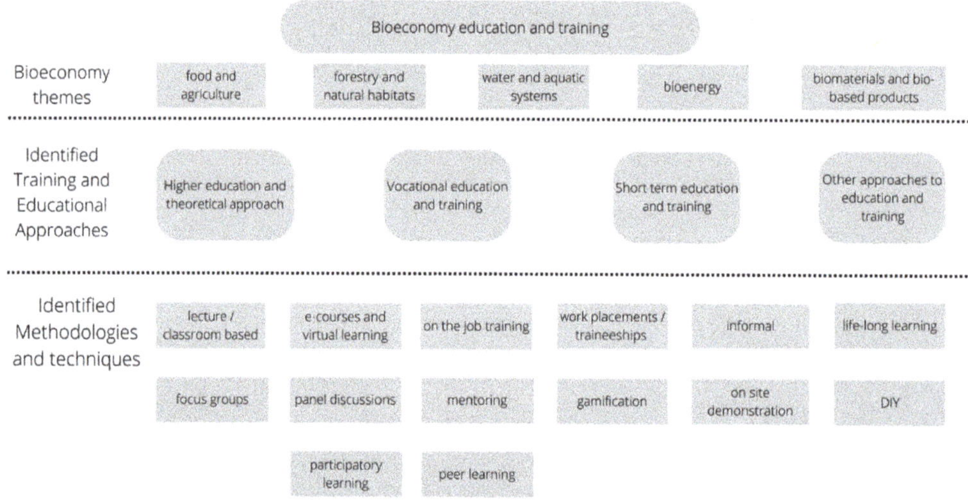

Figure 1. Conceptual Educational Approach.

3.2. Data Sources, Selection Criteria, Search Strategy and Data Collection

This study conducts a review of the current practices of bioeconomy education and training. Data was collected from a range of sources including peer-reviewed publications, reports, results from previous projects, national policies. Our search strategy followed a number of key steps: on the one hand, all partners of the RELIEF consortium were asked to participate in the data collection process by providing relevant information and data on educational approaches in their respective countries as well as the EU. On the other hand, relevant studies were identified through key word searches through google scholar and SCOPUS and desk-based research. Keywords used in these searches included 'bioeconomy education', 'bioeconomy training', and 'bioeconomy learning', these searches identified a total of 259 publications (175 in Scopus). The publications identified show considerable increase in recent years until October 2022 (Figure 2).

The studies included here, and in particular journal articles, were screened for relevance and applicability, the reports and projects used came from respected publication sources, while the journal articles had to be published in well respected and peer-reviewed journals. 72 studies were found relevant and selected for further analysis. These studies were chosen as they contained a clear thematic focus on the bioeconomy and education and training approaches in the EU. Information and data were extracted, compiled and categorised along a number of categories for each bioeconomy theme. Due to the descriptive nature of the data and the goal of this study, data was analysed in a qualitative manner whereby different educational approaches were compared and contrasted.

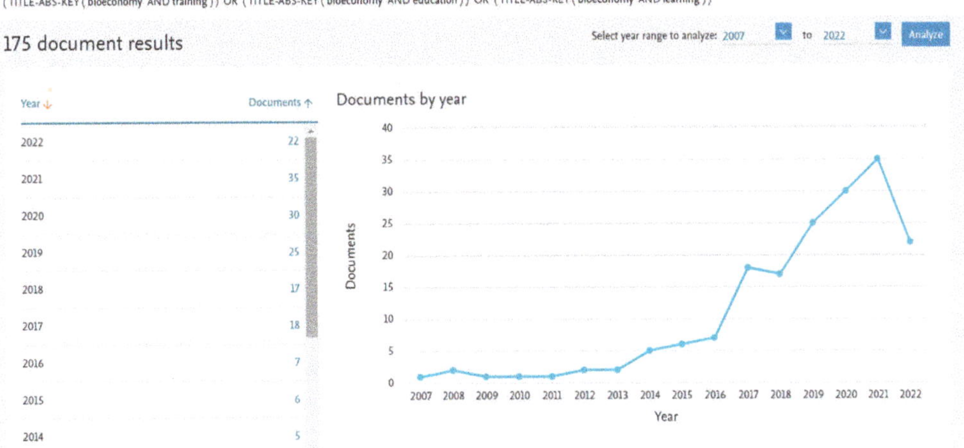

Figure 2. Number of bioeconomy education publications in Scopus by year.

3.3. Bias Risk and Limitations

There are a number of key limitations to the present study. Bioeconomy is a general concept with varying interpretations. In this sense, there is a risk of losing specific answers/results when discussing the bioeconomy [4] as they might be found under different terms. In addition, and due to these variations, authors and studies mean different things when they refer to 'bioeconomy education'. Moreover, as the bioeconomy is a relatively new concept there are relatively few studies and approaches that focus solely on the bioeconomy or use the term 'bioeconomy'. This is a limitation as some educational approaches adopt lenses other than the bioeconomy. Due to this newness, there is relatively little research available on existing bioeconomy education approaches and techniques and there are considerable data gaps. Future research is needed on more specific aspects of each education approach in the bioeconomy. There are also a number of bias risks when conducting a review. As the study is based on data and knowledge that has been shared with the authors and on keyword searches in google scholar and Scopus, the inclusion of knowledge and data is at risk of bias.

4. Results

4.1. Higher Education and Academic Approaches

In recent years, many higher education programs centred around the bioeconomy and its components have been developed throughout the EU. These exist at a range of higher education levels, including at bachelor, master/post graduate and PhD level as categorised in Table 1. Common characteristics amongst these programs include: (i) interdisciplinary design, often bridging STEM and SSH disciplines, (ii) focus on the bioeconomy as a whole and/or on one or more bioeconomy-related themes, (iii) priority is given on academic approach, and (iv) teaching and learning approaches are predominantly classroom/lecture-based formats. These programs tend to be geared around supporting the development of a knowledge economy and have a strong sustainability-related dimension. This sustainability dimension appears in many forms throughout the available courses, for example there could be a focus on the study of the circular economy, the design of eco/efficient bio-products or on sustainable practices in general [32].

Further, there is a variety of designs and focus attached to these programs and they can generally be categorised along three main lines: (i) programs focused on a scientific discipline that emphasise on specific bioeconomy themes (e.g., Bio-technology) [33–36] (ii) general Bioeconomy programs that provide a general overview of the Bioeconomy and all its components [32,37], and (iii) programs that give equal emphasis in two disciplines, usually combining a STEM and a SSH discipline [38], and that focus on the relevant aspects of the bioeconomy (e.g., a course combining Bio-technology with Economics or with Ethics). In recent years, there has been a rapid increase in all of these programs, and specifically the interdisciplinary design, and this trend is likely to continue [24]. Such study designs are increasingly accepted within academia and higher education programs, while inclusive and interdisciplinary approaches are favoured as being paramount to the creation of a circular and sustainable bioeconomy, which is inherently interdisciplinary. Such courses approach the subject with a holistic and critical understanding in order to accommodate the multidisciplinary character of the bio-economy which, as explained above, requires a re-designing of the traditional educational methodologies which are fragmental.

Regarding bachelor level programs, several were identified in several countries and can broadly be split into either general bioeconomy programs or specific theme-based programs. The design of these general bioeconomy programs has a focus on learning and education across the entire bioeconomy value chain, while some of them may include an additional focus on particular aspects of the bioeconomy. These programs generally adopt an academic approach; although the teaching approach somewhat varies, they appear to use traditional approaches to higher education centred around an on-campus lecture format combined with problem-based learning with some or limited access to field-work and laboratory programs [32,37,39].

Regarding the bachelor programs that are focused on specific bioeconomy themes, this study located several bachelors of science, such as biotechnology, biomaterials, bioenergy, agricultural sciences and agricultural engineering, forestry and bioeconomy. These programs are likely to incorporate more of a hands-on approach than the general programs discussed above, with access and use of laboratory facilities and research infrastructure. In many instances, these programs also have close links with industry and can provide relevant work placements for the students. Course descriptions often mention approaches centred around problem-based or challenge-based learning or experiential learning, as well as hands-on learning through lab experiments, field trips, and a capstone project [33,34,36,40].

Regarding Masters programs, these tend to have a closer focus on specific themes of the bioeconomy and generally adopt a mixed education approach that combines academic and practical methodologies [11]. The course descriptions of such post-graduate programs often appear to include a more general, foundational module focused on the bioeconomy/bio-circular economy as a whole, followed by more focused modules on, for instance, biobased industries, management of biobased feedstocks, etc. Often, in addition to the course modules which are generally lecture-based and emphasise on theory, there is a more practical component as part of the program. This practical component may include carrying out experiments, either in lab or real-life conditions, or a work placement, or some independent research [6]. Similarly, to bachelor programs course descriptions are also often centred around problem-based or challenge-based learning [38,41–46].

Despite these commonalities bioeconomy Masters programs and their development generally follow country specific developments. For instance, in Italy, Master degrees are intended as in-depth thematic studies offered independently by universities. The first example of a Master program related to the Bioeconomy was the "IMES Master in Bioenergy and Environment", set up in 2004 with the support of the "EU/US Programme for Cooperation in Higher Education and Vocational Education and Training" [47]. The 60 ECTS equivalent course run in three US Universities and in two EU Universities and was mainly focussed on the Biomass to energy chain. It was held in ten editions until 2019 at the University of Florence, while at the Universidade Nova de Lisboa, it was transformed into a Master Degree at the end of the 2010's. In this Master, the multidisciplinary approach was developed all along five different axes: Biomass Production (Agriculture, Forestry, and Energy crops), Energy conversion (Renewable Energies and Bioenergy Generation), Biofuels (Conventional and Advanced) and Biorefineries, Environmental Impact (modelling and LCA), and Legislation & Economy. Another important example, started in 2016 and presently running, is the 2nd level Master "Bioeconomy in the Circular Economy

BIOCIRCE". It is an interdisciplinary program jointly offered by four Italian Universities (University of Bologna, University of Milano-Bicocca, University of Naples Federico II, and University of Turin), by 4 non-academic partners (Intesa Sanpaolo, Novamont SpA, GFBiochemicals SpA, and PTP Science Park di Lodi), and 2 Italian Technological Cluster (Cluster SPRING and Cluster CLAN agrifood) providing skills and expertise necessary to deal with the full range of issues in the complex bioeconomy field [45].

Regarding PhD and further higher education programs, these are not prescribed and they are variable depending on the topic, however, the general trend shows that there are an increasing number of PhD programs around the bioeconomy [48]. Such programs can be practical, academic or hybrid, including a mix of educational approaches. They are research-focused with the goal of advancing bioeconomy-related research and development, and they often go in-depth on one more specific topic within the sectors of the bioeconomy [48–50].

Table 1. Current higher education practices for the bioeconomy.

Type of Program	Description	Typical Learning Method	Time Period (Years)	Geographical Locations	Prevalence	Qualifications
General Bioeconomy Bachelor	Theory-based learning with some practical elements.	lecture/classroom based, e-courses and virtual learning, participatory learning, some practical elements/on site demonstration	3–4	Germany [37], Finland [51], Norway [32], Poland [39]	Several	Undergraduate degrees
Bachelor on specific bioeconomy themes	Mixed learning approach	lecture/classroom based, e-courses and virtual learning, participatory learning, many practical elements/on site demonstrations	3–4	Germany [33,34], Spain [40], Finland [36], UK [35]	Several	Undergraduate degrees
Masters	Mixed learning approach	lecture/classroom based, e-courses and virtual learning, participatory learning, many practical elements/on site demonstrations	1–2	Greece [38,41], UK [42], Austria, Ireland, France [31], Germany [43], Netherlands [44], Italy [45], Sweden [46]	Many	Postgraduate certificates and degrees
PhD, Post-doc	Research oriented	Self-learning, some lecture/classroom based	2+	Ireland [49], Switzerland, Spain, Italy, Sweden, Netherlands, Germany, Austria, Belgium [48], France [50]	Several	Degree

4.2. VET and Practical Approaches

The EC defines VET as "the training in skills and teaching of knowledge related to a specific trade, occupation or vocation in which the student or employee wishes to participate. Vocational education may be undertaken at an educational institution, as part of secondary or tertiary education, or may be part of initial training during employment, for example as an apprentice, or as a combination of formal education and workplace learning" [52].

VET programs that are focused on specific themes of the bioeconomy are prevalent across the EU. This study has located a multitude of agricultural courses, forestry courses, energy/electrician courses. Relevantly, the EU ERASMUS + project FIELDS has created a database on VET related programs in the bioeconomy [31]. VET programs are generally not standardised across the EU in terms of education and training methodologies and there is a wide range of approaches and learning methods used. In practice, different countries have different approaches to VET education and the training is linked differently with secondary and tertiary education systems. For instance, it is common for VET education to be centred around education institutions but in some countries VET education and certification can also be achieved through apprenticeships, which is the case in France.

VET courses are predominantly focused on training and teaching of skills and knowledge for employment. In this sense, in VET courses on the several bioeconomy themes, the typical learning approach is centred around the teaching of practical skills, achieved through experiential learning and hands-on practice [53]. A common VET course description is likely to include a short theoretical overview of the relevant topic/bioeconomy theme, followed by a range of practical training directed towards the specific topic and the practical skills attached to it. There is a large variation in the types of practical training offered, which can come in the form of lab work, traineeship, apprenticeship, etc.

Common thematic focuses for the courses located (Table 2) are agricultural technicians, digital/technological expertise combined with a bioeconomy theme and forest management expertise. A commonality amongst all VET courses located is that they all involve some level of practical training, usually done through on site/farm demonstration and learning activities and students receive a certificate at the end of their studies. This may also be combined with a work placement, generally towards the end of the program. Amongst the VET courses located there are considerable differences in terms of the time period of all courses ranging from a couple of weeks and/or hundreds of hours to 2+ years and/or 2000 h [53–59]. Clearly rural extension services are also important methods for providing training and education for rural stakeholders on the bioeconomy, however, during this review no extension programs with a focus on the bioeconomy were located. The development and integration of these is an important area of future research.

Table 2. Approaches and Methodologies in VET Education and Training for the Bioeconomy.

Course Title	Description	Typical Learning Methods	Time Period	Location	Qualifications
Smart Farming and Bioeconomy Technician [53]	Training to become a smart farming technician following a range of courses, field learning and cv related training	Lecture/classroom and on-farm demonstration activities	800 h (400 on site, 400 classroom)	Italy	Certificate
Professional Higher Technician in Agrotechnology [60]	Agrotechnology technician training for the management of a small-medium agricultural enterprise	Lecture/classroom and on-farm demonstration activities	2 years	Portugal	Certificate
Technicians in rural areas [54]	Focused on creating entrepreneur with thematic focuses on innovation in agriculture, the viability of farms, irrigation	Lecture/classroom and on-farm demonstration activities	1 year	Spain	Certificate
Data-Driven Agri-Food Business [61]		Online learning	10 weeks	Netherlands	Certificate
Agrogardening [56]	Vocational training with job placements	On-farm demonstration activities and job Placements	2000 h	Spain	Certificate
Forest Harvesting [55]	Training on forestry use and management with associated job placements	Lecture/classroom, on-site demonstration activities and job placements	2 years	Spain	Certificate

Table 2. *Cont.*

Course Title	Description	Typical Learning Methods	Time Period	Location	Qualifications
Environmental Technician [57]	Technical training as an environmental technician through training and job placements	On-site demonstration activities and job placements	2 years	Belgium	Certificate
Technician in agricultural production [62]	Technical training on producing agricultural products	Lecture/classroom and on-farm demonstration activities	2000 h	Spain	Certificate
Technician in Forest Management and the natural environment [63]	Technical studies for forestry and environmental work in mountainous areas and nurseries	Lecture/classroom and on-farm demonstration activities		Spain	Certificate
Entrepreneur Biodynamic agriculture [59]	Training for sustainable agricultural production technician integrated with high school	Lecture/classroom and on-farm demonstration activities		Netherlands	Certificate
Agricultural Technician [58]	Technical training as an agricultural technician through training and job placements	Classroom/lecture and participatory learning	Up to 6 years	Belgium	Certificate
Agricultural Production Technician [64]	Training for Agricultural Production Technician Integrated with high school	Lecture/classroom and on-farm demonstration activities	1 year	Portugal	Certificate

Table 3. Short term education and training approaches for the bioeconomy.

Category	Typical Learning Approach	Methods	Description	Example Courses
Workshops	Mainly academic, possible some practical	Panel Discussions Focus groups Lecture based	Short events ranging from a few hours to a few days. Focused on specific themes around the bioeconomy or general bioeconomy. Condensed knowledge transfer. Often are organised in the context of EU and/or national bioeconomy related projects	Unlocking Regional bioeconomy transitions. State of the art and ways forward [65] Sustainable Production of Biobased Products in the Bioeconomy Era [66]
Short Courses	Mix between academic and practical	Lecture based Tutorials Practical teaching E-learning courses	Range from a few days to Some assessment. Students with bioeconomy background or not. Generally, a mix of lectures, lab and field visits. Online courses	Training course for farmers: non-food crops (NFC) for bioeconomy in Italy [67] Summer School: Towards a Biobased Economy [68] Bioeconomy school [69] ELLS Summer School on Bioeconomy [70]
SME Training	-	Mentoring	Generally centred around capacity building and supporting SME developments in the bioeconomy	Bioeconomy Ventures [71] MPowerBio [72] DigiCirc [73]

With regards to connections between these more practical VET courses and the academic higher education courses described above, there seems to be a considerable overlap between the two; the VET courses usually include a theoretical aspect, albeit smaller than the practical part of the course, often in the introductory modules, while in the higher education courses there often is a module that requires more practical skills, possibly in the form of lab experimentation or work placement.

4.3. Short-Term Training and Education Approaches

Several short-term education and training approaches for the bioeconomy have been located and are summarised in Table 3. They have been categorised according to workshops, short courses and SME training. These categories generally adopt varying and mixed learning approaches and methods, depending on the goal and the scope of the training, they last for a short period of time and are generally knowledge-intensive. These courses are run by several bodies and are supported through various funding modes, including national and EU projects, universities, research institutes, NGOs and for-profit enterprises. Most of these courses either focus on general knowledge sharing on the bioeconomy or the development of specific skills for specific topics within the subject.

4.4. Other Approaches

A range of other bioeconomy education approaches is also mentioned by various actors. These approaches are often informal and/or are not attached to educational institutes and are centred around individual and lifelong learning schemes, though are often dependent on open-source data published by educational institutes. These approaches are generally characterised as self-motivated and voluntary, for personal and professional reasons, and can be practiced by a range of methods, including both academic and practical, depending on the relevant scope and goals [24,27,74]. A range of education techniques are practiced but are generally based around on the job training and participatory learning techniques, mentoring and self-learning through audio-visual material but can also include learning appropriated through informal means such as through discussions with other relevant stakeholders. Multiple authors argue that these self-motivated processes are crucial for supporting a society wide transition towards a circular bioeconomy and need to be supported by educational material and scientific research that is available to the general public [20,27].

5. Discussion

The findings presented in this report illustrate that education approaches attached to the subject of the bioeconomy vary considerably across the EU, with a range of academic, practical, hybrid, short term and other approaches. There are some important commonalities in the methodologies used within these approaches; the education offered is generally problem-based, interdisciplinary and combines academic and experiential learning. Furthermore, the teaching methods vary, traditional lecture based and lab based formats are popular while in recent years online-learning has also become popular [31] replacing and/or adding to the more traditional lecture formats.

Courses focused around academic and higher education have proliferated, especially in recent years. These have also been centred around creating increased collaborations between existing institutions, for instance, In 2022 the European Bioeconomy University was launched as a collaboration between 6 European universities to promote bioeconomy education [75]. The main learning approaches used in these educational systems are based around traditional, campus-based lectures and tutorials, sometimes with slight variations in the format [24]. There is some focus on problem-based and experiential learning within relevant courses. More specifically, there is a relatively high selection of masters' courses that combine bioeconomy with a thematic focus. Pubule et al. (2020) highlight that bioeconomy master programs are designed around thematic focuses that aim to facilitate long-term employment in the bioeconomy sector. They also highlight that most of these programs are currently concentrated in Western Europe though they predict a likely spread to other areas around the world as the bioeconomy becomes more and more prevalent [6].

Regarding VET programs, in the research process of this study, it was relatively easy to locate higher education courses but harder to locate practical VET courses. This is in line with the study by Ciriminna et al. (2022) who highlight that in recent years university courses on the bioeconomy have proliferated but that there is a need for more practical courses [12]. However, the distinction between the various education approaches is not always clear when it comes to VET programs, while the categorisations between VET and HE programs appear increasingly blurred. On the one hand, this is likely to be due to the newness of the bioeconomy as a concept but also due to the inherent interdisciplinary nature of bioeconomy education that generally requires and includes a mix of academic and practical approaches to education. The lack of data around specific design of VET courses is a major data limitation this study, and as more practical courses become available, our understanding of the approaches used and their effectiveness will become clearer. A key recommendation coming out of various studies is that there is a need for considerably more practical bioeconomy education approaches, especially vocational ones, that support the development of relevant skills across a variety of economic sectors [76].

Given that a thriving bioeconomy is the foundation for the transformation towards a circular economy, there is a need for more bioeconomy programs [3]. It is clear, that on a policy level, the importance of bioeconomy education is now widely recognised [12], however our understanding of what transformative bioeconomy training approaches are and how they fit into supporting the transition towards a circular economy remains limited. There is a need for bioeconomy experts across the economy, within research, the public sector and private sector [12]. Considerable attention is now being put towards mapping and creating new approaches to training for the bioeconomy and these can build on the research presented in this paper. Such programs, courses and modules are necessary to support knowledge-wise this transition and the training approaches and methodologies need to also be transformative and need to combine higher education, VET and industry in each of the themes of the bioeconomy, in order to approach the concept in an interdisciplinary manner that supports an understanding of the complexities of a sustainable bioeconomy. Indeed, various studies [3,11,27,77,78] argue that bioeconomy educational programs need to be designed to create a knowledge-based economy and to provide the new skills needed for the new and upcoming bioeconomy [17]. To fulfil this, relevant programs need to be innovative, interdisciplinary, holistic and open to advancements.

This review indicates multiple areas for future research on bioeconomy training and education. Overall, it is clear that, considering the size of the EU and the bioeconomy, bioeconomy education programs remain relatively limited. As the development and transformation to a sustainable bioeconomy is depended on new and relevant skills and competences a unified EU bioeconomy training and education program is needed that is both multidisciplinary and dedicated to the development of a sustainable bioeconomy. For this to happen further research is needed on what skills are required for such a transition. Such research can be strengthened by EU wide initiatives that monitor and analyse new bioeconomy learning approaches and identify and disseminate best practices. A standardised methodology that carries this out could focus on first 'identifying emerging skills needs', then 'updating existing content and teaching methods' and then 'adopting a modular approach' that allows for the development of tailor made education programs [79].

This review also suggests that there is a lack of current VET programs focused on the bioeconomy, further research on how to develop and integrate them into existing structures would facilitate their development. There are various ways to support this including research on what long term skills and competences for the bioeconomy are needed and how to better embed educational approaches in career guidance, this could be then be followed by a process that integrates relevant educational recommendations in relevant national and EU strategies. In addition, Kuckertz 2020 recommends integrating and facilitating entrepreneurship, research on how entrepreneurship can aid bioeconomy training programs could be valuable [80]. This review also suggests that a focus on how to integrate extension and advisory services would be beneficial for bioeconomy stakeholders. Extensive research is needed on how bioeconomy extension services can be integrated in existing extension services. This is particularly important as the remit of the bioeconomy is considerably larger than existing services.

A considerable drawback of current bioeconomy related research is that there is no concrete accepted definition of what bioeconomy education entails. A recent study found that some programs still have a tendency to be discipline oriented and that this can hinder the capacity of students to dealt with complex issues [81]. In the design of educational programs and pedagogical practices ways need to be found to overcome learning boundaries [81] whilst ensuring an interdisciplinary approach to bioeconomy education [82]. Moreover, drawing upon D'Adamo's et al. [83] remarks, it is important the education aspects in bioeconomy to be investigated in relation with the EU Next Generation Fund (as VET and skills consist major priorities) and its funding initiatives and outcomes.

6. Conclusions

This paper has conducted a review of the current practices of bioeconomy education and training in the EU, as well as the associated methodologies, techniques and approaches. It has provided an overview of education approaches for the bioeconomy, including higher education and academic approaches, VET and practical approaches, short term training and education approaches, and other approaches. The main commonalities amongst these approaches are that they are generally problem based and interdisciplinary, and combine academic and experiential. Higher education approaches are generally based around traditional lecture/campus-based formats together with some integrated experiential approaches. In contrast, VET approaches often combine academic and practical learning methods while focusing on developing practical skills. Various areas for future research are identified in this study. In recent years, considerable efforts have been made towards developing appropriate bioeconomy education and training programs. However, due to the scale and complexity required for the bioeconomy transition it is clear that considerably more bioeconomy education programs are required, especially focused on specific bioeconomy themes.

Author Contributions: Conceptualization, B.P., D.M., A.T.B., E.A., D.P. (Duarte Pimentel) and N.A.; methodology, B.P., D.M. and D.P. (Duarte Pimentel); validation, B.P., D.M. and A.T.B.; formal analysis, B.P.; investigation, B.P.; data curation, B.P.; writing—original draft preparation, B.P., D.M., A.T.B., L.N., J.S., H.L. and D.P. (Duarte Pimentel); writing—review and editing, B.P., D.M., A.T.B., L.N., J.S., H.L., D.P. (Duarte Pimentel), C.d.S., E.A., D.P. (Dimitrios Petropoulos) and N.A.; visualization, B.P.; supervision, E.A., D.P. (Dimitrios Petropoulos) and N.A.; project administration, E.A., D.P. (Dimitrios Petropoulos) and N.A.; funding acquisition, E.A., D.P. (Dimitrios Petropoulos) and N.A. All authors have read and agreed to the published version of the manuscript.

Funding: This study received funding from the European Union's Erasmus+ programme under Agreement Number 101056181.

Institutional Review Board Statement: Not applicable.

Informed Consent Statement: Not applicable.

Data Availability Statement: No new data were created or analyzed in this study. Data sharing is not applicable to this article.

Acknowledgments: This study has been developed as part of the Erasmus + RELIEF project (www. relief.uop.gr) co-funded by the European Union, Agreement Number 101056181. We would like to thank all partners for contributions and insight into developing this article.

Conflicts of Interest: The authors declare no conflict of interest.

References

1. Kalnbalkite, A.; Pubule, J.; Blumberga, D. Education for Advancing the Implementation of the Green Deal Goals for Bioeconomy. *Environ. Clim. Technol.* **2022**, *26*, 75–83. [CrossRef]
2. Hakovirta, M.; Lucia, L. Informal STEM education will accelerate the bioeconomy. *Nat. Biotechnol.* **2019**, *37*, 103–104. [CrossRef] [PubMed]
3. Urmetzer, S.; Lask, J.; Vargas-Carpintero, R.; Pyka, A. Learning to change: Transformative knowledge for building a sustainable bioeconomy. *Ecol. Econ.* **2020**, *167*, 106435. [CrossRef]
4. Takala, T.; Tikkanen, J.; Haapala, A.; Pitkänen, S.; Torssonen, P.; Valkeavirta, R.; Pöykkö, T. Shaping the concept of bioeconomy in participatory projects—An example from the post-graduate education in Finland. *J. Clean. Prod.* **2019**, *221*, 176–188. [CrossRef]
5. Pascoli, D.U.; Aui, A.; Frank, J.; Therasme, O.; Dixon, K.; Gustafson, R.; Kelly, B.; Volk, T.A.; Wright, M.M. The US bioeconomy at the intersection of technology, policy, and education. *Biofuels Bioprod. Biorefining* **2022**, *16*, 9–26. [CrossRef]
6. Pubule, J.; Blumberga, A.; Rozakis, S.; Vecina, A.; Kalnbalkite, A.; Blumberga, D. Education for advancing the implementation of the bioeconomy goals: An analysis of MASTER study programmes in bioeconomy. *Environ. Clim. Technol.* **2020**, *24*, 149–159. [CrossRef]
7. BRANCHES. Boosting Rural Bioeconomy Networks BRANCHES. Available online: https://www.branchesproject.eu/ (accessed on 4 August 2022).
8. BE-Rural Home. BE-Rural. Available online: https://be-rural.eu/ (accessed on 4 August 2022).
9. MPowerBIO. MPowerBIO. Available online: https://mpowerbio.eu/ (accessed on 4 August 2022).
10. EBL European Bioeconomy Library—The Bioeconomy Knowledge Base Meeting Point! Available online: https://www.bioeconomy-library.eu/ (accessed on 4 August 2022).
11. Watkinson, I.I.; Bridgwater, A.V.; Luxmore, C. Advanced education and training in bioenergy in Europe. *Biomass Bioenergy* **2012**, *38*, 128–143. [CrossRef]
12. Ciriminna, R.; Albanese, L.; Meneguzzo, F.; Pagliaro, M. Educating the managers of the bioeconomy. *J. Clean. Prod.* **2022**, *366*, 132851. [CrossRef]
13. European Commission. *A sustainable Bioeconomy for Europe: Strengthening the Connection between Economy, Society and the Environment*; European Commission: Brussels, Belgium, 2018; ISBN 9789279941450.
14. European Commission. *Farm to Fork Strategy*; European Commission: Brussels, Belgium, 2020.
15. European Commission. *Circular Economy Action Plan*; European Commission: Brussels, Belgium, 2020.
16. European Commission. Knowledge Centre for Bioeconomy. Available online: https://knowledge4policy.ec.europa.eu/bioeconomy_en (accessed on 4 August 2022).
17. Deloitte; Empirica; FGB. *Promoting Education, Training and Skills Across the Bioeconomy—Final Report*; European Commission: Brussels, Belgium, 2022.
18. Bejinaru, R.; Hapenciuc, C.V.; Condratov, I.; Stanciu, P. The university role in developing the human capital for a sustainable bioeconomy. *Amfiteatru Econ. J.* **2018**, *20*, 583–598. [CrossRef]
19. Masiero, M.; Secco, L.; Pettenella, D.; Da Re, R.; Bernö, H.; Carreira, A.; Dobrovolsky, A.; Giertlieova, B.; Giurca, A.; Holmgren, S.; et al. Bioeconomy perception by future stakeholders: Hearing from European forestry students. *Ambio* **2020**, *49*, 1925–1942. [CrossRef]
20. Ray, S.; Srivastava, S.; Diwakar, S.; Nair, B.; özdemir, V. Delivering on the promise of bioeconomy in the developing world: Link it with social innovation and education. *Biomark. Discov. Dev. World Dissecting Pipeline Meet. Challenges* **2016**, 73–81. [CrossRef]
21. Baptista, F.; Lourenço, P.; Fitas da Cruz, V.; Silva, L.L.; Silva, J.R.; Correia, M.; Picuno, P.; Dimitriou, E.; Papadakis, G. Which are the best practices for MSc programmes in sustainable agriculture? *J. Clean. Prod.* **2021**, *303*, 126914. [CrossRef]
22. Sacchi, S.; Lotti, M.; Branduardi, P. Education for a biobased economy: Integrating life and social sciences in flexible short courses accessible from different backgrounds. *N. Biotechnol.* **2021**, *60*, 72–75. [CrossRef] [PubMed]
23. Onpraphai, T.; Jintrawet, A.; Keoboualapha, B.; Khuenjai, S.; Guo, R.; Wang, J.; Fan, J. Sustaining Biomaterials in Bioeconomy: Roles of Education and Learning in Mekong River Basin. *Forests* **2021**, *12*, 1670. [CrossRef]
24. Lask, J.; Maier, J.; Tchouga, B.; Vargas-Carpintero, R. The bioeconomist. *Bioeconomy Shap. Transit. A Sustain. Biobased Econ.* **2017**, 341–354. [CrossRef]
25. VET4BioECONOMY. VET4BioECONOMY. Available online: https://vet4bioeconomy.sumins.hr/ (accessed on 4 August 2022).
26. Nibbi, L.; Chiaramonti, D.; Palchetti, E. Project BBChina: A new master program in three Chinese universities on bio-based circular economy; from fields to bioenergy, biofuel and bioproducts. *Energy Procedia* **2019**, *158*, 1261–1266. [CrossRef]
27. Global Bioeconomy Summit. How to shape education for a sustainable circular bioeconomy? In Proceedings of the Conclusions from the GB2020 Workshop on Education, training and capacity building, Berlin, Germany, 16–20 November 2020.
28. European Commission. Bioeconomy. Available online: https://ec.europa.eu/info/research-and-innovation/research-area/environment/bioeconomy_en (accessed on 27 July 2022).
29. Eurostat Farms and Farmland in the European Union-statistics-Statistics Explained. Available online: https://ec.europa.eu/eurostat/statistics-explained/index.php?title=Farms_and_farmland_in_the_European_Union_-_statistics#The_evolution_of_farms_and_farmland_from_2005_to_2016 (accessed on 11 April 2022).
30. Eurostat Forests, Forestry and Logging—Statistics. Explained. Available online: https://ec.europa.eu/eurostat/statistics-explained/index.php?title=Forests,_forestry_and_logging#Forests_in_the_EU (accessed on 13 September 2021).

31. FIELDS Database-Erasmus Fields. Available online: https://www.erasmus-fields.eu/database/ (accessed on 4 August 2022).
32. NORD University. Study Plan Bachelor in Circular Bioeconomy. 2022. Available online: https://www.nord.no/en/Student/study-plans/basirk (accessed on 1 September 2022).
33. Rhine-Waal University of Applied Sciences Biomaterials Science, B.Sc. Available online: https://www.hochschule-rhein-waal.de/en/faculties/technology-and-bionics/degree-programmes/biomaterials-science-bsc (accessed on 20 October 2022).
34. University of Hohenheim Biobased Products and Bioenergy (Bachelor's). Available online: https://www.uni-hohenheim.de/en/biobased-products-and-bioenergy-bachelors (accessed on 19 October 2022).
35. Imperial College London BSc Biotechnology. Available online: https://www.imperial.ac.uk/study/ug/courses/life-sciences-department/biotechnology-bsc/ (accessed on 20 October 2022).
36. HAMK Information and Communication Technology, Bioeconomy-Häme University of Applied Sciences. Available online: https://www.hamk.fi/dp-bachelor/bioeconomy-engineering/?lang=en (accessed on 19 October 2022).
37. University of Hohenheim Sustainability & Change Bachelor. Available online: https://www.uni-hohenheim.de/en/sustainability-and-change-bachelors-studies (accessed on 19 October 2022).
38. International Hellenic University MSc in Bioeconomy: Biotechnology and Law—University Center of International Programmes of Studies. Available online: https://www.ihu.gr/ucips/postgraduate-programmes/bioeconomy (accessed on 19 October 2022).
39. Politechnika Łódzka Bioeconomy-I Stopnia-Faculty of Biotechnology and Food Sciences. Available online: https://rekrutacja.p.lodz.pl/en/biogospodarka-first-cycle-faculty-biotechnology-and-food-sciences (accessed on 19 October 2022).
40. ETSIAAB BSc Degree in Agricultural Sciences and Bioeconomy. Available online: https://www.etsiaab.upm.es/Internacional/Studies?fmt=detail&prefmt=articulo&id=fa45d4183f117610VgnVCM10000009c7648a____ (accessed on 19 October 2022).
41. University of Piraeus Bioeconomy, Circular Economy and Sustainable Development. Available online: https://www.unipi.gr/unipi/en/oik-spoudes-en/metapt-spoudes-en/bioeconomy-and-sustainable-development.html (accessed on 12 October 2022).
42. FindAMasters Masters Degrees (Bioeconomy). Available online: https://www.findamasters.com/masters-degrees/?Keywords=bioeconomy (accessed on 19 October 2022).
43. TUM Bioeconomy-Master of Science (M.Sc.)-TUM. Available online: https://www.tum.de/studium/studienangebot/detail/bioeconomy-master-of-science-msc (accessed on 19 October 2022).
44. Wageningen University Master's Biobased Sciences-WUR. Available online: https://www.wur.nl/en/education-programmes/master/msc-programmes/msc-biobased-sciences.htm (accessed on 19 October 2022).
45. BIOCIRCE Master BioCIRCE. Available online: https://masterbiocirce.com/ (accessed on 19 October 2022).
46. University of Borås Master Programme in Resource Recovery-Biotechnology and Bioeconomy. Available online: https://www.hb.se/en/international-student/program/programmes/master-programme-in-resource-recovery---biotechnology-and-bioeconomy/ (accessed on 19 October 2022).
47. University of Florence IMES Master in Bioenergy and Environment. Available online: https://docplayer.net/15146972-The-master-is-managed-by-the-crear-research-centre-on-renewable-energies-of-the-university-of-florence.html (accessed on 5 November 2022).
48. EURAXESS 11 PhD Positions Available in Bioeconomy. Available online: https://euraxess.ec.europa.eu/worldwide/australia-nz/11-phd-positions-available-bioeconomy (accessed on 18 October 2022).
49. University College Dublin PhD in Planning and the Bio-economy. Available online: https://www.findaphd.com/phds/project/phd-in-planning-and-the-bio-economy/?p145296 (accessed on 19 October 2022).
50. ESP Postdoctoral Position in Sustainable Bioeconomy Strategies. Available online: https://www.es-partnership.org/postdoctoral-position-in-sustainable-bioeconomy-strategies/ (accessed on 20 October 2022).
51. EUNICAS BSc Bioeconomy Enginering | Hame University of Applied Sciences (HAMK) | Search Programmes | European Universities Information. Available online: https://www.eunicas.ie/index.php/eunicas/course/bsc-bioeconomy-enginering-1442.html (accessed on 19 October 2022).
52. Eurostat Glossary: Vocational Education and Training (VET)-Statistics Explained. Available online: https://ec.europa.eu/eurostat/statistics-explained/index.php?title=Glossary:Vocational_education_and_training_(VET) (accessed on 4 August 2022).
53. CEFAP Smart Farming & Bioeconomy Technician. Available online: https://www.cefap.fvg.it/it/31220/tecnico-smart-farming-bioeconomy (accessed on 19 October 2022).
54. MAPA Plan de Formación Continua Para Técnicos del Medio Rural. Available online: https://www.mapa.gob.es/es/desarrollo-rural/formacion/cursos/ (accessed on 19 October 2022).
55. EFA Casagrande Aprovechamientos Forestales. Available online: https://www.efacasagrande.org/programas-formativos/fp-basico/ (accessed on 19 October 2022).
56. EFA Centro AGROJARDINERÍA-Formación Profesional Dual en EFAS. Available online: https://efa-centro.org/fp-basica-agrojardineria/ (accessed on 19 October 2022).
57. EPASC Qualification en Environnement. Available online: https://www.epasc-ciney.be/epasc/enseignement/technique-de-qualification/environnement/ (accessed on 19 October 2022).
58. EPASC Qualification en Agriculture. Available online: https://www.epasc-ciney.be/epasc/enseignement/technique-de-qualification/agriculture/ (accessed on 19 October 2022).
59. Warmonderhof Voltijd Biologisch-Dynamische Landbouw. Available online: https://aereswarmonderhof.nl/studeren/voltijd-leren-wonen-werken (accessed on 19 October 2022).

60. Politécnico de Coimbra Técnico Superior Profissional em Agrotecnologia. Available online: https://www.ipc.pt/pt/estudar/cursos/18197 (accessed on 19 October 2022).
61. Wageningen University Course Data-Driven Agri-Food Business-WUR. Available online: https://www.wur.nl/en/education-programmes/wageningen-academy/what-we-offer-you/courses/show/online-course-data-driven-agri-food-business.htm (accessed on 19 October 2022).
62. CIFP CICLO FORMATIVO DE GRADO MEDIO: Producción Agroecológica. Available online: http://www.cifpaguasnuevas.es/2012-10-08-14-46-23/actividades-agrarias/59-ciclo-formativo-de-grado-medio-produccion-agroecologica-tecnico-en-produccion-agroecologica (accessed on 19 October 2022).
63. CIPF CICLO FORMATIVO DE GRADO SUPERIOR: Gestión Forestal y del Medio Natural. Available online: http://www.cifpaguasnuevas.es/2012-10-08-14-46-23/actividades-agrarias/61-ciclo-formativo-de-grado-superior-gestion-forestal-y-del-medio-natural (accessed on 19 October 2022).
64. EPADRC Técnico/a de Produção Agropecuária | Profissionais | EPADRC-Escola Profissional de Agricultura e Desenvolvimento Rural de Cister. Available online: https://www.epadrc.pt/pt/profissionais/tecnicoa-de-producao-agropecuaria (accessed on 19 October 2022).
65. European Forest Institute Bioregions Workshop: Unlocking Regional Bioeconomy Transitions. Available online: https://efi.int/events/bioregions-workshop-unlocking-regional-bioeconomy-transitions-2022-06-28 (accessed on 20 October 2022).
66. Agricultural University of Athens Online Workshop Entitled "Sustainable Production of Biobased Products in the Bioeconomy Era". Available online: https://www2.aua.gr/en/news-events/ekdiloseis/online-workshop-entitled-sustainable-production-biobased-products-bioeconomy (accessed on 20 October 2022).
67. Panacea Training Course for Farmers: Non-Food Crops (NFC) for Bioeconomy in Italy. Available online: http://www.panacea-h2020.eu/2020/10/01/training-course-for-farmers-non-food-crops-nfc-for-bioeconomy-in-italy/ (accessed on 20 October 2022).
68. BICOCCA TOWARDS A BIO-BASED ECONOMY: Science, Innovation, Economics, Education. Available online: https://www.summerschoolbicocca.com/017-10-bio-based-economy-course.php (accessed on 20 October 2022).
69. Lake Como School of Advanced Studies Bioeconomy School: From Basic Science to a New Economy. Available online: https://bsne.lakecomoschool.org/ (accessed on 20 October 2022).
70. Euroleague for Life Sciences ELLS Summer School on Bioeconomy 2021. Available online: https://www.euroleague-study.org/en/bioeconomy-2021 (accessed on 20 October 2022).
71. Bioeconomy Ventures Home-BioeconomyVentures. Available online: https://www.bioeconomyventures.eu/ (accessed on 20 October 2022).
72. CTA MPowerBIO Project is Calling Bioeconomy SMEs to Participate in a Business Support Programme. Available online: https://www.corporaciontecnologica.com/en/sala-de-prensa/noticias/MPowerBIO-project-is-calling-bioeconomy-SMEs-to-participate-in-a-business-support-programme/ (accessed on 20 October 2022).
73. DigiCirc DigiCirc–Accelerator for Circular Economy. Available online: https://digicirc.eu/ (accessed on 20 October 2022).
74. Mustalahti, I. The responsive bioeconomy: The need for inclusion of citizens and environmental capability in the forest based bioeconomy. *J. Clean. Prod.* **2018**, *172*, 3781–3790. [CrossRef]
75. EBU European Bioeconomy University: Six Universities Form International Alliance with Focus on Bioeconomy—European Bioeconomy University. Available online: https://european-bioeconomy-university.eu/european-bioeconomy-university-six-universities-form-international-alliance-with-focus-on-bioeconomy-21-11-18/ (accessed on 19 September 2022).
76. Sakellaris, G. Bioeconomy Education. *Bio#Futures* **2021**, 489–506. [CrossRef]
77. Stegmann, P.; Londo, M.; Junginger, M. The circular bioeconomy: Its elements and role in European bioeconomy clusters. *Resour. Conserv. Recycl. X* **2020**, *6*, 100029. [CrossRef]
78. *Bio-Based Industries Consortium Report about the Analysis of Educational Gaps Identified in the Different Regional Contexts and Action Fields*; Bio-Based Industries Consortium: Madrid, Spain, 2020.
79. European Commission. *Promoting Education, Training and Skills across the Bioeconomy-Policy Brief*; European Commission: Brussels, Belgium, 2022.
80. Kuckertz, A. Bioeconomy Transformation Strategies Worldwide Require Stronger Focus on Entrepreneurship. *Sustainability* **2020**, *12*, 2911. [CrossRef]
81. Vuojärvi, H.; Vartiainen, H.; Eriksson, M.; Ratinen, I.; Saramäki, K.; Torssonen, P.; Vanninen, P.; Pöllänen, S. Boundaries and boundary crossing in a multidisciplinary online higher education course on forest bioeconomy. *Teach. High. Educ.* **2022**. [CrossRef]
82. Výbošťok, J.; Navrátilová, L.; Dobšinská, Z.; Dúbravská, B.; Giertliová, B.; Aláč, P.; Suja, M.; Šálka, J. Bioeconomy perception by students of different study programs-study from Slovakia. *Cent. Eur. For. J.* **2022**, *68*, 91–100. [CrossRef]
83. D'Adamo, I.; Gastaldi, M.; Morone, P.; Rosa, P.; Sassanelli, C.; Settembre-Blundo, D.; Shen, Y. Bioeconomy of Sustainability: Drivers, Opportunities and Policy Implications. *Sustainability* **2022**, *14*, 200. [CrossRef]

Article

New or Traditional Approaches in Argentina's Bioeconomy? Biomass and Biotechnology Use, Local Embeddedness, and Sustainability Outcomes of Bioeconomic Ventures

Jochen Dürr [1,*] and Marcelo Sili [2]

1   Center for Development Research (ZEF), University of Bonn, 53113 Bonn, Germany
2   CONICET-Centro de Investigación ADETER, Universidad Nacional del Sur, Bahía Blanca 8000, Argentina
*   Correspondence: jduerr@uni-bonn.de

Abstract: The bioeconomy continues to be a contested field in the political debate. There is still no consensus on how a bioeconomy should be designed and anchored in society. Alternative bioeconomy concepts that deviate from the mainstream discourse and are based on small-scale, agro-ecological models are usually underrepresented in the debate. This also applies to Argentina, where the diversity of bioeconomic approaches has not yet been documented and analyzed. The objective of this paper is to identify bioeconomic approaches in Argentina, and characterize alternative, more socio-ecological and locally embedded approaches in order to make them more visible for the political debate. Based on literature research, categories were extracted that can be used to distinguish different types of the bioeconomy. Subsequently, these categories were used in an online survey of 47 enterprises representing different sectors of Argentina's bioeconomy. Using cluster analysis, three groups can be distinguished: a biomass, a biotechnology, and a bioembedded cluster. Argentina's bioeconomy seems to follow a path dependency logic, but new development paths are also opening up. The bioeconomic approaches discovered in Argentina are partly consistent with contemporary bioeconomy typologies, but there is also great diversity within the groups. All bioeconomic approaches have local connections, but are locally embedded in different ways. In addition to the differences between the bioeconomic approaches, two common elements could also be detected: an interest in sustainable use of natural resources and in building networks using synergies with other actors in the territory. These two elements mean that bioeconomic initiatives could pave the way for a new rural development model in Argentina.

Keywords: biomass; biotechnology; agro-ecology; territorial development; sustainability; cluster analysis

Citation: Dürr, J.; Sili, M. New or Traditional Approaches in Argentina's Bioeconomy? Biomass and Biotechnology Use, Local Embeddedness, and Sustainability Outcomes of Bioeconomic Ventures. *Sustainability* **2022**, *14*, 14491. https://doi.org/10.3390/su142114491

Academic Editors: Idiano D'Adamo and Massimo Gastaldi

Received: 19 September 2022
Accepted: 31 October 2022
Published: 4 November 2022

Publisher's Note: MDPI stays neutral with regard to jurisdictional claims in published maps and institutional affiliations.

1. Introduction

The importance of locally available resources has again come back into focus as a result of the current crises, and in this context also the bioeconomy, which focuses in particular on a more sustainable and efficient use of local resources [1]. However, the bioeconomy seems to continue a "contested field" [2] in the political debate. Two of the most prominent bioeconomy visions, concepts, and strategies come from the OECD [3], which is strongly biotechnology focused, and from the EU [4], which is more biomass oriented. There are also alternative concepts emerging, such as that from the European Technology Platform TP Organics, which follows a more agro-ecological vision, and stresses the inclusion of different stakeholders from science, politics, business, and civil society [5–7]. There is still no consensus on how a bioeconomy should be designed and how it could be anchored in society. Some authors even argue that these approaches are fundamentally unsuitable for achieving a societal transformation towards a truly sustainable bioeconomy when viewed in the context of global inequalities [8]. In general, alternative bioeconomy concepts that deviate from the mainstream discourse and are based on small-scale, agro-ecological models are usually underrepresented in the debate and marginalized [6].

This is also the case in Argentina, where the bioeconomy is mainly linked to genetically modified (GM) monoculture crops, intensive use of inputs, and export orientation, with a biotechnological and agro-industrial focus [9]. Argentine agriculture has been driven by the soybean model since 1996, when GM crops were first introduced, and has since expanded greatly in terms of acreage and production levels. This production model induced biotechnology research and innovations, such as drought-tolerant seeds and no-tillage systems [10], but has also had negative consequences on air and water quality, land use changes, land distribution, health and employment [11], and on deforestation [12].

However, official documents of the Argentine government stress the potential of the bioeconomy for regional development, for new industrial developments and local value added, for institutional frameworks, and for sustainable, decentralized, renewable energy supply [13,14]. Yet, as Tittor [14] (p. 325) points out for the case of Argentina: "Agroecological initiatives, a solidarity economy or de-centralized energy systems are not part of the debate." However, the author also mentions that "there are interesting and innovative small-scale projects developing as part of the bioeconomy framework" (p. 324), and that sustainability aspects are gaining in importance in the debate. Moreover, "the agricultural commodity production is still the mainstay of the Argentinian bioeconomy, although small-scale local initiatives, which also include socio-institutional and agro-ecological innovations, are coming up", as Sili and Dürr [15] (p.19) noted. These authors observe the coexistence of two bioeconomic development models, which, following Priefer et al. [16], can be categorized into the technological-based approach, which builds mainly on biotechnologies and genetic engineering for higher biomass production and is oriented towards international markets, and the socio-ecological approach, which envisions a decentralized, ecological agriculture for the development of rural areas through the creation of regional value chains.

This diversity of bioeconomic approaches have not yet been documented and analyzed in Argentina. There is an extensive bibliography at the international level, but only a few studies on the Argentine case, mostly on biotechnological approaches, especially case studies of companies and bioeconomic conglomerates of this sector. For example, on the seed industry that has developed locally adapted soybean seeds [17], on the biotechnology sector and its characteristics, strengths, and weaknesses [18], and on the factors which influence the development of some specific bioeconomic ventures [19]. These studies show that Argentina has caught up in the biotech sector and is now becoming more competitive internationally. Moreover, Argentina has a high biomass potential for developing its bioeconomy [20]. However, there is still no clear description and characterization of a socio-ecological approach in Argentina. Moreover, it is unclear whether and how the bioeconomic approaches can be clearly distinguished from each other in practice, or whether there are also mixed forms.

We have chosen Argentina as a potentially interesting example because, on the one hand, the bioeconomy in this country offers new development opportunities which can overcome the constraints of the previous prevailing models, which were characterized by the contrast between agricultural and industrial development and which can now overcome this in order to contribute to a more balanced territorial development [21]. On the other hand, the private sector has been crucial for the development of the bioeconomy, and only recently have public policies gained importance [22], so bioeconomic models might look different compared to countries where public strategies have been more prominent.

As a hypothesis, we suppose that the socio-ecological approach in Argentina is much less visible, because it has not the size, the capacity for political lobbying, or the same importance for exports, as the technology-based approach. We also hypothesize that there might be combinations of these ideal types of bioeconomic approaches. Despite its low visibility, the socio-ecological approach may have certain characteristics that could make it strategically important for more equitable territorial development, namely: the use of advanced scientific and technological knowledge, together with knowledge based on local experience, for the generation of dense local linkages, and the possibility of generating local employment and income. Different bioeconomic approaches in rural areas might follow different logics and generate different outcomes for local development, benefiting varying actors, such as small- or large-scale producers. In particular, the socio-ecological approach can be expected to be highly locally embedded, using mainly local resources from low-intensive, small-scaled farming [5,16], and the entrepreneurs to be highly committed to their local communities, a phenomenon that Korsgaard et al. [22] named "entrepreneurship in the rural". To our knowledge, the concept of local embeddedness has not been taken into account in the characterization of bioeconomic approaches so far, which we aim to change in this article.

The main purpose of this paper is to identify the diversity of bioeconomic approaches in Argentina, and characterize the socio-ecological approach, in order to make it more visible for the political debate, which then could also lead to more targeted support policies. So far, to our knowledge, there have been, no attempts to group bioeconomic enterprises into the different ideal types of the bioeconomy described in the literature. Based on some recent literature that categorize such bioeconomic approaches, we extracted categories that can be used to distinguish different bioeconomy types. Then, we used these categories in an online survey with 47 bioeconomic enterprises representing different sectors of the Argentine bioeconomy. The novelty of this approach lies in the operationalization of categories that characterize bioeconomic types and their application to the real business world. In this way, we aim to answer three important research questions: First, can different bioeconomic approaches be clearly distinguished in the case of Argentina? Second, what are the characteristics of the bioeconomic approaches that could be identified? Third, what kind of linkages do these approaches maintain with rural territories, and what impact do they have on sustainability there?

The paper is structured as follows: First, we describe the main bioeconomy typologies discussed in the literature, and the concept of local embeddedness of enterprises. Second, we explain the methods used. Third, we describe the three bioeconomic groups which emerged from the cluster analysis. Fourth, we discuss the bioeconomic approaches and their linkage with territorial development. Finally, we provide recommendations for further research on the bioeconomy in Argentina.

2. Bioeconomy Typologies and Local Embeddedness

2.1. Bioeconomy Typologies

Bugge et al. [5] identified three ideal types of the bioeconomy: (1) a biotechnology vision, (2) a bio-resource vision, and (3) bio-ecology vision. These visions can be characterized by some key variables: aims & objectives, value creation, drivers & mediators of innovation, and the spatial focus. The aims and objectives of vision 1 are mainly focused on economic growth and job creation. This also applies to vision 2, where sustainability also plays a central role. Aspects such as sustainability, biodiversity, conservation of ecosystems, and avoiding soil degradation are crucial for vision 3. The value creation of vision 1 is based on biotechnologies, and the commercialisation of research & technology, while vision 2 focuses on the conversion and upgrading of bio-resources, and vision 3 on the development of integrated production systems and high-quality products with territorial identity. The innovation process of vision 1 follows a linear model of transforming biotechnological research into new products and processes, stresses the importance of cooperation with universities and research centers, and is based on patents, whereas the innovation drivers of vision 2 relate to optimizing the use of land, bio-resources, and waste, and is more interdisciplinary and network-oriented. Vision 3 drivers are based on the search for sustainable agro-ecological practices, re-use and recycling of waste, and efficiency in land use, and research and innovation activities are related to transdisciplinary sustainability issues. Finally, the spatial focus of vision 1 is on global biotechnology centers and regions, while vision 2 highlights the potentials for rural development, also in peripheral regions, as does vision 3, but with a stronger emphasis on the development of territorial identities and locally embedded economies.

The scientific and social debate on the bioeconomy was analyzed by Priefer et al. [16]. Using several key categories, the authors identified two different approaches, which they call the technology-based vs. the socio-ecological approach. The technology-based approach stresses the importance of increasing biomass production through intensification of agriculture, but increasingly also through laboratories, and the major role played in this process by biotechnologies and patents, multinational companies and global value chains, international competitiveness and innovations, centralized solutions and economies of scale, the partnership between politics, science, and companies, as well as the promotion of life sciences. In contrast, the socio-ecological approach emphasizes multifunctional, ecological agriculture, natural cycles and reduced resource consumption, the promotion of social innovations, the use of local knowledge, the strengthening of rural areas, the creation of regional value chains, a more localized food and energy supply based on small-scale, region-specific biomass production with greater participation by civil society, and inter- and transdisciplinary research. According to the authors, these approaches are not necessarily mutually exclusive, and some features might be connected or otherwise combined.

By doing literature research, expert interviews, and conference participations, Vivien et al. [7] characterized three ideal-types of bioeconomy narratives that revolve around the notions of socio-technical relations, governance, sustainability, and tensions/paradoxes. Bioeconomy type I, based on the works of Georgescu-Roegen, is defined as an ecological economy, respecting the limits of the biosphere, whereas bioeconomy type II is defined as a biotechnology based economy driven by science, and type III as a bio-based economy that replaces fossil fuels with biomass. Type I takes a strong sustainability approach, promotes an economy of prudence and sharing, and favors democratic, ecological planning, while criticizing pure technical solutions. Type II, on the other hand, sees the techno-scientific promises of the bioeconomy, is based on commodification of knowledge (patents), but has a weak sustainability approach. Bio-refineries are at the heart of bioeconomy type III, which pursues mission-driven policies to identify ecological transitions, by substituting products and processes. However, this is still done within a weak sustainability concept, and with the problem that pressure on resources and land could increase.

Through discourse analysis of official policy documents and stakeholder interviews, Hausknost et al. [6] propose two dimensions by which the different visions of the bioeconomy can be located in a continuum: on the one hand the technological dimension, ranging from visions of agroecology to industrial biotechnology; and the political–economic dimension ranging from notions of sufficiency to capitalist growth. This allows the authors to distinguish four different areas, namely (1) Sustainable Capital, founded on the belief of bio-technologies and industrial innovations for further, sustained and sustainable economic growth; (2) Eco-Growth, based on the narrative of agro-ecological innovations for intensification and efficiency gains; (3) Eco-Retreat, characterized by the combination of ecological practices and socio-economic sufficiency; and (4) Planned Transition, constituted by a high-tech vision together with a sufficiency approach. Different actors (state, business, academia, civil society) not only have different visions, but also their roles for the transition to the bioeconomy need to be critically assessed.

In analyzing the European bioeconomy agenda, Levidow et al. [23] describe what they call the (dominant) life science agenda versus the (marginalized) agro-ecological agenda. The former aims to modify plants and animals for greater productivity or new uses, and to convert biomass into various inputs and outputs that can be de- and recomposed for different industrial products, and relies on laboratory knowledge and bio-refinery plants. The latter looks for agro-ecological systems that minimize the use of external inputs, emphasize product identity with territorial characteristics that can be recognized by consumers and therefore add local value, and is based on small-scale farming units and knowledge of agro-ecological methods.

Although this is a short literature review, it presents key issues in the current discussion on the bioeconomy. The typologies elaborated by the authors cited, referred to as bioeconomic "visions", "approaches", "agendas", "narratives", or "models", reflect official documents, research publications, public discourses, etc. Yet, these typologies have been developed using different categories that are not necessarily compatible with each other. Moreover, none of the authors listed above analyzed whether in the real world of bioeconomic ventures, the different bioeconomy types can be found and clearly distinguished. Furthermore, the territorial dimension of the bioeconomy is mentioned by some authors, but not described in detail. If the bioeconomy is to foster territorial development, it seems crucial that the territorial dimension of the bioeconomy is considered more thoroughly (see the following section). As our aim was to apply the concepts to business companies, we extracted three main dimensions, which were then further used for the development of variables that could be utilized in a cluster analysis to characterize bioeconomic approaches (see Section 3.2): (1) biomass production and use; (2) technology, research, and innovations; and (3) sustainability impacts and territorial linkages. Table 1 shows in highly summarized form how the authors describe the main characteristics of the different approaches in relation to (1)–(3).

Table 1. Characteristics of different typologies of the bioeconomy.

	Biotechnological Approach <————————————————————————————> Socio-Ecological Approach			
Bugge et al. [5]	Biotechnology Vision (1) biomass transformation into marketable products (2) biotechnologies based on R&D (3) global markets	Bio-Resource Vision (1) upgrading bio-resources and optimizing land use and waste (2) engineering and science (3) rural development, but weak sustainability		Bio-Ecology Vision (1) sustainable agro-ecological practices (2) transdisciplinary (3) territorial identity, strong sustainability
Hausknost et al. [6]	Sustainable Capital (1) eco-efficient use of renewable resources (2) biotechnologies and industrial innovations (3) global economic growth	Planned transition (1) reduced resource use (2) high biotech vision (3) sufficiency approach, global trade	Eco-Growth (1) organic farming (2) agro-ecological innovations (3) regional, small-scale	Eco-retreat (1) ecological practices (2) small-scale, democratic control over technologies (3) socio-economic sufficiency
Vivien et al. [7]	Science-based economy (1) industrial biotechnologies, cell factories (2) commodification of knowledge, patents (3) weak sustainability	Bio-based economy (1) replacing fossil resources by biomass (2) heterogeneous knowledge base (3) weak sustainability		Ecological economy (1) respecting the limits of the biosphere (2) prudence, against "promethean technologies" (3) strong sustainability
Priefer et al. [16]	Technology Based-Approach (1) intensive production, efficiency gains (2) biotechnologies, competitiveness, technology leadership, patents (3) multinational companies and global value chains		Socio-Ecological Approach (1) multifunctional, ecological agriculture, reduced resource consumption (2) social innovations, local knowledge, transdisciplinary research (3) regional value chains, autarchy, local stakeholders	
Levidow et al. [23]	Life science trajectories (1) modifying plants and animals, conversion of biomass (2) lab knowledge and bio-refineries (3) competition in global markets		Agro-ecological trajectories (1) systems that minimize external input use (2) knowledge systems for agroecology (3) territorial identity and distribution systems	

Source: own elaboration.

Different visions translate into various paths to follow for a transition to the envisioned bioeconomy: for example, a technology-based transition based on research and on technological innovations that can be transformed into competitive bio-based products, opposed to a socio-ecological transition that follows sustainability concerns of the biosphere's supply and regeneration capacities. However, there is also a need and possibility to integrate the different perspectives of the transition [24,25]. The bioeconomic transition pathways proposed by Dietz et al. [26] are not so much influenced by visions, but take into account the different roles that techno-economic mechanisms can play in the development of the bioeconomy, especially for factor substitutions and for efficiency gains, and differentiate the (1) fossil-fuel substitution, (2) primary sector productivity enhancement, (3) new and more efficient biomass uses, and (4) low-bulk and high value applications pathway. Sili and Dürr [15] proposed, for the case of Argentina, a fifth pathway defined as the "generation of new, innovative products and services with local value added."

A value chain approach to differentiate bioeconomic types and a potential upgrading process is proposed by Mac Clay and Sellare [27]. The authors distinguish six value chain models with different characteristics of biomass and biotechnology use and innovation processes. The concept of bioeconomic upgrading from high-volume, low value to low-volume, high value chains describes possible trajectories towards lower environmental impacts with higher economic opportunities. Finally, Bröring et al. [28] differentiate four innovation types (IT) in the bioeconomy, namely (I) substitute products, (II) new processes, (III) new products, and (IV) new behavior. These innovation types are associated with specific challenges, related to markets, value chains, resources, innovation capacities, consumers, and sustainability. For example, for IT I, the integration of bio-based substitutes into existing value chains is a particular challenge, and there is market competition with the fossil-based industry. For IT II, technology adoption and diffusion as well as knowledge transfer are important challenges for establishing new processes and value chains. One of the biggest challenges of IT III is innovation capacity, as new products require long development periods, high investments, and the transfer of biotechnologies to users. Finally, for IT IV, behavioral changes are confronted with different problems such as lack of acceptance (of new products or processes), knowledge gaps, insufficient public communication, and unwillingness to change. The typology could be used to analyze the dynamics and impact of policies and other factors on innovations in the bioeconomy, as well as their sustainability performance [28].

2.2. Local Embeddedness

One of the advantages of the bioeconomy is seen in the promotion of rural development and stimulation of local economic and social development, with the possibility to adapt to local characteristics and based on knowledge of local stakeholders [29]. Especially the agro-ecological version of the bioeconomy is associated in the literature with local knowledge, small-scale production units, shorter supply chains, territorial identity, and rural development [30].

Bioeconomic ventures in rural areas may follow different logics and therefore achieve different outcomes for local development. One important differentiation in this respect might be what Korsgaard et al. [22] idealized as "entrepreneurship in the rural" vs. "rural entrepreneurship". The authors argue that the former is weakly embedded in the rural space, i.e., follows a profit-oriented choice of location, which can also vary depending on the (economic) framework conditions. Entrepreneurs in the rural react to economic incentives (such as low land prices or labor costs) in their location decision, and the location is considered as a space for profit-making, but they are not very engaged in local communities and have no specific interest in rural development (which is not to say that these ventures cannot have positive development impacts). In contrast, the latter represents activities that rely on specific local resources that cannot be easily exchanged, and involves a particular commitment of the company to its location and is heavily place-related, which entails a different logic regarding the location decision and makes a spatial relocation more difficult or unlikely. Rural entrepreneurs see the location also as a space for social life, and they create new values with local resources that contribute to the development of the communities from which these resources originate, not only in economic, but also in socio-cultural terms [22].

Rural entrepreneurial processes are influenced by their spatial context such as the resource base and market outlets, which also determine whether these processes are more or less embedded locally. On the one hand, this depends on the resource endowment and the extent to which rural enterprises use local resources. On the other hand, embeddedness also hinges on whether and how rural entrepreneurs connect their local place to non-local spaces, i.e., whether they link (or not) the local to the national or global economy. "Bridging" of localized resources and products to non-local spaces, i.e., market outlets outside the territory, can lead to dynamics and opportunities for the local economy [31]. This also means that some enterprises might be locally embedded through their resource base, but not through their customers (which are mainly non-local), and vice versa, leading to four types of rural entrepreneurs and their embeddedness, depending on the extent to which local resources are used and the extent to which there is "bridging" to non-local costumers: (1) low embeddedness, but high bridging of local resources to non-local customers, (2) high embeddedness and high bridging to non-local customers, (3) high embeddedness and low bridging to non-local customers, and (4) low embeddedness and low bridging.

This means that enterprises with high local embeddedness might share many characteristics of the socio-ecological approaches (see Table 1), for example, minimizing the use of external inputs and showing local identity, which are mentioned by Bugge et al. [5] and Levidow et al. [23]. The importance of regional value chains, autarchy, and the connection to local stakeholders [16] also aims in this direction. However, by bridging local resources to non-local costumers, there might also be cases that belong more to the approach of biomass transformation into marketable products [5] by replacing fossil resources by biomass [7], and cases where enterprises are active on international markets, a characteristic rather belonging to the biotech model [16]. We have therefore included local embeddedness as an additional category to characterize bioeconomic approaches.

3. Methods

3.1. Selection of Cases and Sample Structure

In the absence of consolidated databases in Argentina, which might be, inter alia, due to lack of a clear definition of the bioeconomy, and thus a clear delineation of which enterprises "belong" to the bioeconomy and which do not, we took an ad hoc approach using three of the existing, but surely incomplete, lists of bioeconomic ventures. First, we consulted the list of the Ministry of Science and Technology and the Ministry of Agriculture of Argentina, which comprises 110 ventures linked to the bioeconomy. Second, we used the list of the web portal "Bioeconomy in Argentina", developed by Argentine agricultural journalists, which includes 30 ventures. Thirdly, we utilized a list of 20 companies interviewed in the documentary on the Argentine bioeconomy, produced by public television. In total, a list of 160 ventures was compiled, but due to the high number of repetitions between the different lists, a final list of 102 cases was consolidated.

Since we do not know the criteria used to create the lists, nothing can be said about possible selection biases. Nevertheless, the companies included are highly diverse in terms of sectors, regions, scale, technological level, etc. They also seem to belong to different bioeconomic approaches (shown in Table 1). For example, there are companies from the biotechnology sector, there are biofuel enterprises, and there are organic producers. Our aim was to include as many as possible of the listed firms, so we tried to contact all of the 102 firms by sending emails and, if no reaction occurred, by reminding them with a telephone call. In the end, 48 of the 102 contacted enterprises filled out the questionnaire (response rate of 47%). One of the enterprises had to be eliminated because it turned out to be a recycling of electronic devices firm, resulting in the following sample structure of 47 enterprises (see Table 2). The distribution of the sectors as well as the regions and the size of the companies (measured by the number of employees) is relatively even. However, the Cuyo region is missing. The map in Appendix C shows the location of the 47 enterprises in the different regions of Argentina.

Table 2. Structure of the sample.

Sector	No. of Cases	Region	No. of Cases	Size	No. of Cases
Food and Beverages	10	Metropolitan	8	Very small	10
Bioenergy	12	Pampa	18	Small	13
Agro-Inputs	11	Patagonia	5	Medium	9
Pharma and Cosmetics	6	Northeast	8	Big	9
Biomaterials	8	Northwest	8	Very big	6

3.2. Data Collection

In order to operationalize the different types of the bioeconomy described above, our aim was to distill the main categories from the literature, find proper variables for each category, and formulate appropriate questions suitable for the use of a Likert-type scale. As mentioned above, we extracted three main topics with opposing views from literature, which refer to (1) biomass production and use (intensive agriculture vs. agro-ecological systems; large scale use of biotechnologies and biomass production vs. small scale, circular systems); (2) technology, research and innovations (technology-based vs. socio-ecology-based; life sciences, R&D and patent-based vs. transdisciplinary and agro-ecological practices-based), and (3) sustainability issues and territorial linkages (weak vs. strong sustainability; high input monocultures vs. conservation of ecosystems, biodiversity and soils; global vs. regional value chains; central regions vs. peripheral regions; international players vs. local stakeholders). Furthermore, we added from the literature of local embeddedness the resource base (local/non-local input and outlet markets), locational choices (a-spatialized vs. sustainable place making), cooperation with local stakeholders, and the role local identity plays for the entrepreneurs.

We grouped these topics into four categories: biomass, scale, technology, and territoriality, and used 19 variables to describe them, see Table 3. The exact questions are listed in the questionnaire in Appendix B. The variables are mainly ordinal (such as "very small" to "very high" or "0%", "1–24%", etc. to "100%"), so that a Likert-type scale could be used. We decided to use a 5-point scale as a sufficient, not too differentiated scale. Score "1" is supposed to be closest to the socio-ecological approach, and score "5" would fully represent the biotech–biomass focused approach, with the exception of the category "biomass", where low values are also to be expected for the biotech approach. Taking the category "biomass" as an example, score "1" stands for low volumes of biomass used, produced by small-scale farms with no intensive production methods, and score "5" would be a highly intensive production of high biomass volumes by large farms. "Size" was measured by three variables (turnover, number of employees, and production volumes compared to other companies). Again, low levels are to be expected for the socio-ecological approach. In the category "technology", the percentages show how much of total production value depends on biotechnologies, local knowledge, and patents. We wanted to know how much companies cooperate with scientific and with private organizations. The socio-ecological approach was expected to fall into low levels of all variables except for the variable "local knowledge". Hence, for this variable, low scores mean high importance. In the category "territoriality", we asked which markets are mainly served, from where inputs mainly come from (excluding biomass, which was asked in the "biomass" category), and how much international prices influence the profitability of the business. Moreover, we wanted to know to what extent products are based on local identity, to what extent business activities contribute to an improved environment and to a more sustainable use of natural resources, and to what extent the company interacts with local stakeholders. Note that some variables are inversely formulated, i.e., the higher the variable outcome, the lower the score. This occurs mainly in the category "territoriality", meaning that higher local embeddedness and higher contribution to the environment are associated with lower scores, which is expected for the socio-ecological approach. Three of the variables are nominal (Biomass 2, Territoriality 1 and 2, which go from "local" to "global"), so that no scale points were used.

Table 3. Variables used to differentiate bioeconomic approaches.

Categories	Scale				
Biomass	1	2	3	4	5
1: Volumes	Zero	<10 t	10–100 t	100–1000 t	>1000 t
2: Origin *	Local	Regional	National	L. America	Global
3: Production Scale	Very small	Small	Medium	High	Very high
4: Intensity	No use	Low	Medium	High	Very high
Size	1	2	3	4	5
1: Turnover (1000$) **	<50	50–250	250–1000	1000–10,000	>10,000
2: Compared to others	Much smaller	Smaller	Average	Bigger	Much bigger
3: No. of Employees	1–5	6–20	21–100	101–500	>500
Technologies	1	2	3	4	5
1: Biotechnologies	0%	1–24%	25–49%	50–74%	75–100%
2: Local knowledge	75–100%	50–74%	25–49%	1–24%	0%
3: Patents	0%	1–24%	25–49%	50–74%	75–100%
4: Scientific coop.	Not	Not Much	Medium	Much	Very much
5: Private sector coop.	Not	Not Much	Medium	Much	Very much
Territoriality	1	2	3	4	5
1: Markets *	Local	Regional	National	L. America	Global
2: Suppliers *	Local	Regional	National	L. America	Global
3: International prices	No influence	Some inf.	Medium inf.	High inf.	Very high inf.
4: Local identity	Very much	Much	Medium	Not much	Not
5: Environment	Very much	Much	Medium	Not much	Not
6: Natural resources	Very much	Much	Medium	Not much	Not
7: Local stakeholders	Very much	Much	Medium	Not much	Not

* no scale used; ** Argentine Pesos (ARS); at the time of the interviews, ARS 100 = USD 1.

In addition to the 19 variables shown in Table 3, an open-ended question on the type of products and services was asked, as well as multiple choice questions with predetermined answers on four key characteristics of the firms, three reasons of their locational choice, three levels of product specialization, and the type(s) of pathways followed (see the questionnaire).

The survey was conducted as an online survey between October and December 2021. An online survey with a standardized questionnaire was considered to be an appropriate tool, firstly to do justice to the ongoing epidemic and secondly to provide the ready-made scales. The answers were formulated in a Likert-type form, for which Google forms was used. The interview questionnaire was developed in Spanish; an English translation is added in Appendix B.

3.3. Data Analysis

In a first analytical step, we divided the highly diverse 47 enterprises by their description of activities, complemented by information given in their websites, into five sectors: (1) Bioenergy, (2) Biomaterials, (3) Food and beverages, (4) Cosmetics and pharmaceutics, (5) Agro-inputs, see Table 2.

A hierarchical cluster analysis (Ward method, using quadratic Euclid distance) was carried out with the values of the ordinal variables. However, we had to exclude two of these variables (Size 1 and Size 2) due to missing data. We decided to exclude the two variables rather than the (three) cases so as not to reduce the total number cases, which were already few. Moreover, the variable Size 1 ("turnover") was biased (64%) towards the upper scale point of "5" (over ARS 10 million), the cut-off of which had been chosen too low, so that the variable was not able to differentiate the enterprises well. In addition, there were four answers of "zero biomass used", so that the following questions on origin, scale, and intensity of biomass use had be automatically scored as "1", even if there was no biomass production at all.

The use of cluster analysis was intended to answer the question of whether there are relatively homogenous groups, as expected the biomass, the biotech, and an agro-ecological group, or whether there are more distinguishable groups with different characteristics, or whether there are no clearly distinguishable groups at all. This could lead to a more differentiated typology, for example, small-scale biotech enterprises with high local embeddedness, or large-scale, international biomass-related enterprises with agro-ecological characteristics, etc.

A first analysis using 14 variables showed no preferred number of clusters. With a predefined number of clusters (3), results showed highly diverse groups of enterprises where no clear pattern could be detected. We then decided to concentrate on only four variables that were considered decisive for the three approaches, namely volumes of biomass, size, use of biotechnologies, and cooperation with local stakeholders. The result for three clusters showed distinguishable groups, but with some enterprises belonging to the "wrong" cluster (two clearly biotech-focused and one clearly biomass-focused companies).

We then decided to use a very simple method, guided by the underlying knowledge and theory of the different types of production models: Firstly, the cluster of enterprises which probably would belong to the biomass-based approach was defined by separating all companies with high and very high biomass utilization volumes (scale 4 and 5), resulting in a group of 21 companies, mainly from the bioenergy and food sectors. Secondly, the biotechnology group was defined as all companies that mainly base their productive processes on biotechnologies (level 5 and 4) plus the companies with medium levels of biotechnology use (level 3) if cooperation with scientific and technological organizations has some importance for these companies (at least level 2). There was an overlapping of six enterprises with the biomass approach that were excluded from the biotech group, resulting in 15 companies, mostly active in pharmaceuticals and agro-inputs. The third bioeconomic group was elaborated as the residual of all companies not belonging to the already defined biomass and biotechnology groups, resulting in 11 companies belonging to different sectors, in particular bio-products and foodstuffs. The result of this simple algorithm was similar to the cluster analysis with four variables, but with the difference that the three biotech companies now belonged to the "right" group. We therefore decided to use the three clusters produced by the simple algorithm, using only two variables (biomass and biotechnology use).

Each of the three clusters was characterized by the number of enterprises belonging to each of the five scale levels of each variable. It was also determined how the other variables not used for cluster analysis were distributed between the clusters. For each variable, analysis of independence was carried out, using Fisher's exact test statistics. A non-parametric correlation analysis (using Spearman's Rho) was performed with the scale points of all ordinary variables for each cluster separately. In this way, it can be investigated whether there are strong and significant relationships between certain variables of the different clusters. For some of the variables that showed such relationships, a second cluster analysis was then taken for each of the three clusters (using Ward method) to detect sub-groups in each cluster.

4. Results

Derived from the literature and the cluster analysis, we have identified three bioeconomy approaches in our material: the biomass, the biotechnological and an alternative cluster we will describe below in further detail. For this cluster, according to the main characteristics of this group, and deviating from the names used in literature so far, we propose an alternative name, i.e., the locally embedded bioeconomy approach, in short, the bioembedded approach.

Table 4 presents a synthesis of each of these clusters according to the different variables of analysis. To differentiate more clearly, only the most frequent scale levels are shown (if there are equally frequent levels, all of them are listed), and used the same colors as in Figure A1 of the Appendix A, where all levels are presented. Dark green stands for the lowest level 1, dark red for the highest level 5, and medium level 3 is painted in yellow. The differences were tested with Fisher–Freeman–Halton exact statistic, and are significant for biomass 1, biomass 3, size, technology 1, and territorial 2.

Table 4. Cluster characteristics of scale variables.

Variables	Fisher Exact	Cluster 1. Biomass (n = 21)	Cluster 2. Biotechnology (n = 15)	Cluster 3. Bioembedded (n = 11)
Biomass 1: Volume	51.3 ** (0.000)	>1000 tn: 71%	<10 tn: 73%	<10 tn: 82%
Biomass 2: Origin	6.3 (0.346)	local: 71%	local: 53%	local: 82%
Biomass 3: Scale	15.9 * (0.025)	medium: 48% very high: 29%	small, very small, medium: 27%	very small: 64%
Biomass 4: Intensity	12.6 (0.092)	medium: 38% low: 24%	no use: 47%	no use: 36% low: 36%
Size 3: No. of Employees	19.5 ** (0.005)	101–500: 33% >500: 24%	1–5: 33% 6–20: 33%	1–5: 45% 6–20: 27%
Technology 1: Biotechnologies	28.4 ** (0.000)	level 2: 43% level 1, 3: 19%	level 5: 60%	level 2: 64%
Technology 2: Local knowledge	8.8 (0.324)	level 2: 38% level 3: 29%	level 1: 33% level 4: 27%	level 3: 45% level 1, 2: 18%
Technology 3: Patents	7.5 (0.490)	level 1: 57%	level 1: 60%	level 1: 45% level 2: 27%
Technology 4: Scient. Cooperation	4.7 (0.848)	level 4: 38% level 3: 29%	level 4: 33% level 2: 20%	level 3: 27% level 4: 27%
Technology 5: Private Cooperation	6.1 (0.682)	level 4: 38% level 2, 3, 5: 19%	33% level 1 27% level 4	45% level 4 18% level 3,5
Territorial 1: Markets	5.3 (0.486)	national: 48% international: 29%	national: 53% international: 27%	national: 73% international: 27%
Territorial 2: Suppliers	12.9 * (0.022)	national: 76% local: 14%	international: 40% national: 40%	national: 36% international: 27%
Territorial 3: Internat. Prices Influence	3.5 (0.790)	very high: 33% high: 29%	high: 47% medium: 33%	medium: 36% high, very high: 27%
Territorial 4: Local Identity	3.3 (0.986)	much: 33% very much: 29%	medium: 33% much: 33%	very much: 36% much: 36%
Territorial 5: Environment	8.7 (0.341)	very much: 38% much: 24%	much: 53%	much: 36% very much, medium: 27%
Territorial 6: Natural resources	8.9 (0.304)	much: 43% very much: 38%	much: 33% not much: 27%	very much: 36% much: 27%
Territorial 7: Local Stakeholders	11.5 (0.115)	much: 52% very much: 33%	much: 47% medium, not much: 20%	much: 45% not much: 27%

* significant ($p < 0.05$) ** significant ($p < 0.01$).

Cluster 1 uses in 71% of the cases more than 1000 t of biomass, whereas in the other two clusters, most enterprises use less than 10 t per year. The scale of biomass production is also medium or very high in Cluster 1, while it is mostly very small in Cluster 3. Cluster 1 consists of mainly larger companies with more than 100 or even 500 employees, while the other two clusters mostly employ only up to 20 people. Of Cluster 2, 60% heavily use biotechnologies, while in Cluster 3 most enterprises (64%) use biotechnologies only to a small extent. Finally, input suppliers (excluding biomass) are overwhelmingly (76%) national in Cluster 1, while in Cluster 2 and 3 the picture is more mixed, including international suppliers. It seems that the equipment necessary for Cluster 1 enterprises can already be produced by national suppliers (provided that they do not import them), while some specialized inputs needed for Cluster 2 and 3 still require imports.

Even if for the other variables no significant relationships with the clusters could be detected, Table 4 demonstrates that the most frequent scale levels differentiate between the clusters. For example, 82% of enterprises of Cluster 3 source their biomass locally, which is only the case for 53% of Cluster 2 enterprises. Scientific cooperation interestingly is higher for Cluster 1 than for Cluster 2, which could mean that biomass-related enterprises nowadays are searching for new knowledge and innovation capacities, while biotech companies might be more independent from public R&D institutions, as they often possess their own laboratories and research departments. Also interestingly, patent use is more common in Cluster 3, where only 45% of enterprises do not use patents, compared to 57% and 60% of Cluster 1 and 3, respectively. Cooperation with the private sector is stronger in Cluster 1 and 3, where 38% and 45% of enterprises fall in level 4, which is only the case for 27% of Cluster 2. 73% of enterprises in Cluster 3 serve the national market, whereas only around half of Cluster 1 and 3 enterprises do this. Local identity is slightly more important for Cluster 3 enterprises (72% high or very high, in contrast to 62% and 60%, respectively, of Cluster 1 and 2). Only 9% of Cluster 3 enterprises do not contribute positively to the local environment, whilst 27% of Cluster 2 and 20% of Cluster 1 stated this. Also, more enterprises (27%) of Cluster 2 contribute only little to sustainable use of natural resources, probably because many of these enterprises do not use any biomass at all. Finally, cooperation with local stakeholders is most common for Cluster 1.

Another feature can be detected in Table 4, and even better in Figure A1 (see Appendix A): there are some intra-cluster differences, which might make subgrouping worthwhile. For example, in Cluster 1, big companies predominate, but there are also some small businesses (24%) involved. Around half of enterprises of Cluster 2 do use local, traditional knowledge, the other half, not (much). Of Cluster 3, 45% cooperate much with local stakeholders, 27% little, etc.

Apart from some variables, clusters are also not independent from the sector they belong to, the pathway they follow, or the type of products they produce, see Table 5: the bioenergy sector predominates (57%) in Cluster 1, whereas in Cluster 2 the agro-input (40%) and the pharmaceutical sector (33%) stand out, and in Cluster 3, the biomaterial and the food sector are, each with 36%, most important. In Cluster 1, 57% of the enterprises follow pathway 1, substituting fossil resources, whereas Cluster 2 enterprises tread pathways 2 and 4, the productivity enhancement and the low volume–high value pathway, and 45% of Cluster 3 take pathway 5, creating new, innovative products with local value added. Finally, Cluster 1 is mainly producing standardized (43%) and specialized products (47%), while Cluster 2 (53% and 47%, respectively) and 3 (45% and 36%, respectively) concentrate on niche and specialized products.

Table 5. Cluster characteristics of qualitative variables.

Variables	Fisher Exact	Cluster 1. Biomass (n = 21)	Cluster 2. Biotechnology (n = 15)	Cluster 3. Bioembedded (n = 11)
Regions	11.4 (0.181)	Pampa 48% NOA 24%	Pampa 48% Metropolitana 24%	Patagonia 27%
Pathways	23.7 ** (0.000)	P1 57%	P2 40% P4 40%	P5 45% P1, P2 18%
Product Characteristics	14.1 ** (0.005)	Specialized 48% Standardized 43%	Niche 53% Specialized 47%	Niche 45% Specialized 36%
Sectors	30.5 ** (0.000)	Bioenergy 57% Food 23% Agro-inputs 14%	Agro-inputs 40% Pharmacy 33% Biomaterials 20%	Food 36% Biomaterials 36% Agro-inputs 18%
Keywords	no results possible	Local development 18% Sustainability 15% Add value 15%	Innovation 20% R&D 20% Biotechnology 15%	Add value 22% Sustainability 20% Innovation 15%
Location criteria	17.9 (0.530)	Raw materials 43% Local connections 21% Social, cultural, and environmental quality 10%	Local connections 24% Social, cultural, and environmental quality 14% Raw materials, skilled labor 14%	Social, cultural, and environmental quality 25% Local connections 21% Raw materials 17%

** significant ($p < 0.01$).

4.1. Cluster 1. Biomass Approach

The companies that make up this cluster are characterized by the use of large quantities of biomass (71% of them use more than 1000 tons), which, due to the high cost of its mobility, comes from the same areas (71%). These companies often integrate primary production with the transformation of biomass into bioenergy or foodstuffs for human consumption, but more especially for animal consumption. In short, these are generally companies that have historically developed their activities in primary and food production (soya, maize, meat, sugar), but which have been able to move up the value chain, taking advantage of their experience in the primary and processing sector, the available infrastructure and equipment, and the scientific and technological facilities in their regions. The biomass is produced mainly in medium (48%) and very big (29%) production systems, with medium (38%) or high and very high (together, 33%) production intensities. The pathway followed by most enterprises (57%) is the substitution of fossil resources, meaning mainly bioenergy production.

They are generally larger companies: 33% of them have more than 100, and 24% more than 500 employees. The technological levels are low to medium, as they use generic, internationally recognized technology, which is why these activities can be replicated in different places without any inconvenience. The companies of this cluster maintain a high level of cooperation with scientific and technological organizations from which they obtain information or with which they build their innovation processes, but more than anything else they maintain a high level of links with other companies and with local stakeholders, especially municipalities and provincial governments with which they maintain cooperation initiatives, especially for the creation and maintenance of infrastructures or for bureaucratic and administrative management.

These companies mainly produce for the national (47%) or, especially in the case of the larger ones, international markets (29%). Meanwhile, their main suppliers are national (76%). The reason for this could be that in the biomass-related sector, the requirement for highly sophisticated equipment is not that high, or that Argentine suppliers have already caught up technologically and specialized their production, so that they are able to offer the required modern technologies. Most companies (62%) stated that they attach great importance to the preservation of the environment and the care of natural resources

(81%), perhaps because they depend on local biomass production. In this sense, these are companies that have a high level of embeddedness, as also suggested by the most often mentioned key words "local development" (18%), "sustainability" (15%), and "added value" (15%). This level of anchoring in the territory is also manifested in the location criteria proposed by the actors: nearly all (95%) of enterprises, or 43% of total answers, mentioned as one of the main location criteria the availability of raw materials. This is one of the reasons why most of these companies are located in the Pampas (48%) and the NOA (24%) region.

Correlation analysis of the variables, using Spearman's Rho, allows for the identification of certain key elements within this bioeconomic approach. Firstly, there is a strong relationship between the volume of biomass used and the size of the companies. (0.523*), meaning that enterprises with higher biomass volumes tend to be larger. This is not really surprising, but it is a peculiarity of the biomass cluster, not relevant for the other clusters, where size is not significantly correlated with biomass use. Secondly, a strong relationship (-0.477*) can also be observed between size and cooperation with the private sector (Technology 5), meaning that enterprises with bigger size tend to have more private sector cooperation. Again, this result is not unexpected, but it differentiates the biomass group from the other two clusters, where no strong and only insignificant correlations exist. Thirdly, the correlation index also allows us to observe that there is a strong correlation between company size and cooperation with local stakeholders (-0.478*), i.e., the larger the companies are, the more links they have with these stakeholders. This correlation is also negative for Cluster 2 and 3, but less strong (-0.378 and -0.162, respectively) and not significant ($p = 0.165$ and $p = 0.635$). Fourth, from a technological point of view, significant correlations can also be observed between Technology 1 and Technology 4 (0.433*), and 5 (0.483*), meaning that the more biotechnologies are used, the more scientific and private sector cooperation exist. This is quite logical as these companies require the support of science and technology centers to drive high-tech processes. These correlation are not strong and insignificant for Cluster 2, the biotech group, probably because the values for Technology 1 are all very high (level 4 or 5) for this cluster, and because these companies have their own R&D teams, with less need to rely on other partners. The same holds true for the correlations between Technology 1 and Territorial 4 (-0.584**) or Territorial 5 (-0.507*), meaning that use of more biotechnologies is associated with more local identity and more environmental preservation. Finally, the biomass cluster has strong and significant correlations between Technology 5 and Territorial 3 (-0.527*), Territorial 4 (-0.438*), Territorial 5 (-0.668**), and Territorial 6 (-0.438*), implying that more private sector cooperation goes along with less influence of international prices, more local identity, more environmental preservation, and sustainable use of natural resources. In sum, it seems that within the biomass-focused group two tendencies could be detected: as size increases, cooperation with the private sector (for example, with local biomass producers) and with local stakeholders (such as local authorities) becomes more important, and this translates into more local embeddedness. Additionally, the more biotechnologies are used, the more important cooperation with other sectors becomes, for example with local R&D centres, which also leads to stronger local embeddedness.

Because of these strong correlations, an intra-group cluster analysis was carried out, using Size, Technology 1, and Biomass 1 as variables, which allowed the observation that there are three sub-groups within this bioeconomic approach: one composed of large companies with a low technological level, especially linked to the basic production of biofuels; another of medium-sized companies with high technology levels, linked to productive integration with various products; and a third subgroup of small companies with low technology, mainly oriented towards the production of food and some type of basic processing. An example of the first group is a large company which produces sugar and its by-products, especially alcohol and bioethanol, and which exploits the demand for biofuels, using tested technology. An example of the second group is a fully integrated company which produces cereals, oilseeds, and meat intensively, and produces energy and waste from maize production, which is used for animal feed, and the waste from animal production is then used to generate electricity. An example of the third group is a company which produces juice concentrates and essential oils on a large scale, but also with a technology that is already well known on the market, using the large volumes of fruit and fruit waste production in their area.

4.2. Cluster 2. Biotechnological Approach

The companies that make up this cluster are characterized by their emphasis on the generation and application of modern biotechnological knowledge, supported also by a strong relationship with scientific and technological organizations present in the country (INTA, CONICET, INTI, among others) and by having international patents. The pathway most companies follow is the low volume–high value (40%) or the productivity-enhancement (40%) pathway. Knowledge is key to this cluster, so the keywords that characterize these companies are innovation, R&D, and biotechnology. In this bioeconomic approach, no (20%) or only few (73% of companies) biomass is used, which is mainly locally sourced (53%) but can also come from different parts of the country. The companies are mainly oriented towards the generation of specialized (43%) and niche (48%) products to improve the productivity of the agricultural sector in Argentina (seeds, liquid fertilizers, biostimulants, etc.), and pharmaceutical products. They are (very) small (66%) or, on the contrary, medium (27%) and very large (7%) companies, depending on their level of development. They are closely linked to national and international markets, both for the purchase of inputs and for the sale of their products.

The importance of innovation, R&D, and the relationship with universities or scientific and technological centers is also reflected in the importance of certain localization factors that these companies consider, such as local connections (24% of total answers) and the availability of skilled labor in the territory (14%). The need for good scientific and technological networks has determined that these companies are located especially in the Pampa region and in the metropolitan area of Buenos Aires.

There are some strong and important correlations in this bioeconomic approach, such as between Size and Territorial 4 (-0.610^*) and Territorial 6 (-0.564^*), meaning that bigger companies tend to have more local identity and more sustainable use of natural resources. Also, there was a strong correlation between Territorial 6 and Territorial 7 (0.705^{**}), meaning that more sustainable use of resources goes along with more cooperation with local stakeholders, or vice versa. In sum, the tendency in Cluster 2 seems to be that bigger companies are more locally embedded.

This led us to do another cluster analysis using the above cited variables with significant correlations, and detecting two sub-groups of companies: one of smaller companies that have a lower level of embeddedness, and another group of larger companies that have a higher level of embeddedness, which is a finding in contrast to the existing assumptions that the larger companies are alien or not linked to the rural territories. An example of the first group is a company which develops and manufactures formulations for pharmaceutical, dermo-cosmetic and nutraceutical industries aiming for the higher quality standards that each industry requires, and another one which develops inoculants, bio-controllers, and growth promoters for the agricultural sector. An example of the second group is a company which develops and commercializes in Argentina and in the international market in vitro diagnostic reagents for human health, biological research, and agro-biotechnology (animal health, plants, and seeds).

4.3. Cluster 3. Bioembedded Approach

The third bioeconomic cluster is clearly different from the biomass and the biotechnology approach. The approach referred to as the locally embedded bioeconomy approach, in short, the bioembedded approach, is characterized by using small amounts of biomass (82%) but of local origin (82%), having low levels of biotechnology use, and being (very) small sized (73%). They are mostly involved in the production of value-added food and bioproducts that make use of local or national biomass, and also make use of or recycle waste from other activities. These are generally niche (45%) or specialized (36%) products, with very clearly identified markets. Two elements that are very important for this approach are the attention to the protection of natural resources and the importance they attach to the local identity of their products or processes, which is why designations of origin or local quality seals are often used. The construction of a new model of local productive development, more respectful of the environment and the circular economy, is also evident in the key words most frequently repeated by these actors, where value addition, local development, and sustainability appear as the main business objectives. This concern for the environment, identity, and the construction of local development processes is also manifested through the criteria for the location of these initiatives, where the social and environmental quality of the place (25%) comes first, followed by the availability of connections with the territory (21%), and only thirdly by the supply of raw materials (17%). Cluster 3 is also distinguished by the main pathway it follows, P5, which is the innovation of products and services that create local value added. These initiatives are distributed throughout the country, although there is a greater concentration in Patagonia, a region that is seen as a fertile territory for the generation of new bioeconomic initiatives, linked to nature and organic production.

However, despite these basic characteristics, certain differences can be found within this group. The correlation analysis carried out on the variables of this cluster allows us to observe that in technological terms there is a strong correlation evident: the companies with more patents have more scientific cooperation (Technology 3 and Technology 4, 0.623*). This allows us to observe that there are two types of companies within this cluster, those that have lower technological levels, such as natural foods and/or their derivatives, and those that operate with higher technological levels, especially those that produce biomaterials. Companies of the first group comprise, for example, a company producing caiman skin and meat in a "ranching system", in partnership with local communities. The high-value skins are exported to different markets, and the meat is consumed in the domestic market. Other companies produce special, totally natural sauces and dressings, and organic food, e.g., high quality hazelnuts. The second group contain companies producing bioplastics from sugar, cellulose and proteins, and a company producing packaging, bags, and other compostable, organic products.

5. Discussion

(1)   Argentina's bioeconomy is path dependent, with a predominance of the biomass–biotech approaches, but new development paths with more socio-ecological traits are opening up.

In relation to the main purpose of this paper, which is to identify the diversity of bioeconomic approaches in Argentina, and make alternative models more visible for policy debate, the analysis has shown that the bioeconomic clusters identified are consistent with the history and production model of Argentina, a country endowed with large biomass resources, an agricultural tradition, and the presence of solid scientific and technological networks [32]. This is manifested in the presence of Clusters 1 and 2. In short, these two clusters mark two key elements, firstly that there is a strong availability of biomass that is beginning to be exploited in a much more comprehensive way by multiple activities through mechanisms for generating added value (production of oils and bioenergy), not only substituting products, but also generating new products and processes [28]. Secondly, there is a strong demand for products to boost agricultural production, in order to make Argentine agriculture much more competitive, which is clearly visible through Cluster 2 innovations of new processes for higher agricultural productivity. Both clusters 1 and 2 have different levels of complexity, one takes direct advantage of biomass resources, the other builds on agriculture but enhances it through knowledge-based innovation of new products and processes [28]. Apart from their differences, it is clear that both clusters indicate a path dependency [33] where synergies between the two lead to increased primary sector productivity, driven by innovations in the primary sector and in the biotech industry.

Cluster 3 emerges, on the one hand, as a product of a new look at natural resources and the environment. The development and growth of this sector of the bioeconomy is directly related to the new demand for organic products, new forms of consumption, and new forms of production that reduce waste and the environmental impact of production processes [34]. However, on the other hand, the emergence of this cluster is an important phenomenon as it indicates that the bioeconomy in Argentina does not necessarily only rely on traditional production sectors linked to the agricultural sector, or on the availability of biomass, but is opening up to new bioproducts or non-traditional or niche foods [5], which means that these activities are often not only located in rural areas, but increasingly in cities of different sizes, closer to consumer markets, or in areas with a stronger environmental protection and which cares about its identity, like the Patagonia region. Although this cluster still has lower levels of development than the other clusters built on the strong availability of biomass and the agricultural tradition, we expect that this bioeconomic approach will expand in the coming years due to the growing demand for bio-products and more specialized or niche foods. Yet, this might face challenges of behavioral innovations described by Bröring et al. [28].

However, there is one key element that has enabled the development of the bioeconomy in Argentina, namely the availability of universities and R&D, which has enabled the development of multiple bioeconomic activities [14]. Although the research did not focus on assessing the importance of Argentina's scientific and technological apparatus, a large part of the companies surveyed maintain different types of links with the scientific and technological system. This is most evident in Cluster 2, which depends on the generation and dissemination of innovative and modern knowledge, but surprisingly also in Cluster 1, which reflects a positive trend in the innovation capacity of traditional biomass-based companies. Even if Cluster 3 has lower values in scientific cooperation than Cluster 1 and 2, more than half of Cluster 3 enterprises use patents. However, despite the differences in the level of knowledge and technologies used, and despite the different scientific cooperation strategies of each cluster, it is important to note that most ventures of all clusters are highly prone and oriented towards the use of new knowledge as a key factor in the bioeconomy [3].

(2) Bioeconomic approaches in Argentina are partly consistent with contemporary conceptual approaches, but there is diversity within the clusters which makes a more differentiated analysis of the approaches worthwhile.

Answering our first and second research question, whether different bioeconomic approaches can be clearly distinguished and what their characteristics are, we found certain concordance between the theoretical models described above and the clusters identified in Argentina. Following Bugge et al. [5], there is a focus on the use and upgrading of biomass in Cluster 1, on biotechnologies and the use of research & technology in Cluster 2, and on high-quality products with territorial identity in Cluster 3. This cluster is also characterized, as the literature points out, by having small-scale production units, shorter supply chains, and strong relations with local stakeholders [30]. The identification of these three clusters were possible by using only two key variables: biomass volumes, and level of biotechnology used. A cluster analysis including all of the 14 variables did not reveal any clear patterns, indicating that in our sample, there is not such a clear continuum from a biotechnological to a socio-ecological way, as depicted in Table 1.

This fact can also be detected in Table 4 (and in Figure A1 of Appendix A), where two things stand out and deserve attention: first, that there is a strong intra-cluster diversity for some variables, and second, that there does not seem to be much difference between the clusters for some variables. For example, intra-cluster analysis showed that in Cluster 1 there are sub-groups with different company size and technological levels, and that this translates into more or less local embeddedness, i.e., more or less cooperation with local business, local R&D centers, or other local stakeholders. For Cluster 2, there seem to be a group of smaller companies with a lower level of embeddedness, and another group of larger companies with a higher level of embeddedness. Finally, also in Cluster 3, there seem to be at least two types of companies, depending on the technological level. One explanation for the intra-cluster diversity could be what Mac Clay and Sellare [27] described as different stages of value chain upgrading, with different levels of biomass use, technological innovation, investments, risks, cooperation, and knowledge-sharing. Some companies of each cluster might be more advanced in the upgrading process than others, leading to different characteristics and also, to different economic and environmental outcomes.

Another reason for the intra-group clustering might be the sectorial affiliation of the enterprises, as described above. It seems that the sector characteristics overlap with the characteristics of the clusters. The sectorial dimension of bioeconomy models is not explicitly discussed by the authors presented in the theoretical background section [5–7,16]; only Müller and Korsgaard [31] mention in their typology for embeddedness some prevalent sectors such as tourism or specialty food and beverages. Our research shows that in Argentina, there seem to be significant differences in terms of sectors dominating the three distinguished clusters: the bioenergy sector prevails in the biomass, the agro-input sector in the biotech, and the food and the biomaterial sectors in the bioembedded approach. The other, less predominant sectors within the clusters might then have different characteristics, be it size, technological level, or embeddedness.

The intra-group diversity also means that for some of the variables, differences between the groups are not very pronounced. For example, no significant differences between the three clusters could be found in the origin of biomass, in the importance of local knowledge, in the use of patents, or in terms of target markets and the influence of international prices. Expressed in another way, this means that in Cluster 3, too, highly technological, not just local and traditional, knowledge is used [5–7,16,23], and that its products also serve and are dependent on the national or international markets, and that Cluster 1 and 2 do not always consist of multinational companies acting on global value chains [5,6,16,23] or are based on the commodification of knowledge and use of patents [16,23], contrasting with some of the bioeconomic typologies described in Table 1. This calls for a more differentiated discussion on the characteristics of the approaches. For example, it would be highly interesting to have a closer look on small-sized enterprises of Cluster 1, or on enterprises that depend on local, traditional knowledge of Cluster 2, or on large sized enterprises of Cluster 3 which use patents, etc.

(3)   All bioeconomic approaches are strongly linked to the territory where the respective companies are present, but the clusters are locally embedded in different ways with implications on possible sustainability outcomes.

Regarding the third research question, what kind of links the different approaches maintain with rural territories, and what sustainability impacts they have, it was possible to observe different relationships established between bioeconomic types and rural areas [35]. All bioeconomic clusters exert local linkages and play a role in local development, albeit to a different extent and for different reasons. This raises the question if local embeddedness strictu sensu can be claimed by all clusters, or if it is, as we argue, a key characteristic of the third cluster.

The first cluster has local linkages through the high volumes of biomass used. The difficulty of mobilizing large volumes of biomass encourages local and regional production and therefore the development of supply chains in the same territory [36], even if it is not clear to which maximum distance biomass can be transported without the costs inhibiting the profitability and development of the enterprises. It can be supposed that the development of the bioeconomic activities of this cluster promote the construction of very dense productive networks, generating new jobs and boosting local development, but also the enrichment of the local productive fabric allows the improvement of the socio-economic conditions of the territory, improving the attractiveness of the territory, and the anchoring of income at local level. In this sense, this cluster can be a generator of virtuous cycles of development in its own territories, provided that the conditions for its development are met, such as, in addition to the presence of biomass resources, the availability of certain basic infrastructures, and the availability of qualified human resources for these activities. However, this model is often linked in Argentina to monoculture production systems, which can lead to land use change, biodiversity loss, and other adverse socio-ecological consequences. However, some of the companies also use organic side streams or waste as inputs, making a more efficient and cascading use of locally available biomass.

The second cluster is, as Korsgaard et al. [22] point out, an active participant in the globalized flow of resources, services and products across multiple locations given that it buys inputs or sells its products in multiple locations, but it also keeps links to the territories, partly because rural territories are the places where biomass is supplied, and also where agro-inputs (seeds, fertilizers) are needed. This cluster is also linked to the territory because the companies build strong relationships with the scientific and technological research centers and networks operating in the nearest cities, as in the model of the Italian industrial districts. These clusters contribute to build a greater exchange and mobility between the countryside and the city, which has allowed numerous small and medium-sized cities in Argentina to grow significantly in the last decades, often driven by the location of these types of companies and other services linked to them [37]. However, this model is often less biomass demanding, and therefore, on the one hand, creates less linkages to local biomass producers, but, on the other hand, may have less impacts on the environment. Moreover, it is often linked to an increase in agricultural productivity, and might lead to reduced land requirements.

The third cluster has a strong relationship with the territory, as it involves smaller companies that require biomass and local inputs, but above all because these companies need strong links with other private actors and with local and provincial governments, which support them through different mechanisms to be viable, i.e., their competitiveness depends on the networks they can build with other local actors to sustain themselves. This cluster also maintains a strong relationship with the territory due to the fact that many of the products generated include quality seals or have a local identity recognition, which expresses the close relationship between products and territory [38]. Following the reflections of Müller and Korsgaard [31], one can affirm that Cluster 3 initiatives have a special capacity to articulate global dynamics (markets, cultural, and consumer trends) with the capacities and characteristics of the territory ("bridging"), due to the fact that many of the products are either linked to international markets, or require global scientific knowledge, or require specific inputs imported from other countries. This articulation can often operate as a key factor in unlocking new opportunities for the development of the territory itself. Moreover, the ventures of this approach often explicitly follow agro-ecological principles, want to preserve the environment, and use only natural or recycled ingredients.

Considering these characteristics, we refer to Cluster 3 as the locally embedded bioeconomic approach, or the bioembedded approach, as it comes closest to the features described in the literature. This is not to say that the other approaches do not have linkages and are somehow locally embedded. On the contrary, despite existing assumptions that bioeconomic activities, especially those most dependent on biotechnology, are activities that tend to operate without relations to the local territory, and do not to generate dynamics of local development, the research shows that they build links that are key to territorial development. However, the three approaches contribute in different ways to the embeddedness and the development of the territories. Even if the analysis of the cases does not allow us to observe differences in the forms of action of the different entrepreneurs in relation to rural territories, which distinguish, as Korsgaard et al. [22] point out, "entrepreneurship in the rural" from "rural entrepreneurship", and further detailed research on the behavior of entrepreneurs is needed to investigate the different territorial logics of entrepreneurs, it can be postulated that the bioembedded approach, based on rather small-scale units, and with a strong local identity, follows a pathway of generating rather low-tech innovations that utilize resources locally available, thereby adding value, bridging local products to non-local customers, and contributing to a more circular, sustainable economy.

6. Concluding Remarks

The bioeconomy in Argentina is still strongly following the biomass and biotechnology approach, a path dependency which has developed over the last three or four decades, and which will certainly continue in the years to come. However, the bioeconomy is becoming more diverse. Different bioeconomic approaches will further develop, and probably coexist, in the short and medium term. Although the clusters identified show clear differences in the use of biomass, in technology, in the size of the companies, among other variables, there are two common elements in all clusters. Firstly, the interest in sustainability, the protection of natural resources, and innovation as a path to development, and secondly, the need of building networks and synergies with other actors or companies in the territory to generate better conditions for their own development and competitiveness. This means that there is a clear will to create greater embeddedness, which is considered a basic pillar of sustainability. These two observations suggest that the different bioeconomic initiatives could be setting the course towards a new model for the development of rural territories in Argentina, given that the model in force in recent decades has been characterized by a logic of little cooperation and articulation between economic agents, very little attention to the sustainability of natural resources, and especially very little concern for the future development of the territories. In fact, the bioeconomy appears as a new opportunity for territorial development in Argentina [38]. This will also require developing and implementing, besides general policies to foster the bioeconomy in Argentina [15], specific programs tailored to the needs of the different bioeconomic types. A good example of how a circular bioeconomy could be strengthened at the local level is the production of biomethane. The installation of biomethane plants requires the involvement and dialogue with local stakeholders, and offers the opportunity to achieve territorial energy self-sufficiency through small-scale systems [39].

However, we have to admit that our sample of 47 cases was rather small, and based on lists that were probably not systematically elaborated. For example, there were very few purely agro-ecological ventures in our sample, a fact that might have biased the characteristics of Cluster 3. We used an online survey with predefined answers, where it was not possible to ask further questions for clarifications or to go into detail. The survey used was opinion-based, so it reflects the subjective estimations of the interviewees on their local embeddedness, etc.

Having said that, we outline further research needs arising from this paper. A new research focused on the forms of linkage between bioeconomy and territory would allow to identify and analyze the type of relationship established between the development of the bioeconomy and the development of rural territories, as this will allow to argue for the need to deepen policies for territorial development through bioeconomic development strategies. In this sense, we put forward a hypothesis for rural development in Argentina: bioeconomic activities can be a clear factor in the development of rural territories, as they have a strong capacity for embedding, and to create virtuous cycles of development in rural areas, superior to the traditional extensive agricultural activity in Argentina, not only because of its capacity to use local resources, but also because of its capacity to link local dynamics with global markets while, at the same time, creating links to local stakeholders and a sense of local identity. This could be especially true for the bioembedded approach. Even if this approach currently might have limited relevance for territorial development in Argentina, it can make a significant contribution in the future and become an opportunity for a country deeply marked by territorial imbalances. Stronger support from public institutions is needed to promote this approach. However, such public support might not be enough. A solid social and productive base in the territories seems to be indispensable for the scaling up of those initiatives [40]. To explore these hypotheses further, it will be important to look more closely at the business models, local value addition, and territorial socio-economic and ecological impacts of the ventures of the different approaches. Our research could only give some qualitative indications of these developments, and more quantitative data are needed. In addition, research is needed on the social impacts of the different bioeconomic approaches, e.g., in relation to aspects such as food security, health, and gender. Future research should also look more closely at agro-ecological initiatives, so that the picture of the Argentine bioeconomy becomes even more diverse, and appropriate support policies can be tailored to the needs of these ventures.

Author Contributions: Conceptualization, J.D. and M.S.; methodology, J.D. and M.S.; software, J.D.; validation, J.D. and M.S.; formal analysis, J.D. and M.S.; investigation, J.D. and M.S.; resources, J.D. and M.S.; data curation, J.D. and M.S.; writing—original draft preparation, J.D. and M.S.; writing— review and editing, J.D.; visualization, M.S.; supervision, J.D.; project administration, J.D. and M.S.; funding acquisition, J.D. All authors have read and agreed to the published version of the manuscript.

Funding: This research has been funded by the German Federal Ministry of Education and Research within the project STRIVE (Sustainable Trade and Innovation Transfer in the Bioeconomy, grant number BMBF: 031B0019, www.strivebioecon.de, accessed on 9 September 2022), and of the Bio-economy Science Center of the Federal State of North Rhine Westfalia within the Transform2Bio project (Integrated Transformation Processes and their Regional Implementations: Structural Change of Fossil Economy to Bioeconomy, grant number BioSC: 53F-50000-00-13050200, https://www.biosc. de/transform2bio, accessed on 9 September 2022). This work was supported by the Open Access Publication Fund of the University of Bonn.

Institutional Review Board Statement: Not applicable.

Informed Consent Statement: Informed consent was obtained from all subjects involved in the study.

Data Availability Statement: Data can be made available upon request.

Acknowledgments: The authors gratefully acknowledge the contributions of team coordinator Jan Börner, and the statistical advice of Andrés Meiller. The authors further thank all the interviewees of the enterprises in Argentina who dedicated their time and gave us valuable information. We are also grateful for the valuable comments from Jorge Sellare, Pablo Mac Clay, and from the anonymous reviewers.

Conflicts of Interest: The authors declare no conflict of interest.

Appendix A

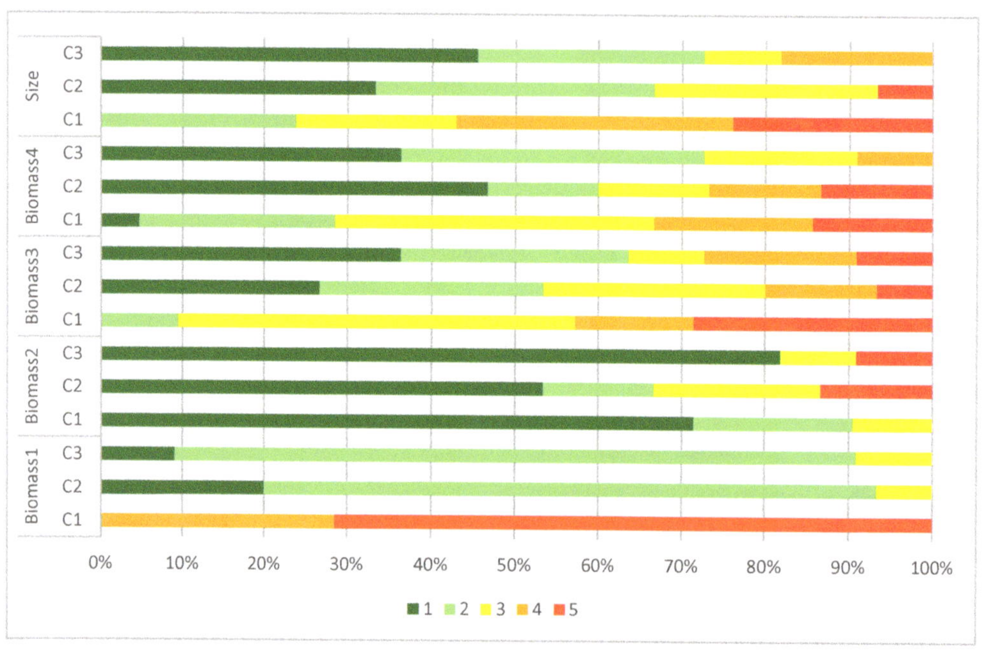

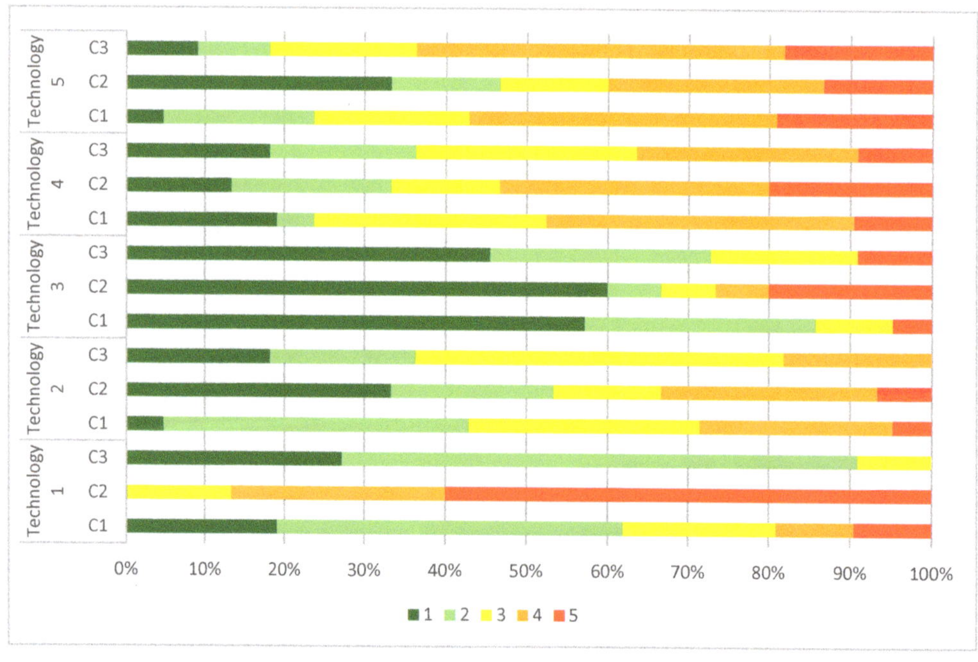

Figure A1. *Cont.*

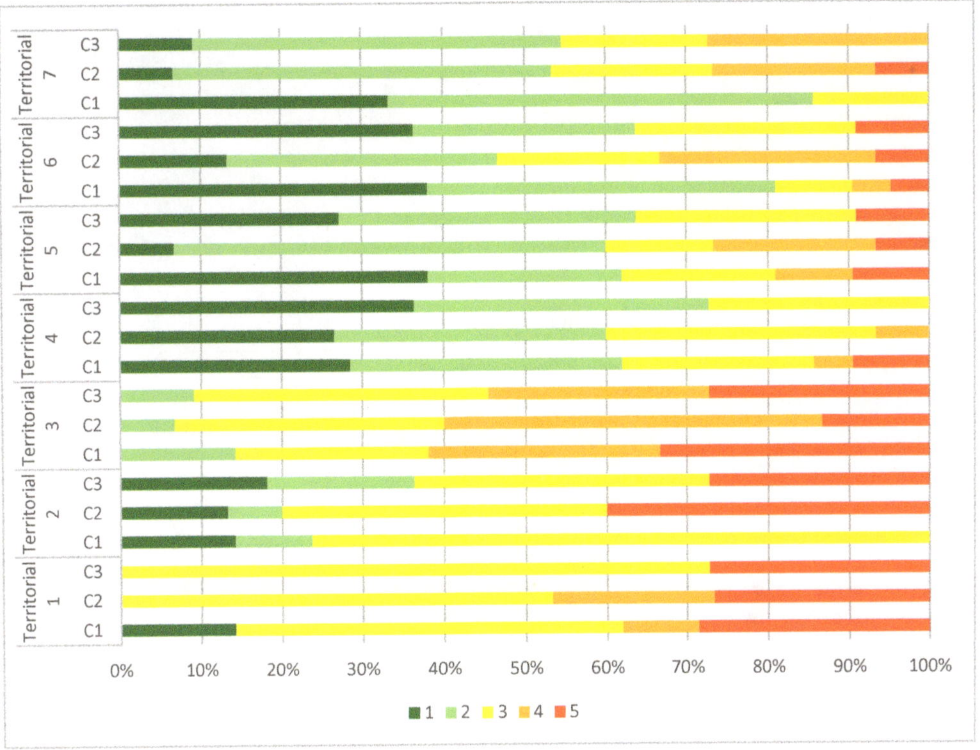

Figure A1. Distribution of enterprises belonging to 5-point scale, per variable and Cluster C1, C2 and C3.

Appendix B. Online Interview Questionnaire (Translation from Spanish)

Enterprise:

Location:

Province:

Web page:

Sector

1. Which products and/or services do you generate in your company? (ordered by importance, free text)
2. Which type of products or services do you produce:

   - Products that substitute fossil fuels
   - Products that enhance the productivity of the primary sector
   - Products that allow for a better and more efficient use of biomass
   - Products of low volume and high value
   - New products with local value added

3. Please, select FOUR key words of the list which best characterize your venture:

   - Local Development
   - Biomass
   - Biotechnology
   - Circular economy
   - Biodiversity
   - Investigation and development
   - Soil fertility

- Competitiveness
- Sustainability
- Patents
- Value addition
- Innovation

4. How are the products that you generate?
   - They are generic products and there are many other producers on the market
   - They are specialized products with little competition in the market
   - They are niche products, very specific, there are no other companies that generate them at a national level

Biomass

1. What total amount of biomass (in tons) do you use per year?

Zero	Less than 10 t	Between 10 and 100 t	Between 100 and 1000 t	More than 1000 t

2. Where does most of the biomass you use come from?

Local	Province	National	Latin America	International

3. What is the size of the agricultural/forestry/marine farms where the biomass that you use in your activity comes from?

Very big	Big	Medium	Small	Very small

4. How intensive is the use of chemical inputs, modern varieties and sophisticated machinery for the production of the biomass that your company produces and/or buys?

Very intensive	Intensive	Medium	Low intensive	Not used

5. What category do your main clients belong to (in percentages)?
   - Industrial enterprises %
   - Food companies %
   - Agricultural service companies %
   - Agricultural/forestry/fisheries producers %
   - Final consumers %
   - Government/State Organizations %
   - Biotech companies %
   - Human health companies/Laboratories %
   - Others: %

6. The products that you generate are mainly for sale and use at which level:

Local	Province	National	Latin America	International

Size

7. What is your annual turnover?

Less than 50.000$	Between 50.000 and 250.000$	Between 250.000 and 1 millon $	Between 1 y 10 millon $	More than 10 millon $

8. How do you consider the production scale of your company compared to most other companies in the same field in Argentina?

Much bigger	Bigger	Average	Smaller	Much smaller

9. How many people work in your company?

Between 1 and 5 persons	Between 6 and 20 persons	Between 21 and 100 persons	Between 101 and 500 persons	More than 500 persons

Innovation and Tecnology

10. How much of the value of your production depends on innovations and processes based on modern and standardized biotechnology?

0%	1–24%	25–49%	50–74%	75–100%

11. How much of your production value depends on innovations and processes based on local experience and tradition?

0%	1–24%	25–49%	50–74%	75–100%

12. How much of the value of your production is based on patents?

0%	1–24%	25–49%	50–74%	75–100%

13. How important has been the cooperation with state scientific and technological organizations for the development of your products? (Universities, INTA, INTI, CONICET, others)?

Very important	Important	Regular	Less important	Not important

14. How important has cooperation with companies or private groups been for the development of your products?

Very important	Important	Regular	Less important	Not important

Territorial Conditions

15. **Excluding** biomass, where are most of your company's suppliers of inputs, goods and services?

Local	Province	National	Latin America	International

16. What were the **THREE** main reasons why you chose the place where your company is located?
- Availability of raw material nearby
- Workforce cost
- Access to international markets
- Availability of public infrastructure
- Personal connections to the area
- Availability of skilled labor
- Accessible and cheap land
- Proximity to research and development centers
- Proximity to Universities
- Subsidies
- Social, cultural and/or environmental quality of the place
- Others:

17. To what extent do international market prices influence the profitability of your business?

Very much	Much	Somehow	Not much	Not at all

18. To what extent are your products based on a local brand, with local identity and cultural recognition?

Very much	Much	Somehow	Not much	Not at all

19. To what extent have your activities or products contributed to improving the environmental quality of the area?

Very much	Much	Somehow	Not much	Not at all

20. To what extent have your activities or products contributed to a more sustainable use of the area's natural resources?

Very much	Much	Somehow	Not much	Not at all

21. How much does your company interact with local organizations and local civil society?

Very much	Much	Somehow	Not much	Not at all

22. Could you identify three companies that operate in the same field as you in Argentina?

Appendix C. Map of Enterprises of the Different Clusters in Argentina

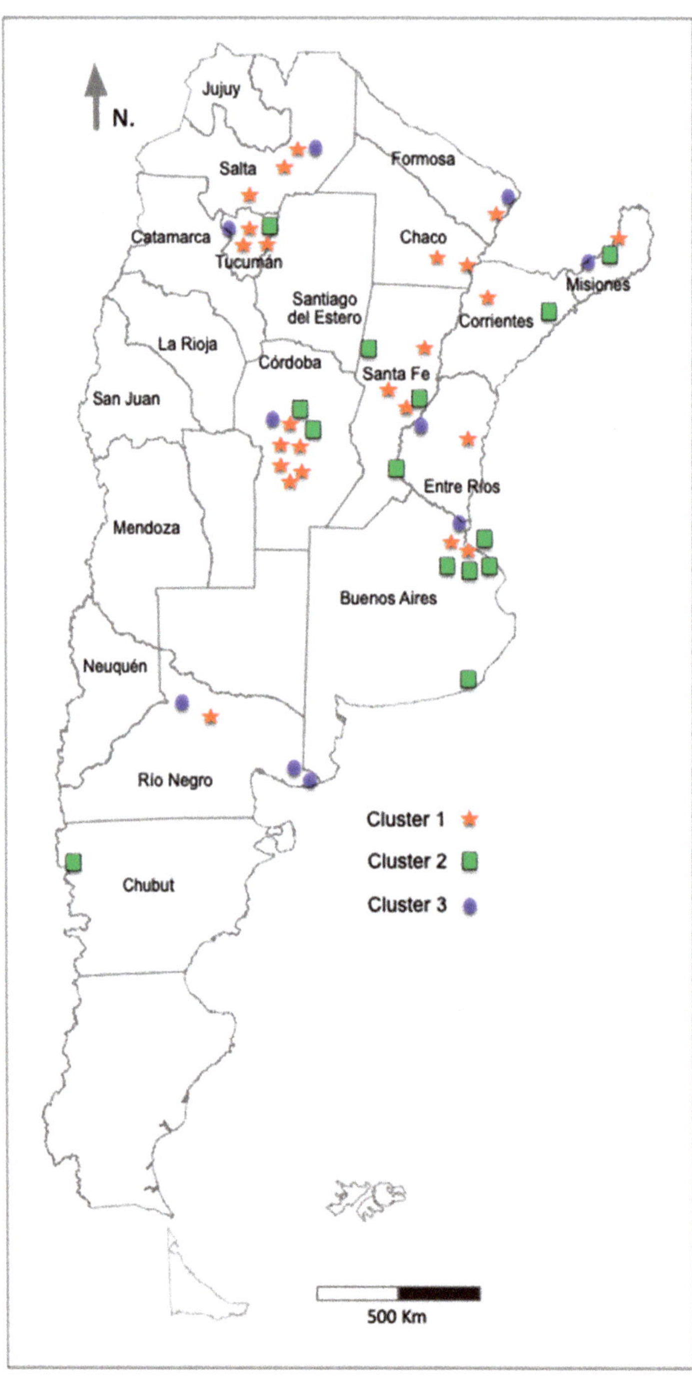

References

1. D'Adamo, I.; Gastaldi, M. Sustainable Development Goals: A Regional Overview Based on Multi-Criteria Decision Analysis. *Sustainability* **2022**, *14*, 9779. [CrossRef]
2. Befort, N. Going beyond definitions to understand tensions within the bioeconomy: The contribution of sociotechnical regimes to contested fields. *Technol. Forecast. Soc. Chang.* **2020**, *153*, 119923. [CrossRef]
3. OECD. *The Bioeconomy to 2030: Designing a Policy Agenda*; OECD Publishing: Paris, France, 2009. [CrossRef]
4. European Commission. Innovation for Sustainably Growth: A Bioeconomy for Europe. (Communication from the Commission to the European Parliament, the Council, the European Economic and Social Committee and the Committee of the Regions, No. SWD, 2012, 11 Final). Publication Office of the EU. Available online: https://op.europa.eu/en/publication-detail/-/publication/1f0d8515-8dc0-4435-ba53-9570e47dbd51 (accessed on 23 February 2022).
5. Bugge, M.; Hansen, T.; Klitkou, A. What Is the Bioeconomy? A Review of the Literature. *Sustainability* **2016**, *8*, 691. [CrossRef]
6. Hausknost, D.; Schriefl, E.; Lauk, C.; Kalt, G.A. Transition to Which Bioeconomy? An Exploration of Diverging Techno-Political Choices. *Sustainability* **2017**, *9*, 669. [CrossRef]
7. Vivien, F.-D.; Nieddu, M.; Befort, N.; Debref, R.; Giampietro, M. The Hijacking of the Bioeconomy. *Ecol. Econ.* **2019**, *159*, 189–197. [CrossRef]
8. Backhouse, M.; Lühmann, M.; Tittor, A. Global Inequalities in the Bioeconomy: Thinking Continuity and Change in View of the Global Soy Complex. *Sustainability* **2022**, *14*, 5481. [CrossRef]
9. Tittor, A. Towards an Extractivist Bioeconomy? The Risk of Deepening Agrarian Extractivism When Promoting Bioeconomy in Argentina. In *Bioeconomy and Global Inequalities*; Backhouse, M., Lehmann, R., Lorenzen, K., Lühmann, M., Puder, J., Rodríguez, F., Tittor, A., Eds.; Springer Nature Switzerland: Cham, Switzerland, 2021; pp. 309–330.
10. Sasson, A.; Malpica, C. Bioeconomy in Latin America. *New Biotechnol.* **2018**, *40*, 40–45. [CrossRef]
11. Phélinas, P.; Choumert, J. Is GM Soybean Cultivation in Argentina Sustainable? *World Dev.* **2017**, *99*, 452–462. [CrossRef]
12. Fehlenberg, V.; Baumann, M.; Gasparri, N.I.; Piquer-Rodriguez, M.; Gavier-Pizarro, G.; Kuemmerle, T. The Role of Soybean Production as an Underlying Driver of Deforestation in the South American Chaco. *Glob. Environ. Chang.* **2017**, *45*, 24–34. [CrossRef]
13. BioökonomieRat. *Bioeconomy Policy (Part III). Update Report of National Strategies around the World*; A Report from the German Bioeconomy Council; BioökonomieRat: Berlin, Germany, 2018.
14. MINAGRO (Ministerio de Agroindustria). Bioeconomía Argentina. Visión desde Agroindustria. 2017. Available online: https://www.magyp.gob.ar/sitio/areas/bioeconomia/_archivos/000000_Bioeconomia%20Argentina.pdf (accessed on 15 September 2021).
15. Sili, M.; Dürr, J. Bioeconomic Entrepreneurship and Key Factors of Development: Lessons from Argentina. *Sustainability* **2022**, *14*, 2447. [CrossRef]
16. Priefer, C.; Jörissen, J.; Frör, O. Pathways to Shape the Bioeconomy. *Resources* **2017**, *6*, 10. [CrossRef]
17. Marín, A.; Stubrin, L. *Innovation in Natural Resources: New Opportunities and New Challenges the Case of the Argentinian Seed Industry*; UNU-MERIT Working Papers; UNU-MERIT: Maastricht, The Netherlands, 2015.
18. Anlló, G.; Corley, E.; Añon, M.C.; Fuchs, M.; Bassó, S.; Genovesi, M.; Bellinzoni, R.; Gutierrez, M.A.; Bisang, R.; Ortiz, I.; et al. Biotecnología argentina al año 2030: Llave estratégica para un modelo de desarrollo tecno-productivo; contribuciones de Ri-cardo Carri... [et al.]; coordinación general de Alicia Balbina Recalde; dirigido por Crisólogo Martín Villanueva; Gustavo Arber.—1a ed.—Buenos Aires: Ministerio de Ciencia, Tecnología e Innovación Productiva, 2016. Available online: https://www.argentina.gob.ar/sites/default/files/est_bio_biotecnologia-argentina-al-2030-sintesis.pdf (accessed on 22 February 2022).
19. Lachman, J.; Bisang, R.; Obschatko, E.S.; Trigo, E. *Bioeconomía: Una Estrategia de Desarrollo Para la Argentina del Siglo XXI.*; Instituto Interamericano de Cooperación para la Agricultura: Buenos Aires, Argentina, 2020.
20. CIECTI (Centro Interdisciplinario de Estudios en Ciencia, Tecnología e Innovación). La Bioeconomía en la Argentina. In *Nuevas Opciones de Desarrollo*; CIECTI: Buenos Aires, Argentina, 2018.
21. Trigo, E.; Regúnaga, M.; Costa, R.; Coremberg, A. Bioeconomy in Argentina: Scope, Current Situation, and Sustainable Development Opportunities. In *Bioeconomy. New Framework for Sustainable Growth in Latin America*; Hodson de Jaramillo, E., Henry, G., Trigo, E., Eds.; Pontificia Universidad Javeriana: Bogotá, Colombia, 2019; pp. 25–48.
22. Korsgaard, S.; Müller, S.; Tanvig, H.V. Rural entrepreneurship or entrepreneurship in the rural–between place and space. *Int. J. Entrep. Behav. Res.* **2015**, *21*, 5–26. [CrossRef]
23. Levidow, L.; Nieddu, M.; Vivien, F.-D.; Béfort, N. Transitions towards a European Bioeconomy: Life Sciences versus agroecology trajectories. In *Ecology, Capitalism and the New Agricultural Economy: The Second Great Transformation*; Allaire, G., Daviron, B., Eds.; Routledge: London, UK, 2019; Volume Chapter 9, pp. 181–203.
24. Priefer, C.; Meyer, R. One Concept, Many Opinions: How Scientists in Germany Think About the Concept of Bioeconomy. *Sustainability* **2019**, *11*, 4253. [CrossRef]
25. Hinderer, S.; Brändle, L.; Kuckertz, A. Transition to a sustainable bioeconomy. *Sustainability* **2021**, *13*, 8232. [CrossRef]
26. Dietz, T.; Börner, J.; Förster, J.J.; von Braun, J. Governance of the Bioeconomy: A Global Comparative Study of National Bioeconomy Strategies. *Sustainability* **2018**, *10*, 3190. [CrossRef]
27. Mac Clay, P.; Sellare, J. *Value Chain Transformations in the Transition to a Sustainable Bioeconomy*; ZEF Discussion Papers on Development Policy No. 319, Center for Development Research: Bonn, Germany, 2022. [CrossRef]

28. Bröring, S.; Laibach, N.; Wustmans, M. Innovation types in the bioeconomy. *J. Clean. Prod.* **2020**, *266*, 121939. [CrossRef]
29. Pfau, S.F.; Hagens, J.E.; Dankbaar, B.; Smits, A.J.M. Visions of sustainability in bioeconomy research. *Sustainability* **2014**, *6*, 1222–1249. [CrossRef]
30. Meyer, R. Bioeconomy Strategies: Contexts, Visions, Guiding Implementation Principles and Resulting Debates. *Sustainability* **2017**, *9*, 1031. [CrossRef]
31. Müller, S.; Korsgaard, S. Resources and bridging: The role of spatial context in rural entrepreneurship. *Entrep. Reg. Dev.* **2018**, *30*, 224–255. [CrossRef]
32. Bisang, R.; Anlló, G.; Fuchs, M.; Lachman, J.; Monasterios, S. Bioeconomía. Cambio Estructural, Nuevos Desafíos y Respuestas globales: Una Ventana de Oportunidad Para las Producciones Basadas en Recursos Naturales Renovables. 2015. Available online: http://www.ucar.gob.ar/images/publicaciones/DocumentoBioeconomia.pdf (accessed on 5 October 2020).
33. Maciejczak, M. What are production determinants of Bioeconomy? *Probl. World Agric.* **2015**, *15*, 137–146.
34. Sili, M. *Por Un Futuro Rural. Innovación, Renacimiento Rural y Nuevos Itinerarios de Desarrollo En La Argentina Pospandemia*; Editorial Biblos CULTURALIA: Buenos Aires, Argentina, 2021.
35. Marsden, T.; Farioli, F. Natural powers: From the bio-economy to the eco-economy and sustainable place-making. *Sustain. Sci.* **2015**, *10*, 331–344. [CrossRef]
36. Romano, L. *Bioeconomía Como Estrategia Para El Desarrollo Argentino*; Buenos Aires, Argentina, 2019; Available online: https://fibamdp.files.wordpress.com/2020/06/la-bioeconomicc81a-como-estrategia-para-el-desarrollo-argentino.pdf (accessed on 20 February 2022).
37. Bruno, M.P.; Viteri, M.L.; Sili, M. El Rol de Las Agronomías y Acopios En La Consolidación Del Modelo de Agronegocios En Balcarce (2000–2019). *Mundo Agrar.* **2021**, *22*, e171. [CrossRef]
38. Sili, M.; Martin, C. *Innovación y Recursos Bioculturales En El Mundo Rural. Lecciones Para Un Desarrollo Sostenible*; Editorial Biblos SOCIEDAD: Buenos Aires, Argentina, 2022; Volume 59.
39. D'Adamo, I.; Gastaldi, M.; Morone, P.; Rosa, P.; Sassanelli, C.; Settembre-Blundo, D.; Shen, Y. Bioeconomy of Sustainability: Drivers, Opportunities and Policy Implications. *Sustainability* **2022**, *14*, 200. [CrossRef]
40. Sili, M.; Haag, M.I.; Nieto, M.B. Constructing the Transitions and Co-Existence of Rural Development Models. *Sustainability* **2022**, *14*, 4625. [CrossRef]

Review

Potential Use of Cow Manure for Poly(Lactic Acid) Production

Ricard Garrido [1], Luisa F. Cabeza [1,*], Víctor Falguera [2] and Omar Pérez Navarro [3]

[1] GREiA Research Group, Universitat de Lleida, Pere de Cabrera s/n, 25001 Lleida, Spain
[2] AKIS International, c/Dr. Robert 33, 25171 Albatàrrec, Spain
[3] Department of Chemical Engineering, Faculty of Chemistry and Pharmacy, Central University "Marta Abreu" of Las Villas, Santa Clara 54830, Cuba
* Correspondence: luisaf.cabeza@udl.cat; Tel.: +349-73003576

Abstract: Cow manure is an abundant residue and poses a problem regarding recycling. Intensive animal farming produces manure, which, if not properly managed, can contaminate nearby water bodies and soils with nutrient excess. There are 1.9 billion cattle worldwide, with a calculated capacity to produce 7.6 billion tons per year. Feeding of these cows is carried out mainly with cellulosic material. Therefore, cow manure contains an important fraction of lignocellulose. Cow manure can be valorized using such lignocellulosic fractions as the raw material of several fermentative processes. This fraction can be transformed into sugar, which can, in turn, be used to feed lactic acid bacteria (LAB). LAB produces lactic acid (LA), which can later be polymerized to poly(lactic acid) (PLA), a bioplastic with promising market forecasts. This review describes the most updated processes for all of the necessary steps to produce lactic acid from lignocellulosic biomass with LAB. Key process parameters to obtain PLA from lignocellulose are reviewed and analyzed herein, including lignocellulosic fraction extraction, sugar transformation, pretreatment, hydrolysis, fermentation, purification, and polymerization. This review highlights the potentiality to obtain lignocellulose from cow manure, as well as its use to obtain PLA.

Keywords: bioplastic; lactic acid; PLA; poly(lactic acid); bioeconomy; cellulose; lignocellulose; cow manure; circular economy

Citation: Garrido, R.; Cabeza, L.F.; Falguera, V.; Pérez Navarro, O. Potential Use of Cow Manure for Poly(Lactic Acid) Production. *Sustainability* **2022**, *14*, 16753. https://doi.org/10.3390/su142416753

Academic Editors: Idiano D'Adamo and Massimo Gastaldi

Received: 19 November 2022
Accepted: 10 December 2022
Published: 14 December 2022

Publisher's Note: MDPI stays neutral with regard to jurisdictional claims in published maps and institutional affiliations.

1. Introduction

Synthetic polymers created from fossil fuels have caused significant environmental issues. Current methods of manufacture, use, and disposal are not environmentally friendly and pose risks to both human and animal health. The accumulation of waste in landfills and natural habitats, physical issues for wildlife brought on by ingesting or becoming entangled in plastic, the leaching of chemicals from plastic products, and the possibility that plastics will transfer chemicals to wildlife and humans are just a few of the many issues surrounding their use and disposal [1]. From coastal areas alone, around 20 megatons (Mt) of improperly disposed plastic debris will reach the oceans by 2025 [2], with an additional 1.15–2.41 Mt brought yearly by rivers from inner areas of the planet [3].

One sustainable source of energy and organic carbon for our industrial society is biomass, because it is a renewable resource [4]. Seventy-five percent of the biomass that photosynthesis generates in nature belongs to the class of carbohydrates. Surprisingly, humans only consume 3–4% of these substances for food and non-food uses [5]. It can be used as an alternative to produce biodegradable and/or biobased plastics.

Lignocellulose, as a carbohydrate source, is an attractive raw material for biotechnological operations because of its renewable nature, global distribution, abundance, and low cost. Plant biomass (lignocellulosic biomass) is an abundant, affordable, and ecologically friendly resource that has great potential to be used in the production of fuels and chemicals. Agricultural waste is a good source of lignocellulosic biomass, which is cheap, renewable, and mostly under-utilized. In the past, these resources have included woody crops, maize stover, sugarcane bagasse, rice hulls, leaves, stems, and stalks from corn fiber. Agricultural and industrial activities generate a variety of types of lignocellulosic waste, such as citrus peel waste, sawdust, paper pulp, industrial trash, municipal solid waste, and paper mill sludge [6]. Cow manure has also a lignocellulosic fraction, which can be used as raw material.

More than 1.4 billion cattle are kept worldwide, and 159 million (11 percent) are in the regions of Europe and Central Asia [7]. Worldwide, Brazil is the country with the most heads [8]. The potential production of cow manure worldwide is 7.6 billion tons per year [9].

The necessity of and opportunity for successful and cost-effective methods that transform lignocellulosic biomass into value-added chemicals, which are now produced from non-renewable resources such as fossil fuels, are highlighted by the abundance of lignocellulosic biomass [1]. Production of LA and PLA from lignocellulosic material has been widely studied.

Chemical synthesis and the fermentation of renewable carbohydrates are both production paths to lactic acid. Using biomass as a carbohydrate source, it is possible to produce LA. LA is an organic acid that occurs naturally and serves as the primary metabolic intermediate in the majority of living bodies, including humans and anaerobic prokaryotes [10]. LA is a versatile organic acid that presents diverse applications, mostly in food and food-related industries. The United States Food and Drug Administration (USFDA) has classed it as Generally Recognized As Safe (GRAS) for general-purpose food additives [11]; it serves a variety of purposes, including flavoring, regulating pH, acting as an acidulant, enhancing microbiological quality, fortifying minerals, and extending shelf life [12]. Ninety percent of LA's worldwide production is achieved through fermentation [13]. Substrates can be renewable and low-cost materials, such as lignocellulosic residues [14]. The production of D-Lactic acid or L-Lactic acid with high optical purity, or a mixture of both with low optical purity, might result from the lactic acid fermentation process, depending on the specific microbe. There are several bacteria that can generate lactic acid, but a competitive commercial process demands a strong, quickly expanding, low-pH, high production strain with affordable nutritional needs. The usual anaerobic fermentation process for *Lactobacillus* uses little energy to run, with the majority of the expense typically coming from medium components such as carbohydrates.

The most widespread biodegradable and biocompatible polymers used today, PLA and poly(lactic-co-glycolic acid) (PLGA), are synthesized and processed using LA [15]. Because of its excellent processing capabilities and strong mechanical characteristics, PLA is one of the most commercially successful bioplastics (at least among the stiff types). Through fermentation, its monomer, lactic acid, is produced from renewable resources such as starch or sugar. After fermentation, lactic acid must be extracted from the broth and purified to the desired requirements in order to produce PLA [16]. The majority of PLA production procedures use ring-opening polymerization (ROP), which is more effectively used to convert lactide (the cyclic dimer of lactic acid) to PLA [17].

PLA is biodegradable and resembles polypropylene (PP), polyethylene (PE), or polystyrene (PS) in terms of its properties. It can be manufactured using currently-in-use manufacturing machinery (those designed and originally used for petrochemical industry plastics). As a result, production is quite inexpensive. In light of this, PLA has the second-highest volume of manufacture of any bioplastic (the most common typically cited as thermoplastic starch). It can be easily converted into molded components, film, or fibers using normal plastic processing machinery [18].

Traditionally, manure waste is used as a fertilizer for agricultural soils or to produce biogas. While biogas is a mixture of methane, CO_2, and other gases, produced by anaerobic digestion of organic materials in an oxygen-free environment, biomethane is a nearly pure source of methane created either by upgrading biogas or by gasifying solid biomass, followed by methanation [19]. However, increasing limitations on its use as a fertilizer [20] or biogas air pollution [21] are key deciding factors.

As stated in [19], data show how much potential for economic growth is being missed by not utilizing renewable sources. The dilemma of having accessible substrates but not using them to create products (or energy) seems to exist. Cow manure lignocellulosic fraction has the potential to be converted into valuable products, falling into the circular bioeconomy field. This paper reviews the most up-to-date knowledge on the unit operations of the process of converting cow manure to PLA, such as the conversion of lignocellulosic material into sugar and its conversion to LA and PLA. To obtain a complete picture of the state of the art, processes from other, extensively researched lignocellulosic material sources are also explored.

2. Cow Manure Composition

Cow manure is an important residue, and that produced in the greatest quantity in rural farms. As a fertilizer, it is suitable for all plants and soils; it gives consistency to sandy and mobile soil and lightness to chalky soil, and also refreshes warm, limestone, and loamy soils. Of all manures, it lasts the longest and is the most uniform. The duration of its strength depends mainly on the kind of feed given to the cattle that produce it.

Production of animal waste worldwide is not counted exactly, but estimations are made based on census and the type of management. Dejection information varies with the species and size of the animal as well as the type of installation, which also has a great influence on the amount and type of waste generated (solid manure or slurry). The European Union (27) generates 1500 million tons of manure annually, mainly from cattle and swine [22]. In Europe, the proportion of the liquid form of the manure (slurry) varies greatly in different countries, from 95% of total production in the Netherlands to 20% in Eastern European countries [23]. Overall, in central Europe, Spain and Portugal, over 65% of the dejections are produced as slurry, with the highest proportion in pig farming [24]. The estimation of Spanish livestock sector's waste production is around 140 million tons per year [25].

The biodegradable fraction is also made up of more complex compounds of slower degradation, such as lipids and relatively stable lignins and tannins, as well as a series of phenolic polymers called melanin, synthesized mainly by fungi, which are characterized by their great similarity to humic acids in terms of composition, structure, and resistance to degradation [26]. In Figure 1, a description of the solid organic fraction transformation is shown.

Livestock manure is classified according to the percentage of dry matter in a solid. Dejections that have approximately more than 20% are characterized as dry matter; those dejections that have around 10–22% dry matter are considered semisolid; and those dejections that have a dry matter content of around 0–15% are liquid or semi-liquid.

The typical composition of both solid and liquid manure is difficult to establish, as it depends on many factors. The fertilizer value of both solid manure and slurry varies greatly from one farm to another, since it depends on the type of farm (breed, species, age, etc.), diet, type of production, type of accommodation, and how the waste is stored. As an example, Table 1 shows the typical nutrient content of manure and slurry studied by different authors.

A cow will consume between 4 and 5 tons of fodder annually on a dry matter basis. Cows are fed a diet that is heavy in lignocellulosic fiber, and following digestion, they generate manure [27]. The most prevalent type of agricultural waste is cow manure, which is also a typical lignocellulosic material [28]. Table 2 shows the dejections produced in a year by a cow.

Cow manure's potential has not yet been fully revealed, because only a small portion of the lignocellulosic fraction has been utilized by microorganisms to produce biogas, leaving huge amounts of the anaerobically digested cow manure unused [29]. Enzymatic hydrolysis into fermentable sugars could effectively disrupt the treated lignocellulosic fraction [30].

Table 3 illustrates the fiber content of the most commonly used cow manure residue. Despite comparing the same type of manure, recorded data display a wide range. This is because different regions of the world have variable levels of animal digestibility, which causes the percentages of lignocellulosic material to vary greatly [31].

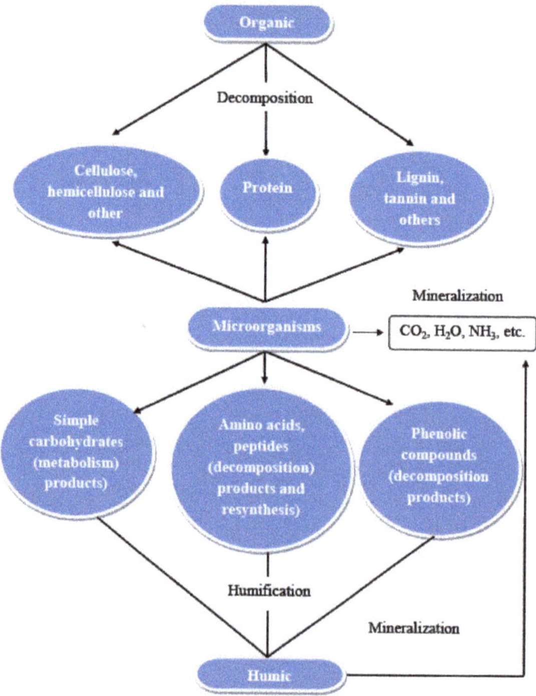

Figure 1. Transformation and evolution of the soil organic fraction. Adapted from [32].

Table 1. General manure composition (g/kg over fresh weight). Adapted from [33].

	Dry Matter		Total Nitrogen		N-Ammonia		Phosphorous	
	Mean	Range	Mean	Range	Mean	Range	Mean	Range
				Liquid manure/slurry				
Cow	65	15–123	3.9	2.0–7.2	2.3	1.0–4.9	2.3	0.2–6.0
				Solid manure				
Cow	223	160–430	4.8	2.0–7.7	1.3	0.5–2.5	3	1.0–3.9

Table 2. Characteristics of four typical livestock farms, and the manure they generate. Adapted from [9].

Parameter	Farm A	Farm B
Type of farm	Fattening Cow	Dairy cow
Capacity (places)	850	400
Cycles/year	1.22	1
Produced dejections (ton/year)	1700	8525
Dejections	Stable/manure heap	Pit/Pond
Nitrogen concentration (kg N/ton)	11	4.8

Table 3. Cow manure's lignocellulosic material percentages.

Cellulose (%)	Hemicellulose (%)	Lignin (%)	Reference
21.2	30.4	11.6	[34]
23.5	12.8	8.0	[31]
17.9	15.7	18.2	[35]
22.9	22.9	8.1	[36]
26.59	11.27	11.24	[37]
23.51	12.82	7.95	[38]
21.89	12.47	13.91	[39]

The compositions of lignocellulose in cow manure and anaerobically digested cow manure have been examined [40], and the composition data for cow manure are presented in Table 4. The analysis revealed that glucan (16.62%) and xylan (15.26%) formed the majority of cow manure. Different feedstock types and seasonality may have a small influence on the overall conversion process design [41]. Due to the lignocellulose fraction's breakdown during biogas generation in the anaerobic reactor, the amount of glucan still present in the anaerobically digested cow manure after biogas production was considerably lower than that in the unfermented cow manure. Thus, glucan (14.50%) and xylan (12.56%) were present in anaerobically digested cow manure. These findings showed that cow manure had a relatively high lignocellulosic content, and that the amount of fiber was somewhat reduced during anaerobic digestion for methane generation.

Table 4. Composition of cow manure and anaerobically digested cow manure [40].

Biomass Components	Cellulose (%)	Hemicellulose (%)
Cow manure	16.62	15.26
Alkali-treated cow manure	35.34	15.48
Acid-treated cow manure	26.62	7.61
Anaerobically digested cow manure	14.5	12.56
Alkali-treated anaerobically digested cow manure	28.94	15.98
Acid-treated anaerobically digested cow manure	22.56	3.22

3. Sugar Production from the Lignocellulosic Portion of Cow Manure

A potentially sustainable method of creating innovative bioprocesses and products is through the biotechnological conversion of lignocellulosic biomass. Due to the lignin-protected, highly crystalline structure of lignocelluloses, which is complicated in structure, this substance has a high degree of recalcitrance, making the process of depolymerization a challenge [1].

Enzymatic hydrolysis could be used to successfully transform cow manure into fermentable sugars [42]. It consists of the partition of cellulose into glucose units and the hemicellulosic fraction in their constituent sugars.

While phenolic monomers can be employed as a chemical intermediary in the chemical industry, sugars in cellulose and hemicellulose are attractive as feedstock for fermentation operations [43]. In addition, lignin can be pyrolyzed to provide an oil fuel that can power combustion engines [44].

The majority of pretreatment techniques combine enzymatic hydrolysis with a thermo-chemical process. Thermo-chemical pretreatment can dissolve or deconstruct (part of) the lignin, deacetylate the hemicellulose to increase accessibility, and/or increase the accessibility of cellulose and hemicellulose polymers. By maximizing process variables, including temperature, pressure, pH, and chemical addition, efficient degradation of lignocellulose during chemical pretreatment is possible [42].

In order to use lignocellulosic biomass for value-added product production from its constituent fractions, such as cellulose and hemicelluloses, barriers that prevent chemical or biological catalysts from contributing to its transformation must first be removed. Recalcitrance or alteration of the crystalline structures of the fraction of interest must also be reduced in order to achieve a higher reaction speed and better product quality. It is also preferred that carbohydrates not be broken down, or that no additional products are developed which could stop enzymes or microbes from fermenting. While physical pretreatment refers to size reduction and steam explosion, in chemical pretreatments, biomass structure is altered with solvents that promote cellulose degradation, hemicellulose, and lignin [45].

Most of the lignocellulose must be hydrolyzed twice to become dextrose, a fermentable sugar, utilizing amylolytic enzymes such as α-amylase and glucoamylase. Usually, the first stage is completed quickly at high temperatures (between 90 and 130 °C), then a longer saccharification to dextrose process is completed at cooler temperatures. This technology has been practiced on the industrial scale for decades. Enzymes for this process are highly developed and efficient, and are available at a relatively low cost from companies specializing in industrial enzymes, such as Novozymes and Genencor. The resulting dextrose from this process can then be used for lactic acid fermentation [46]. Figure 2 presents a representation of the lignocellulosic structure, before and after pretreatment.

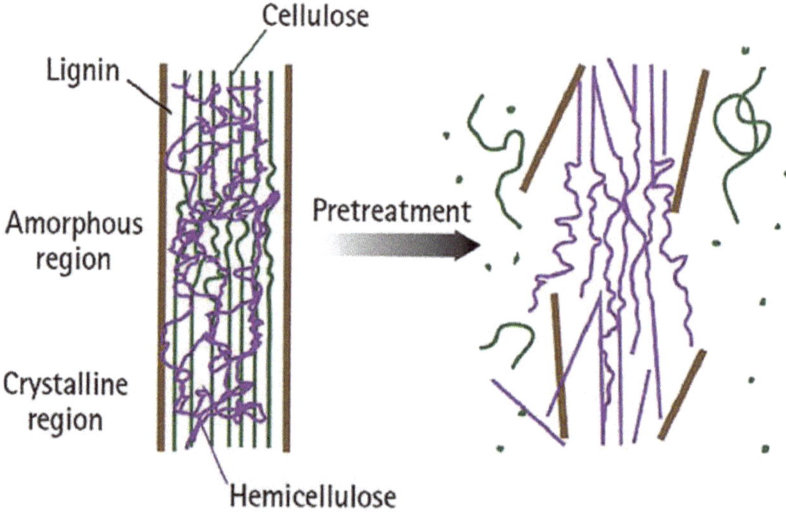

Figure 2. Lignocellulose material structure before and after pretreatment. Adapted from [47].

3.1. Pretreatment

To convert lignocellulosic hemicellulose and cellulose fraction into value-added products, first the elimination of barriers which make accessing its catalysts difficult (biological or chemical) is required. The main purpose of the pretreatment process is to increase the porosity, lower the quantity of crystalline cellulose, and remove lignin and hemicelluloses [37]. Related through the use of physical, chemical, physicochemical, and biological procedures, we can classify different types of pretreatment processes [48]. Figure 3 presents the classification of pretreatments according to their nature.

In terms of byproduct formation, the most critical pretreatments are the chemical ones. Depending on the decomposition method or the lignocellulosic source, different byproducts could be formed. There are three groups of byproducts, namely phenolics, furans, and organic acids [49].

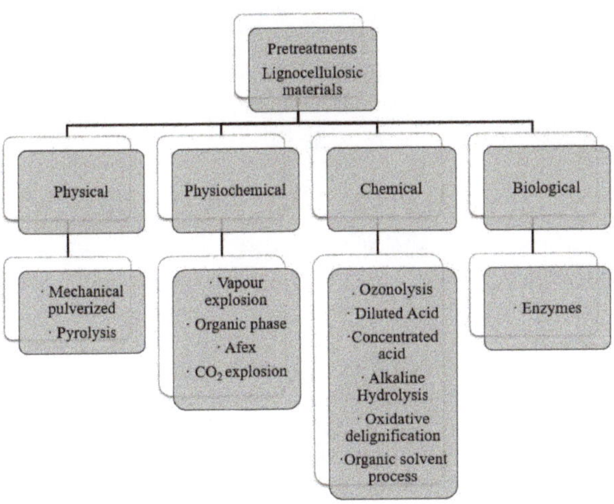

Figure 3. Pretreatment process for lignocellulosic material. Adapted from [45].

3.2. Hydrolysis

An important step is the cellulose partition into glucose units and the hemicellulosic fraction in their constituent sugars. Among several processes for this purpose, two are best-known: acid hydrolysis and enzymatic hydrolysis [50].

3.2.1. Acid Hydrolysis

The acid hydrolysis of polysaccharides comprises the processes in which acid is added at the beginning of the process (acid hydrolysis) and those in which acid is generated during the process (hydrothermal or autohydrolytic). These procedures have the same chemical principle, but differ in their operating conditions, mainly in temperature and acid concentration. The acid hydrolysis processes are mainly classified into concentrated processes and diluted processes. Table 5 presents the pros and cons of this procedure.

Table 5. Advantages and disadvantages of acid hydrolysis. Adapted from [51].

Type of Acid Hydrolysis	Advantages	Disadvantages
Concentrated acid process	Low temperature operation; high sugar performance.	High acid consumption; high energetic cost; long reaction time (2–6 h).
Diluted acid process	Low acid consumption; lower permanence time.	High temperatures; low sugar performance; equipment corrosion; formation of non-desired products (degradation).

Among the chemical pretreatment techniques, dilute-acid hydrolysis is perhaps the most frequently used. It can be used either to prepare lignocellulose for enzymatic hydrolysis or to carry out the actual hydrolysis process, yielding fermentable sugars. In order to pretreat or hydrolyze lignocellulosic materials using dilute-acid procedures, many reactor designs, including batch, percolation, plug flow, countercurrent, and shrinking-bed reactors, have been used. Recent reviews have examined these procedures, as well as many facets of dilute-acid hydrolysis and pretreatment [51,52]. The dilute-acid treatment can achieve high reaction rates and greatly enhance cellulose hydrolysis at high temperatures (e.g., 140–190 °C) and low acid concentrations (e.g., 0.1–1% sulfuric acid). By using diluted acid as a pretreatment, hemicellulose can be removed almost completely. Although the pretreatment does not effectively dissolve lignin, it can disrupt it and make cellulose more accessible to enzymatic hydrolysis [53].

High contents of lignin and glucose have an important inhibitory response to the activity of the enzymatic cocktail, with the final glucose yield decreasing as the addition of the initial concentration of these compounds increases. Other inhibitory compounds (cellobiose, xylose, arabinose, furfural, hydromethylfurfural, and acetic acid) only have a slight effect on cellulose-to-glucose enzymatic conversion, at least at the concentration levels studied. However, it is likely that the synergistic effect of mixtures of these compounds could have a significant negative impact on enzymatic saccharification [54].

Yan et al. [40] studied lignocellulose extracted from cow manure. Cow manure and samples of anaerobically digested cow manure were milled using a micromiller prior to material processing, and they were then thoroughly dried at 45 °C. A one-gram sample was pretreated with 10 milliliters of either 2% sulfuric acid or 2% sodium hydroxide (m/v) in a 500-milliliter conical flask. The lignocellulosic fraction was extracted by two methods. (i) Treated with NaOH: The surface area of the air-dried cow manure materials was increased physically by milling, and the resistant structure was then disrupted by soaking the cow manure in an alkaline NaOH solution. Then, a further detoxification procedure, including water washing, was carried out to remove any remaining chemicals that prevented further microbial fermentation. (ii) Treated with H_2SO_4: lignin and hemicelluloses were further removed during the dilute-acid pretreatment of biomass materials, which enhanced the enzymatic hydrolysis of the biomass to cellulose [55].

3.2.2. Enzymatic Hydrolysis

Several authors have presented theories to explain the total degradation of cellulose. These involve the three enzymatic components of cellulase and the synergism between these, where endoglucanases attack to the amorphous regions of cellulose fibers, creating sites for exoglucanases that would be directed towards the crystalline fiber region. The β-glucosidases would execute the last step of the hydrolysis and would prevent the accumulation of cellobiose, which would inhibit exoglucanases [56].

Other authors have suggested cellulose degradation according to the following steps: (i) adsorption and formation of the substrate enzyme complex; (ii) formation of the product; and (iii) desorption and re-adsorption of the enzyme, or movement of the enzyme along the cellulose molecule. On a string cellulose model, extended hydrolysis of the outer chains would expose the non-terminal internal chain reducers.

The performance of the enzymatic hydrolysis stage depends on the pretreatment used. Alkaline and acid pretreatment methods present performances of less than 85% after hydrolyzing with enzymes, contrary to pretreatments where only water is used in its explosion forms, steam and hot liquid water [57], where performance exceeds 90%. Table 6 summarizes process yields for different types of biomass.

Table 6. Process yields summary for different biomasses [45].

Raw Material	Pretreatment	Enzymes	Hydrolysis Conditions	%	Reference
Corn bran	AFEX	Specyme Cp Accellerase 1000	T = 50, t = 48 h, 15 mg/g glucose	40%	[41]
Cassava bagasse	—	Termamyl 120 L AMG 200 L	T = 90 °C, t = 1 h, pH = 6.5 T = 60 °C, t = 24 h, pH = 4.5 T = 50 °C, t = 96 h, pH = 4.8	97.3%	[58]
Palm oil logs	AFEX	Accellerase 1000	N = 170 rpm	95.4%	[59]
Corn		Celluclast 1.5 L Novozyme 188	T = 50 °C, t = 72h, N = 150 rpm, pH = 5	80%	[60]
Cane bagasse	Diluted organosolv acid	Celluclast 1.5 L Novozyme 188 Xilanasa	T = 50 °C, t = 24 h, pH = 4.8 N = 150 rpm	48–76%	[61]
Cane bagasse	Diluted phosphoric acid	Biocellulase W Novozyme-188	T = 50 °C, t = 96 h, pH = 5 N = 100rpm	74%	[62]
Banana tree	Gelatinization	Celluclast 1.5 L Novozyme-188 Pectinasa (P-2611)	T = 50 °C, t = 9 h, pH = 5	80%	[63]

For acid-pretreated corn stover (PCS), the outcome of fluid dynamic parameters on enzymatic hydrolysis was assessed. When enzymatic PCS biomass saccharification is carried out in a stirred tank batch reactor, with low rotating speeds (<100 rpm) and final Reynolds number (Re) values (<10), a low glucose yield is obtained, with no effect due to higher amounts of enzymes. With stirring speeds over 300 rpm (final Re > 2000), maximal and similar glucose yields are retrieved. Estimated mass transfer coefficients and rates increase with agitation and reaction time. Low stirring speed involves overall rate control by mass transfer, while higher values rapidly lead to low viscosity, high Re, and enhanced mass transfer, with enzymatic reactions as the overall process-controlling step [64].

4. Microbial Fermented Lactic Acid Production

The common term for 2-hydroxypropanoic acid is lactic acid. L(+)-lactic acid and D(-)-lactic acid are the two optical isomers of lactic acid. D(-)-lactic acid can occasionally be hazardous to human metabolism, and can cause acidosis and decalcification [11].

Chemical synthesis and microbiological fermentation can both produce lactic acid. The biotechnological approach for producing lactic acid has various benefits over chemical synthesis, including lower substrate prices, lower production temperatures, and less energy usage [10].

Lactic acid fermentation and product recovery and/or purification are typically included in biotechnological methods for lactic acid production. Numerous studies have been conducted on the creation of biotechnological techniques for the production of lactic acid, with the ultimate goal being to make the process more effective and affordable [12].

From agricultural waste, byproducts, and residues, LA can be synthesized biotechnologically and used to create biodegradable, biocompatible LA polymers. It can also be made from the lignocellulosic component of cow manure, and then turned into sugars which feed LAB. These polymers are often utilized in high-end applications, but with lower production costs, they could find use in a far wider range of applications [65].

The primary method for creating LA is based on the fermentation of various raw materials by microorganisms that produce it, such as bacteria, fungi, and yeast [66]. Group of wild LA producers includes bacteria (LAB: *Carnobacterium, Enterococcus, Lactobacillus, Lactococcus, Leuconostoc, Pediococcus, Streptococcus, Tetragenococcus, Vagococcus*, and members of the genus *Bacillus*) and fungi (several species belonging to the genera *Mucor, Monilia*, and *Rhizopus*) [65].

Microorganisms can be either homofermentative or heterofermentative. Homofermentative LAB use the Embden–Meyerhof–Parnas (EMP) route to convert 1 mol of glucose into 2 mol of LA. Depending on the available substrates, ambient factors, etc., heterofermentative LAB can use either the EMP or phosphoketolase (PKP) route, with the end products being a mixture of LA, ethanol, acetate, and CO_2, or even mannitol in the case of fructose catabolism [66,67].

The majority of the world's commercially manufactured lactic acid is created by using homolactic organisms from the genus *Lactobacillus*, which only produce lactic acid. The following organisms are the main producers of the L(+)-isomer: *Lactobacillus amylophilus, L. bavaricus, L. casei, L. maltaromicus*, and *L. salivarius*. Strains such as *L. delbrueckii, L. jensenii*, or *L. acidophilus* produce either the D-isomer or mixtures of both. Under typical fermentation conditions, such as relatively low to neutral pH, temperatures around 40 °C, and low oxygen concentrations, these strains exhibit significant carbon conversions from feedstock [1]. Among the members of the genus *Lactobacillus*, *Lactobacillus Cassei* has frequently appeared in studies on the generation of lactic acid [68].

Depending on the source of the lignocellulosic material, the liquid phase produced by hydrolyzing cellulose and hemicellulose contains six-carbon sugars (hexoses) and five-carbon sugars (pentoses). The syrup will primarily contain glucose, xylose, arabinose, galactose, mannose, and rhamnose [69].

The optimal settings for the lactic acid formation route including bacteria (*Lactobacillus* sp.) are pH between 5–7, temperature between 40–45 °C [70–75], a nutrient-rich environment, and sterile conditions. To maintain the pH, the lactic acid created during fermentation must be neutralized, which increases the cost of lactic acid production and recovery. Using fungi, lactic acid fermentation is another option (*Rhizopus* sp.). Compared to bacteria, fungi can grow in nutrient-limited conditions and efficiently ferment both hexose and pentose carbohydrates [76]. However, because other products (such ethanol and fumaric acid) are produced during fungal fermentation, the amount of lactic acid obtained is reduced [77]. Aeration is also necessary during the fungal fermentation process for larger lactic acid yields, which raises the price of lactic acid production. Yeast can also be used to make lactic acid, and because it can ferment at low pH levels, there is no longer a need to neutralize and recover lactic acid [46]. An overview of the performance of different processes studied for LAB can be found in Table 7.

Table 7. Microorganism productivity for LA.

Organism	Substrate	Lactic Acid Production [g/L]	Yield [g/g]	Productivity [g/(L·h)]	Reference
Enterococcus faecalis RKY1	Glucose	144	0.96	5.1	[78]
	Molasses	95.7	0.95	4.0	[79]
	Corn starch, potato, and wheat	129.9	1.04	1.5	[80]
	Wood hydrolyzate	93	0.93	1.7	[81]
E. mundtii QU 25	Modified Rogosa and Sharpe (mMRS)	119	0.83	1.12	[82]
	Xylose	86.7	0.84	0.9	[83,84]
Lactobacillus rhamnosus ATCC 10863	Biomass pellets and glucose	67	0.84	2.5	[85]
Lactobacillus rhamnosus ATCC 7469	Paper sludge	73	0.97	2.9	[86]
Lactobacillus rhamnosus CECT-288	Yeast and meat extract, peptone	32.5	0.88	5.4	[87]
	Cellulosic biosludges	42	0.38	0.87	[88]
L. casei subsprhamnosus	Softwood	21.1–23.75	0.74–0.83	0.15–0.23	[89]
L. rhamnosus and *L. brevis*	Cornstover	20.95	0.7	0.58	[90]
Lactobacillus rhamnosus strain CASL	Cassava powder	175.4	0.71	1.8	[91]
L. rhamnosus ATCC 7469	Distillery stillage	97.1	–	1.80	[92]
L. rhamnosus ATCC 7469	Malting, brewing, and oil production byproducts	58.01	–	1.19	[93]
Lactobacillus helveticus ATCC 15009	Whey permeate and yeast extract	65.5	0.66	2.7	[94]
Lactobacillus bulgaricus NRRL B-548	Lactose, glucose, and galactose	38.7	0.9	3.5	[95]
L. bulgaricus CGMCC 1.6970	Dairy waste	113.18	–	2.36	[96]
Lactobacillus casei NRRL B-441	Barley malt sprouts	82	0.91	5.6	[97]
E. casseli flavus and *L. casei*	Xylose and glucose by co-cultivation	95	–	–	[98]
L. casei NCIMB 3254	Cassava bagasse	83.8	0.96	1.4	[99]
L. casei and *L. lactis*	Date juice extract	60.3	–	3.2	[100]
L. casei ATCC 10863	Ram horn waste	–	0.08	–	[101]

Table 7. *Cont.*

Organism	Substrate	Lactic Acid Production [g/L]	Yield [g/g]	Productivity [g/(L·h)]	Reference
L. casei TISTR 390	Sugarcane bagasse	21.3	–	0.18	[102]
Lactobacillus pentosus ATCC 8041	Trimmings of vine shoots	21.8	0.77	0.8	[103]
L. pentosus CECT-4023T (ATCC-8041)	Corncobs	24	0.76	0.51	[104]
L. brevis and *L. pentosus*	Wheat straw	7.1	0.95	–	[105]
L. pentosus	*L. pentosus*	74.8	0.65	–	[106]
Lactobacillus amylophilus GV6	Wheat flour	76.2	0.7	0.8	[107]
Bacillus sp. strain	Corncob molasses	74.5	0.5	0.38	[14]
Bacillus coagulans strains 36D1	Paper sludge	92	0.77	0.96	[70]
Bacillus coagulans DSM 2314	Lime-treated straw	40.7	0.43	–	[71]
Bacillus coagulans DSM 2314	Sugarcane bagasse	58.7	0.73	1.81	[72]
B. coagulans IPE22	Lignocellulosic hydrolysates	50.48	–	3.16	[73]
B. coagulans	Coffee pulp hydrolysate	48.0	–	1.20	[74]
Lactobacillus sp. RKY2	Glucose, corn steep liquor, and yeast extract	27	0.9	6.7	[108]
Lactobacillus sp. RKY2	Amylase-treated rice and wheat brans	129	0.95	2.9	[109]
Lactobacillus bifermentans DSM 20003	Wheat bran syrup	62.8	0.83	1.2	[110]
L. lactis RM2-24	α-cellulose	73	0.73	1.52	[111]
L. lactis RM2-24	Molasses and cellbiose	70	0.88	1.45	[112]
Lactococcus lactis IO-1	Sugarcane bagasse	10.9	0.36	0.17	[113]
Lactococcus lactis sp. *lactis* IFO 12007	Raw cassava starch	90	0.76	1.6	[114]
Lc. Lactis IO-1	Yeast extract, polypeptone, and xylose	33.26	0.68	–	[115]
L. lactis sub sp. *lactis* AS211	Wheat flour	–	0.77	–	[116]
L. lactis sub sp. *lactis* ATCC 19435	Wheat starch	–	–	1.5	[117]
Sporolactobacillus sp. CASD	Peanut meal	207	0.93	3.8	[118]
Sporolactobacillus laevolacticus DSM442	Agricultural waste cottonseed	144.4	–	4.13	[119]
L. brevis	Hydrolysate of lignocellulosic	39.1	0.7	0.81	[120]
L. coryniformis ATCC 25600	Yeast and meat extract	54	0.89	0.5	[121]
L. coryniformis spp. Torquens ATCC 25600	Waste cardboard	23.4	0.56	0.48	[122]
Lactobacillus coryniformis ATCC 25 600	Pretreated cardboard	23	0.56	0.49	[123]
L. coryniformis sub sp. Torquens ATCC 25600	Pulp mill residue	55.7	–	2.80	[124]

Table 7. *Cont.*

Organism	Substrate	Lactic Acid Production [g/L]	Yield [g/g]	Productivity [g/(L·h)]	Reference
Lactobacillus plantarum ATCC 21028	Lactobacilli Man–Rogosa–Sharpe (MRS) broth (ATCC formula 416, DIFCO 0881)	41	0.97	1	[125]
L. plantarum	Alfalfa fiber	46.4	0.46	0.64	[126]
L. plantarum (Recombinant)	Cellooligosaccharides and β-glucan	1.47	–	–	[127]
	Arabinose	38.6	0.82	3.78	[128]
	Xylose	41.2	0.89	1.6	[129]
Leuconostoc lactis SHO-47 and SHO-54	Xyloo-ligosaccharide	2.3	–	–	[130]
Bacillus sp. Strain 36D1	Solka floc	40	0. 65	0.22	[131]
L. salivarious NRRL B-1950	Soy molasses	–	0.76	–	[132]
Lactobacillus sp.	Sugarcane juice	8.1	–	–	[133]
L. amylovorus	Cassava starch substrate	4.8	–	–	[134]
Leuconostoc mesenteroides NRRL B 512	Yeast extract and sugarcane juice	60.2	–	1.25	[135]
Mixed culture of *B. coagulans* LA1507 and engineered and adapted *E. coli* WYZ-L	Sweet sorghum juice	118.0	–	1.84	[136]
	Molasses and corn steep liquor	75.0	–	0.48	[137]
Bacillus sp. XZL9	Corncob molasses	74.7	–	0.38	[14]
L. paracasei	Sweet sorghum	88	0.79	–	[138]
Adapted *L. paracasei* A-22	Agro-industrial substrate	169.9	–	1.42	[139]
L. paracasei LA104/*L. coryniformis* ATCC 25600	Waste Curcuma longa biomass	97.1	–	2.70	[140]
Lactobacillus sp. B2	Crab (Callinectes bellicosus) wastes	19.5	–	0.81	[141]
R. oryzae TS-61	Chicken feather protein hydrolysate and sugar beet molasses	38.5	–	0.92	[142]
R. oryzae NLX-M-1	Xylo-oligosaccharides manufacturing waste residue	60.3	–	1.0	[143]
R. oryzae As 3.819	Tobacco waste extract	173.5	–	1.45	[144]
Delbrueckii IFO 3202	De-fatted rice bran	28	0.78	0.28	[145]
Delbrueckii NCIMB 8130	Molasses	90	0.97	3.8	[146]
Lactobacillus delbrueckii Uc-3	Cellobiose and cellotriose	90	0.9	2.3	[147]
L. delbrueckii mutant Uc-3	Molasses	166	0.87	4.2	[79]
	Waste sugarcane bagasse	90	0.9	2.25	[148]
	Molasses	166	0.87	4.2	[79]
Delbrueckii	Sugarcane juice	118	0.95	1.7	[149]
L. delbreuckii	Alfalfa fiber	35.4	0.35	0.75	[126]
L. delbreuckii NRRL-B445	Cellulosic material	65	0.18	–	[150,151]
L. delbrueckii NCIM 2025	Cassava bagasse	81.9	0.94	1.36	[99]
L. delbrueckii UFV H2B20	Brewer's spent grain	35.5	0.99	0.59	[152]

Table 7. *Cont.*

Organism	Substrate	Lactic Acid Production [g/L]	Yield [g/g]	Productivity [g/(L·h)]	Reference
L. delbrueckii ZU-S2	Cellulosic material	48.7/44.2	0.95/0.92	1.01/5.7	[153]
	Cellbiose and celltriose	90	0.9	2.3	[147]
	α-cellulose	67	0.83	0.93	[148]
Lactobacillus delbrueckii HG 106	Unpolished rice	90	0.73	1.5	[154]
L. delbrueckii sub sp. Lactis	Starch	–	1.0	–	[155]
L. delbrueckii NRRL B-445	Molasses	–	0.81	–	[156]
L. delbrueckii sp. lactis NCDC290/*L. delbrueckii* sp. Delbrueckii NBRC3202	Kodo millet (Paspalum scrobiculatum) bran residue	10.53	–	0.44	[157]
Engineered and adapted *Pediococcus Acidilactici*	Yeast extract, peptone, and glucose	130.8	–	1.82	[158]
S. inulinus YBS1-5	Wheat bran, corn steep liquor, and yeast extract	70.7	–	0.65	[159]

Note: production (g/L): grams of LA produced by liter (biomass concentration). Yield (g/g): product yield per gram of substrate. Productivity (g/L·h): grams of LA produced every hour per liter (biomass concentration).

4.1. Simultaneous Saccharification and Fermentation

By combining the enzymatic hydrolysis of carbohydrate substrates and the microbial fermentation of the produced sugars into a single process, also referred to as "Simultaneous Saccharification And Fermentation" (SSAF), the bioconversion of carbohydrate sources to lactic acid can be made significantly more effective [1]. Due to lactic acid's low added value as a bulk chemical, it is necessary to convert lignocellulosic material into lactic acid in an effective manner with high productivities and yields [72,160].

Enzymatic hydrolysis should proceed significantly faster when fermentation and enzymatic hydrolysis are coupled in an SSAF process, because the microbe can directly absorb the monomerized sugars, reducing product inhibition. Thus, the processing time of SSAF can be drastically decreased [161]. Moreover, the complete hydrolysis of the carbon substrates before fermentation is not necessary with SSAF. Enzymatic hydrolysis, cell development, and microbial generation all take place concurrently throughout the SSAF process. SSAF's ability to lessen the inhibition brought on by mono- or disaccharide buildup raises the saccharification rate, which will, in turn, increase productivity and minimize reactor volume and capital costs [1].

To increase process effectiveness, SSAF is combined with dilute acid or hot water pretreatments. Cellulases and xylanases are responsible for converting carbohydrate polymers into fermentable sugars, which are susceptible to inhibition by the products (glucose, xylose, cellobiose, and other oligosaccharides) [162].

There is an important fraction of minor sugars in lignocellulosic biomass. From an economical and production perspective, the ability to ferment minor sugars has to be verified. LAB laboratory testing shows the sugar consumption order of the microorganisms: glucose first, then mannose, followed by xylose and galactose almost at the same time [89].

Some thermophilic *Lactobacilli* can ferment pentoses (arabinose and ribose) homofermentatively [163]. It has also been reported that strains related to *L. salivatrius*, further described as *L. murinus* [164], exhibited the same property.

The yield of *B. coagulans IPE22* was enhanced with glucose as substrate, but xylose as substrate enhanced cell proliferation. This demonstrated how the characteristics of simulated mixed sugar would affect the efficiency of LA fermentation, but not necessarily the amount of LA that would be produced in the end [73].

The reduction in end-product inhibition of the enzymatic hydrolysis, as well as the decreased investment costs, are the main advantages of conducting the enzymatic hydrolysis concurrently with the fermentation, rather than in a separate phase. This process has a higher hydrolysis rate, and requires a low enzyme load, higher product yield, lower risk of contamination, shorter time process, and smaller reactor volume. On the other hand, the major disadvantages include the necessity to find optimal conditions (for example, temperature and pH) for both enzymatic hydrolysis and fermentation, as well as the challenge of recycling the fermenting organism and the enzymes. The temperature is typically kept below 37 °C to meet the first condition [165].

Simultaneous Saccharification and Co-Fermentation (SSCF) has been demonstrated to be an efficient bioconversion strategy for LA production. With regard to yield and productivity, fermentation of LA from lignocellulosic feedstock has been investigated extensively [166]. *Lactobacillus pentosus* metabolizes hexose (glucose) through the EMP pathway under anaerobic conditions, producing LA as the sole product (homofermentation). It also ferments pentoses (xylose and arabinose) through the PKP pathway, generating equal mole fractions of LA and acetic acid (heterofermentation) [106].

4.2. Lactic Acid Recovery

LA recovery processes are still not viable for industrial application. The principal disadvantages are equipment costs, solvent recovery, and energy consumption.

The purification and separation of LA involve several processes, which make up 50% of the process cost [167,168]. The principal drawbacks to achieving a high purity LA are due to its decomposition at elevated temperatures, its high energy consumption, and its affinity with water [169]. Table 8 compares different methods available for the separation and recovery of LA from final broth.

Membrane recovery processes are very effective, but their drawbacks include high membrane cost, polarization, and fouling problems. The solvent method requires a high exchange area, and it also has high solvent costs and high toxicity of extractants. There are some promising advances in terms of novel extractants. The precipitation method is the most widely used, but it creates lactic acid with low purity and solid waste residue, and is economically competitive [170].

Table 8. LA separation and recovery methods after SSAF. Adapted from [170–172].

Process	Advantage	Disadvantage	Reference
Precipitation	Can be applied in industrial plants Easy to operate Elevates product yield	Sulfuric acid consumption is high Generation of gypsum, which requires landfill disposal Product purity is low	[170,173–175]
Liquid extraction	Gypsum is not produced Thermal decomposition risk decreased	Regeneration by distillation or back-extraction (stripping) of the extractants is needed Purity is low Disadvantageous distribution coefficients of the extraction agents	[170,175–177]
Membrane processes	Scale of production flexibility High differentiation Good performance at purification Possibly to integrate with current fermenters	High membrane cost Membrane fouling Polarization problems Retention of lactic acid Difficult to increase production	[170,178,179]

Table 8. *Cont.*

Process	Advantage	Disadvantage	Reference
Molecular distillation	Low thermal decomposition risk High differentiation No solvents needed No need for purification	Difficult to increase production High vacuum needed	[171,179,180]
Reactive distillation	Reaction and separation realized together Elevates purification values Energy consumption is low	Mechanism is complex Reversible chemical reactions in the liquid phase applicability Applications are restricted to systems with reasonably fast reaction rates and no mismatch between the temperatures suitable for reaction and separation Homogeneous catalyst usage could cause corrosion and separation issues	[170]
Ion exchange adsorption	No waste generated Easy and simple to operate Elevate operational stability and selectivity Accelerated product recovery Low maintenance expenses and energy usage Not toxic to microorganisms Possibility to be integrated in an heterogeneous system Reusable resin	Large liquor waste as a result of widespread fluent use Only applicable with low temperatures and short-/mid-term production Co-extraction of other compounds is difficult	[181–188]

4.3. Purification of Lactic Acid

The purification of LA from the fermentation broth is a vital step for its commercialization, since high-purity LA is required as a building block for the synthesis of high value-added products, such as PLA and other chemicals. When both isomers are created during fermentation, the purification stage might become even more challenging, as some applications require optically pure D-Lactic acid or L-Lactic acid [189].

Usually, two methods are used: (i) the crude lactic acid is obtained by acidifying the cleared fermented liquor with sulfuric acid, with a concentration of 32%, which is substantially above the crystallization point; (ii) using the method outlined in [190,191], the crude calcium salt that precipitates from the concentrated fermented liquor is crystallized, filtered, dissolved, and then acidified with sulfuric acid. This method uses filtering and washing to separate the precipitated calcium lactate from the dissolved contaminants. However, certain contaminants are still present in the cake. Along with the washes, calcium lactate is lost. The calcium lactate method is currently the most frequently used strategy for separation. This technique, which is based on calcium hydroxide precipitation, is already being used commercially by NatureWorks and Purac in starch-based LA production methods [177].

The traditional method produces enormous amounts of calcium sulphate cake. It is difficult to eliminate cake that still contains organic pollutants. Filtered fermented broth primarily contains contaminants including color, various organic acids, and remaining sugar compounds [13]. Reactive extraction, adsorption, electrodialysis, and esterification–hydrolysis with distillation are methods for removing these contaminants.

Calcium carbonate is first used to neutralize the fermentation broth using the conventional chemical separation procedure. After filtering the broth containing calcium lactate, in order to remove cells, sulfuric acid is added to acidify it and to create LA and insoluble calcium sulfate [10]. In addition, hydrolysis, esterification, and distillation are used to produce pure lactic acid. The creation of a significant amount of calcium sulfate (gypsum) as a byproduct, in addition to the high sulfuric acid consumption, are drawbacks of this method [192]. Adsorption [193], reactive distillation [194], ultrafiltration and electrodialysis [10,195–198], and nanofiltration [199,200] have all been investigated as alternative lactic acid separation technologies that do not produce salt waste. Compared to conventional chemical separation methods, these processes are more economical and energy-efficient. Additionally, they have a number of benefits, such as the absence of costly solvents, adsorbents, or energy-intensive phase transitions, as well as the potential for the simultaneous separation and concentration of lactic acid [201].

4.4. Byproduct Formation

The fermentability of substrates obtained from lignocellulose can be negatively impacted by byproducts created during the pretreatment of lignocellulose. The majority of the lignocellulosic byproducts produced can significantly prolong lag phases, reduce productivity, and/or restrict the growth of microorganisms. There are three main categories of lignocellulosic byproducts that can be identified: phenolics, furans, and small organic acids [49].

As a way to deal with the issues caused by byproducts produced during the fermentation of pretreated lignocellulose into LA, three solutions have been presented: (i) It is possible to enhance thermochemical pretreatment techniques to limit the production of byproducts. Since a decrease in byproduct synthesis should not affect the accessibility of hemicellulose to enzymes, achieving this goal can be challenging. Furthermore, since some of the byproducts are inherent to the hemicellulosic structure and are produced during the monomerization of hemicellulosic sugars, their synthesis cannot be completely avoided. Acetic acid, ferulic acid, and coumaric acid are a few examples. The correct pretreatment conditions can considerably limit the formation of additional chemicals that are not a part of the lignocellulose structure, including furfural, without compromising sugar monomerization or accessibility [103]. (ii) After lignocellulose has completed thermochemical pretreatment, byproduct removal can be implemented immediately. Examples include utilizing microbial detoxification, extracting byproducts using active charcoal or lime, or washing pretreated lignocellulose with extremely hot water. Byproduct removal can be an efficient way to limit the number of byproducts in the substrates made from lignocellulose, but this complicates the process, raising costs and perhaps causing a (slight) material loss [202,203]. (iii) Microorganisms can be reinforced to survive greater byproduct concentrations. Methods such as genetic engineering, evolutionary engineering, or mutagenesis can be used to make this improvement. Although the use of genetic engineering and evolutionary engineering may be challenging and time-consuming in some circumstances, and few targets for genetic engineering have been identified, it may be a potent tool to lessen the impact of byproducts on the microorganism.

5. PLA Production by Lactic Acid Polymerization

PLA with variable molecular weight can be produced using LA, although typically only high molecular weight (Mw) PLA has significant economic use in the fiber, textile, plastics, additive manufacturing [204], printed circuit board [205], and packaging industries [206].

The indirect way of obtaining PLA from the lactide, known as ring opening, is the most widely employed technique. It has been shown to be the most efficient method in the industry for producing high molecular weight polymers, and it has the highest implantation. Two of the largest PLA makers use it: Corbion-Purac (Amsterdam, the Netherlands; www.corbion.com, accessed on 7 July 2021) and NatureWorks LLC (Minnetonka, Minnesota; www.natureworksllc.com, accessed on 7 July 2021). Lactic acid is first oligomerized and then depolymerized in the indirect process to create lactide, a cyclic dimer of lactic acid. Following this, ROP transforms lactide into PLA. The three primary techniques for producing high Mw PLA from LA are shown in Figure 4: (1) direct condensation polymerization; (2) direct polycondensation in an azeotropic solution; and (3) polymerization through lactide formation [18].

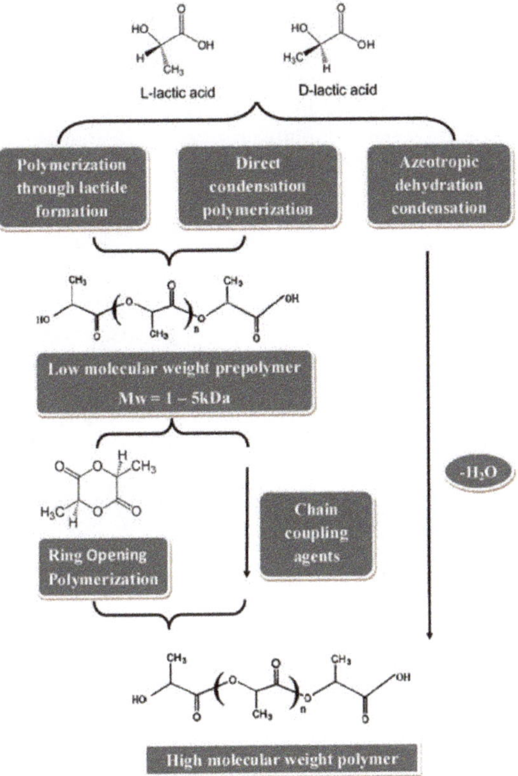

Figure 4. The manufacturing process to produce high molecular weight PLA, from [18].

ROP is the most used technique on the industrial scale due to its benefits, which include light process conditions, shorter residence durations, the absence of side products, and high molecular weights. The most common catalyst is 2-ethylhexanoic tin(II) salt, also known as stannous octoate [Sn(Oct)2], which is authorized by the USFDA, and is typically used in conjunction with alcohol as a cocatalyst. Since impurities have a negative impact on material qualities due to the reaction's sensitivity to residual non-cyclic monomers, the real bottleneck of ROP is the availability of cyclic monomers, as well as their optical and chemical purity. Figure 5 illustrates the three stereo-isomeric forms of cyclic dimer lactide, the cyclic raw material for PLA.

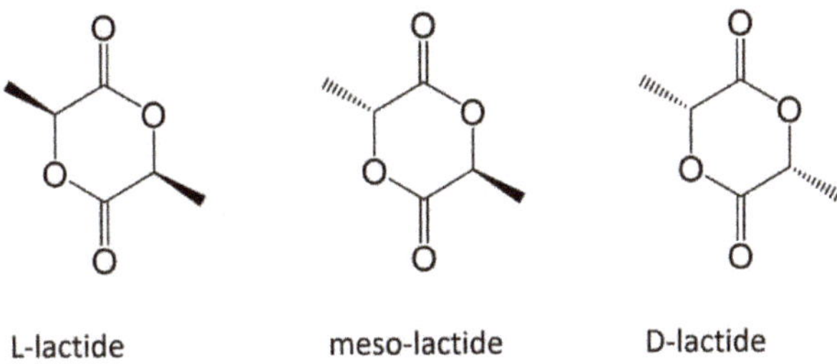

Figure 5. Cyclic dimers for ROP process.

The amount of L- and D-lactide used to create PLA depends on the procedure. Water, meso-lactide, contaminants, LA, and LA oligomers are all products of the lactide reactor [207]. It is necessary to purify the mixture, in this case by vacuum distillation via a set of columns. The highest Mw PLA is formed from L-lactide plus a tiny quantity of meso-lactide, due to the difference in the boiling temperatures of lactide and meso-lactide. The stereochemical purity of the PLA increases with the lactide mixture's stereochemical purity. Because a significant quantity of meso-lactide is produced during this process, the qualities of the PLA resin that is produced can vary depending on the amount of meso-lactide in the mixture. PLA that includes a significant amount (93%) of L-LA crystallizes.

Meso-lactide content in PLA monomers should ideally be as low as possible. NatureWorks LLC has made it feasible to refine meso-lactide into different functionalities, despite the fact that its manufacture is undesired and frequently accompanied by contaminants. The meso-lactide byproduct is used in a variety of surfactants, coatings, and copolymers as chemical intermediates [208]. The manufacture of lactide and PLA with cheap manufacturing and production costs, as well as improved characteristics, has also been the subject of extensive investigation [209]. To obtain high Mw and high optical purity, PLA has been polymerized using a variety of catalysts, including metal, cationic, and organic ones [210]. According to reports, metal complexes are one of the most effective catalysts for the ROP of the rac-lactide-based method of producing stereoblock isotactic PLA, because they can regulate factors such as molecular and chain microstructure [211].

PLA Processing

The procedures for processing PLA are tried-and-true polymer production processes utilized for other commercial polymers, including PS and PET [212]. The primary method for mass producing high Mw PLA is called melt processing, and it involves transforming the resulting PLA resin into finished goods for use in packaging [213], consumer goods, and other applications. The technique of molding molten polymers into desired shapes while they are still in a liquid state is known as melt processing. The polymer is then cooled to stabilize its final dimensions [214].

6. Discussion

It is possible to extract lignocellulosic material from cow manure, which can later be converted into glucose. This glucose can feed microorganisms that produce lactic acid, which can be polymerized into PLA. Thus, this means that it is possible to create a valuable product from waste origin. Production of bioethanol from waste manure has been studied, as has PLA production from a variety of lignocellulosic materials.

Cows are fed a high percentage of lignocellulosic fiber in their diets, and following digestion, they produce manure. Enzymatic hydrolysis into fermentable sugars could successfully disrupt the treated lignocellulosic fraction from those dejections.

Anaerobic digestion of cow manure for methane production resulted in a moderate reduction in the proportion of lignocellulosic fiber, according to research on the composition of lignocellulosic materials, in cow manure and anaerobically digested cow manure. The first step in using lignocellulosic biomass for the production of added-value products from its constituent fractions, such as hemicelluloses and cellulose, is to remove barriers that prevent chemical or biological catalysts from aiding in its transformation. Size reduction and steam explosion are examples of physical pretreatments, whereas solvents that promote cellulose, hemicellulose, and lignin degradation are used in chemical pretreatments to change the structure of biomass. The partition of cellulose into glucose units and of the hemicellulosic biomass into its constituent sugars are necessary for the conversion of lignocellulosic biomass into value-added goods. Two main approaches have been developed: acid and enzymatic hydrolysis. Enzymatic hydrolysis is the biotechnological process, while acid hydrolysis is the chemical path.

On agricultural waste, byproducts, and residues, LA can be synthesized biotechnologically and used to create biodegradable, biocompatible LA polymers. The majority of lactic acid that is commercially generated around the world is created through the fermentation of carbohydrates by homolactic organisms from the genus *Lactobacillus*, which only produce lactic acid. By combining the enzymatic hydrolysis of carbohydrate substrates and microbial fermentation of the resulting sugars into a single phase, the bioconversion of carbohydrate materials to lactic acid can be made considerably more effectively (SSAF process).

To achieve the LA product, purification is needed after fermentation. Separation and the final purification process incur almost 50% of the production costs. Due to water's high affinity, breakdown at high temperatures, and significant energy consumption, obtaining high purity LA is problematic. Variable molecular weight PLA can be produced using LA, but typically only high Mw PLA has significant economic use in the fiber, textile, plastics, and packaging industries. The indirect way of obtaining PLA from ROP is the most widely employed approach.

Although PLA degradation has been studied for a while, knowledge of the underlying mechanisms is still lacking. Methods for the chemical recycling of these materials, together with the obtained chemical products, have been reviewed [215]. According to several studies, PLA breakdown only happens by hydrolysis and does not require any enzyme activity [216]. According to some reports, enzymes play a crucial part in the breakdown of PLA [217].

7. Conclusions

There is concern about the impact of cow manure waste and fossil-based plastic accumulation. Cow manure decomposition leads to gases such as methane and nitrogen oxide, both of these being greenhouse emissions that have an impact on the environment. Combining both problems, it is possible to manufacture a value-added product such as PLA from cow manure, and thus solve emission problems.

Lignocellulosic biomass contains a variety of minor sugars in addition to xylose and glucose. From an economic standpoint, it is highly beneficial if all sugar substrates are utilized. Hexoses' and pentoses' transformation ratios from lignocellulosic fraction into LA would have an important impact on process viability.

It is possible to extract the lignocellulosic fraction from cow manure and convert it into glucose. This glucose can be used to feed the microorganisms that produce LA. The obtained lactic acid can be polymerized, achieving the final product: PLA. The obtained PLA would be a replacement for petro-chemical polymers due to its biodegradability properties, improving the environmental problems.

Existing pretreatments are not adequate, and existing nonthermal advances combined with organic mild chemical process are likely to increase the conversion yields' released inhibitory compounds potential. It is fundamental to enhance LA generation without sacrificing the optical quality or adding extra costs into LA polymers. Cow manure is a possible source for PLA production.

The separation and purification processes are critical for LA production viability: 50% of the costs are incurred by these processes, and they have many drawbacks, such as water affinity, decomposition at high temperatures, and elevated energy consumption.

As a result of the review, to obtain PLA form cow manure, the required technological sequence consists of manure treatment, pretreatment to prepare for hydrolysis process, saccharification, and fermentation. Pretreatment may be performed with a diluted acid treatment if lignin content is low enough. Saccharification and fermentation could be carried out together or separately, depending on the microorganism, affecting the degradation of hexoses and pentoses. After the LA process, the sequence is similar to current industrial processes, but it is necessary to consider the effect of the substrate on the purification, polymerization, and final product characteristics. Further research is needed to validate the entire process yield, as well as the properties of the obtained PLA.

Author Contributions: Conceptualization, L.F.C., R.G. and V.F.; methodology, L.F.C. and V.F.; investigation, R.G. and O.P.N.; resources, L.F.C.; data curation, L.F.C.; writing—original draft preparation, R.G.; writing—review and editing, L.F.C., O.P.N. and V.F.; visualization, R.G.; supervision, L.F.C. and V.F.; project administration, L.F.C.; funding acquisition, L.F.C. All authors have read and agreed to the published version of the manuscript.

Funding: This work has received financial support from the Doctorat Industrial grant (2021 DI 22) from the AGAUR through the Secretariat of Universities and Research of the Department of Business and Knowledge of the Generalitat de Catalunya. This work is partially supported by ICREA under the ICREA Academia programme.

Informed Consent Statement: Not applicable.

Data Availability Statement: Data are available upon request to the corresponding author.

Acknowledgments: The authors at the University of Lleida would like to thank the Catalan Government for the quality accreditation given to their research group GREiA (2017 SGR 1537). GREiA is a certified agent TECNIO in the category of technology developers from the Government of Catalonia.

Conflicts of Interest: The authors declare no conflict of interest.

References

1. Singhvi, M.; Gokhale, D. Biomass to biodegradable polymer (PLA). *RSC Adv.* **2013**, *3*, 13558–13568. [CrossRef]
2. Jambeck, J.R.; Geyer, R.; Wilcox, C.; Siegler, T.R.; Perryman, M.; Andrady, A.; Narayan, R.; Law, K.L. Plastic waste inputs from land into the ocean. *Science* **2015**, *347*, 768–771. [CrossRef] [PubMed]
3. Lebreton, L.C.M.; van der Zwet, J.; Damsteeg, J.W.; Slat, B.; Andrady, A.; Reisser, J. River plastic emissions to the world's oceans. *Nat. Commun.* **2017**, *8*, 15611. [CrossRef]
4. Chheda, J.N.; Huber, G.W.; Dumesic, J.A. Liquid-Phase Catalytic Processing of Biomass-Derived Oxygenated Hydrocarbons to Fuels and Chemicals. *Angew. Chem. Int. Ed.* **2007**, *46*, 7164–7183. [CrossRef] [PubMed]
5. Röper, H. Renewable Raw Materials in Europe-Industrial Utilisation of Starch and Sugar. *Starch-Stärke* **2002**, *54*, 89–99. [CrossRef]
6. Maki, M.; Leung, K.T.; Qin, W. The prospects of cellulase-producing bacteria for the bioconversion of lignocellulosic biomass. *Int. J. Biol. Sci.* **2009**, *5*, 500–516. [CrossRef]
7. FAOSTAT. *Statistical Yearbook of 2012: Europe and Central Asia*; FAO: Rome, Italy, 2012.
8. Beef2Live. *World Cattle Inventory Rankng*; Beef2live: Overland Park, AR, USA, 2020.
9. Boldú, F.P.; Pous, J.P. *Guia de les Tecnologies de Tractament de les Dejeccions Ramaderes a Catalunya*; Departament d'Agricultura, Ramaderia, Pesca i Alimentació (DARP): Barcelona, Spain, 2020.
10. Datta, R.; Henry, M. Lactic acid: Recent advances in products, processes and technologies—A review. *J. Chem. Technol. Biotechnol.* **2006**, *81*, 1119–1129. [CrossRef]
11. Datta, R.; Tsai, S.-P.; Bonsignore, P.; Moon, S.-H.; Frank, J.R. Technological and economic potential of poly(lactic acid) and lactic acid derivatives. *FEMS Microbiol. Rev.* **1995**, *16*, 221–231. [CrossRef]
12. Wee, Y.J.; Kim, J.N.; Ryu, H.W. Biotechnological production of lactic acid and its recent applications. *Food Technol. Biotechnol.* **2006**, *44*, 163–172.
13. Joglekar, H.G.; Rahman, I.; Babu, S.; Kulkarni, B.D.; Joshi, A. Comparative assessment of downstream processing options for lactic acid. *Sep. Purif. Technol.* **2006**, *52*, 1–17. [CrossRef]
14. Wang, L.; Zhao, B.; Liu, B.; Yu, B.; Ma, C.; Su, F.; Hua, D.; Li, Q.; Ma, Y.; Xu, P. Efficient production of l-lactic acid from corncob molasses, a waste by-product in xylitol production, by a newly isolated xylose utilizing *Bacillus* sp. strain. *Bioresour. Technol.* **2010**, *101*, 7908–7915. [CrossRef]
15. Anderson, J.M.; Shive, M.S. Biodegradation and biocompatibility of PLA and PLGA microspheres. *Adv. Drug Deliv. Rev.* **2012**, *64*, 72–82. [CrossRef]
16. Jem, K.J.; Tan, B. The development and challenges of poly (lactic acid) and poly (glycolic acid). *Adv. Ind. Eng. Polym. Res.* **2020**, *3*, 60–70. [CrossRef]
17. Henton, D.E.; Gruber, P.; Lunt, J.; Randall, J. Polylactic Acid Technology. In *Natural Fibers, Biopolymers, and Biocomposites*; CRC Press: Boca Raton, FL, USA, 2005; pp. 527–578.
18. Hartmann, M.H. High Molecular Weight Polylactic Acid Polymers. In *Biopolymers from Renewable Resources*; Springer: Berlin/Heidelberg, Germany, 1998; pp. 367–411. [CrossRef]
19. D'Adamo, I.; Gastaldi, M.; Morone, P.; Rosa, P.; Sassanelli, C.; Settembre-Blundo, D.; Shen, Y. Bioeconomy of Sustainability: Drivers, Opportunities and Policy Implications. *Sustainability* **2022**, *14*, 200. [CrossRef]
20. Ministerio de la Presidencia, Real Decreto 506, de 28 de Junio, Sobre Productos Fertilizantes, Boletín Oficial del Estado. 2013; pp. 1–24. Available online: https://www.boe.es/boe/dias/2013/07/10/pdfs/BOE-A-2013-7540.pdf (accessed on 7 May 2021).
21. Seppälä, M.; Laine, A.; Rintala, J. Screening of novel plants for biogas production in northern conditions. *Bioresour. Technol.* **2013**, *139*, 355–362. [CrossRef] [PubMed]

22. Holm-Nielsen, J.B.; Al Seadi, T.; Oleskowicz-Popiel, P. The future of anaerobic digestion and biogas utilization. *Bioresour. Technol.* **2009**, *100*, 5478–5484. [CrossRef]

23. Burton, C.H. Manure Management–Treatment Strategies for Sustainable Agriculture, second edition. *Livest. Sci.* **2006**, *102*, 256–257. [CrossRef]

24. Burton, C.H.; Beck, J.; Bloxham, P.F.; Derikx, P.J.L.; Martinez, J. *Manure Management. Treatment Strategies for Sustainable Agriculture*; Silsoe Research Institute: Bedford, UK, 1997; Available online: https://research.wur.nl/en/publications/manure-management-treatment-strategies-for-sustainable-agricultur (accessed on 7 January 2021).

25. Rufete Sáez, A.B. Caracterización de Residuos Ganaderos del Sureste Español: Implicaciones Agronómicas y Medioambientales, Universidad Miguel Hernández de Elche. 2015. Available online: http://dspace.umh.es/bitstream/11000/2099/1/TD.pdf (accessed on 6 August 2021).

26. Valmaseda, M.; Martínez, A.T.; Almendros, G. Contribution by pigmented fungi to P-type humic acid formation in two forest soils. *Soil Biol. Biochem.* **1989**, *21*, 23–28. [CrossRef]

27. Hassanat, F.; Gervais, R.; Benchaar, C. Methane production, ruminal fermentation characteristics, nutrient digestibility, nitrogen excretion, and milk production of dairy cows fed conventional or brown midrib corn silage. *J. Dairy Sci.* **2017**, *100*, 2625–2636. [CrossRef]

28. Ashekuzzaman, S.M.; Poulsen, T.G. Optimizing feed composition for improved methane yield during anaerobic digestion of cow manure based waste mixtures. *Bioresour. Technol.* **2011**, *102*, 2213–2218. [CrossRef]

29. Díaz, I.; Figueroa-González, I.; Miguel, J.Á.; Bonilla-Morte, L.; Quijano, G. Enhancing the biomethane potential of liquid dairy cow manure by addition of solid manure fractions. *Biotechnol. Lett.* **2016**, *38*, 2097–2102. [CrossRef] [PubMed]

30. Zhao, X.-Q.; Zi, L.-H.; Bai, F.-W.; Lin, H.-L.; Hao, X.-M.; Yue, G.-J.; Ho, N.W.Y. Bioethanol from lignocellulosic biomass. *Adv. Biochem. Eng. Biotechnol.* **2015**, *128*, 25–51. [CrossRef]

31. Li, K.; Liu, R.; Sun, C. Comparison of anaerobic digestion characteristics and kinetics of four livestock manures with different substrate concentrations. *Bioresour. Technol.* **2015**, *198*, 133–140. [CrossRef] [PubMed]

32. Casanellas, J.P. Edafología: Uso y Protección de Suelos. 2019. Available online: https://www.mundiprensa.com/catalogo/9788484767503/edafologia--uso-y-proteccion-de-suelos (accessed on 7 January 2021).

33. Burton, C.H.; Turner, C. (Eds.) Treatment strategies for livestock manure for sustainable livestock agriculture. In *Manure Management: Treatment Strategies for Sustainable Agriculture*, 2nd ed.; International Society for Animal Hygiene: Bedford, UK, 2003; p. 490.

34. Li, R.; Chen, S.; Li, X. Anaerobic co-digestion of kitchen waste and cattle manure for methane Production. *Energy Sources Part A Recover. Util. Environ. Eff.* **2009**, *31*, 1848–1856. [CrossRef]

35. Shen, J.; Zhao, C.; Liu, Y.; Zhang, R.; Liu, G.; Chen, C. Biogas production from anaerobic co-digestion of durian shell with chicken, dairy, and pig manures. *Energy Convers. Manag.* **2019**, *198*, 110535. [CrossRef]

36. Zhao, Y.; Sun, F.; Yu, J.; Cai, Y.; Luo, X.; Cui, Z.; Hu, Y.; Wang, X. Co-digestion of oat straw and cow manure during anaerobic digestion: Stimulative and inhibitory effects on fermentation. *Bioresour. Technol.* **2018**, *269*, 143–152. [CrossRef]

37. Kumar, P.; Barrett, D.M.; Delwiche, M.J.; Stroeve, P. Methods for pretreatment of lignocellulosic biomass for efficient hydrolysis and biofuel production. *Ind. Eng. Chem. Res.* **2009**, *48*, 3713–3729. [CrossRef]

38. Liao, W.; Liu, Y.; Liu, C.; Wen, Z.; Chen, S. Acid hydrolysis of fibers from dairy manure. *Bioresour. Technol.* **2006**, *97*, 1687–1695. [CrossRef]

39. Wen, Z.; Liao, W.; Chen, S. Hydrolysis of animal manure lignocellulosics for reducing sugar production. *Bioresour. Technol.* **2004**, *91*, 31–39. [CrossRef]

40. Yan, Q.; Liu, X.; Wang, Y.; Li, H.; Li, Z.; Zhou, L.; Qu, Y.; Li, Z.; Bao, X. Cow manure as a lignocellulosic substrate for fungal cellulase expression and bioethanol production. *AMB Express* **2018**, *8*, 190. [CrossRef]

41. van Dyk, J.S.; Pletschke, B.I. A review of lignocellulose bioconversion using enzymatic hydrolysis and synergistic cooperation between enzymes-Factors affecting enzymes, conversion and synergy. *Biotechnol. Adv.* **2012**, *30*, 1458–1480. [CrossRef] [PubMed]

42. Howard, R.L.; Abotsi, E.; van Rensburg, E.L.J.; Howard, S. Lignocellulose biotechnology: Issues of bioconversion and enzyme production. *Afr. J. Biotechnol.* **2003**, *2*, 602–619. [CrossRef]

43. Gosselink, R.J.A.; Teunissen, W.; van Dam, J.E.G.; de Jong, E.; Gellerstedt, G.; Scott, E.L.; Sanders, J.P.M. Lignin depolymerisation in supercritical carbon dioxide/acetone/water fluid for the production of aromatic chemicals. *Bioresour. Technol.* **2012**, *106*, 173–177. [CrossRef] [PubMed]

44. Czernik, S.; Bridgwater, A.V. Overview of applications of biomass fast pyrolysis oil. *Energy Fuels* **2004**, *18*, 590–598. [CrossRef]

45. Castro, Y.P. *Aprovechamiento de Biomasa Lignocelulósica: Algunas Experiencias de Investigación en Colombia*; UTadeo: Bogotá, Colombia, 2014.

46. Miller, C.; Fosmer, A.; Rush, B.; McMullin, T.; Beacom, D.; Suominen, P. *Industrial Production of Lactic Acid*, 2nd ed.; Elsevier B.V.: Amsterdam, The Netherlands, 2011. [CrossRef]

47. Maican, E.; Coz, A.; Ferdeş, M. Continuous Pretreatment Process For Bioethanol Production. In Proceedings of the 4th International Conference on Thermal Equipment, Renewable Energy and Rural Development, Arge County, Romania, 6 June 2015.

48. Kabel, M.A.; Bos, G.; Zeevalking, J.; Voragen, A.G.J.; Schols, H.A. Effect of pretreatment severity on xylan solubility and enzymatic breakdown of the remaining cellulose from wheat straw. *Bioresour. Technol.* **2007**, *98*, 2034–2042. [CrossRef] [PubMed]

49. Palmqvist, E.; Hahn-Hägerdal, B. Fermentation of lignocellulosic hydrolysates. II: Inhibitors and mechanisms of inhibition. *Bioresour. Technol.* **2000**, *74*, 25–33. [CrossRef]

50. Alvira, P.; Tomás-Pejó, E.; Ballesteros, M.; Negro, M.J. Pretreatment technologies for an efficient bioethanol production process based on enzymatic hydrolysis: A review. *Bioresour. Technol.* **2010**, *101*, 4851–4861. [CrossRef]

51. Taherzadeh, M.J.; Karimi, K. Enzyme-based hydrolysis processes for ethanol from lignocellulosic materials: A review. *BioResources* **2007**, *2*, 707–738. [CrossRef]

52. Taherzadeh, M.J.; Karimi, K. Acid-based hydrolysis processes for ethanol from lignocellulosic materials: A review. *BioResources* **2007**, *2*, 472–499.

53. Yang, B.; Wyman, C.E. Effect of Xylan and Lignin Removal by Batch and Flowthrough Pretreatment on the Enzymatic Digestibility of Corn Stover Cellulose. *Biotechnol. Bioeng.* **2004**, *86*, 88–98. [CrossRef]

54. Wojtusik, M.; Villar, J.C.; Zurita, M.; Ladero, M.; Garcia-Ochoa, F. Study of the enzymatic activity inhibition on the saccharification of acid pretreated corn stover. *Biomass Bioenergy* **2017**, *98*, 1–7. [CrossRef]

55. Taherzadeh, M.; Karimi, K. Pretreatment of Lignocellulosic Wastes to Improve Ethanol and Biogas Production: A Review. *Int. J. Mol. Sci.* **2008**, *9*, 1621–1651. [CrossRef] [PubMed]

56. Garzón, H.; Orozco, D. Hidrólisis Enzimática del Material Lignocelulósico de la Planta de Banano y su Fruto. Bachelor's Thesis, Escuela de Procesos y Energía, Universidad Nacional de Colombia-Sede Medellín, Medellín, Colombia, 2006.

57. Hamelinck, C.N.; van Hooijdonk, G.; Faaij, A.P.C. Ethanol from lignocellulosic biomass: Techno-economic performance in short-, middle- and long-term. *Biomass Bioenergy* **2005**, *28*, 384–410. [CrossRef]

58. Pi, Y.; Lozano, J.T. *Aprovechamiento de Biomasa Lignocelulósica, Algunas Experiencias de Investigación en Colombia*, 2014th ed.; Fundación Universidad de Bogotá Jorge Tadeo Lozano Carrera: Bogotá, Colombia, 2016; Available online: https://www.researchgate.net/publication/279448880%0AAprovechamiento (accessed on 7 April 2021).

59. Jung, Y.H.; Kim, I.J.; Kim, J.J.; Oh, K.K.; Han, J.-I.; Choi, I.-G.; Kim, K.H. Ethanol production from oil palm trunks treated with aqueous ammonia and cellulase. *Bioresour. Technol.* **2011**, *102*, 7307–7312. [CrossRef]

60. Kahar, P.; Taku, K.; Tanaka, S. Enzymatic digestion of corncobs pretreated with low strength of sulfuric acid for bioethanol production. *J. Biosci. Bioeng.* **2010**, *110*, 453–458. [CrossRef]

61. Mesa, L.; González, E.; Romero, I.; Ruiz, E.; Cara, C.; Castro, E. Comparison of process configurations for ethanol production from two-step pretreated sugarcane bagasse. *Chem. Eng. J.* **2011**, *175*, 185–191. [CrossRef]

62. Geddes, C.C.; Peterson, J.J.; Roslander, C.; Zacchi, G.; Mullinnix, M.T.; Shanmugam, K.T.; Ingram, L.O. Optimizing the saccharification of sugar cane bagasse using dilute phosphoric acid followed by fungal cellulases. *Bioresour. Technol.* **2010**, *101*, 1851–1857. [CrossRef]

63. Oberoi, H.S.; Sandhu, S.K.; Vadlani, P.V. Statistical optimization of hydrolysis process for banana peels using cellulolytic and pectinolytic enzymes. *Food Bioprod. Process.* **2012**, *90*, 257–265. [CrossRef]

64. Wojtusik, M.; Zurita, M.; Villar, J.C.; Ladero, M.; Garcia-Ochoa, F. Influence of fluid dynamic conditions on enzymatic hydrolysis of lignocellulosic biomass: Effect of mass transfer rate. *Bioresour. Technol.* **2016**, *216*, 28–35. [CrossRef]

65. Djukić-Vuković, A.; Mladenović, D.; Ivanović, J.; Pejin, J.; Mojović, L. Towards sustainability of lactic acid and poly-lactic acid polymers production. *Renew. Sustain. Energy Rev.* **2019**, *108*, 238–252. [CrossRef]

66. Gao, C.; Ma, C.; Xu, P. Biotechnological routes based on lactic acid production from biomass. *Biotechnol. Adv.* **2011**, *29*, 930–939. [CrossRef] [PubMed]

67. Wilhelm, B.J.B.W.; Holzapfel, H. *Lactic Acid Bacteria: Biodiversity and Taxonomy*; Wiley: New York, NY, USA, 2014.

68. Chaisu, K.; Charles, A.L.; Guu, Y.-K.; Yen, T.-B.; Chiu, C.-H. Optimization Lactic Acid Production from Molasses Renewable Raw Material through Response Surface Methodology with Lactobacillus Casei M-15. *APCBEE Procedia* **2014**, *8*, 194–198. [CrossRef]

69. Keshwani, D.R.; Cheng, J.J. Switchgrass for bioethanol and other value-added applications: A review. *Bioresour. Technol.* **2009**, *100*, 1515–1523. [CrossRef]

70. Budhavaram, N.K.; Fan, Z. Production of lactic acid from paper sludge using acid-tolerant, thermophilic Bacillus coagulan strains. *Bioresour. Technol.* **2009**, *100*, 5966–5972. [CrossRef]

71. Maas, R.H.W.; Bakker, R.R.; Jansen, M.L.A.; Visser, D.; de Jong, E.; Eggink, G.; Weusthuis, R.A. Lactic acid production from lime-treated wheat straw by Bacillus coagulans: Neutralization of acid by fed-batch addition of alkaline substrate. *Appl. Microbiol. Biotechnol.* **2008**, *78*, 751–758. [CrossRef]

72. van der Pol, E.C.; Eggink, G.; Weusthuis, R.A. Production of l(+)-lactic acid from acid pretreated sugarcane bagasse using Bacillus coagulans DSM2314 in a simultaneous saccharification and fermentation strategy. *Biotechnol. Biofuels* **2016**, *9*, 248. [CrossRef]

73. Wang, Y.; Cao, W.; Luo, J.; Wan, Y. Exploring the potential of lactic acid production from lignocellulosic hydrolysates with various ratios of hexose versus pentose by Bacillus coagulans IPE22. *Bioresour. Technol.* **2018**, *261*, 342–349. [CrossRef]

74. Pleissner, D.; Neu, A.K.; Mehlmann, K.; Schneider, R.; Puerta-Quintero, G.I.; Venus, J. Fermentative lactic acid production from coffee pulp hydrolysate using Bacillus coagulans at laboratory and pilot scales. *Bioresour. Technol.* **2016**, *218*, 167–173. [CrossRef]

75. Åkerberg, C.; Hofvendahl, K.; Zacchi, G.; Hahn-Hägerdal, B. Modelling the influence of pH, temperature, glucose and lactic acid concentrations on the kinetics of lactic acid production by *Lactococcus lactis* ssp. lactis ATCC 19435 in whole-wheat flour. *Appl. Microbiol. Biotechnol.* **1998**, *49*, 682–690. [CrossRef]

76. Soccol, C.R.; Stonoga, V.I.; Raimbault, M. Production of l-lactic acid by Rhizopus species. *World J. Microbiol. Biotechnol.* **1994**, *10*, 433–435. [CrossRef] [PubMed]

77. Tay, A.; Yang, S.T. Production of L(+)-lactic acid from glucose and starch by immobilized cells of Rhizopus oryzae in a rotating fibrous bed bioreactor. *Biotechnol. Bioeng.* **2002**, *80*, 1–12. [CrossRef] [PubMed]

78. Yun, J.S.; Wee, Y.J.; Ryu, H.W. Production of optically pure L(+)-lactic acid from various carbohydrates by batch fermentation of *Enterococcus faecalis* RKY1. *Enzym. Microb. Technol.* **2003**, *33*, 416–423. [CrossRef]

79. Dumbrepatil, A.; Adsul, M.; Chaudhari, S.; Khire, J.; Gokhale, D. Utilization of molasses sugar for lactic acid production by *Lactobacillus delbrueckii* subsp. delbrueckii mutant Uc-3 in batch fermentation. *Appl. Environ. Microbiol.* **2008**, *74*, 333–335. [CrossRef]

80. Wee, Y.-J.; Reddy, L.V.A.; Ryu, H.-W. Fermentative production of L(+)-lactic acid from starch hydrolyzate and corn steep liquor as inexpensive nutrients by batch culture of *Enterococcus faecalis* RKY1. *J. Chem. Technol. Biotechnol.* **2008**, *83*, 1387–1393. [CrossRef]

81. Wee, Y.J.; Yun, J.S.; Park, D.H.; Ryu, H.W. Biotechnological production of L(+)-lactic acid from wood hydrolyzate by batch fermentation of *Enterococcus faecalis*. *Biotechnol. Lett.* **2004**, *26*, 71–74. [CrossRef]

82. Abdel-Rahman, M.A.; Tashiro, Y.; Zendo, T.; Shibata, K.; Sonomoto, K. Isolation and characterisation of lactic acid bacterium for effective fermentation of cellobiose into optically pure homo l-(+)-lactic acid. *Appl. Microbiol. Biotechnol.* **2011**, *89*, 1039–1049. [CrossRef]

83. Abdel-Rahman, M.A.; Tashiro, Y.; Zendo, T.; Sonomoto, K. Effective (+)-Lactic Acid Production by Co-fermentation of Mixed Sugars. *J. Biotechnol.* **2010**, *150*, 347–348. [CrossRef]

84. Abdel-Rahman, M.A.; Tashiro, Y.; Zendo, T.; Sonomoto, K. Isolation and characterization of novel lactic acid bacterium for efficient production of l (+)-lactic acid from xylose. *J. Biotechnol.* **2010**, *150*, 347. [CrossRef]

85. Berry, A.R.; Franco, C.M.M.; Zhang, W.; Middelberg, A.P.J. Growth and lactic acid production in batch culture of *Lactobacillus rhamnosus* in a defined medium. *Biotechnol. Lett.* **1999**, *21*, 163–167. [CrossRef]

86. Marques, S.; Santos, J.A.L.; Gírio, F.M.; Roseiro, J.C. Lactic acid production from recycled paper sludge by simultaneous saccharification and fermentation. *Biochem. Eng. J.* **2008**, *41*, 210–216. [CrossRef]

87. Gullón, B.; Yáñez, R.; Alonso, J.L.; Parajó, J.C. l-Lactic acid production from apple pomace by sequential hydrolysis and fermentation. *Bioresour. Technol.* **2008**, *99*, 308–319. [CrossRef] [PubMed]

88. Romaní, A.; Yáñez, R.; Garrote, G.; Alonso, J.L. SSF production of lactic acid from cellulosic biosludges. *Bioresour. Technol.* **2008**, *99*, 4247–4254. [CrossRef]

89. Iyer, P.V.; Thomas, S.; Lee, Y.Y. High-Yield Fermentation of Pentoses into Lactic Acid. *Appl. Biochem. Biotechnol.* **2000**, *84–86*, 665–678. [CrossRef] [PubMed]

90. Cui, F.; Li, Y.; Wan, C. Lactic acid production from corn stover using mixed cultures of *Lactobacillus rhamnosus* and *Lactobacillus brevis*. *Bioresour. Technol.* **2011**, *102*, 1831–1836. [CrossRef] [PubMed]

91. Wang, L.; Zhao, B.; Liu, B.; Yang, C.; Yu, B.; Li, Q.; Ma, C.; Xu, P.; Ma, Y. Efficient production of l-lactic acid from cassava powder by *Lactobacillus rhamnosus*. *Bioresour. Technol.* **2010**, *101*, 7895–7901. [CrossRef]

92. Djukić-Vuković, A.P.; Mojović, L.V.; Vukašinović-Sekulić, M.S.; Nikolić, S.B.; Pejin, J.D. Integrated production of lactic acid and biomass on distillery stillage. *Bioprocess Biosyst. Eng.* **2013**, *36*, 1157–1164. [CrossRef]

93. Pejin, J.; Radosavljević, M.; Pribić, M.; Kocić-Tanackov, S.; Mladenović, D.; Djukić-Vuković, A.; Mojović, L. Possibility of L-(+)-lactic acid fermentation using malting, brewing, and oil production by-products. *Waste Manag.* **2018**, *79*, 153–163. [CrossRef]

94. Schepers, A.W.; Thibault, J.; Lacroix, C. Lactobacillus helveticus growth and lactic acid production during pH-controlled batch cultures in whey permeate/yeast extract medium. Part I. Multiple factor kinetic analysis. *Enzym. Microb. Technol.* **2002**, *30*, 176–186. [CrossRef]

95. Burgos-Rubio, C.N.; Okos, M.R.; Wankat, P.C. Kinetic Study of the Conversion of Different Substrates to Lactic Acid Using *Lactobacillus bulgaricus*. *Biotechnol. Prog.* **2000**, *16*, 305–314. [CrossRef]

96. Liu, P.; Zheng, Z.; Xu, Q.; Qian, Z.; Liu, J.; Ouyang, J. Valorization of dairy waste for enhanced D-lactic acid production at low cost. *Process. Biochem.* **2018**, *71*, 18–22. [CrossRef]

97. Hujanen, M.; Linko, Y.-Y. Effect of temperature and various nitrogen sources on L (+)-lactic acid production by *Lactobacillus casei*. *Appl. Microbiol. Biotechnol.* **1996**, *45*, 307–313. [CrossRef]

98. Taniguchi, M.; Tokunaga, T.; Horiuchi, K.; Hoshino, K.; Sakai, K.; Tanaka, T. Production of l-lactic acid from a mixture of xylose and glucose by co-cultivation of lactic acid bacteria. *Appl. Microbiol. Biotechnol.* **2004**, *66*, 160–165. [CrossRef] [PubMed]

99. John, R.P.; Madhavan Nampoothiri, K.; Pandey, A. Simultaneous Saccharification and Fermentation of Cassava Bagasse for L-(+)-Lactic Acid Production Using Lactobacilli. *Appl. Biochem. Biotechnol.* **2006**, *134*, 263–272. [CrossRef] [PubMed]

100. Nancib, A.; Nancib, N.; Boudrant, J. Production of lactic acid from date juice extract with free cells of single and mixed cultures of *Lactobacillus casei* and *Lactococcus lactis*. *World J. Microbiol. Biotechnol.* **2009**, *25*, 1423–1429. [CrossRef]

101. Kurbanoglu, E.B.; Kurbanoglu, N.I. Utilization for lactic acid production with a new acid hydrolysis of ram horn waste. *FEMS Microbiol. Lett.* **2003**, *225*, 29–34. [CrossRef]

102. Oonkhanond, B.; Jonglertjunya, W.; Srimarut, N.; Bunpachart, P.; Tantinukul, S.; Nasongkla, N.; Sakdaronnarong, C. Lactic acid production from sugarcane bagasse by an integrated system of lignocellulose fractionation, saccharification, fermentation, and ex-situ nanofiltration. *J. Environ. Chem. Eng.* **2017**, *5*, 2533–2541. [CrossRef]

103. Bustos, G.; Moldes, A.B.; Cruz, J.M.; Domínguez, J.M. Production of fermentable media from vine-trimming wastes and bioconversion into lactic acid by *Lactobacillus pentosus*. *J. Sci. Food Agric.* **2004**, *84*, 2105–2112. [CrossRef]

104. Moldes, A.B.; Torrado, A.; Converti, A.; Domínguez, J.M. Complete bioconversion of hemicellulosic sugars from agricultural residues into lactic acid by *Lactobacillus pentosus*. *Appl. Biochem. Biotechnol.* **2006**, *135*, 219–228. [CrossRef]

105. Garde, A.; Jonsson, G.; Schmidt, A.S.; Ahring, B.K. Lactic acid production from wheat straw hemicellulose hydrolysate by *Lactobacillus pentosus* and *Lactobacillus brevis*. *Bioresour. Technol.* **2002**, *81*, 217–223. [CrossRef]

106. Zhu, Y.; Lee, Y.Y.; Elander, R.T. Conversion of aqueous ammonia-treated corn stover to lactic acid by simultaneous saccharification and cofermentation. *Appl. Biochem. Biotechnol.* **2007**, *137–140*, 721–738. [CrossRef]

107. Vishnu, C.; Seenayya, G.; Reddy, G. Direct fermentation of various pure and crude starchy substrates to L(+) lactic acid using Lactobacillus amylophilus GV6. *World J. Microbiol. Biotechnol.* **2002**, *18*, 429–433. [CrossRef]

108. Wee, Y.-J.; Ryu, H.-W. Lactic acid production by *Lactobacillus* sp. RKY2 in a cell-recycle continuous fermentation using lignocellulosic hydrolyzates as inexpensive raw materials. *Bioresour. Technol.* **2009**, *100*, 4262–4270. [CrossRef] [PubMed]

109. Yun, J.-S.; Wee, Y.-J.; Kim, J.-N.; Ryu, H.-W. Fermentative production of dl-lactic acid from amylase-treated rice and wheat brans hydrolyzate by a novel lactic acid bacterium, *Lactobacillus* sp. *Biotechnol. Lett.* **2004**, *26*, 1613–1616. [CrossRef] [PubMed]

110. Givry, S.; Prevot, V.; Duchiron, F. Lactic acid production from hemicellulosic hydrolyzate by cells of Lactobacillus bifermentans immobilized in Ca-alginate using response surface methodology. *World J. Microbiol. Biotechnol.* **2008**, *24*, 745–752. [CrossRef]

111. Singhvi, M.; Joshi, D.; Adsul, M.; Varma, A.; Gokhale, D. d-(−)-Lactic acid production from cellobiose and cellulose by Lactobacillus lactis mutant RM2-24. *Green Chem.* **2010**, *12*, 1106–1109. [CrossRef]

112. Joshi, D.S.; Singhvi, M.S.; Khire, J.M.; Gokhale, D.V. Strain improvement of Lactobacillus lactis for d-lactic acid production. *Biotechnol. Lett.* **2010**, *32*, 517–520. [CrossRef]

113. Laopaiboon, P.; Thani, A.; Leelavatcharamas, V.; Laopaiboon, L. Acid hydrolysis of sugarcane bagasse for lactic acid production. *Bioresour. Technol.* **2010**, *101*, 1036–1043. [CrossRef]

114. Roble, N.D.; Ogbonna, J.C.; Tanaka, H. L-lactic acid production from raw cassava starch in a circulating loop bioreactor with cells immobilized in loofa (*Luffa cylindrica*). *Biotechnol. Lett.* **2003**, *25*, 1093–1098. [CrossRef]

115. Tanaka, K.; Komiyama, A.; Sonomoto, K.; Ishizaki, A.; Hall, S.; Stanbury, P. Two different pathways for D-xylose metabolism and the effect of xylose concentration on the yield coefficient of L-lactate in mixed-acid fermentation by the lactic acid bacterium *Lactococcus lactis* IO-1. *Appl. Microbiol. Biotechnol.* **2002**, *60*, 160–167. [CrossRef]

116. Hofvendahl, K.; Hahn-Hägerdal, B. l-lactic acid production from whole wheat flour hydrolysate using strains of *Lactobacilli* and *Lactococci*. *Enzym. Microb. Technol.* **1997**, *20*, 301–307. [CrossRef]

117. Hofvendahl, K.; Hahn-Hägerdal, B.; Åkerberg, C.; Zacchi, G. Simultaneous enzymatic wheat starch saccharification and fermentation to lactic acid by *Lactococcus lactis*. *Appl. Microbiol. Biotechnol.* **1999**, *52*, 163–169. [CrossRef]

118. Wang, L.; Zhao, B.; Li, F.; Xu, K.; Ma, C.; Tao, F.; Li, Q.; Xu, P. Highly efficient production of d-lactate by *Sporolactobacillus* sp. CASD with simultaneous enzymatic hydrolysis of peanut meal. *Appl. Microbiol. Biotechnol.* **2011**, *89*, 1009–1017. [CrossRef]

119. Li, Y.; Wang, L.; Ju, J.; Yu, B.; Ma, Y. Efficient production of polymer-grade d-lactate by Sporolactobacillus laevolacticus DSM442 with agricultural waste cottonseed as the sole nitrogen source. *Bioresour. Technol.* **2013**, *142*, 186–191. [CrossRef] [PubMed]

120. Guo, W.; Jia, W.; Li, Y.; Chen, S. Performances of *Lactobacillus brevis* for producing lactic acid from hydrolysate of Lignocellulosics. *Appl. Biochem. Biotechnol.* **2010**, *161*, 124–136. [CrossRef] [PubMed]

121. Yáñez, R.; Moldes, A.B.; Alonso, J.L.; Parajó, J.C. Production of D(-)-lactic acid from cellulose by simultaneous saccharification and fermentation using *Lactobacillus coryniformis* subsp. *torquens*. *Biotechnol. Lett.* **2003**, *25*, 1161–1164. [CrossRef]

122. Yáñez, R.; Alonso, J.L.; Parajó, J.C. D-Lactic acid production from waste cardboard. *J. Chem. Technol. Biotechnol.* **2005**, *80*, 76–84. [CrossRef]

123. Berlowska, J.; Cieciura-Wloch, W.; Kalinowska, H.; Kregiel, D.; Borowski, S.; Pawlikowska, E.; Binczarski, M.; Witonska, I. Enzymatic Conversion of Sugar Beet Pulp: A Comparison of Simultaneous Saccharification and Fermentation and Separate Hydrolysis and Fermentation for Lactic Acid Production. *Food Technol. Biotechnol.* **2018**, *56*, 188–196. [CrossRef]

124. de Oliveira Moraes, A.; Ramirez, N.I.B.; Pereira, N. Evaluation of the Fermentation Potential of Pulp Mill Residue to Produce d(−)-Lactic Acid by Separate Hydrolysis and Fermentation Using *Lactobacillus coryniformis* subsp. *torquens*. *Appl. Biochem. Biotechnol.* **2016**, *180*, 1574–1585. [CrossRef]

125. Fu, W.; Mathews, A.P. Lactic acid production from lactose by *Lactobacillus plantarum*: Kinetic model and effects of pH, substrate, and oxygen. *Biochem. Eng. J.* **1999**, *3*, 163–170. [CrossRef]

126. Sreenath, H.K.; Moldes, A.B.; Koegel, R.G.; Straub, R.J. Lactic acid production by simultaneous saccharification and fermentation of alfalfa fiber. *J. Biosci. Bioeng.* **2001**, *92*, 518–523. [CrossRef] [PubMed]

127. Okano, K.; Zhang, Q.; Yoshida, S.; Tanaka, T.; Ogino, C.; Fukuda, H.; Kondo, A. D-lactic acid production from cellooligosaccharides and β-glucan using l-LDH gene-deficient and endoglucanase-secreting *Lactobacillus plantarum*. *Appl. Microbiol. Biotechnol.* **2010**, *85*, 643–650. [CrossRef] [PubMed]

128. Okano, K.; Yoshida, S.; Tanaka, T.; Ogino, C.; Fukuda, H.; Kondo, A. Homo-D-Lactic Acid Fermentation from Arabinose by Redirection of the Phosphoketolase Pathway to the Pentose Phosphate Pathway in L-Lactate Dehydrogenase Gene-Deficient *Lactobacillus plantarum*. *Appl. Environ. Microbiol.* **2009**, *75*, 5175–5178. [CrossRef] [PubMed]

129. Okano, K.; Yoshida, S.; Yamada, R.; Tanaka, T.; Ogino, C.; Fukuda, H.; Kondo, A. Improved production of homo-D-lactic acid via xylose fermentation by introduction of xylose assimilation genes and redirection of the phosphoketolase pathway to the pentose phosphate pathway in L-lactate dehydrogenase gene-deficient *Lactobacillus plantarum*. *Appl. Environ. Microbiol.* **2009**, *75*, 7858–7861. [CrossRef] [PubMed]

130. Ohara, H.; Owaki, M.; Sonomoto, K. Xylooligosaccharide fermentation with Leuconostoc lactis. *J. Biosci. Bioeng.* **2006**, *101*, 415–420. [CrossRef] [PubMed]
131. Patel, M.A.; Ou, M.S.; Ingram, L.O.; Shanmugam, K.T. Simultaneous saccharification and co-fermentation of crystalline cellulose and sugar cane bagasse hemicellulose hydrolysate to lactate by a thermotolerant acidophilic *Bacillus* sp. *Biotechnol. Prog.* **2005**, *21*, 1453–1460. [CrossRef] [PubMed]
132. Montelongo, J.-L.; Chassy, B.M.; McCORD, J.D. *Lactobacillus salivarius* for Conversion of Soy Molasses into Lactic Acid. *J. Food Sci.* **1993**, *58*, 863–866. [CrossRef]
133. Timbuntam, W.; Sriroth, K.; Tokiwa, Y. Lactic acid production from sugar-cane juice by a newly isolated *Lactobacillus* sp. *Biotechnol. Lett.* **2006**, *28*, 811–814. [CrossRef]
134. Xiaodong, W.; Xuan, G.; Rakshit, S.K. Direct fermentative production of lactic acid on cassava and other starch substrates. *Biotechnol. Lett.* **1997**, *19*, 841–843. [CrossRef]
135. Coelho, L.F.; de Lima, C.J.B.; Bernardo, M.P.; Contiero, J. d(−)-Lactic Acid Production by *Leuconostoc mesenteroides* B512 Using Different Carbon and Nitrogen Sources. *Appl. Biochem. Biotechnol.* **2011**, *164*, 1160–1171. [CrossRef] [PubMed]
136. Wang, Y.; Chen, C.; Cai, D.; Wang, Z.; Qin, P.; Tan, T. The optimization of L-lactic acid production from sweet sorghum juice by mixed fermentation of Bacillus coagulans and *Lactobacillus rhamnosus* under unsterile conditions. *Bioresour. Technol.* **2016**, *218*, 1098–1105. [CrossRef] [PubMed]
137. Wang, Y.; Li, K.; Huang, F.; Wang, J.; Zhao, J.; Zhao, X.; Garza, E.; Manow, R.; Grayburn, S.; Zhou, S. Engineering and adaptive evolution of Escherichia coli W for l-lactic acid fermentation from molasses and corn steep liquor without additional nutrients. *Bioresour. Technol.* **2013**, *148*, 394–400. [CrossRef] [PubMed]
138. Richter, K.; Träger, A. L(+)-Lactic acid from sweet sorghum by submerged and solid-state fermentations. *Acta Biotechnol.* **1994**, *14*, 367–378. [CrossRef]
139. Mladenović, D.; Pejin, J.; Kocić-Tanackov, S.; Djukić-Vuković, A.; Mojović, L. Enhanced Lactic Acid Production by Adaptive Evolution of *Lactobacillus paracasei* on Agro-industrial Substrate. *Appl. Biochem. Biotechnol.* **2019**, *187*, 753–769. [CrossRef] [PubMed]
140. Nguyen, C.M.; Kim, J.S.; Nguyen, T.N.; Kim, S.K.; Choi, G.J.; Choi, Y.H.; Jang, K.S.; Kim, J.C. Production of l- and d-lactic acid from waste Curcuma longa biomass through simultaneous saccharification and cofermentation. *Bioresour. Technol.* **2013**, *146*, 35–43. [CrossRef]
141. Flores-Albino, B.; Arias, L.; Gómez, J.; Castillo, A.; Gimeno, M.; Shirai, K. Chitin and L(+)-lactic acid production from crab (*Callinectes bellicosus*) wastes by fermentation of *Lactobacillus* sp. B2 using sugar cane molasses as carbon source. *Bioprocess Biosyst. Eng.* **2012**, *35*, 1193–1200. [CrossRef]
142. Taskin, M.; Esim, N.; Ortucu, S. Efficient production of l-lactic acid from chicken feather protein hydrolysate and sugar beet molasses by the newly isolated Rhizopus oryzae TS-61. *Food Bioprod. Process.* **2012**, *90*, 773–779. [CrossRef]
143. Zhang, L.; Li, X.; Yong, Q.; Yang, S.T.; Ouyang, J.; Yu, S. Simultaneous saccharification and fermentation of xylo-oligosaccharides manufacturing waste residue for l-lactic acid production by Rhizopus oryzae. *Biochem. Eng. J.* **2015**, *94*, 92–99. [CrossRef]
144. Zheng, Y.; Wang, Y.; Zhang, J.; Pan, J. Using tobacco waste extract in pre-culture medium to improve xylose utilization for l-lactic acid production from cellulosic waste by Rhizopus oryzae. *Bioresour. Technol.* **2016**, *218*, 344–350. [CrossRef]
145. Tanaka, T.; Hoshina, M.; Tanabe, S.; Sakai, K.; Ohtsubo, S.; Taniguchi, M. Production of d-lactic acid from defatted rice bran by simultaneous saccharification and fermentation. *Bioresour. Technol.* **2006**, *97*, 211–217. [CrossRef]
146. Kotzamanidis, C.; Roukas, T.; Skaracis, G. Optimization of lactic acid production from beet molasses by *Lactobacillus delbrueckii* NCIMB 8130. *World J. Microbiol. Biotechnol.* **2002**, *18*, 441–448. [CrossRef]
147. Adsul, M.; Khire, J.; Bastawde, K.; Gokhale, D. Production of Lactic Acid from Cellobiose and Cellotriose by *Lactobacillus delbrueckii* Mutant Uc-3. *Appl. Environ. Microbiol.* **2007**, *73*, 5055–5057. [CrossRef] [PubMed]
148. Adsul, M.G.; Varma, A.J.; Gokhale, D.V. Lactic acid production from waste sugarcane bagasse derived cellulose. *Green Chem.* **2007**, *9*, 58–62. [CrossRef]
149. Calabia, B.P.; Tokiwa, Y. Production of d-lactic acid from sugarcane molasses, sugarcane juice and sugar beet juice by *Lactobacillus delbrueckii*. *Biotechnol. Lett.* **2007**, *29*, 1329–1332. [CrossRef] [PubMed]
150. Iyer, P.V.; Lee, Y.Y. Product inhibition in simultaneous saccharification and fermentation of cellulose into lactic acid. *Biotechnol. Lett.* **1999**, *21*, 371–373. [CrossRef]
151. Iyer, P.V.; Lee, Y.Y. Simultaneous Saccharification and Extractive Fermentation of Lignocellulosic Materials into Lactic Acid in a Two-Zone Fermentor-Extractor System. In *Twentieth Symposium on Biotechnology for Fuels and Chemicals*; Humana Press: Totowa, NJ, USA, 1999; pp. 409–419. [CrossRef]
152. Mussatto, S.I.; Fernandes, M.; Mancilha, I.M.; Roberto, I.C. Effects of medium supplementation and pH control on lactic acid production from brewer's spent grain. *Biochem. Eng. J.* **2008**, *40*, 437–444. [CrossRef]
153. Shen, X.; Xia, L. Lactic acid production from cellulosic material by synergetic hydrolysis and Fermentation. *Appl. Biochem. Biotechnol.* **2006**, *133*, 251–262. [CrossRef]
154. Lu, Z.; Lu, M.; He, F.; Yu, L. An economical approach for d-lactic acid production utilizing unpolished rice from aging paddy as major nutrient source. *Bioresour. Technol.* **2009**, *100*, 2026–2031. [CrossRef]
155. Tsai, T.S.; Millard, C.S. Improved Pre-treatment Process for Lactic Acid Production. WO/1994/013826, 23 June 1994. Available online: https://patentscope.wipo.int/search/en/detail.jsf?docId=WO1994013826 (accessed on 6 April 2021).

156. Aksu, Z.; Kutsal, T. Lactic acid production from molasses utilizing *Lactobacillus delbrueckii* and invertase together. *Biotechnol. Lett.* **1986**, *8*, 157–160. [CrossRef]

157. Balakrishnan, R.; Tadi, S.R.R.; Sivaprakasam, S.; Rajaram, S. Optimization of acid and enzymatic hydrolysis of kodo millet (*Paspalum scrobiculatum*) bran residue to obtain fermentable sugars for the production of optically pure D (−) lactic acid. *Ind. Crops Prod.* **2018**, *111*, 731–742. [CrossRef]

158. Qiu, Z.; Gao, Q.; Bao, J. Engineering *Pediococcus acidilactici* with xylose assimilation pathway for high titer cellulosic L-lactic acid fermentation. *Bioresour. Technol.* **2018**, *249*, 9–15. [CrossRef] [PubMed]

159. Bai, Z.; Gao, Z.; He, B.; Wu, B. Effect of lignocellulose-derived inhibitors on the growth and d-lactic acid production of Sporolactobacillus inulinus YBS1-5. *Bioprocess Biosyst. Eng.* **2015**, *38*, 1993–2001. [CrossRef] [PubMed]

160. John, R.P.; Nampoothiri, K.M.; Pandey, A. Fermentative production of lactic acid from biomass: An overview on process developments and future perspectives. *Appl. Microbiol. Biotechnol.* **2007**, *74*, 524–534. [CrossRef] [PubMed]

161. Van Der Pol, E.C. Development of a Lactic Acid Production Process Using Lignocellulosic Biomass as Feedstock. 2016. Available online: https://edepot.wur.nl/374060 (accessed on 5 March 2021).

162. Balat, M. Production of bioethanol from lignocellulosic materials via the biochemical pathway: A review. *Energy Convers. Manag.* **2011**, *52*, 858–875. [CrossRef]

163. Barre, P. Identification of Thermobacteria and Homofermentative, Thermophilic, Pentose-utilizing Lactobacilli from High Temperature Fermenting Grape Musts. *J. Appl. Bacteriol.* **1978**, *44*, 125–129. [CrossRef]

164. Hemme, D.; Raibaud, P.; Ducluzeau, R.; Galpin, J.V.; Sicard, P.; van Heyenoort, J. *Lactobacillus murinus'* n.sp., a new species of the autochthonous dominant flora of the digestive tract of rat and mouse. *Ann. Inst. Pasteur* **1980**, *131*, 297–308. Available online: https://hal.inrae.fr/hal-02732519 (accessed on 11 September 2021).

165. Olofsson, K.; Bertilsson, M.; Lidén, G. A short review on SSF—An interesting process option for ethanol production from lignocellulosic feedstocks. *Biotechnol. Biofuels* **2008**, *1*, 7. [CrossRef]

166. Abdel-Rahman, M.A.; Tashiro, Y.; Sonomoto, K. Lactic acid production from lignocellulose-derived sugars using lactic acid bacteria: Overview and limits. *J. Biotechnol.* **2011**, *156*, 286–301. [CrossRef]

167. Biddy, M.J.; Scarlata, C.; Kinchin, C. *Chemicals from Biomass: A Market Assessment of Bioproducts with Near-Term Potential*; NREL: Golden, CO, USA, 2016. [CrossRef]

168. Komesu, A.; Martins, P.F.; Lunelli, B.H.; Rocha, J.O.; Filho, R.M.; Maciel, M.R.W. The Effect of Evaporator Temperature on Lactic Acid Purity and Recovery by Short Path Evaporation. *Sep. Sci. Technol.* **2014**, *50*, 1548–1553. [CrossRef]

169. Järvinen, M.; Myllykoski, L.; Keiski, R.; Sohlo, J. Separation of lactic acid from fermented broth by reactive extraction. *Bioseparation* **2000**, *9*, 163–166. [CrossRef]

170. Komesu, A.; Maciel, M.R.W.; Filho, R.M. Separation and Purification Technologies for Lactic Acid—A Brief Review. *BioResources* **2017**, *12*, 6885–6901. [CrossRef]

171. Komesu, A.; Maciel, M.R.W.; de Oliveira, J.A.R.; da Silva Martins, L.H.; Maciel Filho, R. Purification of Lactic Acid Produced by Fermentation: Focus on Non-traditional Distillation Processes. *Sep. Purif. Rev.* **2017**, *46*, 241–254. [CrossRef]

172. Din, N.A.S.; Lim, S.J.; Maskat, M.Y.; Mutalib, S.A.; Zaini, N.A.M. Lactic acid separation and recovery from fermentation broth by ion-exchange resin: A review. *Bioresour. Bioprocess.* **2021**, *8*, 31. [CrossRef]

173. Yankov, D.; Molinier, J.; Kyuchoukov, G.; Albet, J.; Malmary, G. Improvement of the lactic acid extraction. Extraction from aqueous solutions and simulated fermentation broth by means of mixed extractant and TOA, partially loaded with HCI. *Chem. Biochem. Eng. Q.* **2005**, *19*, 17–24.

174. Daful, A.G.; Haigh, K.; Vaskan, P.; Görgens, J.F. Environmental impact assessment of lignocellulosic lactic acid production: Integrated with existing sugar mills. *Food Bioprod. Process.* **2016**, *99*, 58–70. [CrossRef]

175. Wasewar, K.L.; Pangarkar, V.G.; Heesink, A.B.M.; Versteeg, G.F. Intensification of enzymatic conversion of glucose to lactic acid by reactive extraction. *Chem. Eng. Sci.* **2003**, *58*, 3385–3393. [CrossRef]

176. Henczka, M.; Djas, M. Reactive extraction of acetic acid and propionic acid using supercritical carbon dioxide. *J. Supercrit. Fluids* **2016**, *110*, 154–160. [CrossRef]

177. López-Garzón, C.S.; Straathof, A.J.J. Recovery of carboxylic acids produced by fermentation. *Biotechnol. Adv.* **2014**, *32*, 873–904. [CrossRef]

178. Kumar, A.; Thakur, A.; Panesar, P.S. Lactic acid and its separation and purification techniques: A review. *Rev. Environ. Sci. Bio/Technology* **2019**, *18*, 823–853. [CrossRef]

179. Wojtyniak, B.; Kołodziejczyk, J.; Szaniawska, D. Production of lactic acid by ultrafiltration of fermented whey obtained in bioreactor equipped with ZOSS membrane. *Chem. Eng. J.* **2016**, *305*, 28–36. [CrossRef]

180. Aqar, D.Y.; Rahmanian, N.; Mujtaba, I.M. Integrated batch reactive distillation column configurations for optimal synthesis of methyl lactate. *Chem. Eng. Process. Process Intensif.* **2016**, *108*, 197–211. [CrossRef]

181. Nielsen, D.R.; Amarasiriwardena, G.S.; Prather, K.L. Predicting the adsorption of second generation biofuels by polymeric resins with applications for in situ product recovery (ISPR). *Bioresour. Technol.* **2010**, *101*, 2762–2769. [CrossRef] [PubMed]

182. Pradhan, N.; Rene, E.R.; Lens, P.N.L.; Dipasquale, L.; D'Ippolito, G.; Fontana, A.; Panico, A.; Esposito, G. Adsorption behaviour of lactic acid on granular activated carbon and anionic resins: Thermodynamics, isotherms and kinetic Studies. *Energies* **2017**, *10*, 665. [CrossRef]

183. Seeber, G.; Buchmeiser, M.R.; Bonn, G.K.; Bertsch, T. Determination of airborne, volatile amines from polyurethane foams by sorption onto a high-capacity cation-exchange resin based on poly(succinic acid). *J. Chromatogr. A* **1998**, *809*, 121–129. [CrossRef]

184. Kumar, S.; Jain, S. History, introduction, and kinetics of ion exchange Materials. *J. Chem.* **2013**, *2013*, 957647. [CrossRef]

185. Aljundi, I.H.; Belovich, J.M.; Talu, O. Adsorption of lactic acid from fermentation broth and aqueous solutions on Zeolite molecular sieves. *Chem. Eng. Sci.* **2005**, *60*, 5004–5009. [CrossRef]

186. Boonmee, M.; Cotano, O.; Amnuaypanich, S.; Grisadanurak, N. Improved Lactic Acid Production by In Situ Removal of Lactic Acid During Fermentation and a Proposed Scheme for Its Recovery. *Arab. J. Sci. Eng.* **2016**, *41*, 2067–2075. [CrossRef]

187. Zhang, Y.; Qian, Z.; Liu, P.; Liu, L.; Zheng, Z.; Ouyang, J. Efficient in situ separation and production of l-lactic acid by Bacillus coagulans using weak basic anion-exchange resin. *Bioprocess Biosyst. Eng.* **2018**, *41*, 205–212. [CrossRef]

188. Quintero, J.; Acosta, A.; Mejía, C.; Ríos, R.; Torres, A.M. Purification of lactic acid obtained from a fermentative process of cassava syrup using ion exchange resins. *Rev. Fac. Ing. Univ. Antioq.* **2012**, *65*, 139–151.

189. Kuo, Y.C.; Yuan, S.F.; Wang, C.A.; Huang, Y.J.; Guo, G.L.; Hwang, W.S. Production of optically pure l-lactic acid from lignocellulosic hydrolysate by using a newly isolated and d-lactate dehydrogenase gene-deficient *Lactobacillus paracasei* strain. *Bioresour. Technol.* **2015**, *198*, 651–657. [CrossRef]

190. Peckham, G.T. The Commercial Manufacture of Lactic Acid. *Chem. Eng. News* **1944**, *22*, 440–443, 469. [CrossRef]

191. Inskeep, G.C.; Taylor, G.G.; Breitzke, W.C. LACTIC ACID FROM CORN SUGAR. *Ind. Eng. Chem.* **1952**, *44*, 1955–1966. [CrossRef]

192. Qin, J.; Wang, X.; Zheng, Z.; Ma, C.; Tang, H.; Xu, P. Production of L-lactic acid by a thermophilic Bacillus mutant using sodium hydroxide as neutralizing agent. *Bioresour. Technol.* **2010**, *101*, 7570–7576. [CrossRef] [PubMed]

193. Chen, C.-C.; Ju, L.-K. Adsorption Characteristics of Polyvinylpyridine and Activated Carbon for Lactic Acid Recovery from Fermentation of *Lactobacillus delbrueckii*. *Sep. Sci. Technol.* **1998**, *33*, 1423–1437. [CrossRef]

194. Kumar, R.; Nanavati, H.; Noronha, S.B.; Mahajani, S.M. A continuous process for the recovery of lactic acid by reactive distillation. *J. Chem. Technol. Biotechnol.* **2006**, *81*, 1767–1777. [CrossRef]

195. Madzingaidzo, L.; Danner, H.; Braun, R. Process development and optimisation of lactic acid purification using electrodialysis. *J. Biotechnol.* **2002**, *96*, 223–239. [CrossRef]

196. Choi, J.H.; Kim, S.H.; Moon, S.H. Recovery of lactic acid from sodium lactate by ion substitution using ion-exchange membrane. *Sep. Purif. Technol.* **2002**, *28*, 69–79. [CrossRef]

197. Hábová, V.; Melzoch, K.; Rychtera, M.; Přibyl, L.; Mejta, V. Application of electrodialysis for lactic acid recovery. *Czech J. Food Sci.* **2013**, *19*, 73–80. [CrossRef]

198. Kim, Y.H.; Moon, S.H. Lactic acid recovery from fermentation broth using one-stage electrodialysis. *J. Chem. Technol. Biotechnol.* **2001**, *76*, 169–178. [CrossRef]

199. Yebo Li, A.S. Lactic Acid Recovery From Cheese Whey Fermentation Broth Using Combined Ultrafiltration and Nanofiltration Membranes. *Appl. Biochem. Biotechnol.* **2006**, *132*, 985–996. [CrossRef]

200. González, M.I.; Alvarez, S.; Riera, F.A.; Álvarez, R. Lactic acid recovery from whey ultrafiltrate fermentation broths and artificial solutions by nanofiltration. *Desalination* **2008**, *228*, 84–96. [CrossRef]

201. Li, Y.; Shahbazi, A.; Williams, K.; Wan, C. Separate and Concentrate Lactic Acid Using Combination of Nanofiltration and Reverse Osmosis Membranes. *Appl. Biochem. Biotechnol.* **2007**, *147*, 1–9. [CrossRef] [PubMed]

202. Zhao, K.; Qiao, Q.; Chu, D.; Gu, H.; Dao, T.H.; Zhang, J.; Bao, J. Simultaneous saccharification and high titer lactic acid fermentation of corn stover using a newly isolated lactic acid bacterium *Pediococcus acidilactici* DQ2. *Bioresour. Technol.* **2013**, *135*, 481–489. [CrossRef] [PubMed]

203. van der Pol, E.; Bakker, R.; van Zeeland, A.; Garcia, D.S.; Punt, A.; Eggink, G. Analysis of by-product formation and sugar monomerization in sugarcane bagasse pretreated at pilot plant scale: Differences between autohydrolysis, alkaline and acid pretreatment. *Bioresour. Technol.* **2015**, *181*, 114–123. [CrossRef] [PubMed]

204. Kaščak, J.; Gašpár, Š.; Paško, J.; Husár, J.; Knapčíková, L. Polylactic Acid and Its Cellulose Based Composite as a Significant Tool for the Production of Optimized Models Modified for Additive Manufacturing. *Sustainability* **2021**, *13*, 1256. [CrossRef]

205. Nassajfar, M.N.; Deviatkin, I.; Leminen, V.; Horttanainen, M. Alternative Materials for Printed Circuit Board Production: An Environmental Perspective. *Sustainability* **2021**, *13*, 12126. [CrossRef]

206. Castro-Aguirre, E.; Iñiguez-Franco, F.; Samsudin, H.; Fang, X.; Auras, R. Poly(lactic acid)—Mass production, processing, industrial applications, and end of life. *Adv. Drug Deliv. Rev.* **2016**, *107*, 333–366. [CrossRef]

207. Groot, W.; van Krieken, J.; Sliekersl, O.; de Vos, S. Production and Purification of Lactic Acid and Lactide. In *Poly(Lactic Acid): Synthesis, Structures, Properties, Processing, and Applications*; Auras, R., Lim, L.-T., Selke, S.E.M., Tsuji, H., Eds.; John Wiley & Sons, Inc.: Hoboken, NJ, USA, 2010; pp. 1–18. [CrossRef]

208. Vink, E.T.H.; Rábago, K.R.; Glassner, D.A.; Gruber, P.R. Applications of life cycle assessment to NatureWorks^{TM} polylactide (PLA) production. *Polym. Degrad. Stab.* **2003**, *80*, 403–419. [CrossRef]

209. Hormnirun, P.; Marshall, E.L.; Gibson, V.C.; Pugh, R.I.; White, A.J. Study of ligand substituent effects on the rate and stereoselectivity of lactide polymerization using aluminum salen-type initiators. *Proc. Natl. Acad. Sci. USA* **2006**, *103*, 15343–15348. [CrossRef]

210. Masutani, K.; Kimura, Y. *PLA Synthesis and Polymerization*; The Royal Society of Chemistry: London, UK, 2014.

211. Yang, Y.; Wang, H.; Ma, H. Stereoselective Polymerization of rac -Lactide Catalyzed by Zinc Complexes with Tetradentate Aminophenolate Ligands in Different Coordination Patterns: Kinetics and Mechanism. *Inorg. Chem.* **2015**, *54*, 5839–5854. [CrossRef]
212. Jamshidian, M.; Tehrany, E.A.; Imran, M.; Jacquot, M.; Desobry, S. Poly-Lactic Acid: Production, Applications, Nanocomposites, and Release Studies. *Compr. Rev. Food Sci. Food Saf.* **2010**, *9*, 552–571. [CrossRef] [PubMed]
213. Michaliszyn-Gabryś, B.; Krupanek, J.; Kalisz, M.; Smith, J. Challenges for Sustainability in Packaging of Fresh Vegetables in Organic Farming. *Sustainability* **2022**, *14*, 5346. [CrossRef]
214. Lim, L.-T.; Vanyo, T.; Randall, J.; Cink, K.; Agrawal, A.K. PROCESSING OF POLY(LACTIC ACID). In *Poly(Lactic Acid): Synthesis, Structures, Properties, Processing, and Applications, and End of Life*, 2nd ed.; John Wiley & Sons, Ltd.: New York, NY, USA, 2022; pp. 231–270. [CrossRef]
215. Siddiqui, M.N.; Redhwi, H.H.; Al-Arfaj, A.A.; Achilias, D.S. Chemical Recycling of PET in the Presence of the Bio-Based Polymers, PLA, PHB and PEF: A Review. *Sustainability* **2021**, *13*, 10528. [CrossRef]
216. Li, S.M.; Garreau, H.; Vert, M. Structure-property relationships in the case of the degradation of massive aliphatic poly-(α-hydroxy acids) in aqueous media. *J. Mater. Sci. Mater. Med.* **1990**, *1*, 123–130. [CrossRef]
217. Hoshino, A.; Isono, Y. Degradation of aliphatic polyester films by commercially available lipases with special reference to rapid and complete degradation of poly(L-lactide) film by lipase PL derived from *Alcaligenes* sp. *Biogeochemistry* **2002**, *13*, 141–147. [CrossRef]

Review

Second-Generation Bio-Fuels: Strategies for Employing Degraded Land for Climate Change Mitigation Meeting United Nation-Sustainable Development Goals

Atreyi Pramanik [1], Aashna Sinha [1], Kundan Kumar Chaubey [1,*], Sujata Hariharan [1], Deen Dayal [2], Rakesh Kumar Bachheti [3], Archana Bachheti [4] and Anuj K. Chandel [5,*]

1 Division of Research and Innovation, School of Applied and Life Sciences, Uttaranchal University, Arcadia Grant, P.O. Chandanwari, Premnagar, Dehradun 248007, Uttarakhand, India
2 Department of Biotechnology, GLA University, Mathura 281406, Uttar Pradesh, India
3 Centre of Excellence in Nanotechnology, Department of Industrial Chemistry, Addis Ababa Sciences and Technology University, Addis Ababa P.O. Box 16417, Ethiopia
4 Department of Environment Science, Graphic Era University, Dehradun 248002, Uttarakhand, India
5 Department of Biotechnology, Engineering School of Lorena (EEL), University of São Paulo (USP), Estrada Municipal do Campinho, N°. 100, Lorena, São Paulo 12602-810, Brazil
* Correspondence: kundan2006chaubey@gmail.com (K.K.C.); anuj10@usp.br (A.K.C.)

Abstract: Increased Greenhouse Gas (GHG) emissions from both natural and man-made systems contribute to climate change. In addition to reducing the use of crude petroleum's derived fuels, and increasing tree-planting efforts and sustainable practices, air pollution can be minimized through phytoremediation. Bio-fuel from crops grown on marginal land can sustainably address climate change, global warming, and geopolitical issues. There are numerous methods for producing renewable energy from both organic and inorganic environmental resources (sunlight, air, water, tides, waves, and convective energy), and numerous technologies for doing the same with biomass with different properties and derived from different sources (food industry, agriculture, forestry). However, the production of bio-fuels is challenging and contentious in many parts of the world since it competes for soil with the growth of crops and may be harmful to the environment. Therefore, it is necessary to use wildlife management techniques to provide sustainable bio-energy while maintaining or even improving essential ecosystem processes. The second generation of bio-fuels is viewed as a solution to the serious issue. Agricultural lignocellulosic waste is the primary source of second-generation bio-fuel, possibly the bio-fuel of the future. Sustainable practices to grow biomass, followed by their holistic conversion into ethanol with desired yield and productivity, are the key concerns for employing renewable energy mix successfully. In this paper, we analyze the various types of bio-fuels, their sources, and their production and impact on sustainability.

Keywords: greenhouse gas; biomass production; second-generation bio-fuels; sustainability; environment; climate change

Citation: Pramanik, A.; Sinha, A.; Chaubey, K.K.; Hariharan, S.; Dayal, D.; Bachheti, R.K.; Bachheti, A.; Chandel, A.K. Second-Generation Bio-Fuels: Strategies for Employing Degraded Land for Climate Change Mitigation Meeting United Nation-Sustainable Development Goals. *Sustainability* **2023**, *15*, 7578. https://doi.org/10.3390/su15097578

Academic Editors: Georgios Archimidis Tsalidis, Idiano D'Adamo and Massimo Gastaldi

Received: 18 February 2023
Revised: 16 April 2023
Accepted: 26 April 2023
Published: 5 May 2023

1. Introduction

Natural resource depletion and finite fossil fuel supplies have put tremendous strain on the world's expanding human population. Global concerns about declining ecosystem services and rising greenhouse gas emissions demand immediate attention. This decade (2021–2030) has been dubbed the "Century on Natural Regeneration" by a number of international organizations, including the United Nations (UN), to quicken the achievement of the UN Sustainable Development Goals (SDG) [1]. Based on the most recent statistics and projections, the SDGs Report 2022 offers a thorough analysis of how the 2030 agenda for Sustainable Development has progressed. It maintains record of regional and global achievement towards the 17 Goals through in-depth analyses of particular indicators for each Goal. According to the report, cascading and related problems are seriously endangering both the 2030 plan for sustainable development and human life. The importance and magnitude of the issues we confront are emphasized throughout the report.

All of the SDGs are impacted by the convergence of crises, which are primarily caused by COVID-19, climate change, and wars as well as issues with food and nutrition, health, learning, the ecosystem, and peace and security SDGs. The study explains how years of gains in eradicating hunger and poverty, enhancing health and education, supplying necessities, and a lot more have been reversed. It also highlights areas that must get critical intervention if the SDGs are to be preserved and significant advancements for people and the environment accomplished by 2030 [2].

For the purpose of addressing climate change challenges at a regional, national, and international level, numerous research efforts are being made to address the present concern of recovering natural resources, such as the degraded land resources and second-generation biomass or bio-energy crop production. Complete investigation and optimization, though, continue to be a major ground obstacle for sustainability. It is necessary to thoroughly investigate the many options and technological aspects for second-generation biomass crop plantation from the degraded areas in order to address these difficulties in bio-energy production [3].

Further, the achievement of the UN-SDGs for global targets (Figure 1) would be research and development of various optimization methodologies for sustainable bio-energy production in industrial systems.

Figure 1. 17 UN-SDG goals.

The scope of important studies focused on forestry techniques, phytoremediation technologies, biomass, and bio-energy crop production for restoring degraded lands and reducing the effects of climate change would be expanded by this holistic approach from the ground to industrial systems [4]. As they are an alternative energy source that is in danger of going extinct, we will analyze the various types of bio-fuels, their sources, and their manufacturing processes in this review study. The circular economy strategy puts forward the idea of trash as a benefit, opening up fresh viewpoints and proving its environmental sustainability [5]. System analysis in conjunction with bio-processing and biomass conversion processes could provide additional opportunities for optimization research. Utilizing various accounting techniques could assist in evaluating the viability of manufacturing systems from the ground, and utilizing various accounting techniques could aid in evaluating both the industrial systems' conversion processes as well as the long-term viability of production methods [6]. However, many pilot and demo facilities have lately been planned, with research accomplishments predominantly happening in North America, Europe, and a few developing countries. International Energy Agency (IEA) forecasts indicate that demand for bio-fuels, particularly second-generation bio-fuels, will increase quickly in an energy industry that aims to stabilize atmospheric CO_2 at 450 parts per million (ppm).

Recently, both the United States (US) and the European Union (EU) passed comprehensive bio-fuel support policies. Due to the size of the two markets and their significant bio-fuel imports, the US and EU mandates could be a major driving force behind the development of second-generation bio-fuels globally. Additionally, current IEA analysis predicts a loss in national production in both the US and EU that would need to be filled by imports [7]. Brazil and China, whose infrastructure permits the export of bio-fuels and whose prototype facilities are already operational, may benefit most from this gap in second-generation bio-fuels. Due to a lack of R&D activity, inadequate infrastructure, and a shortage of skilled personnel, other countries, such as Cameroon and Tanzania, may face considerable difficulties in meeting the demand for second-generation bio-fuels in the EU and US in the imminent future [1]. In this review paper, we discuss the different types of bio-fuels, their sources, and their method of production as an alternative source of energy that is on the verge of extinction.

2. Methodology

This article presents a comprehensive literature review with the aim to critically analyze the methods of bio-fuel and bioenergy production, and sustainable development goals, within the scope of renewable fuel production.

For data collection, first, a search was conducted through database search engines like PubMed, Scopus, Google Scholar, and Directory of Open Access Journal (DOAJ) using keywords such as (a) "Biomass" AND "Pretreatment", (b) "Saccharification" AND "Ethanol fermentation", (c) "Second generation bio-fuel production", (d) "Forestry practices for biomass production" (e) "SDGs", (f) "Food and fuel debate" AND "Climate change mitigation", among others. This yielded 150 studies, and this number was reduced to 100 studies after removing duplicates. Following this, 61 articles were selected depending on the criteria and scope (Figure 2).

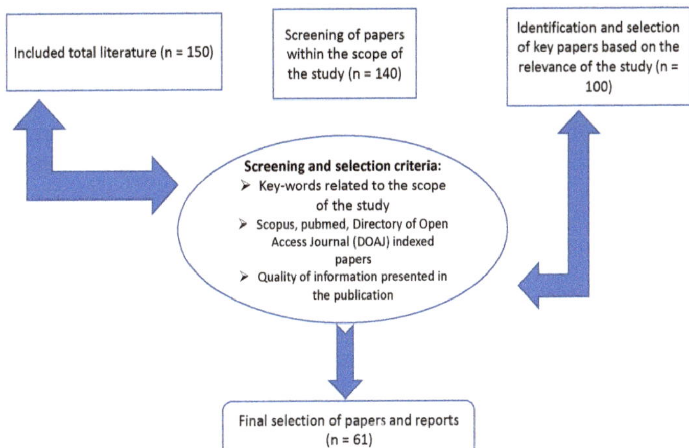

Figure 2. Flow diagram showing the selection criteria and procedure of second-generation bio-fuels production.

3. Bioenergy from Feedstocks

Utilization of marginal land for bio-fuel production has drawn a lot of interest due to its potential to produce bioenergy feedstock while reducing the use of fertile agricultural land for food/feed crop production. According to Gopalakrishnan et al. (2011), at least two criteria allow for the classification of 1.6 million ha, or 4 million acres, of land (or around 8% of the total land area) as marginal. On this land, second-generation lignocellulosic bioenergy crops like switch-grass (*Panicum virgatum* L.), miscanthus (*Miscanthus giganteus*), natural prairie grasses, and short-rotation woody crops might be cultivated without substantially affecting the production of food or feed [8].

Forest weed *Lantana camara* (*L. camara*) is poisonous. In many nations throughout the world, it is known as an invasive weed. The plant *L. camara* is extremely competitive. It swiftly covers an open area. It seriously harms the local biodiversity and environment. Although it is an ecologically resilient species, the infestation density rather than the species' vast distribution is a potential future threat to ecosystems. Due to its invasiveness, capacity for spread, and negative effects on the economy and ecology, it is considered one of the worst weeds in approximately 50 countries. The weed engulfs native vegetation in dense, impenetrable thickets. A dominant weed species has been implicated with threats to ecology and biodiversity. However, *Lantana* generates a significant amount of woody biomass used for energy purposes [9]. The invasive species *Prosopis juliflora* (*P. juliflora*) is another one. Millions of hectares of farmland and forest have been blanketed nationwide. The ecological effects of *P. juliflora* expansion are numerous. The loss of local flora and animals as well as diminishing water bodies are associated with its existence. The weed has been used for a variety of purposes. Its wood is used to make activated carbon and charcoal. Its wood is highly calorific, extensively dispersed throughout India's many agro-climatic zones, and anticipated to play a significant part in upcoming bioenergy programs. The habitat for wildlife is negatively impacted by these two invasive species. *L. camara* has less ash than other plants. *P. juliflora* biomass (2.3%) and *L. camara* biomass (3.8%) provide the additional benefit of these species' briquettes. Another crucial factor for contrasting the fuel characteristics of various feedstock is the fixed carbon concentration. A fuel's high energy value is correlated with its high fixed carbon content [9,10].

4. Eucalyptus as a Source of Paper and Pulp Production

One of the necessities for daily life is paper, and the paper industry is the foundation of the economy in many Eucalyptus-growing countries. Other industries, including education, communication, and product packaging, benefit from the use of paper and paper-based products. It was assumed that India's consumption of paper and paperboard is projected to double from 10 Mt yr^{-1} (WWF, 2010) [11]. Pulp and paper mills will need to decrease their consumption of forest products, and counties need to increase their forest plantations, because of the execution of national and state government policies toward forest protection and afforestation. Additionally, the government is promoting the establishment of plantations on unused and degraded land. The paper industry will have to rely more and more on imports of pulp or finished paper goods due to the generally limited availability of raw resources. In India, eucalyptus has been cultivated in about 170 different species, varieties, and provenances; the most notable and popular of them is the eucalyptus hybrid, also known as Mysore gum, which is a variety of *Eucalyptus tereticornis*. Eucalyptus hybrids play a major role in providing wood, and house-hold applications in Indian environments, due to its: (a) fast growth, (b) the ability to overtake weeds, (c) a fire-resistant character, and (d) the capacity to adapt to a wide range of edaphoclimatic conditions [11].

5. Bio-Fuels

Green Fuel refers to clean fuels, which are also known as bio-fuels. Substrates that produce heat when combined with oxygen, i.e., during the combustion reaction, form fuel. Green fuels originated from green or biological sources; therefore, they add less load on the environment; i.e., they are eco-friendly and also bio-renewable [12]. Biofuels are the kind of energy materials which are obtained from the renewable resource materials (biomass) produced in agricultural land and natural aquatic systems as the primary products of photosynthesis (Tables 1 and 2) [13]. Interest in producing bio-fuels is enormously increasing for the following reasons:

- Cutting down the dependence on petroleum (crude and products)
- Growing environmental concerns
- Economic concerns.

Table 1. Different feedstocks contributing to production of bio-fuel.

Feedstock	Condition	Bio-Fuel Production (L/ha)	Reference
Corn	Hydrolysis/fermentation	3800	[14–16]
Sugarcane	Fermentation	7200	[14,17–19]
Sugar beet	Hydrolysis/fermentation	7900	[14,20–22]
Wheat	Hydrolysis/fermentation	1700	[14,23,24]
Cassava	Hydrolysis/fermentation	137	[14,25,26]

Table 2. Raw materials used for second-generation bio-fuel production.

Second Generation Fuel (Raw Material)	Strategy	Reference
Soybean oil	Dominant biomass for the manufacturing of biodiesel	[27]
Palm oil	Dominant biomass for the manufacturing of biodiesel	[27]
Rapeseed	Dominant biomass for the manufacturing of biodiesel	[27]
Crude glycerol	Electrochemically converted (uses in the pharmaceutical, cosmetics, food, etc.), thermochemical conversion (Biomass gasification, Biomass pyrolysis, Biomass combustion)	[28,29]
Sunflower stalk (saccharification)	Residue is liquefied to produce bio polyol. Sunflower stalk waste products (strongly condemn) and biodiesel (crude glycerol) are combined to create biopolyol, which can result in the manufacture of polyurethane.	[30]

There are two types of bio-fuels:

- Gaseous fuels like biomethane and biohydrogen
- Liquid fuels like bio-ethanol, biobutanol, bio-gasoline, biokerosene, and biodiesel.

An overview of bio-fuel production is given below. Tables 2 and 3 gives an overview of different types of bio-fuels and method of production.

5.1. Municipal Solid Waste (MSW) as a Source of Biomass

In addition to certain commercial and industrial compost that is comparable in type to home waste and has been dumped in municipal landfill sites, municipal solid waste (MSW) is predominantly waste produced by households. In addition to being a potential liability if it needs to be disposed of, MSW is a sizable resource that can be profitably recovered: for example, through the recycling of commodities like aluminum cans, metals, glass, fibers, etc., or by recovery processes like energy conversion and composting. Countries should follow the waste hierarchy as depicted in Figure 3.

Table 3. Method of production of different bio-fuels.

Types of Bio-Fuels	Method of Production	Reference
Biodiesel In-vivo In-vitro Semi In-vitro/In-vivo	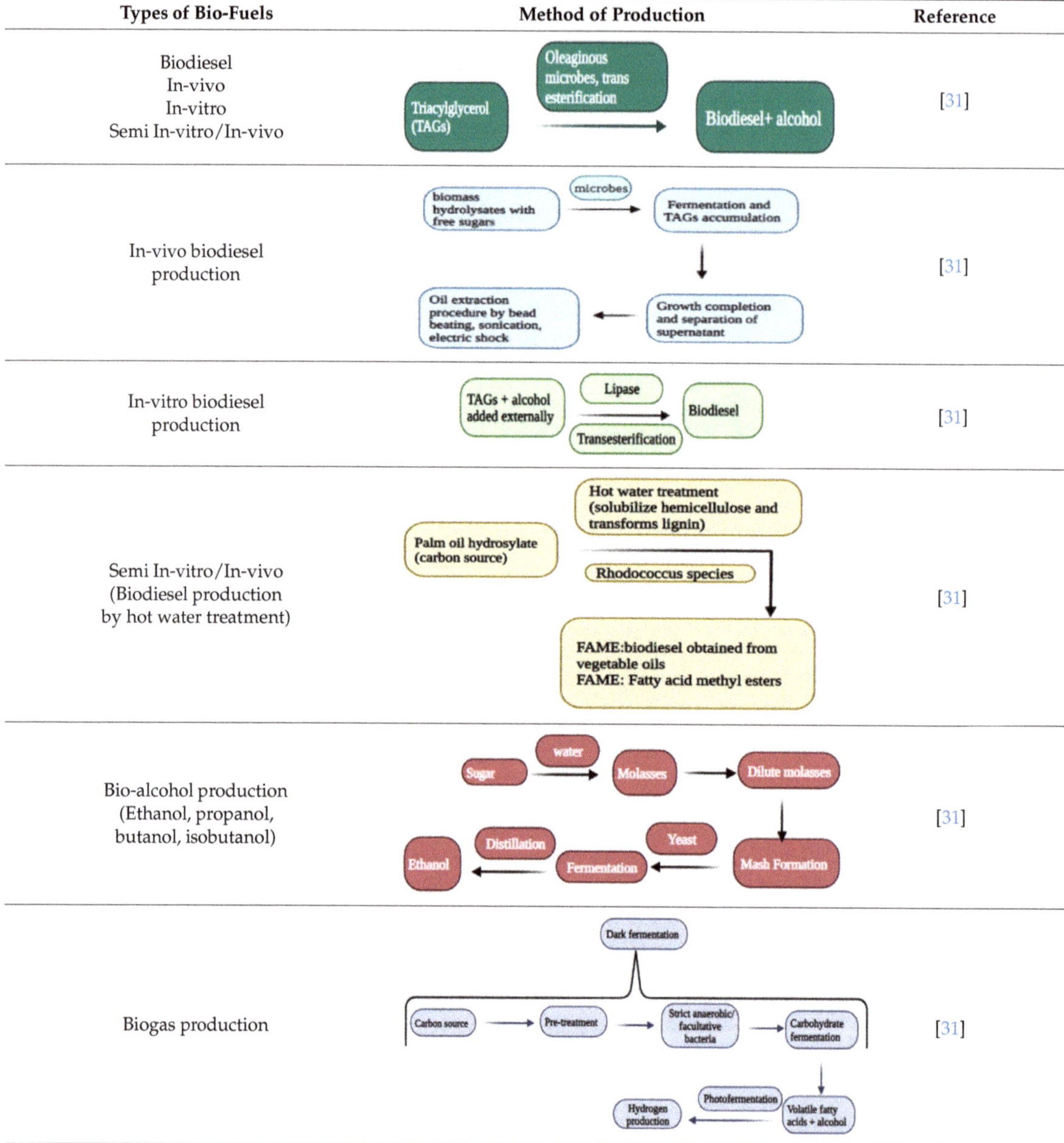	[31]
In-vivo biodiesel production		[31]
In-vitro biodiesel production		[31]
Semi In-vitro/In-vivo (Biodiesel production by hot water treatment)		[31]
Bio-alcohol production (Ethanol, propanol, butanol, isobutanol)		[31]
Biogas production		[31]

Table 3. *Cont.*

Types of Bio-Fuels	Method of Production	Reference
Biomethane production	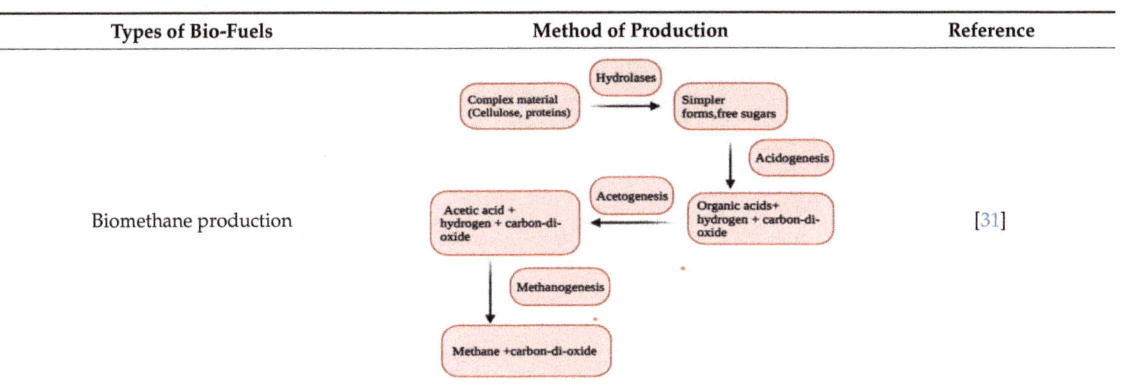	[31]

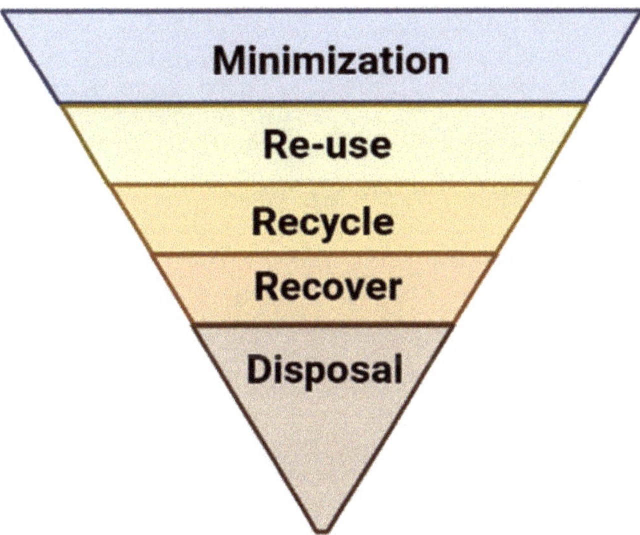

Figure 3. Waste energy inverse pyramid.

The various methods for turning MSW into energy include those shown in Figure 4. These mostly consist of thermo-chemical processes (such combustion, gasification, and pyrolysis) and biological processes (such as anaerobic digestion). All other process paths make use of an improved fuel, with the exception of mass burning or incineration systems. Either separation at the source followed by a straightforward mechanical treatment, like size reduction, or intensive mechanical treatment of MSW to generate Solid Recovered Fuel, can be used to achieve this. Major environmental consequences (or sustainability indicators) of MSW have been subject to life-cycle-based evaluations, which have demonstrated the advantages of MSW energy recovery. These benefits come in the form of decreased leaching into waterways, decreased acid gas emissions, decreased depletion of natural resources (fossil fuels and materials), and decreased soil contamination [32].

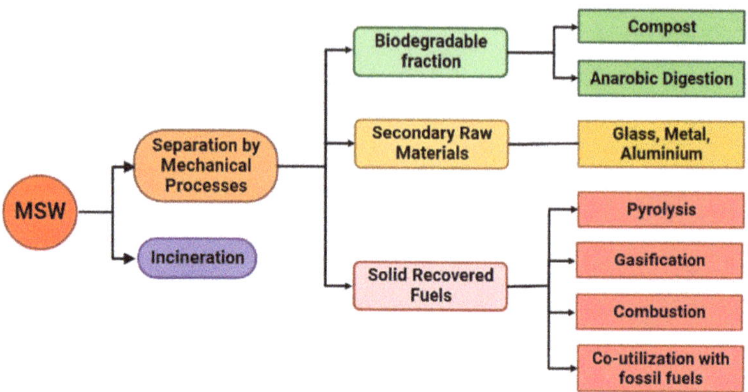

Figure 4. Methods and turning MSW into energy.

5.2. Biomethane Production Technology

Anaerobic processes involve different anaerobic bacteria, utilizing hydrolytic bacteria, acetogenic bacteria, and methanogenic bacteria. Hydrolytic bacteria break down biopolymers like cellulose and proteins to form H_2, CO_2, and low molecular organic acids. Acetogenic bacteria convert the above products into acetate while methanogenic bacteria use acetate and H_2/CO_2 etc. to produce methane.

5.3. Biohydrogen Production Technology

Bio-photolysis refers to the production of hydrogen from water with the help of sunlight by photosynthetic microbes.

$$\text{Water} + \text{Sunlight} \rightarrow \text{Hydrogen}$$

5.4. Biodiesel

The term "biodiesel" refers to a non-petroleum-based diesel fuel created by transesterifying vegetable oil or animal fat (tallow) and having long-chain alkyl (methyl, propyl, or ethyl) esters (Figure 5). In automobiles with stock diesel engines, biodiesel may be utilized either by itself or in combination with regular petro-diesel. Biodiesel is distinct from straight vegetable oil (SVO), which is utilized as fuel in certain converted diesel cars (either alone or in a mix). It can be made from vegetable oil or from microbial oil by trans-esterification process. The trans-esterification process could be chemical catalyzed or lipase catalyzed. The current concept of biodiesel is created by converting oils and fats to alkyl esters of monohydric alcohols to solve problems with high viscosity, high boiling point, and reactivity.

Biodiesel is now only defined as the "Monoalkyl" esters of long-chain fatty acids produced from the oils/fats of vegetable and animal sources that meet nearly all the standards of diesels generated from petroleum. Non-edible oils, like Jatropha, Karanja, and minor oils like mowrah, neem, undinahor, waste cooking oils, microbial oil and fatty acid distillates, are potential raw materials for biodiesel. Algal biomass could serve as an ideal feedstock for a spectrum of bio-fuels (biodiesel, bioethanol, biohydrogen), food (single cell protein or food ingredients), and nutraceuticals (vitamins, amino acids, biopigments, and others) [33].

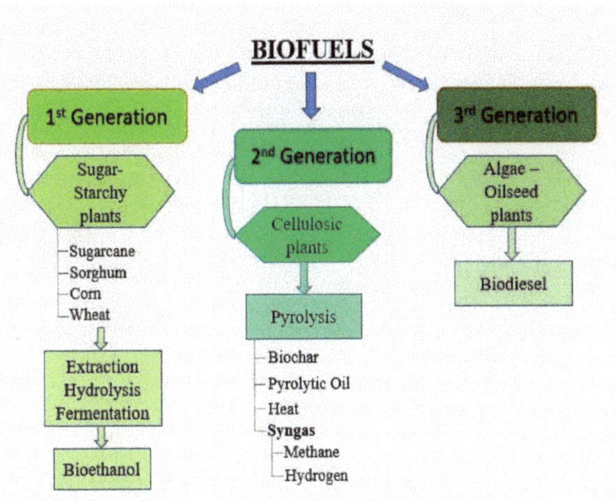

Figure 5. Bio-fuel products from lignocellulosic biomass.

6. Second Generation Bio-Fuels

Using inedible raw materials for the production of 2G bio-fuels, often referred to as "advanced bio-fuels", solves the issue of the "food versus fuels" competition. Furthermore, burning 2G bio-fuels results in balance or even lower correlation carbon emissions (Table 3).

Among the non-food feed stocks employed in novel processes that produced 2G bio-fuels were food waste, manure, spent cooking oil, wood, sawdust, garbage, leftovers from agriculture and food processing techniques (abandoned fuelwood), and energy crops (Figures 5 and 6) [34]. The lignocellulosic remains of crops would be the most plausible options among these sources due to their quantity, widespread availability around the world and throughout the year, and low cost. 2G bio-fuels are not industrially viable due to heterogeneity in structure in their leftovers' composition, which necessitates additional complicated production procedures. Currently, it is projected that 0.4 billion liters of 2G bio-fuels are produced annually: that is, 0.4% of all ethanol produced. Cellulosic ethanol production is predicted to increase to 0.8 billion liters in 2023 by the IEA for advanced bio-fuels [35].

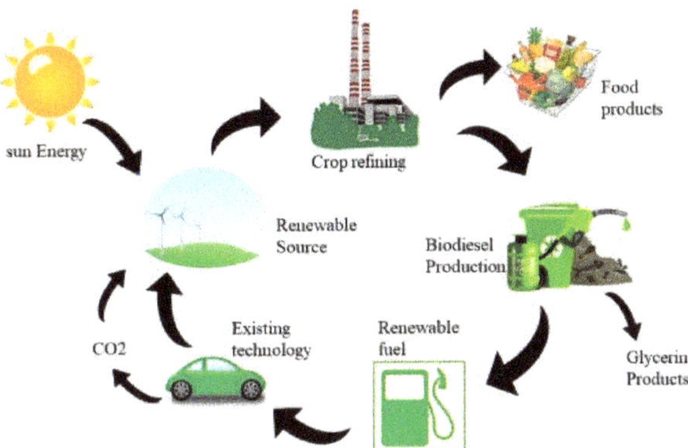

Figure 6. An overview of biodiesel production from algal feedstock.

7. Food vs. Fuel Debate

The dispute around the trade-offs involved in using grains and oilseeds for the production of bio-fuels vs. animal feed and human nourishment is a part of this problem. The relevance of evaluating the sustainability of bio-fuel production and use has increased recently, particularly in light of the need to reduce GHG emissions, the argument over whether to use fuel or food, and the growing need to uphold environmental and social norms [36]. The second generation switched to the development of specialized energy crops like Jatropha in response to the dispute over whether food or fuel should come first. *Jatropha curcas* is a low-cost biodiesel feedstock that has more oil than other species and strong fuel characteristics. It is a feedstock made from inedible oilseeds. Thus, it will not have an effect on food costs or the argument between food and fuel. Jatropha can be used in diesel engines to get a similar level of performance while emitting fewer pollutants than diesel. The crop has a significant positive impact on rural living as well. Additionally, the plant can produce up to 40% more oil per seed depending on weight [37].

Jatropha seed is a crop that is gaining popularity due to a number of factors, including the manufacturing of biodiesel. This is due to its seeds' high oil levels, which can range from 25% to 35%. Jatropha is appealing for more reasons than only its high oil content, however. It is a highly adaptable crop that thrives in dry or semiarid climates and is perennial, drought-resistant, and nutrient-restricted tolerant. The fact that this extremely adaptable crop is not seen as a food crop and does not have to compete for agricultural land offers a potential resolution to the ongoing food vs. fuel conflict. This makes Jatropha an appealing choice for bio-fuel conversion; however, there are still many questions regarding this plant [35]. Figure 7 shows a transition to green fuels from conventional fossil fuels, which will result in the development of clean energy and a carbon-neutral economy empowering employment creation at various levels.

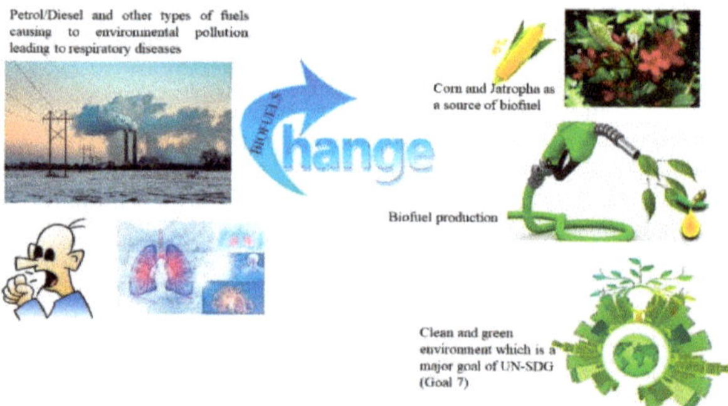

Figure 7. Figure shows that a transition to green fuels will result in a green environment resulting in clean energy.

Sugarcane's strong biomass output is an important biological component that helps the plant create bioethanol with a high positive LCEB (life cycle energy balance) and a positive GHGE (greenhouse gas emission) balance [38]. Low net greenhouse gas (GHG) emissions from sugarcane bio-fuel result in a diminished negative environmental impact as measured by pollution indices. In 2007, it was estimated that Brazil's net emissions of greenhouse gases decreased by 25.8 million tons CO_2 equivalent. With 9.8% fermentable sugars in its juice, sugarcane is the most cost-effective source of bioethanol, whereas sweet sorghum has 11.8%. While sweet sorghum has 21% lignin, 27% hemicellulose, and 45% cellulose with potential ethanol output of 12,938 and 5804 kg per ha, respectively, sugarcane bagasse has 22% lignin, 24% hemicellulose, and 43% cellulose. Although sugarcane is more productive than its rivals (sweet sorghum, sugar beet), more fiber and sugar must still be added to the plant for it to become energy cane. About 50% of the dry mass of sugarcane is made up of sucrose, whereas 14% is made up of fiber. Additionally, the following characteristics make sugarcane a superior option as an energy crop: less flower output, erect growth habit, ratooning ability, drought tolerance, cold tolerance, pest and disease resistance, and rapid early growth. Changing plant cellulose content and total biomass through the control of growth hormones and biomass synthesis pathways may be essential for the development of energy cane [14].

8. Forestry Practices for Biomass Production and Tackling Climate Change Issues

Bio-fuels of the second generation are created from lignocellulosic crops. The technology of this generation makes it possible to separate the plant's cellulose and lignin so that the former may be fermented into alcohol. Many types of biomass can be utilized to create these bio-fuels because biomass is defined as any source of organic carbon. Algae or multiple other products are other possibilities. Biomass energy is produced using a variety of feedstocks, including landfill gases, garbage, and plants like perennial grasses and trees. Wood pellets from forests can be used to produce electricity, heat, and possibly even liquid fuels. The key benefit of biomass is that it cannot be drained like fossil fuels. The fundamental benefit of biomass is that, unlike fossil fuels, it is not exhaustible. Due to the abundance of plants on Earth, biomass has the potential to be a significant way for renewable energy that must be utilized as a long-term replacement for fossil fuels. While burning fossil fuels creates emissions into the environment that leads to climate change, sustainably managed biomass is regarded as being carbon-neutral. This fuel source has had fewer harmful effects on the environment, such as global warming and water and air pollution [39].

It is clear that the loss of fertile land and the requirement to secure the food security of a growing global population are mutually exclusive. It is necessary to continue plants' net carbon uptake in order to lessen climate change. As a result, creating resilient ecosystems depends on both preventing additional land degradation as well as facilitating land rehabilitation. Training, research, dissemination, and slightly higher compared advanced technologies are necessary to enable efficient biomass and bio-energy manufacturing. Recent studies give the following examples of technologies for producing bio-energy from various types of biomass:

- Although old-fashioned biomass practices like firewood and charcoal are highly prevalent in many areas, their effectiveness might be significantly increased by using controlled charcoal manufacture units and capturing solid, liquid, and gaseous fuel portions. Although these bio-energy methods need significant capital to create and maintain, it is challenging to go from traditional to scientifically higher specialized biomass power technologies (combined heat and power schemes, bio-fuels, etc.). This demonstrates the urgent need for eco-friendly, easy integration of low-cost bio-energy technologies with present biomass supplies and technology (i.e., wood chips or pellets from wastes, to replace coal).
- To determine (i) the level and nature of energy demand (such as electricity, liquid fuel, etc.), as well as (ii) the available renewable production to fuel bio-energy units, it is also necessary to examine more complicated systems designed to replace fossil fuels. Although several biological and/or chemical processes can convert biomass into energy, the economic sustainability of these methods in underdeveloped nations is still unknown.
- To enable the development of better energy systems, new bio-energy technologies should be carefully evaluated for their socioeconomic benefits and drawbacks. The merging of biomass resources with low-impact forestry to support the creation of green technologies was a significant result. Boosting energy output while keeping the environment green is difficult, and there are several different approaches to this problem (Figure 7).
- Using methods to cultivate deciduous biomass plants on barren or damaged terrain. Although the technique relies on cautious variety identification across all domains, it

may lead to initially poor yields due to the creation of varied stands, cautious usage, and long-term yield studies.

- Bio-energy leftovers (slurries, burn, and smoke) should be used to stop the degradation of soil nutrients; these can be recycled into soils directly for soil improvement. However, during the generation of ethanol and bioenergy, nitrogen and other important macronutrients for plants are typically lost through chemical activation or exhaust fumes. To maximize regulated absorption and long-term ammonium nutrition while minimizing environmental impact, slow-release fertilizers made from bio-energy wastes (or modified biochar) are a potential option.
- However, the impact of the biomass energy part of the economy on different societies has received little attention. The deployment of bio-energy technology is intended to increase socioeconomic wellbeing in countries where forests are maintained by people (such as in Indonesia and Nepal) [40].

A plan to counteract the damaging effects of tropical habitat destruction is land rehabilitation. Lands are deemed to be degraded when they have poor biotic diversity and edaphic conditions are decreased to such an extent that the land is unable to serve any particular use. In other words, degradation is destruction of ecosystem reserves such biomass, seed banks, soil minerals, and organic matter to a point from where returning to original land quality is not possible. On the other hand, reclamation is the process by which degraded lands are returned to productivity with biotic functions. However, this is a difficult stage as there are limitations of biota. On the contrary, rehabilitation is a process where a degraded land is returned to a fully functional state [41]. Land rehabilitation systems involves reversal effects and a couple of actions:

- Removal of the stress source such as high frequency forest fire, over-grazing, or removal of biomass.
- Addition of species (plants and/or animals) or materials (fertilizers, organic residues or water).
- Managing the soil quality to control the synchronization of release of nutrients and plant uptake.

One of the most important components of land rehabilitation is planting trees. Forests can be brought back to their old form only by tree planting, which henceforth increases biomass production. By planting trees, soil nutrition, soil fertility and soil organic matters are re-established [41].

Borchard et al. (2016) introduced advanced technology and bio-energy techniques to incorporate solid, liquid, and gaseous bio-fuels into current methods for producing green and clean energy (Figure 6) [40]. Two different kinds of bio-fuels were covered. (a) Usually, species with rapid growth rates are chosen for biomass production. Although commonly utilized in the tropics, *Calliandracalothyrsus* and *Gliricidiasepium* do not do well in conditions of water-logged soil. Large tracts of peat land exist in Southeast Asia, and during the rainy season, high water tables are necessary to protect them. Therefore, it is imperative to find woody crops that can produce valuable biomass outputs and are compatible with semi-terrestrial soils. (b) *Calophylluminophyllum*, *Pongamiapinnata*, and *Reutealistrisperma* are a few of the species that have been discovered and assessed for their capacity to generate oil and convert it into usable liquid fuels.

9. Phytoremediation Technologies for Effective Land Restoration and Biomass Production

A striking method for purifying soils contaminated with a variety of hazardous organic and inorganic substances is phytoremediation. Typically, phytoremediation entails restoring the soil's microbiota, physicochemical qualities, and fertility in order to promote the growth of the suitable climax cover plant. In order to accomplish these goals for land damaged by coal mining, South Africa created a myco-phytoremediation method known as Fungcoal. In the late 1990s, the practice of using eco-friendly plants to remove or make harmless ecological contaminants was known as phytoremediation. Both organic and inorganic pollutants, present in either solid or liquid substrates, or air, were addressed utilizing one or a mixture of phytoextraction, phytodegradation, phyto-stabilization, rhizo-filtration, and phyto-volatilization.

In 2005, phytoremediation was expanded to encompass both plants and the utilization of microorganisms associated with plants for restoring the environment. It was viewed as a practical and non-intrusive bioprocess remediation approach at the time. The development of modelling frameworks, according to its proponents, offers a possibility to capitalize commercially on a phytoremediation strategy by producing value-added goods from biomass. Nevertheless, the cycling of soil organic carbon (SOC) for plant growth is mostly dependent on soil microbes [42]. Recent years have seen growing evaluation of biomass integration in energy production processes to boost the positive effects on the economy and environment. This combination can lessen metal remobilization and assist in removing the financial barriers to phytoremediation. In addition to assisting in meeting the world's energy needs, the commercialization of residual biomass from phytoremediation that can be converted into bioethanol, biodiesel, biogas, and heat, and thus, would also offer a way to promote a bio-based economy for long-term growth.

A process called phyto-mining aims to extract valuable metals from plant biomass. Because these metals are used as raw materials in contemporary industries like batteries and electrochemistry, metals are released into the environment. Therefore, it is paradoxical that they end up as soil contaminants at the conclusion of their life cycle. By retrieving the metals, a process called phyto-mining seeks to break this cycle. The crucial step is leaching the polluted plant biomass with acids in order to recover the adsorbed metal and produce biomass with metal concentrations within the limits of environmental guidelines [43]. The buildup of heavy metals in soil has increased dramatically because of several human (industrial) actions in addition to natural processes.

Due to their non-biodegradability, heavy metals linger in the environment, pose a risk of contaminating crop plants, and may eventually build up in people's bodies due to bio-magnification. Since heavy metals are harmful, they provide a serious risk to both public health and the biosphere. The importance of cleaning up land pollution is therefore crucial. Phytoremediation makes it feasible to ecologically friendly replant soil that has been contaminated by heavy metals. Phytoremediation is the use of plants to remove hazardous substances from the environment or to lessen their bioavailability in soil [39]. Plants can take up ionic compounds from the soil through their root systems. Plants extend their root systems into the underlying ground to form rhizosphere ecosystems that absorb heavy metals and manage their bioavailability, clean up contaminated soil, and preserve soil fertility.

10. Optimization Strategies in Bioprocessing Industries for Efficient Second-Generation Bio-Fuel Production

Given the effects of global warming as well as the depletion of non-renewable resources, there is an urgent need for alternative forms of transportation fuel. The cellulose, hemicellulose, and lignin structural elements of lignocellulosic biomass are presented, together with the technological unit stages of pre-treatment, enzymatic hydrolysis, fermentation, distillation, and dehydration. The goal of the pre-treatment step is to reduce the amount of inhibitors present and increase the amount of carbohydrate surface area available for enzymatic saccharification. Enzymatic hydrolysis yields fermentable sugars, which are then transformed into ethanol by microbial catalysts. Lignocellulose-derived carbohydrates (hemicellulose and cellulose) pave a sustainable platform for the production of not only bio-fuel but also bulk and specialty chemicals in addition to functional oligosaccharides, eventually benefitting lignocellulose bio-refineries [44,45]. Energy demand rises because of population increase and expanding industrialization, yet conventional fossil fuels, especially petroleum, are limited resources that release GHG when burned. Future global energy demands must be met using environmentally responsible and sustainable energy sources.

Researchers, business partners, and governments are consequently very interested in bio-fuels, specifically cellulosic bioethanol, butanol, and biodiesel. Particularly, bioethanol is seen as a possible drop-in fuel that could serve as a substitute for gasoline in the transportation industry [46]. Many people have characterized the growth of bio-fuels as essential for addressing two sides of such an "energy trilemma"—by lowering greenhouse gases through the development of diesel substitutes, and also (especially in the USA) by permitting countries to break their improved energy security by reducing reliance on the global trade of fossil fuels. Only the first generation of bio-fuels has so far seen broad acceptance. Bio-fuels are typically divided into each of two (or rarely up to four) decades or phases of development. The majority of first-generation bio-fuels come from plants that are converted into ethanol, such as sugar cane, corn, or soybeans. Second-generation bio-fuels, which have not yet been extensively accepted commercially, convert lignocellulosic biomass— typically, "deciduous tree" agricultural waste, but sometimes specifically grown plants like switch grass or *Miscanthus*—into fuels via a range of physical, biological, or chemical methods (and other useful chemicals) [47].

11. Highlights from COP26 Agenda towards Net Zero Emissions through Bio-Fuels Implementation

The Conference of the Parties (COP) to the UN Framework Convention of climate change, alternatively known as the climate summit event, brings together governments to unanimously discuss and review how climate change and global warming can be managed. The 26th meeting called COP26 was held at Glasgow, Scotland. The following are the four goals of COP26 that need to be achieved [48]:

Goal 1: "Secure global net zero by mid-century and keep 1.5 degrees within reach". Carbon neutrality or zero carbon emission is to be reached by 2050 and the global warming temperature should be below +1.5 degree centigrade. To achieve this deliverable, countries are required to fast-track the phase-out of coal and encourage investments in renewable energy. This can be done by restricting deforestation and gearing up to electric vehicles.

Goal 2: "Adapt to protect communities and natural habitats". The climate is changing due to human behavior and deforestation. It will continue to change even if there is carbon emission reduction. COP26 aimsto encourage countries affected by climate change to protect and restore the ecosystem. It encourages building defenses and installing warning systems in areas, helping infrastructure and agriculture to be more resilient to avoid loss of homes, lives, and livelihood.

Goal 3: "Mobilize finance". To achieve the first two agendas, approximately $100 billion is required per year. As a helping hand, international financial institutions must play their role and work towards the releasing of funds in private and public sector finance. This step is essential to secure global net zero.

Goal 4: "Working together to deliver". COP26 looks forward to achieving the challenges discussed only when countries work together. It urges to turn ambitions into action by speeding up collaboration between governments, businesses, and civil society to deliver goals faster than stipulated time.

Concisely, COP26 urges every country to come forward and help hand-in-hand with the vision of net zero carbon emission, low deforestation, and renewable economy.

12. Sustainability Assessment of Bio-Fuels

Sustainability is showing a growing trend in terms of paper publications, which reflects the fact that people are interested in improving the environment. Nevertheless, COP26 also demonstrates that not all nations are eager to assist this transformation, and that it will ultimately be difficult to do so given the stark inequalities in growth, emission levels, economic production factors, efficiency, and social repercussions [49].

- Bio economy: A bio economy is defined by the European Commision as an economy which uses renewable biological resources [50]. A bio-economy involves several sectors like agriculture, forestry, fishing and aquaculture, and the manufacture of food, beverages, and tobacco, etc. It is categorized under three main standpoints: (i) the bioecology vision; (ii) the biotechnology vision, and (iii) the bioresource vision [51].

Bioeconomy sectors and sub-sectors investigated in some countries showed vast variations in their objectives and strategies. Table 4 summarizes the objectives and priorities of different countries.

Table 4. Objectives and priorities of some countries for their bioeconomic growth.

Name of Country	Priorities/Objectives	Reference
Argentina	Bio economy is seen as tool for sustainable development Recognized as a positive substitute for new behavior generation Source of employment to face the stern challenge of climate change	[52] [53]
Germany	Country has primacy for progressing towards a knowledge-based bio economy The priorities include: the creation of a reliable supply of high-quality food; the conversion of an economy based on fossil fuels to one that is more resource- and raw-material-efficient while preserving biodiversity and soil fertility	[52] [54]
Malaysia	Significant donor to economic growth, that provide benefit to society via innovations in agricultural productivity, inventions in healthcare and implementation of sustainable industrial processes.	[52]
Netherlands	The utilization of renewable natural resources and wastes offers Dutch enterprises economic opportunity, as do CO_2 emissions, the circular economy, and knowledge of the limited nature of fossil fuels.	[52,55]
South Africa	Objective is to make the country economically sound using renewable feedstock, particularly in industrial and agriculture sectors The goal is to have a low-carbon economy	[56] [52]

- Environment: The pollution caused by petroleum hydrocarbons, oil, and heavy metals is becoming an increasingly significant problem because of the increased demand for crude oil and products related to crude oil in many fields of application. Due to the ecological harm it causes to terrestrial, aquatic, and marine ecosystems, this pollution has attracted a great deal of attention. Recently, biosurfactant compounds have drawn a lot of attention since they are seen as a viable solution and environmentally friendly material for remediation technology. The unique trait of biosurfactants is that they minimize and lower the interfacial tension of liquids. These qualities make it possible for biosurfactants to be used in a range of industrial situations, including emulsification, de-emulsification, biodegradability, foam generation, cleaning efficiency, surface activity, and detergent composition [57].
- Surfactants' primary job is to reduce interfacial tension; emulsifiers, on the other hand, progressively bind to the surface of the droplet and provide longer-term stability. Many synthetic surfactants and emulsifiers have significant levels of toxicity and ecological effect, which has sparked interest in more natural compounds like bio-surfactants and bio-emulsifiers. The primary sectors connected to human health, including pharmaceuticals, food, and cosmetics, are interested in these bio-based surface-active compounds, many of which have previously been discovered and used extensively [58].
- Bio-fuel productions offer several employment opportunities at various levels. Beside job creation, this sector offers unique economic support to farmers, in turn strengthening the rural economy [59]. A bio-fuel driven economy will give a big impetus to the circular economy, eventually developing a renewable economy. Microalgae have the potential to deliver a value-based product using waste from dairy industries. Microalgae cultivation using dairy waste can provide biolipids, carotenoids, aminoacids, enzymes, and other high value products. Recently, Gramegna et al. [60] obtained lipids (12% to 21% (*w/w*)) from *Auxenochlorella protothecoides* cultivated on dairy-wastes, showing a high lipid productivity of ~0.16 g/L/d.

- The governmental sector along with the private sector should make concerted efforts to develop the circular economy globally, offering large-scale employment for the development of a better society and finally addressing the UN-SDG goals of poverty alleviation, employment, and climate change.

13. Limitations

The production of bio-fuel is a complex process involving many process configurations. First-generation (1G) ethanol production is a well-established process. It is produced directly from sugars from sugarcane molasses, corn grains, and other starchy and sugar feedstock. Brazil and USA are global leaders in production of 1G ethanol from sugarcane molasses and corn grains, respectively. Blending of 1G ethanol in petroleum is being employed in many countries. However, 2G ethanol production needs diversified process configurations involving pretreatment, enzymatic hydrolysis, and fermentation, making the process costly and time consuming. Agricultural residues such as sugarcane biomass (bagasse and straw), rice straw, and corn stover, among others are principal feedstock for 2G ethanol production. However, technical maturity and economic viability of the 2G ethanol process are two major hindrances for the successful deployment of lignocellulosic bio refineries [61,62].

Another promising bio-fuel is biodiesel, which also has significant momentum in world. Vegetable oil, oil from seed yielding crops, and microalgae are principal feedstock explored for biodiesel production. However, the cost of production and regular availability of feedstock with affordable price, and food preference over to bio-fuel production, are the major limitations to making biodiesel a successful renewable fuel alternative. Beside the technical maturity and cost of the production process, the use of sugars and oil as raw material for bioethanol and biodiesel has an important role in the success of bio-fuel production. More research towards the development of a robust and simplified process employing low-cost feedstock, preferably grown on marginal land proven at an industrial scale, is required for bio-fuel production at commercial scale.

14. Future Prospects

To assure the creation of effective forest-based bio-energy systems, a variety of groups of research and development for the bio-energy area are needed:

(a) Creating strategies for forest-based bio-energy that are tailored to: (a) regional energy needs; (b) socioeconomic and environmental factors to support successful implementation; and (c) delivering advantageous groupings of renewable energy kinds (e.g., bioenergy, water energy, etc.).

(b) Creating straightforward evaluation tools to facilitate the creation of strategies for the production of sustainable and financially viable forest-based bio-energy (covering biomass manufacture, bio-energy technologies, efficient recycle of deposits and effective policies).

(c) Evaluating forest community-based bio-energy methods for their socioeconomic and environmental effects as solutions to advance energy security and alleviate poverty in emerging nations.

(d) Inspiring administrations to create laws and guidelines that encourage the growth of forestry-based bio-energy in training, economic incentives, and the advancement of bio-energy activities.

With increasing pollution and global warming, icebergs are melting at an alarming rate. Scientists fear that if the current condition persists, the water body of the globe, which covers 70% of the planet, will increase to a far greater extent, which will be a matter of concern for the human race. Therefore, it is high time that people should be aware and take care of the environment. Deforestation is another concern for increasing global warming. With phytoremediation being a solution, people should plant more trees and decrease the use of fossil fuels to reduce air pollution.

15. Conclusions

Bio-fuels area promising renewable and sustainable energy alternatives that can help in addressing the issues of environmental pollution. There are different types of bio-fuels that are used, the production of first-generation bio-fuels has significant constraints, and second-generation bio-fuel methods have been developed. These issues can be resolved, and second-generation bio-fuels can produce a higher percentage of ethanol sustainably and inexpensively with greater environmental advantages. In order to increase the quantity of bio-fuel that can be produced sustainably utilizing biomass, second-generation bio-fuel methods are used. Numerous techniques exist to generate renewable energy from organic and inorganic environmental assets (such as sunlight, air, water, tides, waves, and convective energy), as well as a number of technologies that do the same with biomass with various characteristics and derived from various sources (e.g., food industry, agriculture, forestry). However, the manufacture of bio-fuels is difficult and controversial in many regions of the world since it competes for soil with the availability of crops, and could harm the ecosystem. As a result, wildlife management practices are required to generate sustainable bio-energy while preserving or even enhancing crucial ecosystem processes.

Author Contributions: Conceptualization and supervision: K.K.C. and A.K.C.; writing—original draft preparation: A.P., A.S. and K.K.C.; review and editing, artwork, and schemes: A.P., A.S., K.K.C., S.H., D.D., R.K.B., A.B. and A.K.C. All authors have read and agreed to the published version of the manuscript.

Funding: This research received no external funding.

Institutional Review Board Statement: Not applicabale.

Informed Consent Statement: Not applicabale.

Data Availability Statement: Not applicabale.

Acknowledgments: The authors acknowledge Uttaranchal University, Dehradun for providing the opportunity to write this review. A. K. Chandel gratefully acknowledges The Brazilian National Council for Scientific and Technological Development (CNPq), Brazil for scientific productivity program (Process number: 309214/2021-1).

Conflicts of Interest: The authors declare no conflict of interest.

References

1. Eisentraut, A. *Sustainable Production of Second-Generation Biofuels: Potential and Perspectives in Major Economies and Developing Countries*; OECD: Paris, France, 2010.
2. Transforming Our World: The 2030 Agenda for Sustainable Development, Department of Economic and Social Affairs. Available online: https://sdgs.un.org/2030agenda (accessed on 21 March 2023).
3. Kabeyi, M.J.B.; Olanrewaju, O.A. Sustainable Energy Transition for Renewable and Low Carbon Grid Electricity Generation and Supply. *Front. Energy Res.* **2022**, *9*, 1032. [CrossRef]
4. Yan, A.; Wang, Y.; Tan, S.N.; Mohd Yusof, M.L.; Ghosh, S.; Chen, Z. Phytoremediation: A Promising Approach for Revegetation of Heavy Metal-Polluted Land. *Front. Plant Sci.* **2020**, *11*, 359. [CrossRef] [PubMed]
5. D'Adamo, I.; Ribichini, M.; Tsagarakis, K.P. Biomethane as an energy resource for achieving sustainable production: Economic assessments and policy implications. *Sustain. Prod. Consum.* **2023**, *35*, 13–27. [CrossRef]
6. Osman, A.I.; Mehta, N.; Elgarahy, A.M.; Al-Hinai, A.; Al-Muhtaseb, A.H.; Rooney, D.W. Conversion of biomass to biofuels and life cycle assessment: A review. *Environ. Chem Lett.* **2021**, *19*, 4075–4118. [CrossRef]
7. World Energy Outlook 2009—Analysis. Available online: https://www.iea.org/reports/world-energy-outlook-2009 (accessed on 30 April 2023).
8. Gopalakrishnan, G.; Cristina Negri, M.; Snyder, S.W. A novel framework to classify marginal land for sustainable biomass feedstock production. *J. Environ. Qual.* **2011**, *40*, 1593–1600. [CrossRef]
9. Rajan, S.S.; Sekar, I.; Divya, M.; Parthiban, K.; Sudhagar, R.J. Suitability of invasive species for briquette production: Lantana camara and Prosopis juliflora. *Pharma Innov.* **2022**, *11*, 530–533. [CrossRef]
10. Tsalidis, G.A.; Kryona, Z.-P.; Tsirliganis, N. Selecting south European wine based on carbon footprint. *Resour. Environ. Sustain.* **2022**, *9*, 100066. [CrossRef]

11. Rana, V.; Joshi, G.; Singh, S.P.; Gupta, P.K. 20 Eucalypts in Pulp and Paper Industry. In *Eucalypts in India*; Bhojvaid, P.P., Kaushik, S., Singh, Y.P., Kumar, D., Thapliyal, M., Barthwal, S., Eds.; ENVIS Centre on Forestry, National Forest Library and Information Centre, Forest Research Institute: Dehradun, India, 2014; pp. 470–506.

12. Zabermawi, N.M.; Alsulaimany, F.A.S.; El-Saadony, M.T.; El-Tarabily, K.A. New eco-friendly trends to produce biofuel and bioenergy from microorganisms: An updated review. *Saudi J. Biol. Sci.* 2022; *in press*. [CrossRef]

13. Fraiture, C.; Giordano, M.; Liao, Y. Biofuels and implications for agricultural water use: Blue impacts of green energy. *Water Policy* **2008**, *10*, 67–81. [CrossRef]

14. Khan, M.S.; Mustafa, G.; Joyia, F.A.; Mirza, S.A.; Khan, M.S.; Mustafa, G.; Joyia, F.A.; Mirza, S.A. *Sugarcane as Future Bioenergy Crop: Potential Genetic and Genomic Approaches*; IntechOpen: London, UK, 2021; ISBN 978-1-83968-936-9.

15. Kauffman, N.; Hayes, D.; Brown, R. A life cycle assessment of advanced biofuel production from a hectare of corn. *Fuel* **2011**, *90*, 3306–3314. [CrossRef]

16. Ramos, M.D.N.; Milessi, T.S.; Candido, R.G.; Mendes, A.A.; Aguiar, A. Enzymatic catalysis as a tool in biofuels production in Brazil: Current status and perspectives. *Energy Sustain. Dev.* **2022**, *68*, 103–119. [CrossRef]

17. Mookherjee, P. *The Implications of India's Revised Roadmap for Biofuels: A Lifecycle Perspective*; Policy Commons, 2022. Available online: https://policycommons.net/artifacts/2273619/the-implications-of-indias-revised-roadmap-for-biofuels/3033480/ (accessed on 27 April 2023).

18. Vandenberghe, L.P.S.; Valladares-Diestra, K.K.; Bittencourt, G.A.; Zevallos Torres, L.A.; Vieira, S.; Karp, S.G.; Sydney, E.B.; de Carvalho, J.C.; Thomaz Soccol, V.; Soccol, C.R. Beyond sugar and ethanol: The future of sugarcane biorefineries in Brazil. *Renew. Sustain. Energy Rev.* **2022**, *167*, 112721. [CrossRef]

19. Raj Singh, A.; Kumar Singh, S.; Jain, S. A review on bioenergy and biofuel production. *Mater. Today Proc.* **2022**, *49*, 510–516. [CrossRef]

20. Panella, L. Sugar Beet as an Energy Crop. *Sugar Tech.* **2010**, *12*, 288–293. [CrossRef]

21. Salazar-Ordóñez, M.; Pérez-Hernández, P.P.; Martín-Lozano, J.M. Sugar beet for bioethanol production: An approach based on environmental agricultural outputs. *Energy Policy* **2013**, *55*, 662–668. [CrossRef]

22. Isler-Kaya, A.; Karaosmanoglu, F. Life cycle assessment of safflower and sugar beet molasses-based biofuels. *Renew. Energy* **2022**, *201*, 1127–1138. [CrossRef]

23. Kaparaju, P.; Serrano, M.; Thomsen, A.B.; Kongjan, P.; Angelidaki, I. Bioethanol, biohydrogen and biogas production from wheat straw in a biorefinery concept. *Bioresour. Technol.* **2009**, *100*, 2562–2568. [CrossRef]

24. Saravanan, A.; Senthil Kumar, P.; Jeevanantham, S.; Karishma, S.; Vo, D.-V.N. Recent advances and sustainable development of biofuels production from lignocellulosic biomass. *Bioresour. Technol.* **2022**, *344*, 126203. [CrossRef]

25. Sivamani, S.; Chandrasekaran, A.P.; Balajii, M.; Shanmugaprakash, M.; Hosseini-Bandegharaei, A.; Baskar, R. Evaluation of the potential of cassava-based residues for biofuels production. *Rev. Environ. Sci. Biotechnol.* **2018**, *17*, 553–570. [CrossRef]

26. Nizzy, A.M.; Kannan, S. A review on the conversion of cassava wastes into value-added products towards a sustainable environment. *Environ. Sci. Pollut. Res.* **2022**, *29*, 69223–69240. [CrossRef]

27. Atabani, A.E.; Silitonga, A.S.; Badruddin, I.A.; Mahlia, T.M.I.; Masjuki, H.H.; Mekhilef, S. A comprehensive review on biodiesel as an alternative energy resource and its characteristics. *Renew. Sustain. Energy Rev.* **2012**, *16*, 2070–2093. [CrossRef]

28. Dou, B.; Song, Y.; Wang, C.; Chen, H.; Xu, Y. Hydrogen production from catalytic steam reforming of biodiesel byproduct glycerol: Issues and challenges. *Renew. Sustain. Energy Rev.* **2014**, *30*, 950–960. [CrossRef]

29. Inayat, A.; Inayat, M.; Shahbaz, M.; Sulaiman, S.; Raza, M.; Suzana, Y. Parametric Analysis and Optimization for the Catalytic Air Gasification of Palm Kernel Shell using Coal Bottom Ash as Catalyst. *Renew. Energy* **2019**, 145. [CrossRef]

30. Kim, K.H.; Yu, J.-H.; Lee, E.Y. Crude glycerol-mediated liquefaction of saccharification residues of sunflower stalks for production of lignin biopolyols. *J. Ind. Eng. Chem.* **2016**, *38*, 175–180. [CrossRef]

31. Bhatia, S.K.; Kim, S.H.; Yoon, J.J.; Yang, Y.H. Current status and strategies for second generation biofuel production using microbial systems. *Energy Convers. Manag.* **2017**, *148*, 1142–1156. [CrossRef]

32. Bello, A.S.; Al-Ghouti, M.A.; Abu-Dieyeh, M.H. Sustainable and long-term management of municipal solid waste: A review. *Bioresour. Technol. Rep.* **2022**, *18*, 101067. [CrossRef]

33. Bhatia, L.; Bachheti, R.K.; Garlapati, V.K.; Chandel, A.K. Third-generation biorefineries: A sustainable platform for food, clean energy, and nutraceuticals production. *Biomass Conv. Bioref.* **2022**, *12*, 4215–4230. [CrossRef]

34. Wang, P.; Lü, X. Chapter 1—General introduction to biofuels and bioethanol. In *Advances in 2nd Generation of Bioethanol Production*; Lü, X., Ed.; Woodhead Publishing Series in Energy; Woodhead Publishing: Sawston, UK, 2021; pp. 1–7; ISBN 978-0-12-818862-0.

35. Dahman, Y.; Syed, K.; Begum, S.; Roy, P.; Mohtasebi, B. 14—Biofuels: Their characteristics and analysis. In *Biomass, Biopolymer-Based Materials, and Bioenergy*; Verma, D., Fortunati, E., Jain, S., Zhang, X., Eds.; Woodhead Publishing Series in Composites Science and Engineering; Woodhead Publishing: Sawston, UK, 2019; pp. 277–325; ISBN 978-0-08-102426-3.

36. Gautam, R.; Nayak, J.K.; Daverey, A.; Ghosh, U.K. Chapter 1—Emerging sustainable opportunities for waste to bioenergy: An overview. In *Waste-to-Energy Approaches Towards Zero Waste*; Hussain, C.M., Singh, S., Goswami, L., Eds.; Elsevier: Amsterdam, The Netherlands, 2022; pp. 1–55; ISBN 978-0-323-85387-3.

37. Indra, T.; Sebayang, A.; Padli, Y.; Rizwanul Fattah, I.M.; Kusumo, F.; Ong, H.C.; Mahlia, T.M.I. Current Progress of Jatropha Curcas Commoditisation as Biodiesel Feedstock: A Comprehensive Review. *Front. Energy Res.* **2022**, *9*, 1019. [CrossRef]

38. Broda, M.; Yelle, D.J.; Serwańska, K. Bioethanol Production from Lignocellulosic Biomass—Challenges and Solutions. *Molecules* **2022**, *27*, 8717. [CrossRef]
39. Biomass: A Sustainable Energy Source for the Future? College of Natural Resources News. Available online: https://cnr.ncsu.edu/news/2021/01/biomass-a-sustainable-energy-source-for-the-future/ (accessed on 23 November 2022).
40. Borchard, N.; Artati, Y.; Lee, S.M.; Baral, H. *Sustainable Forest Management for Land Rehabilitation and Provision of Biomass-Energy*; Center for International Forestry Research (CIFOR): Bogor, Indonesia, 2017.
41. Brown, S.; Lugo, A. Rehabilitation of Tropical Lands: A Key to Sustaining Development. *Restor. Ecol.* **1994**, *2*, 97–111. [CrossRef]
42. Edgar, V.-N.; Fabián, F.-L.; Mario, P.-C.J.; Ileana, V.-R. Coupling Plant Biomass Derived from Phytoremediation of Potential Toxic-Metal-Polluted Soils to Bioenergy Production and High-Value by-Products—A Review. *Appl. Sci.* **2021**, *11*, 2982. [CrossRef]
43. Sekhohola-Dlamini, L.M.; Keshinro, O.M.; Masudi, W.L.; Cowan, A.K. Elaboration of a Phytoremediation Strategy for Successful and Sustainable Rehabilitation of Disturbed and Degraded Land. *Minerals* **2022**, *12*, 111. [CrossRef]
44. Bhatia, L.; Sharma, A.; Bachheti, R.K.; Chandel, A.K. Lignocellulose derived functional oligosaccharides: Production, properties, and health benefits. *Prep. Biochem. Biotechnol.* **2019**, *49*, 744–758. [CrossRef] [PubMed]
45. Chandel, A.; Silveira, M. *Advances in Sugarcane Biorefinery Technologies, Commercialization, Policy Issues and Paradigm Shift for Bioethanol and By-Products*; Elsevier: Amsterdam, The Netherlands, 2018.
46. Robak, K.; Balcerek, M. Review of Second Generation Bioethanol Production from Residual Biomass. *Food Technol. Biotechnol.* **2018**, *56*, 174–187. [CrossRef]
47. Groves, C.; Sankar, M.; Thomas, P.J. Second-generation biofuels: Exploring imaginaries via deliberative workshops with farmers. *J. Responsible Innov.* **2018**, *5*, 149–169. [CrossRef]
48. COP26 Explained pdf. Available online: https://ukcop26.org/wp-content/uploads/2021/07/COP26-Explained.pdf (accessed on 30 January 2023).
49. D'Adamo, I.; Gastaldi, M.; Morone, P.; Rosa, P.; Sassanelli, C.; Settembre-Blundo, D.; Shen, Y. Bioeconomy of Sustainability: Drivers, Opportunities and Policy Implications. *Sustainability* **2022**, *14*, 200. [CrossRef]
50. A sustainable bioeconomy for Europe—Strengthening the Connection between Economy, Society and the Environment: Updated Bioeconomy Strategy | Knowledge for Policy. Available online: https://knowledge4policy.ec.europa.eu/publication/sustainable-bioeconomy-europe-strengthening-connection-between-economy-society_en (accessed on 30 January 2023).
51. Johnson, F.X.; Canales, N.; Fielding, M.; Gladkykh, G.; Aung, M.T.; Bailis, R.; Ogeya, M.; Olsson, O. A comparative analysis of bioeconomy visions and pathways based on stakeholder dialogues in Colombia, Rwanda, Sweden, and Thailand. *J. Environ. Policy Plan.* **2022**, *24*, 680–700. [CrossRef]
52. Bracco, S.; Calicioglu, O.; Gomez San Juan, M.; Flammini, A. Assessing the Contribution of Bioeconomy to the Total Economy: A Review of National Frameworks. *Sustainability* **2018**, *10*, 1698. [CrossRef]
53. Puder, J.; Tittor, A. Bioeconomy as a promise of development? The cases of Argentina and Malaysia. *Sustain. Sci.* **2023**, *18*, 617–631. [CrossRef]
54. Richter, S.; Szarka, N.; Bezama, A.; Thrän, D. What Drives a Future German Bioeconomy? A Narrative and STEEPLE Analysis for Explorative Characterisation of Scenario Drivers. *Sustainability* **2022**, *14*, 3045. [CrossRef]
55. Sanders, J.; Langeveld, H. A biobased economy for The Netherlands. In *The Biobased Economy: Biofuels, Materials and Chemicals in the Post-Oil Era*; Routledge: Abingdon, UK, 2009.
56. Hanekom, M.D. Bio-Economy Strategy. Available online: https://www.gov.za/sites/default/files/gcis_document/201409/bioeconomy-strategya.pdf (accessed on 27 April 2023).
57. Jimoh, A.A.; Lin, J. Biosurfactant: A new frontier for greener technology and environmental sustainability. *Ecotoxicol. Environ. Saf.* **2019**, *184*, 109607. [CrossRef] [PubMed]
58. Sałek, K.; Euston, S.R. Sustainable microbial biosurfactants and bioemulsifiers for commercial exploitation. *Process Biochem.* **2019**, *85*, 143–155. [CrossRef]
59. Leistritz, F.L.; Hodur, N.M. Biofuels: A major rural economic development opportunity. *Biofuels Bioprod. Biorefining* **2008**, *2*, 501–504. [CrossRef]
60. Gramegna, G.; Scortica, A.; Scafati, V.; Ferella, F.; Gurrieri, L.; Giovannoni, M.; Bassi, R.; Sparla, F.; Mattei, B.; Benedetti, M. Exploring the potential of microalgae in the recycling of dairy wastes. *Bioresour. Technol. Rep.* **2020**, *12*, 100604. [CrossRef]
61. Chandel, A.; Sukumaran, R. *Sustainable Biofuels Development in India*; Springer: Berlin/Heidelberg, Germany, 2017; p. 557.
62. Chandel, A.K.; Garlapati, V.K.; Jeevan Kumar, S.P.; Hans, M.; Singh, A.K.; Kumar, S. The role of renewable chemicals and biofuels in building a bioeconomy. *Biofuels Bioprod. Biorefining.* **2020**, *14*, 830–844. [CrossRef]

 *sustainability*

Review

Shaping the Knowledge Base of Bioeconomy Sectors Development in Latin American and Caribbean Countries: A Bibliometric Analysis

Maria Lourdes Ordoñez Olivo [1,*] and Zoltán Lakner [2]

1   Doctoral School of Economy and Regional Planning, Hungarian University of Agriculture and Life Sciences, 2100 Gödöllő, Hungary
2   Department of Agricultural Business and Economics, Institute of Agricultural and Food Economics, Hungarian University of Agriculture and Life Sciences, 1118 Budapest, Hungary; lakner.zoltan.karoly@uni-mate.hu
*   Correspondence: mordoniez0584@gmail.com or ordonez.lourdes.maria@uni-mate.hu; Tel.: +36-202209432

Abstract: Academic research on bioeconomy sectors in Latin American and Caribbean countries has developed exponentially over the last few years. Based on the Web of Science (WOS) database and statistical analysis of more than 18.9 thousand documents, the current article offers a bibliometric analysis of these datasets. The main bioeconomy sector identified in the results was biofuel production and all the background terms related to the primary processes of bioenergy. The other segments of the bioeconomy in the Latin America and Caribbean (LAC) region have not yet been studied with the same relevance as biofuels. Since 2008, researchers from Latin American and Caribbean countries have participated significantly in the scientific production of the field studied. However, the most relevant scientific journals belong to European countries or the United States. Journals from Latin American and Caribbean countries have very low representation, although the search topics are directly related to this region. Based on the co-occurrence of keywords, eight clusters with different levels of importance can be distinguished: (1) agriculture; (2) climate change; (3) biodiversity; (4) bioremediation; (5) bioenergy; (6) biofuels; (7) energy efficiency; and (8) bioeconomy. The above results highlight the significant research gap between biofuels and other types of bioeconomy sectors in the region. This is despite the immense biodiversity potential of the LAC countries, which can generate innovative products with bioeconomic added value that can stimulate scientific research in the sustainable bioeconomy.

Keywords: bioeconomy; LAC region; bibliometric analysis; biofuels; sustainability; R software

Citation: Ordoñez Olivo, M.L.; Lakner, Z. Shaping the Knowledge Base of Bioeconomy Sectors Development in Latin American and Caribbean Countries: A Bibliometric Analysis. *Sustainability* **2023**, *15*, 5158. https://doi.org/10.3390/su15065158

Academic Editors: Idiano D'Adamo and Massimo Gastaldi

Received: 29 January 2023
Revised: 9 March 2023
Accepted: 10 March 2023
Published: 14 March 2023

1. Introduction

According to the Global Bioeconomy Summit in 2018, bioeconomy refers to "the production, utilization, and conservation of biological resources, including related knowledge, science, technology, and innovation, to provide information, products, processes, and services across all economic sectors aiming towards a sustainable economy [1]".

One of the main goals of the bioeconomy is the reduction of non-renewable fossil energy use and its replacement by renewable resources [2–4]. Nevertheless, other objectives include linking all economic and industrial sectors that use biological resources and process them to produce food, feed, bio-based products, and services [5] or, in some cases, the optimization of the life cycle of the products and the creation of secondary markets for bio-based products [6].

According to Linser, S. (2020), many bioeconomy strategies are relevant to several SDGs (14 out of 17), making it a sustainable pathway to achieving the UN Sustainable Development Goals [7,8]. Furthermore, bioeconomy can be seen as a response to at least four emerging and converging global challenges: (a) growing global population; (b) increasing global demand for biomass (at least 60% above current rates), exacerbating the scarcity of natural resources; (c) growing evidence that the era of oil and cheap energy is coming to an end; and (d) concerns about climate change [4,8]. In summary, the relationship of the bioeconomy to SDGs and the global challenges can be grouped into three dimensions: socio-economic, environmental, industrial, and economic drivers [9,10].

These drivers are directly related to the sustainability aspects of the bioeconomy, towards which progress can be made when certain conditions are met [11]: "(i) sustainability of the resource base; (ii) sustainability of processes and products; and (iii) circular processes of material flows" [12]. In addition, the environmental and production components of bioeconomy development approaches need to be closely linked to how bio-resources are supplied, produced, and consumed [13].

The potential of the bioeconomy needs to be steered in the right direction to ensure that it works for people, food and nutrition security, and sustainable economic growth while preventing climate change and not harming the environment [14,15]. Therefore, some countries around the world contributed with significant knowledge, policy, and institutional efforts to develop bioeconomy strategies.

According to the German Bioeconomy Council [1], the bioeconomy has gained strength worldwide and is a certainty in many developed countries such as Germany, France, Finland, the Netherlands, Russia, and Japan [16]. At the beginning of 2018, nearly fifty countries have included a defined bioeconomy policy or strategy in their development plans or in their sub-regional procedures.

Nowadays, the bioeconomy has also been adopted by many low and middle-income countries as a new development concept and as part of their commitments to the Paris Climate Agreement [17]. In the case of Latin America and the Caribbean, a sustainable bioeconomy could open up new opportunities for economic development and industrialization and support economic and social goals [8,18].

Latin American and Caribbean countries have the most significant global endowments of natural capital because of their great diversity and natural resources, which are primarily the basis of their economies [19]. The region possesses the highest biomass production related to the availability of soil, water, and land [20]. Due to its high level of biodiversity, it tends to make a more significant contribution to the quality of life of people on average than other regions of the world [21,22].

In this context, the bioeconomy in Latin America and Caribbean countries has two main sets of objectives. On a global level, the region plays a critical role in contributing to global food, fiber, and energy balances, while improving environmental sustainability. Within the region's boundaries, the bioeconomy is a new source of opportunities for equitable growth through improved agricultural and biomass production [23,24].

Considering the comparative advantages and experiences in the countries of the Latin America and Caribbean region, Trigo et al. identified six distinct pathways that offer a holistic approach to the bioeconomy initiatives in the region. These six pathways include "(a) biodiversity resources exploitation; (b) eco intensification of agriculture, (c) biotechnology applications; (d) bio-refineries and bio-products, (e) value chain efficiency improvement; and (f) ecosystem services" [23].

The aim of this paper is to present a bibliometric analysis of the bioeconomy sectors developed in Latin American and Caribbean countries in recent years, based on the authors' evaluation criteria to determine the final products obtained from biomass processing, which in this study are biofuels, bioenergy, biotextiles, biocosmetics, and biopharmaceuticals. The countries considered in the analysis are those that have relevant bioeconomic approaches according to the revised bibliography (in the case of South America, Brazil, Argentina, Uruguay, Colombia, and Chile and in the case of the Caribbean, Mexico, Cuba, and Costa Rica).

2. Materials and Methods

In the present research, we applied the traditional bibliometric analysis. Figure 1 presents the workflow of the essential steps used in the dataset.

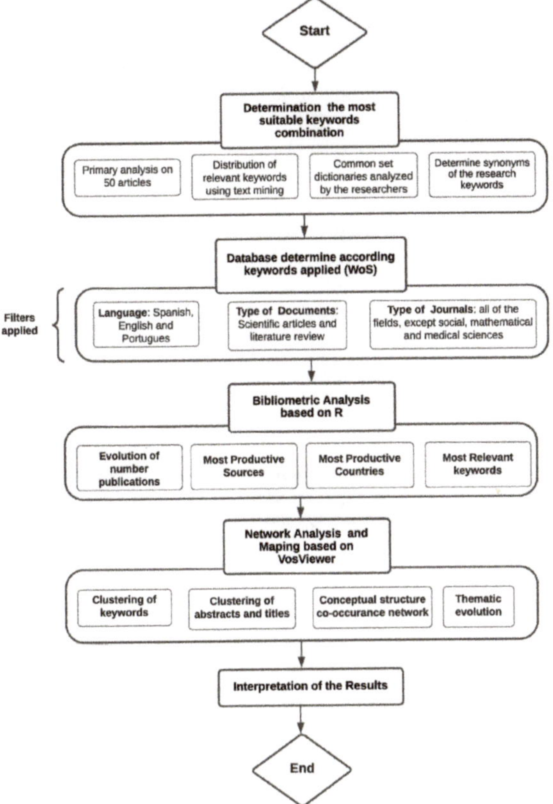

Figure 1. Flowchart of the research.

2.1. Data Sources and Collection

The bibliometric research has been carried out on the basis of the Web of Science database. A total of 18,971 documents were the subject of the analysis, and the time span of the publications under consideration was between 1977 and 2021.

To determine the most suitable keyword combination, the authors applied the parti pris concept [25]. In the first phase, we downloaded 50 articles using the simplest keyword combinations and then analyzed the distribution of relevant keywords using text mining methods. Subsequently, we set up a preliminary dictionary of potential keywords, which was carried out separately by each of the two co-authors. In the second phase, we determined the standard set of dictionaries and analyses the proportion of such words in the lexicon of each of the authors. At the end of this phase, we applied "Roget's Thesaurus" [26] to determine the potential synonyms of the research terms. After all these stages, we consider that we have achieved the type of keyword combination, which proved to be quite solid and robust, and which includes the main sectors of the bioeconomy developed in Latin American and Caribbean countries, with relevant bioeconomic initiatives related to the final products obtained from biomass.

The present research had the most reliable and interpretable results with the following keyword combination: TS = (((("bioet*") OR ("bioenergy*") OR ("biodies*") OR ("biogas*") OR ("short rotation crop*") OR ("biofuel") OR ("energytree*") OR ("energygaps") OR ("energyplantation") OR ("energy plantation") OR ("energy forest*") OR ("biomass*") OR ("biocosmetic*") OR ("bio-cosmetic*") OR ("biopharma*") OR ("biofiber*") OR ("biofibrer*")) AND ((("Brazil*") OR ("Brasil*")) OR ("Argentin*") OR ("Chile*") OR ("Uruguay*") OR ("Mexic*") OR ("Cuba*") OR ("Colombia*") OR ("Costa Rica*"))).

Given the large size of the corpus to be analyzed, it was decided to apply structural breaks according to the number of publications that had changed over time for the bioeconomy sectors developed in Latin America and the Caribbean. The structural breaks in the time series were determined using the algorithms of the Strucchange R-package [27], the specialist estimation Z.L, and the econometric time-series analysis carried out by the econometric software Gretl (ver. 2022.c-64) [28]. On this basis, four periods were identified: 2000, 2001, and 2007, 2008 and 2014, and from 2015 to 2021. The interpretation and justification of these four periods are based on the milestone years in various historical datasets that have been researched for this purpose (Table 1), in addition to the results of the mathematical and statistical methods carried out using R and Gretl software. The mentioned sources provide statistical data regarding the established breakpoints and the period variance (Appendix A, Figures A1–A3).

Table 1. Historical records of three different variables which support the breakpoints established in the study.

Milestones Years	Public Investment in Bioenergy Sector (Million Dollars)	Renewable Share (Modern Renewables) in Final Energy Consumption (Percentage)	Primary Energy Supply from Clean Renewable Sources in LAC and the Caribbean (Thousands of Barrels of Oil Equivalent)
Up to 2000	13.24	6.86	431,710.1
2001–2007	126.52	7.48	551,923.1
2008–2014	7682	9.04	650,628.1
2015–2021	4046.48	10.88	740,115.5

Prior to the year 2000, the bioeconomy sectors in the world were taking their first steps towards development, in particular in regions such as Latin America and the Caribbean. According to the International Renewable Agency [29], public investment in the bioenergy sector did not exceed 13.6 million. This indicator is also related to the consumption of final energy from renewable sources, which was 0.62% lower than the next established breakpoint [30,31]. Since 2001, there has been evidence of significant changes in the level of consumption of biofuels at the global level and in the clean supply of renewable sources in the countries of Latin America and the Caribbean [32,33]. Finally, a significant increase in bioeconomy sectors worldwide occurred between 2008 and 2014; this could also be directly related to the policies and strategies proposed in EEUU, Canada, Germany, Austria, and Finland [34,35].

2.2. Data Analysis

In order to analyze and visualize the corpus data, a detailed bibliometric analysis was carried out using the Bibliometrix R-package. This program provides a wide variety of statistical functions (linear and non-linear modeling, classical statistical tests, time-series analysis, classification, clustering) and graphical techniques [36,37]. To complement the statistical analysis of the research, we also used the VOSviewer 1.6.18 software [38–40]. Table 2 shows the statistical indicators applied in the present research.

It is important to emphasize that the data obtained are the result of a global search for scientific production in the periods indicated and follow the keywords defined as the most appropriate for this research.

Table 2. Software tools applied to the corpus for statistical analysis.

Type of Software	Type of Statistical Analysis
Gretl econometric software	• Econometric analysis of the structural break points of the main dataset base based on the article's years of production
Bibliometric (biblioshiny)	• Yearly academic production. • Most relevant sources in the field of Bioeconomy in relation to the Latin America and Caribbean region. • Evolution of authors' keywords in the bioeconomy literature [41].
VOSviewer	• Co-occurrence of different keywords and expressions applying a combination of clusters in the four periods to analyze [42].

2.3. Limitations

During the keyword search process, after following the guidelines in Figure 1, we noted that we had achieved the type of keyword combinations that were robust and firm enough to achieve the results to be displayed. The previous statement indicates that subtracting or adding a less relevant keyword could not considerably influence the number of results.

It is important to note that, as described in the methodology, the definition of the search terms was based on the previous review of 50 scientific articles that showed a priority focus on one of the most technologically and economically developed sectors in the region, i.e., biofuels. A few initiatives related to other biomass-derived products could also be considered relevant sectors. Therefore, only those initiatives identified in the keyword search were included.

However, as authors, we are aware that our research only covers some sectors currently considered part of the bioeconomy, which may have been further developed and researched. Similarly, we know this research only covers some relevant articles on the subject, as it is extensive and has grown exponentially in recent years.

3. Results

3.1. General Characteristics of the Corpus

The corpus of this research contains 18,971 documents. It represents a global search of scientific production in the four periods indicated, according to the keywords used. The published documents are analyzed according to the authors, the institutions linked to the corresponding countries, and the average number of citations of the articles at the world level [43]. In this set of data, the average number of citations per article per year is relatively high (2.068).

Figure 2 shows the global scientific publications between 1977 and 2021 in relation to the bioeconomy sectors developed in the LAC region. The high level of interest is reflected between the years 2000 and 2021, with 91% of the scientific production concentrated within the evaluation period of 44 years. This interest is also directly related to the investments made in LAC countries in the final products derived from biomass processing, such as biofuels and others. According to the International Renewable Energy Agency [41], there has been an 11 percentage point increase in investment in renewable energy in LAC since 2004, compared to a 6 percentage point increase worldwide. Countries such as Brazil, Argentina, Mexico, and Chile have joined the list of the world's top 10 renewable energy markets.

Table 3 shows an analysis of the cumulative share of publications in the corpus, based on the global search of scientific output in the analyzed period. This cumulative share is a measure constructed from the publication frequency of the corresponding authors (intra- and inter-CCP country) during the four periods analyzed. In the early years of bioeconomy development in LAC, it was dominated by the United States and four representative South

American countries: Brazil, Mexico, Argentina, and Chile. The European countries of Germany and the United Kingdom appear in sixth and seventh place among the producing countries in the following period analyzed. From 2008 onwards, there is a rapid increase in the participation of other South American countries, such as Colombia and Cuba, and the appearance of major Asian countries, such as China and India. In recent years, the share of Latin American countries has grown even faster: five countries are among the 20 most productive.

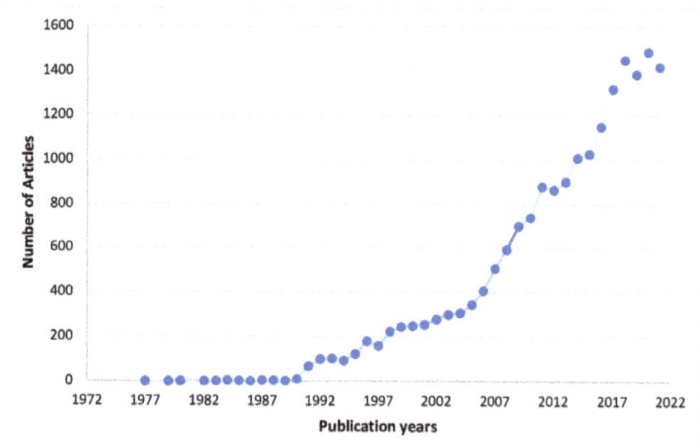

Figure 2. Scientific publications between 1977 and 2021 related to bioeconomy sectors developed in Latin America and Caribbean region.

Table 3. The cumulative share of the top 15 countries in publications with corresponding authors' contribution to bioeconomy sectors (countries indicated by 3-digit ISO codes).

1977–2000		2001–2007		2008–2014		2015–2021	
Country	Cum. Share of Publications (%)	Country	Cum. Share of Publications (%)	Country	Cum. Share of Publications (%)	Country	Cum. Share of Publications (%)
USA	0.28	USA	0.24	BRA	0.37	BRA	0.42
BRA	0.44	BRA	0.46	USA	0.52	MEX	0.54
MEX	0.57	MEX	0.57	MEX	0.61	USA	0.63
ARG	0.68	ARG	0.68	ARG	0.70	ARG	0.69
CHL	0.74	CHL	0.74	CHL	0.76	CHL	0.74
DEU	0.78	DEU	0.79	COL	0.79	COL	0.79
GBR	0.81	GBR	0.80	DEU	0.81	DEU	0.81
FRA	0.84	CAN	0.82	GBR	0.82	CHN	0.83
CAN	0.85	CUB	0.83	ESP	0.84	GBR	0.84
CRI	0.86	FRA	0.85	FRA	0.85	ESP	0.86
COL	0.88	ESP	0.86	CHN	0.87	CRI	0.87
NLD	0.89	COL	0.88	CAN	0.88	IND	0.88
ESP	0.90	AUT	0.89	NTD	0.89	NTD	0.89
IND	0.91	CRI	0.90	IND	0.90	URY	0.90
AUT	0.92	ITA	0.90	CRI	0.91	CUB	0.91

The above data are consistent with González, C. et al., (2016) findings. They show that developing countries are narrowing their science gap, with R&D investment and scientific impact growing at more than twice the rate of the developed world. However, among the countries assessed, the scientific output and impact are relative to their level of investment and the resources available to them and are not necessarily being carried out in an efficient manner [44].

Figure 3 shows the temporal changes during the four implementation periods of the bioeconomy sectors in Latin America and the Caribbean. It is based on the frequency of publications per country over time and interpreted in the territorial maps. In all the periods analyzed, Brazil and the United States were in the lead in terms of scientific production, followed by Central and South American countries such as Mexico and Argentina. The main difference lies in the frequency fluctuation of publications (expressed as a percentage on the map) among these key countries. Another interesting process observed in the last ten years is the active role of countries such as Brazil, which is the leading producer, accounting for almost 42% of the total number of publications, followed by the United States (17%), Mexico (13%), Argentina (9%), and Chile (7%). All these data highlight the global nature of bioeconomy sectors and the importance of Latin American countries in scientific production on this subject.

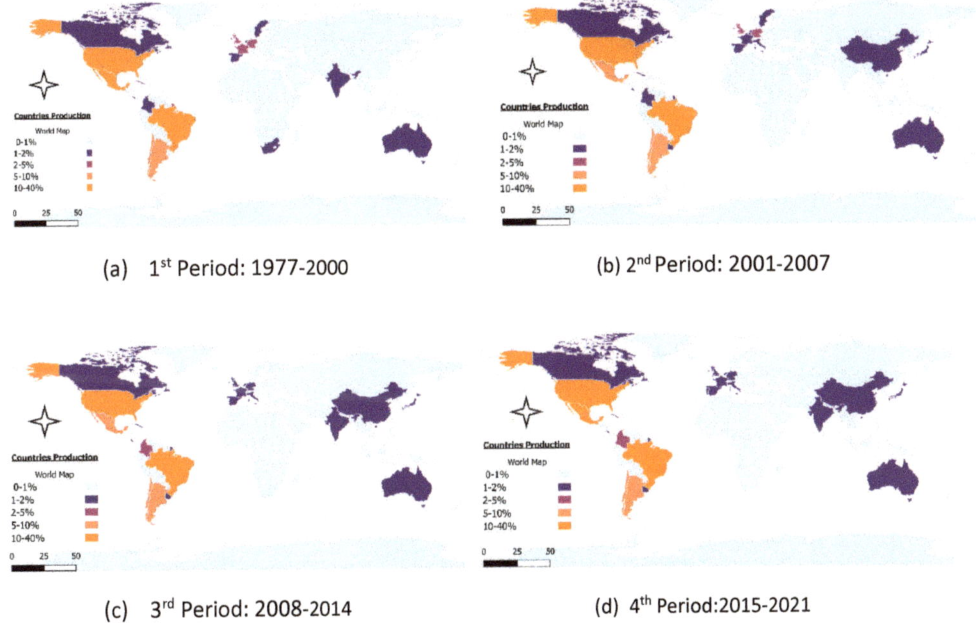

Figure 3. Temporal changes in the spatial distribution of bioeconomy-related publications measured by countries' production percentage (Map created in R with base map courtesy of OpenStreetMap).

Based on the number of scientific articles published per journal, Table 4 shows the most relevant academic journals. For the purposes of this analysis, the descendant rank of the journal and the country of origin are taken into account accordingly. It is interesting to note that developed countries account for a higher proportion of top scientific research articles and have a robust research impact in this field. European countries such as the Netherlands, the United Kingdom, and Germany, together with the United States, are in the top 10 of the journal spectrums. Contrary to the previous analysis of country production, in the case of top journals, only a few belong to South American countries, such as Brazil, with the highest participation, followed by Chile and Costa Rica.

Table 4. The 14 most relevant academic journals in the 4 periods evaluated in the field of bioeconomy sectors in the Latin America and Caribbean region.

| 1977–2000 | | 2001–2007 | | 2008–2014 | | 2015–2021 | |
Name Journal	Country	Name Journal	Country	Name Journal	Country	Name Journal	Country
Journal of Geophysical Research-Atmospheres	USA	Forest Ecology and Management	NLD	Forest Ecology and Management	NLD	Journal of Cleaner Production	UK
Hydrobiologia	NLD	Journal of Geophysical Research-Atmospheres	USA	Atmospheric Chemistry and Physics	DEU	Forest Ecology and Management	NLD
Marine Ecology Progress Series	DEU	Hydrobiologia	NLD	Revista Brasileira de Ciencia do Solo	BRA	Plos One	USA
Revista de Biologia Tropical	CRI	Marine Ecology Progress Series	DEU	Biomass and Bioenergy	UK	Renewable and Sustainable Energy Reviews	UK
Forest Ecology and Management	NLD	Revista de Biologia Tropical	CRI	Energy Policy	UK	Science of the Total Environment	NLD
Pesquisa Agropecuaria Brasileira	BRA	Revista Brasileira de Ciencia do Solo	BRA	Plos One	USA	Biomass and Bioenergy	UK
Biotropica	USA	Atmospheric Environment	UK	Revista de Biologia Tropical	CRI	Industrial Crops and Products	NLD
Oecologia	USA	Ecological Applications	USA	Renewable and Sustainable Energy Reviews	UK	Renewable Energy	UK
Revista Chilena de Historia Natural	CHL	Biotropica	USA	Hydrobiologia	NLD	Remote Sensing	USA
Plant And Soil	NDL	Ecological Modeling	NLD	Brazilian Journal of Biology	BRA	Sustainability	CHE
Journal of Range Management	USA	Plant And Soil	NLD	Latin American Journal of Aquatic Research	CHI	Environmental Science and Pollution Research	DEU
Journal of Tropical Ecology	UK	Field Crops Research	NLD	Atmospheric Environment	UK	Forests	CHE
Field Crops Research	NDL	Journal ff Arid Environments	USA	Pesquisa Agropecuaria Brasileira	BRA	Revista de Biologia Tropical	CRI
Soil Biology and Biochemistry	UK	Global Change Biology	UK	Plant And Soil	NLD	Fuel	NLD

This phenomenon can also be explained in terms of the gross domestic expenditure on R&D that countries invest on an annual basis. In 2018, North America and Western Europe invested around 2.5% of their GDP, while Latin America and the Caribbean invested only 0.7%, according to the Unesco Global R&D Investment Report [45]. If we analyze the countries of Latin America and the Caribbean, Brazil is the country that invests the most, with 1.7% of GDP, and is ranked 9th among the top 10 countries in the world for investing in R&D. Several studies show that there is a strong positive correlation between R&D expenditure and scientific production [46,47]. Melo et al. conclude that countries that spend more on R&D have more universities and ISI-indexed journals and produce a significant volume of research papers.

3.2. Analysis of Keywords and Co-Keywords

Figure 4 shows the frequency of the different keywords over time. Aspects related to "biomass" and its different variants and concepts such as "biodiversity" reflect permanent growth. Terms related to climate change factors, such as land use change, deforestation, and degradation, are another group of terms identified in the corpus. Keywords associated with renewable fuels reflect their growth in importance over the last 15 years. Finally, terms such as "sustainability" and "life cycle assessment" show less growth. However, these latter terms have a direct cross-cutting relationship with most of the topics analyzed and are fundamental concepts for compliance with Sustainable Development Goals.

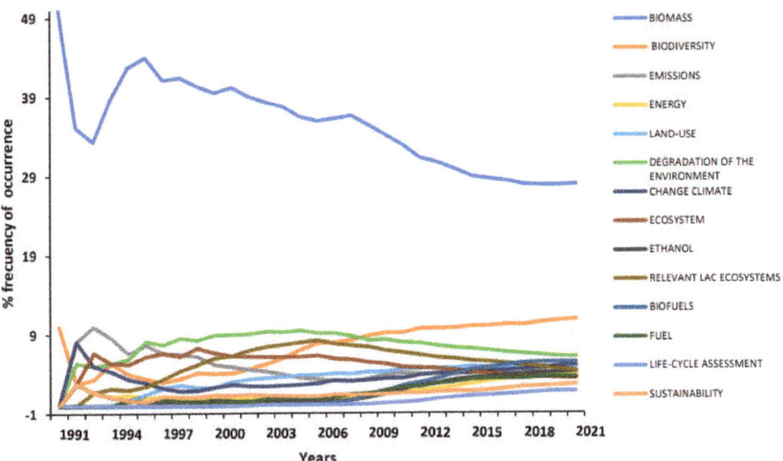

Figure 4. Frequency of occurrences of keywords on the corpus " −1".

The analysis of the second body is based on the frequency of the relevant words used by the authors during the period described (Figure 5). These words are directly related to the bioeconomy sectors developed in Latin America and the Caribbean. Some keywords appear exponentially as "biofuels" or "biogas", which shows the growing sensitivity of the academic environment towards renewable energy as the most important bioeconomy sector that has been established in this region.

For the purposes of this study, "biomass production" includes other sectors of the bioeconomy (e.g., biocosmetics, biopharmaceuticals) which do not have the same relevance and scientific production compared with biofuels.

Interestingly, terms such as "bioethics" appear with greater frequency from 2006 to 2016 as part of the glossary used by the authors, demonstrating the importance of including moral principles and values in all scientific research. It emphasizes the balance to be struck between ethical principles, technological possibilities, and several conflicting human needs, such as producing food and, in particular, renewable energy based on first-generation biofuels [48].

Finally, the term "REDD" appears with minority participation, although it is a relevant concept for the sustainable management of ecosystems, especially for developing countries such as those in the Latin America and Caribbean region.

3.3. Clustering of Research Topics Based on Co-Occurrence of Keywords

As mentioned in the previous sections, there have been significant changes in the research area in recent years. For this reason, a co-occurrence-based analysis was carried out in the VOSviewer software between the years 2015 and 2020. Figure 6 shows a summary of the results obtained.

The analysis performed makes it possible to differentiate eight-dimensional coordinate clusters. These clusters are interconnected and have different levels of importance according to the number of items they contain.

The largest cluster in terms of word number includes Agriculture and Soil Research (No. 1, shown by red color). In this cluster, soil management and soil properties are key factors for agriculture productivity. The same applies to biomass production systems, which deal with the use of agricultural land for bioenergy production. In terms of the relationship between sustainability and agriculture, the cluster contains important keywords such as conservative, sustainable, and functional diversity.

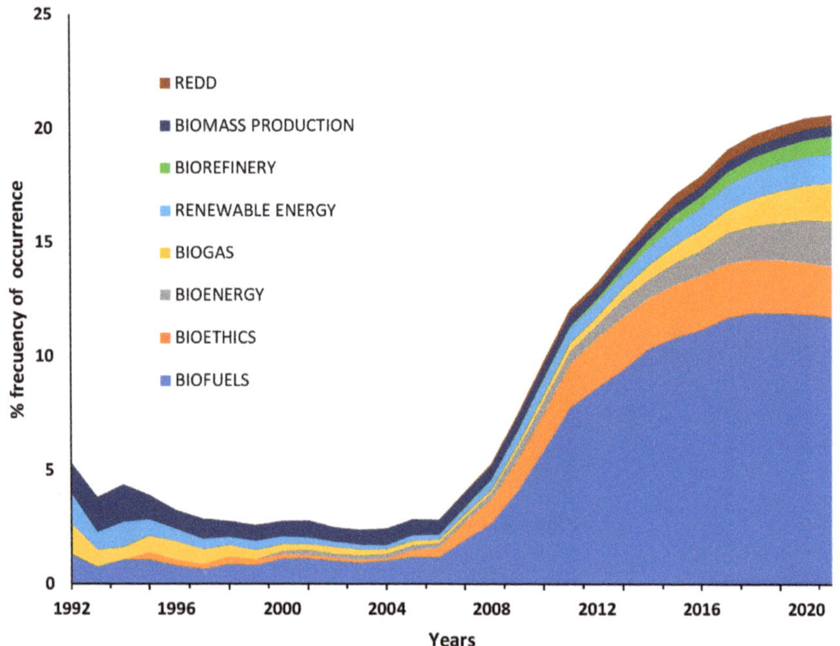

Figure 5. Frequency of occurrences of specific authors' keywords relevant to the corpus " −2".

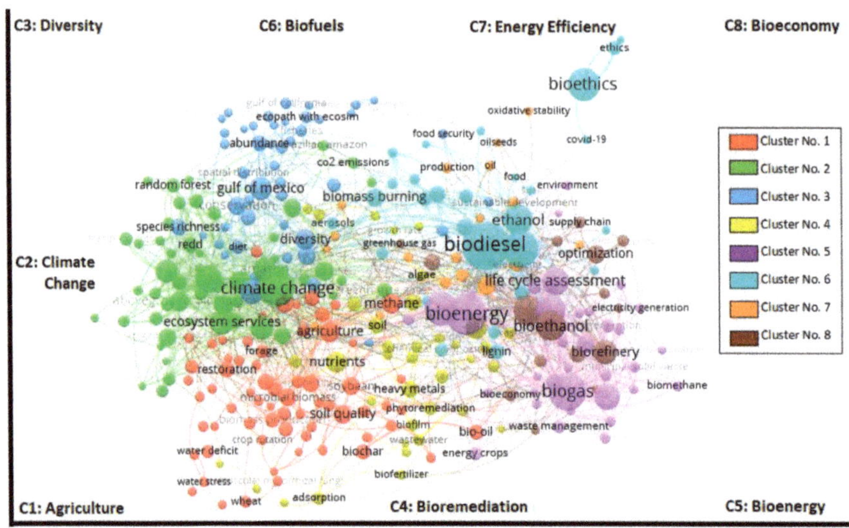

Figure 6. Key clusters of bioeconomy sectors developed in Latin America and the Caribbean.

The second cluster, indicated by green color (No. 2) consists of 37 items, most of which are related to factors that affect climate change. This cluster highlights the relationship between biomass production strategies, particularly in the context of bioenergy and the impact of climate change.

The third cluster (No. 3, colored blue) deals with diversity and conservation aspects, including important ecosystems located in Latin America and the Caribbean, where high biodiversity indicators are one of the main features. The fourth cluster, shown in yellow, represents the different aspects of bioremediation, highlighting the current techniques used and how they are linked to sustainable agriculture and ecological restoration.

The fifth and sixth clusters are the highest in terms of word frequency and group the most related bioeconomy keywords. Containing a total of 64 words, these items focus on bioenergy and biofuel production as the most important bioeconomy sector developed in Latin American and Caribbean countries. The fifth cluster highlights the environmental impact and the life cycle assessment as evaluation methodologies for this type of model. Bioethics, food security, and COVID-19 appear in cluster six as relevant and topical issues.

The seventh cluster, marked in orange, refers to energy efficiency, the production of oilseeds, and concepts related to the sustainable development of the bioeconomy in Latin America and the Caribbean.

Finally, the eighth cluster brings together in a single word ("bioeconomy") a holistic approach to the previous clusters and summarizes, in a few keywords, the concepts of biomass production (such as biofuels), which is considered the most important sector identified in the region in this study.

In conclusion, it can be said that the bioeconomic sectors developed in Latin America, according to the scientific articles evaluated in this research, are concentrated in the production of bioenergy. The backbone of the field studied is the production of various biofuels, with leading countries such as Brazil and Argentina, which are considered the largest producers in the region. Although other types of biomass production in LAC were included in the keyword search, no words or clusters were found in the results.

3.4. Mapping of Topic-Evolution

Figure 7 examines the evolution of the topic map of the main research directions in recent years. Among the motor themes, three basic directions can be observed: firstly, the growing importance of bioenergy as a general category, including biofuels and other types of renewable energy; secondly, biomass as a primary source for bioenergy production; and thirdly, climate change as a cross-cutting theme of the previous ones since it is directly related to the production of biofuels as a strategy to mitigate the climate impact of fossil fuels. No basic topics are reported in this period of the research.

According to Plaza-Delgado E. et al., an alternative to reduce the consumption of fossil fuels is the use of biomass as a source of energy, especially in the Latin American and Caribbean region, which has great potential due to its diverse sources of biomass [49]. The study by Bailis R. et al. points out that this region is a world leader in the production of biofuels, accounting for 27% of the world's supply [50]. However, it is important to consider that the production of biofuels could mean an expansion of the production frontier, which poses a serious challenge to the region's environment and biodiversity [51].

Finally, bioethics is still present as an emerging theme in the analysis; according to Gutierrez-Prieto, H in the last thirty years, Bioethics, as a developing discipline, has obtained gradual and increasing worldwide recognition, not only for its novelty but also for the connection with the future research topics [39]. In the same vein, the Nuffield Council on Bioethics highlighted the ethical issues raised by current and future approaches to biofuel development, as global biofuel production indirectly has serious negative impacts on agricultural and food sustainability [52].

In conclusion, scientific research on the region's bioeconomic sectors in the period 2015 to 2021 has focused on bioenergy as a fundamental strategy in both ways, on the one hand in relation to the production of biofuels as an energy source, and on the other as a transitional energy model for some Latin American and Caribbean countries.

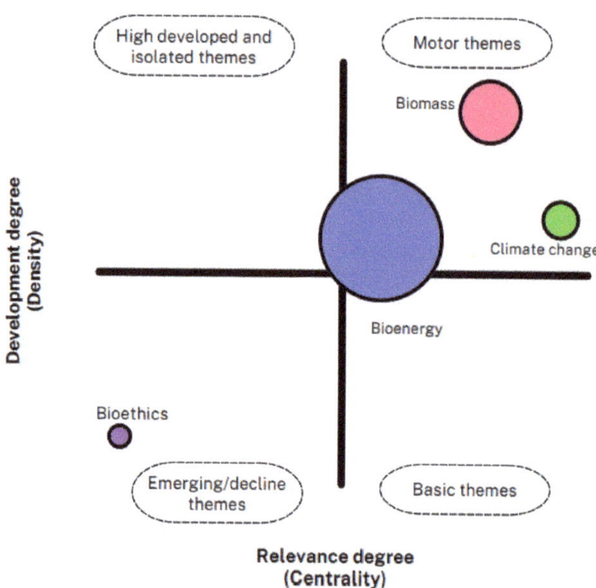

Figure 7. Science map of research topics from 2015 to 2021.

4. Discussion

The results of the bibliometric analysis have highlighted the importance of biofuels as the most important sector of the bioeconomy developed in the Latin America and Caribbean countries evaluated. This finding is in support of the fact that the number of relevant publications in this area has been growing exponentially. Aydogan H. et al. pointed out that biofuels have been rapidly gaining prominence due to their continuous increase in economic value and, at the same time, less harmful effects on the environment [6,30,53].

According to the IICA study, by 2020 Brazil will be the world's second-largest producer of liquid biofuels, with a 23 percent share, behind the United States [54]. On the other hand, Argentina has a significant share in world biodiesel production, with around 7 percent, followed by Thailand, Colombia, and Paraguay [55,56].

Our research has also highlighted the relationship between biofuels and land use, particularly the crops used as feedstock for their production, and their transversal link to the region's ongoing concerns about food security and the sustainable development of such products. The World Bank's 2008 Development Report of Agriculture Development notes that the major challenge for governments in developing countries, such as those in Latin America and the Caribbean, is to "implement regulations and to develop certification systems that reduce the environmental and food security risks of biofuel production" [57].

The sustainability of agricultural land use for biofuel production is one of the priority issues to be discussed in the future, given the importance of biofuels as a major sector in the region. According to UNCTAD, biofuels compete directly with existing arable and grazing land for food production [58]. Moreover, bioenergy crops can lead to agricultural expansion, competition for water, and threats to biodiversity, especially in rural LAC areas with high ecological and social vulnerability [59].

Climate change was another cluster identified within the keyword analysis that is directly related to biofuels. Jeswani H et al. [60] point out that biofuels do not exist in isolation and, like other production systems, have an impact on various ecosystem services such as land, water, and food. In addition, authors such as Prasad, S. et al. [61] point out that producing biofuels from biomass has the potential to promote sustainable development and mitigate climate change while providing socio-economic benefits.

In terms of scientific publications, Latin American and Caribbean countries such as Brazil, Argentina, Mexico, and Chile are among the top twenty countries in terms of scientific production over the last period. This reflects a positive evolution in the concentration of developing countries as producers of scientific publications, compared to previous years when developed countries such as the United States had a significant and majority participation. However, the research shows that in terms of the most relevant scientific journals, the majority of them are from European countries or the United States. The representativeness of journals from Latin American and Caribbean countries is very low, even though the topics covered are directly related to this region.

With regard to the other types of bioeconomy sectors based on the final products obtained from biomass, we did not find any significant scientific publications in the Latin American and Caribbean countries evaluated during the period under review. However, the region has a wide and diverse range of renewable natural resources that could provide the essential basis for the development of a competitive bioeconomy [62] and the production of innovative products with added bioeconomic value [63].

Within the clustering of research topics, we could identify important keywords such as 'bioethics', 'food security', and 'COVID', which are transversally related to the bioeconomy sectors and also represent current topics in the scientific fields. According to Woźniak E et al., the "COVID crisis may be the driving force for the global integration related to bioeconomy, especially in implementing the SGD goals, development of national and regional bioeconomy strategies, ensuring food security and protecting biodiversity" [64]. Regarding bioethics as an emerging term, several bibliographical references indicate its importance over time and its close relationship with bioeconomy sectors, especially those environmental and sustainable aspects that seek social agreements to support human well-being while preserving the natural environment [65].

Finally, the structural breakpoints in the research database have been able to indicate the importance of bioeconomy sectors over time. It is noteworthy that since 2013, several European countries, including Germany, Spain, and Finland, proposed major policies to develop their bioeconomy, followed by several public policy documents and research papers covering different aspects of bioeconomy in developing countries at local and national levels [66].

5. Conclusions

The study highlights the importance of biofuels as the most important bioeconomy sector developed in the Latin American and Caribbean countries evaluated. Brazil and Argentina are the region's main producers and rank first in the world. Even though the region has a wide and diverse range of renewable natural resources, we were not able to find any significant scientific publications on other types of bioeconomy sectors that are based on the final products obtained from biomass.

Based on the co-occurrence keywords, our research has also shown the relationship between the eight identified clusters: (1) agriculture; (2) climate change; (3) diversity; (4) bioremediation; (5) bioenergy; (6) biofuels; (7) energy efficiency; and (8) bioeconomy. The first seven are linked to the region's current concerns for food security and the sustainable development of bioenergy production, while the eighth relates to the holistic approach of the research. In terms of scientific production in recent years, Brazil, Argentina, Mexico, and Chile are at the top positions. However, the research shows that the most relevant scientific journals belong to developed European countries or the United States.

In conclusion, in the LAC countries under review, there has been a significant increase in scientific production in relation to bioeconomy sectors over the last 15 years, with the main focus of research on biofuel production as the main source of bioenergy. The above results highlight the significant research gap on other types of innovative products with bioeconomic added value that could be generated in the region, given its immense biodiversity potential.

Author Contributions: Conceptualization, Z.L. and M.L.O.O.; methodology, Z.L.; software, Z.L. and M.L.O.O.; validation, Z.L.; formal analysis, M.L.O.O.; investigation, M.L.O.O.; data curation, Z.L. and M.L.O.O.; writing—original draft preparation, M.L.O.O.; writing—review and editing, Z.L. and M.L.O.O.; supervision, Z.L. All authors have read and agreed to the published version of the manuscript.

Funding: This research received no external funding.

Institutional Review Board Statement: Not applicable.

Informed Consent Statement: Not applicable.

Data Availability Statement: The data presented in this study are available on request from the corresponding author. The data are not publicly available due to the intellectual property of the ISI enterprise.

Acknowledgments: The authors are particularly grateful to the Hungarian University of Agriculture and Life Sciences for the outstanding informatics support in the access to different databases requested for this research article, as well as the use of statistical software, need it to analyze the data presented.

Conflicts of Interest: The authors declare no conflict of interest.

Appendix A

Statistical analysis completed by using "Gretl software" to corroborate the structural breakpoints established by the R-package of the main dataset and the historical records mentioned in Table 1.

Model 1: OLS using observations 1997-2021 (T=45)

Depended Variable: Years Production

	Coefficient	std. error	t-radio	p-value
Const	-357.033	59.4080	-6.010	3.54×10^{7} ***
Time	33.8536	2.24916	15.05	9.54×10^{19}***
Mean depend variable	421.6000		S.D. depend var	484.9947
Sum squared resid	1651018		S.E. of regression	195.9484
R- squared	0.840476		Adjusted R-squared	0.836766
F (1,43)	226.5524		P- values (F)	9.54×10^{19}
Long likelihood	-300.3326		Akaike criterion	604.6653
Schwarz criterion	608.2786		Hannan-Quinn	606.0126
Rho	0.949934		Durbin-Watson	0.072092

Chow test for structural break at observation 2000 –
  Null Hypothesis: no structural break
  Test Statistic: F (2, 41) = 208.784
  with p-value= P (F (2, 41) > 208.784) = 3.18626×10^{22}
Chow test for structural break at observation 2007 –
  Null Hypothesis: no structural break
  Test Statistic: F (2, 41) = 272.063
  with p-value= P (F (2, 41) > 272.063) = 2.15505×10^{24}
Chow test for structural break at observation 2014 –
  Null Hypothesis: no structural break
  Test Statistic: F (2, 41) = 36.8912
  with p-value= P (F (2, 41) > 36.8912) = 6.83324×10^{10}

Figure A1. Article year's production of the main dataset (***: $p \leq 0.001$).

Model 1: OLS using observations 1997-2020 (T=44)

Depended Variable: PrimaryEnergySupplyLAC

	Coefficient	std. error	t-radio	p-value
Const	124266	7989.66	15.55	4.92×10^{19} ***
Time	13.993.4	309.245	45.55	2.88×10^{37} ***

Mean depend variable	439117.9	S.D. depend var	181582.8
Sum squared resid	2.85×10^{10}	S.E. of regression	26248.26
R- squared	0.979900	Adjusted R-squared	0.979422
F (1,43)	2847.587	P- values (F)	2.88×10^{37}
Long likelihood	-507.7889	Akaike criterion	1021.578
Schwarz criterion	1025.146	Hannan-Quinn	1022.901
Rho	0.794095	Durbin-Watson	0.387197

Chow test for structural break at observation 2000 –
  Null Hypothesis: no structural break
  Test Statistic: F (2, 40) = 28.029
  with p-value= P (F (2, 40) > 28.029) = 2.48453×10^{08}
Chow test for structural break at observation 2007 –
  Null Hypothesis: no structural break
  Test Statistic: F (2, 40) = 30.0038
  with p-value= P (F (2, 40) > 30.0038) = 1.097084×10^{08}
Chow test for structural break at observation 2014 –
  Null Hypothesis: no structural break
  Test Statistic: F (2, 40) = 3.6204
  with p-value= P (F (2, 40) > 3.6204) = 0.0358802

Figure A2. Primary energy supply from renewable sources in Latin American and Caribbean countries. (***: $p \leq 0.001$).

Model 1: OLS using observations 1997-2019 (T=30)

Depended Variable: GlobalRenewableConsumption

	Coefficient	std. error	t-radio	p-value
Const	5.66437	0.187839	30.16	6.83×10^{23} ***
Time	0.163804	0.0105807	15.48	2.96×10^{15} ***

Mean depend variable	8.203333	S.D. depend var	1.523943
Sum squared resid	7.045141	S.E. of regression	0.501610
R- squared	0.895395	Adjusted R-squared	0.891659
F (1,43)	239.6725	P- values (F)	2.96×10^{15}
Long likelihood	-20.83527	Akaike criterion	45.67053
Schwarz criterion	48.47293	Hannan-Quinn	46.56704
Rho	0.942290	Durbin-Watson	0.175627

Chow test for structural break at observation 2000 –
  Null Hypothesis: no structural break
  Test Statistic: F (2, 26) = 36.5613
  with p-value= P (F (2, 26) > 36.5613) = 2.78235×10^{08}
Chow test for structural break at observation 2007 –
  Null Hypothesis: no structural break
  Test Statistic: F (2, 26) = 141.826
  with p-value= P (F (2, 26) > 141.826) = 1.03109×10^{14}
Chow test for structural break at observation 2014 –
  Null Hypothesis: no structural break
  Test Statistic: F (2, 26) = 26.554
  with p-value= P (F (2, 26) > 26.554) = 5.2215×10^{07}

Figure A3. Renewable share (modern renewables) in final energy consumption, word wide (***: $p \leq 0.001$).

References

1. Global Bioeconomy Summit 2018. Communiqué Global Bioeconomy Summit 2018 Innovation in the Global Bioeconomy for Sustainable and Inclusive Transformation and Wellbeing. 2018. Available online: http://gbs2018.com/resources/ (accessed on 5 May 2022).
2. Hodson De Jaramillo, E.; Henry, G.; Trigo, E. *Bioeconomy New Framework for Sustainable Growth in Latin America*; Editorial Pontificia Universidad Javeriana: Bogota, Colombia, 2019.
3. Papadopoulou, C.-I.; Loizou, E.; Melfou, K.; Chatzitheodoridis, F. The knowledge based agricultural bioeconomy: A bibliometric network analysis. *Energies* **2021**, *14*, 6823. [CrossRef]
4. Perišić, M.; Barceló, E.; Dimic-Misic, K.; Imani, M.; Brkić, V.S. The Role of Bioeconomy in the Future Energy Scenario: A State-of-the-Art Review. *Sustainability* **2022**, *14*, 560. [CrossRef]
5. European Commission. Mainstreaming the Bioeconomy. EU Rural Review. 2019. Available online: https://enrd.ec.europa.eu (accessed on 25 May 2022).
6. Dorokhina, E.Y.; Kharchenko, S.G. Business models of the circular economy as mechanism of sustainable development achievement. *Ecol. Ind. Russ.* **2017**, *21*, 58–61. [CrossRef]
7. Linser, S.; Lier, M. The Contribution of Sustainable Development Goals and Forest-Related Indicators to National Bioeconomy Progress Monitoring. *Sustainability* **2020**, *12*, 2898. [CrossRef]
8. Food and Agriculture Organization of the United Nations. Assessing the Contribution of Bioeconomy to Countries Economy. 2018. Available online: www.fao.org/publications (accessed on 25 May 2022).
9. Diemer, A.; Batisse, C.; Gladkykh, G.; Bennich, T. Role of Bioeconomy in the Achievement of Sustainable Development Goals. In *Partnerships for the Goals: Encyclopedia of the UN Sustainable Development Goals*; Leal Filho, W., Marisa Azul, A., Brandli, L., Lange Salvia, A., Wall, T., Eds.; Springer: Cham, Switzerland, 2021; pp. 1054–1067.
10. Heimann, T. Bioeconomy and SDGs: Does the Bioeconomy Support the Achievement of the SDGs? *Earths Future* **2019**, *7*, 43–57. [CrossRef]
11. D'Adamo, I.; Gastaldi, M.; Morone, P.; Rosa, P.; Sassanelli, C.; Settembre-Blundo, D.; Shen, Y. Bioeconomy of Sustainability: Drivers, Opportunities and Policy Implications. *Sustainability* **2021**, *14*, 200. [CrossRef]
12. Gawel, E.; Pannicke, N.; Hagemann, N. A Path Transition Towards a Bioeconomy—The Crucial Role of Sustainability. *Sustainability* **2019**, *11*, 3005. [CrossRef]
13. Bracco, S.; Calicioglu, O.; Gomez San Juan, M.; Flammini, A. Assessing the Contribution of Bioeconomy to the Total Economy: A Review of National Frameworks. *Sustainability* **2018**, *10*, 1698. [CrossRef]
14. Olivier, D.; Martha, G.S.J. How Sustainability Is Addressed in Official Bioeconomy Strategies at International, National, and Regional Leveles—An Overview. 2016. Available online: https://landportal.org/node/50172 (accessed on 7 March 2023).
15. Federal Ministry of Food and Agriculture. National Policy Strategy on Bioeconomy. 2014. Available online: www.bmel.de (accessed on 7 March 2023).
16. Iriarte, L. *Sustainability Governance of Bioenergy and the Broader Bioeconomy*; IINAS—International Institute for Sustainability Analysis and Strategy: Pamplona, Spain, 2021.
17. Delzeit, R.; Heimann, T.; Schuenemann, F.; Söder, M.; Zabel, F.; Hosseini, M. Scenarios for an impact assessment of global bioeconomy strategies: Results from a co-design process. *Res. Glob.* **2021**, *3*, 100060. [CrossRef]
18. Schröder, P.; Albaladejo, M.; Ribas, A.; Macewen, M.; Tilkanen, J. *The Circular Economy in Latin America and the Caribbean Opportunities for Building Resilience*; Chatham House: London, UK, 2020.
19. Boeri, P.; Piñuel, L.; Dalzotto, D.; Sharry, S. Native Biodiversity: A Strategic Resource to Accelerate Bioeconomy Development in Latin America and the Caribbean. In *Agricultural, Forestry and Bioindustry Biotechnology and Biodiscovery*; Springer: Cham, Switzerland, 2020; pp. 163–174.
20. Interamerican Institute for Agriculture Cooperation. The Outlook for Agriculture and Rural Development in the Americas: A Perspective on Latin America and the Caribbean. 2019. Available online: www.fao.org/americas (accessed on 26 May 2022).
21. Sasson, A.; Malpica, C. Bioeconomy in Latin America. *New Biotechnol.* **2018**, *40*, 40–45. [CrossRef] [PubMed]
22. Rodríguez, A.G.; Mondaini, A.O.; Hitschfeld, M.A. *Bioeconomía en América Latina y el Caribe Contexto Global y Regional y Perspectivas*; Economic Commission for Latin America and the Caribbean: Santiago, Chile, 2017.
23. Trigo, E.J.; Henry, G.; Sanders, J.; Schurr, U.; Ingelbrecht, I.; Revell, C.; Santana, C.; Rocha, P. Towards a Latin America and Caribbean Knowledge Based Bio-Economy in Partnership with Europe towards Bioeconomy Development in Latin America and the Caribbean. 2013. Available online: www.bioeconomy-alcue.org (accessed on 26 May 2022).
24. Chavarría, H.; Trigo, E.; Martínez, J.F. Policies and Business for the Bioeconomy in LAC: An Ongoing Process. 2021. Available online: https://www.uco.es/ucopress/ojs/index.php/bioeconomy/article/download/13150/11977/v (accessed on 26 May 2022).
25. Steedman, J. Longitudinal Survey Research into Progress in Secondary Schools, Based on the National Child Development Study. In *Doing Sociology of Education*, 1st ed.; Walford, G., Ed.; Routledge: Oxford, UK, 2014; pp. 177–213.
26. Jarmasz, M.; Szpakowicz, S. Roget's Thesaurus: A Lexical Resource to Treasure. *arXiv* **2012**, arXiv:1204.0258.
27. Zeileis, A.; Leisch, F.; Hornik, K.; Kleiber, C. strucchange: An R Package for Testing for Structural Change in Linear Regression Models. *J. Stadistical. Softw.* **2002**, *7*, 1–38. Available online: http://www.R-project.org/ (accessed on 27 May 2022). [CrossRef]
28. Baiocchi, G.; Distaso, W. GRETL: Econometric Software for the GNU Generation. *J. Appl. Econom.* **2003**, *18*, 105–110. [CrossRef]

29. International Renewable Energy Agency. Renewable Energy Financial Flows. 2019. Available online: https://www.irena.org/Data/View-data-by-topic/Finance-and-Investment/Renewable-Energy-Finance-Flows (accessed on 27 May 2022).

30. Aydogan, H.; Hirz, M.; Brunner, H. The use and future of biofuels alternative propulsion technologies, new mobility concepts. *Int. J. Soc. Sci.* **2014**, *3*, 1–10. Available online: https://www.researchgate.net/publication/278300294 (accessed on 27 May 2022).

31. IAE. Energy Statistics Data Browser. 2022. Available online: https://www.iea.org/data-and-statistics/data-tools/energy-statistics-data-browser?country=WORLD&fuel=Renewables%20and%20waste&indicator=SDG72modern (accessed on 23 September 2022).

32. Pistonesi, H.; Nadal, G.; Bravo, V.; Bouille, D. *The Contribution of Biofuels to the Sustainability of Development in Latin America and the Caribbean: Elements for Formulating Public Policy*; Economic Commission for Latin America and the Caribbean: Santiago, Chile, 2008.

33. CEPALSTAT. Primary Energy Supply from Renewable (Combustible and Non-Combustible) and Non-Renewable Sources by Energy Resource. 2020. Available online: https://statistics.cepal.org/portal/cepalstat/dashboard.html?theme=3&lang=en (accessed on 20 May 2022).

34. Mougenot, B.; Doussoulin, J.P. Conceptual evolution of the bioeconomy: A bibliometric analysis. *Environ. Dev. Sustain.* **2022**, *24*, 1031–1047. [CrossRef] [PubMed]

35. Staffas, L.; Gustavsson, M.; McCormick, K. Strategies and policies for the bioeconomy and bio-based economy: An analysis of official national approaches. *Sustainability* **2013**, *5*, 2751–2769. [CrossRef]

36. Dervis, H. Bibliometric analysis using bibliometrix an R package. *J. Sci. Res.* **2019**, *8*, 156–160. [CrossRef]

37. Aria, M.; Cuccurullo, C. Bibliometrix: An R-tool for comprehensive science mapping analysis. *J. Informetr.* **2017**, *11*, 959–975. [CrossRef]

38. Van Eck, N.J.; Waltman, L. Software survey: VOSviewer, a computer program for bibliometric mapping. *Scientometrics* **2010**, *84*, 523–538. [CrossRef]

39. Moral-Muñoz, J.A.; Herrera-Viedma, E.; Santisteban-Espejo, A.; Cobo, M.J. Software tools for conducting bibliometric analysis in science: An up-to-date review. *Prof. Inf.* **2020**, *29*, e290103. [CrossRef]

40. Lakner, Z.; Plasek, B.; Kasza, G.; Kiss, A.; Soós, S.; Temesi, Á. Towards understanding the food consumer behavior—Food safety—Sustainability triangle: A bibliometric approach. *Sustainability* **2021**, *13*, 12218. [CrossRef]

41. Mulet-Forteza, C.; Martorell-Cunill, O.; Merigó, J.M.; Genovart-Balaguer, J.; Mauleon-Mendez, E. Twenty five years of the Journal of Travel & Tourism Marketing: A bibliometric ranking. *J. Travel Tour. Mark.* **2018**, *35*, 1201–1221.

42. Rajeswari, S.; Saravanan, P.; Kumaraguru, K.; Jaya, N.; Rajeshkannan, R.; Rajasimman, M. The scientometric evaluation on the research of biodiesel based on HistCite and VOSviewer (1993–2019). *Biomass Convers. Biorefin.* **2021**. [CrossRef]

43. OECD and SCImago Research Group. Compendium of Bibliometric Science Indicators. 2016. Available online: https://lostandtaken.com/ (accessed on 27 September 2022).

44. Gonzalez-Brambila, C.N.; Reyes-Gonzalez, L.; Veloso, F.; Perez-Angón, M.A. The Scientific Impact of Developing Nations. *PLoS ONE* **2016**, *11*, e0151328. [CrossRef]

45. UNESCO. Global Investments in R&D A Snapshot of R&D Expenditure. 2020. Available online: http://uis.unesco.org (accessed on 15 February 2023).

46. Prodan, I. Influence of Research and Development Expenditures on Number of Patent Applications: Selected Cases Studies in OECD Countries and Central Europe, 1981–2001. *Appl. Econom. Int. Dev.* **2005**, *5*, 5–22.

47. Meo, S.A.; Al Masri, A.A.; Usmani, A.M.; Memon, A.N.; Zaidi, S.Z. Impact of GDP, Spending on R&D, Number of Universities and Scientific Journals on Research Publications among Asian Countries. *PLoS ONE* **2013**, *8*, e66449.

48. Gutiérrez-Prieto, H. Bioethics and Ecology: Towards "Sustainable Bioethics". *Vniversitas* **2008**, *117*, 275–294. Available online: http://www.wwf.org (accessed on 27 May 2022).

49. Delgado-Plaza, E.; Carrillo, A.; Valdés, H.; Odobez, N.; Peralta-Jaramillo, J.; Jaramillo, D.; Reinoso-Tigre, J.; Nuñez, V.; Garcia, J.; Reyes-Plascencia, C.; et al. Key Processes for the Energy Use of Biomass in Rural Sectors of Latin America. *Sustainability* **2022**, *15*, 169. [CrossRef]

50. Bailis, R.; Solomon, B.D.; Moser, C.; Hildebrandt, T. Biofuel sustainability in Latin America and the Caribbean—A review of recent experiences and future prospects. *Biofuels* **2014**, *5*, 469–485. [CrossRef]

51. Ludena, C. Biofuels Potential in Latin America and the Caribbean: Quantitative Considerations and Policy Implications for the Agricultural Sector. 2005. Available online: http://ageconsearch.umn.edu (accessed on 7 March 2023).

52. Nuffield Council on Bioethics. Biofuels: Ethical Issues. Nuffield Council on Bioethics. 2011. Available online: https://www.nuffieldbioethics.org/assets/pdfs/Biofuels_ethical_issues_A5_Guide.pdf (accessed on 2 October 2022).

53. Subramaniam, Y.; Masron, T.A. The impact of economic globalization on biofuel in developing countries. *Energy Convers. Manag. X* **2021**, *10*, 100064. [CrossRef]

54. Torroba, A. *Atlas de los Biocombustibles Líquidos*; Interamerican Institute for Agriculture Cooperation: San Jose, CA, USA, 2020.

55. CEPAL. *Brazil, Argentina and Colombia Lead Biofuels Production in the Region*; CEPAL: Santiago, Chile, 2011.

56. OECD. *Food and Agriculture Organization of the United Nations. Biofuels: OCDE-FAO Agricultural Perspectives 2020–2029*; OECD: Paris, France, 2021.

57. Falck-Zepeda, J.; Mangi, S.; Sulser, T.; Zambrano, P.; Falconí, C. *Biofuels and Rural Development in Latin America and the Caribbean*; Cooperative Programme FAO: Rome, Italy, 2010.

58. United Nations Conference on Trade and Development. Biofuels Controversy. 2007. Available online: https://unctad.org (accessed on 15 February 2023).
59. Castiblanco Rozo, C.; Hortúa Romero, S. Biofuels' Energetic Paradigm and Its Implications: A Global Overview and the Colombian case. Environment and Development. 2012. Available online: https://revistas.unal.edu.co/index.php/gestion/article/view/33718 (accessed on 15 February 2023).
60. Jeswani, H.K.; Chilvers, A.; Azapagic, A. Environmental sustainability of biofuels: A review. *Proc. Math. Phys. Eng. Sci.* **2020**, *476*, 20200351. [CrossRef] [PubMed]
61. Prasad, S.; Yadav, A.N.; Singh, A. Impact of Climate Change on Sustainable Biofuel Production. In *Biofuels Production—Sustainability and Advances in Microbial Bioresources*; Yadav, A.N., Rastegari, A.A., Yadav, N., Gaur, R., Eds.; Springer: Cham, Switzerland, 2020; pp. 79–97.
62. Bucaram-Villacis, S.; Trabacchi, C.; Schneider, D.A.C.M.E.N.; Watson, G. A Call for an Integrated Framework for the Bioeconomy in Latin America and the Caribbean Region. Available online: https://blogs.iadb.org/sostenibilidad/en/a-call-for-an-integrated-framework-for-the-bioeconomy-in-latin-america-and-the-caribbean-region/ (accessed on 1 January 2023).
63. Instituto Interamericano de Cooperación para la Agricultura. *Bioeconomía: Potencial y Retos para su Aprovechamiento en America Latina y el Caribe*; Chavarria, H., Blanco, M., Eds.; IICA: San José, CA, USA, 2020; Available online: http://www.iica.int (accessed on 29 May 2022).
64. Woźniak, E.; Tyczewska, A. Bioeconomy during the COVID-19 and perspectives for the post-pandemic world: Example from EU. *EFB Bioecon. J.* **2021**, *1*, 100013. [CrossRef]
65. Barañano, L.; Garbisu, N.; Alkorta, I.; Araujo, A.; Garbisu, C. Contextualization of the bioeconomy concept through its links with related concepts and the challenges facing humanity. *Sustainability* **2021**, *13*, 7746. [CrossRef]
66. Rodríguez, A.G.; Rodrigues, M.; Sotomayor, O. Hacia una Bioeconomía Sostenible en América Latina y el Caribe Elementos para una Visión Regional. 2003. Available online: www.cepal.org/apps (accessed on 1 January 2023).

Article

Chemo-Sonic Pretreatment Approach on Marine Macroalgae for Energy Efficient Biohydrogen Production

Shabarish Shankaran [1], Tamilarasan Karuppiah [1,*] and Rajesh Banu Jeyakumar [2]

1   Department of Civil Engineering, Vel Tech Rangarajan Dr. Sagunthala R&D Institute of Science and Technology, Chennai 600062, India
2   Department of Biotechnology, Central University of Tamil Nadu, Neelakudi, Thiruvarur 610005, India
*   Correspondence: tamilkaruppiah@gmail.com; Tel.: +91-9486536474

Abstract: The core objective of this analysis is to implement a combination of alkaline (NaOH) and sonication pretreatment techniques to produce energy-efficient biohydrogen from the marine macroalgae *Chaetomorpha antennina*. Anaerobic fermentation was implemented in control, sonic solubilization (SS) and sonic alkali solubilization (SAS) pretreatment for 15 days. In control, a biohydrogen production of 40 mL H_2/gCOD was obtained. The sonicator intensities varied from 10% to 90% for a period of 1 h during SS pretreatment. About 2650 mg/L SCOD release with a COD solubilization of 21% was obtained at an optimum intensity of 50% in a 30 min duration, in which 119 mL H_2/gCOD biohydrogen was produced in the anaerobic fermentation. SAS pretreatment was performed by varying the pH from 8 to 12 with the optimum conditions of SS where a SCOD release of 3400 mg/L, COD solubilization efficiency of 26% and a maximum biohydrogen production of 150 mL H_2/gCOD was obtained at a high pH range of 11 in the fermentation. The specific energy required by SS (9000 kJ/kgTS) was comparatively higher than SAS (4500 kJ/kg TS). SAS reduced half of the energy consumption when compared to SS. Overall, SAS pretreatment was found to be energetically favorable in a field application.

Keywords: COD solubilization; chemo sonic pretreatment; biohydrogen; specific energy

Citation: Shankaran, S.; Karuppiah, T.; Jeyakumar, R.B. Chemo-Sonic Pretreatment Approach on Marine Macroalgae for Energy Efficient Biohydrogen Production. *Sustainability* **2022**, *14*, 12849. https://doi.org/10.3390/su141912849

Academic Editors: Idiano D'Adamo and Massimo Gastaldi

Received: 1 September 2022
Accepted: 3 October 2022
Published: 9 October 2022

Publisher's Note: MDPI stays neutral with regard to jurisdictional claims in published maps and institutional affiliations.

1. Introduction

Recently, a lot of environmental issues have been raised owing to the usage of fossil fuels. It motivates researchers and scientists to predict prompt remedial action to create a proper substitute for fossil fuels [1]. Furthermore, most countries extract energy from many natural resources such as wind, hydropower and solar power. Biomass is a significant potential source of energy among these energy resources [2]. As a photosynthetic organism, marine macroalgae has the promising potential to act as a bioresource for biofuel production. Since it is associated with the green color type of marine macroalgae autotrophs, it is a rich source of biopolymers such as protein, carbohydrates and lipids, which are responsible for more biofuel production [3]. Furthermore, the lack of lignin content makes the marine macroalgae even more appropriate for an effective anaerobic fermentation process [4]. Marine macroalgae is a collection of rapidly growing plant organisms that can grow to substantial sizes in marine environments such as rock surfaces. The median photosynthetic activity of this marine macroalgae was 6–8%, much higher than that of earthbound biomass (1.8–2.2%) [5].

The circular economy involves energy recovery from trash and residues, which can fulfil the material and energy cycle. A very promising pathway toward sustainability is the biogas–biohydrogen chain. It can be transmitted into the natural gas grid, used as a vehicle fuel, or transformed into electricity-generating units. It is produced from a variety of different substrates, such as crop leftovers, algae, animal wastes, organic portion of municipal solid wastes and sludge [6,7]. Anaerobic fermentation is the sustainable way of extracting or generating bioenergy from macroalgae since it is adaptable to this process [8]. Hydrogen (H_2) is one of the various fuel sources that evolved from igniting hydrogen-holding elements such as natural gas, oil and coal. However, with regards to excessive energy content, hydrogen has an enormous energy density than other surviving sources of fuels such as methane and ethanol [9]. Figure 1 shows the global hydrogen production in the last ten years and its market value based on the report of GHR, 2021 [10].

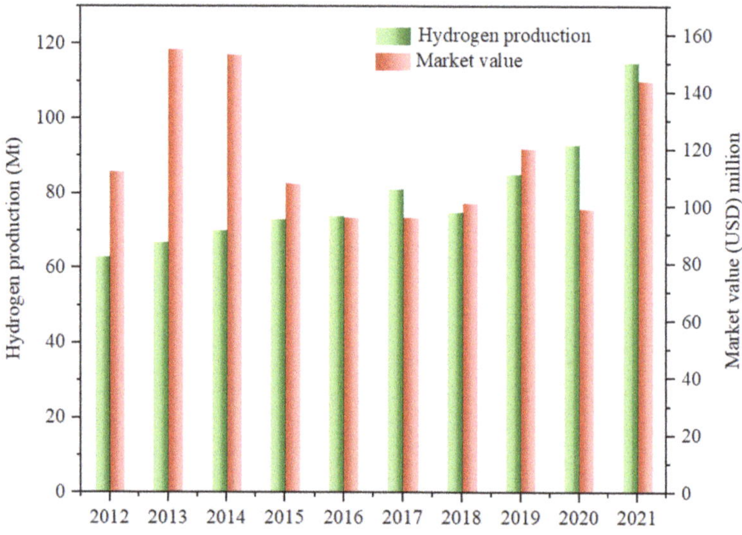

Figure 1. Global hydrogen production in last ten years and its market value.

Additionally, hydrogen is a high potential energy supplier for various utilities such as an ignition engine fuel for vehicles, rocket propellant, fuel cells for power supply and reactor coolant [11]. Furthermore, hydrogen generation makes way for the ecological crisis in the surrounding environment, such as global warming, melting glaciers and icebergs, rising ocean water levels and air pollution. Therefore, the bio-based hydrogen generation method is considered more appropriate than the standard fuel generation source. Moreover, the output of biohydrogen using substrate biomass has quite a few benefits such as uncomplicated performance, outspread accessibility of energy resources such as residues of food, vegetables, fruits, fauna manure such as cow chip, unbiased carbon source and cost worthy in its functions [12]. When specific requirements are met, the bioeconomy can advance toward sustainability: (i) the resource base is sustainable; (ii) processes and products are sustainable and (iii) transport of materials is viable [13]. Furthermore, customer satisfaction with bio-based products is also necessary to assess the influence of green premiums and the significance of sustainability certification [14]. However, these factors affect various industries because they are essential when considering sustainability as a factor that facilitates market success and a competitive advantage source [15]. In this perspective, the use of biological resources to replace non-renewable resources, escalating the use of biomass and reducing biowaste are excellent examples of a circular bioeconomy, which can be crucial in achieving sustainable development goals (SDGs) [16]. Marine macroalgae and biohydrogen satisfy all the mentioned criteria of the bioeconomy. Therefore, both marine macroalgae and biohydrogen are bio-economically feasible. Hydrolysis is the primary step of the anaerobic fermentation at which the cell cleavage occurs, a rate-limiting factor of the fermentation process. It is a difficult stage in anaerobic fermentation since the cell wall of the biomass may be more vital to rupture [17]. To augment the hydrolysis phase, the structural integrity of the biomass can be degraded through various pretreatment techniques and biopolymers such as proteins, carbohydrates, lipids and starch present in the biomass come out for the oxygen-free fermentation process [18]. Therefore, for this purpose, various pretreatment techniques, such as physical, chemical, mechanical and biological ones, are incorporated [19]. Ultrasonication is a method of generating acoustic waves used to disrupt the cell wall of the biomass [20]. The sonicator gives rise to high-intensity ultrasound waves through a probe over the substrate kept in a beaker with water inside the apparatus. These high-intensity sonic waves are generated with the help of an intensity generator during the sonication process. These high-intensity sonic waves initiate the pressure wave formation and due to these pressure waves, cavitation develops. This cavitation collapses the cell wall of marine macroalgae species and disrupts it [21]. Energy exhaustion is a primary concern because the mechanical (sonication) pretreatment consumes much energy (electric current) to disrupt the biomass cell wall [22]. In order to overcome this problem, additives such as alkali and surfactants can be added, making the operation process of sonication energetically feasible [23]. Microwave–surfactant, microwave–acidic and disperser–ozone were the combinative pretreatment techniques used to solubilize marine macroalgae until now [24]. However, there are no published studies on marine macroalgae (*Chaetomorpha antennina*) solubilization using the sonication and alkali (NaOH) combination. Therefore, the marine macroalgae were solubilized in this study using a novel technique called alkali-assisted sonication. The objectives of this research are (1) to optimize the solubilization conditions for SAS for energy-effective performance; (2) to perform kinetic analysis for SS and to analyze its efficiency; (3) to assess the beneficial impact of this SAS pretreatment; (4) to evaluate the effect of this SAS pretreatment on the production of biohydrogen; (5) to perform an energy analysis of SAS in terms of field applicability.

2. Materials and Methods

2.1. Marine Macroalgae Sample

The marine macroalgae biomass species *Chaetomorpha antennina* was collected from ennore, a marine area of chennai (13°12′23.4864″ N, 80°19′38.0100″ E), Tamil Nadu, India. The marine macroalgae were entirely washed with water to detach the residue particles. The cleaned sample was shade-dried and sliced into pieces of less than 2 cm in size for the convenience of pretreatment. This biomass was kept in a refrigerator for the subsequent study [24].

2.2. Biomass Pretreatment

2.2.1. Sonic Solubilization (SS)

SS pretreatment was implemented to rupture the cell wall of the biomass. The operation mechanism utilized a sonicator (Model VCX130, New Town, CT, USA) instrument with a frequency of 20 kHz and a maximum power input of 130W. A beaker of 1L capacity volume filled with water and substrate sample was taken for this pretreatment. The substrate and water ratio taken for pretreatment was 1:50. The sonication power intensity and the time duration varied from 10 to 90% and from 1 to 60 min, respectively. The sonic probe produces the combined effect of pressure waves and cavitation. This effect results in the marine macroalgae cell wall weakening for enhanced solubilization and biopolymers release. The only drawback of this SS pretreatment was that it consumed more electrical energy to solubilize the marine macroalgae. The samples were taken and examined for a regular period.

2.2.2. Sonic Alkali Solubilization (SAS)

The solubilization of the substrate by SAS was carried out by adding alkali "sodium hydroxide (NaOH)" with an optimum condition obtained from SS pretreatment. The pH of the sample varied from 8 to 12. SAS pretreatment is appropriate for the following reasons: (1) the mechanical (sonication) pretreatment gives high and efficient output within a quick session compared to physical and biological pretreatment methods. (2) Alkali (NaOH), when added to sonication, are divided into cations (Na^+) and anions (OH^-). Cations transform into bubbles, clash with the cell wall of marine macroalgae and break it. Anions settle over the marine macroalgae cell wall surface and weaken it. This phenomenon accelerates the solubilization process of marine macroalgae and more biopolymers are released in a short duration. The samples are taken at a regular time interval and subjected to analysis.

2.3. Anaerobic Fermentation Study

Anaerobic fermentation was performed for control, SS and SAS, into which anaerobically digested sludge (inoculum) taken from a wastewater treatment plant was added at a ratio of 9:1 in serum bottles of 250 mL volume capacity for three days. To suppress the fermentation within the acetogenic phase and to compute the volatile fatty acids (VFA) produced, a methanogenic phase obstructor 50 mM of 2- Bromo ethane sulphonic acid (BESA) was added to each bottle. The computation of VFA was performed to substantiate the pretreatment and biohydrogen production efficiency [25]. To remove O_2, nitrogen gas was introduced into all serum bottles. The bottles were firmly sealed by stoppers and positioned in an orbital shaker under agitation at a speed and temperature of about 150 rpm and 35 °C [26]. VFA analysis was performed through the distillation method [27].

2.4. Biohydrogen Potential Assessment (BPA)

BPA analysis was applied for control, SS and SAS to evaluate the biohydrogen production capability under moderate temperatures. The process of BPA was performed in serum bottles with a functioning volume of 150 mL. In all three serum bottles, the marine macroalgae sample (70%), inoculum (25%) and the nourishment food (5%) were taken [28]. As a point of expelling methanogens in the inoculum and enriching the microbes for hydrogen production, the inoculum was subjected to calefaction for 30 mins at 100 °C [29]. To maintain an oxygen-free environment, nitrogen (N_2) gas was filled in the remaining bottle area for 10 mins [30]. Rubber stoppers were used to seal the bottles. Finally, the bottles were kept in a shaker and incubated at 37 °C at 130 rpm. A gas chromatograph with a thermal conductivity detector and stainless column packed with Porapak Q (3.25 mm diameter, 2 cm length and 80/100 mesh) was used to calculate hydrogen production [31]. The experiments were triplicated. To estimate the cumulative H_2 yield, the modified Gompertz Equation (1) was used.

$$AH = Hl * \exp\left(-\exp\left(-pr(Hc - Hfb)\right)\right) \tag{1}$$

where:

AH—Increased H_2 production (mL);
Hl—H_2 production (mL H_2/g COD);
pr—Peak H_2 generation rate (mL H_2/g COD d);
Hc—Commencing phase of hydrogen production (days);
Hfb—Lag phase of hydrogen production (days).

2.5. Analytical Methods

The biopolymers proteins, carbohydrates and lipids released as a result of pretreatment were measured based on the method prescribed by Kavitha et al. (2016) [32]. In addition, total chemical oxygen demand (TCOD), soluble chemical oxygen demand (SCOD), and VFA were analyzed with the help of standard methods as per APHA (2005) [33].

2.6. Statistical Analysis

One way analysis of variance (ANOVA) ($\alpha = 0.05$) approach was made to assess the deliverables of the experiment. The differences between experimental deliverables during the pretreatment could be subjected to statistical significance analysis if the p-values were less than 0.05. To be precise, for p-values < 0.05, the difference between SCOD release averages was statistically significant. On the contrary, for p-values > 0.05, the difference between SCOD release averages was not statistically significant [34].

2.7. Specific Energy for Sonication (SES)

Specific energy (SE) is considered for the measure of vital energy required by the sonicator to solubilize the cell wall of marine macroalgae. The SE was calculated using the subsequent Equation (2):

$$SES\ (kJ/kg\ TS) = (P_D \times S_T)/(Vs \times TS) \tag{2}$$

where:

SES—Specific energy for sonication;
P_D—Power used for disruption of the biomass cell wall (kW);
S_T—Sonication treatment time (s);
V_S—Volume of the sample (L);
TS—Total solids (kg).

2.8. Energy Analysis

One prominent contemplation in the massive scale biofuel production is the energy employed in the entire process. From an economic angle, minimum input energy should exhibit the uttermost output energy, which will be profitable [35]. This investigation studied the energy required to treat 1 kilogram of marine macroalgae biomass sample to produce H_2 gas. The total net energy that has been dominated was calculated using Equation (3).

$$N_E = O_E - I_E \tag{3}$$

where:

N_E—Net energy (kWh);
O_E—Output energy (kWh);
I_E—Input energy (kWh).

The solubilization energy taken by the sonicator is the input energy as shown in Equation (4).

$$I_E = P_S * T_S * V_R * B \tag{4}$$

where:

I_E—Input energy (kWh);
P_S—Power utilized for the sonication process (kW/kg);
T_S—Time consumed for solubilization (h);
V_R—Reactor volume (m^3);
B—Biomass (kg/m^3).

The output energy was calculated based on various parameters such as biomass biodegradability, organic load, the volume of the reactor and hydrogen yield, as mentioned in Equation (5).

$$O_E = B_{SB} * L_{COD} * H_Y * V_R * B_{CF} \tag{5}$$

where:

O_E—Output energy (kWh);
B_{SB}—Biodegradability of marine macroalgae biomass (g COD/g COD);
L_{COD}—COD load (g COD/m^3);
H_Y—Hydrogen yield (m^3 /g COD);
V_R—Reactor volume (m^3);
B_{CF}—Biohydrogen conversion factor.

By determining the optimistic and pessimistic amount of net energy, the profit and loss in the energy are confirmed in the SS and SAS processes.

The energy ratio is given in Equation (6),

$$Er = O_E / I_E \tag{6}$$

where:

Er—Energy ratio;

O_E—Output energy (kWh);

I_E—Input energy (kWh).

3. Results and Discussion

3.1. Sequel of SS in the Liberation of Soluble Organics Release

Solubilization potential was estimated by the release of soluble organics during the SS process. Figure 2 shows the release of the soluble organics for sonication intensity and period. The sonicator was operated by varying its power intensities from 10% to 90% for 1 h. During SS operation, the marine macroalgae that had to be solubilized was kept under the sonicator probe. It was subjected to the impact of high-power ultrasonic waves, which resulted in the emergence of pressure waves and cavity bubbles. This simultaneous evolution of cavity bubbles and pressure waves weakened the marine macroalgae cell wall. At each intensity, the solubilized marine macroalgae sample was taken and analyzed. In Figure 2, it was observed that the release of the soluble organics was classified into two phases, namely, the faster phase (1–30 min) and the slower phase (30–60 min). The figure shows that when the sonication pretreatment time increases, there is an increment found in the soluble organics release. In the faster phase, 1–30 min, the release of soluble organics was high up to 30 mins, but in the slower phase, beyond 30 min, the minor release was found. A steady trend was spotted in the slower phase after 30 min. This trend indicates that most of the soluble organics got unleashed within 30 min in the faster phase. For a sonication process, the pretreatment time was recognized as an ideal parameter [36]. Hence, the sonication pretreatment time of 30 min was acknowledged as an optimum pretreatment time for SS. Furthermore, the sonication intensity for pretreatment also plays an indispensable part in SS. When the release of the soluble organics was reasoned against the intensity of SS, an extraneous behavior was noticed in the release of the soluble organics. In the intensity range (10–40%), there was a minimum release found in soluble organics and the release range was 1750–2320 mg/L. This provided authentic evidence that the marine macroalgae were partially solubilized [37]. When the intensity is further increased to 50%, drastic enhancement in the soluble organics release of 2650 mg/L was obtained due to the combined effort of high-power ultrasonic waves, pressure waves and increased formation of cavity bubbles. This caused the marine macroalgae cell wall to smash and become solubilized. Increasing the intensity beyond 50%, there was no excess improvement obtained. The soluble organics release found between 50–90% was in the SCOD release range of only 2819–3010 mg/L. This marginal release was found because most of the soluble organics got released at up to 50% intensity. Hence, 50% was considered to be optimum for SS. For the soluble organics released during SS, statistical analyses were carried out via ANOVA. Table 1 represents the one-way ANOVA of variance for various intensities of sonicator on the SCOD release basis. When the intensity varied from 10% to 40%, the probability value was found to be 0.46, which was greater than 0.05. This signifies that there is no statistical difference. For intensities between 40% and 50%, the probability value of 0.013 obtained was less than 0.05. This shows that there was a considerable difference found between 40% and 50%. The mean values of SCOD release from 50% to 90% imply a lack of significant difference between them, with a probability value of 0.84, which was greater than 0.05. Therefore, considering all these outcomes, a power intensity of 50% with a duration of 30 min was considered as optimum.

3.2. Response of SE over COD Solubilization

Significant attention is given to SE regarding the economy of the process for enormous biofuel production. Figure 3 represents the solubilization of SS concerning SE. It was noticed that the solubilization trend increases with an increase in SE input for all sonic intensities. The solubilization tendency can be divided into three phases: X, Y and Z. Slower solubilization was represented by phase X, which corresponds to intensities of 10% to 40%.

Phase Y represents a faster solubilization rate, ranging from 40% to 50%. Finally, phase Z extends from 60% to 90%. At a sonicator, SE input of around 1800–7200 kJ/kg TS, solubilization of about 13.46–17.84% was achieved during phase X. The amount of solubilization obtained was insignificant and can be ignored for further analysis. An effective rise in solubilization was observed in phase Y, with a maximum of 21% reached at a sonicator SE input of about 9000 kJ/kg TS for an intensity 50%. Even though the sonicator intensity and SE were increased from 60% to 90% and 9000 kJ/kg TS to 10,800 kJ/kg TS in phase Z, there was no significant increase in solubilization. To increase solubilization from 21% to 22%, for example, a sonicator SE input of 10,800 kJ/kg TS was required. As a result, it can be concluded that simply raising sonicator intensities during the SS process may waste energy. Instead, SS was found to benefit from an optimum sonicator SE input of 9000 kJ/kg TS.

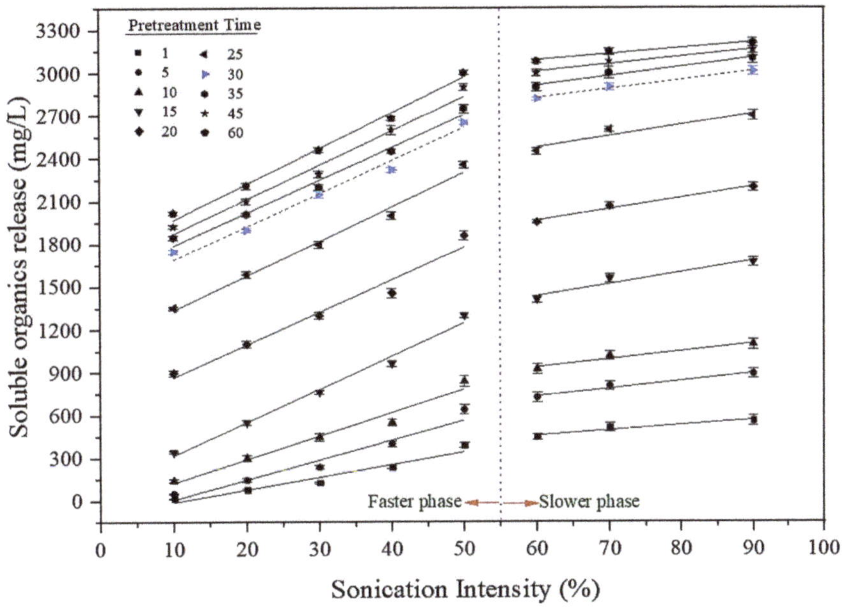

Figure 2. Soluble organics release with respect to sonication intensity.

Table 1. One-way analysis of variance for various intensities of sonicator on the SCOD release basis.

Variation Source	Sum of Squares	Degrees of Freedom	Mean Square	F Value	*p*-Value Prob > F	Results
10–40%	56,512	3	977,011	0.5	0.46	Not significant
40–50%	52,040,402	1	6,900,710	7.5	0.013	Significant
50–90%	40,951	3	1,052,333	0.031	0.84	Not significant

3.3. Impact of SAS in the Discharge of Organic Biopolymers

Sodium hydroxide (NaOH), as an alkaline solution, has the massive potential to fracture the cell wall's ester bond, resulting in increased cellulose decrystallization [38]. During SAS, alkali, when added to the sample, gets split into cations (Na^+) and anions (OH^-). Due to saponification, the cations get transmuted into bubble form, clash with the marine macroalgae cell wall and break it; and due to solvation, it settles in the bottom of the beaker as salts. On the other hand, the anions settle over the cell wall and make it squashy, which makes the sonication process even more rapid and comfortable. This results in the reduction of energy consumption by the sonicator. Thus, the alkali (NaOH) acts as an excellent energy-saving additive and intensifies the sonication pretreatment even more effectively [39]. In the present study, alkaline (NaOH) was combined with the SS process to enhance the solubilization capability of the previous certainties. Figure 4 signifies the soluble organics and biopolymer release at various pH levels. The alkali was added by differing its pH from 8 to 12. The sonicator was operated at 50% of power intensity and 30 min of duration, which was optimized in SS, and the sample's pH was varied. During the operational time of SAS, for every 5 min, the solubilized biomass sample was taken and examined for each pH from 8 to 12. From the figure, it was understood that the patterns of soluble organics (SCOD) and biopolymers (protein, carbohydrates and lipids) show two divergent phases: an accelerated and a slow phase. The accelerated phase occurs from pH 8 to pH 11, where a soluble organics release (2900–3400 mg/L) was obtained. This proves that the combinative pretreatment was very effective as more SCOD were released in SAS (3400 mg/L) compared to SS (2650 mg/L), as presented in Figure 2. This massive increase in the release of soluble organics during the accelerated phase could be due to the combined action of SAS, which prompts the fracturing of marine macroalgae cell walls and the release of intercellular components. It is similar to the work of Kumar et al. (2017) [40], where the SCOD release of 1603 mg/L was obtained from microalgae via combined pretreatment of sonication and electrolysis. The slow phase lies from pH 11 to pH 12, where a soluble organics release (3400–3450 mg/L) was obtained. A significant hike was found in the release of soluble organics between pH 8 and 11, but in the slow phase beyond pH 11, a minimum rise was noted in the release of soluble organics. This makes it evident that almost all the soluble organics got released within pH 11 and it was adequate to solubilize the marine macroalgae cell wall. Therefore, increasing the pH level beyond 11 will increase chemical cost rather than marine macroalgae solubilization. From Figure 4, it is evident that at optimum solubilization of 21%, SAS consumed less SE (4500 KJ/kg TS) compared to SS (9000 kJ/kg TS), which shows that SAS is more energetically feasible than SS.

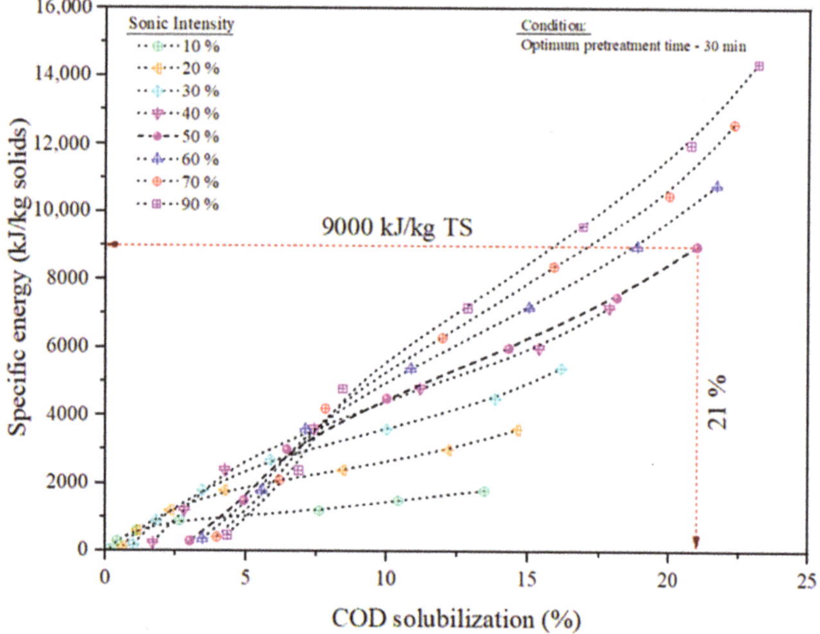

Figure 3. Solubilization efficiency of SS with respect to specific energy.

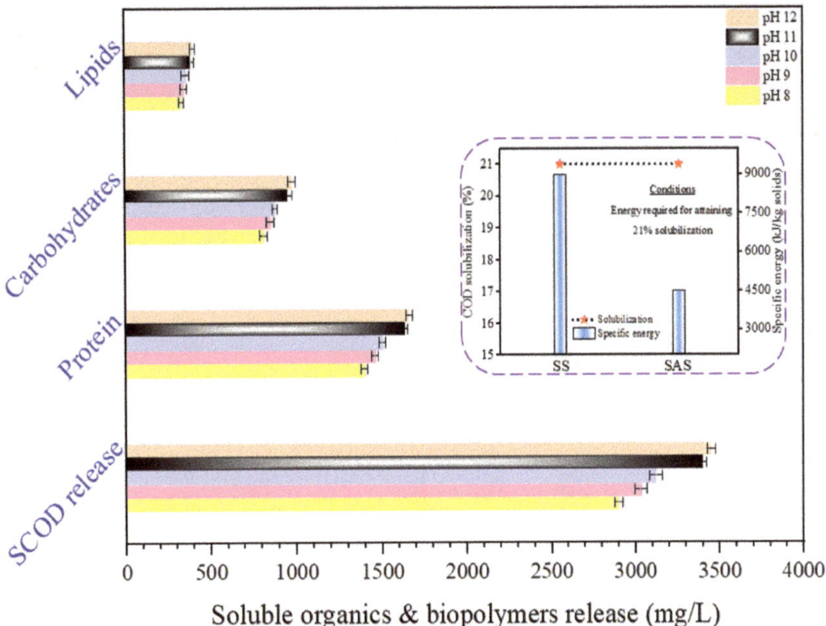

Soluble organics & biopolymers release (mg/L)

Figure 4. Soluble organics and biopolymers release in SAS.

The biopolymer's existence in marine macroalgae boosts hydrogen generation. Figure 4 elucidates the biopolymer release from pH 8 to pH 12. Indisputably, the biopolymers trend is similar to the SCOD trend and could be grouped into two phases: active and inactive. The active phase begins at pH 8 and ends at pH 11. A moderate increase in the biopolymers release was observed in this active phase up to a pH of 11, where a protein, carbohydrate and lipid release of 1637, 957 and 390 mg/L was obtained. The inactive phase begins beyond 11 where a protein, carbohydrate and lipid release of 1660, 978 and 402 mg/L, respectively, were obtained and there is no sturdy increase in biopolymers release after that, which signifies that the majority of the biopolymers got released in the pH 11. The collaborative effect of chemo sonic pretreatment makes way for effective solubilization of marine macroalgae cell wall and the liberation of biopolymers into the liquid phase of marine macroalgae. Hence from the facts mentioned earlier, it was concluded that SAS is more effective in solubilization and biopolymers release.

3.4. VFA Production in SS and SAS

The VFA investigation done for control, SS and SAS pretreated samples during anaerobic fermentation was analyzed and conveyed in Figure 5. In the commencing hydrolysis stage, the complicated hydrolytic components released during pretreatment got converted into sugars, amino acids and fatty acids. In the peripheral stage of acetogenesis, the simple monomers got transmuted into VFA [41]. Due to the biological action of microbes in the inoculum, the biopolymers got transformed into VFA [42]. Anaerobic fermentation was carried out for 72 h. At the end of 72 h, as predicted, SAS showed an enormous decrement in protein, and carbohydrate concentration from 1637, 957 mg/L to 623, 364 mg/L, which denotes the hydrolysis competence. On the other hand, SS showed a slight protein and carbohydrate concentration reduction from 1300, 760 mg/L to 498, 289 mg/L. The depletion in the concentration of biopolymers was found to be a lot less in SS compared to SAS. This made authentic evidence that highly solubilized biopolymers are easily accessible by fermentative microbes, which defines the effectiveness of combinative pretreatment [43]. It is similar to the combinative pretreatment strategy suggested by Tamilarasan et al. (2017) [44].

In contrast, the untreated control sample did not manifest a major decrement; instead, a build-up was spotted in the biopolymer's concentration. In control protein, the carbohydrate concentration was increased from 160,110 mg/L to 180,130 mg/L respectively. The reason behind this is that the biopolymers are not solubilized since there was no pretreatment in control; hence, the microbes try to break the marine macroalgae cell wall and release the biopolymers. This release was found only using disintegration instead of fermentation. The increased VFA production should have a higher hydrogen yield at the end of fermentation process. The VFA production analysis was performed to validate the effectiveness of biohydrogen production in the fermentation process. The utmost liberation of VFA during fermentation intensifies biohydrogen production [45]. Figure 5 clearly states that among control (110 mg/L), SS (860 mg/L) and SAS (1800 mg/L) after 72 h of anaerobic fermentation, SAS showed higher VFA production compared to SS and control due to the alkali sonication impact and effective utilization of pretreated and hydrolyzed biopolymers by acetogenic microbes. From the findings, SAS presents effectiveness in VFA production, hence proving that SAS will yield more hydrogen at the end of the fermentation process.

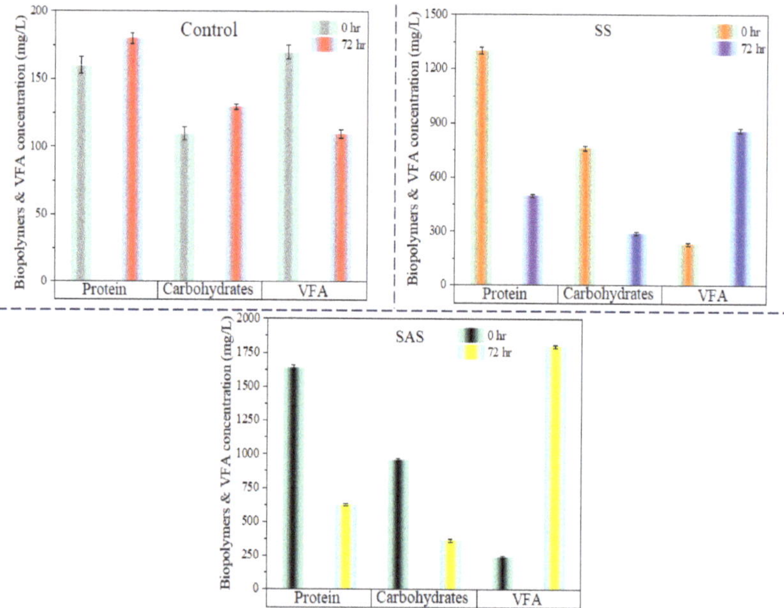

Figure 5. VFA production in control, SS and SAS.

3.5. Biohydrogen Potential Assay (BPA)

Figure 6 signifies the biohydrogen production in control, SS and SAS. From Figure 6, it was unquestionably understood that the biohydrogen generation got varied with control, SS and SAS. Biohydrogen analysis was done for 15 days. Regardless of augmentation in biohydrogen generation concerning increasing days of fermentation, the generation rate of biohydrogen was less in control (40 mL H_2/g COD) in comparison with SS (119 mL H_2/g COD) and SAS (150 mL H_2/g COD) on the eighth day of fermentation. This is due to the certainty that the microbes in inoculum are more comfortable in the biological degradation of marine macroalgae to generate hydrogen when the biomass is in soluble form than solid form. SAS sample has more effectiveness in biohydrogen production than control and SS because the alkali and sonication gave an impressive hydrolysis effect. Hydrogen-producing microbes' subsequent utilization of acetogenic elements enhances biohydrogen generation [46]. Owing to the combined pretreatment method imposed over the marine macroalgae, the anaerobic culture media had a very suitable approach to liberating biohydrogen. At the same time, depending upon the composition, solubilization efficiency and pretreatment conditions, the biohydrogen production potential may vary for different substrates. In this condition, the released solubilized compounds, especially proteins and carbohydrates, declined as there was a rise in VFA and biohydrogen production due to the effective hydrolysis and consumption of biopolymers by the microbes. The commencement of biohydrogen fermentation starts with the biopolymer's biodegradation. The proceedings of biopolymers degradation by the fermentative and hydrogen-producing microbes resulted in the emergence of biohydrogen. The biopolymers which were solubilized got exploited by fermentative microbes as a source of energy and electrons [47]. Then, the hydrogen-generating microbes use these compounds and transmute them into biohydrogen. Anaerobic microbes in the inoculum can easily access the biopolymers in the marine macroalgae via this combinative pretreatment. The inoculum (anaerobic sludge) comprises microbes that effectively utilize the solubilized biopolymers and convert them into monosaccharides, thus escalating biohydrogen production [48]. In the preliminary stage, the third day of the operation, the biohydrogen production was low for all samples. This may be due to the instantaneously unadaptable condition of the microbes in the environment. After the third day in the augmented stage, there was a steady increase in the biohydrogen generation where control, SS and SAS showed a biohydrogen production of 5, 75 and 106 mL H_2/g COD, respectively. This rising scenario of biohydrogen in the augmented stage guarantees an effective proliferation and fermentative action of microbes. The eighth day of fermentation begins with the sound stage where control, SS and SAS showed a biohydrogen production of 40, 119 and 150 mL H_2/g COD respectively, beyond which there was no rise in biohydrogen production since a stable range was observed. The summary of this stable stage shown in Figure 6 shows that the biohydrogen producers have unreservedly exploited the solubilized substrates. A maximum biohydrogen yield of 150 mL H_2/g COD was obtained in SAS than SS 119 mL H_2/g COD and control 40 mL H_2/g COD. This is due to the chemo sonic pretreatment that makes the biopolymers in the marine macroalgae easily approachable to the anaerobic microbes in the inoculum sludge, which is essential for biohydrogen production. Table 2 signifies the kinetics constants accomplished through Gompertz modeling of control, SS and SAS samples. SAS shows an uttermost hydrogen production potential and rate (150 mL H_2/g COD and 0.91 mL/d) in correlation with SS (119 mL H_2/g COD and 0.67 mL/d) and control (40 mL H_2/g COD and 0.47 mL/d) expressing the combinative potency of sonication and alkali [49]. It is witnessed that the SAS has a very short preliminary stage (1.5 days) in comparison with control (3.7 days) and SS (2.6 days). An excellent fit was observed in exploratory data as the correlation coefficient of 0.995 was obtained. A similar range of fit was obtained in the work of Tamilarasan et al. (2018) [50]. Based on the above points, it was proved that SAS is more effective in biohydrogen generation than control and SS. Table 3 shows biohydrogen production from different species of marine macroalgae with various combinative pretreatments. From a sustainability point of view, marine macroalgae have emerged as prospective sources for biobased products and biofuel.

3.6. Energy Interpretation

The overall energy consumed for the operation of the marine macroalgae (1 kg) accounted for energy interpretation. Figure 7 depicts the overall energy interpretation between SS and SAS, which includes optimum condition, total energy spent, energy gained through biohydrogen production, net energy, and energy ratio [28,45,51,52]. For effective pretreatment accomplishment, the exhausted input energy should be compensated by the output biohydrogen production. In the evaluation aspect, the output biohydrogen and input sonication energy of SS and SAS observed at an optimum setup were considered. Solubilization efficiency of 21% was kept as an indicator to derive the energy constants for the appraisal of SS and SAS pretreatment energy efficiency [51,52]. The energy consumed by SS (0.1 kWh/kg solids) and SAS (0.05 kWh/kg solids) was determined based on all these specifications. The output biohydrogen production energy of (0.09 kWh/kg solids) was obtained for both SS and SAS since the SCOD solubilization efficiency was taken as

21% to derive the energy parameters. Net energy and energy ratio are the two fundamental factors that conclude the energy competence and pretreatment efficiency [2,45,53,54]. The net energy (−0.01 kWh) and energy ratio (0.8) for SS were less compared to SAS, where the net energy (0.04 kWh) and an energy ratio of (1.8) were obtained. It is proclaimed that the SAS pretreatment process would benefit when there is an energy ratio greater than 1. This is similar to the work of Rajesh Banu et al. (2020) [55]. This certifies that combinative pretreatment of SAS was a more energy valuable pretreatment than SS.

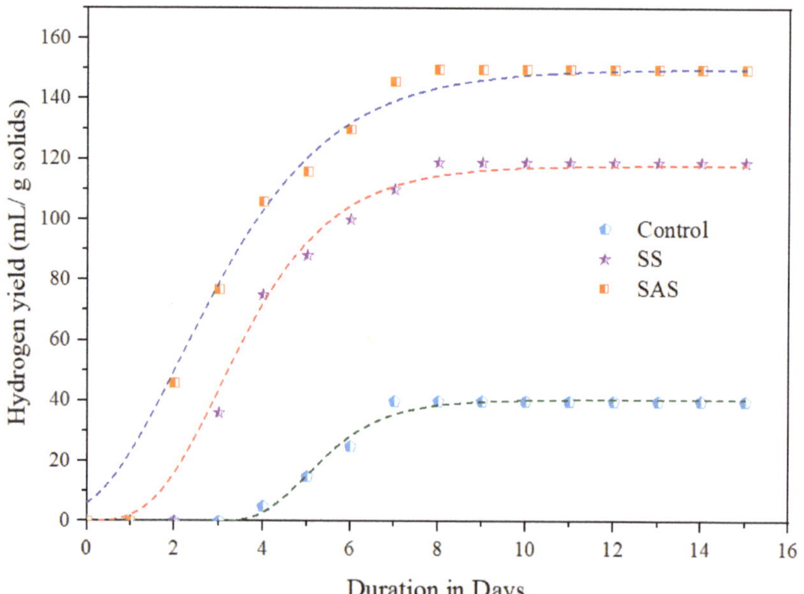

Figure 6. Biohydrogen production in control, SS and SAS.

Table 2. Kinetic analysis for various solubilized samples through Gompertz modelling.

S. No.	Samples	K (mL/d)	H_P (mL)	H_l (days)	R^2
1	SAS	0.99	150.1	1.5	0.995
2	SS	0.67	119	2.6	0.986
3	Control	0.47	40	3.7	0.983

Table 3. Biohydrogen production from different species of marine macroalgae with various combinative pretreatments.

S.no	Marine *Macroalgae Species*	Pretreatment	Operational *Parameters*	Hydrogen *Yield*	Reference
1	Ulva reticulata	Surfactant coupled with disperser pretreatment	Disperser—10,000 rpm, time—30 min, Surfactant—80 dosage (21.6 mg/L)	ΔY: 63 mL H_2/g COD	[30]
		Acidic-hydrogen peroxide coupled with microwave pretreatment	Microwave power—40%, time—10 min, pH—5, H_2O_2 concentration—0.024 g/g TS,	ΔY: 63 mL H_2/g COD	[28]

Table 3. *Cont.*

S.no	Marine *Macroalgae Species*	Pretreatment	Operational *Parameters*	Hydrogen *Yield*	Reference
2	Laminania Japonica	Heat pretreatment using autoclave	Temperature—121 °C, Duration—30 min	ΔY: 83.45 ± 96 mL/g	[51]
		Sonication pretreatment	Frequency—20Khz	ΔY: 23.56 ± 4.56 mL/g	[51]
		Thermal pretreatment	Temperature—170 °C Duration—20 min	ΔY: 109.6 mL/g	[52]
		Microwave combined with acidic pretreatment	Temperature—140 °C, Duration—15 min, H_2SO_4—1%	ΔY: 28 mL/g	[45]
3	*Padina tetrastromatica*	Acidic pretreatment	Sulphuric acid—1% *v/v* of H_2SO_4	ΔY: 78 ± 2.9 mL/ 0.05 gVS	[53]
4	*Chaetomorpha antennina*	Surfactant coupled with microwave pretreatment	Microwave power-0.36 KW, Duration—15 min, Surfactant dosage—0.0035 g /g TS	ΔY: 74.5 mL H_2/g COD	[2]
		Alkali (NaOH) combined with sonication pretreatment	Sonication intensity—50%, Duration—30 min, pH—11	ΔY: 150 mL H_2/g COD	This study

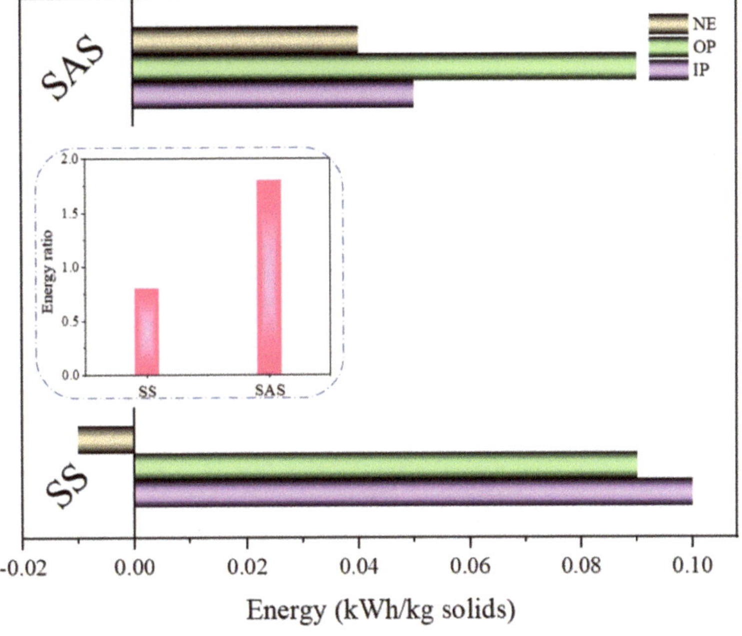

Figure 7. Energy interpretation in SS and SAS.

4. Conclusions and Future Areas of Research

An exploration was made to generate energy-efficient biohydrogen from marine macroalgae by utilizing chemo sonic pretreatment. SS liberated a SCOD release of 2650 mg/L and COD solubilization of 21%, which was lesser than SAS in which a SCOD release of

3400 mg/L and COD solubilization of about 26% was obtained. In comparison with control (40 mL H_2/gCOD) and SS (119 mL H_2/gCOD), SAS (150 mL H_2/gCOD) showed maximum biohydrogen production. pH 11 was the appropriate range for alkali with 50% sonication intensity and 30 min duration for energy-efficient biohydrogen production. VFA production was higher in SAS (1800 mg/L) when compared to SS (860 mg/L) and control (110 mg/L). SAS stated net energy of 0.04 kWh/kg of marine macroalgae biomass and an energy ratio of 1.8, which was effective when compared to SS, in which net energy of -0.01 kWh/kg and an energy ratio of 0.8 was obtained. Hence, chemo sonic pretreatment was regarded as a promising pretreatment approach for biohydrogen generation from marine macroalgae.

Marine macroalgae have emerged as prospective sources for biobased products and biofuel, making them the most viable and desirable biofuel sources. The development of commercial bio-refinery technologies, which primarily utilize marine macroalgae as feed, may be restricted by a distinct lack of practical concepts that must be addressed before its prototype can be successfully sold. Numerous lab-scale experiments are currently being performed, however, it is uncertain whether these technologies could be implemented in the near future. The efficacy of the bioprocess and output of the bioproduct should be reviewed as a result of the scale-up process in order to keep records of losses that happened. Other challenges include species selection as well as conventional microorganisms' role in hydrolysis, conversion and utilization of particular polysaccharides. The development of marine macroalgal biorefineries may be limited by its inability to scale up the biotechnologies which is now being used to conduct ongoing research. Freshwater utilization rises as the biorefinery process progresses, which leads to a freshwater shortage worldwide. The feasibility of using saltwater in a specific biorefinery process has been demonstrated in some research, but it has not yet been verified in a comprehensive marine macroalgal biorefinery process, which entails a number of interrelated processes and activities.

It is essential to identify the spectrum of potential bioproducts and biofuels for each marine macroalgae variety that may be grown sustainably, as well as the best, most comprehensive and unified bioprocessing methods. This information can depend on the long-term sustainability and financial benefit of the green economy. The marine macroalgal sector develops if all bioprocessing steps and the range of potential bioproducts are maintained in a centralized system that can be accessed globally. A strong collaboration between academics and industries which comprises environmental engineers, marine scientists, skillful laborers and economists should yield effective methods for biofuel production from marine macroalgae. The organization of the bioeconomy in a particular nation could undergo a dramatic change in the following decades due to the effects of global warming. As a result of the rising temperatures brought on by climate change, research has revealed potential changes in the geographical distribution of marine macroalgae in diverse coastal environments. Shifts in marine macroalgal distribution affect the infrastructure, locations, employment opportunities and overall viability of marine macroalgal biorefineries in the bioeconomy. Therefore, it is essential to model and predict the transformation of the commercially significant marine macroalgal species under climate change. Better macroalgae collection techniques to have a high yield of biofuels via genetic alteration will be the future of algal biology. The marine macroalgae chosen for biofuel production should suit all the environmental requirements so that they can be considered as a sustainable feedstock. The bioprocessing characteristics of each marine macroalgal species, such as life cycle evaluation, energy and energy-based modeling, should be accurately examined using various eco-friendly techniques. This could help for better sustainable biorefinery development.

Author Contributions: Conceptualization, supervision, T.K.; writing—original draft preparation, S.S.; data curation, R.B.J. All authors have read and agreed to the published version of the manuscript.

Funding: This research received no external funding.

Institutional Review Board Statement: Not applicable.

Informed Consent Statement: Not applicable.

Data Availability Statement: Not applicable.

Conflicts of Interest: The authors declare no conflict of interest.

References

1. Banu, J.R.; Kavitha, S.; Kannah, R.Y.; Usman, T.M.; Kumar, G. Application of chemo thermal coupled sonic homogenization of marine macroalgal biomass for energy efficient volatile fatty acid recovery. *Bioresour. Technol.* **2020**, *303*, 122951. [CrossRef] [PubMed]
2. Kumar, D.; Eswari, A.P.; Park, J.-H.; Adishkumar, S.; Banu, J.R. Biohydrogen Generation from Macroalgal Biomass, Chaetomorpha antennina Through Surfactant Aided Microwave Disintegration. *Front. Energy Res.* **2019**, *7*, 78. [CrossRef]
3. Praseptiangga, D. Development of Seaweed-based Biopolymers for Edible Films and Lectins. *IOP Conf. Ser. Mater. Sci. Eng.* **2017**, *193*, 012003. [CrossRef]
4. Wahlström, N.; Edlund, U.; Pavia, H.; Toth, G.; Jaworski, A.; Pell, A.J.; Choong, F.X.; Shirani, H.; Nilsson, K.P.R.; Richter-Dahlfors, A. Cellulose from the green macroalgae Ulva lactuca: Isolation, characterization, optotracing, and production of cellulose nanofibrils. *Cellulose* **2020**, *27*, 3707–3725. [CrossRef]
5. Naik Kishore, K.; Naik Prarameshwara, T. Preparation and Characterisation of Biodiesel from Marine Macro-algae Chaetomorpha spp. *Int. J. Res. Appl. Sci. Biotechnol.* **2020**, *7*, 242–247. [CrossRef]
6. Lindfors, A.; Feiz, R.; Eklund, M.; Ammenberg, J. Assessing the Potential, Performance and Feasibility of Urban Solutions: Methodological Considerations and Learnings from Biogas Solutions. *Sustainability* **2019**, *11*, 3756. [CrossRef]
7. Baena-Moreno, F.; Malico, I.; Marques, I. Promoting Sustainability: Wastewater Treatment Plants as a Source of Biomethane in Regions Far from a High-Pressure Grid. A Real Portuguese Case Study. *Sustainability* **2021**, *13*, 8933. [CrossRef]
8. Milledge, J.J.; Nielsen, B.V.; Maneein, S.; Harvey, P.J. A Brief Review of Anaerobic Digestion of Algae for Bioenergy. *Energies* **2019**, *12*, 1166. [CrossRef]
9. Rahman, S.; Masdar, M.; Rosli, M.; Majlan, E.; Husaini, T.; Kamarudin, S.; Daud, W. Overview biohydrogen technologies and application in fuel cell technology. *Renew. Sustain. Energy Rev.* **2016**, *66*, 137–162. [CrossRef]
10. Global Hydrogen Review 2021. Available online: https://www.iea.org/reports/global-hydrogen-review-2021 (accessed on 18 September 2022).
11. Sekoai, P.T.; Awosusi, A.; Yoro, K.O.; Singo, M.; Oloye, O.; Ayeni, A.O.; Bodunrin, M.O.; Daramola, M. Microbial cell immobilization in biohydrogen production: A short overview. *Crit. Rev. Biotechnol.* **2017**, *38*, 157–171. [CrossRef] [PubMed]
12. Sarangi, P.K.; Nanda, S. Biohydrogen production through dark fermentation. *Chem. Eng. Technol.* **2020**, *43*, 601–612. [CrossRef]
13. Gawel, E.; Pannicke, N.; Hagemann, N. A Path Transition Towards a Bioeconomy—The Crucial Role of Sustainability. *Sustainability* **2019**, *11*, 3005. [CrossRef]
14. D'Adamo, I.; Gastaldi, M.; Morone, P.; Rosa, P.; Sassanelli, C.; Settembre-Blundo, D.; Shen, Y. Bioeconomy of Sustainability: Drivers, Opportunities and Policy Implications. *Sustainability* **2022**, *14*, 200. [CrossRef]
15. Majer, S.; Wurster, S.; Moosmann, D.; Ladu, L.; Sumfleth, B.; Thrän, D. Gaps and Research Demand for Sustainability Certification and Standardisation in a Sustainable Bio-Based Economy in the EU. *Sustainability* **2018**, *10*, 2455. [CrossRef]
16. D'Adamo, I.; Gastaldi, M.; Imbriani, C.; Morone, P. Assessing regional performance for the Sustainable Development Goals in Italy. *Sci. Rep.* **2021**, *11*, 24117. [CrossRef]
17. Kucharska, K.; Rybarczyk, P.; Hołowacz, I.; Łukajtis, R.; Glinka, M.; Kamiński, M. Pretreatment of Lignocellulosic Materials as Substrates for Fermentation Processes. *Molecules* **2018**, *23*, 2937. [CrossRef]
18. Kumar, A.K.; Sharma, S. Recent updates on different methods of pretreatment of lignocellulosic feedstocks: A review. *Bioresour. Bioprocess.* **2017**, *4*, 7. [CrossRef]
19. Aarthy, A.; Kumari, S.; Turkar, P.; Subramanian, S. An insight on algal cell disruption for biodiesel production. *Asian J. Pharm. Clin. Res.* **2018**, *11*, 21–26. [CrossRef]
20. Naveena, B.; Armshaw, P.; Pembroke, J.T. Ultrasonic intensification as a tool for enhanced microbial biofuel yields. *Biotechnol. Biofuels* **2015**, *8*, 140. [CrossRef] [PubMed]
21. SriBala, G.; Chennuru, R.; Mahapatra, S.; Vinu, R. Effect of alkaline ultrasonic pretreatment on crystalline morphology and enzymatic hydrolysis of cellulose. *Cellulose* **2016**, *23*, 1725–1740. [CrossRef]
22. Mamvura, T.A.; Iyuke, S.E.; Paterson, A.E. Energy changes during use of high-power ultrasound on food grade surfaces. *South Afr. J. Chem. Eng.* **2018**, *25*, 62–73. [CrossRef]
23. Wahid, R.; Romero-Guiza, M.; Moset, V.; Møller, H.B.; Fernández, B. Improved anaerobic biodegradability of wheat straw, solid cattle manure and solid slaughterhouse by alkali, ultrasonic and alkali-ultrasonic pre-treatment. *Environ. Technol.* **2020**, *41*, 997–1006. [CrossRef]
24. Tamilarasan, K.; Banu, J.R.; Kumar, M.D.; Sakthinathan, G.; Park, J.-H. Influence of Mild-Ozone Assisted Disperser Pretreatment on the Enhanced Biogas Generation and Biodegradability of Green Marine Macroalgae. *Front. Energy Res.* **2019**, *7*, 89. [CrossRef]
25. Kumar, A.; Kamra, D.; Agarwal, N.; Chaudhary, L. Effect of Graded Levels of Bromoethanesulfonic Acid Supplementation on Methane Production, Rumen Microbial Diversity and Fermentation Characteristics. *Anim. Nutr. Feed Technol.* **2019**, *19*, 15. [CrossRef]

26. Vasiliadou, I.; Berná, A.; Manchon, C.; Melero, J.A.; Martinez, F.; Esteve-Nuñez, A.; Puyol, D. Biological and Bioelectrochemical Systems for Hydrogen Production and Carbon Fixation Using Purple Phototrophic Bacteria. *Front. Energy Res.* **2018**, *6*, 107. [CrossRef]

27. Woo, H.C.; Kim, Y.H. Eco-efficient recovery of bio-based volatile C2–6 fatty acids. *Biotechnol. Biofuels* **2019**, *12*, 92. [CrossRef]

28. Kumar, M.D.; Kaliappan, S.; Gopikumar, S.; Zhen, G.; Banu, J.R. Synergetic pretreatment of algal biomass through H2O2 induced microwave in acidic condition for biohydrogen production. *Fuel* **2019**, *253*, 833–839. [CrossRef]

29. Magrini, F.E.; de Almeida, G.M.; Soares, D.D.M.; Fuentes, L.; Ecthebehere, C.; Beal, L.L.; da Silveira, M.M.; Paesi, S. Effect of different heat treatments of inoculum on the production of hydrogen and volatile fatty acids by dark fermentation of sugarcane vinasse. *Biomass Convers. Biorefinery* **2020**, *11*, 2443–2456. [CrossRef]

30. Kumar, M.D.; Tamilarasan, K.; Kaliappan, S.; Banu, J.R.; Rajkumar, M.; Kim, S.H. Surfactant assisted disperser pretreatment on the liquefaction of Ulva reticulata and evaluation of biodegradability for energy efficient biofuel production through nonlinear regression modelling. *Bioresour. Technol.* **2018**, *255*, 116–122. [CrossRef]

31. Weijun, Y. Analytical accuracy of hydrogen measurement using gas chromatography with thermal conductivity detection. *J. Sep. Sci.* **2015**, *38*, 2640–2646. [CrossRef]

32. Kavitha, S.; Banu, J.R.; Subitha, G.; Ushani, U.; Yeom, I.T. Impact of thermo-chemo-sonic pretreatment in solubilizing waste activated sludge for biogas production: Energetic analysis and economic assessment. *Bioresour. Technol.* **2016**, *219*, 479–486. [CrossRef]

33. APHA; AWWA. *WEF, Standard Methods for the Examination of Water and Wastewater*, 21st ed.; American Public Health Association/American Water Works Association/Water Environment Federation: Washington, DC, USA, 2005.

34. Banu, J.R.; Tamilarasan, K.; Rani, R.U.; Gunasekaran, M.; Cho, S.-K.; Al-Muhtaseb, A.H.; Kumar, G. Dispersion aided tenside disintegration of seagrass Syringodium isoetifolium: Towards biomethanation, kinetics, energy exploration and evaluation. *Bioresour. Technol.* **2019**, *277*, 62–67. [CrossRef]

35. Karlsson, H.; Ahlgren, S.; Sandgren, M.; Passoth, V.; Wallberg, O.; Hansson, P.-A. A systems analysis of biodiesel production from wheat straw using oleaginous yeast: Process design, mass and energy balances. *Biotechnol. Biofuels* **2016**, *9*, 229. [CrossRef]

36. Shojaeiarani, J.; Bajwa, D.; Holt, G. Sonication amplitude and processing time influence the cellulose nanocrystals morphology and dispersion. *Nanocomposites* **2020**, *6*, 41–46. [CrossRef]

37. Karray, R.; Hamza, M.; Sayadi, S. Evaluation of ultrasonic, acid, thermo-alkaline and enzymatic pre-treatments on anaerobic digestion of Ulva rigida for biogas production. *Bioresour. Technol.* **2015**, *187*, 205–213. [CrossRef]

38. Soontornchaiboon, W.; Kim, S.M.; Pawongrat, R. Effects of alkaline combined with ultrasonic pretreatment and en zymatic hydrolysis of agricultural wastes for high reducing sugar production. *Sains Malays.* **2016**, *45*, 955–962.

39. Jákói, Z.; Lemmer, B.; Hodúr, C.; Beszédes, S. Microwave and Ultrasound Based Methods in Sludge Treatment: A Review. *Appl. Sci.* **2021**, *11*, 7067. [CrossRef]

40. Kumar, G.; Sivagurunathan, P.; Zhen, G.; Kobayashi, T.; Kim, S.-H.; Xu, K. Combined pretreatment of electrolysis and ultrasonication towards enhancing solubilization and methane production from mixed microalgae biomass. *Bioresour. Technol.* **2017**, *245*, 196–200. [CrossRef]

41. Li, W.; Leong, T.S.H.; Ashokkumar, M.; Martin, G.J.O. A study of the effectiveness and energy efficiency of ultrasonic emulsification. *Phys. Chem. Chem. Phys.* **2018**, *20*, 86–96. [CrossRef]

42. Yin, Y.; Wang, J. Isolation and characterization of a novel strain Clostridium butyricum INET1 for fermentative hydrogen production. *Int. J. Hydrogen Energy* **2017**, *42*, 12173–12180. [CrossRef]

43. Sambusiti, C.; Bellucci, M.; Zabaniotou, A.; Beneduce, L.; Monlau, F. Algae as promising feedstocks for fermentative biohydrogen production according to a biorefinery approach: A comprehensive review. *Renew. Sustain. Energy Rev.* **2015**, *44*, 20–36. [CrossRef]

44. Tamilarasan, K.; Kavitha, S.; Banu, J.R.; Arulazhagan, P.; Yeom, I.T. Energy-efficient methane production from macroalgal biomass through chemo disperser liquefaction. *Bioresour. Technol.* **2017**, *228*, 156–163. [CrossRef]

45. Yin, Y.; Wang, J. Pretreatment of macroalgal Laminaria japonica by combined microwave-acid method for biohydrogen production. *Bioresour. Technol.* **2018**, *268*, 52–59. [CrossRef]

46. Yang, G.; Wang, J. Pretreatment of grass waste using combined ionizing radiation-acid treatment for enhancing fermentative hydrogen production. *Bioresour. Technol.* **2018**, *255*, 7–15. [CrossRef]

47. Xia, A.; Jacob, A.; Tabassum, M.R.; Herrmann, C.; Murphy, J.D. Production of hydrogen, ethanol and volatile fatty acids through co-fermentation of macro- and micro-algae. *Bioresour. Technol.* **2016**, *205*, 118–125. [CrossRef]

48. Hatamoto, M.; Kaneko, T.; Takimoto, Y.; Ito, T.; Miyazato, N.; Maki, S.; Yamaguchi, T.; Aoi, T. Microbial Community Structure and Enumeration of Bacillus species in Activated Sludge. *J. Water Environ. Technol.* **2017**, *15*, 233–240. [CrossRef]

49. Kannah, R.Y.; Kavitha, S.; Sivashanmugham, P.; Kumar, G.; Nguyen, D.D.; Chang, S.W.; Banu, J.R. Biohydrogen production from rice straw: Effect of combinative pretreatment, modelling assessment and energy balance consideration. *Int. J. Hydrogen Energy* **2019**, *44*, 2203–2215. [CrossRef]

50. Tamilarasan, K.; Arulazhagan, P.; Rani, R.U.; Kaliappan, S.; Banu, J.R. Synergistic impact of sonic-tenside on biomass disintegration potential: Acidogenic and methane potential studies, kinetics and cost analytics. *Bioresour. Technol.* **2018**, *253*, 256–261. [CrossRef]

51. Liu, H.; Wang, G. Fermentative hydrogen production from macro-algae Laminaria japonica using anaerobic mixed bacteria. *Int. J. Hydrogen Energy* **2014**, *39*, 9012–9017. [CrossRef]

52. Jung, K.-W.; Kim, D.-H.; Shin, H.-S. Fermentative hydrogen production from Laminaria japonica and optimization of thermal pretreatment conditions. *Bioresour. Technol.* **2011**, *102*, 2745–2750. [CrossRef]
53. Radha, M.; Murugesan, A. Enhanced dark fermentative biohydrogen production from marine macroalgae Padina tetrastromatica by different pretreatment processes. *Biofuel Res. J.* **2017**, *4*, 551–558. [CrossRef]
54. Sharmila, V.G.; Tamilarasan, K.; Kumar, M.D.; Kumar, G.; Varjani, S.; Kumar, S.A.; Banu, J.R. Trends in dark biohydrogen production strategy and linkages with transition towards low carbon economy: An outlook, cost-effectiveness, bottlenecks and future scope. *Int. J. Hydrogen Energy* **2022**, *47*, 15309–15332. [CrossRef]
55. 55. Banu, J.R.; Tamilarasan, T.; Kavitha, S.; Gunasekaran, M.; Gopalakrishnankumar; Al-Muhtaseb, A.H. Energetically feasible biohydrogen production from sea eelgrass via homogenization through a surfactant, sodium tripolyphosphate. *Int. J. Hydrogen Energy* **2020**, *45*, 5900–5910. [CrossRef]

 *sustainability* 

Article

Participatory Planning for the Drafting of a Regional Law on the Bioeconomy

Elvira Tarsitano [1,2,*], Simona Giordano [3], Gianluigi de Gennaro [4], Annalisa Turi [5], Giovanni Ronco [6] and Lucia Parchitelli [7]

1   Sustainability Center, University of Bari Aldo Moro, 70100 Bari, Italy
2   ABAP-APS, 70100 Bari, Italy
3   Department of Research and Humanistic Innovation, University of Bari Aldo Moro, 70100 Bari, Italy
4   Department of Biology, University of Bari Aldo Moro, 70100 Bari, Italy
5   U.O. Center of Third Mission, Research Section, University of Bari Aldo Moro, 70100 Bari, Italy
6   Confindustria Puglia, 70125 Bari, Italy
7   Council of the Puglia Region, 70100 Bari, Italy
*   Correspondence: elvira.tarsitano@uniba.it; Tel.: +39-080-467-9838

Abstract: In an increasingly complex global economic scenario, sustainability represents a fundamental compass aimed to guide actions of institutions and individuals. A nondissipative use of Earth's resources is feasible through a common effort that reconsiders the actual development system according to the key principles of the bioeconomy. It is vital to start from local contexts to reach the global dimension by exploiting the opportunities available in each territory. Starting from these assumptions, the participatory process activated in the Apulia region has represented the first step towards an intervention strategy in the panorama of the bioeconomy, and has made it possible to increase the awareness of a development based on the adoption of bioeconomy models and, therefore, circular economy ones through an effective inclusion process. A process has given rise to a project allowing all involved actors to reflect on the double economy–environment system, to share good practices and promote the adoption of lifestyles and consumption styles more compatible with the principles of the bioeconomy and to elaborate a proposal for a participatory regional law for the bioeconomy in the Apulia region as an expression of the collaboration between different bodies and institutions (universities, Confindustria and the council of the Puglia region).

Keywords: bioeconomy; sustainable development; 2030 Agenda; natural resources; participation

Citation: Tarsitano, E.; Giordano, S.; de Gennaro, G.; Turi, A.; Ronco, G.; Parchitelli, L. Participatory Planning for the Drafting of a Regional Law on the Bioeconomy. *Sustainability* **2023**, *15*, 7192. https://doi.org/10.3390/su15097192

Academic Editors: Idiano D'Adamo and Massimo Gastaldi

Received: 23 February 2023
Revised: 22 April 2023
Accepted: 23 April 2023
Published: 26 April 2023

1. Introduction

In line with the communication to the European parliament, the council, the European economic and social committee and the committee of the regions of 11 March 2020 [1], the European commission has defined a new action plan for the circular economy, entitled "For a cleaner and more competitive Europe", establishing a future-oriented program to reach the cited objective in cocreation with different actors [2]. Furthermore, the plan aims to accelerate the profound changes required by the European Green Deal, based on actions to the circular economy implemented since 2015. This plan aims to rationalize the regulatory framework, making it suitable for a sustainable future, ensuring the optimization of new opportunities arising from the transition and minimizing the burden on people and businesses. The same plan embeds a series of interconnected initiatives designed to establish a strategic framework for sustainable products, services and business models with the goal to help transform consumption patterns so as to avoid, in the first place, waste generation. In fact, the new regulatory framework has the potential to allow for the achievement of the objectives set out by the new directives on waste prevention, recycling and reduction in landfill disposal. At the same time, the same framework needs to support the transition to the circular economy by removing those administrative and procedural criticalities that too often hinder and slow down its development, aiming to overcome the strong territorial inhomogeneities currently existing in the management of the waste cycle in Italy as a whole, as well as through the construction of necessary systems and infrastructures. As a consequence, it appears necessary that the process of drafting legislative decrees be accompanied by extensive discussions with stakeholders and that the deadline set for the transposition of the new directives into national law is respected. In this scenario, the bioeconomy [3–6] represents the answer to a large part of the current global challenges, from global warming to all the issues related to climate change, to smart agriculture limiting the adoption of pesticides. The bioeconomy [7–10], including the mentioned principles of the circular economy, fosters the adoption of a model of sustainable development, not only devoted to mere profits and profitability, but also to social progress, considered the driving force for achieving the objectives of the 2015 Paris Agreement, as well as the United Nations 2030 Agenda for Sustainable Development [11–13]. By virtue of this, Europe, as well as Italy, recognizing its key role, and has strived to implementing a sound strategy for the bioeconomy. As far as Italy is concerned, in May 2019, the update of the "National Bioeconomy Strategy" [14–17] was presented, with the related implementation program in view of the new "European Bioeconomy Strategy", strongly emphasizing the need to orient all sectors of the bioeconomy towards circularity and environmental, economic and social sustainability.

As for Italy, the European Green Deal [18,19] plays an extraordinary role and constitutes a precious opportunity towards development along a path of ecological transition; this necessarily requires that Italy be able to define its own coherent strategic framework and develop actions to effectively increase and use the financial resources made available by the European plan. The start of the process for a National Green Deal constitutes an essential reference from the point of view of the transition to a circular economy. However, this project needs to be significantly strengthened both from the point of view of public and private investments and from the point of view of a more comprehensive and coherent reorientation of all public policies towards the ecological transition and the circular economy, all within the framework of the European Green Deal. The different regions can play a decisive strategic role in the transition to a circular economy, as they have the necessary regulatory skills and responsibilities, in addition to the knowledge and experience on the different territories, capable of defining realistic objectives, to be pursued on a local territorial and differentiated scale, as "the regions are large enough to make a difference and small enough to make it happen". The OECD [12], through "The Bioeconomy to 2030: designing a policy agenda", defines a true industrial revolution capable of innovating mature sectors, such as those of raw materials, waste, energy production and of guaranteeing long-term environmental, economic and social sustainability within the global economic system. Taking into account the territorial processes included in the annual "Program for the participation of the Puglia Region pursuant to LR N.28/2017—Law on Participation" [20], into which the Manifesto for the Bioeconomy in Puglia (MaBiP) project [21] was inserted, it is vital to highlight, with regard to the issue at stake, the importance of participation from all stakeholders vital to combine innovation and environmental protection. Change is a collective action: the public expresses the needs and governs; the private sector provides skills and financial resources. Without collaboration and partnership, sustainability cannot be achieved (Goal 17, Agenda 2030) [11]. In order for the bioeconomy to win the challenge of "re-integrating economy, society and the environment", it is not enough to simply use biomass for industrial applications or to use renewable raw materials instead of fossil ones. It cannot all be considered a mere question of integrating biological knowledge into existing technology; to overcome the described challenge, the transition must also take place at a social level, stimulating awareness and dialogue, as well as supporting innovation in social structures in order to promote more conscious and aware behaviours.

It is, moreover, fundamental to enhance knowledge related to what is consumed (in particular food products and related processes) to favour the improvement of people's health and lifestyle, thus, stimulating a demand that pushes companies towards sustainable innovation. This process of transition in the economy and society, in order to truly benefit from it, requires a systemic approach according to which citizens must become the real protagonists of the social transformation that the bioeconomy can produce [9,10]. Social dialogue and an understanding of the challenges and opportunities related to the bioeconomy both play a decisive role in the level of demand for new products and services, and in the innovations and technological developments associated with them. Activities such as public procurement should be placed in the context of participatory processes, so as to foster involvement, understanding and the potential for replication. Consequently, the bioeconomy also represents a challenging playground for reconnecting with the environment, economy and society, generating economic value together with new social values and a new cultural approach [22–24].

This takes renewed skills in building consensus for both the public and private sectors, and the opening of a social dialogue.

The challenge at stake requires the following:

- For private economic actors to provide business models that involve customers, workers, users and subjects interested in their activities (primarily citizens) in a common vision of sustainability; while new products, services and investments connected to the bioeconomy are created, new economic value, employment, relationships and interactions are created, thus, making it clear that the bioeconomy is able to meet social needs and improve the wellbeing of the community by also enhancing individual participation and involvement;
- On a public level, the widespread adoption of both a participatory approach to local development and of a new concept of territory, understood as a localized set of tangible and intangible assets and relationships between different public and private entities present in each region. Being aware of the territorial distribution of renewable resources, of the strengths and weaknesses, of the needs and of the barriers to development allows to recompose fragmented skills and knowledge into new stocks and flows of productive knowledge, forming an innovation matrix for the bioeconomy and contributing to creating a new territorial identity.

Starting from the above-described issues, the present contribution aims at deepening the analysis of the participatory process that led to the involvement, in a context such as the south of Italy, of various actors in sharing good practices in line with the principles of the bioeconomy; the final objective, through the same process that is detailed in the following paragraphs, consists of elaborating a proposal for a participatory regional law for the bioeconomy in the Apulia region.

2. Materials and Methods

2.1. Preliminary Considerations and Scenario Analysis: The "MaBiP" Project

As part of the public notice for the selection of participatory processes to be admitted to provide regional support within the annual program of participation of the Puglia region, pursuant to LR N.28/2017—Law on Participation-AD n.28 of 21.11.2018 [20]—the University Centre of Excellence for Sustainability of the University of Bari Aldo Moro, in partnership with the University Centre of Excellence for Innovation and Creativity and Confindustria Puglia presented the "Manifesto for the Bioeconomy in Puglia (MaBiP)" proposal [21,25], the winning result with resolution no. 238 of 16 December 2019 of the head of the special institutional communication structure. The MaBiP project was conceived as a continuation of the subscription on 20 March 2019 of the Manifesto for the Bioeconomy in Puglia by the presidency of the Puglia region, research bodies of the territory (including the University of Bari) and Confindustria, thus, involving all business world, a partnership extended to all stakeholders interested in what the OECD [12], through "The Bioeconomy to 2030: designing a policy agenda", defines a true industrial revolution capable of innovating mature sectors, such as raw materials, waste and energy, ensuring long-term environmental, economic and social sustainability within the economic system. As to the analysis conducted up to this point, it is fundamental, in order to increase awareness of the importance to promote the definition of a new economic model based on the principles of the bioeconomy, especially in industrial areas that have a strong impact on the territory, to favour the promotion, transition, creation and adoption of bioeconomy models, and, therefore, the circular economy. All regional stakeholders need to be involved at various levels in order to: facilitate connection and dialogue between stakeholders belonging to different value chains; promote and disseminate the principles of the bioeconomy at all levels; frame the regional context in the field of bioeconomy for subsequent mapping; draw up a roadmap for the strategic development of the bioeconomy; promote the drafting of a regional law proposal on the bioeconomy [26].

2.2. Phases of the Process and Activities Carried Out

The entire participation process consisted of four main steps and took a total of six months, from June to December 2020. The activities of the participatory process were carried out in a mixed way: in presence and remotely. Despite the obvious difficulties in carrying out most of the activities foreseen by the project in person, due to the concomitant pandemic caused by the SARS-CoV-2 virus [27], the technological and multimedia support and the various video-calling applications managed to ensure that all the activities foreseen from the project could be realized. The expected methodology for reaching the objectives was to achieve learning content, the effectiveness of the interventions with an integrated assessment system and the active and participatory assessment of learning. The laboratories were carried out with a small group mode with support from expert facilitators.

The working method used was design thinking (DT). The DT approach is characterized by tools and methodologies that support the generation of ideas, such as the "How Might We", in which prototyping plays a very important role. The method is not limited to a mere definition of the steps aimed at conceiving an idea, a solution, but also allows for the work team to reach its realization by drafting a prototype (Table 1).

Table 1. Design thinking.

Steps	Activities	Thematic Working Groups N.04	Target Categories
1° Exploration	OBSERVE, UNDERSTAND, DEFINE (Open innovation design thinking) Set a track for the interview • The person (target category) • The map PLENARY Each group, through its representative, talks about the group's output	What they do: Each working group designs its own interview track using target categories. One facilitator for each group.	
2° Definition	DESIGNING • Identifying the opportunities • Benchmarking with other ideas and experiences, similar and distant • Brainstorming on possible solutions PLENARY Each group, through its representative, tells the group's outputs	What they do: Benchmarking Brainstorming	Citizens Enterprises Third sector Public institutions
3° Ideation and creation	PROTOTYPE and TEST The question "How can we … " to answer to meet the needs of our target PLENARY Each group, through its representative, tells the group's outputs	What they do: Define the "How Might We" question (HMW) Each work group designs and manufactures at least 3 prototypes to solve needs. The prototype can be composed of any means (real drawing, software, web, etc.)	
4° Sharing and validation	TEST AND ORGANIZATIONAL CHANGE PLENARY Each group, through its representative, tells the group's outputs	What to do: They share the prototype in plenary Acquire feedback with a shared word file, with chat and with direct intervention OUTPUT: Each group produces a work report	

2.3. The Hackathon

The "Circular Economy Action" [28] Hackathon, a "rally call" to map the best practices of the bioeconomy in Apulia, has represented a positive example of an effective methodology. It aimed at searching through different actions resulting from start-up or company initiatives, from associations or individual citizens, with the objective of narrating the practices capable of generating experiences of new production and consumption models. The process envisaged the following phases: a launch of call; registration on the platform; evaluation; identification of models and mapping of best practices; drafting and processing of documents.

Private and public actors were not mere spectators of the process but, indeed, protagonists within the entire project through moments of discussion, sharing of ideas and good practices already present in the region, with a particular view of innovation and highlighting the essential dimensions of circularity. The training activity, developed in a modular way for a total of 72 h, had the purpose of providing participants with in-depth knowledge related to the bioeconomy with the ultimate goal of activating specialized offices of Confindustria dedicated to the bioeconomy. The awarded operators were granted a free participation in the training course in management systems for sustainable development in the communities, a specialized module of the ISO 37101:2019 standard [29]. Thirty-four organizations participated in the "Circular Economy Action" [28] Hackathon award ceremony; despite the restrictions imposed by the COVID-19 pandemic situation [27], over three hundred people could participate and fruitfully share the experience.

2.4. Participating Laboratories

The workshops (4), led by expert facilitators, were delivered online through webinars; approximately two hundred people actively took part in them. In the course of the four participatory workshops, four themes considered as fundamental were addressed:

- Circularity, waste and climate change; circularity, food, health and lifestyles; circularity and new business models; circularity and the sustainable development of the territory;
- Each of the workshops, lasting four hours, included the following moments:
    1. Opening plenary, during which the organizers presented the methodology to conduct each workshop;
    2. Working groups divided by categories around the target themes in four virtual rooms, one for each of the themes in which the bioeconomy in Apulia had declined;
    3. Output: in this phase, each working group was asked to draw up a report embedding the main results that emerged;
    4. Closing plenary, during which each of the four working groups gave feedback on what was discussed and defined within the same working group.

The detailed reports for each laboratory were uploaded to the Puglia Partecipa platform.

3. Results

3.1. Results of the Participating Laboratories

The participants in the described workshops totalled 202; out of these, 52% were female and had an average age of forty-six. As shown in Figure 1, more than 50% of the participants were between thirty-six and fifty-five years old. Rather marginal was the presence of young people under the age of twenty-five (only 1.2%). However, the youth segment of the Apulian population was still represented, taking into account that almost 18% of the participants were under the age of thirty-five.

Although the participants born in the province of Bari constituted 44% of the total participants, the data collected showed a representativeness of all the Apulian provinces (Figure 2).

The educational level of participants in the workshops was particularly high (Figure 3), with 86% having at least a bachelor's degree; it is worth noting the data related to those who declared to have the title of PhD (22%). Only 14% of participants declared that they held a high school diploma.

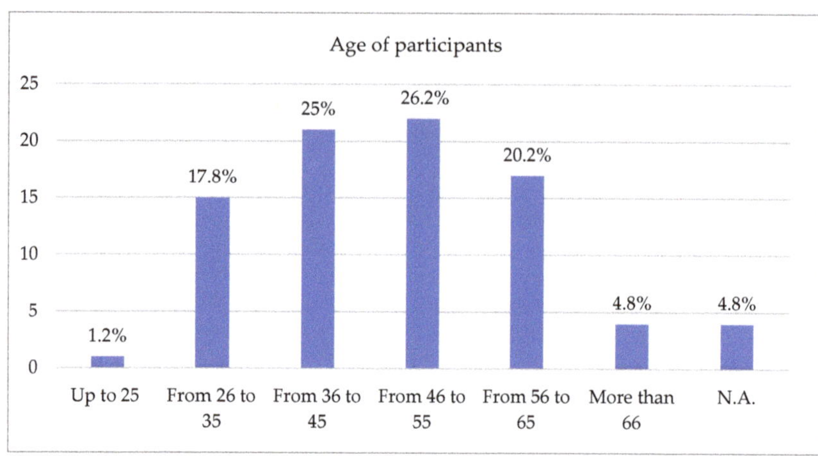

Figure 1. Age of participants.

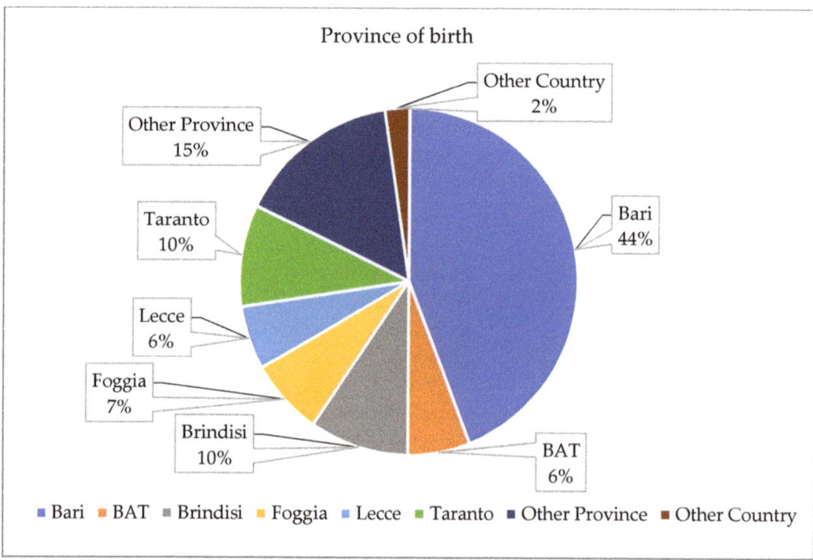

Figure 2. Province of birth.

Almost 36% of those enrolled in the participatory process took part in all four workshops included in the course. Approximately one point higher was the percentage of those who enrolled in a single laboratory. In total, 17.8% of the participants enrolled in two laboratories, and 9.5% in three laboratories (Figure 4).

Regarding the preference for the themes of each workshop (Figure 5), the recorded data showed that the percentage of those who enrolled in the workshop "Circularity and sustainable development" was slightly higher (29%). However, there was no particularly high percentage difference between this last topic and that of the other laboratories, namely, "Circularity, food, health and lifestyles", "Circularity, waste and climate change" (both at 24%) and "Circularity and new business models" (23%).

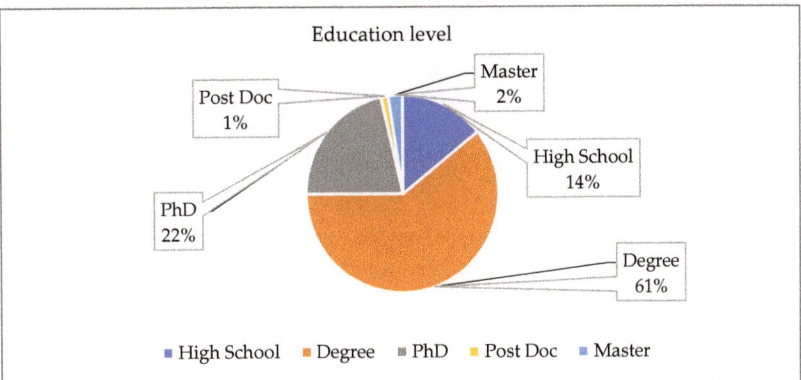

Figure 3. Education level.

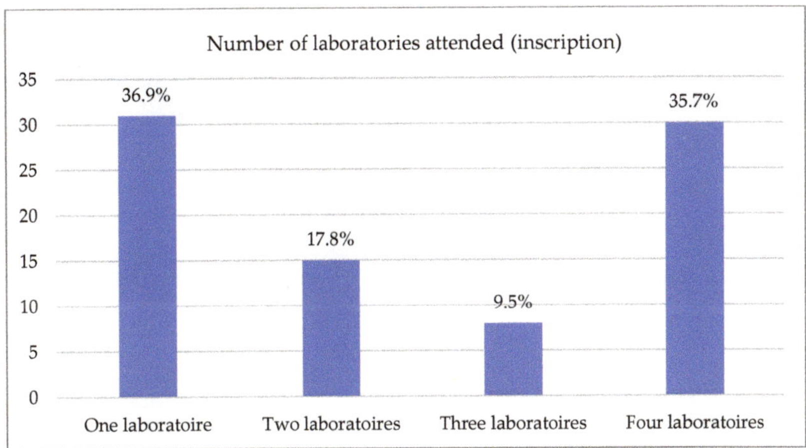

Figure 4. Number of laboratories.

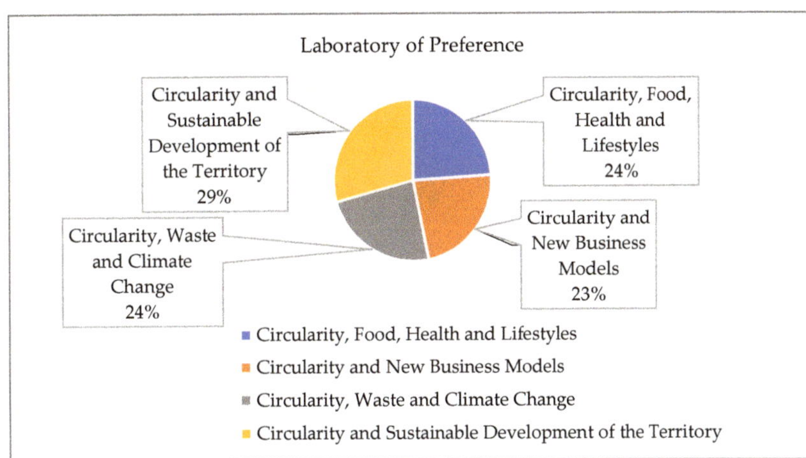

Figure 5. Laboratory of preference.

Participants in the workshops reflected a clear expression of the great variety of organizational structures present in the Apulia region. In addition to the 33% composed of citizens involved in the process, the rest was represented by the following: almost 24% came from the business world, 13.2% represented the world of associations and 10.7% belonged to public research bodies (in particular the ENEA [17] and CNR). Furthermore, albeit in a more limited percentage, the presence of cooperatives with a percentage of 5.9% and public bodies, at 2.4% (representing the Environment Council of the Municipality of Bari) was noted. On the other hand, 1.2% belonged to voluntary organizations.

3.2. Results of the Hackathon

As above-described, after the Hackathon, the award ceremony was attended by 34 organizations from the region, distributed at a prevalence of those based in Bari or the cities and towns of the same province (53%), followed in percentage by the organizations located in Taranto (17%) and Lecce (16%). Less than 10 was the percentage of the organizations coming from Foggia (9%) and from Brindisi and BAT (3% in both cases) (Figure 6).

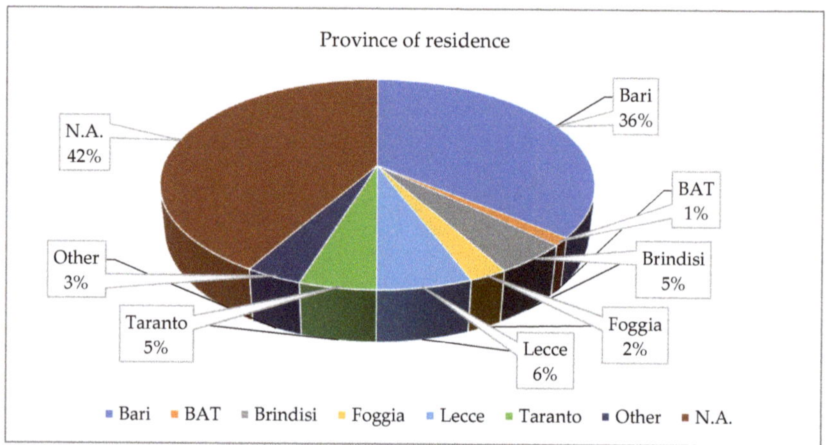

Figure 6. Province of residence.

The legal form of the organizations that took part in the Hackathon also varied (Figure 7). Companies with limited liability were obviously the most common legal forms among the participants, with a percentage that stood at 41%. Percentages greater than or equal to 15% were reported for cooperative enterprises (17%) or for associations (15%). Within the record "Other", including 18% of participants, social promotion associations, general partnerships and sole proprietorships in less significant percentages were included. In total, 9% of Hackathon participants declared they represented a natural person.

The data collected related to the sectors in which the organizations operated showed the great vivacity of the Apulia region in the field of the bioeconomy. The most represented sector was that of "Recovery, reuse and recycling", with a percentage of participating organizations equal to 32%. This was followed by the sectors of "Culture, Education and Information", with a percentage of organizations equal to 23%, that of "Technologies and solutions for the environment and the territory", with a percentage of 15%, and that of companies in the "Agrifood sector", with a percentage of 12%. Less than 10% were organizations belonging to sectors such as "Fashion and design" and "Food and fight against food waste", both at 6%, and "Sustainable mobility" and "Research and innovation", with a percentage of 3%.

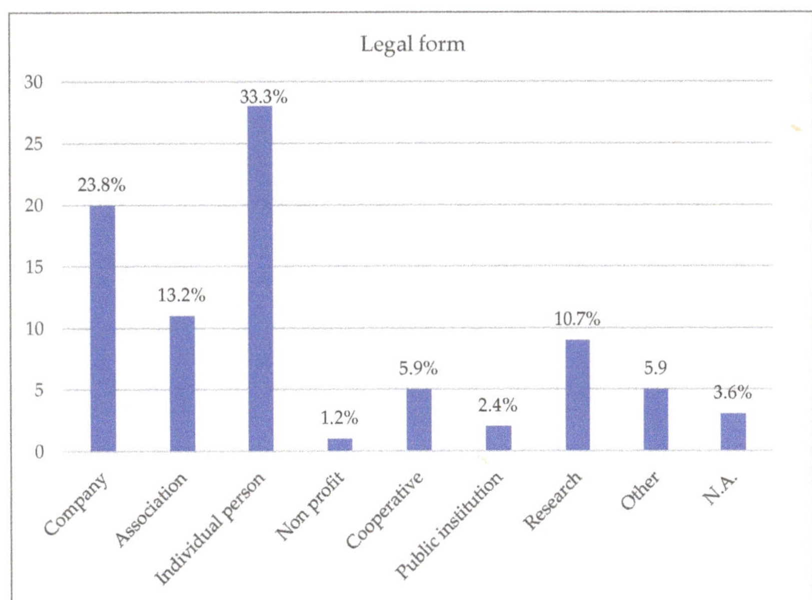

Figure 7. Legal forms of participants.

3.3. Outcomes of the Participatory Process

With regard to the outcomes of the participatory process, it emerged that during the participated workshops, each of the participants in the different working groups, divided into the cited four thematic areas, highlighted the requests/needs that the regional law on bioeconomy should possess, as summarized in Table 2.

The "Circularity, Waste and Climate Change" group carried out a reflection on how it would be possible to overcome the culture of waste, highlighting the need to define a new economic model capable of combining both the health of the environment and that of citizens, to focus on renewable energy and on the circularity of production, especially in the agrifood sector, one of the strengths of the Apulian economic system.

The "Circularity, Food, Health and Lifestyles" group sought to reflect on the promotion of a culture capable of generating new lifestyles aimed at improving the health and wellbeing of citizens, as well as through the enhancement of small production chains.

The "Circularity and New Business Models" group reflected on the need to encourage the transition to a new model of production and sustainable management of businesses, enhancing their role in reaching an effective growth of the territory.

The "Circularity and Sustainable Development of the Territory" group wanted to reflect on the need to promote a widespread and shared awareness of sustainable development, highlighting the complexity of the issue and the need to address it in a multidimensional and multidisciplinary way.

The presentation of the results of the project with the delivery of the participatory proposal document, embedding the law proposal, which took place on 27 November 2020 [11], during the final workshop. A proposal for a participatory regional law on the bioeconomy that was presented during the final meeting was the result of a development vision that should permeate the regional political strategy in order to fully achieve all the described objectives.

Table 2. Results of participating laboratories.

Target	Thematic Groups			
	Circularity, Waste and Climate Change	Circularity, Food, Health and Lifestyles	Circularity and New Business Models	Circularity and Sustainable Development of the Territory
Citizens	Enhancement through reward systems of recycling and reuse practices; greater control in the management of waste services; widespread training and involvement of citizens as a vehicle for the improvement of products and processes.	Introduction of control mechanisms that can limit food waste and other unsustainable behaviours; definition and implementation of information and training courses also in view of the recovery of the sense of active and sustainable citizenship.	Involvement and information of citizens for the definition of sustainability problems and solutions, as well as through the design of apps for measuring the impacts of behaviour; implementation of training courses for schools of all types and levels in the field of circular economy.	Promotion of the use of participation tools (consultations, forums and civic networks) focusing on the bioeconomy by recovering the sense of active citizenship and belonging to the community; providing incentives, including economic ones, that favour a change in mentality and the adoption of new lifestyles; implementation of information and training courses on the bioeconomy from primary school.
Companies	Improvement in research and technological developments in the sector, as well as through specific funding and the regeneration of regional production districts; economic support for the development of MOCAs (mitigation of obsolescence cost analyses) for a reduction in production costs; creation of supply chains for the recovery and enhancement of waste and reduction in energy use, favouring a transition in line with the Green Deal.	Improvement in research and developments in the sector, with a particular focus on packaging; recovery of the ethical sense and CSR, applying the logic of fair trade to indigenous productions and limiting the use of natural capital as much as possible; promoting dialogue between small local businesses and large retailers in order to promote the visibility of the former by making them protagonists of the regional economy; attention to waste reduction by promoting the redistribution of leftovers and reusable waste.	Improvement in research and developments in the sector to anticipate the market; promoting corporate social responsibility and increasing the social return on investment (SROI) by collaborating with local associations and creating synergy with the nonprofit sector; to provide the presence of a resource manager in each company.	Improvement in research and technological and managerial innovations for the definition of a new economic model; to encourage the creation of business networks that collaborate with universities and research institutions; promotion of tax incentives for companies that provide for the presence of green jobs (reconversion such as reskilling, upskilling and/or new hires); introduction of ad hoc managerial figures, for example, sustainability managers; promoting dialogue between small and large companies, including through the creation of environmental consultancy services by the business confederations in support of small- and medium-sized enterprises.
Third sector	Promotion of the culture of sustainable consumption through information and communication campaigns, with particular attention paid to the issue of reducing plastics and waste disposed.	Creation of networks and associations between organizations and associations of the third sector and to promote greater awareness of citizens regarding the logic of the economy and the market; to support the fight against food waste and to promote sharing economy initiatives.	Promotion of dialogue between businesses and the third sector so that there is a mutual and territorial improvement; implementation of information, training and education courses for schools and citizens.	Improvement in the relationship and communication between citizens and institutions by providing greater opportunities for participation and the creation of bodies that favour and guarantee the above.
Public institutions	Investment in research and innovation, as well as by financing the creation of innovative start-ups and encouraging the intervention of business accelerators that make large investments possible, and the creation of "shared technology halls" in order to allow investments that would otherwise be impractical; adjustment in the reference regulatory framework in order to guarantee a "new life" to the greatest possible number of waste.	Establishment of reward systems rather than sanctions to promote waste disposal, the use of renewable resources, improvement in CO_2 compensation mechanisms; creation, involvement and enhancement of organic and local production chains, especially if attentive to reuse and recycling; promotion of information and training courses on the issues in question that also involve schools of all levels; identification of certifiable rating protocols on sustainable production models, identifiable with a logo that can be used on the labels.	Encouraging the forms and practices of a sharing economy; financial support for businesses to become sustainable and aim for continuous improvements; facilitation from a regulatory point of view of reuse and recycling by small, medium and large enterprises, simplifying the bureaucracy as much as possible.	Establishment of a control room to connect the human resources of companies, universities and research bodies to involve different skills in the management of circularity to create a new regulatory framework for sustainability; definition and calling of ad hoc calls to promote and encourage the green conversion of businesses; for the conception of tax relief tools (to de-tax citizens and businesses that implement actions with reduced environmental impacts).

In order to achieve a circular and sustainable bioeconomy, it is vital that each political and strategic structure absorbs its principles and declines them in its own activities. The joint commitment of politics and citizenship prompted to elaborate, as part of the "Manifesto for the Bioeconomy" project in Apulia (MaBiP) [21], the following recommendations to the presidency of the Puglia region:

- The creation of a regional observatory on the bioeconomy under the guidance of the presidency of the Puglia region, through the participation office, with the objective to take care of relations and dialogue with the various departments and sectors involved in bioeconomy processes;

- Hinge the aforementioned observatory within the structures of the presidency and, in particular, of the participation office;
- The creation of a dedicated office on the bioeconomy to support companies;
- The activation of a participatory process that leads to the definition of a regional strategy for the bioeconomy that integrates with the regional forum for sustainable development and with the regional strategy for sustainable development;
- Promote the drafting of a roadmap that identifies regional models and best practices in the field of the bioeconomy.

Starting from the principles of evidence-based policy and participation, the proposed law aims at defining the regulatory principles for the establishment of a place of synergy and institutional capacity capable of facilitating the sustainable development of Apulia, structuring the collaboration between stakeholders. The participatory process produced the draft text entitled "Participatory Proposal Document", containing the proposal for a participatory regional law on the bioeconomy.

4. Conclusions

Through the described participatory process, it was possible to initiate a path of shared reflection on the double economy–environment system, with related intersections and implications. Economic systems always require positive growth rates and shun both stabilization and immobility; the environment, instead, requires balance and stability. Since there was no spontaneous convergence between the needs of the two systems, the real issue at stake was which of the two should give way to the other, whereas economy and nature should recognize the need for common subsistence and the necessary balance between themselves. A sound answer must be sought in the different degrees of modifiability in order to reach the objective to create an equilibrium in which both experience life and good health. The economy, as a human product, is, by its nature, modifiable through cultural, social, technological and design innovations, including possible changes in lifestyles to such an extent that it is possible to rely on an elasticity factor that is not only economic–technological, but also cultural–behavioural.

As to the case of the environment, it is worth noting that it is different, as natural balances have their own rules (including limits in the carrying capacity of each system) that cannot be modified or neglected by human activities. The natural equilibrium can "endure" up to a certain point, and the permitted threshold level cannot be shifted. There is no elasticity in natural balances with respect to human actions. This implies having to put aside prejudices, interpretations and absolute values, and devote time and energy to the critical and positive rediscovery of the distinctive characteristics that animate the two systems. For these reasons, recognizing the need and the potential that participation can have in the dynamics of sustainability, also in light of the contents of the United Nations 2030 Agenda (in particular Goals 4, 16 and 17) [11], the described process aimed at being innovative and multidisciplinary in order to promote the definition and enhancement of the economic and cultural model of the bioeconomy that was launched.

The participatory process retraced a creative path and a local, collective and inclusive reflection in the different contexts that experienced the same reflection. The "map" created was at the same time a participatory census, a business plan, a self-portrait and a collective biography. As a consequence, a participatory, innovative, inclusive and multidisciplinary methodological process was launched, designed to build the participation path around the four previously analysed themes.

This approach made it possible to favour the identification and sharing of development policies at a territorial level and disseminate success stories that constitute a fundamental example of how to activate bioeconomy processes, starting from existing good practices and outlining new horizons and projects that could contribute to the sustainable development of both the territory of belonging and of the entire regional area, respecting the vocations and specificities of the territories themselves.

The participated events and workshops involved companies, organizations, institutions and representative associations of all the six Apulian provinces, starting from the analysis of the different elements that contributed to the cited process.

The proposed participatory process promoted throughout the Apulian territory [26] the engagement of the main stakeholders and privileged observers in a path with particularly innovative effects; the result of the identification and sharing of new local production and consumption models strongly oriented towards sustainability in order to promote a business model that puts different and complementary sectors of the economy into a mutual dialogue, also in the context of urban policies. Not surprisingly, there was a growing consensus on the idea that to implement sustainable development paths, learning through experience and community-centred approaches is necessary.

By focusing on participation, it was possible to encourage the promotion and identification of effective and replicable bioeconomy models in the entrepreneurial and cooperative institutions that took part in the project, and in view of the setting of a regional strategic development model linked to the bioeconomy.

From an environmental point of view, the bioeconomy contributes both opportunities and challenges. Opportunities are connected to the gradual transition in the context of production processes, from the use of nonrenewable resources to renewable ones, so as to limit the environmental pressure on ecosystems and enhance their value for the purpose of their conservation, not merely considering their intrinsic value or the connection with ecosystem services that are "natural" solutions to combat climate change and hydrogeological risk, but also as a source of relevant services for the whole economy. Furthermore, the bioeconomy implies the possibility of reducing dependence on resources scarcely available in Italy. The strengthening of production activities deriving from renewable sources holds the potential to facilitate waste management, as these sources can be more easily assimilated.

However, the bioeconomy can also amplify a series of challenges as well as highlight the numerous examples of unsustainable management for the environment and human health, particularly in the food and fish industries. Furthermore, it is evident that it is often not necessary to increase the production of raw materials, but rather to increase their added value to society, improving the quality of products (e.g., in agriculture) and processes in response to the requirements of Objective 12 of the 2030 Agenda [11].

As a consequence, it is vital to proceed towards a sustainable economic system that assumes economic growth limited conditionally to the sustainability of material resources and leading to the valorisation of the new economic and cultural model of the bioeconomy in Apulia [26].

This "new" economy, despite being an interconnected whole on a conceptual level, can be divided into two parts. The first, measurable in material and energetic terms, is necessarily limited in its expansion within the natural carrying capacity, which is constant. The second component, on the other hand, being immaterial, keeps its virtually "unlimited" peculiarity. It is based on information in the availability of services in the required times and methods, as well as in the quality, in particular of relationships, both on a global level and on a territorial one.

It necessarily requires institutional and regulatory interventions with respect to the current market. Accelerating the transition towards the bioeconomy is fundamental to increase not only the competitiveness of regional industry, research and training to strengthen the position Apulia deserves in the national and international context, but above all to safeguard the environmental and sociocultural heritage of the territories. Through the dynamics of debate and comparison, the participatory process aimed at simplifying relations between regional actors on the subject of the bioeconomy, favouring transversal connections and allowing for the dissemination and use of good practices and ready-to-use technologies on the territory in order to reach a sound exploitation of the resources that the Apulian context offers.

The described participatory process enabled us to obtain a strategic vision on how to intervene in the main macroareas of the bioeconomy in Apulia (the environment, economic development and agrifood chain) with an approach devoted to effective sustainability and based on a circular logic, one that does not subtract resources from the territory, but maximizes the opportunities for reuse through technological innovation.

The fundamental objective is to generate a change in the mindset and generate a value of all the actors involved, from companies to institutions to individual citizens to such an extent that it is possible, through the participatory process, to implement a shared strategy of the development of the territory not connected to profit but, instead, to the protection of the Apulian context from an environmental and social point of view. A real industrial revolution that, from below, contribution by contribution, had as its objective the drafting of a law on the bioeconomy through a participatory process [13].

As a result, following the participatory process on 20th May at the headquarters of the Puglia regional council, the draft law "Provisions on the Bioeconomy" was presented; a proposal, originating from the described process carried out by the Centres of Excellence for Sustainability of the University of Bari Aldo Moro, in collaboration with Confindustria Puglia, had the aim of recognizing, for the Apulia region, the importance of fostering a territorial development inspired by the principles of the bioeconomy, in line with the objectives of the 2030 Agenda and the NRRP (National Recovery and Resilience Plan) [11].

Europe and Italy, recognizing their key roles, have proceeded to implement a strategy for the bioeconomy. However, being the achievement of global challenges necessarily based on the active involvement of territories and strategic levers for a sustainable revolution, it is fundamental to commit to create an alliance between institutions, research and the industry. A partnership extended to all stakeholders interested in what the OECD, "The Bioeconomy to 2030: designing a policy agenda" [11], defines a true industrial revolution capable of innovating mature sectors such as raw materials, waste, energy ensuring long-term environmental, economic and social sustainability within the economic system.

A participatory process linked to the bioeconomy [30–32], in view of its enormous innovative potential, can be a response to most of the regional and global challenges to be faced in the coming years, from environmental remediation to the problems of climate change, to the invention of new medicines, to the need to feed a world in which food needs are predicted increase by 70% between now and 2050, reconciling the economy, the environment and society.

The "transversal" nature of the bioeconomy offers a unique opportunity to face, in a comprehensive and systemic way, the mentioned cogent social challenges [33,34], as envisaged by the EU The communication "Innovation for sustainable growth: a Bioeconomy for Europe" [35,36].

In the described scenario, it is of particular interest to carry on a reflection on how bioeconomy intertwines with EU policies related to actual cogent challenges.

Among them, it is important to mention climate change; as a matter of fact, the council and the European Parliament set specific goals as to the climate for the near future. In line with these goals, the fit for 55% relates to the objective of cutting down net greenhouse gas emissions by at least 55% by 2030. As a consequence, the fit for the 55% package contains a whole set of legislative proposals to ensure that EU legislation are in line with the cited 2030 reduction goal.

As the described package deals with a comprehensive series of sectors, from agriculture to industry and the energy sector, in the framework of the present contribution, it is of particular relevance, as it addresses all aspects at the core of bioeconomy and, moreover, represents a crucial witness of how participation (at the core of this article) plays a vital role in reaching a sound and effective legislation at different levels, including the regional one [37].

Furthermore, the actual global scenario has been deeply affected by the Russia–Ukraine conflict; consequent challenges relate, as easily understood, to the supplies of gas being weaponised from Russia. The manipulation of energy markets has led to skyrocketing energy prices in the EU. In addition, unpredictable events and connected risks of the discontinuation or even the interruption of supply holds the potential to create additional pressure on energy markets.

The alternative option proposed by the renewable energy technology field has been strongly supported by means of a series of recent policies in other regions, leading to a weak outlook on the competitiveness of the European renewable energy technology industries and value chains.

In the described context, it is vital to address the exposure of consumers and businesses within the EU to increasing and volatile energy prices; this objective could be achieved by means of fostering supplies from renewable sources, thus, as well, increasing the security of the supply itself.

As a matter of fact, regulation 2022/2577 aims at accelerating the deployment of renewable energy sources through the adoption of ad hoc urgent measures mostly effective in the short term. The time frame is connected to the importance of allowing member states to adopt these same measures rapidly and to ease the permit-granting process applicable to renewable energy projects without requiring burdensome changes to their national procedures and legal systems, and ensuring a positive acceleration of the deployment of renewables in the short term. This reflects the important role that renewable energy can play in the decarbonisation of the European Union's energy system, by offering immediate solutions to replace fossil-fuel-based energy and by addressing the aggravated situation in the market [38].

The issue of sustainability is, of course, a huge challenge; it is difficult to promote sustainability, as it implies a broad vision, a strong determination and a great balance. These three characteristics of vision, determination and balance are necessary, and the open challenges appear epochal and require deeper, faster and more ambitious responses and integrated solutions, to initiate the social and economic transformation necessary to achieve the Sustainable Development Goals (SDGs) of the 2030 Agenda [11]. The present contribution, through the described process, carried out a comprehensive analysis of the participatory process that led to the development of a proposal for a participatory regional law for the bioeconomy in the Apulia region through the involvement, in a context such as the south of Italy, of various actors in sharing reflections and good practices. The outlined path represents an important case study both in the local described context and with a broader perspective.

Author Contributions: Conceptualization, E.T.; methodology, E.T. and S.G.; validation, L.P. and G.R.; formal analysis, E.T.; investigation, L.P.; resources, A.T.; data curation, G.d.G.; writing—original draft preparation, A.T.; writing—review and editing, S.G.; supervision, G.R. and E.T.; project administration, G.d.G.; funding acquisition, E.T. All authors have read and agreed to the published version of the manuscript.

Funding: This research was funded by the Apulia region—AVVISO DD 28/2018, BURP n.1250/2018.

Institutional Review Board Statement: Not applicable.

Informed Consent Statement: Not applicable.

Data Availability Statement: Not applicable.

Acknowledgments: We acknowledge the financial support of Apulia region—AVVISO DD 28/2018, BURP n.1250/2018.

Conflicts of Interest: The authors declare no conflict of interest.

References

1. A Sustainable Bioeconomy for Europe: Strengthening the Link between Economy, Society and the Environment. Communication from the Commission to the European Parliament, the Council, the European Economic and Social Committee and the Committee of the Regions. 2018. Available online: https://eur-lex.europa.eu/legal-content/IT/TXT/PDF/?uri=CELEX:52018DC0673&from=EN (accessed on 11 October 2021).
2. Van Langen, S.K.; Vassillo, C.; Ghisellini, P.; Restaino, D.; Passaro, R.; Ulgiati, S. Promoting circular economy transition: A study about perceptions and awareness by different stakeholders groups. *J. Clean. Prod.* **2021**, *316*, 128166. [CrossRef]

3. Velenturf, A.P.M.; Purnell, P.; Tregent, M.; Ferguson, J.; Holmes, A. Co-Producing a Vision and Approach for the Transition towards a Circular Economy: Perspectives from Government Partners. *Sustainability* **2018**, *10*, 1401. [CrossRef]

4. Genovese, A.; Pansera, M. The circular economy at a crossroad: Technocratic eco-modernism or convivial technology for social revolution? *Capital. Nat. Social.* **2021**, *32*, 95–113. [CrossRef]

5. European Commission—Directorate General for Research and Innovation. *Innovating for Sustainable Growth*; A Bioeconomy for Europe: Brussels, Belgium, 2012.

6. Research and Innovation. 2021. Available online: https://ec.europa.eu/info/research-and-innovation_en (accessed on 10 December 2022).

7. D'Adamo, I.; Gastaldi, M.; Morone, P.; Rosa, P.; Sassanelli, C.; Settembre-Blundo, D.; Shen, Y. Bioeconomy of Sustainability: Drivers, Opportunities and Policy Implications. *Sustainability* **2022**, *14*, 200. [CrossRef]

8. Majer, S.; Wurster, S.; Moosmann, D.; Ladu, L.; Sumfleth, B.; Thrän, D. Gaps and research demand for sustainability certification and standardisation in a sustainable bio-based economy in the EU. *Sustainability* **2018**, *10*, 2455. [CrossRef]

9. Appolloni, A.; Chiappetta Jabbour, C.J.; D'Adamo, I.; Gastaldi, M.; Settembre-Blundo, D. Green recovery in the mature manufacturing industry: The role of the green-circular premium and sustainability certification in innovative efforts. *Ecol. Econ.* **2022**, *193*, 107311. Available online: https://www.sciencedirect.com/science/article/abs/pii/S0921800921003700?via%3Dihub (accessed on 2 February 2023). [CrossRef]

10. Bioeconomy. Research and Innovation on the Bioeconomy, Funding, Collaboration and Job Opportunities, Projects and Results, Events and News. 2020. Available online: https://ec.europa.eu/info/research-and-innovation/research-area/environment/bioeconomy_en (accessed on 6 December 2022).

11. United Nations Department of Economic and Social Affairs (UNDESA)-Division for Sustainable Development Goals (DSDG) (2015) Transforming Our World: The 2030 Agenda for Sustainable Development. pp. 1–41. Available online: https://sustainabledevelopment.un.org/post2015/transformingourworld (accessed on 15 December 2022).

12. OECD. *The Bioeconomy to 2030: Designing a Policy Agenda*; OECD: Paris, France, 2009.

13. European Commission—Directorate General for Research and Innovation. *Bioeconomy Strategy Review*; A Commission Staff Working Document: Summary; European Commission: Brussels, Belgium, 2019.

14. D'Adamo, I.; Gastaldi, M.; Imbriani, C.; Morone, P. Assessing regional performance for the Sustainable Development Goals in Italy. *Sci. Rep.* **2021**, *11*, 24117. Available online: https://www.nature.com/articles/s41598-021-03635-8 (accessed on 2 February 2023). [CrossRef] [PubMed]

15. Appelstrand, M. Participation and societal values: The challenge for lawmakers and policy practitioners. *For. Policy Econ.* **2002**, *4*, 281–290. [CrossRef]

16. Circular Economy Network & ENEA. Report on the Circular Economy in Italy. 2020. Circular Economy Report 2021 in Italy-Focus on the Role of Circular Economy in the Transition to Climate Neutrality | European Circular Economy Stakeholder Platform (europa.eu). Available online: https://circulareconomy.europa.eu/platform/en/knowledge/circular-economy-report-2021-italy-focus-role-circular-economy-transition-climate-neutrality (accessed on 10 November 2022).

17. Re, B.; Magnani, G.; Zucchella, A. The future of sustainability. Value co-creation processes in the circular economy. In *The Palgrave Handbook of Corporate Sustainability in the Digital Era*; Palgrave Macmillian: London, UK, 2020.

18. Marchetti, M.; Palahì, M. Perspectives in Bioeconomy: Strategies, Green Deal and COVID-19. *J. Silvic. For. Ecol.* **2020**, *17*, 52–55.

19. Presidency of the Council of Ministers. Report "The Bioeconomy in Italy". 2020. Available online: https://cnbbsv.palazzochigi.it/it/bioeconomia/strategia-italiana/ (accessed on 5 November 2022).

20. Official Bulletin of Puglia Region—n. 84 of 17-7-2017. Participation Act. REGIONAL LAW 13 July 2017, n. 28. Notice of Selection of Participatory Process Proposals to be Admitted pursuant to L.R. No. 28/2017—Law on Participation. 2017. Available online: http://nodopp.regione.puglia.it/avviso-selezione.html (accessed on 11 October 2022).

21. MaBiP—Manifesto for the Bioeconomy in Puglia. 2021. Available online: https://partecipazione.regione.puglia.it/processes/bioeconomy-in-puglia?locale=it (accessed on 11 October 2021).

22. Nancy, B.; Grove, J.M.; Pickett, S.T.A.; Redman, C.L. Integrated approach to Long-Term studies of urban ecological systems. *BioScience* **2000**, *50*, 571–584.

23. Stephens, J.C.; Hernandez, M.E.; Roman, M.; Graham, A.C.; Scholz, R.W. Higher education as a change agent for sustainability in different cultures and contexts. *Int. J. Sustain. High. Educ.* **2008**, *9*, 317–338. [CrossRef]

24. Tarsitano, E. Interaction between the environment and animals in urban settings: Integrated and Participatory Planning. *Environ. Manag.* **2006**, *38*, 799–809. [CrossRef] [PubMed]

25. Zilahy, G.; Huising, D.; Melanen, M.; Phillips, V.D.; Sheffy, J. Roles of academia in regional sustainability initiatives: Outreach for more sustainable future. *J. Clean. Prod.* **2009**, *17*, 1053–1056. [CrossRef]

26. Lepore, A.; Palermo, S.; Pomella, A. Dalla green economy alla bioeconomia circolare. Un nuovo paradigma di crescita per il Sud e per il Paese. *Riv. Econ. Mezzog.* **2021**, *35*, 523–539. [CrossRef]

27. Giudice, F.; Caferra, R.; Morone, P. COVID-19, the Food System and the Circular Economy: Challenges and Opportunities. *Sustainability* **2020**, *12*, 7939. [CrossRef]

28. Stelpstra, T. New Circular Economy Action Plan, CDR 1265/2020, Adopted on 14th October 2020. Available online: https://cor.europa.eu/en/our-work/Pages/OpinionTimeline.aspx?opId=CDR-1265-202 (accessed on 10 December 2022).

29. *ISO 37101:2016*; Sustainable Development In Communities—Management System For Sustainable Development—Requirements with Guidance For Use. ISO: Geneva, Switzerland, 2016.
30. Sterling, S. Higher education, sustainability and the role of systemic learning. In *Higher Education and Challenge of Sustainability: Problematics, Promise, and Practice*; Corcoran, P.B., Ed.; Kluwer Academic Press: Dordrecht, The Netherlands, 2005.
31. Brown, P.; Baldassarre, B.; Konietzko, J.; Bocken, N.; Balkenende, R. A tool for collaborative circular proposition design. *J. Clean. Prod.* **2021**, *297*, 126354. [CrossRef]
32. Plattner, Meinel, Leifer. Design Thinking: Understand—Improve—Apply (PDF). 2011. Available online: https://hpi.de/ (accessed on 20 December 2022).
33. Newman, P.; Jennings, I. Cities as Sustainable Ecosystems. In *Principles and Practices*; Island Press: Washington, DC, USA, 2008.
34. Steiner, G. Higher education for sustainability by means of transdisciplinary case studies: An innovative approach for solving complex, real world problems. *J. Clean. Prod.* **2006**, *14*, 877–890. [CrossRef]
35. UNESCO Rethinking Education. Towards a Global Common Good? Paris. 2015. Available online: http://www.unesco.org/new/fileadmin/MULTIMEDIA/FIELD/Cairo/images/rethinkingeducation.pdf (accessed on 10 December 2022).
36. Wals, A.E.J. Review of Context and Structure for Education for Sustainable Development: Learning for Sustainable World UNDESD 2005-2014, UNESCO, Paris, 2009. Available online: https://unesdoc.unesco.org/ark:/48223/pf0000184944 (accessed on 10 December 2022).
37. Pronti per il 55%. Available online: https://www.consilium.europa.eu/it/policies/green-deal/fit-for-55-the-eu-plan-for-a-green-transition/ (accessed on 5 April 2023).
38. Council Regulation (EU) 2022/2577 of 22 December 2022 Laying down a Framework to Accelerate the Deployment of Renewable Energy. Available online: https://eur-lex.europa.eu/eli/reg/2022/2577/oj (accessed on 4 April 2023).

 *sustainability*

Review

Integrating Multi-Criteria Techniques in Life-Cycle Tools for the Circular Bioeconomy Transition of Agri-Food Waste Biomass: A Systematic Review

Felipe Romero-Perdomo [1,2] and **Miguel Ángel González-Curbelo** [1,*]

[1] Departamento de Ciencias Básicas, Facultad de Ingeniería, Universidad EAN, Bogotá 110221, Colombia
[2] Corporación Colombiana de Investigación Agropecuaria-AGROSAVIA, Mosquera 250047, Colombia
* Correspondence: magonzalez@universidadean.edu.co

Abstract: Agri-food waste biomass (AWB) is consolidating as a relevant bioresource for supplying material products and energy in a circular bioeconomy. However, its recovery and sustainable processing present trade-offs that must be understood. The integration of multi-criteria decision analysis (MCDA) into life-cycle assessment (LCA) tools has emerged as a novel way to address this challenge. This paper aims to conduct a systematic literature review to critically synthesize how MCDA has been integrated into LCA in an assessment framework and how helpful it is in AWB's circular bioeconomy transition. The literature shows that the most studied AWBs are rice husk, sugarcane bagasse, and household food waste. These are processed through the technologies of composting, anaerobic digestion, and pyrolysis for applications such as biofuels, bioenergy, and soil amendment. Environmental LCA (E-LCA) is the most widely used LCA tool, while both the analytical hierarchy process (AHP) and the technique for ordering preference by similarity to the ideal solution (TOPSIS) are the most applied techniques for MCDA. The current trend of integrating MCDA into LCA does not fully cover the LCA phases, favoring solely the impact assessment phase and indicating that the other phases are overlooked. The potential and involvement of the stakeholders are partially explored. Although there are holistic sustainability assessments, the social implications are rarely considered. The number of MCDA/LCA studies is expected to increase, assessments at the micro-, meso-, and macro-scales to become more articulated, and the impact of the results to become more aligned with government and company goals.

Keywords: circular economy; sustainable agriculture; multi-criteria decision making; TOPSIS; stakeholders; social LCA; life-cycle costing; bioenergy; biorefinery

Citation: Romero-Perdomo, F.; González-Curbelo, M.Á. Integrating Multi-Criteria Techniques in Life-Cycle Tools for the Circular Bioeconomy Transition of Agri-Food Waste Biomass: A Systematic Review. *Sustainability* **2023**, *15*, 5026. https://doi.org/10.3390/su15065026

Academic Editors: Idiano D'Adamo and Massimo Gastaldi

Received: 2 February 2023
Revised: 8 March 2023
Accepted: 10 March 2023
Published: 12 March 2023

1. Introduction

The circular bioeconomy is an emerging field that is booming in the scholarly and political communities. The most widely accepted definition of circular bioeconomy stems from the intersection of bioeconomy and circular economy [1,2]. The circular bioeconomy focuses on the sustainable and efficient recovery of biomass and biowaste resources in integrated, multi-output production chains, optimizing their value over time and taking the three sustainability pillars into account [3]. The term "circular bioeconomy" appeared around 2015 and has been the focus of many scientific publications since 2016 [1]. The European Commission's 2018 updated bioeconomy strategy emphasized that the *"European Bioeconomy needs to have sustainability and circularity at its heart"* [4]. The circular bioeconomy is envisioned as one of the approaches that will make the greatest contribution to addressing sustainability challenges.

Properly providing biomass-based feedstock is a multipurpose objective of the circular bioeconomy [5]. Agri-food waste biomass (AWB), a type of biomass, is produced and used in large quantities around the world, with the amount increasing more than threefold in the last 50 years [6]. The AWB amounts are inputs for the development of biorefineries and biotechnologies that are part of the circular bioeconomy. Nevertheless, they are affecting the sustainability of the planet by emitting greenhouse gases and causing inefficient water use. AWB management is a widespread issue that poses a challenge for food safety, environmental pollution, and economic stability [7].

Several circular-bioeconomy-based strategies for recovering AWB and producing food in a less-polluting manner have been proposed. While findings with immediate applications of AWB have been abundantly studied, the sustainable implications of the technologies have been questioned [8,9]. Some circular bioeconomy strategies and research consider the circular bioeconomy to be inherently sustainable [10]. Yet, there is emerging evidence that highlights potential obstacles and trade-offs [11,12]. Therefore, a sustainability check should not be overlooked in this research context.

Analyzing the sustainability of AWB management framed in a circular bioeconomy is pivotal for the definition and implementation of economic activities and investments, technological development plans, and policies. This action requires a robust measurement and interpretation of indicators associated with decision-making processes that impact society [13]. To fulfill this purpose, methodological tools that integrate the social, economic, and environmental dimensions as pillars of sustainability are necessary [14].

Environmental life-cycle assessment (E-LCA) is a well-known and widely applied method in AWB [15]. E-LCA is a tool focused on the environmental impacts of a product, service, or process based on the life-cycle perspective [16]. This perspective represents the processes that converge from raw materials to waste management [17]. The E-LCA has been used not only in AWB but also in other economic sectors, translating environmental science into useful knowledge in business and regulatory areas [18,19]. However, the E-LCA by itself does not cover sustainability as a whole [20]. Over time, efforts have been made to broaden the approach to the economic and social spheres, thus developing life-cycle costing (LCC) and social life-cycle analysis (S-LCA). The integration of these three tools later constituted the life-cycle sustainability assessment (LCSA) [21].

Multiple-criteria decision analysis (MCDA) is a recognized approach to complex decision making that has also been applied to AWB management issues [22]. It offers a systematic way for selecting, ordering, assigning, and weighting various indicators via processes that can involve a diverse range of stakeholders [23]. Some of the most popular MCDA techniques are analytical network processing (ANP), analytical hierarchical processing (AHP), technique for order performance by similarity to the ideal solution (TOPSIS), and multi-criteria optimization and compromise solution (VIKOR) [24].

Combining LCA and MCDA as a holistic sustainability assessment framework is a prominent stream of research. Its origins are not recent, but its popularity is growing. The purpose of this synergistic framework is to achieve more complete and conclusive results for decision makers. Research efforts have been reported in renewable energy systems, the automotive industry, and agricultural practices [25–27]. However, what is the current state of the MCDA/LCA framework's application in AWB's circular bioeconomy transition? A synthesis of previous research on this study context in the literature is currently lacking. This paper is an attempt to close this gap by conducting a systematic review on the usefulness and operability of the MCDA/LCA framework in assessing sustainability in the transition to AWB's circular bioeconomy. Trends and intrinsic aspects of biomass, technologies, applications, spatial scales, stakeholder participation, tools, and impact categories used are identified and discussed. Additionally, strengths and weaknesses of the methodological interaction and crucial issues for future work are examined.

The rest of this paper is structured as follows: The second and third sections represent the theoretical background of AWB as well as the rationale for integrating MCDA and LCA. The fourth section is about the methodology used. The fifth section deals with the MCDA/LCA research landscape in AWB evaluation, while the sixth section discusses the synergies and trade-offs of combining the two tools. The seventh section shows possible steps forward to improve the impact of the MCDA/LCA framework. The eighth section provides the conclusions.

2. AWB Characteristics in the Circular Bioeconomy

Biomass is the term used to describe all organic materials that represent an alternative feedstock to crude oil and natural gas. Biomass is used in a variety of applications, including wood-based materials, pulp and paper production, biomass-derived fibers, and biofuels [28]. Biomass sources can be classified as animal waste, municipal waste, forest waste, industrial waste, and AWB (Figure 1A). AWB is divided into two main types: agricultural waste and food-processing waste. Agricultural waste is cropping field waste that remains in the fields as a by-product of post-harvesting activities, whereas food-processing waste is solid or liquid waste from the industrial processing of agricultural products [29].

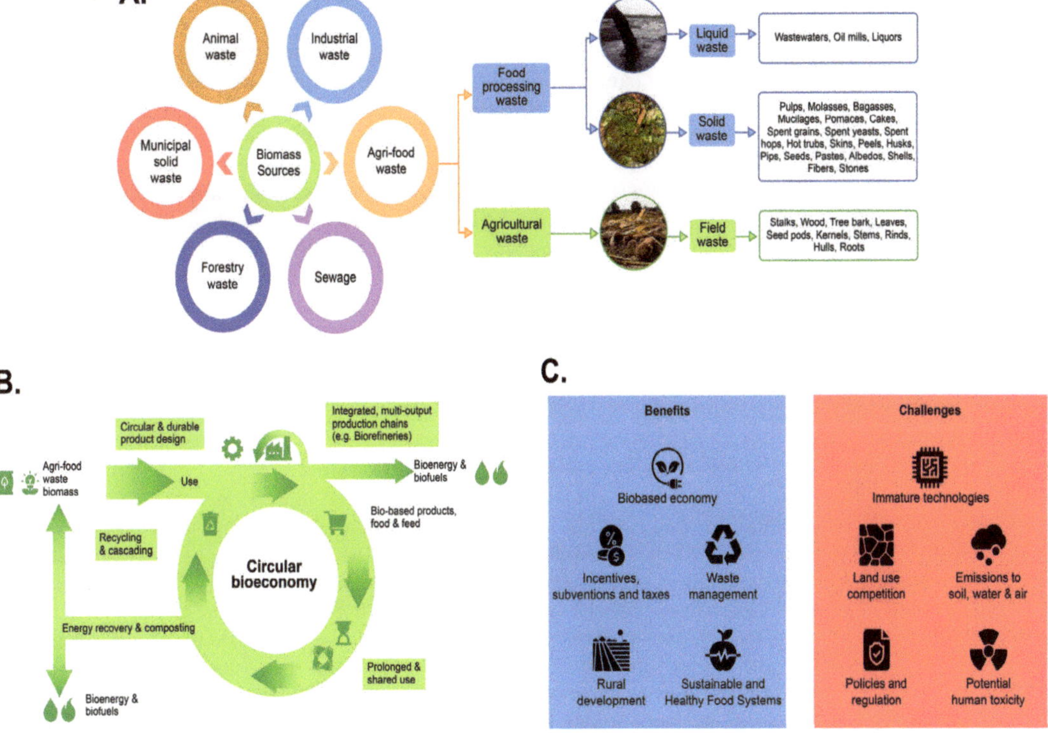

Figure 1. Features of AWB's transition to a circular bioeconomy. Classification of biomass sources and AWB types (**A**). A circular bioeconomy conceptual model for AWB (**B**). Benefits and challenges of AWB development in the circular bioeconomy (**C**). (Own elaboration with modifications from [3,29–32].)

It was estimated that more than 9 billion tons of crops were produced in 2017, with 1.3 billion tons of food wasted [33]. Roughly 1000 tons of food are wasted every minute, and 50% of food is lost at the production stage alone. By 2025, the agri-food sector could generate approximately 44% of global waste. AWB varies greatly between developed and developing countries, not only in quantity but also in location. North America, Oceania, Europe, and East Asia waste the most food in terms of volume [7]. AWB occurs primarily at the consumer level in developed countries, whereas food waste occurs at the production level in developing countries [34]. There are three major stages of AWB: agricultural production (33%), post-harvest and storage (54%), and processing, consumption, and distribution (46%). The primary AWB sources are beverages (26%), fruits and vegetables (14.8%), cereals and seeds (12.9%), and edible oils (3.9%), among others [35].

Accumulation and non-recovery of AWB have a negative impact on sustainability. In the social dimension, food waste is facing world hunger. A quarter of the world's total food waste could be used to feed the hungry. In terms of economics, developed countries waste USD 680 billion in food each year, while developing countries waste USD 310 billion [7]. The environmental impact of AWB is largely ignored. Every year, AWBs emit approximately 3.3 billion tons of carbon dioxide into the atmosphere, or approximately 1000 tons of carbon dioxide per minute [36]. AWBs strain water resources by consuming 250 square kilometers of fresh water; in other words, a quarter of the world's fresh water is wasted. In land use, approximately 1.4 billion hectares of arable land, or the size of the United States, India, and Egypt combined, has been designated for AWB [37].

A circular bioeconomy requires the use of sustainable biomass to ensure that the restoration cycle is completed and continues. As illustrated in Figure 1B, integrated and multi-output production chains can extend the life of biobased production to become a source of biomass and bioenergy [38]. The pillars of the circular bioeconomy are resource efficiency, optimizing the value of biomass over time, and sustainability, which push practitioners toward an energy-efficient and renewable management of AWB [3]. Some of the technologies highlighted in AWB to generate applications in bioenergy, biomaterials for construction, soil amendments, animal feed, and biopolymer production are composting, anaerobic digestion, transesterification, pyrolysis, and gasification [38,39].

AWB's transition to a circular bioeconomy has the potential to deliver several benefits, for example, the consolidation of a biobased economy to produce goods, services, or energy through the promotion of incentives, taxes, and subsidies. Other benefits include improving waste management, promoting rural development, and promoting sustainable and healthy food systems [27,32]. However, several impediments to achieving these benefits remain, such as immature technological development; competition for land use; soil, air, and water pollution; and indirect toxicity to humans. Furthermore, there are no legal guidelines in most countries for the disposal of unrecoverable AWB [27,36] (Figure 1C).

3. The Need to Integrate MCDA into LCA

LCA is based on the collection of inputs and outputs associated with the environmental, economic, and social impacts of a system of products and services throughout its life cycle [17]. LCA results are often difficult to interpret due to trade-offs between the impact category results of the scenarios under consideration [19]. These findings may even contradict one another, making decision making difficult. In response to this issue, two actions have been reported. The first action is to weight and add the LCA results from each impact category to produce a single scoring indicator. The second is to use a small number of impact categories to make interpretation of the results easier. These two decisions are contentious issues among LCA practitioners [40,41].

MCDA is based on mathematical protocols that are applied to inputs in structure analysis and interpretation processes that lead to decision making in numerous fields of knowledge [24]. However, there is a remarkable limitation in how the impact criteria are measured, as decision makers typically use qualitative scales [42]. Although qualitative scales are more practical for social aspects, it has been demonstrated that they are not always an accurate and precise way of representing the environmental and economic performance of the alternatives under consideration [43]. It is critical to achieve sustainability by using structured tools to assess the performance of alternatives in terms of multiple criteria [42].

LCA is composed of four phases, and MCDA generally has three, but their operation follows the same logical sequence (Figure 2). LCA is objective, reproducible, and standardized. MCDA is subjective on many occasions and can capture various perspectives that provide a broader view of the study context [44].

According to the descriptions above, LCA and MCDA are decision support tools that present different approaches [45]. LCA quantifies impact indicators that must be properly interpreted, while MCDA interprets real-world contexts for decision making that must be effectively based on indicators [44]. Furthermore, the two tools have complementary properties. Therefore, LCA and MCDA can be used in tandem.

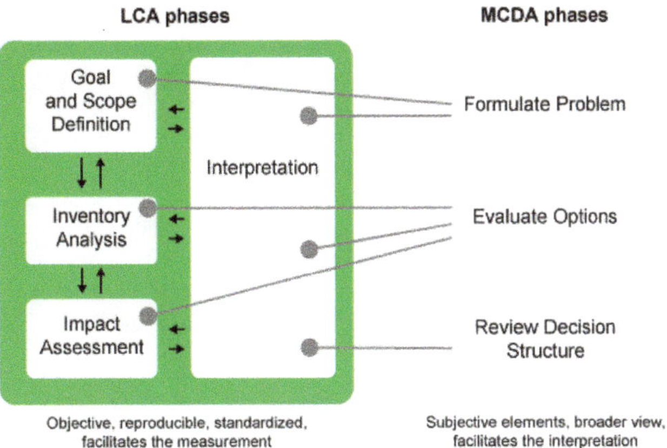

Figure 2. Complementary features between LCA and MCDA—own elaboration with modifications from [44,46].

The first applications of the MCDA/LCA framework emerged between the 1990s and 2000s. The work of Miettinen et al. [47] is an example of this, where they combined the use of E-LCA and the hierarchical analysis of processes (AHP) to weigh the impact categories of industrial beverage packaging systems. Hence, the methodological association of MCDA and LCA has been applied and discussed for more than 25 years.

The joint use of MCDA and LCA can be carried out in two ways: by integrating LCA into MCDA or by integrating MCDA into LCA [40]. The first is for LCA to provide indicators for the MCDA process. The second seeks for the robust interpretation of MCDA to be included in the life-cycle perspective. This review focuses on the second way: integration of MCDA into LCA. It is necessary to mention that the acronym LCA can be used as an umbrella concept encompassing life-cycle-based tools (i.e., E-LCA, S-LCA, LCC, and LCSA, among others), which is used in this paper.

4. Materials and Methods

This review article was structured as a systematic literature review to make the research replicable in other study areas or even updating the findings shown here in the future. The procedures described by Tranfield et al. [48] were adopted as follows:

- Step 1: Identify the opportunity for research.
- Step 2: Define the steps to consolidate the reported literature.
- Step 3: Select the aspects that will be analyzed to extract the information.

The first step, identifying the research opportunity, was already justified in this paper's introduction and background. The relevance of multi-criteria and life-cycle tools for AWB's transition to a circular bioeconomy was emphasized, as was the lack of state-of-the-art on the integration of these two tools in the scholarly literature.

The second step was carried out using the PRISMA (Preferred Reporting Items for Systematic Reviews and Meta-Analyses) protocol [49]. This protocol establishes the following four phases: identification, screening, eligibility, and inclusion (Table 1). The identification phase consists of defining the parameters for the literature search in the databases and thus identifying the set of publications. The search required the definition of associated keywords using Boolean operators (" ", OR, AND, *) (Table S1). Keywords were meticulously selected from the reported bibliometric analysis [23,40,50]. The search was not limited to a period to cover the evidence published to date and was carried out in a single day to avoid bias caused by daily database updates. The search query was applied to the SCOPUS and Web of Science databases. These databases have a high incidence of access in research fields and contain peer-reviewed academic publications [51].

Table 1. Phases of the PRISMA protocol applied to obtain the set of publications to be reviewed.

Steps	Criterion	Effect
Identification	Search query Time horizon Search date Database	in Tittle, Abs, Key No limit 9 January 2023 Scopus and ISI Web of Science
	Finding publications by searching databases	Scopus: $n = 43$ ISI WOS: $n = 117$
Screening	Inclusion criteria: 1. Research articles 2. English publications 3. No duplicate publications	Records included Scopus: $n = 41$ ISI WOS: $n = 109$ Scopus: $n = 41$ ISI WOS: $n = 109$ Full-text publications consolidated: $n = 120$
Eligibility	Inclusion criteria: Publications related to the topic (Review of the first reading)	Full-text publications included: $n = 40$
Included	Review of the second reading, critical reading, and scrutiny	Final sample of reviewed publications: $n = 23$

The screening phase entailed developing criteria such as the type and language of publications required to conduct the review process as thoroughly as possible. The types of publications included were research articles; non-English publications were excluded. Moreover, duplicate publications between the two databases were eliminated to consolidate a single set of papers. The eligibility phase represented an exhaustive review of the content of the publications to confirm their intrinsic association with the research context. Since the scope of this paper is the integration of MCDA into LCA, only publications that applied both tools in this integration way were considered. In the included phase, the final sample of publications was established as an input for the next step.

The third step involved a critical reading, scrutiny, and detailed synthesis of the publications from two analytical perspectives that define the scope of this review. The first was based on the general characteristics of this field of research, such as biomass type, recovery technologies and applications, stakeholder contribution, spatial scales, and the techniques and indicators implemented (Table S2). The categories of these aspects were based on reported works and were defined iteratively, i.e., the reviews of the publications allowed the categories to be confirmed, added, or reconsidered [52,53]. The second perspective was a more in-depth interpretation that focused on neglected aspects of technique development, synergies and trade-offs in methodological association, and key issues for future sustainability assessments in AWB recovery.

5. Overview of MCDA/LCA Studies in AWB Recovery

MCDA/LCA studies on AWB are a relatively new but increasingly common approach as a sustainability assessment framework. Evidence collected and explored in this review paper covers several sectors (i.e., agri-food, construction, manufacturing, and energy) and geographic contexts (i.e., North America, South America, Europe, West Africa, and Eastern Asia).

5.1. Biomass, Technologies, Applications, and Spatial Scales Used

The MCDA/LCA framework implementation has been useful for a variety of applications involving conversion and processing technologies utilizing different AWBs (Table 2).

The production of biofuels from rapeseed, sugarcane bagasse, rice straw, wheat straw, moringa, maize, and triticale has been the most notable application.

Rapeseed (*Brassica napus*) research has sought to propose feasible scenarios for producing second-generation biofuels. Comparisons of biodiesel chains with different geographic feedstock origins have been conducted [54]. Locally produced feedstocks were chosen over imported feedstocks based on job creation and gross value added [55]. The establishment of environmental impact assessment processes has identified aspects that are affected but are not contemplated in the LCA standard, such as biodiversity [56].

Sugarcane bagasse research has pursued multiple purposes, including investigating value-added product processing routes and selecting feedstock sources. Joglekar et al. [57] studied the influence of six sugarcane bagasse processing routes, highlighting that the first- and second-generation ethanol-based routes with biogas promote both environmental preservation and productivity. Ramesh et al. [58] compared five lignocellulosic biomasses for second-generation bioethanol production and found that vetiver grass favors sustainability, followed by moringa, rice straw, wheat straw, and sugarcane bagasse. *Vetiver zizanioides* is cultivated in many countries, such as India, China, and Brazil, for its essential oil. This grass is drawing attention for exhibiting biorefinery potential, growing in conditions of drought, flood, high temperature, and contaminated soil [59].

Biofuels derived from maize have been used to standardize methodologies for evaluating the sustainability performance of fossil and renewable fuel chains. Ekener et al. [60] developed a life-cycle sustainability assessment methodology by applying value-based sustainability weights. These authors have shown that the maize-based biofuel chain leads to sustainable benefits compared to fossil fuels.

Table 2. Overview of all studies reviewed in terms of applications, recovery, and processing technologies, AWB, and spatial scales.

Applications	Recovery/Processing Technologies	AWB	Spatial Scales	Reference
Biofuel (Transportation)	Pre-treatment, saccharification, fermentation, and purification	Sugarcane bagasse, rice straw, wheat straw, moringa, and vetiver	Nation	[58]
	Extraction and transesterification	Rapeseed (*Brassica napus*)	Supply chain	[54]
	pre-treatment, saccharification, fermentation, and purification	Sugarcane bagasse	Process	[57]
	Extraction and transesterification	Rapeseed (*Brassica napus*) and soybean	Nation	[55]
	Husky process, gasification, pre-treatment with lime, saccharification, co-fermentation, dry milling, extrusion, and pelletizing	Triticale (X *Triticosecale* Wittmack)	Process	[61]
	Mechanical compressing, purifying, and refinement of biodiesel	Rapeseed (*Brassica napus*) and oil palm	Supply chain	[56]
	Pyrolysis, gasification, and methanol synthesis	Rice straw	Supply chain	[62]
	Biofuel production processes	Sugarcane and maize	Supply chain and farm-based	[60]

Table 2. *Cont.*

Applications	Recovery/Processing Technologies	AWB	Spatial Scales	Reference
Bioenergy (Bioelectricity and bioheating)	Collection, incineration, centralized composting, anaerobic digestion, biogas upgrading, and post-composting	Household food waste	City	[63]
	Anaerobic digestion, in-vessel composting, incineration, and landfilling	Household food waste	World regions	[64]
	Anaerobic digestion	A mixture of grape marc and cow manure	World regions	[65]
	Direct-combustion power generation, gasification power generation, and briquette fuel	Urban food waste	Resources	[66]
	Bioenergy systems based mainly on combustion, gasification, and pyrolysis	Lignocellulosic biomass	Resources	[67]
Soil amendments	Fertilizer production	Oil palm	Product and Farm-based	[68]
	Composting	Coffee residue	Farm-based	[69]
	Planting, pre-harvesting, harvesting, straw recovery	Sugarcane straw	Farm-based	[70]
	Anaerobic digestion	Household food waste	City	[41]
Construction biomaterials	Manufacturing, construction, and demolition	Rice husk ash and carbon nanotubes	Product and process	[71]
	Manufacturing processes	Rice husk ash and cotton mill waste	Product and process	[72]
Food waste recovery manufacturing strategy	Extraction and anaerobic digestion	Urban food waste	Process	[73]
Biopolymers	Anaerobic digestion, booster technology, polyhydroxybutyrate technology	Sugarcane straw, sugarcane bagasse, rice straw, rice husk ash	World regions, nation, city	[74]
Biochemicals	Polyphenol extraction methods	Red wine pomace	Process	[75]
Animal feed	Landfilling, incineration, and production process	Urban food waste	Nation	[76]

Triticale (*X Triticosecale Wittmack*) has been cataloged as a preferred non-food energy crop for biorefineries. The advantages of triticale lie in its ability to grow on marginal land and its higher yields compared to cereal crops [77]. These properties have driven decade-long research on triticale-based biorefinery processes from the MCDA/LCA perspective. For example, Liard et al. [61] analyzed the technical, economic, and environmental risks of three triticale-based biorefinery platforms incorporating two technological options. The platforms were ethanol, polylactic acid, and a mixture of thermoplastic starch and polylactic acid, while the technologies were ultrafiltration and cogeneration. These researchers noted that ultrafiltration technology substantially mitigates the environmental effects of the polylactic acid platform.

MCDA/LCA studies with AWB highlight bioenergy generation in terms of bioelectricity and bioheating. Here, the main source has been household food waste. Quantifications of the environmental and socioeconomic impacts of household food waste management in the Amsterdam Metropolitan Area for biogas production have been consolidated [65]. Findings by Wang et al. [66] have pointed out that direct-combustion power generation, gasification power generation, and briquette fuel are recognized as sustainable bioenergy technologies for household food waste. Slorach et al. [64] present a methodology to assess environmental performance in the food–energy–water–health nexus using four treatment options: anaerobic digestion, in-vessel composting, incineration, and landfill. They found that anaerobic digestion is the most environmentally sustainable option with the lowest overall impact on the nexus and that in-vessel composting is the worst option, even though it dominates circular economy waste hierarchies.

The production of biogas and the co-production of polyhydroxyalkanoates and biogas from grape pomace and cow manure have been investigated. Vega et al. [65] determined that the environmental performance of a biorefinery with polyhydroxyalkanoates and biogas co-production is preferable to a biorefinery only producing biogas at the territorial level, both in southern France and the western United States. Lignocellulosic biomass has also been associated with bioelectricity and bioheating. Von Doderer and Kleynhans [67] studied 37 plausible lignocellulosic bioenergy systems based on economic-financial, socioeconomic, and environmental indicators. These researchers indicated that a feller buncher for harvesting, a forwarder for biomass extraction, mobile comminution at the roadside, secondary transport in truck–container–trailer combinations, and an integrated gasification system for conversion to electricity are feasible and sustainable options.

To achieve sustainable production in a circular bioeconomy, bioenergy management faces challenges such as decreasing competition with food production that supports food security, mitigating the environmental impact of crop production, increasing feedstock options, incentivizing the use of AWB, and reducing production costs [78,79]. AWB represents a renewable energy source with a high growth potential for the circular bioeconomy.

Oil palm, coffee residue, and sugarcane straw are AWBs that have aroused research interest as soil amendments. The strategy of converting AWB to organic fertilizer is a common practice that has been applied to oil palm. In that vein, Lim et al. [68] carried out a fertilizer formulation process that incorporated organic and chemical compounds based on sustainability indices. The formulation obtained consisted of 0.96 wt% urea, 1.14 wt% monoammonium phosphate, 0.10 wt% kieserite, and 97.81 wt% palm-based organic fertilizer for oil palm. The circulation of resources that recovers biowaste has also been applied in horticulture systems, with promising results that are making another goal possible: replacing substrates obtained from processes under the perspective of the linear economy [80].

Composting coffee residues was used as a circular practice to propose the LCA4CSA method, which promotes climate-smart agriculture at the farm and cropping system levels. This work quantified the climate change mitigation potential of the use of compost, which ranged between 22% and 41%, by considering operations that occur on and upstream of the farm [69]. They also noted contamination transfers between impact categories such as climate change indicators, acidification, and terrestrial eutrophication. One emerging biotechnological strategy is the combination of AWB and microbial biostimulants to reduce contamination transfer, improve plant growth, reduce fossil fertilizer doses, and mitigate abiotic stress [39,81].

Sugarcane straw has been studied for the economic, social, and environmental impacts caused by manual and mechanical technologies for its recovery, as well as for the planting and harvesting of sugarcane. Cardoso et al. [70] showed that there are clear differences in the sustainable performance of manual and mechanized technologies. Manual technologies encourage job creation but negatively affect internal rates of return, ethanol production costs, and the environmental effect. By contrast, mechanized technologies present lower ethanol production costs, higher internal rates of return, and better environmental impact, confirming that mechanized scenarios are more sustainable.

The MCDA/LCA literature on soil amendments as biofertilizers for agricultural management practices so far is industrially focused for commercialization. However, no studies were found that sought to strengthen the economy of rural areas through the valorization of AWB. An investigation that promoted this social purpose is the one carried out by Juanpera et al. [82]. This work deduced that vermifiltration is a feasible alternative for the post-treatment of digestates from low-tech digesters implemented in small-scale farms. Vermifiltration produces a high-quality biofertilizer, is easy to use, and is implemented with local materials.

Biobased building materials are gaining ground in this research context. The circularity strategy to produce construction biomaterials is to combine virgin materials and materials recovered from AWB. For example, partially replacing the cement in the green concrete mix with rice hull ash nanowaste particles has been shown to be a promising alternative [71]. Its use reduces carbon emissions and requires less energy during the cement manufacturing phase. Cotton mill waste is another AWB used as an input for alternative building materials. Joglekar et al. [72] suggest that bricks based on this waste use fewer natural resources and use moderate conditions in the manufacturing process compared to burnt clay bricks used for masonry. Their study underlines that the serious handling and disposal problems of this waste are converted into an opportunity for green construction.

There are applications in food waste recovery manufacturing strategy, biopolymer production, biochemical synthesis, and animal feed with less dominance. The implementation of the MCDA/LCA framework, combined with the cost–benefit analysis has shed light on optimal strategies for citrus waste management by companies [73]. Strategies identified included the wholesale of imperfect but still edible waste as well as investments in facilities to extract higher-value pectin using a microwave-assisted pectin extraction process. Technological scenarios associated with the production of bioplastics have been evaluated to determine the benefits that they produce at territorial scales [74]. The environmental and economic performance at both laboratory and industrial scales of the extraction of polyphenols from red wine pomace has been crucial for making decisions on process standardization [75]. Lastly, some authors have indicated that using urban food waste as animal feed is environmentally feasible if the safe recovery rate exceeds 48% compared to the use of sanitary landfills and incineration [76].

The reported findings reveal that the macro-, meso-, and micro-scales have been addressed. Most of the case studies have been conducted at the meso-scale (i.e., supply chain and farm base), followed by the micro-scale (i.e., resources, processes, products), and the macro-scale (i.e., national, regional, and world). The integration of studies at the three scales would provide a complete picture of the circular bioeconomy debate in AWB. Following this direction, research can strongly contribute to larger-scale policymaking to foster circular and sustainable production patterns [83].

Three approaches must be promoted to move the circular bioeconomy toward sustainability: sustainability of the bioresource base, sustainability of processes and products, and sustainability of circular processes of material flows [84]. All the research summarized above indicated that the use of the MCDA/LCA framework has been aligned with these three approaches, mainly the second and third, suggesting that it has contributed to the sustainable advancement of the circular bioeconomy. Although the prevailing narrative in the literature is optimistic, there are critical positions on the barriers and limits, even suggesting that "circular" does not necessarily mean "environmentally friendly". The rationale of this statement is associated with the rebound effect, the risk of greenwashing strategies, and the development of new technologies without sufficient knowledge of their consequences [85,86]. These aspects must be addressed on a case-by-case basis to guide the academic community and determine the real benefits.

5.2. Stakeholder Engagement

The role, worldview, and values held by stakeholders strongly influence sustainability motivation and performance [60]. The relevance of communication and cooperation between governments, organizations, and actors are essential elements for decision making on sustainability [87]. The AWB literature associated with the MCDA/LCA framework revealed that stakeholders are not considered in most studies. In fact, less than half of the investigations included them (Figure 3A).

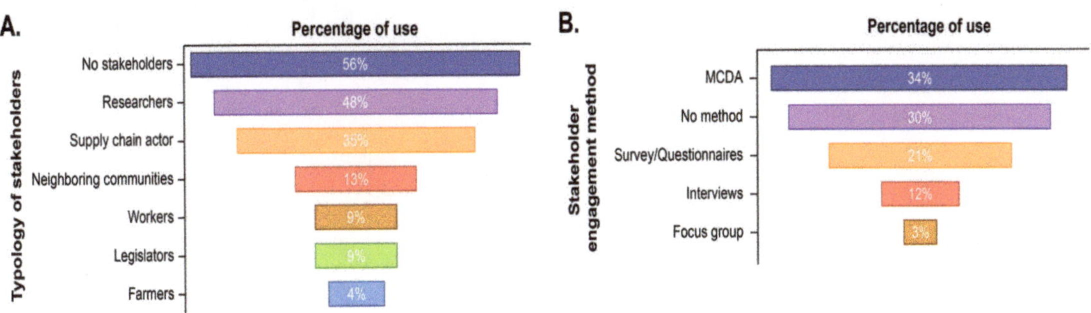

Figure 3. Role of stakeholders in the joint use of MCDA and LCA for AWB's circular bioeconomy transition. (**A**) Typology of stakeholders. (**B**) Stakeholder engagement method—own elaboration.

The stakeholders included in the studies consisted mainly of researchers and supply chain actors. Neighboring communities, workers, legislators, and farmers were less prominent. Some works underline the involvement of researchers for their ability to objectively compare sustainability impacts rather than relying on subjective opinions [57,58]. The frequently observed reason for choosing stakeholders is experience and knowledge. However, few studies have clearly documented a systematic selection of stakeholders, which could affect fully encompassing the multifaceted connotation of AWB management.

The selection of stakeholders may be based on their legitimate participation. Although there are divergent answers on how to define legitimate participation, Kruetli et al. [88] suggest addressing the following two questions: *Does the stakeholder affect or are they affected by the decision? And what participatory approach does the stakeholder play?* Brandt et al. [89] contribute to this debate by proposing four degrees of intensity of legitimate stakeholder engagement: information, consultation, collaboration, and empowerment. These grades are distinguished according to the attitudes of the stakeholders toward the evaluation process.

The reviewed studies elicited that various stakeholder engagement methods have been implemented. The popularly used method is to integrate the stakeholders into the development of the MCDA technique, which is performed with face-to-face questionnaires or via email (Figure 3B). Other methods used are interviews, focus groups, and surveys. These methods are carried out when the role of the stakeholders is multifunctional and independent of MCDA, thus contributing to various actions throughout LCA. De Luca et al. [27] suggest that stakeholder engagement methods need to be carefully elucidated as they determine the intensity of stakeholder engagement. Unfortunately, few studies have reflected a comprehensive understanding of the significance of the participatory method for stakeholders.

The literature showed that stakeholders have participated in all four phases of LCA (Figure 4). The impact assessment phase is the most impacted, followed by the inventory analysis phase. The interpretation phase is approached lightly, and the goal and scope definition phases are substantially overlooked.

The most common reason for involving stakeholders is to weigh indicators during the impact assessment phase. This trend, which was noted in AWB's circular bioeconomy transition, has also been seen in research promoting sustainable agriculture and renewable energy technologies [42,43]. The major benefit of this practice is ranking impacts measured by capturing multiple perspectives provided by stakeholders. Consequently, decision making is less complex in the interpretative phase and is based on the real interests of society.

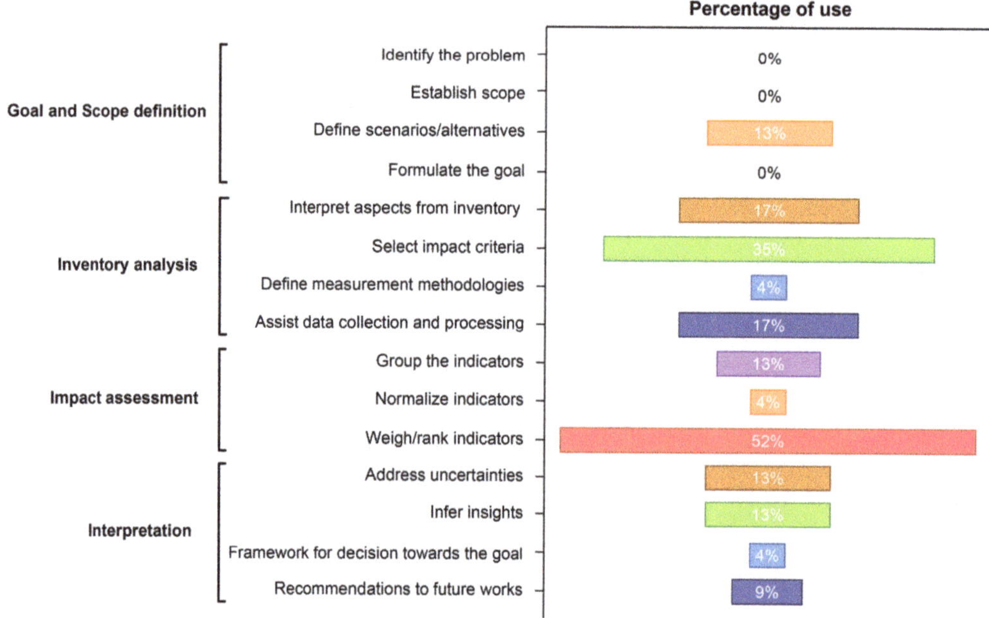

Figure 4. Contribution of stakeholders by LCA phases in the joint use of MCDA and LCA for AWB's circular bioeconomy transition. The percentage of use corresponds to the total number of publications consolidated for this review—own elaboration.

As mentioned above, there are authors of studies that did not involve stakeholders. They argue that the decision makers' analysis is time- and resource-intensive, which may hinder their application in practice [90]. They also expose that it is preferable to use stochastic methods that avoid subjective value judgments about which indicators are most important. Under this approach, different types of weighting, such as equal weights for all criteria evaluated and higher criteria values, and methodologies to define weighting are used [18]. Dias et al. [54] propose an analysis strategy based on stochastic multi-criteria acceptability analysis and variable interdependent parameter analysis to provide an exact limit of how much better each alternative is compared to another, from a perspective of pairwise comparison.

While these studies are methodologically sound, Thokala and Madhavan [91] take a different stance. They claim that the outputs will not be translated into tangible actions because decision makers do not trust them due to their limited or no involvement in the process. Iofrida et al. [92] describe that the consideration of stakeholder values is challenging but should not be a scientific weakness to interpretivism to the point that it is avoided. This review emphasizes that a methodological synergy can be achieved between these two approaches. Scientific validity must be promoted and can strongly support both the reconciliation of different objectives and the resolution of conflicting ambitions in social contexts.

The grouping of indicators is the second action carried out by stakeholders during the impact assessment phase, for which MCDA served as a methodological bridge. The two most common options are grouping by sustainable dimension and grouping by relevance to the identified problem. Both options lead to a decrease in the number of indicators that favor the following actions in LCA; yet, grouping by sustainability dimension has been recommended to encourage balanced approaches to sustainability [45].

Stakeholders have supported the inventory phase. The action preferably carried out by them has been the selection of impact categories. This observation possibly responds to previously reported suggestions to mitigate the bias of the research authors themselves [93]. Stakeholder selection of impact categories reflected alignment with the LCA goal but not the adoption of holistic sustainability perspectives. In line with this observation, Wang et al. [94] suggest using the Sustainable Development Goals as a guide to choose and structure the impact categories.

Stakeholders also perform actions such as interpreting aspects of the inventory and assisting with data collection and processing. While several theoretical frameworks have been proposed at a theoretical level, few works have documented comprehensive data collection with stakeholders. Contrasting studies illustrating the specific benefits of including stakeholders in the establishment of inputs have not been reported, but their influence may promote a more accurate analysis of the study object. Regarding data management, stakeholders have been reported to have contributed to the measurement of semi-quantitative indicators, mainly in the social dimension [76].

Stakeholders have been involved in the interpretation phase to address uncertainties and gain insights. A less considerate action for stakeholders is to propose recommendations for future work. Interestingly, living stakeholder labs have been created to follow up on the implementation of the findings obtained and thus establish a roadmap [63]. Living laboratories, known as living labs, are open innovation ecosystems in real-life settings whose main goal is to solve societal challenges through iterative feedback processes between stakeholders for collective ideation and collaboration [95].

Stakeholders have been weakly linked to the scope and objective definition phase. The sole action carried out by them is the establishment of alternatives and scenarios to be evaluated through the MCDA/LCA framework. Stakeholders have not been considered to identify the problem, establish the scope, and formulate the goal. Huttunen et al. [96] reported that stakeholders can improve the definition of system boundary elements, goal orientation, and the scope of the social dimension in the first phase of LCA. Stakeholders can even be empowered by learning about the consequences of their decisions and actions.

In the scope and objective definition phase, it is key to make sure that the set of stakeholders is comprehensively composed and that they represent all types of stakeholders. It is relevant to consider whether there are any value chain stakeholders who are indirectly dependent on the AWB. Garcia-Garcia et al. [73] exemplify this issue by asking the following question: *if local farmers are collecting waste for use as free animal feed, what would they do if this supply became unavailable?*

Taken together, stakeholder engagement reflected a routine development. The degree of integration of the stakeholders was not at the process level but at the level of the final findings. Their potential and development are partially exploited and detailed in AWB's circular bioeconomy transition when the MCDA/LCA framework is applied. The need for inter- and transdisciplinary approaches that combine, interpret, and communicate scientific and local knowledge is still increasing [97]. The processes of selection and involvement of the stakeholders depend on the study context, but being aware of the different ways of carrying out these processes contributes to the design of more meaningful, effective, and practical processes in the MCDA/LCA framework.

5.3. Techniques Applied

In promoting AWB's circular bioeconomy, the MCDA/LCA framework has shown the integration of various multi-criteria techniques and life-cycle tools. As expected, the E-LCA is the predominant tool in LCA; however, S-LCA and LCSA have rarely been applied, and LCC is not reported (Figure 5A). This pattern has been described in both the standalone implementation of LCA and its combination use with other tools [98–100]. Therefore, it is not due to an incompatibility of LCA and MCDA.

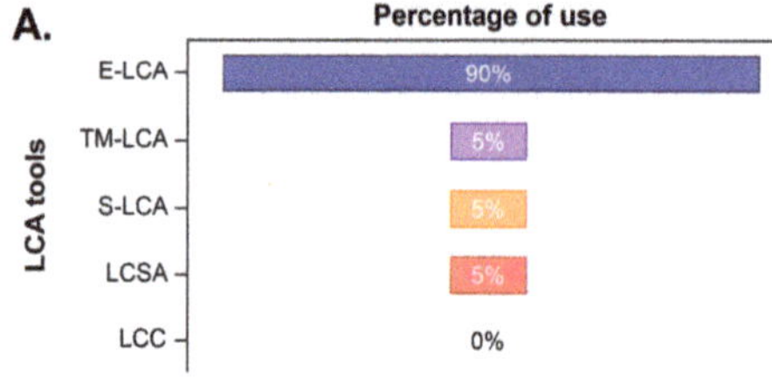
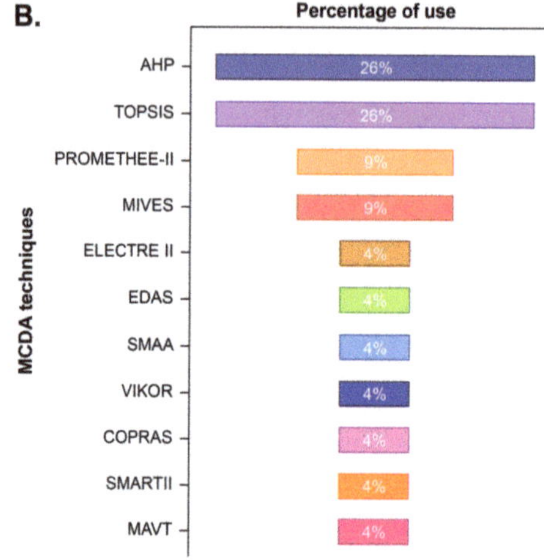

Figure 5. Use of tools in the MCDA/LCA framework for the transition to AWB's circular bioeconomy. (**A**) LCA tools. (**B**) MCDA tools. The frequency of LCSA use was not included in the S-LCA and LCC tools. In other words, the percentage of application of S-LCA and LCC was based on their individual use and was independent of LCSA—own elaboration.

The observed trend is to use the E-LCA and, at the same time, quantify various economic and social parameters independently. The authors do not expose reasons to avoid the implementation of S-LCA and LCC. However, three issues have been identified: (i) The comprehensive sustainable assessment method based on life-cycle theory for AWB circular management is still immature [6]; (ii) the E-LCA, LCC, and S-LCA analyses do not have the same level of maturity [101]; and (iii) there is a lack of harmonization of LCA tools that makes comparative analyses difficult [102].

E-LCA is a method that ISO 14044 standardized in 2006 and is widely used to investigate the environmental impacts of products and processes [16]. LCC is not yet standardized, but it considers all the costs and revenues attributable to cost objects from invention to abandonment [103]. S-LCA is a methodology that has shown a lack of consensus, but it has recently received standardization guidelines [104]. It evaluates the social impacts of stakeholder groups throughout the product life cycle [105].

LCSA can be extremely useful when used in conjunction with MCDA, contributing to a more comprehensive assessment and ensuring that all stakeholder concerns are included in the analyses [44]. Despite these virtues, LCSA is rarely used. Its application has consisted of the three life-cycle tools' results evaluated separately, concluding the fence of sustainability in a comparative way. This LCSA development path is due to the complex synergy of many of the existing methods in LCA tools [106]. As an alternative option to LCSA, triple-bottom-line evaluation has been used. The environmental bottom line has been calculated using LCA, the economic bottom line has been determined using net present value, and the social bottom line has been evaluated using SWOT analysis [76].

A study that is outside the context of this review but that represents a relevant reference for the use of LCSA is the one published by Hildebrandt et al. [107]. They assessed three scenarios involving existing and future wood-based value-added networks in Germany. The framework implemented in this work includes a set of 55 calibrated categories to sustainably monitor regional bioeconomy clusters. According to Visentin et al. [18], future research on LCSA should concentrate on three directions: standardizing analytical methodologies, establishing and measuring indicators, and applying LCSA in case studies. These priorities must be considered in the MCDA/LCA framework.

The territorial metabolism–life-cycle assessment (TM-LCA) framework is booming for circular AWB management. Its use allows for the study of the environmental and economic consequences of AWB management chains on a macro-scale, addressing their complexity, seasonality, and regionality [108]. Interestingly, the joint use of the multi-criteria approach to TM-LCA has been highlighted for the simulation and prediction of the environmental performance of future systems [109]. Its integration further extends the approach with scenario analysis, including regional and seasonal aspects, several product life cycles, and comparing these across a wide range of impact categories simultaneously [110].

The substantially used MCDA tools are AHP, TOPSIS, PROMETHEE-II, and MIVES (Figure 5B). Other tools, less commonly reported, are ELECTRE II, EDAS, SMAA, VIKOR, COPRAS, SMART, and MAVT. This trend indicated that the MCDA techniques used are aligned with two typologies: the theory of multi-attribute value and the prioritization and classification method, where the former is more prevalent. It is noteworthy that some researchers stated that they were using a multi-criteria approach but did not refer to any specific MCDA technique [27,111]. Some researchers describe the pros and cons of MCDA models regarding the system and objective under study to determine the most feasible model [41].

AHP is a hierarchical MCDA with levels of objectives, criteria, possible sub-criteria, and alternatives. This method compares the criteria pairwise to determine a preference ranking [112]. AHP has presented a broad area of application spanning the social, natural, and economic sciences [113,114]. AHP is used in the LCA/MCDA framework to establish the weight and ranking of the categories. The AHP method was applied either individually or in combination with other MCDAs (i.e., AHP + TOPSIS + Entropy and AHP + VIKOR). The most frequent application of AHP in the reviewed literature was to select AWB, the conversion and processing technology, and the supply chain scenario that would lead to the best sustainable beneficial impacts. The benefits of integrating AHP with LCA are that it is a robust, flexible, and well-known method that clearly illustrates the pillars of sustainability, measures the consistency of decision makers' judgments, can be easily combined with other methods, and presents an algorithmic structure that facilitates the communication of results to decision makers [42,115].

The TOPSIS method is based on the idea that the chosen alternative should be the shortest geometric distance from the best solution and the furthest away from the worst solution [116]. It requires little input data and produces results that are both understandable and reliable. TOPSIS's benefits include its simplicity, ability to measure the relative performance of each alternative, and high computational efficiency [117].

PROMETHEE is a method that classifies and selects a finite set of alternative actions between criteria that are often contradictory [118]. MIVES is a tool especially aligned to sustainability that combines features of multi-level requirement aggregation, inclusion of a weighting process, and indicator value utility functions [119]. MCDA technique comparisons have been reported to support appropriate selection based on study context [120,121]. Lastly, a potential strategy is to implement different MCDA techniques to compare the weights obtained and consolidate them into a final ranking of the study alternatives. This alternative provides a broader picture for decision making.

5.4. Impact Categories Assessed

AWB's circular bioeconomy transition has been measured using 57 impact categories spanning six dimensions via the MCDA/LCA association. Environmental categories account for 39% of the total, followed by social categories (26%), economic categories (18%), and technical categories (7%). Both the business strategy and government categories received 5% (Table S3).

In many cases, the impact categories were established without regard for selection criteria. Some studies base their impact criteria on previous studies conducted with similar conditions [57]. In macro-scale studies, selection criteria such as reliability, measurability, and relevance to the territory's situation have been used [122].

Global warming potential is the most measured indicator in the environmental area, to the point that all studies included it (Figure 6). However, few authors measure it in more specific terms, such as sulfur dioxide mitigation, nitrous oxide mitigation, and chemical oxygen demand discharge, to provide a more detailed environmental description. Overall, the findings showed global warming potential decreases through circular strategies, indicating that it is possible to improve the current response to the global climate crisis [123]. Acidification showed a measurement frequency of 65%, while eutrophication showed a frequency of 57%. These three indicators have been vital to studying the implications of circular AWB flows. Contamination transfers from AWB to the system have been found in them [41].

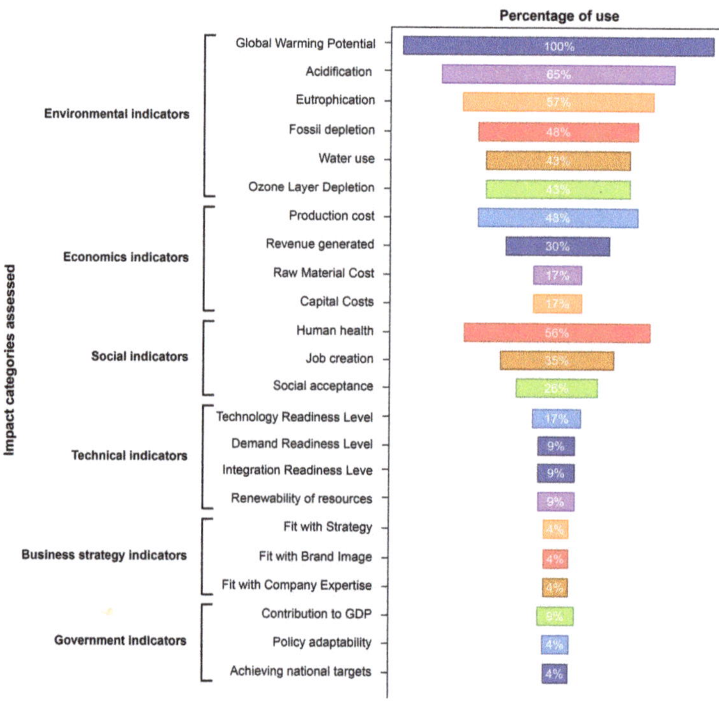

Figure 6. Impact categories most reported in the literature on the use of the MCDA/LCA framework in AWB's circular bioeconomy transition. The other categories are found in Table S3—own elaboration.

Fossil depletion was the fourth most reported indicator, covering 48% of the literature. This is a highly discussed impact category in LCA as it is a problem crossing the economy–environment system boundary [124]. Water use and ozone layer depletion registered 43% each. Moderately frequent environmental indicators in the studies are ecotoxicity potential, land use, photochemical oxidation, and particulate matter. The above impact categories are addressed in nearly every E-LCA. Rarely measured environmental indicators are residual waste, macronutrient (N, P, and K) content recovery, recycled contents, and undesirable substances.

Several authors have mentioned two limitations concerning environmental indicators. The first is that the estimates of LCA impacts were limited to impact categories for which inventory data were available [125]. The second is linked to the inaccurate interpretations that can occur when the databases are not regionalized [126]. In response to these barriers, computational frameworks have been proposed to strengthen the efficiency of data collection in LCA, and government efforts to promote and finance this goal have been highlighted on global agendas [127,128].

Studying the social dimension to measure its feasibility in the circular bioeconomy has always been a challenge [105]. The social indicators outnumber the economic indicators when compared to the total number of indicators; however, there is considerable variation in the number of indicators between studies. Consequently, current studies usually capture indicators of environmental impact, economic indicators, and, on occasion, social indicators.

The most contemplated social indicator is human health, with its presence in more than half of the literature. Previous works have emphasized the relevance of this indicator, which is a long-lasting endeavor of mankind [129]. This category implies the evaluation and comparison of both positive and negative human health impacts along product life cycles [130]. Job creation is included in a third of the publications, and it is an indicator that has received considerable research attention. Methodologies aimed at characterizing the social dimension through the evaluation of potential job creation have been reported [131]. Social acceptance stands out with 26% usage. The focus that has predominated in the categories of social impact is health and socioeconomic repercussions.

The participation of the workforce in each sector, occupational accident levels, education levels, skilled labor, safety, odor generation, noise creation, participation of associations, and stakeholder support are minimally measured. The researchers who included these indicators acknowledge that they were useful but that they have not been adequately disseminated in the scholarly literature; therefore, they recommend them for use in future studies [63,132]. Bartzas and Komnitsas [133] underline that stakeholder support is crucial for the success of any change and that its measurement is linked to the social sphere. They quantified this indicator as the total number of training programs followed by the farmer. The measurement of most social indicators is qualitative and does not follow the guidance provided by S-LCA.

Analyzing the social aspects of biomass management, such as AWB, in a circular bioeconomy can have several important benefits for participation, equity, effective policy development, sustainability, and social acceptance. However, evaluating it can present significant challenges, including complexity, a lack of data, different perspectives, and cultural change [105]. Social aspects are inherently complex and multifaceted. Biomass management can have implications for different social groups, cultures, and values, which can complicate the identification of suitable solutions and strategies [134,135]. The circular bioeconomy may require a significant cultural change and a new focus on the relationship between society and natural resources, which requires novel approaches to promote it [12]. Although the reviewed studies did not state that strategies based on the circular bioeconomy on AWB can have negative social consequences, it should not be assumed that these strategies generate added social value automatically. Recent research has found that circular economy practices can have negative social consequences, implying that each practice should be evaluated individually for its social implications [136].

There are several calls that the academic community has promoted so that the social sphere in sustainability is not neglected [137,138]. As LCA, MCDA, and their combined use are involved, every practitioner must be certain that the social categories require more attention. Murphy [139] states that elucidating how the social pillar relates to the environmental pillar will allow a clearer understanding of its significance in sustainable development. He poses the following two questions: *How might the goal of global equity be made compatible with environmental objectives? And how might participatory mechanisms incorporate the aspirations of vulnerable groups, current and future?* The answers to these questions are key inputs for formulating an alternative set of social impact categories. Gutowski [140] reported a criticism of LCA, questioning where the people are. He provided a framework and examples to illustrate how human behavior can alter the environmental outcomes suggested by LCA. In response to this cross-sectoral challenge, Leipold et al. [141] suggest that social policies and initiatives should be at the heart of circular bioeconomy narratives, highlighting collective equity issues over individual opportunities for consumers and employees.

The economic dimension has had an impact on categories related to financial costs. The cost of production leads the list of categories, followed by the revenue generated.

Raw material costs and capital costs share the third position. Most studies ambiguously describe production costs, without specifying whether feedstock costs, common goods, utility costs, and maintenance costs are measured. The detail of the estimates has been previously suggested [142]. Since economic activities can have a wide range of positive and negative consequences, the need to broaden the scope of economic evaluations in LCA has been suggested [143]. This criticism applies to the circular management of AWB, since the impact categories used are limited to current economic structures and indicators that do not denote the search for resilient economies. Valdivia et al. [144] invite us to rethink the approach to the economic field in LCA based on inclusive paths that do not neglect social stability.

The studies' environmental, economic, and social developments revealed that there are holistic assessments of sustainability. The impact categories that cover the three pillars were balanced at the beginning of the studies, but when carrying out the discussion with the stakeholders, the indicators decreased due to their level of association with the defined problem.

Technology and government impact categories were included in the MCDA/LCA analyses. The technological aspect was described by the level of technological readiness, the level of readiness for demand, the level of readiness for integration, and the renewability of resources. The aspects of government were analyzed in relation to the strategy, the brand image, and the experience of the company [145]. The inclusion of these two aspects played a key role in the decision-making process for the studies. However, its low prevalence in the reviewed literature indicated that there is a negligent alignment between the purposes of MCDA/LCA investigations and the objectives of business and government. Implications arise when LCA findings are scaled up to make claims about possible future outcomes within a scenario that ignores public and private sector goals.

6. Synergies and Trade-Offs in the Methodological Association

The complex links and trade-offs between AWB, the environment, and people under a transformative circular vision are now sharply in focus. The growing understanding of this three-way interaction has prompted the development of sustainable assessment frameworks. This review reports that the MCDA/LCA framework has benefited AWB's circular bioeconomy. The primary motivation for its use has been to maximize economic performance while minimizing environmental footprints. The predominance of eco-efficiency and technical-environmental approaches demonstrate this. Debates have been published about this motivation, since it does not represent the essence of sustainability [146]. Nevertheless, the social sphere's connotations are gradually being captured.

The relevance of the MCDA/LCA framework lies in promoting a multiple approach based on dimensions, indicators, and stakeholders to assess and make decisions from the perspective of the life cycle. The methodological synergy that the integration of MCDA techniques into LCA tools has shown lies in the impact assessment phase. MCDA has not been integrated into the processes or actions of the other three phases of LCA (Figure 7). MCDA has traditionally been used to group and weight the outputs of LCA. As previously stated, the weighting methods used were classified as "objective weighting methods" using mathematical methods and "subjective weighting methods" using stakeholders. Although the ISO standards on LCA do not require normalization, grouping, or weighting, LCA practitioners see these as beneficial in simplifying the analysis of the overview for decision making. This influence of MCDA is crucial to resolving trade-offs between impact categories. It has been notably applied in different areas of sustainability [27,42].

The methodological trade-offs of incorporating MCDA into LCA have revolved around three aspects: poor integration in the three phases of LCA, uncertainties, and operability. Unfortunately, few authors either mentioned or applied recommendations and limitations exposed in other works.

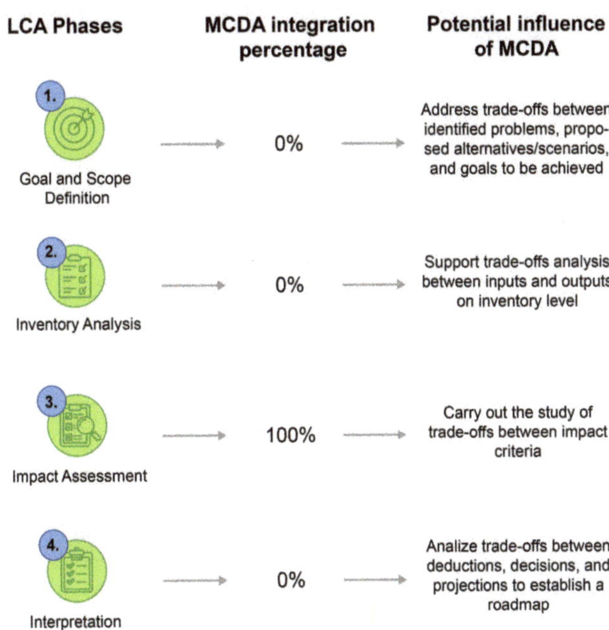

Figure 7. Integration of MCDA into the LCA phases in the studies framed in the circular bioeconomy of AWB and the benefits that MCDA can trigger if its complete integration is carried out—own elaboration.

The absence of MCDA in the first, second, and fourth stages of LCA diminishes its potential impact. The benefits provided by MCDA in the objective and scope definition phase are the structuring of the problem and the mitigation of uncertainty in methodological decisions. The description of the study context can present various challenges and numerous solution alternatives that are not dimensioned and projected by the authors of the paper. Likewise, recently proposed methodological innovations can be incorporated, or methodological gaps can be identified and strengthened.

Including MCDA in the inventory phase supports the analysis of trade-offs between inputs and outputs. Depending on the objective of the study, it is necessary to select different typologies and numbers of categories. A greater diversity of typologies and number of categories does not translate into a greater validation of the work. Tedious processes can be simplified by including a small step based on MCDA. Furthermore, some indicators are measured qualitatively, and for this, MCDA is suitable.

The primary benefit of incorporating MCDA in the interpretation phase is that it strengthens decisions and projections. The establishment of roadmaps is uncommon, which limits the dissemination of concatenated publications that can be used to justify policy changes. Many more research efforts are needed to determine the acceptability of various approaches and their adequacy to inform decision making in real-world situations [44]. Integrating MCDA into these three phases of LCA may represent small efforts compared to the impact assessment phase.

The possibility that MCDA could introduce uncertainty through a loss of information by adding subjective, value-laden data is an identified trade-off. To assess the effect of these changes, for example, on the overall sustainability index weighting, a sensitivity analysis is performed. Unfortunately, more than half of the reviewed studies did not have sensitivity analyses. A robust treatment of uncertainty is often omitted or partially accounted for by assuming selected key scenarios [147]. Several researchers have mentioned that the MCDA/LCA framework is not operationally feasible as another trade-off. It implies a considerable alignment of efforts that does not depend entirely on the scope of the researchers.

To improve sustainability assessments under LCA, one or more MCDA processes may be needed throughout its phases. This decision, however, must be carefully considered on a case-by-case basis. In line with Dias et al. [44], two expectations of this review are that LCA and MCDA practitioners collaborate more closely to investigate the potential of this framework and that its use serves as a bridge between science and practice.

7. Key Issues for a Comprehensive AWB Recovery Sustainability Assessment through the MCDA/LCA Framework

Research on AWB and sustainability issues has been prolific, and it relates to the call to become sustainable for a cleaner planet [148–150]. It has produced an amount of information that was not previously available, and the MCDA/LCA association has contributed to this. Based on the shown potential of this framework and its adoption by researchers and practitioners, it is expected that its use will continue to grow and be implemented in other types of biomass. To enrich the methodological innovation of MCDA/LCA research and thus support the emerging challenges and opportunities of AWB's circular bioeconomy transition in society, this review establishes the following five suggestions:

First, explore new roles for MCDA throughout the phases of LCA, identifying which techniques are feasible and which factors are necessary and deducing pros and cons. This consideration will break with the routine adoption of the MCDA process.

Second, design an appropriate process for stakeholder identification and engagement. Holistically articulating the stakeholders in the phases of LCA leads to benefits that depend on the study contexts. It will also require novel forms of interaction and consensus building to address bottlenecks and map out options for improvement among stakeholders.

Third, strengthen social and technological categories. The social dimension can be analyzed through S-LCA and thus investigate the social consequences of a given change, for example, the adoption of a new technology [70]. Both the United Nations Environment Development Program's life-cycle initiative and social hotspot database provide detailed guidelines for the S-LCA of products [84], which can be adapted to the AWB recovery context to advance in measuring their social feasibility. Likewise, it is necessary to deepen technological implications that in some instances can be a limitation.

Fourth, examine what relevant legislation applies to the AWB in question in terms of permits, taxes, and relevant financial incentives that might be available. Against this background, Stone et al. [146] propose a qualitative impact indicator called political/regulatory compatibility that can be considered and diversified into more indicators for future studies. The impact of MCDA/LCA results must be aligned with government and company goals.

Fifth, promote the articulation of studies at the micro-, meso-, and macro-scales that allow moving from quantification to action. Processes of articulation and discussion of the perspectives that unite the environmental, economic, social, technical, managerial, and political spheres to create a shared vision and a road map will be necessary later [151]. The work published by Lopes et al. [152], which offers a description of how to perform a collaborative process involving stakeholders to support the circular transition with a projection to 2035 in the food and beverage packaging sector in Portugal, can be used as a reference guide.

This review paper considered the literature within the parameters of the search criteria used to find publications in databases. This implies that the search criteria (keywords and Boolean operators) used do not allow for results to be generalized beyond the scope of the study. Moreover, the final version of the published journal articles was considered. There is the option of searching for additional articles that are not yet ready for publication. Although these two factors may be limitations of this study, these are inherent in database operation.

8. Conclusions

The use of the MCDA/LCA framework has aided in the advancement of AWB's circular bioeconomy. It has been versatile due to the various applications, technologies, and AWB reports at micro-, meso-, and macro-scales.

The operability of the MCDA/LCA framework has presented divergences in different aspects. Stakeholders are partially considered in the studies, and neither their role nor the characteristics of their selection are described in sufficient detail. Its potential is limitedly exploited, which could reduce the impact, effectiveness, and practicality of processes. The MCDA/LCA framework has shown the integration of various multi-criteria techniques and life-cycle tools. The predominant methodological approach of life-cycle tools is to use E-LCA and measure independently of various economic and social parameters. Barriers still exist to using S-LCA, LCC, and LCSA. In contrast, the TM-LCA methodology is gaining ground. The AHP and TOPSIS techniques are frequently used in MCDA. Its integration in the LCA tools has focused solely on the impact evaluation phase, with the main purpose of grouping and weighting the outputs of LCA. MCDA has the potential to generate benefits per the LCA phase that need to be examined and proven.

The MCDA/LCA framework has supported holistic assessments of sustainability. Environmental, economic, and social factors have been studied together with technical, business strategy, and governmental aspects. The assessment of these last three categories is still immature. The social dimension has received interest, but it must be strengthened, while the economic dimension has not reflected innovation. Future studies are expected to address and answer the questions discussed here.

Supplementary Materials: The following supporting information can be downloaded at https://www.mdpi.com/article/10.3390/su15065026/s1, Table S1: Search equation applied to the Scopus and Web of Science databases to consolidate the reviewed literature, Table S2: Description of aspects and categories scrutinized in the publications, Table S3: Indicators found by dimension in the publication set.

Author Contributions: All authors have made a direct and intellectual contribution to the manuscript. Conceptualization, F.R.-P. and M.Á.G.-C.; validation, F.R.-P. and M.Á.G.-C.; data analysis, F.R.-P.; writing—original draft preparation, F.R.-P.; writing—review and editing, F.R.-P. and M.Á.G.-C.; supervision, M.Á.G.-C. All authors have read and agreed to the published version of the manuscript.

Funding: This research received no external funding.

Institutional Review Board Statement: Not applicable.

Informed Consent Statement: Not applicable.

Data Availability Statement: The data presented in this study are available on request from the corresponding author.

Acknowledgments: Not applicable.

Conflicts of Interest: The authors declare no conflict of interest.

References

1. Tan, E.C.; Lamers, P. Circular bioeconomy concepts—A perspective. *Front. Sustain.* **2021**, *2*, 701509. [CrossRef]
2. D'Amato, D.; Droste, N.; Allen, B.; Kettunen, M.; Lähtinen, K.; Korhonen, J.; Leskinen, P.; Matthies, B.D.; Toppinen, A. Green, circular, bio economy: A comparative analysis of sustainability avenues. *J. Clean. Prod.* **2017**, *168*, 716–734. [CrossRef]
3. Stegmann, P.; Londo, M.; Junginger, M. The circular bioeconomy: Its elements and role in European bioeconomy clusters. *Resour. Conserv. Recycl. X* **2020**, *6*, 100029. [CrossRef]
4. European Commission. Available online: https://knowledge4policy.ec.europa.eu/publication/sustainable-bioeconomy-europe-strengthening-connection-between-economy-society_en (accessed on 9 January 2023).
5. Muscat, A.; de Olde, E.M.; Ripoll-Bosch, R.; Van Zanten, H.H.E.; Metze, T.A.P.; Termeer, C.J.A.M.; van Ittersum, M.K.; de Boer, I.J.M. Principles, drivers and opportunities of a circular bioeconomy. *Nat. Food* **2021**, *2*, 561–566. [CrossRef]

6. Duque-Acevedo, M.; Belmonte-Ureña, L.J.; Cortés-García, F.J.; Camacho-Ferre, F. Recovery of agricultural waste biomass: A sustainability strategy for moving towards a circular bioeconomy. In *Handbook of Solid Waste Management: Sustainability through Circular Economy*, 1st ed.; Baskar, C., Ramakrishna, S., Baskar, S., Sharma, R., Chinnappan, A., Sehrawat, R., Eds.; Springer: Singapore, 2021; Volume 1, pp. 1–30.

7. United Nations Environment Programme. Available online: https://www.unep.org/resources/report/unep-food-waste-index-report-2021 (accessed on 10 January 2023).

8. Sarangi, P.K.; Subudhi, S.; Bhatia, L.; Saha, K.; Mudgil, D.; Shadangi, K.P.; Srivastava, R.K.; Pattnaik, B.; Arya, R.K. Utilization of agricultural waste biomass and recycling toward circular bioeconomy. *Environ. Sci. Pollut. Res.* **2022**, *30*, 8526–8539. [CrossRef] [PubMed]

9. D'Adamo, I.; Gastaldi, M.; Morone, P.; Rosa, P.; Sassanelli, C.; Settembre-Blundo, D.; Shen, Y. Bioeconomy of Sustainability: Drivers, Opportunities and Policy Implications. *Sustainability* **2021**, *14*, 200. [CrossRef]

10. Pfau, S.F.; Hagens, J.E.; Dankbaar, B.; Smits, A.J.M. Visions of Sustainability in Bioeconomy Research. *Sustainability* **2014**, *6*, 1222–1249. [CrossRef]

11. Salvador, R.; Barros, M.V.; Donner, M.; Brito, P.; Halog, A.; De Francisco, A.C. How to advance regional circular bioeconomy systems? Identifying barriers, challenges, drivers, and opportunities. *Sustain. Prod. Consum.* **2022**, *32*, 248–269. [CrossRef]

12. Giampietro, M. On the Circular Bioeconomy and Decoupling: Implications for Sustainable Growth. *Ecol. Econ.* **2019**, *162*, 143–156. [CrossRef]

13. Angouria-Tsorochidou, E.; Teigiserova, D.A.; Thomsen, M. Limits to circular bioeconomy in the transition towards decentralized biowaste management systems. *Resour. Conserv. Recycl.* **2021**, *164*, 105207. [CrossRef]

14. Singh, R.K.; Murty, H.R.; Gupta, S.K.; Dikshit, A.K. An overview of sustainability assessment methodologies. *Ecol. Indic.* **2009**, *9*, 189–212. [CrossRef]

15. Lam, C.-M.; Yu, I.K.; Hsu, S.-C.; Tsang, D.C. Life-cycle assessment on food waste valorisation to value-added products. *J. Clean. Prod.* **2018**, *199*, 840–848. [CrossRef]

16. Bjørn, A.; Owsianiak, M.; Molin, C.; Hauschild, M.Z. LCA history. In *Life Cycle Assessment: Theory and Practice*, 1st ed.; Hauschild, M., Rosenbaum, R., Olsen, S., Eds.; Springer: Cham, Switzerland, 2018; pp. 17–30.

17. Hauschild, M.Z. Introduction to LCA Methodology. In *Life Cycle Assessment: Theory and Practice*, 1st ed.; Hauschild, M., Rosenbaum, R., Olsen, S., Eds.; Springer: Cham, Switzerland, 2018; pp. 59–66.

18. Visentin, C.; da Silva Trentin, A.W.; Braun, A.B.; Thomé, A. Life cycle sustainability assessment: A systematic literature review through the application perspective, indicators, and methodologies. *J. Clean. Prod.* **2020**, *270*, 122509. [CrossRef]

19. Owsianiak, M.; Bjørn, A.; Laurent, A.; Molin, C.; Ryberg, M.W. LCA Applications. In *Life Cycle Assessment: Theory and Practice*, 1st ed.; Hauschild, M., Rosenbaum, R., Olsen, S., Eds.; Springer: Cham, Switzerland, 2018; pp. 31–41.

20. Moltesen, A.; Bjørn, A. LCA and Sustainability. In *Life Cycle Assessment: Theory and Practice*, 1st ed.; Hauschild, M., Rosenbaum, R., Olsen, S., Eds.; Springer: Cham, Switzerland, 2018; pp. 43–55.

21. Kloepffer, W. Life cycle sustainability assessment of products. *Int. J. Life Cycle Assess.* **2008**, *13*, 89–95. [CrossRef]

22. Vlachokostas, C.; Michailidou, A.; Achillas, C. Multi-Criteria Decision Analysis towards promoting Waste-to-Energy Management Strategies: A critical review. *Renew. Sustain. Energy Rev.* **2021**, *138*, 110563. [CrossRef]

23. Ben Amor, S.; Belaid, F.; Benkraiem, R.; Ramdani, B.; Guesmi, K. Multi-criteria classification, sorting, and clustering: A bibliometric review and research agenda. *Ann. Oper. Res.* **2022**, *316*, 1–23. [CrossRef]

24. Esmail, B.A.; Geneletti, D. Multi-criteria decision analysis for nature conservation: A review of 20 years of applications. *Methods Ecol. Evol.* **2018**, *9*, 42–53. [CrossRef]

25. Onat, N.C.; Gumus, S.; Kucukvar, M.; Tatari, O. Application of the TOPSIS and intuitionistic fuzzy set approaches for ranking the life cycle sustainability performance of alternative vehicle technologies. *Sustain. Prod. Consum.* **2016**, *6*, 12–25. [CrossRef]

26. Väisänen, S.; Mikkilä, M.; Havukainen, J.; Sokka, L.; Luoranen, M.; Horttanainen, M. Using a multi-method approach for decision-making about a sustainable local distributed energy system: A case study from Finland. *J. Clean. Prod.* **2016**, *137*, 1330–1338. [CrossRef]

27. De Luca, A.I.; Iofrida, N.; Leskinen, P.; Stillitano, T.; Falcone, G.; Strano, A.; Gulisano, G. Life cycle tools combined with multi-criteria and participatory methods for agricultural sustainability: Insights from a systematic and critical review. *Sci. Total Environ.* **2017**, *595*, 352–370. [CrossRef]

28. Scarlat, N.; Dallemand, J.F.; Monforti-Ferrario, F.; Nita, V. The role of biomass and bioenergy in a future bioeconomy: Policies and facts. *Environ. Dev.* **2015**, *15*, 3–34. [CrossRef]

29. Tursi, A. A review on biomass: Importance, chemistry, classification, and conversion. *Biofuel Res. J.* **2019**, *6*, 962–979. [CrossRef]

30. Castro-Muñoz, R.; Díaz-Montes, E.; Gontarek-Castro, E.; Boczkaj, G.; Galanakis, C.M. A comprehensive review on current and emerging technologies toward the valorization of bio-based wastes and by products from foods. *Compr. Rev. Food Sci. Food Saf.* **2022**, *21*, 46–105. [CrossRef] [PubMed]

31. Sadh, P.K.; Duhan, S.; Duhan, J.S. Agro-industrial wastes and their utilization using solid state fermentation: A review. *Bioresour. Bioprocess.* **2018**, *5*, 1. [CrossRef]

32. Angulo-Mosquera, L.S.; Alvarado-Alvarado, A.A.; Rivas-Arrieta, M.J.; Cattaneo, C.R.; Rene, E.R.; García-Depraect, O. Production of solid biofuels from organic waste in developing countries: A review from sustainability and economic feasibility perspectives. *Sci. Total Environ.* **2021**, *795*, 148816. [CrossRef]

33. Sherwood, J. The significance of biomass in a circular economy. *Bioresour. Technol.* **2020**, *300*, 122755. [CrossRef]

34. Ishangulyyev, R.; Kim, S.; Lee, S.H. Understanding Food Loss and Waste—Why Are We Losing and Wasting Food? *Foods* **2019**, *8*, 297. [CrossRef]

35. Hodaifa, G.; Garcìa, C.A.; Rodroguez-Perez, S. Revalorization of agro-food residues as bioadsorbents for wastewater treatment. In *Aqueous Phase Adsorption—Theory, Simulations and Experiments*, 1st ed.; Singh, J.K., Verma, N., Eds.; Taylor & Francis Group: New York, NY, USA, 2018; Volume 1, pp. 249–282.

36. Nayak, A.; Bhushan, B. An overview of the recent trends on the waste valorization techniques for food wastes. *J. Environ. Manag.* **2019**, *233*, 352–370. [CrossRef] [PubMed]

37. Mehta, N.; Shah, K.; Lin, Y.-I.; Sun, Y.; Pan, S.-Y. Advances in Circular Bioeconomy Technologies: From Agricultural Wastewater to Value-Added Resources. *Environments* **2021**, *8*, 20. [CrossRef]

38. Yaashikaa, P.; Kumar, P.S.; Varjani, S. Valorization of agro-industrial wastes for biorefinery process and circular bioeconomy: A critical review. *Bioresour. Technol.* **2022**, *343*, 126126. [CrossRef]

39. Cuadrado-Osorio, P.D.; Ramírez-Mejía, J.M.; Mejía-Avellaneda, L.F.; Mesa, L.; Bautista, E.J. Agro-industrial residues for microbial bioproducts: A key booster for bioeconomy. *Bioresour. Technol. Rep.* **2022**, *20*, 101232. [CrossRef]

40. Torkayesh, A.E.; Rajaeifar, M.A.; Rostom, M.; Malmir, B.; Yazdani, M.; Suh, S.; Heidrich, O. Integrating life cycle assessment and multi criteria decision making for sustainable waste management: Key issues and recommendations for future studies. *Renew. Sustain. Energy Rev.* **2022**, *168*, 112819. [CrossRef]

41. Angelo, A.C.M.; Saraiva, A.B.; Clímaco, J.C.N.; Infante, C.E.; Valle, R. Life Cycle Assessment and Multi-criteria Decision Analysis: Selection of a strategy for domestic food waste management in Rio de Janeiro. *J. Clean. Prod.* **2017**, *143*, 744–756. [CrossRef]

42. Campos-Guzmán, V.; García-Cáscales, M.S.; Espinosa, N.; Urbina, A. Life Cycle Analysis with Multi-Criteria Decision Making: A review of approaches for the sustainability evaluation of renewable energy technologies. *Renew. Sustain. Energy Rev.* **2019**, *104*, 343–366. [CrossRef]

43. Tziolas, E.; Bournaris, T.; Manos, B.; Nastis, S. Life cycle assessment and multi-criteria analysis in agriculture: Synergies and insights. In *Multicriteria Analysis in Agriculture: Current Trends and Recent Applications*, 1st ed.; Berbel, J., Bournaris, T., Manos, B., Matsatsinis, N., Viaggi, D., Eds.; Springer: Cham, Switzerland, 2018; Volume 1, pp. 289–321. [CrossRef]

44. Dias, L.C.; Freire, F.; Geldermann, J. Perspectives on multi-criteria decision analysis and life-cycle assessment. In *New Perspectives in Multiple Criteria Decision Making*, 1st ed.; Doumpos, M., Figueira, J., Greco, S., Zopounidis, C., Eds.; Springer: Cham, Switzerland, 2019; Volume 1, pp. 315–329.

45. Zanghelini, G.M.; Cherubini, E.; Soares, S.R. How Multi-Criteria Decision Analysis (MCDA) is aiding Life Cycle Assessment (LCA) in results interpretation. *J. Clean. Prod.* **2018**, *172*, 609–622. [CrossRef]

46. Geldermann, J.; Rentz, O. Multi-criteria Analysis for Technique Assessment: Case Study from Industrial Coating. *J. Ind. Ecol.* **2005**, *9*, 127–142. [CrossRef]

47. Miettinen, P.; Hämäläinen, R.P. How to benefit from decision analysis in environmental life cycle assessment (LCA). *Eur. J. Oper. Res.* **1997**, *102*, 279–294. [CrossRef]

48. Tranfield, D.; Denyer, D.; Smart, P. Towards a Methodology for Developing Evidence-Informed Management Knowledge by Means of Systematic Review. *Br. J. Manag.* **2003**, *14*, 207–222. [CrossRef]

49. Moher, D.; Liberati, A.; Tetzlaff, J.; Altman, D.G.; PRISMA Group. Preferred Reporting Items for Systematic Reviews and Meta-Analyses: The PRISMA Statement. *Ann. Intern. Med.* **2009**, *151*, 264–269. [CrossRef]

50. Ranjbari, M.; Esfandabadi, Z.S.; Quatraro, F.; Vatanparast, H.; Lam, S.S.; Aghbashlo, M.; Tabatabaei, M. Biomass and organic waste potentials towards implementing circular bioeconomy platforms: A systematic bibliometric analysis. *Fuel* **2022**, *318*, 123585. [CrossRef]

51. Delaney, A.; Tamás, P.A. Searching for evidence or approval? A commentary on database search in systematic reviews and alternative information retrieval methodologies. *Res. Synth. Methods* **2018**, *9*, 124–131. [CrossRef] [PubMed]

52. Gésan-Guiziou, G.; Alaphilippe, A.; Andro, M.; Aubin, J.; Bockstaller, C.; Botreau, R.; Buche, P.; Collet, C.; Darmon, N.; Delabuis, M.; et al. Annotation data about multi criteria assessment methods used in the agri-food research: The french national institute for agricultural research (INRA) experience. *Data Brief* **2019**, *25*, 104204. [CrossRef] [PubMed]

53. Liu, Y.; Lyu, Y.; Tian, J.; Zhao, J.; Ye, N.; Zhang, Y.; Chen, L. Review of waste biorefinery development towards a circular economy: From the perspective of a life cycle assessment. *Renew. Sustain. Energy Rev.* **2021**, *139*, 110716. [CrossRef]

54. Dias, L.C.; Passeira, C.; Malça, J.; Freire, F. Integrating life-cycle assessment and multi-criteria decision analysis to compare alternative biodiesel chains. *Ann. Oper. Res.* **2016**, *312*, 1359–1374. [CrossRef]

55. Fernández-Tirado, F.; Parra-López, C.; Romero-Gámez, M. A multi-criteria sustainability assessment for biodiesel alternatives in Spain: Life cycle assessment normalization and weighting. *Renew. Energy* **2021**, *164*, 1195–1203. [CrossRef]

56. Myllyviita, T.; Holma, A.; Antikainen, R.; Lähtinen, K.; Leskinen, P. Assessing environmental impacts of biomass production chains—Application of life cycle assessment (LCA) and multi-criteria decision analysis (MCDA). *J. Clean. Prod.* **2012**, *29–30*, 238–245. [CrossRef]

57. Joglekar, S.N.; Dalwankar, G.; Qureshi, N.; Mandavgane, S.A. Sugarcane valorization: Selection of process routes based on sustainability index. *Environ. Sci. Pollut. Res.* **2022**, *29*, 10812–10825. [CrossRef]

58. Ramesh, P.; Selvan, V.A.M.; Babu, D. Selection of sustainable lignocellulose biomass for second-generation bioethanol production for automobile vehicles using lifecycle indicators through fuzzy hybrid PyMCDM approach. *Fuel* **2022**, *322*, 124240. [CrossRef]

59. Raman, J.K.; Alves, C.M.; Gnansounou, E. A review on moringa tree and vetiver grassPotential biorefinery feedstocks. *Bioresour. Technol.* **2018**, *249*, 1044–1051. [CrossRef]

60. Ekener, E.; Hansson, J.; Larsson, A.; Peck, P. Developing Life Cycle Sustainability Assessment methodology by applying values-based sustainability weighting—Tested on biomass based and fossil transportation fuels. *J. Clean. Prod.* **2018**, *181*, 337–351. [CrossRef]

61. Liard, G.; Lesage, P.; Samson, R.; Stuart, P.R. Systematic assessment of triticale-based biorefinery strategies: Environmental evaluation using life cycle assessment. *Biofuels Bioprod. Biorefin.* **2018**, *12*, S60–S72. [CrossRef]

62. Im-Orb, K.; Arpornwichanop, A. Process and sustainability analyses of the integrated biomass pyrolysis, gasification, and methanol synthesis process for methanol production. *Energy* **2020**, *193*, 116788. [CrossRef]

63. Tonini, D.; Wandl, A.; Meister, K.; Unceta, P.M.; Taelman, S.E.; Sanjuan-Delmás, D.; Dewulf, J.; Huygens, D. Quantitative sustainability assessment of household food waste management in the Amsterdam Metropolitan Area. *Resour. Conserv. Recycl.* **2020**, *160*, 104854. [CrossRef]

64. Slorach, P.C.; Jeswani, H.K.; Cuéllar-Franca, R.; Azapagic, A. Environmental sustainability in the food-energy-water-health nexus: A new methodology and an application to food waste in a circular economy. *Waste Manag.* **2020**, *113*, 359–368. [CrossRef]

65. Vega, G.C.; Sohn, J.; Bruun, S.; Olsen, S.I.; Birkved, M. Maximizing Environmental Impact Savings Potential through Innovative Biorefinery Alternatives: An Application of the TM-LCA Framework for Regional Scale Impact Assessment. *Sustainability* **2019**, *11*, 3836. [CrossRef]

66. Wang, B.; Song, J.; Ren, J.; Li, K.; Duan, H.; Wang, X. Selecting sustainable energy conversion technologies for agricultural residues: A fuzzy AHP-VIKOR based prioritization from life cycle perspective. *Resour. Conserv. Recycl.* **2019**, *142*, 78–87. [CrossRef]

67. von Doderer, C.; Kleynhans, T. Determining the most sustainable lignocellulosic bioenergy system following a case study approach. *Biomass Bioenergy* **2014**, *70*, 273–286. [CrossRef]

68. Lim, J.Y.; How, B.S.; Teng, S.Y.; Leong, W.D.; Tang, J.P.; Lam, H.L.; Yoo, C.K. Multi-objective lifecycle optimization for oil palm fertilizer formulation: A hybrid P-graph and TOPSIS approach. *Resour. Conserv. Recycl.* **2021**, *166*, 105357. [CrossRef]

69. Acosta-Alba, I.; Chia, E.; Andrieu, N. The LCA4CSA framework: Using life cycle assessment to strengthen environmental sustainability analysis of climate smart agriculture options at farm and crop system levels. *Agric. Syst.* **2019**, *171*, 155–170. [CrossRef]

70. Cardoso, T.F.; Watanabe, M.D.; Souza, A.; Chagas, M.F.; Cavalett, O.; Morais, E.R.; Nogueira, L.A.; Leal, M.R.L.; Braunbeck, O.A.; Cortez, L.A.; et al. Economic, environmental, and social impacts of different sugarcane production systems. *Biofuels Bioprod. Biorefin.* **2018**, *12*, 68–82. [CrossRef]

71. Garas, G.; Sayed, A.M.; Bakhoum, E.S.H. Application of nano waste particles in concrete for sustainable construction: A comparative study. *Int. J. Sustain. Eng.* **2021**, *14*, 2041–2047. [CrossRef]

72. Joglekar, S.N.; Kharkar, R.A.; Mandavgane, S.A.; Kulkarni, B.D. Sustainability assessment of brick work for low-cost housing: A comparison between waste based bricks and burnt clay bricks. *Sustain. Cities Soc.* **2018**, *37*, 396–406. [CrossRef]

73. Garcia-Garcia, G.; Woolley, E.; Rahimifard, S.; Colwill, J.; White, R.; Needham, L. A Methodology for Sustainable Management of Food Waste. *Waste Biomass Valorization* **2017**, *8*, 2209–2227. [CrossRef]

74. Vega, G.C.; Voogt, J.; Sohn, J.; Birkved, M.; Olsen, S.I. Assessing New Biotechnologies by Combining TEA and TM-LCA for an Efficient Use of Biomass Resources. *Sustainability* **2020**, *12*, 3676. [CrossRef]

75. Vega, G.C.; Sohn, J.; Voogt, J.; Birkved, M.; Olsen, S.I.; Nilsson, A.E. Insights from combining techno-economic and life cycle assessment—A case study of polyphenol extraction from red wine pomace. *Resour. Conserv. Recycl.* **2021**, *167*, 105318. [CrossRef]

76. Alsaleh, A.; Aleisa, E. Triple Bottom-Line Evaluation of the Production of Animal Feed from Food Waste: A Life Cycle Assessment. *Waste Biomass Valorization* **2022**, *13*, 1–27. [CrossRef] [PubMed]

77. Sanaei, S.; Stuart, P.R. Systematic assessment of triticale-based biorefinery strategies: Techno-economic analysis to identify investment opportunities. *Biofuels Bioprod. Biorefin.* **2018**, *12*, S46–S59. [CrossRef]

78. Sadh, P.K.; Chawla, P.; Kumar, S.; Das, A.; Kumar, R.; Bains, A.; Sridhar, K.; Duhan, J.S.; Sharma, M. Recovery of agricultural waste biomass: A path for circular bioeconomy. *Sci. Total Environ.* **2023**, *870*, 161904. [CrossRef] [PubMed]

79. Jain, A.; Sarsaiya, S.; Awasthi, M.K.; Singh, R.; Rajput, R.; Mishra, U.C.; Chen, J.; Shi, J. Bioenergy and bio-products from bio-waste and its associated modern circular economy: Current research trends, challenges, and future outlooks. *Fuel* **2022**, *307*, 121859. [CrossRef]

80. Salinas-Velandia, D.A.; Romero-Perdomo, F.; Numa-Vergel, S.; Villagrán, E.; Donado-Godoy, P.; Galindo-Pacheco, J.R. Insights into Circular Horticulture: Knowledge Diffusion, Resource Circulation, One Health Approach, and Greenhouse Technologies. *Int. J. Environ. Res. Public Health* **2022**, *19*, 12053. [CrossRef] [PubMed]

81. Mendoza-Labrador, J.; Romero-Perdomo, F.; Abril, J.; Hernández, J.P.; Uribe-Vélez, D.; Buitrago, R.B. Bacillus strains immobilized in alginate macrobeads enhance drought stress adaptation of guinea grass. *Rhizosphere* **2021**, *19*, 100385. [CrossRef]

82. Juanpera, M.; Ferrer-Martí, L.; Diez-Montero, R.; Ferrer, I.; Castro, L.; Escalante, H.; Garfí, M. A robust multicriteria analysis for the post-treatment of digestate from low-tech digesters. Boosting the circular bioeconomy of small-scale farms in Colombia. *Renew. Sustain. Energy Rev.* **2022**, *166*, 112638. [CrossRef]

83. Onat, N.C.; Kucukvar, M. A systematic review on sustainability assessment of electric vehicles: Knowledge gaps and future perspectives. *Environ. Impact Assess. Rev.* **2022**, *97*, 106867. [CrossRef]

84. Gawel, E.; Pannicke, N.; Hagemann, N. A Path Transition Towards a Bioeconomy—The Crucial Role of Sustainability. *Sustainability* **2019**, *11*, 3005. [CrossRef]

85. Ncube, A.; Sadondo, P.; Makhanda, R.; Mabika, C.; Beinisch, N.; Cocker, J.; Gwenzi, W.; Ulgiati, S. Circular bioeconomy potential and challenges within an African context: From theory to practice. *J. Clean. Prod.* **2022**, *367*, 133068. [CrossRef]

86. Blum, N.U.; Haupt, M.; Bening, C.R. Why "Circular" doesn't always mean "Sustainable". *Resour. Conserv. Recycl.* **2020**, *162*, 105042. [CrossRef]

87. Kujala, J.; Sachs, S.; Leinonen, H.; Heikkinen, A.; Laude, D. Stakeholder Engagement: Past, Present, and Future. *Bus. Soc.* **2022**, *61*, 1136–1196. [CrossRef]

88. Kruetli, P.; Stauffacher, M.; Flueeler, T.; Scholz, R.W. Functional-dynamic public participation in technological decision-making: Site selection processes of nuclear waste repositories. *J. Risk Res.* **2010**, *13*, 861–875. [CrossRef]

89. Brandt, P.; Ernst, A.; Gralla, F.; Luederitz, C.; Lang, D.J.; Newig, J.; Reinert, F.; Abson, D.J.; von Wehrden, H. A review of transdisciplinary research in sustainability science. *Ecol. Econ.* **2013**, *92*, 1–15. [CrossRef]

90. Chen, W.; Holden, N.M. Tiered life cycle sustainability assessment applied to a grazing dairy farm. *J. Clean. Prod.* **2018**, *172*, 1169–1179. [CrossRef]

91. Thokala, P.; Madhavan, G. Stakeholder involvement in Multi-Criteria Decision Analysis. *Cost Eff. Resour. Alloc.* **2018**, *16*, 1–3. [CrossRef] [PubMed]

92. Iofrida, N.; De Luca, A.I.; Strano, A.; Gulisano, G. Can social research paradigms justify the diversity of approaches to social life cycle assessment? *Int. J. Life Cycle Assess.* **2016**, *23*, 464–480. [CrossRef]

93. Souza, R.; Rosenhead, J.; Salhofer, S.; Valle, R.; Lins, M. Definition of sustainability impact categories based on stakeholder perspectives. *J. Clean. Prod.* **2015**, *105*, 41–51. [CrossRef]

94. Wang, J.; Maier, S.D.; Horn, R.; Holländer, R.; Aschemann, R. Development of an Ex-Ante Sustainability Assessment Methodology for Municipal Solid Waste Management Innovations. *Sustainability* **2018**, *10*, 3208. [CrossRef]

95. Hossain, M.; Leminen, S.; Westerlund, M. A systematic review of living lab literature. *J. Clean. Prod.* **2019**, *213*, 976–988. [CrossRef]

96. Huttunen, S.; Manninen, K.; Leskinen, P. Combining biogas LCA reviews with stakeholder interviews to analyse life cycle impacts at a practical level. *J. Clean. Prod.* **2014**, *80*, 5–16. [CrossRef]

97. Marttunen, M.; Mustajoki, J.; Dufva, M.; Karjalainen, T.P. How to design and realize participation of stakeholders in MCDA processes? A framework for selecting an appropriate approach. *EURO J. Decis. Process.* **2015**, *3*, 187–214. [CrossRef]

98. Stillitano, T.; Falcone, G.; Iofrida, N.; Spada, E.; Gulisano, G.; De Luca, A.I. A customized multi-cycle model for measuring the sustainability of circular pathways in agri-food supply chains. *Sci. Total Environ.* **2022**, *844*, 157229. [CrossRef]

99. Bareschino, P.; Mancusi, E.; Urciuolo, M.; Paulillo, A.; Chirone, R.; Pepe, F. Life cycle assessment and feasibility analysis of a combined chemical looping combustion and power-to-methane system for CO_2 capture and utilization. *Renew. Sustain. Energy Rev.* **2020**, *130*, 109962. [CrossRef]

100. Miah, J.; Koh, S.; Stone, D. A hybridised framework combining integrated methods for environmental Life Cycle Assessment and Life Cycle Costing. *J. Clean. Prod.* **2017**, *168*, 846–866. [CrossRef]

101. Grubert, E. The Need for a Preference-Based Multicriteria Prioritization Framework in Life Cycle Sustainability Assessment. *J. Ind. Ecol.* **2017**, *21*, 1522–1535. [CrossRef]

102. Escobar, N.; Laibach, N. Sustainability check for bio-based technologies: A review of process-based and life cycle approaches. *Renew. Sustain. Energy Rev.* **2021**, *135*, 110213. [CrossRef]

103. Degieter, M.; Gellynck, X.; Goyal, S.; Ott, D.; De Steur, H. Life cycle cost analysis of agri-food products: A systematic review. *Sci. Total Environ.* **2022**, *850*, 158012. [CrossRef] [PubMed]

104. United Nations Environment Programme. Available online: https://www.unep.org/resources/report/guidelines-social-life-cycle-assessment-products (accessed on 14 January 2023).

105. Rebolledo-Leiva, R.; Moreira, M.T.; González-García, S. Progress of social assessment in the framework of bioeconomy under a life cycle perspective. *Renew. Sustain. Energy Rev.* **2023**, *175*, 113162. [CrossRef]

106. Pesonen, H.-L.; Horn, S. Evaluating the Sustainability SWOT as a streamlined tool for life cycle sustainability assessment. *Int. J. Life Cycle Assess.* **2013**, *18*, 1780–1792. [CrossRef]

107. Hildebrandt, J.; Bezama, A.; Thrän, D. Insights from the Sustainability Monitoring Tool SUMINISTRO Applied to a Case Study System of Prospective Wood-Based Industry Networks in Central Germany. *Sustainability* **2020**, *12*, 3896. [CrossRef]

108. Sohn, J.; Vega, G.C.; Birkved, M. A Methodology Concept for Territorial Metabolism—Life Cycle Assessment: Challenges and Opportunities in Scaling from Urban to Territorial Assessment. *Procedia CIRP* **2018**, *69*, 89–93. [CrossRef]

109. Harris, S.; Martin, M.; Diener, D. Circularity for circularity's sake? Scoping review of assessment methods for environmental performance in the circular economy. *Sustain. Prod. Consum.* **2021**, *26*, 172–186. [CrossRef]

110. Gontard, N.; Sonesson, U.; Birkved, M.; Majone, M.; Bolzonella, D.; Celli, A.; Angellier-Coussy, H.; Jang, G.-W.; Verniquet, A.; Broeze, J.; et al. A research challenge vision regarding management of agricultural waste in a circular bio-based economy. *Crit. Rev. Environ. Sci. Technol.* **2018**, *48*, 614–654. [CrossRef]

111. Recchia, L.; Boncinelli, P.; Cini, E.; Vieri, M.; Garbati Pegna, F.; Sarri, D. Energetic use of biomass and biofuels. In *Multicriteria Analysis and LCA Techniques. With Applications to Agro-Engineering Problems*, 1st ed.; Recchia, L., Boncinelli, P., Cini, E., Vieri, M., Garbati Pegna, F., Sarri, D., Eds.; Springer: London, UK, 2011; Volume 1, pp. 27–56.

112. Saaty, T.L. How to make a decision: The analytic hierarchy process. *Eur. J. Oper. Res.* **1990**, *48*, 9–26. [CrossRef]

113. Vaidya, O.S.; Kumar, S. Analytic hierarchy process: An overview of applications. *Eur. J. Oper. Res.* **2006**, *169*, 1–29. [CrossRef]
114. Ho, W.; Ma, X. The state-of-the-art integrations and applications of the analytic hierarchy process. *Eur. J. Oper. Res.* **2018**, *267*, 399–414. [CrossRef]
115. Ishizaka, A.; Labib, A. Analytical hierarchy process and expert choice: Benefits and limitations. *Oper. Res. Insight* **2009**, *22*, 201–220. [CrossRef]
116. Olson, D. Comparison of weights in TOPSIS models. *Math. Comput. Model.* **2004**, *40*, 721–727. [CrossRef]
117. Zyoud, S.H.; Fuchs-Hanusch, D. A bibliometric-based survey on AHP and TOPSIS techniques. *Expert Syst. Appl.* **2017**, *78*, 158–181. [CrossRef]
118. Behzadian, M.; Kazemzadeh, R.; Albadvi, A.; Aghdasi, M. PROMETHEE: A comprehensive literature review on methodologies and applications. *Eur. J. Oper. Res.* **2010**, *200*, 198–215. [CrossRef]
119. Boix-Cots, D.; Pardo-Bosch, F.; Blanco, A.; Aguado, A.; Pujadas, P. A systematic review on MIVES: A sustainability-oriented multi-criteria decision-making method. *Build. Environ.* **2022**, *223*, 109515. [CrossRef]
120. Zlaugotne, B.; Zihare, L.; Balode, L.; Kalnbalkite, A.; Khabdullin, A.; Blumberga, D. Multi-Criteria Decision Analysis Methods Comparison. *Environ. Clim. Technol.* **2020**, *24*, 454–471. [CrossRef]
121. Sałabun, W.; Wątróbski, J.; Shekhovtsov, A. Are MCDA Methods Benchmarkable? A Comparative Study of TOPSIS, VIKOR, COPRAS, and PROMETHEE II Methods. *Symmetry* **2020**, *12*, 1549. [CrossRef]
122. Nzila, C.; Dewulf, J.; Spanjers, H.; Tuigong, D.; Kiriamiti, H.; van Langenhove, H. Multi criteria sustainability assessment of biogas production in Kenya. *Appl. Energy* **2012**, *93*, 496–506. [CrossRef]
123. Romero-Perdomo, F.; Carvajalino-Umaña, J.D.; Moreno-Gallego, J.L.; Ardila, N.; González-Curbelo, M. Research Trends on Climate Change and Circular Economy from a Knowledge Mapping Perspective. *Sustainability* **2022**, *14*, 521. [CrossRef]
124. Van Oers, L.; Guinée, J. The Abiotic Depletion Potential: Background, Updates, and Future. *Resources* **2016**, *5*, 16. [CrossRef]
125. Cucurachi, S.; Scherer, L.; Guinée, J.; Tukker, A. Life Cycle Assessment of Food Systems. *One Earth* **2019**, *1*, 292–297. [CrossRef]
126. van der Werf, H.M.G.; Knudsen, M.T.; Cederberg, C. Towards better representation of organic agriculture in life cycle assessment. *Nat. Sustain.* **2020**, *3*, 419–425. [CrossRef]
127. Donke, A.C.G.; Novaes, R.M.L.; Pazianotto, R.A.A.; Moreno-Ruiz, E.; Reinhard, J.; Picoli, J.F.; Folegatti-Matsuura, M.I.D.S. Integrating regionalized Brazilian land use change da-tasets into the ecoinvent database: New data, premises and uncertainties have large effects in the results. *Int. J. Life Cycle Assess.* **2020**, *25*, 1027–1042. [CrossRef]
128. Vázquez-Rowe, I.; Kahhat, R.; Sánchez, I. Perú LCA: Launching the Peruvian national life cycle database. *Int. J. Life Cycle Assess.* **2019**, *24*, 2089–2090. [CrossRef]
129. de Araujo, J.B.; Frega, J.R.; Ugaya, C.M.L. From social impact subcategories to human health: An application of multivariate analysis on S-LCA. *Int. J. Life Cycle Assess.* **2021**, *26*, 1471–1493. [CrossRef]
130. Arvidsson, R.; Hildenbrand, J.; Baumann, H.; Islam, K.M.N.; Parsmo, R. A method for human health impact assessment in social LCA: Lessons from three case studies. *Int. J. Life Cycle Assess.* **2018**, *23*, 690–699. [CrossRef]
131. Pillain, B.; Viana, L.R.; Lefeuvre, A.; Jacquemin, L.; Sonnemann, G. Social life cycle assessment framework for evaluation of potential job creation with an application in the French carbon fiber aeronautical recycling sector. *Int. J. Life Cycle Assess.* **2019**, *24*, 1729–1742. [CrossRef]
132. Stone, J.; Garcia-Garcia, G.; Rahimifard, S. Selection of Sustainable Food Waste Valorisation Routes: A Case Study with Barley Field Residue. *Waste Biomass Valorization* **2020**, *11*, 5733–5748. [CrossRef]
133. Bartzas, G.; Komnitsas, K. An integrated multi-criteria analysis for assessing sustainability of agricultural production at regional level. *Inf. Process. Agric.* **2020**, *7*, 223–232. [CrossRef]
134. D'Adamo, I.; Mazzanti, M.; Morone, P.; Rosa, P. Assessing the relation between waste management policies and circular economy goals. *Waste Manag.* **2022**, *154*, 27–35. [CrossRef] [PubMed]
135. Lim, C.H.; Ngan, S.L.; Ng, W.P.Q.; How, B.S.; Lam, H.L. Biomass supply chain management and challenges. In *Value-Chain of Biofuels*, 1st ed.; Yusup, S., Rashidi, N.A., Eds.; Elsevier: Amsterdam, The Netherlands, 2022; Volume 1, pp. 429–444.
136. Luthin, A.; Traverso, M.; Crawford, R.H. Assessing the social life cycle impacts of circular economy. *J. Clean. Prod.* **2023**, *386*, 135725. [CrossRef]
137. Silvestre, B.S.; Țîrcă, D.M. Innovations for sustainable development: Moving toward a sustainable future. *J. Clean. Prod.* **2019**, *208*, 325–332. [CrossRef]
138. Kristjanson, P.; Harvey, B.; Van Epp, M.; Thornton, P.K. Social learning and sustainable development. *Nat. Clim. Chang.* **2014**, *4*, 5–7. [CrossRef]
139. Murphy, K. The social pillar of sustainable development: A literature review and framework for policy analysis. *Sustain. Sci. Pract. Policy* **2012**, *8*, 15–29. [CrossRef]
140. Gutowski, T.G. A Critique of Life Cycle Assessment; Where Are the People? *Procedia CIRP* **2018**, *69*, 11–15. [CrossRef]
141. Leipold, S.; Weldner, K.; Hohl, M. Do we need a 'circular society'? Competing narratives of the circular economy in the French food sector. *Ecol. Econ.* **2021**, *187*, 107086. [CrossRef]
142. Wulf, C.; Werker, J.; Ball, C.; Zapp, P.; Kuckshinrichs, W. Review of Sustainability Assessment Approaches Based on Life Cycles. *Sustainability* **2019**, *11*, 5717. [CrossRef]
143. Neugebauer, S.; Forin, S.; Finkbeiner, M. From Life Cycle Costing to Economic Life Cycle Assessment—Introducing an Economic Impact Pathway. *Sustainability* **2016**, *8*, 428. [CrossRef]

144. Valdivia, S.; Backes, J.G.; Traverso, M.; Sonnemann, G.; Cucurachi, S.; Guinée, J.B.; Goedkoop, M. Principles for the application of life cycle sustainability assessment. *Int. J. Life Cycle Assess.* **2021**, *26*, 1900–1905. [CrossRef]
145. Stone, J.; Garcia-Garcia, G.; Rahimifard, S. Development of a pragmatic framework to help food and drink manufacturers select the most sustainable food waste valorisation strategy. *J. Environ. Manag.* **2019**, *247*, 425–438. [CrossRef] [PubMed]
146. Kuhlman, T.; Farrington, J. What is sustainability? *Sustainability* **2010**, *2*, 3436–3448. [CrossRef]
147. Di Maria, F.; Sisani, F. A sustainability assessment for use on land or wastewater treatment of the digestate from bio-waste. *Waste Manag.* **2019**, *87*, 741–750. [CrossRef] [PubMed]
148. Kamal, H.; Habib, H.M.; Ali, A.; Show, P.L.; Koyande, A.K.; Kheadr, E.; Ibrahim, W.H. Food waste valorization potential: Fiber, sugar, and color profiles of 18 date seed varieties (*Phoenix dactylifera*, L.). *J. Saudi Soc. Agric. Sci.* **2022**, *22*, 133–138. [CrossRef]
149. Narisetty, V.; Zhang, L.; Zhang, J.; Lin, C.S.K.; Tong, Y.W.; Show, P.L.; Bathia, S.K.; Misra, A.; Kumar, V. Fermentative production of 2, 3-Butanediol using bread waste–A green approach for sustainable management of food waste. *Bioresour. Technol.* **2022**, *358*, 127381. [CrossRef] [PubMed]
150. D'Adamo, I.; Gastaldi, M. Perspectives and Challenges on Sustainability: Drivers, Opportunities and Policy Implications in Universities. *Sustainability* **2023**, *15*, 3564. [CrossRef]
151. Leipold, S.; Petit-Boix, A.; Luo, A.; Helander, H.; Simoens, M.; Ashton, W.S.; Babbitt, C.W.; Bala, A.; Bening, C.R.; Birkved, M.; et al. Lessons, narratives, and research directions for a sustainable circular economy. *J. Ind. Ecol.* **2023**, *27*, 6–18. [CrossRef]
152. Lopes, R.; Santos, R.; Videira, N.; Antunes, P. Co-creating a Vision and Roadmap for Circular Economy in the Food and Beverages Packaging Sector. *Circ. Econ. Sustain.* **2021**, *1*, 873–893. [CrossRef]

 *sustainability*

Review

Providing a Roadmap for Future Research Agenda: A Bibliometric Literature Review of Sustainability Performance Reporting (SPR)

Oluyomi A. Osobajo [1], Adekunle Oke [1,*], Ama Lawani [1], Temitope S. Omotayo [2], Nkeiruka Ndubuka-McCallum [1] and Lovelin Obi [3]

[1] Aberdeen Business School, Robert Gordon University, Aberdeen AB10 7QE, UK; o.osobajo@rgu.ac.uk (O.A.O.); a.lawani1@rgu.ac.uk (A.L.); n.ndubuka1@rgu.ac.uk (N.N.-M.)
[2] School of Built Environment, Engineering and Computing, Leeds Beckett University, Leeds LS2 8AG, UK; t.s.omotayo@leedsbeckett.ac.uk
[3] School of Architecture and the Built Environment, University of Wolverhampton, Wolverhampton WV1 1LY, UK; l.obi@wlv.ac.uk
* Correspondence: a.oke1@rgu.ac.uk; Tel.: +44-(0)-1224263974

Citation: Osobajo, O.A.; Oke, A.; Lawani, A.; Omotayo, T.S.; Ndubuka-McCallum, N.; Obi, L. Providing a Roadmap for Future Research Agenda: A Bibliometric Literature Review of Sustainability Performance Reporting (SPR). *Sustainability* **2022**, *14*, 8523. https://doi.org/10.3390/su14148523

Academic Editors: Idiano D'Adamo and Massimo Gastaldi

Received: 1 June 2022
Accepted: 11 July 2022
Published: 12 July 2022

Abstract: The concept of sustainability reporting is now an essential tool through which organisations demonstrate accountability to their stakeholders. The increasing market pressure coupled with the awareness of the consequences of organisations' activities suggests the need for organisations to report their sustainability credentials. Sustainability performance reports should provide adequate information on organisations' social, economic, and environmental performance. However, the current process through which organisations communicate their sustainability performance to stakeholders is questionable and remains a significant concern. This study assessed the current state and direction of research on sustainability performance reporting by conducting a bibliometric literature review of peer-reviewed studies on sustainability performance reporting published between 1987 and 2022. The findings highlight the misconceptions between sustainability and CSR when reporting organisations' sustainability performance. Furthermore, businesses and scholars prioritise reporting instead of communication with stakeholders. The observed lack of engagement with stakeholders indicates that the reported performance may not reflect the impact of business activities on the three dimensions of sustainability. Rather than adopting a one-way information dissemination approach, this study concludes that the desired performance can only be achieved through two-way communication with stakeholders.

Keywords: sustainability; communication; stakeholders; reporting; bibliometric review; sustainability reporting; sustainability performance

1. Introduction

Sustainability is inherently complex, involving many stakeholders with different interests and expectations [1]. Organisations across different sectors, including energy and bioeconomy, demonstrate their sustainability performance (SP) by evaluating the social, economic, and environmental effects of their business activities [1–3]. Bioeconomy, like many other issues, is increasingly important and relevant in achieving sustainable goals [1], necessitating the need for SP. On the one hand, SP shows organisations' commitment to the idea of sustainability to their business stakeholders and shareholders [4,5]. On the other hand, companies now realise the potential of sustainability in creating business value, increasing their market share while adding value to their customers, including shareholders and stakeholders [1]. Although sustainability performance reports allow organisations to address their stakeholders' concerns, it also allows business executives and shareholders to understand the impact of organisations' business activities [6]. As a result, organisations need to publish their sustainability performance report (SPR) to demonstrate how their policies and practices align with stakeholders' expectations while ensuring the effective and efficient utilisation of resources [7,8].

Relatively new, SPR emerged due to increased expectations for public disclosure about the role of businesses in society, including how they contribute to the social, economic, and environmental wellbeing of their business environment [9]. While different terminologies such as citizenship reporting, corporate social responsibility reporting, corporate sustainability reporting, and corporate accountability reporting have been used by practitioners [3–5], the overarching goal of these concepts is to allow organisations to be accountable for the consequences of their activities. These concepts highlight the contribution and importance of business activities to the triple bottom line, economic, social and environment, which are considered the vital performance areas of an organisation [10]. While these concepts are different, particularly regarding what they are designed to achieve [1,2], SPR is a primary tool or platform for businesses to communicate sustainability performance and achieve sustainable certifications/compliance [11,12]. Many businesses include sustainability performance in their annual reports to report and demonstrate their commitment to shareholders' wants.

Identifying and effectively communicating sustainability objectives to business stakeholders is crucial in achieving successful sustainable practices. When ecological problems and sustainability-related awareness are not communicated, it becomes non-existent and socially irrelevant [13]. These views emphasise the essence of communicating sustainability performance with concerned stakeholders, which may legitimise business activities within the business operating environment. Even though there are misconceptions in the literature that SPR represents a means for organisations to communicate sustainability performance to their stakeholders, the current approach negates the concept/theory of communication. Companies provide information on their websites and corporate annual reports regarding their sustainability performance; however, little or no attention is given to the communication of sustainability performance in research and practice [13,14].

Despite the benefits of SPR in revealing the challenges and achievements relating to sustainable activities to business stakeholders, it fails to engage stakeholders in dialogue about the threats and opportunities associated with decision-making and strategies relating to sustainability. The complexity of SPR is heightened by the lack of appropriate and effective methodology and governance for organisations to communicate sustainability performance to stakeholders across the three sustainability dimensions. The diverse methods, including the lack of clarity in the international standards for sustainability reporting, indicate the difficulty for organisations to make operational improvements informed by holistic assessments of their business activities' consequences [15]. Therefore, this study examines sustainability performance reporting as a concept through a bibliometric account of peer-reviewed literature to provide a reference point for further research and organisations to address the challenges of communicating their sustainability performance to stakeholders. It also highlights areas for future research on SPR. As a result, relevant peer-reviewed articles on SPR were retrieved from different databases and systematically reviewed in this study. Consistent with Tranfield et al. [16], this review seeks to map, consolidate, and evaluate published studies on SPR to determine the focus of research, including the extent to which sustainability communication rather than reporting is addressed in research and practice. By addressing these fundamental thematic areas and highlighting the evolution of sustainability performance reporting research over time, this bibliometric analysis provides a roadmap for future research agenda and practice of SPR.

The paper begins with an overview of SPR literature to provide a context for the study. This is followed by a theoretical perspective on sustainability reporting and the material and method section. The findings of this review are presented under different themes, and finally, the authors offer suggestions for future research.

2. Overview of Sustainability Performance Reporting (SPR)

The global reporting initiative (GRI) is known for identifying, developing, and disseminating globally applicable guidelines for sustainability reporting. However, the concept of SPR remains a contemporary global concern, resulting in different ways in which reporting is performed by various organisations [17]. Stakeholders are becoming more outspoken on how organisations align their activities and operations with sustainable development principles [5,6], reinforcing Kolk's [18] assertion that different stakeholders are now much more interested in SPR. Hence, organisations are under pressure to disclose their sustainability performance due to their stakeholders' concerns [7,19,20].

SPR is a means to appraise the economic, social and environmental impacts of the business's products, operations, and gross contribution to sustainable development. Acknowledging stakeholders' importance, GRI [21] defines SPR as the method of assessing, disclosing, and being accountable to external and internal stakeholders regarding how businesses contribute to sustainable development goals (SDGs). Furthermore, Fonseca et al. [22] referred to SPR as a framework comprising indices, indicators, principles, conceptual models, criteria, policies, and goals. Likewise, Kocamiş and Yildirim [23] defined SPR as a report that provides information concerning an organisation's social, economic, and environmental performance. While SPR is perceived as a method or framework, it provides an informative analysis of the organisation's approach, progress, and issues in achieving the goals of its sustainable development and strategy [24]. These views mirror Yılmaz and Nuri İne's [25] claim that SPR represents a means via which organisations provide traceability of their sustainability operations or activities in terms of indicators. The existing conceptualisation of SPR suggests it as an instrument for organisations to present their overall social, economic, and environmental impacts to their stakeholders. Arguably, SPR should foster the exchange of sustainability-related information between organisations and diverse stakeholders.

Therefore, organisations' focus should be beyond making profits for their shareholders as they must consider the impacts of their operations on their stakeholders [26]. Organisations should have structured and formal performance indicators to assess their performance as sustainable development agents [27]. Performance indicators have been considered the most effective way of evaluating sustainability performance to present information for management and decision-making purposes [20]. Furthermore, Singh et al. [28] added that performance indicators are used to condense and summarise data to produce a report. Even though scholars have argued that the selection of performance indicators is influenced by the business activities of organisations, sustainability reports should focus on social, economic, and environmental dimensions [29]. For example, biomethane plants could provide economic and social benefits [1]; however, their sustainability impacts and how they affect stakeholders should be examined and communicated with stakeholders.

2.1. Environmental Sustainability

All organisations have an impact on environmental resources. As a result, environmental sustainability has been the focus of many studies compared to other dimensions of sustainability. The consensus from the available studies suggests that organisations must develop plans to monitor and measure such impacts and design strategies to ensure that the environmental resources are used sustainably both now and for future generations [4].

2.2. Economic Sustainability

A sustainable economy focuses only on increasing the stock of man-made capital. However, this study perceives "economic sustainability" as how business activities increase man-made capital without having negative impacts on the environmental, social, and human capital. In other words, economic sustainability refers to the consistent long-term growth of business activities without jeopardising the environmental, social, and cultural value of the community where businesses operate [30]. This view suggests that economic performance indicators should address the organisation's economic impacts on different stakeholders, demonstrating the contribution of businesses to the economic prosperity of their local community.

2.3. Social Sustainability

Social sustainability is a complex concept with practitioners, including businesses, often conflate the process with corporate social responsibility partly due to the lack of a coherent and precise definition of social sustainability [31]. Despite the lack of consensus in the literature, social sustainability addresses intra- and inter-generational equity and emphasises the relationships between human activities and stakeholders, including communities. Using Elkington's triple bottom line model, this study defines social sustainability as economic activities with minimal or no negative short/long-term effects on people and society. From a business perspective, the dimension establishes decisions and priorities that ensure the achievement of stakeholders' needs and expectations, suggesting that the social performance element focuses on organisations' contribution to stakeholders' wellbeing.

Sustainability performance reports (SPRs) offer organisations the opportunity to incorporate sustainable thinking into their planning, implementation, control, and decision-making activities. Organisations must provide SPRs because it plays a fundamental role in implementing sustainable development [27]. As organisations start acknowledging the importance of SPR, the need for sustainable business practices becomes increasingly apparent [29]. According to Alon and Vidovic [32] and Comyns et al. [33], SPR enhances organisations' reputations and strengthens their legitimacy, particularly through public perceptions. Arguably, organisations that actively report their sustainability practices gain a positive reputation from the stakeholders and promote transparency. However, organisations must provide a feedback mechanism to allow suggestions and contributions from stakeholders on how organisations could improve. This SPR approach reduces information asymmetry [34], decreasing organisations' risk exposure [35]. The feedback loop allows organisations to be transparent in their sustainability reporting, legitimising their business activities and enhancing their reputations with stakeholders. The observed positive relationship between SPR and transparency [4,36] positions SPR as a legitimate way of elevating an organisation's reputation [37]. Therefore, publishing SPR habitually allows businesses to maintain and increase stakeholders' trust [38] and loyalty [39], providing the opportunity for businesses to attract talented human resources and maximise corporate and stakeholders' wealth [11,40,41]. SPR could help promote a harmonious relationship between a company and its stakeholders while fulfilling stakeholders' expectations, reinforcing the need for active stakeholder involvement in SPR [41].

Scholars argue that sustainability performance reports are useful for policymaking and public communication because they provide information on organisations' performance in social, economic, and environmental development areas [28]. However, how organisations communicate sustainability performance to their stakeholders remains a significant concern. Borga et al. [42] emphasised the need for a comprehensive framework to communicate and manage initiatives related to organisations' environmental and social aspects. SPRs are expected to bring about a balanced and complete picture of an organisation's sustainability performance; however, they are prone to a different interpretation from stakeholders [43], possibly because the communication of sustainability results/efforts is mostly unregulated [44]. This view further suggests the disparity in how information concerning organisations' sustainability performance is gathered, written, and disseminated.

3. Theoretical Perspectives on Sustainability Reporting

The need for organisations to be transparent and accountable in their activities and operations has received the attention of scholars and practitioners in recent years [15]. This awareness has resulted in increased disclosure by organisations of their performances due to external influences [4,45]. This interplay makes stakeholder and communication theory relevant in explaining why organisations should report their sustainability performance.

Based on stakeholder theory, organisations have obligations towards different stakeholder groups other than their shareholders [46]. The theory offers a unique approach to understanding business responsibility by suggesting that it is imperative to meet several stakeholders' requirements while satisfying shareholders' needs. Investors, employees, suppliers, customers, shareholders, non-governmental organisations, trade associations, the media, and other interest groups are different stakeholder groups identified within the literature. In addition, Mitchell et al. [47] stated that the relevance of stakeholders is determined by possessing one or more attributes of legitimacy, power, and urgency. Lee [45] added that salient stakeholders' pressure has a significant influence on an organisation's social behaviour. Stakeholders can influence an organisation's actions and decisions based on these attributes. This influence, therefore, compels organisations to yield to stakeholders' expectations on sustainability performance reporting [48]. Stakeholders' potential to exercise influence on an organisation's behaviour has been an inherent part of the classic stakeholder definition that stakeholders are any individual or group affected or can affect by organisations' activities when fulfilling their goals [49]. It echoes Guzman and Becker-Olsen's [50] assertion that organisations made significant changes to their activity and operation mode due to consumer actions. Arguably, integrating different organisations' salient stakeholders' needs into the decision-making process to create sustainability performance reports calls for effective communication with stakeholders.

Ziemann [13] referred to communication as a technologically and human-based activity of the reciprocal interpretation of signs and the reciprocal use of signs for successful coordinating action, understanding, and shaping reality. This view suggests that communication involving at least two actors is a social process and contributes positively to obtaining buy-in, mobilisation, and agreeing on a consensus between parties [51]. Communication, therefore, plays a significant role within and outside the organisation's environment. The stakeholder theory supports the importance of communicating organisations' sustainable development. Communication transpires when sustainability-related matters and performance are conceived, defined, discussed, planned, and initiated between an organisation and its stakeholders [52].

4. Materials and Methods

Systematic and bibliometric literature review, which has drawn the attention of various scholars from different fields of study, is a way of collecting and synthesising previous research [16]. This review approach characterises research studies to address particular issues and identify trends in research efforts [53]. Govindan et al. [54] added that conducting a systematic literature review involves four sequential stages; this process (Table 1) was adopted for the current bibliometric literature review.

Table 1. Systematic Review Process.

Four Sequential Stages	Adopted Article Selection Stages
1. Perform a literature review of the available studies on the topic.	1. Identifying journal articles that relate to sustainable performance reporting.
2. Based on pre-determined criteria, develop a classification framework.	2. Journal articles were coded into seven themes.
3. Tabulate and segregate the literature based on the framework.	
4. Using the classification framework, present and organise the review.	3. Present the findings of the literature review using the coding framework.
5. Review analysis and presentation of suggestions for future work.	4. Discussions and proposed framework to address the current gap in knowledge.

A bibliometric literature review is useful to quantify and highlight the pattern and direction of research efforts on emerging issues while identifying the challenges and need for future research. Many scholars, such as Fahimnia et al. [55], have successfully applied the approach to summarise research findings on similar themes based on predefined criteria. This approach is consistent with Snyder's [56] argument that an in-depth review, such as a bibliometric literature review and systematic literature review, effectively provides evidence of the effect that can inform practice and policy by synthesising the collection of studies addressing a similar topic or theme. As a result, this study was conducted to inform designing an effective communication framework by establishing the current knowledge in SPR through published studies between 1987 and 2022 on sustainability performance reporting. This period is considered necessary because sustainability became prominent among researchers and practitioners due to the emergence of the Brundtland Report in 1987 [57].

Data Sources

To achieve the goal of this study, a search was conducted electronically, through the Web of Science and Scopus, for relevant articles on sustainability reporting published from 1987 to 2022 (see Figure 1). The databases were selected as they are considered comprehensive and cover many fields of study and disciplines [58].

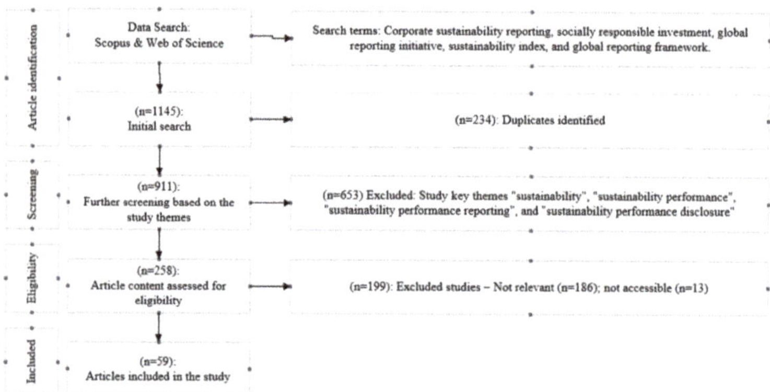

Figure 1. Article Search Scheme.

Figure 1 presents a step-by-step process to select the journal articles considered for this bibliometric review. The initial search resulted in 1145 articles using the following keywords: corporate sustainability reporting, socially responsible investment, global reporting initiative, sustainability index, and global reporting framework. At this point, a two-stage process for selecting and identifying relevant and appropriate studies was used. First, the authors checked the journal articles generated through the search terms and/or phrases for any duplicate records and relevance. This step was imperative as it is impossible to include all the journal articles obtained from the search. We identified and removed 234 duplicates from the 1145 retrieved articles, resulting in 911 relevant articles. In the second phase, we applied search themes such as "sustainability", "sustainability performance", "sustainability performance reporting", and "sustainability performance disclosure" to screen the identified articles. After that, the authors checked the relevance of the remaining articles by reading through the abstract and contents to establish that all the articles addressed sustainability performance. At this stage, an analysis was performed to verify that all selected journal articles' discussion was about sustainability performance reporting or disclosure. Hence, 653 journal articles that did not integrate reporting or disclosure as a theme were excluded from the study. Out of the remaining 258 articles, 199 articles were not considered in this bibliometric analysis because they are non-English journal articles and reviews, while 13 articles were not accessible and were subsequently excluded from this review. In total, 59 peer-reviewed journal articles were considered relevant for this bibliometric review.

Appendix A shows the journals that published the reviewed articles and the number of published articles per year. It should be noted that the focus of this bibliometric literature review is sustainability performance reporting rather than sustainability which could influence the number of articles retrieved from each journal. The subsequent step, which focused on coding the relevant articles in this study, entails data extraction and synthesis to discuss the selected peer-reviewed studies. Information such as the year of publication, authors' name, study focus, study context, the industry of analysis, data collection methods, study type, analytical tool, and continent were subsequently recorded. These data put together formed the basis of the study analysis. The findings of the bibliometric literature review were presented using the coding framework.

After retrieving relevant articles for this bibliometric literature review, classification and coding were performed using letters and numbers (see Table 2). The following coding procedures were applied in this study:

- Study focus, identified as item 1, is coded A to B. This coding focuses on whether the study focuses on sustainability performance reporting or has common themes with sustainability performance reporting.
- The study context, classified as item 2, is coded on a scale of A to C.
- The industry is classified as item 3 and is coded on a scale of A to E.
- The method of data collection, identified as item 4, is coded on a scale of A to G.
- Likewise, the study type identified as item 5 is coded on a scale of A to B.
- The analytical tool, identified as item 6, is coded on a scale of A to D.
- The study's continent is classified as item 7 and coded on a scale of A–F.

Table 2. Journal articles classification and coding framework.

Classification	Description	Codes
Study Focus	Sustainability performance reporting as the central theme	1A
	Sustainability performance reporting as a supporting theme	1B
Study Context	Developing countries	2A
	Developed countries	2B
	Mixed	2C
Industry of Analysis	Extraction (Mining, and Oil and gas)	3A
	Education/Public Sector	3B
	Manufacturing	3C
	Financial Service/Banking	3D
	Others	3E
Method of Data Collection	Observation	4A
	Surveys	4B
	Case Study	4C
	Interviews	4D
	Case study and Interviews	4E
	Literature review	4F
	Case study and Focus Groups	4G
Study Type	Empirical	5A
	Theoretical	5B
Analytical Tool	Qualitative	6A
	Quantitative	6B
	Mixed	6C
	Not applicable	6D
Continent	Europe	7A
	America	7B
	Africa	7C
	Asia	7D
	Australia	7E
	Mixed	7F

5. Results and Discussion

Only fifty-nine peer-reviewed studies were considered relevant for the review based on the selection criteria, and the selected articles were included in the classification and coding process (Table 3).

Table 3. Included articles classification and coding.

Author(s).	Study Focus	Study Context	Industry of Analysis	Data Collection Method	Study Type	Analytical Tool	Continent
Brown et al. [4]	1B	2B	3E	4D	5A	6D	7F
Günther [11]	1B	2C	3C	4B	5A	6A	7F
Ramos et al. [19]	1A	2B	3E	4B	5A	6A	7A
Fonseca et al. [22]	1A	2C	3A	4D	5A	6A	7B
Kocamiş and Yildirim [23]	1A	2A	3E	4F	5B	6D	7D
Alon and Vidovic [32]	1B	2C	3E	4C	5A	6B	7F
Borga et al. [42]	1A	2B	3C	4E	5A	6A	7A
Hahn and Lülfs [43]	1A	2B	3E	4F	5A	6A	7F
Fonseca [59]	1A	2B	3A	4F	5B	6D	7B
Fonseca et al. [60]	1A	2B	3B	4G	5A	6A	7B
Chang et al. [61]	1B	2B	3E	4C	5A	6B	7D
Scagnelli et al. [62]	1A	2B	3E	4C	5A	6D	7A
Fernandez-Feijoo et al. [63]	1A	2C	3E	4F	5A	6B	7F
Lodhia and Hess [64]	1A	2B	3A	4F	5B	6A	7E
Maubane et al. [65]	1A	2A	3E	4C	5A	6A	7C
Hinson, Gyabea and Ibrahim [66]	1A	2A	3B	4F	5B	6A	7C
Husgafvel et al. [67]	1B	2B	3C	4B	5A	6B	7A
Ng and Rezaee [68]	1B	2B	3E	4C	5A	6A	7F
Diaz-Sarachaga et al. [69]	1B	2B	3E	4F	5B	6D	7F
Herremans, Nazari and Mahmoudian [70]	1B	2B	3A	4A	5A	6A	7B
Long et al. [71]	1B	2B	3E	4D	5A	6A	7A
Manetti and Bellucci [72]	1B	2B	3E	4F	5B	6A	7A
Maas et al. [73]	1B	2B	3E	4F	5B	6D	7A
Seele [74]	1B	2B	3E	4F	5B	6A	7A
Thaslim and Antony [75]	1A	2A	3E	4F	5B	6B	7D
Amoako, Lord and Dixon [76]	1A	2C	3A	4C	5A	6A	7F
Anusornnitisarn et al. [77]	1B	2A	3C	4B	5A	6B	7D
Arthur et al. [78]	1A	2A	3A	4C	5A	6A	7C
Aziz, and Bidin [79]	1A	2A	3E	4F	5B	6A	7D
Diouf and Boiral [80]	1A	2B	3E	4D	5A	6A	7B
Domingues et al. [81]	1A	2B	3B	4B	5A	6A	7F
Mickovski and Thomson [82]	1A	2B	3C	4E	5A	6A	7A
Hannibal and Kauppi [83]	1B	2B	3C	4D	5A	6A	7A
Kaur and Lodhia [84]	1A	2B	3E	4E	5A	6A	7E
Laskar and Gopal Maji [85]	1A	2C	3E	4C	5A	6D	7D
Niemann and Hoppe [86]	1A	2B	3B	4C	5A	6A	7A
Watson et al. [87]	1B	2B	3E	4F	5B	6D	7A
Calabrese et al. [88]	1A	2B	3E	4F	5B	6A	7A
Carp et al. [89]	1A	2A	3E	4C	5A	6B	7A
Dissanayake et al. [90]	1A	2A	3E	4C	5A	6B	7D
Semuel et al. [91]	1A	2A	3E	4F	5B	6B	7D
Kouloukoui et al. [92]	1A	2A	3E	4C	5A	6C	7B
Silva et al. [93]	1B	2B	3E	4F	5B	6D	7A
Poon and Law [94]	1B	2B	3E	4F	5B	6D	7D
Sari et al. [95]	1A	2A	3B	4C	5A	6B	7D
Saeed and Kersten [96]	1B	2B	3E	4F	5A	6B	7A
Khan et al. [97]	1A	2A	3D	4F	5A	6A	7D
Ionaşcu et al. [98]	1A	2B	3E	4F	5A	6B	7F
Ceesay [99]	1A	2A	3E	4F	5B	6D	7C
Journeault et al. [100]	1A	2B	3A	4C	5A	6A	7B
Park and Krause [101]	1A	2B	3B	4B	5A	6B	7B
Salehi and Arianpoor [102]	1A	2A	3E	4B	5A	6B	7D
Kumar et al. [103]	1A	2A	3A	4C	5A	6A	7D
Bananuka et al. [104]	1A	2A	3D	4B	5A	6B	7C
Ardiana [105]	1A	2B	3E	4C	5A	6C	7F
Raji and Hassan [106]	1A	2B	3B	4D	5A	6A	7A
Fennell and de Grosbois [107]	1A	2C	3E	4C	5A	6C	7F
Afolabi et al. [108]	1A	2B	3E	4F	5B	6D	7A
Tumwebaze et al. [109]	1A	2A	3D	4B	5A	6B	7C

5.1. Overview of Studies

Although sustainability as a concept came into the limelight in the 1980s [57], SPR only received attention about a decade ago. Figure 2 revealed that SPR gained the attention of scholars from 2009, with the concept gaining more popularity from 2013 onwards, during which an average of four peer-reviewed articles were published. This finding could be explained by Mussari and Monfardini's [17] assertion that SPR remains a contemporary global concern as business stakeholders become more interested and outspoken on how organisations align their activities and operations with sustainable development principles. According to Ardiana [105], stakeholders are increasing pressure on organisations to disclose their sustainability performance.

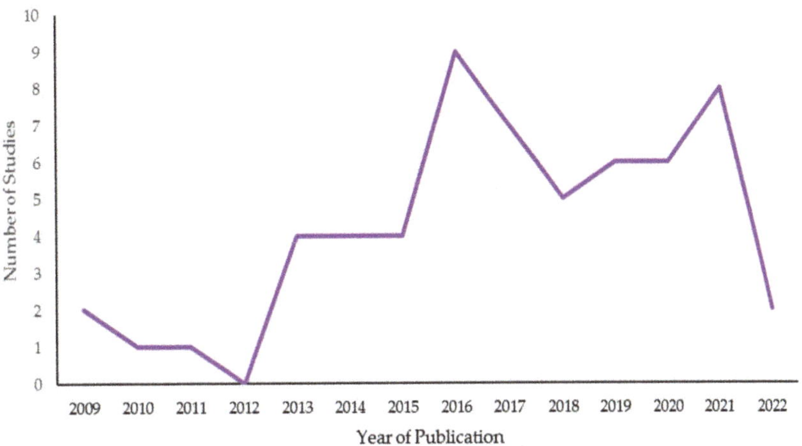

Figure 2. Published Journal Articles by Year.

5.2. Studies Focusing on SPR

As presented in Figure 3, a considerable number of reviewed articles considered SPR a central theme within their study. In contrast, only 18 peer-reviewed articles examined SPR as a corroborative theme to other themes such as CSR, stakeholder engagement, and transparency within their study [72,75].

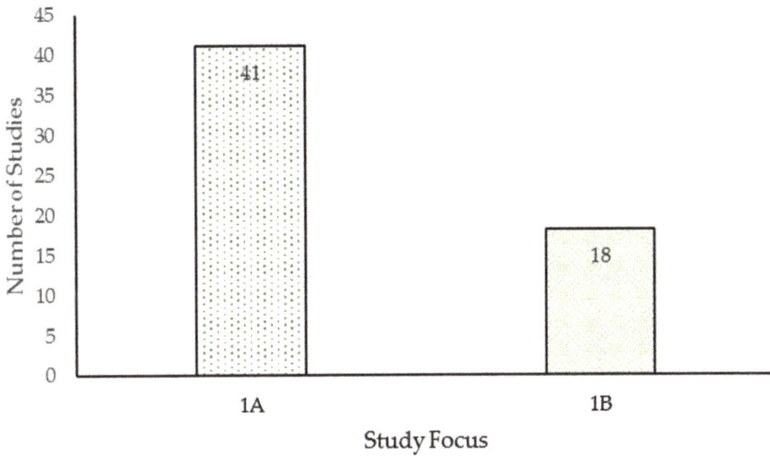

Figure 3. Articles distribution by study focus.

This situation is probably because sustainability performance has increasingly become a central concept among diverse disciplines. This observation is consistent with Lozano's [15] assertion that there is a continuous emphasis by scholars and practitioners alike on organisational accountability and transparency in SPR. The dominance of SPR as the central theme in the review could be due to organisations' response to their stakeholder expectations [7,8,105]. These views emphasise the value and acceptance of SPR among diverse disciplines in reporting the impacts of business activities on sustainability and the risks they face.

Furthermore, while the articles reviewed have considered it a key objective to emphasise the importance of SPR, none attempt to focus on the need for communication. This supports the argument that there is a lack of research efforts on communicating sustainability performance [43,44]. Hence, Herremans et al. [70] concluded that direct communication with stakeholders should be an essential characteristic of sustainability reporting.

5.3. Study Context

Most (i.e., 33) of the studies reviewed were carried out in developed economies (Figure 4). Only 19 studies considered developing economies, while 7 focused on comparing developing and developed economies. From Figure 4, it is obvious that SPR is receiving more attention from scholars within developed economies than in emerging or developing economies. One possible explanation is the importance of regulatory compliance of SPR in fostering accountability and transparency in the most developed economy [59]. Arguably, emerging or developing economies are presented with unique sustainability challenges different from those experienced by developed economies.

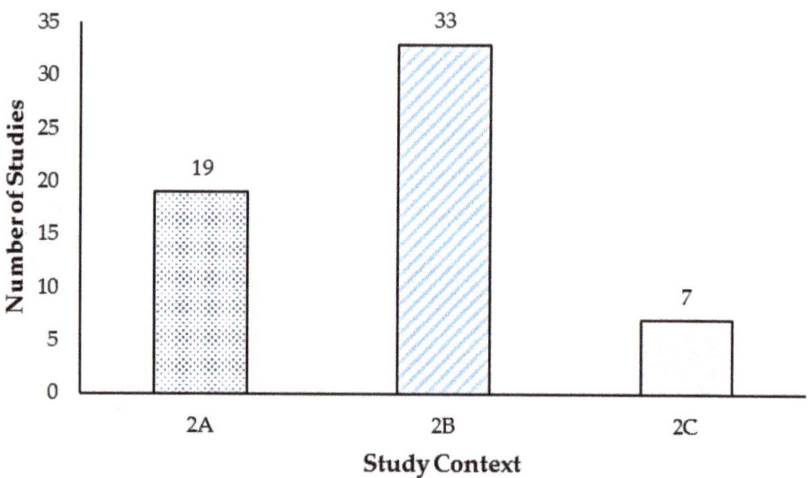

Figure 4. Journal classification based on the study context.

Therefore, it is imperative for scholars and practitioners to further explore how sustainability performance is assessed, monitored, and reported, including how SPR is evolving in developing or emerging economies.

5.4. Industry of Analysis

This review (Figure 5) shows that previous studies examined different industries, although over 55% (i.e., 35 studies) failed to identify a specific industry. Arguably this is because the need to identify, measure and report sustainability performance is, to the same extent, important for all industries irrespective of their business activities and impacts. However, 14 studies were carried out within the extraction and manufacturing industry. It could be because of the impact of these industries on the environment [110]. A total of seven studies and three studies were carried out within the education and financial/banking industries, respectively, while 35 studies were conducted in other industries such as IT, real estate, NGO and tourism. These findings further suggest the need for scholars to examine how organisations in other business sectors measure and report sustainability performance, including the extent of stakeholders' involvement in the process. Moreover, this clarification is necessary considering that disclosures positively impact a company's growth and financial performance [92].

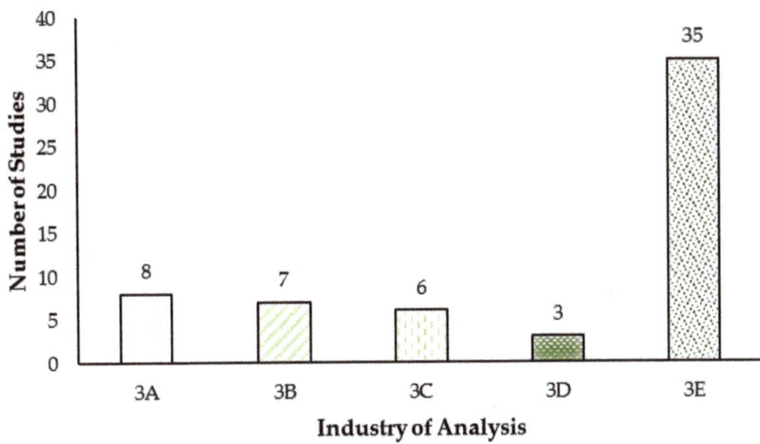

Figure 5. Journal classification based on the industry of analysis.

5.5. Method of Data Collection

All the reviewed journal articles provided information on how research samples were chosen, as shown in Figure 6. However, most of the reviewed studies (i.e., 37) use primary data collection such as observation, surveys, case studies, interviews, and focus groups. This trend could be attributed to the understanding that primary data collection allows researchers to better understand, elaborate, and explain a subject matter in detail.

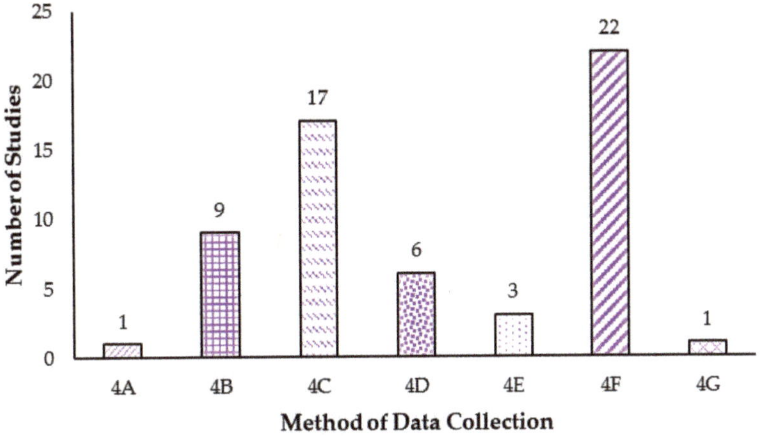

Figure 6. Journal classification based on the method of data collection.

Furthermore, 22 studies employed secondary data collection, such as company reports and information from Bloomberg. While secondary data could be useful, there is a possibility of misrepresentation and information distortion when relying on secondary data. As a result, scholars need to obtain original and first-hand data on organisations' sustainability performance to avoid information asymmetry. However, none of the studies considered for review employed a mix of quantitative and qualitative data collection methods. There is a need for further research studies employing a mixed-methods approach to improve the analysis and findings of any SPR evaluation. According to Wisdom and Creswell [111], combining quantitative and qualitative data in a study can enrich the rigour of the research process, including data analysis and findings.

5.6. Study Type

Out of the 59 studies considered for this bibliometric review, 42 are classified as empirical studies, consistent with Emerald Group Publishing's [112] definition of research studies that focus on observation and measurement of phenomena based on the researcher's direct experience. This finding (Figure 7) suggests that studies are applied primary data, and results are based on the researcher's first-hand and real-life experience. In contrast, 17 studies focused on explaining and formulating a theory to better understand the deeper philosophical issue of the concept. These findings suggest the need for scholars to show further interest in both theoretical and empirical studies. This is because empirical research cannot be separated from theoretical studies, as consideration for theory forms the foundation of most research studies. In addition, theory avails empirical studies as the lead way to replicate and test the results of a study in different contexts [112].

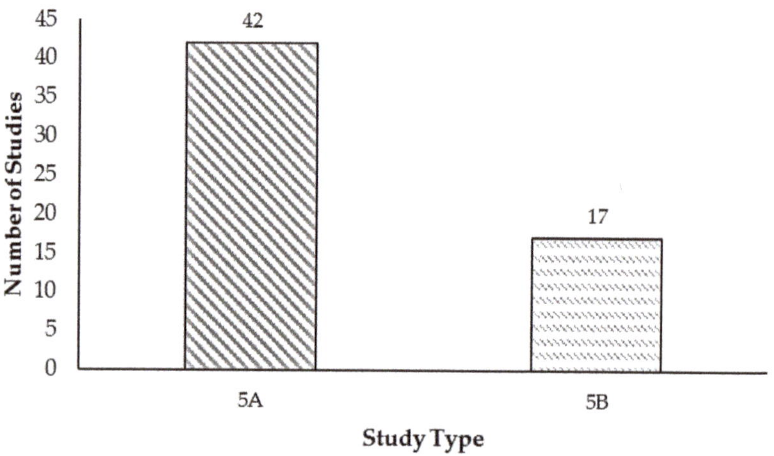

Figure 7. Journal classification based on study type.

5.7. Study Analytical Method

It is methodologically important to understand how scholars analysed the collected data in their studies. Our review shows that 47 studies provided information on how data were analysed, while 12 studies were silent concerning the data analysis tool. As shown in Figure 8, different methods, including qualitative (28 studies), quantitative (16 studies) and a combination of the two methods (3 studies), were used across the identified peer-reviewed articles. This finding revealed that content analysis and thematic analysis are mostly adopted to examine sustainability reporting more in-depth using participants' views. On the contrary, quantitative analysis, such as correlation analysis, regression analysis, and econometric analysis, is another method used by authors, suggesting that only a couple of the reviewed articles are broadly using models/theories to understand and analyse the what of SPR. However, only one of the journal articles reviewed used a combination of the two methods. While it is difficult to argue in support of one analytical tool against the other, future studies are encouraged to employ mixed-method methods, where the weaknesses/strengths of qualitative and quantitative methods supplement each other.

5.8. Continent of Study

All journal articles (N = 59) reviewed for this study were conducted across five continents (Figure 9). The breakdown shows Europe (18 studies), Asia (13 studies), America (8 studies), Africa (6 studies), and Australia (2 studies). However, 12 studies involved researchers from two or more continents. The analysis shows that SPR has gained more attention from researchers in developed economies than in underdeveloped or developing continents such as Africa. This outcome suggests there is room for research activities in developing continents such as Africa, but future research design should consider promoting research activities between two or more continents.

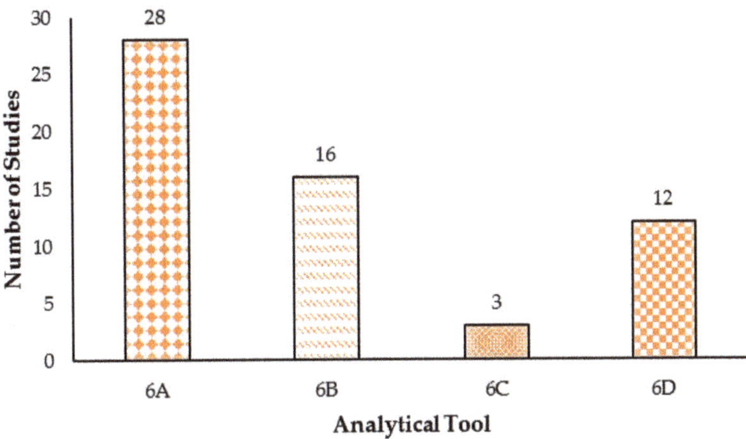

Figure 8. Articles classification based on the study's analytical method.

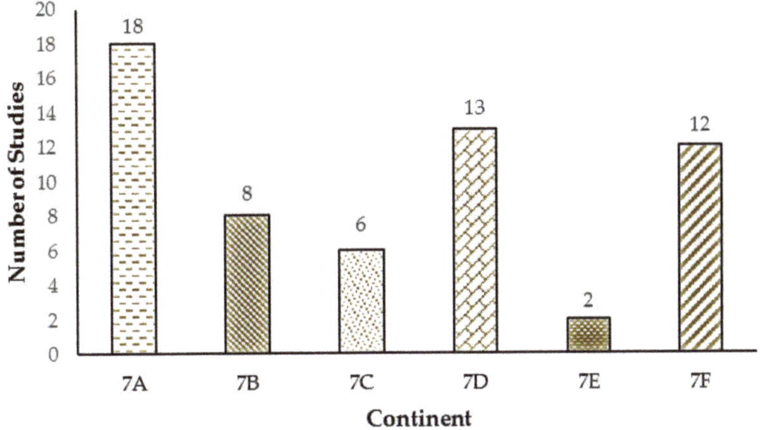

Figure 9. Articles classification based on the study continent.

5.9. SPR Indicators

Sustainability performance indicators are useful in assessing and optimising business activities by determining inadequacies that could be removed or prevented [20]. Scholars have strived to identify and understand several related SPR indicators in the last decade. Most studies reviewed stated that organisations should provide information on their social, economic, and environmental performances. However, scholars arguing from the perspective of public relations [23], economics [65], and accounting [80] disciplines added that organisations should also provide information on their governance performance. Likewise, Kouloukoui et al. [92] asserted the need to disclose corporate climate risk within the sustainability performance report. This is because corporate climate risk disclosures have a positive and significant relationship with firm financial performance, size, and country origin. Their respective industry or sector influences the type of sustainability performance information reported by organisations.

While Chang et al. [61] noted that sustainability performance varies across industries as it is drifting towards continuous improvement in corporate sustainability performance, Romero et al. [12] noted that comparing and evaluating organizations' sustainability-related performance is likely impossible due to the lack of sustainability reporting standards. These views emphasise the need to investigate the type of information in sustainability reporting to clearly understand the sustainability reporting phenomenon. Alghamdi [113] added a need to justify the significance of such reporting by developing and regularly reviewing the reporting of their sustainability performance. This aligns with Ionașcu et al. [98] argument that organizations presenting both quantitative and qualitative key performance indicators is imperative to reveal the degree of achievement of the SDGs. These views emphasise the need to provide transparent and complete sustainability performance reporting.

The majority of the journal articles argued that sustainability performance reports are the kind of corporate reports that foster the transfer of social, economic, and environmental impact of organisation activities to their stakeholders [23]. Consistent with Niemann and Hoppe's [86] assertion, there is a need to develop an effective approach or a single document to engage all stakeholders through SPR due to the lack of a "magic tool" to achieve communication and management functions. These views suggest that SPR cannot fulfil the role of communication between organisations and their respective stakeholders. Hence, Borga et al. [42] emphasised the essence of a comprehensive policy to communicate and manage sustainability performance initiatives achieved by enterprises. Likewise, Herremans et al. [70] argued that direct communication with stakeholders is essential for sustainability reporting. Silva et al. [93] concluded that without clear and definitive consideration of stakeholder expectations in measuring and assessing sustainability performance, results often result in stakeholder dissatisfaction.

6. Conclusions and Direction for Future Research

This study presents the current state and direction of research on sustainability performance reporting considering stakeholders' increasing concerns for organisations to be sustainable. It is, therefore, imperative to evidence the lack of research efforts on communicating sustainability performance [43,44].

According to the results, there are misconceptions between sustainability and CSR when reporting organisations' sustainability performance; however, businesses and scholars prioritise reporting instead of communication with stakeholders. The observed lack of engagement with stakeholders indicates that the reported performance may not reflect the actual impacts of business activities on sustainability. Effective communication is necessary considering that achieving the balance point between economic prosperity, environmental improvement, and social equity [1] without stakeholders' engagement is complex and unrealistic. This review, therefore, argues for a need to consider sustainability performance communication when reporting or investigating organisations' sustainability performance. Organisations need to include a feedback mechanism when reporting their sustainability performance to establish their stakeholders' satisfaction with their performance and understand areas of improvement. Although there is a consensus that organisations should report their economic, social and environmental performance to stakeholders, many scholars argue for the inclusion of corporate governance performance.

Future studies should examine the sustainability performance of organisations in developing economies to reduce the negative consequences of business activities. Moreover, cross-continent comparison is another important research area that future studies could address. Furthermore, this work has not addressed the impact of SP on specific stakeholders, such as consumers, value chain actors, general society, local community and workers; future studies should evaluate this aspect [114], considering the increasingly complex sustainability challenges facing stakeholders. While this review observed that scholars generally adopted mono-method when investigating SPR, future studies should employ a combination of quantitative and qualitative methods to offer a robust explanation of organisations' sustainability performance.

Despite the contributions of this review, some limitations should be addressed in future bibliometric and systematic review studies. Firstly, future studies should consider different parameters, including non-peer-reviewed articles, when selecting journal articles for review. Hence, published books, non-referenced or peer-reviewed journal articles, and conference papers should be included in future studies. Secondly, using language as one of the selection criteria suggests that this review might have excluded many studies not published in English.

This study argues that stakeholders' roles should be considered when reporting sustainability performance. It proposes that stakeholders' needs should be incorporated into the decision-making process when creating a sustainability performance report. The current study contributes to both sustainability performance and communication literature. Identifying research studies addressing sustainability performance reporting is a means to justify the complex concerns involved in communicating organisations' sustainability performance. This approach, therefore, creates an avenue for both empirical and theoretical research studies to understand how organisations should communicate their sustainability performance to their stakeholders. The need for two-way communication, including its impacts on sustainability performance, requires further investigation.

Author Contributions: Conceptualization, O.A.O. and A.O.; methodology, O.A.O., A.L. and A.O.; software, O.A.O., N.N.-M., and T.S.O.; validation, A.O., A.L. and N.N.-M.; formal analysis, O.A.O., A.O., and T.S.O.; investigation, O.A.O. and A.O; resources, O.A.O. and L.O; data curation, O.A.O. and A.L.; writing—original draft preparation, O.A.O., A.O., A.L. and N.N.-M.; writing—review and editing, O.A.O., A.O., A.L., and N.N.-M.; visualization, T.S.O., N.N.-M. and L.O.; supervision, A.O. and L.O.; project administration, N.N.-M. and L.O. All authors have read and agreed to the published version of the manuscript.

Funding: This research received no external funding.

Data Availability Statement: Not applicable.

Conflicts of Interest: The authors declare no conflict of interest.

Appendix A

Table A1. Journals of the reviewed articles.

Journal Name	Number of Articles per Year													Total
	2009	2010	2011	2013	2014	2015	2016	2017	2018	2019	2020	2021	2022	
Business Strategy and the Environment	1													1
Environmental politics	1													1
Corporate Social Responsibility and Environmental Management		1								1				2
International Journal of Sustainability in Higher Education			1											1
Journal of Cleaner Production				2	2		3			1				8
Accounting and Control for Sustainability				1										1
Journal of Business Ethics					2		1							3
European Journal of Economics and Business Studies							1							1
Accounting, Auditing & Accountability Journal							1	1	1		1			4
Public Relations Review					1									1
Corporate Reputation Review						1								1
Communication						1								1

Table A1. *Cont.*

Journal Name	Number of Articles per Year													Total
	2009	2010	2011	2013	2014	2015	2016	2017	2018	2019	2020	2021	2022	
International Journal of Sustainable Engineering						1								1
Journal of Corporate Finance						1								1
Ecological Indicators							1							1
Journal of Applied Leadership and Management							1							1
World Scientific News							1							1
Meditari Accountancy Research								1						1
International Journal of Innovation and Learning								1						1
The International Journal of Business in Society								1						1
Journal of Human Capital Development								1						1
Journal of Environmental Management								1				1		2
Ecological engineering								1						1
International Journal of Production Economics									1					1
Asian Review of Accounting									1					1
Public Management Review									1					1
Journal of Product Innovation Management									1					1
Technological and Economic Development of Economy										1				1
Sustainability										1	1	1	1	4
Pacific Accounting Review										1				1
Indonesian Journal of Sustainability Accounting and Management										1				1
Human Resource Management Review											1			1
International Journal of Innovation, Creativity and Change											1			1
International Journal of Sustainable Development & World Ecology											1			1
Jindal Journal of Business Research											1			1
Critical Perspectives on Accounting												1		1
The TQM Journal												1		1
Business and Society Review												1		1
Journal of Intellectual Capital												1		1
Meditari Accountancy Research												1		1
Tourism Recreation Research												1		1
Journal of Global Responsibility													1	1
Total														59

References

1. D'Adamo, I.; Sassanelli, C. Biomethane community: A research agenda towards sustainability. *Sustainability* **2022**, *14*, 4735. [CrossRef]
2. Roca, L.C.; Searcy, C. An analysis of indicators disclosed in corporate sustainability reports. *J. Clean. Prod.* **2012**, *20*, 103–118. [CrossRef]
3. Gualandris, J.; Golini, R.; Kalchschmidt, M. Do supply management and global sourcing matter for firm sustainability performance? *Supply Chain. Manag. Int. J.* **2014**, *19*, 258–274. [CrossRef]
4. Brown, H.S.; De Jong, M.; Lessidrenska, T. The rise of the Global Reporting Initiative: A case of institutional entrepreneurship. *Environ. Politics* **2009**, *18*, 182–200. [CrossRef]
5. Schaltegger, S.; Burritt, R.L. Sustainability accounting for companies: Catchphrase or decision support for business leaders? *J. World Bus.* **2010**, *45*, 375–384. [CrossRef]

6.	Clarkson, P.M.; Overell, M.B.; Chapple, L. Environmental reporting and its relation to corporate environmental performance. *Abacus* **2011**, *47*, 27–60. [CrossRef]
7.	Valdivia, S.; Bajaj, S.; Sonnemann, G.; Quiros, A.; Ugaya, C.M.L. Mainstreaming life cycle sustainability management in rapidly growing and emerging economies through capacity-building. In *Life Cycle Management*; Springer: Dordrecht, The Netherlands, 2015; pp. 263–277.
8.	Siano, A.; Conte, F.; Amabile, S.; Vollero, A.; Piciocchi, P. Communicating sustainability: An operational model for evaluating corporate websites. *Sustainability* **2016**, *8*, 950. [CrossRef]
9.	Schaltegger, S.; Wagner, M. Integrative management of sustainability performance, measurement and reporting. *Int. J. Account. Audit. Perform. Eval.* **2006**, *3*, 1–19. [CrossRef]
10.	Choudhuri, A.; Chakraborty, J. An insight into sustainability reporting. *ICFAI J. Manag. Res.* **2009**, *8*, 46–53.
11.	Günther, K. Key Factors for Successful Implementation of a Sustainability Strategy. *J. Appl. Leadersh. Manag.* **2016**, *4*, 1–20.
12.	Romero, S.; Fernandez-Feijoo, B.; Ruiz, S. Perceptions of Quality of Assurance Statements for Sustainability Reports. *Soc. Responsib. J.* **2014**, *10*, 480–499. [CrossRef]
13.	Ziemann, A. Communication Theory and Sustainability Discourse. In *Sustainability Communication–Interdisciplinary Perspectives and Theoretical Foundations*; Godemann, J., Michelsen, G., Eds.; Springer: Berlin/Heidelberg, Germany, 2011.
14.	Schaltegger, S.; Burritt, R. Measuring and managing sustainability performance of supply chains. *Supply Chain. Manag. Int. J.* **2014**, *19*, 232–241. [CrossRef]
15.	Lozano, R. The State of Sustainability Reporting in Universities. *Int. J. Sustain. High. Educ.* **2011**, *12*, 67–78. [CrossRef]
16.	Tranfield, D.; Denyer, D.; Smart, P. Towards a methodology for developing evidence-informed management knowledge by means of systematic review. *Br. J. Manag.* **2003**, *14*, 207–222. [CrossRef]
17.	Mussari, R.; Monfardini, P. Practices of Social Reporting in Public Sector and Non-Profit Organizations. *Public Manag. Rev.* **2010**, *12*, 487–492. [CrossRef]
18.	Kolk, A. More Than Words? An Analysis of Sustainability Reports. *New Acad. Rev.* **2004**, *3*, 59–75.
19.	Ramos, T.B.; Cecílio, T.; Douglas, C.H.; Caeiro, S. Corporate sustainability reporting and the relations with evaluation and management frameworks: The Portuguese case. *J. Clean. Prod.* **2013**, *52*, 317–328. [CrossRef]
20.	Arbačiauskas, V.; Staniškis, J. Sustainability performance indicators for industrial enterprise management. *Environ. Res. Eng. Manag.* **2009**, *48*, 42–50.
21.	Global Reporting Initiative (GRI). *Sustainability Reporting Guidelines*; Global Reporting Initiative: Amsterdam, The Netherlands, 2006.
22.	Fonseca, A.; McAllister, M.L.; Fitzpatrick, P. Sustainability reporting among mining corporations: A constructive critique of the GRI approach. *J. Clean. Prod.* **2014**, *84*, 70–83. [CrossRef]
23.	Kocamiş, T.U.; Yildirim, G. *Sustainability* reporting in Turkey: Analysis of companies in the BIST sustainability index. *Eur. J. Econ. Bus. Stud.* **2016**, *2*, 41–51. [CrossRef]
24.	Crouch, E. Chartered Secretary: The governance evolution. *Gov. Dir.* **2017**, *69*, 138.
25.	Yılmaz, G.; Nuri İne, M. Assessment of sustainability performances of banks by TOPSIS method and balanced scorecard approach. *Int. J. Bus. Appl. Soc. Sci.* **2018**, *4*, 62–75.
26.	Rezaee, Z. Supply chain management and business sustainability synergy: A theoretical and integrated perspective. *Sustainability* **2018**, *10*, 275. [CrossRef]
27.	Baumgartner, R. Corporate Sustainability Performance: Methods and Illustrative Example. *Int. J. Sustain. Dev. Plan.* **2008**, *3*, 117–131. [CrossRef]
28.	Singh, R.K.; Murty, H.R.; Gupta, S.K.; Dikshit, A.K. An Overview of Sustainability Assessment Methodologies. *Ecol. Indic.* **2012**, *15*, 281–299. [CrossRef]
29.	Caraiani, C.; Lungu, C.I.; Dascălu, C.; Cimpoeru, M.V.; Dinu, M. Social and environmental performance indicators: Dimensions of integrated reporting and benefits for responsible management and sustainability. *Afr. J. Bus. Manag.* **2012**, *6*, 4990–4997.
30.	Van Niekerk, A.J. Inclusive Economic Sustainability: SDGs and Global Inequality. *Sustainability* **2020**, *12*, 5427. [CrossRef]
31.	Åhman, H. Social sustainability-Society at the intersection of development and maintenance. *Local Environ.* **2013**, *18*, 1153–1166. [CrossRef]
32.	Alon, A.; Vidovic, M. Sustainability performance and assurance: Influence on reputation. *Corp. Reput. Rev.* **2015**, *18*, 337–352. [CrossRef]
33.	Comyns, B.; Figge, F.; Hahn, T.; Barkemeyer, R. Sustainability reporting: The role of "Search", "Experience" and "Credence" information. *Account. Forum* **2013**, *37*, 231–243. [CrossRef]
34.	Cormier, D.; Ledoux, M.; Magnan, M. The Informational Contribution of Social and Environmental Disclosures for Investors. *Manag. Decis.* **2011**, *49*, 1276–1304. [CrossRef]
35.	El Ghoul, S.; Guedhami, O.; Kwok, C.C.Y.; Mishra, D.R. Does Corporate Social Responsibility Affect the Cost of Capital? *J. Bank. Financ.* **2011**, *35*, 2388–2406. [CrossRef]
36.	De Villiers, C.; Marques, A.W. Corporate Social Responsibility, Country-Level Predispositions, and the Consequences of Choosing a Level of Disclosure. *Account. Bus. Res.* **2016**, *46*, 167–195. [CrossRef]
37.	Cho, C.H.; Patten, D.M. The role of environmental disclosures as tools of legitimacy: A research note. *Account. Organ. Soc.* **2007**, *32*, 639–647. [CrossRef]

38. Reddy, K.; Gordon, L.W. The Effect of Sustainability Reporting on Financial Performance: An Empirical Study using Listed Companies. *J. Asia Entrep. Sustain.* **2010**, *6*, 19–42.

39. Hohnen, P. *The Future of Sustainability Reporting*; EEDP Programme Paper; Chatham House: London, UK, 2012.

40. Jizi, M. The Influence of Board Composition on Sustainable Development Disclosure. *Bus. Strategy Environ.* **2017**, *26*, 640–655. [CrossRef]

41. Kurniawan, P.S. An Implementation Model of Sustainability Reporting in Village-Owned Enterprise and Small and Medium Enterprise. *Indones. J. Sustain. Account. Manag.* **2018**, *2*, 90–106. [CrossRef]

42. Borga, F.; Citterio, A.; Noci, G.; Pizzurno, E. Sustainability report in small enterprises: Case studies in Italian furniture companies. *Bus. Strategy Environ.* **2009**, *18*, 162–176. [CrossRef]

43. Hahn, R.; Lülfs, R. Legitimizing negative aspects in GRI-oriented sustainability reporting: A qualitative analysis of corporate disclosure strategies. *J. Bus. Ethics* **2014**, *123*, 401–420. [CrossRef]

44. Legrand, W.; Huegel, E.B.; Sloan, P. Learning from best practices: Sustainability reporting in international Hotel Chains. In *Advances in Hospitality and Leisure*; Emerald Group Publishing Limited: Bingley, UK, 2013; pp. 119–134.

45. Lee, M.D.P. Configuration of External Influences: The Combined Effects of Institutions and Stakeholders on Corporate Social Responsibility Strategies. *J. Bus. Ethics* **2011**, *102*, 281–298. [CrossRef]

46. Hillenbrand, K.; Money, K. Corporate Responsibility and Corporate Reputation: Two Separate Concepts or Two Sides of the Same Coin? *Corp. Reput. Rev.* **2007**, *10*, 261–277. [CrossRef]

47. Mitchell, R.K.; Agle, B.R.; Wood, D.J. Toward a Theory of Stakeholder Identification and Salience: Defining the Principle of Who and What Really Count. *Acad. Manag. Rev.* **1997**, *22*, 853–886. [CrossRef]

48. Hahn, R.; Kühnen, M. Determinants of Sustainability Reporting: A Review of Results, Trends, Theory and Opportunities in An Expanding Field of Research. *J. Clean. Prod.* **2013**, *59*, 5–21. [CrossRef]

49. Freeman, R. *Strategic Management: A Stakeholder's Approach*; Pitman: Boston, MA, USA, 1984.

50. Guzman, F.; Becker-Oslen, K. Strategic Corporate Social Responsibility: A Brand–Building Tool. In *Innovative CSR: From Risk Management to Value Creation*; Louche, C., Idowu, S.O., Filho, L.W., Eds.; Greenleaf: Sheffield, UK, 2010; pp. 197–219.

51. Okereke, C.; Wittneben, B.; Bowen, F. Climate change: Challenging business, transforming politics. *Bus. Soc.* **2012**, *51*, 7–30. [CrossRef]

52. Allen, M. *Strategic Communication for Sustainable Organizations. Theory and Practice*; University of Arkansas: Fayetteville, AR, USA, 2016.

53. Mariano, E.B.; Sobreiro, V.A.; do Nascimento Rebelatto, D.A. Human development and data envelopment analysis: A structured literature review. *Omega* **2015**, *54*, 33–49. [CrossRef]

54. Govindan, K.; Rajendran, S.; Sarkis, J.; Murugesan, P. Multi-criteria decision making approaches for green supplier evaluation and selection: A literature review. *J. Clean. Prod.* **2015**, *98*, 66–83. [CrossRef]

55. Fahimnia, B.; Sarkis, J.; Davarzani, H. Green supply chain management: A review and bibliometric analysis. *Int. J. Prod. Econ.* **2015**, *162*, 101–114. [CrossRef]

56. Snyder, H. Literature review as a research methodology: An overview and guidelines. *J. Bus. Res.* **2019**, *104*, 333–339. [CrossRef]

57. Fernández, E.F.; Malwé, C. The emergence of the' planetary boundaries' concept in international environmental law: A proposal for a framework convention. *Rev. Eur. Comp. Int. Environ. Law* **2019**, *28*, 48–56. [CrossRef]

58. Falagas, M.E.; Pitsouni, E.I.; Malietzis, G.A.; Pappas, G. Comparison of PubMed, Scopus, web of science, and Google scholar: Strengths and weaknesses. *FASEB J.* **2008**, *22*, 338–342. [CrossRef]

59. Fonseca, A. How credible are mining corporations' sustainability reports? A critical analysis of external assurance under the requirements of the international council on mining and metals. *Corp. Soc. Responsib. Environ. Manag.* **2010**, *17*, 355–370. [CrossRef]

60. Fonseca, A.; Macdonald, A.; Dandy, E.; Valenti, P. The state of sustainability reporting at Canadian universities. *Int. J. Sustain. High. Educ.* **2011**, *12*, 22–40. [CrossRef]

61. Chang, D.S.; Kuo, L.C.R.; Chen, Y.T. Industrial changes in corporate sustainability performance–an empirical overview using data envelopment analysis. *J. Clean. Prod.* **2013**, *56*, 147–155. [CrossRef]

62. Scagnelli, S.D.; Corazza, L.; Cisi, M. How SMEs disclose their sustainability performance. Which variables influence the choice of reporting guidelines? In *Accounting and Control for Sustainability*; Emerald Group Publishing Limited: Bingley, UK, 2013; pp. 77–114.

63. Fernandez-Feijoo, B.; Romero, S.; Ruiz, S. Effect of stakeholders' pressure on transparency of sustainability reports within the GRI framework. *J. Bus. Ethics* **2014**, *122*, 53–63. [CrossRef]

64. Lodhia, S.; Hess, N. Sustainiability accounting and reporting in the mining industry: Current literature and directions for future research. *J. Clean. Prod.* **2014**, *84*, 43–50. [CrossRef]

65. Maubane, P.; Prinsloo, A.; Van Rooyen, N. Sustainability reporting patterns of companies listed on the Johannesburg securities exchange. *Public Relat. Rev.* **2014**, *40*, 153–160. [CrossRef]

66. Hinson, R.; Gyabea, A.; Ibrahim, M. Sustainability reporting among Ghanaian universities. *Communication* **2015**, *41*, 22–42. [CrossRef]

67. Husgafvel, R.; Pajunen, N.; Virtanen, K.; Paavola, I.L.; Päällysaho, M.; Inkinen, V.; Heiskanen, K.; Dahl, O.; Ekroos, A. Social sustainability performance indicators–experiences from process industry. *Int. J. Sustain. Eng.* **2015**, *8*, 14–25. [CrossRef]

68. Ng, A.C.; Rezaee, Z. Business sustainability performance and cost of equity capital. *J. Corp. Financ.* **2015**, *34*, 128–149. [CrossRef]
69. Diaz-Sarachaga, J.M.; Jato-Espino, D.; Alsulami, B.; Castro-Fresno, D. Evaluation of existing sustainable infrastructure rating systems for their application in developing countries. *Ecol. Indic.* **2016**, *71*, 491–502. [CrossRef]
70. Herremans, I.M.; Nazari, J.A.; Mahmoudian, F. Stakeholder relationships, engagement, and sustainability reporting. *J. Bus. Ethics* **2016**, *138*, 417–435. [CrossRef]
71. Long, T.B.; Blok, V.; Coninx, I. Barriers to the adoption and diffusion of technological innovations for climate-smart agriculture in Europe: Evidence from the Netherlands, France, Switzerland and Italy. *J. Clean. Prod.* **2016**, *112*, 9–21. [CrossRef]
72. Manetti, G.; Bellucci, M. The use of social media for engaging stakeholders in sustainability reporting. *Account. Audit. Account. J.* **2016**, *29*, 985–1011. [CrossRef]
73. Maas, K.; Schaltegger, S.; Crutzen, N. Integrating corporate sustainability assessment, management accounting, control, and reporting. *J. Clean. Prod.* **2016**, *136*, 237–248. [CrossRef]
74. Seele, P. Digitally unified reporting: How XBRL-based real-time transparency helps in combining integrated sustainability reporting and performance control. *J. Clean. Prod.* **2016**, *136*, 65–77. [CrossRef]
75. Thaslim, K.M.; Antony, A.R. Sustainability reporting–Its then, now and the emerging next! *World Sci. News* **2016**, *42*, 24–40.
76. Amoako, K.O.; Lord, B.R.; Dixon, K. Sustainability reporting: Insights from the websites of five plants operated by Newmont Mining Corporation. *Meditari Account. Res.* **2017**, *25*, 186–215. [CrossRef]
77. Anusornnitisarn, P.; Chindavijak, C.; Rassameethes, B.; Meeampol, S.; Kess, P.; Hidayanto, A.N. Development of sustainability's performance framework: Learning from executive viewpoints. *Int. J. Innov. Learn.* **2017**, *22*, 304–321. [CrossRef]
78. Arthur, C.L.; Wu, J.; Yago, M.; Zhang, J. Investigating performance indicators disclosure in sustainability reports of large mining companies in Ghana. *Corp. Gov. Int. J. Bus. Soc.* **2017**, *17*, 643–660. [CrossRef]
79. Aziz, N.S.A.; Bidin, R.H. A Review on The Indicators Disclosed in Sustainability Reporting of Public Listed Companies in Malaysia. *J. Hum. Cap. Dev.* **2017**, *10*, 1–14.
80. Diouf, D.; Boiral, O. The quality of sustainability reports and impression management: A stakeholder perspective. *Account. Audit. Account. J.* **2017**, *30*, 643–667. [CrossRef]
81. Domingues, A.R.; Lozano, R.; Ceulemans, K.; Ramos, T.B. Sustainability reporting in public sector organisations: Exploring the relation between the reporting process and organisational change management for sustainability. *J. Environ. Manag.* **2017**, *192*, 292–301. [CrossRef]
82. Mickovski, S.B.; Thomson, C.S. Developing a framework for the sustainability assessment of eco-engineering measures. *Ecol. Eng.* **2017**, *109*, 145–160. [CrossRef]
83. Hannibal, C.; Kauppi, K. Third party social sustainability assessment: Is it a multi-tier supply chain solution? *Int. J. Prod. Econ.* **2018**, *217*, 78–87. [CrossRef]
84. Kaur, A.; Lodhia, S. Stakeholder engagement in sustainability accounting and reporting: A study of Australian local councils. *Account. Audit. Account. J.* **2018**, *31*, 338–368. [CrossRef]
85. Laskar, N.; Gopal Maji, S. Disclosure of corporate sustainability performance and firm performance in Asia. *Asian Rev. Account.* **2018**, *26*, 414–443. [CrossRef]
86. Niemann, L.; Hoppe, T. Sustainability reporting by local governments: A magic tool? Lessons on use and usefulness from European pioneers. *Public Manag. Rev.* **2018**, *20*, 201–223. [CrossRef]
87. Watson, R.; Wilson, H.N.; Smart, P.; Macdonald, E.K. Harnessing difference: A capability-based framework for stakeholder engagement in environmental innovation. *J. Prod. Innov. Manag.* **2018**, *35*, 254–279. [CrossRef]
88. Calabrese, A.; Costa, R.; Ghiron, N.L.; Menichini, T. Materiality analysis in sustainability reporting: A tool for directing corporate sustainability towards emerging economic, environmental and social opportunities. *Technol. Econ. Dev. Econ.* **2019**, *25*, 1016–1038. [CrossRef]
89. Carp, M.; Păvăloaia, L.; Afrăsinei, M.B.; Georgescu, I.E. Is Sustainability Reporting a Business Strategy for Firm's Growth? Empirical Study on the Romanian Capital Market. *Sustainability* **2019**, *11*, 30658.
90. Dissanayake, D.; Tilt, C.; Qian, W. Factors influencing sustainability reporting by Sri Lankan companies. *Pac. Account. Rev.* **2019**, *31*, 84–109. [CrossRef]
91. Semuel, H.; Hatane, S.E.; Fransisca, C.; Tarigan, J.; Dautrey, J.M. A Comparative Study on Financial Performance of the Participants in Indonesia. *Indones. J. Sustain. Account. Manag.* **2019**, *3*, 95–108.
92. Kouloukoui, D.; Sant'Anna, Â.M.O.; da Silva Gomes, S.M.; de Oliveira Marinho, M.M.; de Jong, P.; Kiperstok, A.; Torres, E.A. Factors influencing the level of environmental disclosures in sustainability reports: Case of climate risk disclosure by Brazilian companies. *Corp. Soc. Responsib. Environ. Manag.* **2019**, *26*, 791–804. [CrossRef]
93. Silva, S.; Nuzum, A.K.; Schaltegger, S. Stakeholder expectations on sustainability performance measurement and assessment. A systematic literature review. *J. Clean. Prod.* **2019**, *217*, 204–215. [CrossRef]
94. Poon, T.S.C.; Law, K.K. Sustainable HRM: An extension of the paradox perspective. *Hum. Resour. Manag. Rev.* **2022**, *32*, 100818. [CrossRef]
95. Sari, M.P.; Hajawiyah, A.; Raharja, S.; Pamungkas, I.D. The report of university sustainability in Indonesia. *Int. J. Innov. Creat. Chang.* **2020**, *11*, 110–124.
96. Saeed, M.A.; Kersten, W. Sustainability performance assessment framework: A cross–industry multiple case study. *Int. J. Sustain. Dev. World Ecol.* **2020**, *27*, 496–514. [CrossRef]

97. Khan, H.Z.; Bose, S.; Mollik, A.T.; Harun, H. Green washing "or" authentic effort? An empirical investigation of the quality of sustainability reporting by banks. *Account. Audit. Account. J.* **2020**, *34*, 338–369. [CrossRef]

98. Ionașcu, E.; Mironiuc, M.; Anghel, I.; Huian, M.C. The Involvement of Real Estate Companies in Sustainable Development—An Analysis from the SDGs Reporting Perspective. *Sustainability* **2020**, *12*, 798. [CrossRef]

99. Ceesay, L.B. Exploring the Influence of NGOs in Corporate Sustainability Adoption: Institutional-Legitimacy Perspective. *Jindal J. Bus. Res.* **2020**, *9*, 135–147. [CrossRef]

100. Journeault, M.; Levant, Y.; Picard, C.F. Sustainability performance reporting: A technocratic shadowing and silencing. *Crit. Perspect. Account.* **2021**, *74*, 102145. [CrossRef]

101. Park, A.Y.; Krause, R.M. Exploring the landscape of sustainability performance management systems in US local governments. *J. Environ. Manag.* **2021**, *279*, 111764. [CrossRef]

102. Salehi, M.; Arianpoor, A. The relationship among financial and non-financial aspects of business sustainability performance: Evidence from Iranian panel data. *TQM J.* **2021**, *33*, 1447–1468. [CrossRef]

103. Kumar, K.; Kumari, R.; Kumar, R. The state of corporate sustainability reporting in India: Evidence from environmentally sensitive industries. *Bus. Soc. Rev.* **2021**, *126*, 513–538. [CrossRef]

104. Bananuka, J.; Tauringana, V.; Tumwebaze, Z. Intellectual capital and sustainability reporting practices in Uganda. *J. Intellect. Cap.* **2022**. *ahead-of-print*. [CrossRef]

105. Ardiana, P.A. Stakeholder engagement in sustainability reporting by Fortune Global 500 companies: A call for embeddedness. *Meditari Account. Res.* **2021**. *ahead-of-print*. [CrossRef]

106. Raji, A.; Hassan, A. Sustainability and stakeholder awareness: A case study of a Scottish university. *Sustainability* **2021**, *13*, 4186. [CrossRef]

107. Fennell, D.A.; de Grosbois, D. Communicating sustainability and ecotourism principles by ecolodges: A global analysis. *Tour. Recreat. Res.* **2021**, 1–19. Available online: https://www.tandfonline.com/doi/abs/10.1080/02508281.2021.1920225 (accessed on 10 July 2022).

108. Afolabi, H.; Ram, R.; Rimmel, G. Harmonization of Sustainability Reporting Regulation: Analysis of a Contested Arena. *Sustainability* **2022**, *14*, 5517. [CrossRef]

109. Tumwebaze, Z.; Bananuka, J.; Orobia, L.A.; Kinatta, M.M. Board role performance and sustainability reporting practices: Managerial perception-based evidence from Uganda. *J. Glob. Responsib.* **2022**, *13*, 317–337. [CrossRef]

110. Amrina, E.; Vilsi, A.L. Key performance indicators for sustainable manufacturing evaluation in cement industry. *Procedia CIRP* **2015**, *26*, 19–23. [CrossRef]

111. Wisdom, J.; Creswell, J.W. Mixed methods: Integrating quantitative and qualitative data collection and analysis while studying patient-centered medical home models. *Agency Healthc. Res. Qual.* **2013**, *13*, 1–5.

112. Emerald Group Publishing. How to Conduct Empirical Research. 2019. Available online: https://www.emeraldgrouppublishing.com/how-to/research-methods/conduct-empirical-research#theoretical-framework (accessed on 5 January 2020).

113. Alghamdi, N. Sustainability reporting in higher education institutions: What, why, and how. In *International Business, Trade and Institutional Sustainability*; Springer: Cham, Switzerland, 2020; pp. 975–989.

114. D'Adamo, I.; Gastaldi, M.; Morone, P.; Rosa, P.; Sassanelli, C.; Settembre-Blundo, D.; Shen, Y. Bioeconomy of sustainability: Drivers, opportunities and policy implications. *Sustainability* **2021**, *14*, 200. [CrossRef]

 *sustainability*

Article

Global Inequalities in the Bioeconomy: Thinking Continuity and Change in View of the Global Soy Complex

Maria Backhouse *, Malte Lühmann and Anne Tittor

BMBF-Junior Research Group "Bioeconomy and Inequalities", Institute of Sociology,
Friedrich Schiller University Jena, Bachstr. 18k, 07743 Jena, Germany; malte.luehmann@uni-jena.de (M.L.);
anne.tittor@uni-jena.de (A.T.)
* Correspondence: maria.backhouse@uni-jena.de

Abstract: As a proposed pathway to societal transformation, the bioeconomy is aimed at providing a sustainable alternative to the fossil-based economy, replacing fossil raw materials with renewable biogenic alternatives. In this conceptual contribution, we argue that it is impossible to transform societies into sustainable bioeconomies considering the narrow boundaries of the bioeconomy as a policy. Drawing on approaches including agro-food studies, cheap food, and agrarian extractivism, we show that the bioeconomy is entangled in a broader context of social relations which call its claim to sustainability into question. Our analysis of the global soy complex, which represents the core of the current agro-food system, demonstrates how the bioeconomy perpetuates global inequalities with regard to trade relations, demand, and supply patterns, as well as power relations between the involved actors from the global to the local level. Against this background, we propose a fundamental rethink of the underlying understanding of transformation in bioeconomy policies. Instead of thinking the bioeconomy only along the lines of ecological modernisation, its proponents should consider studies on social-ecological transformation, which would entail radical structural change of the prevailing food regime to cope with the social-ecological crisis.

Keywords: agrarian extractivism; bioeconomy; cheap food; food regime; Latin America; social-ecological transformation; soy

Citation: Backhouse, M.; Lühmann, M.; Tittor, A. Global Inequalities in the Bioeconomy: Thinking Continuity and Change in View of the Global Soy Complex. *Sustainability* **2022**, *14*, 5481. https://doi.org/10.3390/su14095481

Academic Editors: Idiano D'Adamo and Massimo Gastaldi

Received: 7 March 2022
Accepted: 28 April 2022
Published: 3 May 2022

Publisher's Note: MDPI stays neutral with regard to jurisdictional claims in published maps and institutional affiliations.

1. Introduction: Bioeconomy, Flexible Biomass, and Societal Transformation

As a proposed pathway to societal transformation, the bioeconomy is aimed at developing and implementing new technologies to produce biomass and transform it into a range of products. Various forms of biomass are to replace fossil resources for energy production and industrial raw materials. Products mostly from agriculture (but also from forestry and aquaculture) are to be turned into flexible biomass for universal usage in "biorefineries" [1] (p. 95) or used to produce biofuels or recently also biomethane [2]. In addition, the bioeconomy agenda claims to provide a sustainable alternative to the fossil-based economy as fossil raw materials are to be replaced with renewable biogenic alternatives [3]. Over 60 countries have adopted bioeconomy strategies or are pursuing bioeconomy-related policies in addition to a growing number of macro-regional bioeconomy strategies such as the one drawn up by the EU [4] (p. 13). This strategy constitutes a multi-faceted global transformation project, as it sets different priorities in research funding and incentives for bioenergy.

However, various national strategies have come under criticism, since they foresee a rise in the agroindustrial production of soy, palm oil, and corn—crops that have come to dominate the contemporary global agro-food system and that are exacerbating climate change, land use change, and land grabbing [5,6]. Therefore, the bioeconomy agenda can aggravate problems already discussed in the food versus fuel debate [6,7]. Consequently, nowadays, many policymakers and expert fora recognise the conflicting socio-ecological objectives and interests that are present in the bioeconomy from the local to the global level.

This acknowledgment is clearly visible in Germany's latest bioeconomy strategy, which emphasises the need to avoid negative socio-ecological impact in order to strengthen the participation of civil society and to achieve the sustainable development goals (SDGs) [8]. Thus, transformation in this context not only involves replacing fossil fuels (the basis of modern societies) with biomass. Rather, this material transformation is now to take place in harmony with the goals of sustainable development worldwide. However, this acknowledgment at the policy level has not led to a reversal in the trend towards expansion of biomass production or even helped mitigate its negative socio-ecological impacts [9]. This, of course, conflicts with implementation of the SDGs.

Against this background, aspects of sustainability in the bioeconomy have become an increasingly important issue for research on bioeconomy policies and in related academic debates [10–12]. In this conceptual contribution, we advance the thesis that it is impossible to transform societies to sustainable bioeconomies that achieve the SDGs, considering the narrow boundaries of the bioeconomy as a policy. Insights from critical agrarian and food studies demonstrate that change and continuity in agricultural biomass production should not be thought of as isolated processes. The understanding and possible transformation of patterns in agricultural biomass production need to include analyses of historical roots, these patterns' embeddedness in wider social relations, and the relevance of power relations. Exploring this argument in detail, we show that critical analyses of the global soy complex provide an understanding of the prospects of the emerging bioeconomy as a societal transformation. The guiding questions for our study are twofold: What are the limits of bioeconomy as a strategy for societal transformation? And more specifically: What does the development of the existing global soy complex tell us about the prospect of bioeconomy transformation?

To answer these questions, we draw on concepts including agro-food-studies [13,14], agrarian extractivism [15–17], and cheap food [18]. These approaches enable us to gain a deeper understanding of historical and structural patterns from the global to the local level that shape all biomass producing industries and attempts to transform them to increase sustainability. In combination, these concepts provide an innovative theoretical framework for our conceptual contribution on the bioeconomy. The theoretical argument is empirically underpinned with data and qualitative studies on the social relations encompassing production, distribution, and consumption of soy and its derivatives. The soy sector is the empirical focus of this study because it is particularly well suited to be used to flesh out a historically rooted, global perspective on the bioeconomy. This view is based on the assumptions that (a) specific social relations, such as those in the soy sector, cannot be understood exclusively in their local context, but are globally embedded in broader social structures; (b) the global soy complex constitutes the core of the current agro-food system—The system that is to evolve into a future bioeconomy; thus, (c) the dynamics of the global soy complex foreshadow issues that would be associated with a full-scale bioeconomy; and (d) the same global social inequalities that shape the soy complex today question the socio-ecological sustainability of the bioeconomy.

Soy is the most important agricultural biomass commodity to date and has globalised production and trade networks. In 2020/2021, soy was grown on 129 million hectares worldwide, almost half of which was in South America [19] (p. 32). This is leading the region to be described as "soylandia" (in English: soy land) [20] (p. 119). The ecological changes are far-reaching, compelling some researchers to speak of the "soy-isation" of agriculture [21,22]. As a flex crop [1], soy can be used in the food, fodder, energy, and other industrial sectors, depending on which form of further processing is more profitable. In addition, the soy sector has been the site of various forms of technological innovation with the aim of improving efficiency and productivity, for instance, by increasing yields through transgenic technologies as well as through better farming techniques such as no-till farming and crop rotation with corn. However, technological innovation has neither led to a halt in the expansion of cultivated areas, nor to more sustainable production [23]. This fact highlights the limitations of a merely technology-driven transformation strategy and the need to analyse the complexity of historical global contexts.

In this study, we analyse the global soy complex and its exemplary role for a bioeconomy with the aim of grasping the transnational social relations it embodies. South America, the main production region, plays a central role in our reflections. Other world regions, which play an important role in bioeconomy visions and in biomass markets beyond soy, are therefore not within the scope of this study. In our understanding, the term "soy complex" encompasses the (mainly transnational) enterprises selling inputs (seeds, pesticides, fertilisers, and machines) necessary to grow soy, as well as (the mainly different) companies buying soy from farmers to process it into food, fodder, or biodiesel, and to store, transport, and export it. Both the upstream and downstream side of the soy complex is dominated by a handful of economic actors—despite regional differences—that have a tendency towards oligopolistic structures. The analysis of the development and prospects of the soy complex at the heart of the global agro-food system and its links to the bioeconomy serves to clarify the role and impact of bioeconomy transformation in a world of inequalities. The evidence leads to the conclusion that a just and truly sustainable transformation of the global agro-food system requires more than what current bioeconomy policies can deliver.

2. Theoretical Analysis of Socio-Ecological Inequalities in the Soy Complex

We use theoretical concepts that capture the global agro-food system in its entirety including its dynamics of transformation to fully grasp the circumstances under which bioeconomy policies operate on a global scale. Concepts embracing this perspective have been developed in the tradition of world systems analysis [24]. More specifically, and for the analysis of agriculture as a central field of biomass production with soy as a pivotal global crop, the concepts of *food regimes* [25], *cheap food* [18], and *agrarian extractivism* [16] are useful.

Food regime analysis is a perspective on global patterns of agriculture, food processing, and consumption that goes back to a seminal article by Friedmann and McMichael [25]. Food regime analysis proposes understanding the political economy of food on a global scale in relation to the process of capital accumulation: "The difference made by food regime analysis is that it prioritises the ways in which forms of capital accumulation in agriculture constitute global power arrangements, as expressed through patterns of circulation of food" [13] (p. 140). Historically different food regimes can be distinguished; food regimes are defined as temporarily stable sets of implicit and formal rules governing the global agro-food system [26] (p. 30).

Coined by Jason Moore [18], the concept of cheap food explains the structural function of the production of cheap surplus food under capitalism as central to the reproduction of the growing working classes in the urban centres. Furthermore, it draws attention to how cheap food is produced through productivity revolutions and commodity frontiers, which appropriate, sometimes dispossess, and exploit natural resources, spaces, and people all over the world. Thus, examining the issue through the lens of cheap food helps to provide an understanding of the current expansion dynamics of soy, as it conceptualises transnational interrelations between consuming and producing classes and regions at the world scale.

The concept of agrarian extractivism has been introduced to describe an economic strategy used by countries in the Global South to generate wealth via the extraction of resources from the ground, as well as by producing large quantities of flex crops for the global market. The term is mainly used to criticise the negative economic, social, and environmental consequences of such strategies. The expansion of soy in the Conosur region has been the first field beyond fossil fuels and mining where the concept of extractivism has been discussed broadly [27,28]. The debate has led to the development of the term "agrarian extractivism" [29–31]. The concept helps provide an understanding of the specific political and economic power relations that hamper socio-ecological transformations.

On the basis of their common lineage from word-systems analysis, the three concepts provide an understanding of the historically developed structures and inequalities of global capitalism and its roots in colonialism. In this understanding, modern capitalism has historically developed with and through the incorporation of the Americas into the world system. Colonialism and the extraction of raw materials from Latin America and the Caribbean (world systems analysis stresses the importance of the colonisation of the Americas for the world system and the emergence of capitalism. Nevertheless, this does not mean that colonialism in Africa and Asia was less brutal or less important. For Wallerstein's analysis of colonialism in Africa, see [32].) has been crucial to the development of wealth in Europe and the constitution of the capitalist world system. On this basis, the three approaches complement each other effectively as part of our analysis: the food regime serves as an overall framework to understand the rules governing agriculture and food production on a global scale. It also highlights the connection to broader social relations in global capitalism. The cheap food perspective helps us gain a deeper understanding of the political economy of consumption patterns by analysing the underlying relations of production. Finally, agrarian extractivism provides an understanding of the specific social relations entailed in the soy complex in South America.

Despite their diverging focus, all three concepts emphasise similar dimensions that are important for our analysis. First, they underline the historical roots of contemporary societies and the weight of past processes (such as colonialism) in shaping social relations. Second, they stress structural inequalities as defining features of society from the global to the local scale. Third, they acknowledge the role of collective actors and the power relations between them in reproducing or transforming social structures. The following analysis is structured by these three dimensions.

In the first chapter, we outline how the food regime (and the soy complex as a part of it) has evolved historically. In the second chapter, we explain the expansion dynamics of soy using Moore's argument about the structural need to produce cheap food. In the third chapter, we show which actor constellations and power relations support the continued expansion of this sector by looking at the main cultivation regions in South America. To this end, we evaluate the current research on soy in the region. At the end of each chapter, we directly link the findings to the emerging bioeconomy and its impact.

3. Historical Contextualisation: Deep-Rooted and Continuous Inequalities

An understanding of the historical roots and trajectory of the global soy complex and the wider agro-food system is fundamental to our perspective. As mentioned above, colonialism and its role in the development of the modern capitalist world system is a central aspect. Food regime analysis, which we use as a framework, emphasises that this historical relationship is not a linear development but needs to be seen as a succession of qualitatively different periods and respective food regimes. Friedmann and McMichael [25] initially identified two food regimes: A first "imperial" regime under British hegemony ranging from 1870 to 1914 (McMichael and other authors later extended the period of the first food regime from 1870 to 1930; see [13,20].) and a second "developmental" regime under U.S. hegemony from 1945 to 1973. A vivid debate is taking place about whether a third food regime (for the period since 1973) is emerging or has already established itself and how it is to be conceptualised [33] (pp. 18–21). McMichael suggests that a "corporate food regime" began in the 1980s, and this seems to be an accurate assessment from the vantage point of the global soy complex. There are substantial differences between the present and earlier food regimes when it comes to the role of soy.

In the first "imperial" food regime under British hegemony, soy was a supplementary, albeit relatively cheap source of protein and fat for the European working classes, and it was produced mainly by Chinese settler families and sold to Europe as part of the British free-trade paradigm and the gold standard [20] (p. 140). As a whole, the imperial food regime was centred around the British state and capital and included two major global food flows [13] (pp. 144–145). Tropical foods such as sugar, coffee, and fruit were imported from plantation colonies to Europe, while temperate foods, mostly wheat and meat, were imported from settler colony states such as the USA, Argentina, Australia, and South Africa. Extracting relatively cheap foodstuffs along with other raw materials in different colonial frontiers at the expense of local populations (over-exploitation of paid labour and appropriation of unpaid labour) as well as nature (appropriation of untilled land, exhaustion, and degradation of soils) enabled British and European capital to provision a growing industrial labour force [13] (p. 145). The two food flows were part of two contemporary dynamics in the world system: "the culmination of European colonialism in Asia and Africa (colonies of 'occupation') and the 'rise of the nation-state system' in which (former) colonies of 'settlement' were now independent" [33] (p. 3). The institutional rules of the imperial food regime were suspended during WW1 and finally crumbled during the 1929 global economic crisis [20] (p. 121).

By the second "developmental" food regime (1945–1973), soy had already become a central ingredient in the transformation of agriculture and the post-WW2 international division of labour [25] (p. 110). In this U.S.-centric regime, soy was produced by an expanding U.S. agro-industrial complex and partially shipped under GATT tariff exemption to Europe as feed for the growing meat production in the post-war era [25] (p. 107). Simultaneously, soy surpluses such as other cheap foods were used politically by the USA during the Cold War to influence the growing number of newly independent nation states in the Global South [20] (p. 122). The expansion of the soy complex in the USA started in the 1930s after the exhaustion of the family-farming model based on wheat during the "Dust Bowl", and this provided it with a dominant position on the world market until the 1970s [14] (p. 252). As the physical expansion of farming in the USA had ended during the previous food regime, the growth of the soy complex took place primarily through the displacement of other crops, as well as through mechanisation and the application of new (agrochemical) technologies [14] (p. 252). This perceived U.S. model of national development based on modernising the farm sector in conjuncture with industry was publicly promoted but also forcibly implemented as an example for the rest of the world under the label of the "Green Revolution" [13] (pp. 145–146). The fictitious picture of national sovereignty conveyed by the U.S. development model stood in contrast to the construction of increasingly transnational commodity chains in agriculture that penetrated national economies under the control of U.S. agribusiness.

As the global economy entered a prolonged crisis in the mid-1970s, including the regulation of global markets for food commodities, a new phase began in the globalisation of soy. The liberalisation of these markets and the increasingly dominant position of TNCs (ABCD group, see below) in the 1980s, marked a shift towards what McMichael calls the "corporate food regime" [34]. In this context, soy has been transformed into an increasingly flexible crop, and cultivated under the control of transnational agribusiness mostly in South America to be channelled through liberalised global markets to Europe and China to provide the growing labour forces with a meat/protein-rich diet [34] (pp. 288–289).

From the 1970s onwards, soy producers in South America posed increasing competition to the soy farmers in the USA [14] (p. 258). In the context of the corporate food regime, Brazil and the other countries in the Southern Cone region became the main producers of soy for the world market. This was achieved through the expansion of land used for soy farming but more importantly through the application of new technologies such as GMO soy, which is resistant to specific pesticides that are intensively used in soy cultivation [14] (pp. 259–263). After a few years, it became clear that the main beneficiaries of this model were the transnational companies that sell the "technological package" of GMO seeds and pesticides [35] (p. 67).

As soy production in South America is primarily directed at the global market under the control of agribusiness TNCs and because these activities contain few linkages to local production and consumption, they constitute an exemplary form of agrarian extractivism [29]. High volumes of raw or semi-processed materials are shipped out of the country to fulfil global demands for resources. This form of production is based on the extraction of nutrients from soils, which are degraded in the long term by soy cultivation [36], as well as increasing land demands, leading to displacement of subsistence farmers and indigenous groups and significant environmental impact such as deforestation, erosion, and contamination of water sources [37] (p. 51).

The soy complex in South America was one of the first fields in which the corporate food regime developed an entirely flexible crop. Only 6% of soy is currently used to feed people; most of it is used for agro-industrial feedstock including fodder as well as for biodiesel and industrial products [23] (p. 252). Through the lens of food system analysis, soy, among other crops, constitutes a biomass resource, which is seen as interchangeable:

> "The corporate food regime has progressively modelled a form of agriculture valuing its product solely as a commodity. The bio-economy represents the highest stage of commodification in the fact of crop substitutability. Here, exchange value erases use value, and crops become fungible investments as the multiple uses of corn, soy, palm oil and sugar, for example, whether as foods, feeds, fuels, cosmetics, stabilizers and so on. For the crops mentioned, their conversion from food to exchange-value is the ultimate fetishization of agriculture, as an input-output process geared to indiscriminate production of commodities for profit." [38] (p. 132)

As the quote shows, the main aim of the bioeconomy—to replace fossil resources for industrial uses and energy production with biogenic resources (see the Introduction)—leads crops such as soy to be conceptualised as interchangeable inputs in a global economy; this strengthens unequal and extractivist relations at the sites of production. Moreover, the emergence of soy as a flex crop rests upon the extractive relations established in Latin America since colonial times and the historic shift from a developmental to a corporate food regime on a world scale. These historical roots are deepened in a bioeconomy that relies on flex crops and other established agro-industrial practices found in the soy complex and throughout the corporate food regime. The next chapter demonstrates that the difficulty of changing such practices is and has been exacerbated by the current and historical structural inequalities that pervade social relations in and beyond the agro-food system.

4. Cheap Food and Structural Inequalities

Contrary to the soy sector's claims to contribute significantly to feeding a growing world population, numerous studies demonstrate that soy is used to feed livestock to provide meat and animal products to the world's growing middle classes and not to feed the poor [23]. Tony Weis argues that soy and other grain and oilseed production is deeply intertwined with the livestock industry within the current food regime, and that this contributes to the "meatification of diets" [39] (p. 127). Today, meat consumption worldwide is twice as high as it was two generations ago, even though there are twice as many people on the planet. However, it would be short-sighted to equate growing meat consumption with the growing world population as meat consumption is highly unequal and even exacerbates social inequalities on a global scale: "People in high-income countries consume over twice as much meat per year as the world average" [40] (p. 562). In 2018, annual per capita meat consumption in the USA was 145 kg; in Nigeria, it was just under 7 kg [41]. Simultaneously, the production of crops and oilseeds for fodder exacerbates food insecurity in poor countries as "nearly one-third of cropland is devoted to producing livestock feed" [40] (p. 564). Hence, growing livestock and meat consumption exacerbates hunger and malnutrition of the poor.

There are many reasons for the growing demand for meat such as taste, beliefs about the need to consume animal protein, cultural veneration, and ideas of masculinity [40] (p. 562). The sole focus on consumption habits, however, obscures the structural background. Therefore, we draw on the political economy of consume patterns to analyse the dialectic relations between demand and supply in capitalist economies [40] (p. 563). Examining the subject through this lens reveals that chronic grain surpluses caused by subsidised production were pivotal in linking the soy complex to the livestock industry, since the crops were absorbed by "fast-rising populations of concentrated livestock, starting with chickens and followed by pigs" [40] (p. 563). At the same time, markets for standardised grain and meat products emerged, making crops and meat the basis of financial instruments, which, in turn, have exacerbated the standardisation, homogenisation, and industrialisation of the respective agricultural products (ibid.). This shift began as part of a transnational restructuring of agricultural sectors in the context of the second food regime [25] (pp. 105–108). Meat developed into a central product-category in the post-war agro-food system:

> "Like the automobile, meat was a key product in the mass production and consumption of standardized products that provide the central dynamic of post-war capitalism in advanced capitalist economies; and like petroleum [...] soy was a critical input to mass production." [25] (p. 106)

Moore provides an additional explanation for the rise of the food regime and its specific consumption patterns with his emphasis on the structural function of *cheap food* in capitalism; the intrinsic logic of capitalism is to extract more value by increasing labour productivity to produce more commodities with less labour. In this understanding, cheap food means that "more calories are produced in less average labour-time in the commodity system" [42] (p. 10). Furthermore, cheap food is essential to keep the wage-bill low. Moore argues that it is a specificity of capitalism that the exploitation of paid labour can be successfully intensified by appropriating unpaid labour (e.g., care work, subsistence farming, slavery) and natural resources (e.g., soil nutrition) for the production and trade of cheap food. In this context, Moore refers to wages, which remained stable while food prices continued to fall in OECD countries, especially during the 1990s [43].

Moore views the development of modern industrial agriculture as decisive for the emergence and development of capitalism: on the one hand, its enclosures have contributed to the continuous displacement of small-scale agriculture over the last five centuries, and this led to large-scale migration to the cities. On the other hand, the combination of major productivity revolutions through technological innovation and the expansion of the frontiers of agro-industrial agriculture (which, in turn, has provoked new enclosures) has provided cheaper food to the growing urban population. In this view, it is not industrialisation in England that enabled the rise of capitalism, but a change in the food system; only the production of surplus food enabled the creation of labour outside of agriculture. The conquest of the Americas, in combination with the emergence of plantation economies, was central to producing this food surplus. As Moore shows in his historical analysis, the importance of the global expansion of capitalist agriculture has become apparent since the conquest of the Americas and the later colonisation of Asian countries, in particular: "problems with English agricultural productivity in the eighteenth century, for example, were never resolved within England, but rather through successive frontier movements, especially in North America" [42] (p. 274). The history of the last few centuries reveals that food security is essential to stabilising great empires [44] (pp. 92–93). The opening up of new frontiers to produce cheap food, therefore, is crucial for social peace and, at the same time, remains an engine of imperial expansion. Soy is a "petrochemical hybrid complex" that combines "new plants, fertilisers, pesticides, and irrigation schemes" ([45] cited in [42] (p. 251)). Consequently, the growing demand for soy pushes forward frontiers in different regions: on the one hand, the horizontal frontiers of the growing plantations appropriate more and more land in South America. On the other hand, vertical frontiers include the growing consumption of underground fossil resources for fertiliser and transport, and potable water [42] (p. 254).

In the early globalised supply chains of cheap food, labour relations in the emerging industrial nations were combined in a completely new way with unpaid labour by slaves in the colonised peripheries. As Sidney Mintz shows, the production of cheap sugar on the sugar cane plantations in the Caribbean was a crucial cheap source of calories for the proletarians in the industrial centres of Europe [46]. Today, the question is how these entanglements of different labour relations and classes are reproduced or reconfigured within and between countries; precarious working conditions and modern slavery are still central pillars of the global production of cheap food. Using the example of cheap chicken in Mexico, Patel and Moore underline the centrality of cheap meat in establishing and maintaining social peace in emerging economies [43]. Patel and Moore demonstrate that chicken became cheap and affordable as a direct consequence of the North American Free Trade Agreement (NAFTA), technological innovation, and the U.S. soy industry [43]. However, Mexican smallholder agriculture paid a very high indirect price; NAFTA plunged smallholder agriculture into crisis and forced many farmers to migrate to the U.S. agricultural sector where they became part of the precarious labour force and consumers of cheap food.

However, soy as the "supercrop" of the green revolution also illustrates the limits of productivity growth and cheap food. The large productivity gains made during the last few decades can no longer be met because soils have been depleted; "superweeds" are threatening plantations; investments for seeds, fertiliser, and pesticides are growing; and climate change is aggravating the socio-ecological crisis. In Brazil, the largest global soy producer, the cost of soy production has been rising by 5 percent annually since 2009 [42] (p. 268). Globally, food prices are rising again, and this led Moore to argue that cheap food, as a key pillar of capitalism, has entered a deep crisis in the 2000s that cannot simply be resolved through technological innovation [42] (pp. 268–276).

Examining the issue through the lens of cheap food draws attention to the political economy of consumption patterns that cannot simply be met by pleas to consumers or more sustainable soy production but that has to involve profound changes to the whole food regime. Even if the change in eating habits envisaged by the bioeconomy, such as meat substitutes, is an important starting point, it will not be enough. Synthetic meat cannot be developed in ahistorical spaces but is based on substances produced within existing structures of the global food regime. A bioeconomy would thus not only have to overcome the food regime's dependence on fossil raw materials, but also to transform the inner logic of cheap food, because it is based on the exploitation of labour as well as the appropriation of unpaid labour and natural resources. Whether or not these relations are addressed in transformation processes such as the bioeconomy is also a question of power. The following looks at the situation in South America—the main production site of the global soy complex—and demonstrates how the power relations in this complex shape social and environmental relations and how this situation transpires into bioeconomy policies.

5. Actors and Power Relations in the Soy Complex and the Bioeconomy

The global soy complex is dominated by a handful of powerful transnational companies that produce most of its inputs (seeds, pesticides, fertilisers, machines), control the export and import relations, and are so powerful that they are able to influence the research agenda and political regulations within the soy sector in diverse countries [47]. The ABCD companies (ADM, Bunge, Cargill, and Louis-Dreyfus) now operate as cross-sectoral "value chain managers" on a truly global scale [48]. In 1996, Argentina became one of the first countries to permit the cultivation of GMO soy. GMO seeds from Argentina have been smuggled to Brazil, Paraguay, and Bolivia and were approved in these countries between 2003 and 2005 [23] (p. 254). This led to the establishment of highly profitable monocultures of soy based on increasing pesticide use and no-till farming. The Argentinian government strongly supported agricultural biotechnology. Both neoliberal (ca. 1990–2002) and developmentalist (2003–2015) administrations enthusiastically endorsed GM crops in Argentina and encouraged farmers to plant soy [49] (p. 706). Empirical studies have shown that soy exports in Argentina disproportionately benefit the elite [50], although they have also generated state revenues. Local farmers and agribusiness aligned with transnational companies such as Monsanto in the 1990s and 2000s to make large profits by using their land for the boom in soy production [49] (p. 701). In the 1970s and 1980s, state-owned agricultural institutes in Brazil and Argentina played a key role in developing a type of soy that is adapted to the specific weather and soil conditions in South America, but these institutes have been displaced by transnational companies and their seeds, which have dominated the region since the 1990s [23] (pp. 253–254), [51]. In the 1990s, Brazil and Argentina were focused on exports, the deregulation of the banking system, and attempts to attract foreign direct investment in trade infrastructure such as ports, warehouses, and crushing facilities [23] (p. 254).

The most important resource for the soy complex is control over land. Industrial actors in the soy complex (e.g., large-scale farmers, corporations) have appropriated the land (independently from those who formally own it) and control land use [37] (p. 50), [52] (p. 62). In Paraguay and Bolivia, soy-isation went hand in hand with processes of land grabbing [31] (p. 65); in Argentina, farmers started leasing land to the emerging "sowing pools", which are investment networks. This attracted all kinds of capital to the agricultural sector, most of it foreign, and generated large profits for the large-scale farmers and investors [35] (p. 68). This practice was partly exported to neighbouring countries [23] (p. 265). As soy production needs very few workers, smallholder farmers living in regions transformed by soy lose their opportunity to work and are often forced to move elsewhere—a process described as "productive exclusion" by McKay and Colque [37] (p. 50). Therefore, the expansion of soy in the Southern Cone has exacerbated existing social inequalities [31] (p. 152). These inequalities are difficult to overcome for multiple reasons: First, as shown above, they build on historical processes that have deeply shaped social structures in South America. Second, the elite is profiting from agrarian extractivism and has no interest in overcoming it and governments strongly support the strategy. This has led to the re-primarisation of soy-exporting countries [53] (p. 10), which are becoming ever more dependent on these exports. On the other side of the spectrum, the poor and marginalised are negatively affected by these strategies. Especially poor rural populations face health risks from agrochemical spraying, (partly violent) displacement, and expropriation [54] (p. 200). Among the peoples who are losing (access and control over) land, indigenous groups are disproportionally affected. Indigenous environmental activists who fight against extractivism face violence more often and more intensively than other people [55] (p. 9), [56] (p. 15).

Some studies have shown that agrarian extractivism also increases gender inequalities. The agribusiness sector and state institutions promoting the large-scale soy model are dominated and led by men. In the last few decades, women have been excluded from commercial agrarian production, a tendency that has been reinforced by the strong entry of financial investors and male-dominated techno-science into commodity production in general and the soy model in particular [54] (p. 206). Agrarian extractivism, therefore, has a gendered structure—an argument also made by other researchers looking into other flex crops such as palm oil production in Colombia, sugarcane, and oil in Ecuador. Diana Ojeda argues that agrarian extractivism in Colombia relies on, and deepens, gender disparities and gender-based violence. Men are often hired to work on the plantations and earn a salary, whereas the reproductive work of women subsidize the plantation model, exacerbating unequal gender relations [57]. A similar argument is made by Landívar García [58] analysing the gendered structure of agrarian extractivism concerning sugarcane in Ecuador. Looking at the chikungunya epidemic in the refinery city of Esmeraldas in Ecuador, a recent study concludes: "Extractivism exacerbates the already heavy burden of women's care work, thus forming a central mechanism of the 'illness-poverty trap' by which ill health is both a consequence of and a contributor to inequalities" [59] (p. 169).

Soy expansion in Argentina and Brazil and its socio-ecological impact has generated discontent and protests among the negatively affected. In both countries, social movements are campaigning against the intensive use of pesticides associated with the soy model, question GMOs, and fight for food/seed sovereignty [49,60,61]. Nevertheless, as most of these protest movements operate on a local scale, they have been unable to stop the entire dynamic, and—at best—have only been able to limit local pesticide use. The soy economy can still count on support from the population because of its use of powerful narratives such as "feeding the world", and because of the employment it generates in the countryside, as well as its contribution to "industrialisation" [37] (p. 45). Especially during the pink tide, left-wing governments argued that soy exports were generating state revenues and that these enabled social welfare programs to be established and therefore helped reduce poverty and inequality. Despite the fact that soy exports do not reduce inequalities [62] (p. 33), [63] (p. 106), this narrative is still powerful and the social movements questioning the soy model have difficulties in gaining a voice.

Additionally, agribusiness has developed a strategy of presenting their activities as sustainable and as contributing to the fight against the climate crisis. Powerful actors from the agribusiness and biotechnology sectors in Argentina have appropriated the narrative of the bioeconomy and play a key role in the development of public policies on bioeconomy in the country [64,65]. This has contributed to a situation in which actors from these sectors have powerful voices in bioeconomy fora and they are the ones with which European bioeconomy proponents collaborate. In contrast, civil society actors that are more critical of the soy economy and its socio-environmental impact are not invited to these dialogues or networks. As such, contentious issues such as pesticides, environmental damage, the lack of decent jobs, and the re-primarisation of South American economies are not being tackled. This is even more dubious given that the bioeconomy is promoted as a strategy of transformation towards sustainable development—a goal which cannot be reached without the participation of civil society and marginalised groups and without addressing the aforementioned critical issues.

6. Results: Continuity and Change in the Global Soy Complex and Its Implications for the Bioeconomy

During this analysis, we have unfolded the conceptual thesis that the transformation of societies towards a sustainable bioeconomy, as envisioned by the German and other bioeconomy strategies, is unachievable in the narrow confines of existing bioeconomy policies. Focusing on the soy complex as a pivotal area in the agro-food system and an unavoidable background for such transformation, we have explored three dimensions of this argument.

1. We have shown how the soy complex developed during the changing historical context of successive food regimes that took place before the emergence of the current corporate food regime. This latest regime has increasingly transformed soy into a flex crop, and it has been instrumental in transnationalising global food markets and placing them under corporate control. The emergence, configuration, and crisis of food regimes shows the ways in which the development of the global agro-food system is embedded in fundamental processes of societal change on a historical scale.

2. Examining the issue through the lens of cheap food reveals that the consumption of cheap meat (produced with cheap soy) is intertwined with the production relations constituted by globalised supply chains. In terms of structural inequalities, we argued how the profits and the socio-ecological costs of the soy complex tend to be unevenly distributed between the centre and the periphery of the world system but also how transnational relations of exploitation and appropriation are not only intertwined but also constitutive of the agro-food system. The case of soy illustrates how the appropriation of unpaid labour and of natural resources in the production regions of this flex crop are entangled with the exploitation of wage labourers in urban centres around the world through the provision of cheap food or more specifically cheap meat. Against the backdrop of global inequalities, more empirical research is needed into the centrality of cheap meat to social peace, and the reproduction of the precarious workforce in the Global South and North, and how these relations are interrelated.

3. Finally, with respect to the power relations in the soy complex, a more detailed analysis of the situation in the central production areas in Latin America shows how transnational agribusiness mostly control the technological inputs and global trade of the soy complex, whereas national agrarian elites control the production of soy on expanding land areas. These groups are powerful political actors and can be contrasted with local farmers and social movements that are side-lined even under

left-wing governments. The power asymmetry in favour of agribusiness actors in countries such as Argentina transpires into the emerging bioeconomy as bioeconomy strategies are shaped under their influence.

These three dimensions describe the current shape of the soy complex that is at the heart of the global agro-food system and locate it in its societal context. It should be noted that other sectors of the global agro-food system have not been a subject of this article, which implies that no general overview of global inequalities across sectors and world regions was intended. Moreover, the dynamics identified in the soy complex and thereof derived findings cannot necessarily be transferred to other crop's production systems and their local and regional contexts. However, the social structures of the global soy complex and their historical trajectory, which we analysed as a key example for the global agro-food system, contradict most of the bioeconomy's claims to aspire towards sustainable development. The limitations of bioeconomy strategies are evident in their avoidance of the fundamental question of how the global interconnections and inequalities constituting the food regime can be addressed and changed. Instead, they focus on technological innovation ranging from the digitalisation of agriculture and the use of residues for bioenergy to artificial meat substitutes. In line with Moore, we have doubts about whether a bioeconomy could succeed in producing cheap food if it were not part of a food regime based on the exploitation of wage labour as well as appropriation of unpaid labour and natural resources.

7. Conclusions and Outlook: Limits and Perspectives of a Socio-Ecological Transformation towards a Sustainable Bioeconomy

In light of these findings, the question arises as to whether references to sustainability provide bioeconomy strategies with anything more than a "selling point" [66]. For example, the German strategy recognises that "Securing a global supply of food is and has always been a priority, and ethical principles and socially recognised goals such as environmental protection, landscape conservation and animal welfare must be accorded similarly high valuation" [8] (p. 10). However, the German strategy and even more so that of the European Commission stress that the bioeconomy is also about strengthening a country's own technological leadership in new biotechnological fields while securing jobs and green growth. Like most other official bioeconomy strategies, the German and EU strategies remain firmly within the framework of green capitalism [67]. In this perspective, the bioeconomy can be seen as an attempt to "green" the food regime without changing the underlying societal and power relations that have led to socio-ecological problems such as hunger and climate change. As has been shown elsewhere [68], a growing bioeconomy in Germany and the EU should be expected to aggravate these problems because of increased import demand for biomass. Bioeconomy policies which continue to rely on fundamental mechanisms of the current food regime and global capitalism more broadly, such as economic growth or extractivism, and which primarily seek innovative technological solutions to complex societal problems seriously limit the room for a transformation towards sustainability. These shortcomings of the bioeconomy as a strategy for societal transformation are recognised not least by many scientists from this field as a recent German survey by Zeug et al. [69] has shown:

> "[M]ost respondents from the stakeholder group science encourage this vision, disagree with current developments, but as active carriers and advocates of ongoing social change hope for a more social and ecological sustainable bioeconomy and societal transformation. We conclude that according to most of the respondents, for a bioeconomy to be socially assertive and a successful sustainability transformation, it needs to go beyond business-as-usual and claim a global responsibility to provide a good life for all within planetary boundaries." [69] (p. 14)

We tend to agree with this view, but we do not expect a more sustainable bioeconomy to grow out of current policies. While other contributions explore ways to increase public involvement and stakeholder engagement in order to foster the democratic legitimacy and thus social sustainability of the bioeconomy [2,70], we propose instead a fundamental rethink of the underlying understanding of socio-ecological transformation in bioeconomy strategies: the bioeconomy is more closely linked to classical ecological modernisation than transformation studies, which would demand radical structural change [71] (pp. 9–10). A democratic social-ecological transformation requires a deep understanding of the economic, political, and cultural structures of the prevailing food regime as part of modern capitalist society [72]. This type of analysis, which is also used in this study, demonstrate that the logics of expansion, appropriation, and growth need to be discarded. These logics form the foundation of—not least—current bioeconomy polices. Furthermore, political action is needed to ensure the decommodification of nature and the democratisation of society and societal nature relations in order to cope with the social-ecological crisis at hand [73] (pp. 169–170). Degrowth and decolonial environmental justice can be further guiding principles for a fundamental social-ecological transformation [74]. Especially in the Global North, a strategy of shrinkage and the reduction of raw material consumption is needed, in line with a perspective of sufficiency, resting on the insight that also bioeconomies need to acknowledge planetary boundaries in terms of land and energy. This would be the kind of radical change of the economy and society implied in the original concept of a bioeconomy as defined by Georgescu-Roegen in the 1970s [75]. From this vantage point, working towards social-ecological transformation means identifying and strengthening starting points for radical societal change on the basis of alternative designs and fields of experimentation, such as those used by social movements and agroecology. The overarching goal should be nothing less than reshaping the economy and the interrelation between society and nature in a democratic and just manner.

Author Contributions: All authors wrote Sections 1, 2, 6 and 7 together. M.L. mainly authored Section 3, M.B. Section 4, and A.T. Section 5. All authors have read and agreed to the published version of the manuscript.

Funding: The research on which the article is based as well as the Open Access Publication were funded by the Federal Ministry of Education and Research (BMBF) as part of the junior research group "Bioeconomy and Inequalities. Transnational Entanglements and Interdependencies in the Bioenergy Sector" (funding number 031B0021).

Institutional Review Board Statement: Not applicable.

Informed Consent Statement: Not applicable.

Data Availability Statement: All relevant data or references can be found in the paper itself.

Acknowledgments: The authors would like to thank three anonymous reviewers as well as Hariati Sinaga, Fabrício Rodríguez, Thomas Vogelpohl, and Axel Anlauf for valuable comments on an earlier draft.

Conflicts of Interest: The authors declare no conflict of interest.

References

1. Borras, S.M.; Franco, J.; Isakson, R.C.; Levidow, L.; Vervest, P. The rise of flex crops and commodities: Implications for research. *J. Peasant Stud.* **2016**, *43*, 93–115. [CrossRef]
2. D'Adamo, I.; Sassanelli, C. Biomethane Community: A Research Agenda towards Sustainability. *Sustainability* **2022**, *14*, 4735. [CrossRef]
3. De Besi, M.; McCormick, K. Towards a Bioeconomy in Europe: National, Regional and Industrial Strategies. *Sustainability* **2015**, *7*, 10461–10478. [CrossRef]
4. International Advisory Council on Global Bioeconomy. *Global Bioeconomy Policy Report (IV): A Decade of Bioeconomy Policy Development around the World*; IACBG: Berlin, Germany, 2020. Available online: https://gbs2020.net/wp-content/uploads/2020/11/GBS-2020_Global-Bioeconomy-Policy-Report_IV_web.pdf (accessed on 2 February 2022).
5. Backhouse, M.; Lehmann, R.; Lorenzen, K.; Lühmann, M.; Puder, J.; Rodríguez, F.; Tittor, A. Contextualizing the Bioeconomy in an Unequal World: Biomass Sourcing and Global Socio-ecological Inequalities. In *Bioeconomy and Global Inequalities: Socio-Ecological Perspectives on Biomass Sourcing and Production*; Backhouse, M., Lehmann, R., Lorenzen, K., Lühmann, M., Puder, J., Rodríguez, F., Tittor, A., Eds.; Palgrave Macmillan: Cham, Switzerland, 2021; pp. 3–22.
6. TNI and Hands on the Land. *The Bioeconomy. A Primer.* 2015. Available online: https://www.tni.org/files/publication-downloads/tni_primer_the_bioeconomy.pdf (accessed on 1 September 2018).

7. Dietz, K.; Engels, B.; Pye, O.; Brunnengräber, A. (Eds.) *The Political Ecology of Agrofuels*, 1st ed; Routledge: London, UK, 2014; ISBN 978-1-315-79540-9.

8. BMBF; BMEL. *National Bioeconomy Strategy*; BMBF: Berlin, Germany, 2020. Available online: https://www.bmbf.de/upload_filestore/pub/BMBF_Nationale_Biooekonomiestrategie_Langfassung_eng.pdf (accessed on 28 September 2020).

9. O'Brien, M.; Schütz, H.; Bringezu, S. The land footprint of the EU bioeconomy: Monitoring tools, gaps and needs. *Land Use Policy* **2015**, *47*, 235–246. [CrossRef]

10. Pfau, S.; Hagens, J.; Dankbaar, B.; Smits, A. Visions of Sustainability in Bioeconomy Research. *Sustainability* **2014**, *6*, 1222–1249. [CrossRef]

11. Priefer, C.; Jörissen, J.; Frör, O. Pathways to Shape the Bioeconomy. *Resources* **2017**, *6*, 10. [CrossRef]

12. D'Adamo, I.; Gastaldi, M.; Morone, P.; Rosa, P.; Sassanelli, C.; Settembre-Blundo, D.; Shen, Y. Bioeconomy of Sustainability: Drivers, Opportunities and Policy Implications. *Sustainability* **2022**, *14*, 200. [CrossRef]

13. McMichael, P. A food regime genealogy. *J. Peasant Stud.* **2009**, *36*, 139–169. [CrossRef]

14. Langthaler, E. Broadening and Deepening: Soy Expansions in a World-Historical Perspective. *HALAC* **2020**, *10*, 244–277. [CrossRef]

15. Gudynas, E. *Extractivismos. Ecología, Economía Y política de un Modo de Entender el Desarrollo y la Naturaleza*; Centro de Documentación e Información Bolivia (CEDIB): La Paz, Bolivia, 2015.

16. McKay, B.; Alonso-Fradejas, A.; Ezquerro-Cañete, A. (Eds.) *Agrarian Extractivism in Latin America*; Routledge: London, UK; New York, NY, USA, 2021.

17. Svampa, M.; Viale, E. *Maldesarrollo. La Argentina del Extractivismo y el Despojo*, 2nd ed.; Katz Editores: Buenos Aires, Argentina, 2015.

18. Moore, J.W. Cheap Food and Bad Climate: From Surplus Value to Negative-Value in the Capitalist World-Ecology. *Crit. Hist. Stud.* **2015**, 1–42. [CrossRef]

19. Foreign Agricultural Service. World Agricultural Production: February 2022; Circular Series WAP 2-22. 2022. Available online: https://apps.fas.usda.gov/psdonline/circulars/production.pdf (accessed on 16 February 2022).

20. Langthaler, E. Ausweitung und Vertiefung: Sojaexpansionen als regionale Schauplätze der Globalisierung. *Österreichische Z. Für Geisteswiss.* **2019**, *30*, 115–147.

21. Delvenne, P.; Vasen, F.; Vara, A.M. The "soy-ization" of Argentina: The dynamics of the "globalized" privatization regime in a peripheral context. *Technol. Soc.* **2013**, *35*, 153–162. [CrossRef]

22. Gras, C.; Hernandez, V.A. Los pilares del modelo agribuisness y sus estilos empresariales. In *El Agro Como Negocio: Producción, Sociedad y Territorios en la Globalización*; Gras, C., Hernandez, V.A., Eds.; Biblos: Buenos Aires, Argentina, 2013; pp. 17–46. ISBN 9876911430.

23. Oliveira, G.; Hecht, S. Sacred groves, sacrifice zones and soy production: Globalization, intensification and neo-nature in South America. *J. Peasant Stud.* **2016**, *43*, 251–285. [CrossRef]

24. Wallerstein, I. *World-Systems Analysis: An Introduction*; 5th print; Duke University Press: Durham, NC, USA, 2007; ISBN 9780822334422.

25. Friedmann, H.; McMichael, P. Agriculture and the state system: The rise and fall of national agricultures, 1870 to the present. *Sociol. Rural.* **1989**, *29*, 93–117. [CrossRef]

26. Friedmann, H. The Political Economy of Food: A Global Crisis. *New Left Rev.* **1993**, *197*, 29–57.

27. Giarracca, N.; Teubal, M. (Eds.) *Actividades Extractivas en Expansión. Reprimarización de la Economía Argentina?* Antropofagia: Buenos Aires, Argentina, 2013.

28. Toledo López, V. La política agraria del kirchnerismo. Entre el espejismo de la coexistencia y el predominio del agronegocio. *Mundo Agrar.* **2017**, *18*, 37. [CrossRef]

29. McKay, B. Agrarian Extractivism in Bolivia. *World Dev.* **2017**, *97*, 199–211. [CrossRef]

30. McKay, B. *The Political Economy of Agrarian Extractivism. Lessons from Bolivia*; Practical Action Publishing: Rugby, UK, 2020.

31. Ezquerro-Cañete, A. The Agrarian Question of Extractive Capital: Political Economy, Rural Change, and Peasant Struggle in 21st Century Paraguay. Ph.D. Thesis, Saint Mary's University, Halifax, NS, Canada, 2020.

32. Wallerstein, I. *Africa and the Modern World*; Africa World Press: Trenton, NJ, USA, 1986.

33. Bernstein, H. *Food Regimes and Food Regime Analysis: A Selective Survey*. Bicas Working Paper No. 2. 2015. Available online: https://www.tni.org/files/download/bicas_working_paper_2_bernstein.pdf (accessed on 12 January 2022).

34. McMichael, P. Global Development and the Corporate Food Regime. *Rural Sociol. Dev.* **2005**, *11*, 269–303.

35. Reboratti, C. Un mar de soja: La nueva agricultura en Argentina y sus consecuencias. *Rev. Geogr. Norte Gd.* **2010**, *45*, 63–76. [CrossRef]

36. Pengue, W.A. Transgenic Crops in Argentina: The Ecological and Social Debt. *Bull. Sci. Technol. Soc.* **2005**, *25*, 314–322. [CrossRef]

37. McKay, B.; Colque, G. Extractive dynamics of agrarian change in Bolivia. In *Agrarian Extractivism in Latin America*; McKay, B., Alonso-Fradejas, A., Ezquerro-Cañete, A., Eds.; Routledge: London, UK; New York, NY, USA, 2021; pp. 45–63.

38. McMichael, P. *Food Regimes and Agrarian Questions: Agrarian Change and Peasant Studies*; Fernwood Publishing: Halifax, NS, Canada; Winnipeg, MB, Canada, 2013.

39. Weis, T. The Meatification of Diets. In *Routledge Handbook of Food and Nutrition Security*; First issued in paperback; Pritchard, B., Ortiz Ríos, R., Shekar, M., Eds.; Routledge: London, UK; New York, NY, USA, 2018; pp. 124–136. ISBN 9781138343498.

40. Weis, T. Meatification. In *Handbook of Critical Agrarian Studies*; Akram-Lodhi, A., Dietz, K., Engels, B., McKay, B., Eds.; Edward Elgar Publishing: Cheltenham, UK, 2021; pp. 561–567. ISBN 9781788972468.

41. Gray, A.; Weis, T. *The Meatifiction and Re-meatification of Diets: The Unequal Burden of Animal Flesh and the Urgency of Plant-Meat Alternatives: Guidance Memo*; Tiny Beam Fund: Boston, MA, USA, 2021.

42. Moore, J.W. *Capitalism in the Web of Life: Ecology and the Accumulation of Capital*; Verso: London, UK; New York, NY, USA, 2015.

43. Patel, R.; Moore, J.W. *A History of the World in Seven Cheap Things: A Guide to Capitalism, Nature, and the Future of the Planet*; University of California Press: Oakland, CA, USA, 2017; ISBN 978-0520293137.

44. Patel, R. *Stuffed and Starved: The Hidden Battle for the World Food System*, 2nd ed.; Melville House: Brooklyn, NY, USA, 2012; ISBN 978-1612191270.

45. Walker, R. *The Conquest of Bread: 150 Years of Agribusiness in California*; New Press, Distributed by Norton: New York, NY, USA, 2004; ISBN 1-56584-877-2.

46. Mintz, S.W. *Sweetness and Power: The Place of Sugar in Modern History*; Penguin Books: New York, NY, USA, 1986; ISBN 9780140092332.

47. Poth, C. The Biotechnological Agrarian Model in Argentina: Fighting against capital with science. In *Agrarian Extractivism in Latin America*; McKay, B., Alonso-Fradejas, A., Ezquerro-Cañete, A., Eds.; Routledge: London, UK; New York, NY, USA, 2021; pp. 21–44.

48. Clapp, J. ABCD and beyond: From grain merchants to agricultural value chain managers. *Can. Food Stud.-La Rev. Can. Des Études Sur L´Aliment.* **2015**, *2*, 126–135. [CrossRef]

49. Lapegna, P.; Perelmuter, T. Genetically modified crops and seed/food sovereignty in Argentina: Scales and states in the contemporary food regime. *J. Peasant Stud.* **2020**, *47*, 700–719. [CrossRef]

50. Gras, C.; Hernández, V. Agribusiness and large-scale farming: Capitalist globalisation in Argentine agriculture. *Can. J. Dev. Stud. / Rev. Can. D'Études Du Développement* **2014**, *35*, 339–357. [CrossRef]

51. Backhouse, M. Global Inequalities and Extractive Knowledge Production in the Bioeconomy. In *Bioeconomy and Global Inequalities: Socio-Ecological Perspectives on Biomass Sourcing and Production*; Backhouse, M., Lehmann, R., Lorenzen, K., Lühmann, M., Puder, J., Rodríguez, F., Tittor, A., Eds.; Palgrave Macmillan: Cham, Switzerland, 2021; pp. 25–44.

52. Veltmeyer, H.; Petras, J. Agro-Extractivism: The Agrarian Question of the 21st Century. In *Extractive Imperialism in the Americas: Capitalism's New Frontier*; Petras, J., Veltmeyer, H., Bowles, P., Canterbury, D.C., Girvan, N., Tetreault, D., Eds.; Brill: Leiden, The Netherlands, 2014; pp. 62–100. ISBN 9789004268869.

53. Teubal, M.; Giarracca, N. Introducción. In *Actividades Extractivas en Expansión. Reprimarización de la Economía Argentina?* Giarracca, N., Teubal, M., Eds.; Antropofagia: Buenos Aires, Argentina, 2013; pp. 9–18.

54. Leguizamón, A. The Gendered Dimensions of Resource Extractivism in Argentina's Soy Boom. *Lat. Am. Perspect.* **2019**, *46*, 199–216. [CrossRef]

55. Global Witness. *Defending Tomorrow. The Climate Crisis and Threats Against Land and Environmental Defenders*; Global Witness: London, UK, 2020.

56. Temper, L.; Avila, S.; Del Bene, D.; Gobby, J.; Kosoy, N.; LeBillon, P.; Martínez Alier, J.; Perkins, P.; Roy, B.; Scheidel, A.; et al. Movements shaping climate futures: A systematic mapping of protests against fossil fuel and low-carbon energy projects. *Environ. Res. Lett.* **2020**, *15*, 12. [CrossRef]

57. Ojeda, D. Social reproduction, dispossession, and the gendered workings of agrarian extractivism in Colombia. In *Agrarian Extractivism in Latin America*; McKay, B., Alonso-Fradejas, A., Ezquerro-Cañete, A., Eds.; Routledge: London, UK; New York, NY, USA, 2021; pp. 85–98.

58. Landívar García, N. Gender inclusion in the sugarcane production of agrofuels in coastal Ecuador: Illusionary promises of rural development within a new agrarian extractivism. In *Agrarian Extractivism in Latin America*; McKay, B., Alonso-Fradejas, A., Ezquerro-Cañete, A., Eds.; Routledge: London, UK; New York, NY, USA, 2021; pp. 117–138.

59. Cielo, C.; Coba, L. Extractivism, Gender, and Disease: An Intersectional Approach to Inequalities. *Ethics Int. Aff.* **2018**, *32*, 169–178. [CrossRef]

60. Motta, R. *Social Mobilization, Global Capitalism and Struggles over Food: A Comparative Study of Social Movements*; Routledge: London, UK; New York, NY, USA, 2016; ISBN 1472479084.

61. Arancibia, F.; Motta, R. Undone Science and Counter-Expertise: Fighting for Justice in an Argentine Community Contaminated by Pesticides. *Sci. Cult.* **2019**, *28*, 277–302. [CrossRef]

62. Svampa, M. *Die Grenzen der Rohstoffausbeutung. Umweltkonflikte und Ökoterritoriale Wende in Lateinamerika*; Bielefeld University Press: Bielefeld, Germany, 2020.

63. Domínguez, R.; Caria, S. Extractivismos andinos y limitantes del cambio estructural. In *Nada dura Para Siempre. Neo-Extractivismo Tras el Boom de las Materias Primas*; Burchardt, H.-J., Domínguez, R., Larrea, C., Peters, S., Eds.; Ediciones Abya-Yala: Quito, Ecuador, 2016; pp. 89–130.

64. Tittor, A. The key role of the agribusiness and biotechnology sectors in constructing the economic imaginary of the bioeconomy in Argentina. *J. Environ. Policy Plan.* **2021**, *23*, 213–226. [CrossRef]

65. Tittor, A. Towards an Extractivist Bioeconomy? The Risk of Deepening Agrarian Extractivism when Promoting Bioeconomy in Argentina. In *Bioeconomy and Global Inequalities: Socio-Ecological Perspectives on Biomass Sourcing and Production*; Backhouse, M.,

Lehmann, R., Lorenzen, K., Lühmann, M., Puder, J., Rodríguez, F., Tittor, A., Eds.; Palgrave Macmillan: Cham, Switzerland, 2021; pp. 309–330.

66. Ramcilovic-Suominen, S.; Pülzl, H. Sustainable development—A 'selling point' of the emerging EU bioeconomy policy framework? *J. Clean. Prod.* **2018**, *172*, 4170–4180. [CrossRef]

67. Hausknost, D.; Schriefl, E.; Lauk, C.; Kalt, G. A Transition to Which Bioeconomy? An Exploration of Diverging Techno-Political Choices. *Sustainability* **2017**, *9*, 669. [CrossRef]

68. Lühmann, M. Sustaining the European Bioeconomy. The Material Base and Extractive Relations of a Bio-based EU-Economy. In *Bioeconomy and Global Inequalities: Socio-Ecological Perspectives on Biomass Sourcing and Production*; Backhouse, M., Lehmann, R., Lorenzen, K., Lühmann, M., Puder, J., Rodríguez, F., Tittor, A., Eds.; Palgrave Macmillan: Cham, Switzerland, 2021; pp. 287–307.

69. Zeug, W.; Kluson, F.R.; Mittelstädt, N.; Bezama, A.; Thrän, D. *Results from a Stakeholder Survey on Bioeconomy Monitoring and Perceptions on Bioeconomy in Germany*. UFZ Discussion Papers 8/2021, Leipzig. 2021. Available online: https://nbn-resolving.org/urn:nbn:de:0168-ssoar-76967-4 (accessed on 17 February 2022).

70. Lynch, D.H.; Klaassen, P.; van Wassenaer, L.; Broerse, J.E. Constructing the Public in Roadmapping the Transition to a Bioeconomy: A Case Study from the Netherlands. *Sustainability* **2020**, *12*, 3179. [CrossRef]

71. Bastos Lima, M.G. *The Politics of Bioeconomy and Sustainability: Lessons from Biofuel Governance, Policies and Production Strategies in the Emerging World*; Springer Nature: Cham, Switzerland, 2021; ISBN 978-3-030-66836-5.

72. Görg, C.; Brand, U.; Haberl, H.; Hummel, D.; Jahn, T.; Liehr, S. Challenges for Social-Ecological Transformations: Contributions from Social and Political Ecology. *Sustainability* **2017**, *9*, 1045. [CrossRef]

73. Brand, U.; Görg, C.; Wissen, M. Overcoming neoliberal globalization: Social-ecological transformation from a Polanyian perspective and beyond. *Globalizations* **2020**, *17*, 161–176. [CrossRef]

74. Ramcilovic-Suominen, S. Envisioning just transformations in and beyond the EU bioeconomy: Inspirations from decolonial environmental justice and degrowth. *Sustain. Sci.* **2022**, 1–16. [CrossRef]

75. Vivien, F.-D.; Nieddu, M.; Befort, N.; Debref, R.; Giampietro, M. The Hijacking of the Bioeconomy. *Ecol. Econ.* **2019**, *159*, 189–197. [CrossRef]

 *sustainability*

Article

Economic Evaluation of Bioremediation of Hydrocarbon-Contaminated Urban Soils in Chile

Roberto Orellana [1,2,3,*,†], Andrés Cumsille [1,†], Paula Piña-Gangas [4], Claudia Rojas [1], Alejandra Arancibia [2], Salvador Donghi [5,6], Cristian Stuardo [1], Patricio Cabrera [1], Gabriela Arancibia [1], Franco Cárdenas [1], Felipe Salazar [1], Myriam González [1], Patricio Santis [1], Josefina Abarca-Hurtado [2], María Mejías [2] and Michael Seeger [1,*]

[1] Laboratorio de Microbiología Molecular y Biotecnología Ambiental, Departamento de Química & Centro de Biotecnología Daniel Alkalay-Lowitt, Universidad Técnica Federico Santa María, Avenida España 1680, Valparaíso 2390123, Chile
[2] Laboratorio de Biología Celular y Ecofisiología Microbiana, Facultad de Ciencias Naturales y Exactas, Universidad de Playa Ancha, Leopoldo Carvallo 270, Valparaíso 2360001, Chile
[3] HUB Ambiental, Universidad de Playa Ancha, Leopoldo Carvallo 270, Valparaíso 2360001, Chile
[4] Escuela de Negocios Internacionales, Universidad de Valparaíso, Valparaíso 2572048, Chile
[5] Instituto de Geografía, Pontificia Universidad Católica de Valparaíso, Valparaíso 2350026, Chile
[6] Simbiosis, Valparaíso 2510117, Chile
* Correspondence: roberto.orellana@upla.cl (R.O.); michael.seeger@usm.cl (M.S.)
† These authors contributed equally to this work.

Abstract: Technical advances have converted bioremediation into a large-scale ecosystem service suitable for the treatment of polluted soils worldwide; however, its application in Chile is scarce. The main hurdles that must be addressed include the capacities of such approaches for the treatment of polluted soils, the lack of knowledge about key factors affecting bioremediation costs and the lack of a legal framework to regulate this activity. In this study, the economic performance of the bioremediation of chronically hydrocarbon-polluted urban soils based on bioaugmentation, biostimulation or the combination of both approaches projected to an industrial scale was evaluated. The cost of bioremediation ranged between USD 50.7 and USD 310.4 per m^3 of contaminated soil. In addition, the items and activities that had the most significant impacts on the final bioremediation cost, such as compost for biostimulation and bacterial growth media for bioaugmentation-based approaches, were identified. The projected costs were compared against an extensive database of 130 soil bioremediation projects. The bioremediation treatment costs fell within the top 60% of the more expensive projects, highlighting the high effort involved in bioremediation of chronically contaminated soils. This framework can facilitate the decision making of entrepreneurs, consultants, researchers and governmental authorities when launching initiatives to develop a local bioremediation industry capable of cleaning up a high number of polluted sites in Chile.

Keywords: microbial bioremediation; cost of bioremediation; bioremediation industry

Citation: Orellana, R.; Cumsille, A.; Piña-Gangas, P.; Rojas, C.; Arancibia, A.; Donghi, S.; Stuardo, C.; Cabrera, P.; Arancibia, G.; Cárdenas, F.; et al. Economic Evaluation of Bioremediation of Hydrocarbon-Contaminated Urban Soils in Chile. *Sustainability* 2022, 14, 11854. https://doi.org/10.3390/su141911854

Academic Editors: Idiano D'Adamo and Massimo Gastaldi

Received: 18 July 2022
Accepted: 29 August 2022
Published: 21 September 2022

Publisher's Note: MDPI stays neutral with regard to jurisdictional claims in published maps and institutional affiliations.

1. Introduction

Petroleum-derived hydrocarbons are the main hazardous compounds causing the contamination of extensive areas of soil and water. The pollution is more frequent at areas near to oil wells and sites devoted to the production, distribution, manipulation, disposal and especially storage of petroleum-derived products. Inadequate management, monitoring and maintenance have led to contaminated sites, generating a potentially permanent source of diffuse pollution that represents a significant health risk for neighboring communities. Historically, conventional techniques for restoring polluted soils have mainly consisted of excavation, removal and disposal of contaminated materials into waste-dumped or hazardous-waste landfills [1,2]. The volatility of the hydrocarbons during removal and management operations constitutes a health risk, especially for workers during excavation, handling and transportation. Microbial communities in soils possess metabolic and physiological flexibility for adaptation to environmental challenges [3]. The input of contaminants usually impacts the structure and function of the soil microbiome, reshaping microbial communities towards the selection of species that can survive and cope with the toxicity of the contaminants [4–10]. Emergent species within the disturbed communities usually share enhanced physiological and substrate degradation capabilities, oxidizing, transforming, immobilizing or binding the contaminants [11–15]. This metabolic flexibility of soil microbiota present in the environment provides a platform for microbial bioremediation as a valuable ecosystem service [16]. In addition, the biodegradable nature of hydrocarbons and the ubiquitous distribution of hydrocarbon-degrading microorganisms have underlined the efficacy of microbially driven bioprocesses for the restoration of polluted ecosystems [17–21]. Currently, there is an increasing need for alternatives in cleaning up contaminated sites worldwide. For instance, it has been estimated that there are 500,000 and 340,000 contaminated sites in the U.S. and Europe, respectively [22,23]. China has over two million hectares of abandoned sites in more than one hundred old industrial cities that require environmental restoration before redevelopment [24]. It is in this context that full-scale bioremediation projects have been successfully applied, mainly in Europe, North America and, more recently, China [25,26]. As a result, the global industry related to bioremediation services has systematically spread, becoming an industry worth USD 35 billion in 2009 [27]. The growth of the global bioremediation market has accelerated in recent years, reaching USD 91.0 billion in 2018, and it is expected to grow to USD 186.3 billion in 2023 [28].

The main sources of soil contamination in Chile are mining activities (30.9% of total sites), including deposition of metals and metalloids, and agro-industrial activities (30.1% of total sites), including petroleum products, pesticides, fertilizers and urban waste (24.2% of total sites) [29]. Regarding contamination by metals and metalloids, the geological evolution specifically associated with the Andes has resulted in an overwhelming proportion of metallic ore deposits being located in the northern part [30]. Chilean porphyry copper deposits have a low ore grade but exist in huge volumes and, together with the lack of regulations, this has encouraged several large-scale mining operations to leave a legacy of contaminated areas [31]. It has been previously reported that these polluted sites include 740 tailings, of which 23.3% are abandoned [32]. Soil contamination associated with agricultural and forestry production systems includes excessive utilization of fertilizers and pesticides, resulting in diffuse contamination across nearby areas [33,34]. The contamination of soils by other chemicals, such as petroleum derivatives and organochlorines, has mainly been associated with industrial activities outside the boundaries of cities. However, with the growth of the urban population, they have been absorbed by urban areas, such as in Valparaíso, Concepción and Santiago, and there are also cities with industrial activities, such as Quintero, Puchuncaví and Talcahuano [1,35–39]. Despite the urgent need for sustainable clean-up processes, applications of microbial bioremediation in Chile are scarce. Recently, the Chilean government evaluated 19 projects designed to restore metal and hydrocarbon-polluted soils, among which only 8 projects (42%) focused on bioremediation. Several factors may explain why bioremediation is not commonly used. Currently, Chile is the only Organization for Economic Co-operation and Development (OECD) country that does not have a soil protection regulation, which would regulate the maximum permissible concentrations of contaminants [40,41]. Bioremediation may entail higher costs in comparison to conventional techniques, such as the confinement of contaminants in authorized places. The high variability in the physical, chemical and microbiological properties of soils requires the design of site-tailored treatments [42,43]. In addition, the lack of directives, regulations, definitions of methodological tools, process standards and requirements for contaminated soils have hindered the application and development of a local bioremediation industry.

The aim of this study was to provide insights into the comparative costs of five approaches for bioremediation of long-term hydrocarbon-polluted urban soil, including biostimulation, bioaugmentation and the combination of both, highlighting the key factors that influence the costs associated with each approach. Since each contaminated soil has specific characteristics, this study is a guide to developing the cost assessments of bioremediation processes. This framework will facilitate the decision making of entrepreneurs and consultants when performing a risk–cost–benefit analysis. This study will help to trigger governmental authorities to generate environmental policies, regulations and standards for contaminated soils, as well as launching initiatives to develop an environmentally safe and robust local bioremediation industry capable of cleaning up different polluted sites in Chile.

2. Materials and Methods

Economic evaluation was based on treating chronically contaminated soil, taking the soil conditions of the Las Salinas site (Viña del Mar, Valparaiso Region, Chile) as a reference. The Las Salinas site is a contaminated brownfield that was subjected to petroleum industrial activity for more than eight decades. During this long period, the contamination pressure not only affected the diversity and community structure of soil, but also hampered the ecosystem function and its natural resilience. In such environments, the remaining contamination is often enriched with heavier and more structurally complex fractions of hydrocarbons, due to volatilization or solubilization of volatile fractions and the degradation of lighter alkanes [44]. The more recalcitrant compounds can also be sorbed to the soil matrix, decreasing their bioavailability and the biodegradation rate and extent [44–47].

In general, three main approaches have been widely used for restoring hydrocarbon-contaminated soils [1,48]. The first technology, called biostimulation, includes the enhancement of the metabolic activity of native microbial communities by providing limiting nutrients, such as phosphorous, nitrogen or oxygen, and further modification of environmental factors [1]. The second approach is based on the addition of hydrocarbon-degrading microorganisms, frequently applied when native microbial communities lack the metabolic capabilities or when their activity is unable to trigger significant biodegradation rates [49]. A third approach is based on the addition of stable organic amendments, such as compost, which has been applied with success across pilot- and full-scale applications [44,50,51]. The analysis presented here was based on the bioremediation of chronically contaminated soils, such as those currently present at Las Salinas, in which there is no evidence of persistent degradation processes over time. Therefore, the economic assessment was based on the addition of compost, the addition of hydrocarbon-degrading microorganisms as well as a mixture of both approaches, for which experimental results were previously published [52]. In this study, the bioremediation was addressed by five treatments. Briefly, the first two were based on bioaugmentation, with the addition of five hydrocarbonoclastic strains (named BA), and also the same treatment with the addition of permanent air venting (BAV). The selected hydrocarbonoclastic bacterial strains for bioaugmentation were *Acinetobacter* sp. DD78, *Acinetobacter* sp. AA64, *Acinetobacter* sp. AF53, *Pseudomonas* sp. DN36 and *Pseudomonas* sp. DN34. These strains were previously isolated from hydrocarbon-contaminated soil (Valparaíso, Chile) and possess the ability to degrade a wide range of hydrocarbons [53]. *Acinetobacter* sp. DD78 possess the ability to produce biosurfactants, which can help improve the bioavailability of the hydrocarbons for biodegradation [54]. The following two treatments were based on biostimulation with the addition of compost in two different ratios, 9:1 (*v/v*) and 3:2 (*v/v*), named BE1 and BE4, respectively. The fifth treatment considered was the combination of bioaugmentation and biostimulation, using a mixture of soil and compost with 3:2 (*v/v*) ratio (BAE).

To determine the feasibility of bioremediation as an industrial activity, an economic evaluation was made based on the results of the bioremediation strategies projected to industrial scale. The projection considered the implementation of an on-site hydrocarbon soil bioremediation process in a square one-hectare field with 10 biopiles of 100 m length each. The biopiles were designed to be trapezoidal with 2 m height, 5 m width (base) and 2.5 m width (top), with a 5 m space between each pile (Figure S1). A potential decrease in the bulk density was assumed to be 25% after construction. Each biopile has a total volume of

750 m^3 of material that was covered with a High-Density Polyethylene (HDPE) membrane to conserve moisture, minimizing leachate production and gas emissions. It was projected that bioremediation process at an industrial scale would take 20 weeks, assuming one week of preparation and one week of dismantling the biopiles and associated materials, as previously observed (unpublished). An additional period of ten weeks was also considered for contingencies and maintenance. Therefore, each bioremediation cycle treats between 3750 and 6250 m^3 of soil, depending on each strategy (Table S1). The proposed equipment and supplies required for the construction of biopiles and soil movement were two front shovel loaders with 2.5 m^3 buckets each. Since water content is one of the most critical factors that regulates microbial activity during bioremediation, all treatments include a system for irrigation that uses one spray truck of 20 m^3 that periodically moistens the soil.

To determine the initial capital for each treatment, we considered that bioaugmentation-based approaches required infrastructure with higher-technology equipment and cost-intensive installation efforts than those required by biostimulation-based approaches. Indeed, a set of bioreactors was included for bioaugmentation and air injection treatments covering two stages. An initial stage of preinoculation where sufficient biomass is grown for inoculating bigger reactors that contain the biomass to be incorporated into the soil. Five jacketed bioreactors of 0.3 m^3 with blowers of 0.0075 m^3 s^{-1} and six jacketed bioreactors of 15 m^3 equipped with blowers of 0.4 m^3 s^{-1} were considered for preinoculation and inoculation stages, respectively. Calculations for culture volume added for bioaugmentation was determined on the basis that each strain reaches a density of 10^6 CFU g^{-1}, value that is 10-fold higher than those levels of cell density of hydrocarbon-degrading microorganisms at which bioremediation will be negligible [55]. Based on bacterial counts made in preliminary laboratory experiments [52], each strain requires one jacketed reactor, except for *Acinetobacter* sp. AA64, which requires two. On the other hand, biostimulation-based approaches require neither biomass reactors nor blowers. Instead, they need machinery to build biopiles, including a front shovel loader and a spray truck. A similar trend was observed when installation services costs were calculated. The cost of installation services was determined by adding the costs of installation of equipment, instrumentation and control, piping, electrical wiring, infrastructure, yard improvements, services, land, engineering and supervision, infrastructure expenses, contractors fees and contingencies (Table S2), and calculated according to suggested values [56].

The determination of operation costs included variable costs, fixed costs, indirect production costs and administrative and sales expenses. Variable and fixed costs enclosed direct raw materials and manpower, respectively. For treatments requiring bioaugmentation, all materials for preparation of Bushnell Haas (BH) medium and monitoring microbial grow were included. The BH broth medium contains (in grams per liter of Milli-Q water): KH$_2$PO$_4$, 1; K$_2$HPO$_4$, 1; NH$_4$NO$_3$, 1; MgSO$_4$, 0.2; CaCl$_2$, 0.020; FeCl$_3$, 0.050. For biostimulation treatments, the supply was assumed to be compost. For every treatment there are several common costs, such as water, diesel and electricity.

The fixed costs considered remuneration for direct and indirect labor, building maintenance, publicity, machinery depreciation and many supplies and services. The incomes, as well as costs, were calculated based on the number of treatments instead of a time scale. Calculations of operating incomes of each treatment were estimated based on the breakeven–total-cost formula as (Fixed costs + Variable costs) = Total revenue = Breakeven, and (Quantity sold × Unit selling price) = Breakeven. Calculations of variable costs per unit produced and the unit selling price were estimated based in the following formula Profit = Unit Sales × (Unit Sales − Variable unit costs)—Total fixed costs. An estimation of contribution margin ratio was made based on ten years life expectancy of each equipment. As every treatment lasts twenty weeks, it was considered that 200 treatments were performed along life expectancy. Revenue estimations were made considering investment costs and the future return of investment in different time lapses (1 to 10 years).

In addition, an extensive cost analysis of 130 soil bioremediation projects was located, reviewed, and evaluated using a collection of peer-reviewed literature, federal and state agency reports (Table S3). This analysis included bioremediation with either hydrocarbons or a mixture of hydrocarbons and other contaminants, such as organic solvents, halogenated organic compounds and heavy metals. Projects were classified according to the bioremediation strategies, in the following categories, "bioventing", "biostimulation", "solvent vapor extraction", "phytoremediation", "thermal treatments", and "bioaugmentation". In addition, two extra categories were included. The category "various treatments" employs a combination of more than one treatment. The second category, "other", is a treatment that was no included in the list.

3. Results

Economic assessment was made based on the results of bioremediation of five treatments to clean-up chronically hydrocarbon-contaminated soils, based on bioaugmentation, biostimulation or combination of both technologies [52]. The first approach was bioaugmentation, with the addition of five hydrocarbonoclastic strains (named as BA). The second technology was bioaugmentation with the addition of permanent air venting (BAV). The following two approaches were based on biostimulation with the addition of compost in two different ratios, 9:1 (*v/v*) and 3:2 (*v/v*), named BE1 and BE4, respectively. The fifth treatment was the combination of bioaugmentation and biostimulation, using a mixture of soil and compost with 3:2 (*v/v*) ratio (BAE). The analysis of the start-up capital, defined as the resources required to acquire the assets needed for each bioremediation approach, resulted in a high difference between bioaugmentation and biostimulation treatments. Among all five treatments, bioaugmentation-based approaches were those with the highest start-up capital. The complete start-up capital of BAE and BA was USD 1,822,159, whereas the capital of BAV was USD2,150,961 (Table S2). In contrast, the complete start-up capital required for biostimulation-based approaches, BE1 and BE4, was an order of magnitude lower than those technologies with bioaugmentation, due to the fact that there is no requirement for equipment, such as reactors and blowers, decreasing costs of installation.

The estimated cost of the bioremediation of chronically hydrocarbon-contaminated soils, assuming that the investment is recovered after five years of operation, ranged between USD 50.7 and USD 310.4 per m^3 of contaminated soil (Table 1). Biostimulation with 10% compost (BE1) was the treatment strategy with the lowest cost (USD 50.7 per m^3), followed by biostimulation with 40% compost (BE4, USD 131.7 per m^3). The estimated costs for treatment per m^3 of soil of both strategies that exclusively involved addition of bacterial cultures, BA and BAV, were USD 141.8 and USD 153.9, respectively. BAE was evaluated as the most expensive technology (USD 310.4 per m^3 of contaminated soil) due to the high technological level required to provide the bacteria to ensure the bioaugmentation of only 3375 m^3 of soil per bioremediation cycle (Figure 1). The analysis encompassed the estimation of product costs, period costs, liabilities and obligations. Projections showed that product costs represent the largest cost for all treatments, reflecting the importance of the selection of raw materials, such as compost and bacterial growth media for biostimulation and bioaugmentation-based approaches (Table S4). The largest item of product cost was the direct material that ranged from USD 17.5 to USD 174 per m^3 of bioremediated soil, contributing to 35% and 56% of total costs of BE1 and BAE, respectively (Figure 1). The direct material cost, mainly based on compost, was USD 12.3 and USD 74.1 for biostimulation with 10% and 40% of compost, respectively (Table 1). Each cycle using biostimulation with 10% was more cost effective due to lower compost addition and the treatment of 5625 m^3 of contaminated soil, whereas biostimulation with 40% of compost applied higher compost concentration (four-fold) and treated only 3750 m^3 per bioremediation cycle (Figure 1). For biostimulation approaches, manufacturing overhead cost was the second largest contribution to the total costs. In contrast, in bioaugmentation approaches, provision and covering assets were the second and the third largest item costs, suggesting that the technological level of bioaugmentation has a significant impact on its economic performance. The provision cost of bioaugmentation treatments is almost two-fold higher than the provision cost of biostimulation, mainly due to an increase in monitoring and quality control. In addition, there is a high level of costs for equipment and machinery for the culture of microbial strains (Tables S2, S4 and S5).

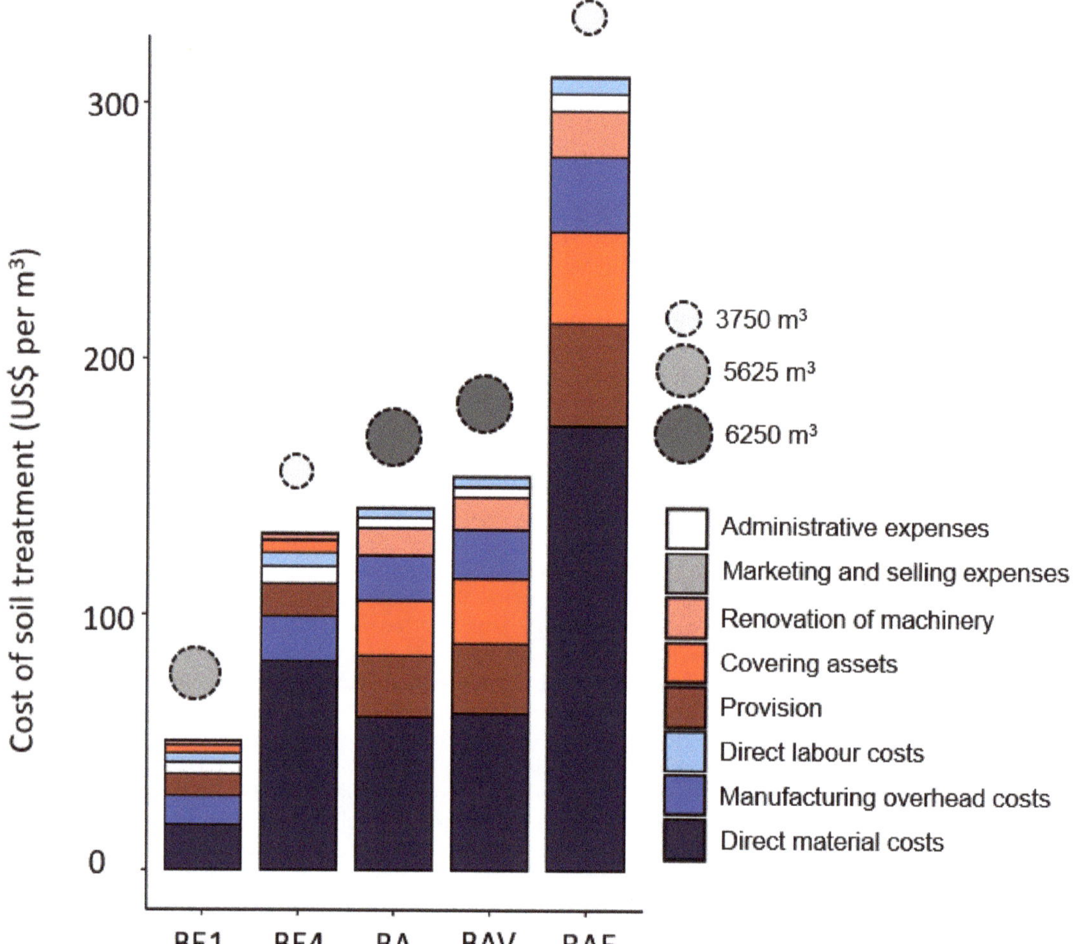

Figure 1. Estimated cost for bioremediation of chronically contaminated soils using different strategies projected to industrial scale. Colors indicate type of cost contributing to the total bioremediation cost. Blue tones show products' costs. Direct material costs are shown in navy blue, manufacturing overhead costs in blue and direct labor costs in light blue. Red tones indicate items included in liabilities and obligations as follows: Marron shows provision, red shows covering assets, and salmon shows renovation of machinery. Marketing and selling expenses are shown in gray. Administrative expenses are shown in white. Dotted line circles shown the quantity (in m³) of soil treated per 10 biopiles of each treatment.

Table 1. Estimated costs per m^3 of contaminated soil treated by different bioremediation approaches.

	Items	BE1	BE4	BAE	BA	BAV
	Direct material costs	USD 17.5	USD 81.9	USD 174.0	USD 59.9	USD 61.4
Products' costs	Direct labor costs	USD 3.5	USD 5.3	USD 6.3	USD 3.8	USD 3.8
	Manufacturing overhead costs	USD 11.6	USD 17.4	USD 29.2	USD 17.5	USD 19.0
Period costs	Administrative expenses	USD 4.5	USD 6.7	USD 6.7	USD 4.0	USD 4.0
	Marketing and selling expenses	USD 0.2	USD 0.4	USD 0.4	USD 0.2	USD 0.2
Liabilities and obligations	Covering assets	USD 3.3	USD 4.9	USD 36.0	USD 21.6	USD 25.5
	Provision	USD 8.4	USD 12.6	USD 39.8	USD 23.9	USD 27.2
	Renewal machinery	USD 1.6	USD 2.5	USD 18.0	USD 10.8	USD 12.7
	Total	USD 50.7	USD 131.7	USD 310.4	USD 141.8	USD 153.9

Furthermore, different scenarios for recovery assets were simulated. When the time length of the start-up investment recovery increased to ten years, the cost per m^3 of treated soil decreased more significantly for the bioaugmentation treatments than for the biostimulation treatments (Table S6). All these aspects have unequal impacts on the economic performance of the different bioremediation technologies, highlighting the relevance of this analysis, especially for evaluating profitability under incipient market conditions, such as those for bioremediation in Chile.

Cost of Bioremediation in Chile and Other Countries

Our results were calculated using average costs in Chile; therefore, they may be highly influenced by domestic dynamics of unrelated sectors rather than those sectors related to the nature of bioremediation. In order to compare our results to worldwide bioremediation operations, we reconstructed an extensive analysis of 130 bioremediation projects of contaminated soils. The analysis included different types of contaminants (e.g., hydrocarbons, organic solvents, halogenated organic compounds, heavy metals), as well as diverse remediation approaches including biostimulation, bioventing, bioaugmentation, solvent vapor extraction and thermal treatments (Table S3).

We determined that the cost of bioremediation was highly variable and ranged between USD 0.5 and USD 1820 per m^3 of treated soil. All the projects that registered costs of < USD 2 per m^3 of treated soil were associated with treatments based exclusively on biostimulation, and these treatments only removed 50% of hydrocarbons (Table S3). In contrast, projects that reported costs higher than USD 700 per m^3 of treated soil involved treatments of a diversity of contaminants, such as BTEX, VOC, PAHs and heavy metals, using a variety of approaches, including biostimulation, bioaugmentation and thermal treatments (Table S3). The distribution of costs of the projects seems to be more influenced by the type of remediation treatment than the type of contaminants in the soil. The 40% of those projects with lower cost (36) were based on bioventing (20), biostimulation (10), solvent vapor extraction (5) and phytoremediation (1). The 60% of projects with higher costs (94) involved approaches such as bioventing (27), thermal (26), solvent vapor extraction (13), biostimulation (12), bioaugmentation (1), other (6) and combined (9) treatments. Costs calculated in our projections fall within this last group, highlighting the important effort involved in bioremediation of chronically contaminated soils (Figure 2). In general, this effort requires reshaping the soil microbiota and dramatically strengthing their biodegradation capabilities towards the most recalcitrant and less bioavailable fractions. In contrast, several projects belonging to the 40% of lower cost are based on bioventing and biostimulation, mainly oriented to metabolize more easily biodegradable pollutants. The cost of our more expensive treatment, BAE (bioaugmentation and biostimulation), was within the top 20% more expensive projects of the dataset, which are enriched by physicochemical aggressive techniques, such as thermal treatments and solvent vapor extraction (Figure 2).

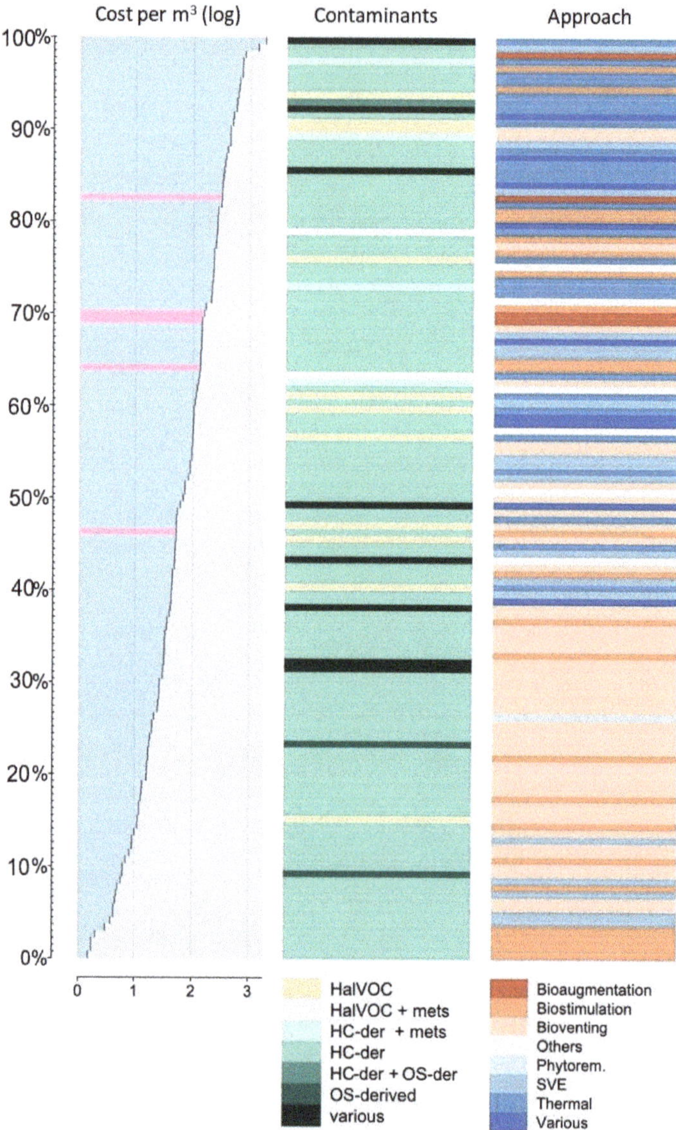

Figure 2. Distribution of the cost of bioremediation across a set of 135 projects around the world. The bar chart in the left shows costs of bioremediation sorted from the cheapest (0%) to the most expensive (100%), and the cost for bioremediation is expressed in USD per m³ of contaminated soils in logarithmic scale. The contaminants column indicates the type of contaminants in the treated soil as the following. "HC-der" indicates soils contaminated with hydrocarbons-derived compounds; "HalVOC" indicates soils contaminated with halogenated volatile organic compounds; "mets" indicates soils contaminated with metals; "OS-der" indicates soils contaminated with organic solvent compounds. The approach column indicates the technique used for soil bioremediation. Abbreviations: Phytorem—phytoremediation; SVE—solvent vapor extraction; Thermal—thermal treatments.

4. Discussion

Far from being a silver bullet, there is not a single bioremediation approach that is useful in all hydrocarbon-contaminated sites [57]. Thus, appropriate characterization of the polluted soils and the adaptation of bioremediation techniques on a case-by-case basis are essential [58]. Adjustments should consider several factors, including technology-specific components, contaminants (type, concentration, aging and distribution), and the physico-chemical properties of the soil. As a result of the variability, the design and implementation of bioremediation projects have a direct impact on their capability to evaluate the economic performance, converting this into a daunting task [43]. An initial guide for the evaluation of costs of different bioremediation approaches, such as the one presented here, is relevant not only to compare these alternatives to projects based on excavation and disposal, which is rather easy to grasp, but to integrate this information through management practices to ensure the technical and economic feasibility of bioremediation projects, where the decision making is empirical rather than knowledge based.

This study indicated the cost of bioremediation of chronically hydrocarbon-contaminated soils by different approaches in a country with neither developed environmental remediation industry, nor economic policies that help to determine the treatments costs. The cost of the bioremediation of chronically hydrocarbon-contaminated soils was estimated to be between USD 50.7 and USD 310.4 per m^3 of soil. Among those, biostimulation with 10% of compost (BE1) was found to be the most cost-effective treatment. Except for BE1, all other bioremediation costs ranged on values that have been previously reported for biopile bioremediation of hydrocarbon-contaminated soils [59]. Our results showed that biostimulation-based treatments have lower costs than their counterparts using bioaugmentation. Specifically, the comparison between all bioaugmented treatments and BE1 revealed that the last treatment, though effective in hydrocarbon biodegradation [52], showed a significant reduction in direct material costs, provision and covering assets. This highlights the properties of compost, which is the mature product of composting, a bioprocess that transform solid organic substrates into relatively stable, organic-rich material via microbial communities [60]. Amendments with compost are considered a blend between the addition of nutrients and microorganisms with enzymatic composition and metabolic capabilities to biodegrade persistent compounds [50]. The first production of compost based on organic urban waste was dated as early as about 6000 years ago, when humans transitioned from being gatherers and hunters to breeders and farmers before establishing themselves in urban settlements [61]. Since then, compost has been widely applied in agricultural fields as a source of limiting nutrients for crops, such as nitrogen and phosphorus [62–65], amendments to reducing soil-borne crop diseases [66,67], and as a way of enhancing soil fertility by increasing natural nutrient cycling [68] in both conventional and organic agriculture [69]. The applications of compost have expanded to other fields, such as control of soil erosion [70], carbon sequestration [71], greenhouse gases biofiltration [72] and enhanced bioremediation [46,73]. Indeed, their application in bioremediation of organic contaminants has exponentially increased over the years [74,75]. However, in Chile, compost currently maintains a low price, since it does not compete with alternative uses besides turf/grass industries and organic-based agriculture. Therefore, the potential impact of higher demand for compost on the cost of biostimulation-based bioremediation remain to be examined in further detail. In contrast, the requirements of equipment and supplies for cultivation of high volumes of microbial biomass, as well as monitoring and quality controls converts bioaugmentation to a more expensive process. These results should be analyzed cautiously as a higher control level of remediation, including addressing the level of success of bioaugmentation and/or biostimulation in supplying hydrocarbon-degrading microorganisms, which may increase the efficiency of the process, but also may raise the bioremediation costs. Further analysis towards analyzing the impact of critical variables on the efficiency and efficacy of bioremediation are beyond the scope of the current work, however, they remain to be examined in future.

The economic evaluation of microbial bioremediation of chronically contaminated soils presented in this study will contribute to improvements in the understanding of how costs vary for each bioremediation approach, and therefore, they may be used as an input for risk–cost–benefit analysis. This is especially relevant in an industry that focuses its efforts on facing environmental remediation liabilities that occurred in the past instead of producing new products or rewarding shareholders [43]. As a whole, the absence of regulation and laws governing the maximum permissible concentrations of contaminants in soils converts the traditional "the polluter pays" to the predominant paradigm in Chile, preventing the advent of alternative paradigms that integrate ecosystem services, such as microbial bioremediation, as a valuable input to support ecological restoration. Furthermore, this framework will help to address similar challenges that other productive sectors have historically faced in Chile, where the controversial tradeoff between economic growth and environmental pollution is still rather frequent [76]. As with many other current sustainability challenges [77], it becomes relevant to establish a debate about regulations and incentive policies to encourage the implementation of local bioremediation industry capable to clean up a high number of polluted sites. Undoubtedly, the technical and scientific dimensions of the debate will have a positive impact on the recent Framework Law for Soils (Ley Marco de Suelos) that, after twenty years of discussion, has advanced to the discussion of several issues, including climate change, land management, land degradation and the prevention of soil contamination [41].

5. Conclusions

During the last few decades, the use of microbial bioremediation as a technology to restore polluted sites has been increasingly applied worldwide. Despite its several technical advances, and the growing need for improved technologies to effectively restore contaminated environments, bioremediation has been scarcely used in Chile. In the present study, the cost of five different bioremediation strategies based on biostimulation and/or bioaugmentation for removing hydrocarbons from chronically contaminated soils in an industrial projected scenario were estimated. The results identified compost and bacterial culture media as the items with the highest cost for biostimulation and bioaugmentation-based approaches, respectively. The comparison of the projected costs with an extensive database of 130 soil bioremediation projects indicate that the treatment costs fall within 60% of the more expensive projects, highlighting the high effort involved in bioremediation of chronically contaminated soils. An initial guide for the evaluation of costs of different bioremediation approaches, such as the one presented here, is relevant not only to compare these alternatives to projects based on excavation and disposal, which is rather easy to grasp, but to integrate this information through management practices to ensure the technical and economic feasibility of bioremediation projects, where the decision making is empirical rather than knowledge based. This framework will also facilitate a debate about regulations and incentive policies to encourage the implementation of local bioremediation industry capable of cleaning up a high number of polluted sites, as well as to improve the decision making of entrepreneurs and consultants, and may help to trigger government-generated environmental policies, regulations and standards for contaminated soils.

Supplementary Materials: The following supporting information can be downloaded at: https://www.mdpi.com/article/10.3390/su141911854/s1, Figure S1: Dimensions of projected soil biopiles for bioremediation; Table S1: Soil volume per hectare treated during each bioremediation cycle; Table S2: Total initial capital of the different bioremediation treatments; Table S3: Cost of 130 bioremediation projects worldwide; Table S4: Products and period costs of the different bioremediation approaches; Table S5: Total machinery depreciation of the different bioremediation treatments; Table S6: Recovery assets of the bioremediation treatments after initial investment recovery. Refs. [44,78–140] are cited in supplementary materials.

Author Contributions: Conceptualization, R.O., A.C. and M.S.; methodology, R.O., A.A., S.D., P.P.-G. and M.S.; formal analysis, A.C., S.D., C.R., C.S., P.C., G.A., F.C., F.S., M.G., P.S., J.A.-H. and M.M.; investigation, A.C., S.D., C.R., C.S., P.C., G.A., F.C., F.S., M.G., P.S., J.A.-H. and M.M.; resources, R.O. and M.S.; data curation, A.C., S.D. and A.A.; writing—original draft preparation, R.O., A.C., A.A., S.D. and M.S.; writing—review and editing, R.O., A.A., S.D. and M.S.; visualization, R.O., A.A. and M.S.; funding acquisition, R.O. and M.S. All authors have read and agreed to the published version of the manuscript.

Funding: This research was funded by Conicyt Programa de Inserción a la Academia "Fortalecimiento de la Investigación y la Docencia en las Áreas de Microbiología Ambiental y Bioinformática de la Universidad de Playa Ancha" grant PAI79170091 (R.O.), "Apoyo a la Formación de Redes Internacionales para Investigadores en Etapa Inicial" grant 170600 (R.O., M.S.), ANID PIA Ring GAMBIO Genomics and Applied Microbiology for Biodegradation and Bioproducts" grant ACT172128 Chile (MS), Fondecyt grants 11190863 (R.O.) and 1200756 (M.S.), and Inmobiliaria Las Salinas grant (M.S.), Beca de Doctorado ANID 21191625 and Programa de Incentivos a la Iniciación Científica, UTFSM (A.C.).

Institutional Review Board Statement: Not applicable.

Conflicts of Interest: The authors declare no conflict of interest.

References

1. Fuentes, S.; Méndez, V.; Aguila, P.; Seeger, M. Bioremediation of petroleum hydrocarbons: Catabolic genes, microbial communities, and applications. *Appl. Microbiol. Biotechnol.* **2014**, *98*, 4781–4794. [CrossRef] [PubMed]
2. Wartell, B.; Boufadel, M.; Rodriguez-Freire, L. An effort to understand and improve the anaerobic biodegradation of petroleum hydrocarbons: A literature review. *Int. Biodeterior. Biodegrad.* **2021**, *157*, 105156. [CrossRef]
3. Orellana, R.; Macaya, C.; Bravo, G.; Dorochesi, F.; Cumsille, A.; Valencia, R.; Rojas, C.; Seeger, M. Living at the Frontiers of Life: Extremophiles in Chile and Their Potential for Bioremediation. *Front. Microbiol.* **2018**, *9*, 2309. [CrossRef] [PubMed]
4. Fuentes, S.; Barra, B.; Caporaso, J.G.; Seeger, M. From Rare to Dominant: A Fine-Tuned Soil Bacterial Bloom during Petroleum Hydrocarbon Bioremediation. *Appl. Environ. Microbiol.* **2016**, *82*, 888–896. [CrossRef]
5. Yergeau, E.; Sanschagrin, S.; Beaumier, D.; Greer, C.W. Metagenomic Analysis of the Bioremediation of Diesel-Contaminated Canadian High Arctic Soils. *PLoS ONE* **2012**, *7*, e30058. [CrossRef]
6. Atlas, R.M.; Hazen, T.C. Oil Biodegradation and Bioremediation: A Tale of the Two Worst Spills in U.S. History. *Environ. Sci. Technol.* **2011**, *45*, 6709–6715. [CrossRef]
7. Hamamura, N.; Ward, D.M.; Inskeep, W.P. Effects of petroleum mixture types on soil bacterial population dynamics associated with the biodegradation of hydrocarbons in soil environments. *FEMS Microbiol. Ecol.* **2013**, *85*, 168–178. [CrossRef]
8. Zheng, J.; Feng, J.-Q.; Zhou, L.; Mbadinga, S.M.; Gu, J.-D.; Mu, B.-Z. Characterization of bacterial composition and diversity in a long-term petroleum contaminated soil and isolation of high-efficiency alkane-degrading strains using an improved medium. *World J. Microbiol. Biotechnol.* **2018**, *34*, 34. [CrossRef]
9. Apul, O.G.; Arrowsmith, S.; Hall, C.A.; Miranda, E.M.; Alam, F.; Dahlen, P.; Sra, K.; Kamath, R.; McMillen, S.J.; Sihota, N.; et al. Biodegradation of petroleum hydrocarbons in a weathered, unsaturated soil is inhibited by peroxide oxidants. *J. Hazard. Mater.* **2022**, *433*, 128770. [CrossRef]
10. Mohapatra, B.; Dhamale, T.; Saha, B.K.; Phale, P.S. Chapter 18—Microbial degradation of aromatic pollutants: Metabolic routes, pathway diversity, and strategies for bioremediation. In *Microbial Biodegradation and Bioremediation*, 2nd ed.; Das, S., Dash, H.R., Eds.; Elsevier: Amsterdam, The Netherlands, 2022; pp. 365–394.
11. Lovley, D.R. Cleaning up with genomics: Applying molecular biology to bioremediation. *Nat. Rev. Microbiol.* **2003**, *1*, 35–44. [CrossRef]
12. Guermouche M'rassi, A.; Bensalah, F.; Gury, J.; Duran, R. Isolation and characterization of different bacterial strains for bioremediation of n-alkanes and polycyclic aromatic hydrocarbons. *Environ. Sci. Pollut. Res.* **2015**, *22*, 15332–15346. [CrossRef] [PubMed]
13. Ehis-Eriakha, C.B.; Chikere, C.B.; Akaranta, O. Functional Gene Diversity of Selected Indigenous Hydrocarbon-Degrading Bacteria in Aged Crude Oil. *Int. J. Microbiol.* **2020**, *2020*, 2141209. [CrossRef] [PubMed]
14. Wu, B.; Deng, J.; Niu, H.; Liang, J.; Arslan, M.; Gamal El-Din, M.; Wang, Q.; Guo, S.; Chen, C. Establishing and Optimizing a Bacterial Consortia for Effective Biodegradation of Petroleum Contaminants: Advancing Classical Microbiology via Experimental and Mathematical Approach. *Water* **2021**, *13*, 3311. [CrossRef]
15. Mangimbulude, J.C.; Lembang, R.K. Biostimulation and Bioaugmentation: An Alternative Strategy for Bioremediation of Ground Water Contaminated Mixed Landfill Leachate and Sea Water in Low Income ASEAN Countries. In *Handbook of Environmental Materials Management*; Hussain, C.M., Ed.; Springer International Publishing: Cham, Switzerland, 2019; pp. 515–533.
16. Dominati, E.; Patterson, M.; Mackay, A. A framework for classifying and quantifying the natural capital and ecosystem services of soils. *Ecol. Econ.* **2010**, *69*, 1858–1868. [CrossRef]
17. Prince, R.C.; Drake, E.N. Transformation and Fate of Polycyclic Aromatic Hydrocarbons in Soil. In *Bioremediation of Contaminated Soils*; American Society of Agronomy, Inc.: Madison, WI, USA, 1999; pp. 89–110.
18. Abed, R.M.M. Interaction between cyanobacteria and aerobic heterotrophic bacteria in the degradation of hydrocarbons. *Int. Biodeterior. Biodegrad.* **2010**, *64*, 58–64. [CrossRef]
19. Almansoory, A.F.; Hasan, H.A.; Abdullah, S.R.S.; Idris, M.; Anuar, N.; Al-Adiwish, W.M. Biosurfactant produced by the hydrocarbon-degrading bacteria: Characterization, activity and applications in removing TPH from contaminated soil. *Environ. Technol. Innov.* **2019**, *14*, 100347. [CrossRef]
20. Prince, R.C.; Amande, T.J.; McGenity, T.J. Prokaryotic Hydrocarbon Degraders. In *Taxonomy, Genomics and Ecophysiology of Hydrocarbon-Degrading Microbes*; McGenity, T.J., Ed.; Springer International Publishing: Cham, Switzerland, 2019; pp. 1–39.

21. Hashmat, A.J.; Afzal, M.; Fatima, K.; Anwar-ul-Haq, M.; Khan, Q.M.; Arias, C.A.; Brix, H. Characterization of Hydrocarbon-Degrading Bacteria in Constructed Wetland Microcosms Used to Treat Crude Oil Polluted Water. *Bull. Environ. Contam. Toxicol.* **2019**, *102*, 358–364. [CrossRef]

22. Zvomuya, F.; Murata, A.P. Soil Contamination and Remediation. In *Encyclopedia of Environmetrics*; John Wiley & Sons: Hoboken, NJ, USA, 2001.

23. Van Liedekerke, M.; Prokop, G.; Rabl-Berger, S.; Kibblewhite, M.; Louwagie, G. *Progress in the Management of Contaminated Sites in Europe*; Publications Office of the European Union: Luxembourg, 2014.

24. The State Council of the People's Republic of China. *National Plan for Adjustment and Reconstruction of Old Industrial Bases (2013–2022)*; The State Council of the People's Republic of China: Beijing, China, 2013.

25. Koshlaf, E.; Ball, A.S. Soil bioremediation approaches for petroleum hydrocarbon polluted environments. *AIMS Microbiol.* **2017**, *3*, 25–49. [CrossRef]

26. Li, X.N.; Jiao, W.T.; Xiao, R.B.; Chen, W.P.; Chang, A.C. Soil pollution and site remediation policies in China: A review. *Environ. Rev.* **2015**, *23*, 263–274. [CrossRef]

27. Singh, A.; Kuhad, R.C.; Ward, O.P. Biological Remediation of Soil: An Overview of Global Market and Available Technologies. In *Advances in Applied Bioremediation*; Singh, A., Kuhad, R.C., Ward, O.P., Eds.; Springer: Berlin/Heidelberg, Germany, 2009; pp. 1–19.

28. BCC Bioremediation: Global Markets and Technologies to 2023. Available online: https://www.bccresearch.com/market-research/environment/bioremediation.html#:~{}:text=The%20global%20bioremediation%20market%20should,15.4%25%20from%202018%20through%202023 (accessed on 5 October 2021).

29. Ministerio del Medio Ambiente. *Quinto Reporte del Estado del Medio Ambiente*; Ministerio del Medio Ambiente: Puente Alto, Chile, 2019; p. 269.

30. Oyarzún, J.; Oyarzún, R. Sustainable development threats, inter-sector conflicts and environmental policy requirements in the arid, mining rich, northern Chile territory. *Sustain. Dev.* **2011**, *19*, 263–274. [CrossRef]

31. Tapia, J.S.; Valdés, J.; Orrego, R.; Tchernitchin, A.; Dorador, C.; Bolados, A.; Harrod, C. Geologic and anthropogenic sources of contamination in settled dust of a historic mining port city in northern Chile: Health risk implications. *PeerJ* **2018**, *6*, e4699. [CrossRef] [PubMed]

32. Lam, E.J.; Montofré, I.L.; Álvarez, F.A.; Gaete, N.F.; Poblete, D.A.; Rojas, R.J. Methodology to Prioritize Chilean Tailings Selection, According to Their Potential Risks. *Int. J. Environ. Res. Public Health* **2020**, *17*, 3948. [CrossRef] [PubMed]

33. Donoso, G.; Cancino, J.; Magri, A. Effects of agricultural activities on water pollution with nitrates and pesticides in the Central Valley of Chile. *Water Sci. Technol.* **1999**, *39*, 49–60. [CrossRef]

34. Melo, O.; Quiñones, N.B.; Acuña, D. Towards Sustainable Agriculture in Chile, Reflections on the Role of Public Policy. *Int. J. Agric. Nat. Resour.* **2021**, *48*, 186–209. [CrossRef]

35. Henriquez, M.; Becerra, J.; Barra, R.; Rojas, J. Hydrocarbons and organochlorine pesticides in soils of the Urban ecosystem of Chillán and Chillán Viejo, Chile. *J. Chil. Chem. Soc.* **2006**, *51*, 938–944. [CrossRef]

36. Barra, R.; Quiroz, R.; Saez, K.; Araneda, A.; Urrutia, R.; Popp, P. Sources of polycyclic aromatic hydrocarbons (PAHs) in sediment of the Biobio River in south central Chile. *Environ. Chem. Lett.* **2008**, *7*, 133–139. [CrossRef]

37. Deelaman, W.; Pongpiachan, S.; Tipmanee, D.; Choochuay, C.; Iadtem, N.; Suttinun, O.; Wang, Q.; Xing, L.; Li, G.; Han, Y.; et al. Source identification of polycyclic aromatic hydrocarbons in terrestrial soils in Chile. *J. S. Am. Earth Sci.* **2020**, *99*, 102514. [CrossRef]

38. Oyarzo-Miranda, C.; Latorre, N.; Meynard, A.; Rivas, J.; Bulboa, C.; Contreras-Porcia, L. Coastal pollution from the industrial park Quintero bay of central Chile: Effects on abundance, morphology, and development of the kelp Lessonia spicata (Phaeophyceae). *PLoS ONE* **2020**, *15*, e0240581. [CrossRef]

39. Fundacion Chile. *Guía Metodológica para la Gestión de Suelos con Potencial Presencia de Contaminantes*; Fundacion Chile: Santiago, Chile, 2015; p. 127.

40. Neaman, A.; Valenzuela, P.; Tapia-Gatica, J.; Selles, I.; Novoselov, A.A.; Dovletyarova, E.A.; Yáñez, C.; Krutyakov, Y.A.; Stuckey, J.W. Chilean regulations on metal-polluted soils: The need to advance from adapting foreign laws towards developing sovereign legislation. *Environ. Res.* **2020**, *185*, 109429. [CrossRef]

41. Salazar, O.; Casanova, M.; Fuentes, J.P.; Galleguillos, M.; Nájera, F.; Perez-Quezada, J.F.; Pfeiffer, M.; Renwick, L.L.R.; Seguel, O.; Tapia, Y. Soil research, management, and policy priorities in Chile. *Geoderma Reg.* **2022**, *29*, e00502. [CrossRef]

42. Bartke, S. Valuation of market uncertainties for contaminated land. *Int. J. Strateg. Prop. Manag.* **2011**, *15*, 356–378. [CrossRef]

43. Steffan, R.J. Developing Bioremediation Technologies for Commercial Application: An Insider's View. In *Consequences of Microbial Interactions with Hydrocarbons, Oils, and Lipids: Biodegradation and Bioremediation*; Steffan, R.J., Ed.; Springer International Publishing: Cham, Switzerland, 2019; pp. 21–32.

44. Brown, D.M.; Okoro, S.; van Gils, J.; van Spanning, R.; Bonte, M.; Hutchings, T.; Linden, O.; Egbuche, U.; Bruun, K.B.; Smith, J.W.N. Comparison of landfarming amendments to improve bioremediation of petroleum hydrocarbons in Niger Delta soils. *Sci Total Environ.* **2017**, *596–597*, 284–292. [CrossRef]

45. Brassington, K.J.; Pollard, S.J.T.; Coulon, F. Weathered Hydrocarbon Biotransformation: Implications for Bioremediation, Analysis, and Risk Assessment. In *Handbook of Hydrocarbon and Lipid Microbiology*; Timmis, K.N., Ed.; Springer: Berlin/Heidelberg, Germany, 2010; pp. 2487–2499.

46. Semple, K.T.; Reid, B.J.; Fermor, T.R. Impact of composting strategies on the treatment of soils contaminated with organic pollutants. *Environ. Pollut.* **2001**, *112*, 269–283. [CrossRef]

47. Trindade, P.V.O.; Sobral, L.G.; Rizzo, A.C.L.; Leite, S.G.F.; Soriano, A.U. Bioremediation of a weathered and a recently oil-contaminated soils from Brazil: A comparison study. *Chemosphere* **2005**, *58*, 515–522. [CrossRef] [PubMed]

48. Dias, R.L.; Ruberto, L.; Calabró, A.; Balbo, A.L.; Del Panno, M.T.; Mac Cormack, W.P. Hydrocarbon removal and bacterial community structure in on-site biostimulated biopile systems designed for bioremediation of diesel-contaminated Antarctic soil. *Polar Biol.* **2015**, *38*, 677–687. [CrossRef]

49. Gentry, T.; Rensing, C.; Pepper, I.A.N. New Approaches for Bioaugmentation as a Remediation Technology. *Crit. Rev. Environ. Sci. Technol.* **2004**, *34*, 447–494. [CrossRef]

50. Kästner, M.; Miltner, A. Application of compost for effective bioremediation of organic contaminants and pollutants in soil. *Appl. Microbiol. Biotechnol.* **2016**, *100*, 3433–3449. [CrossRef]

51. Antizar-Ladislao, B.; Lopez-Real, J.; Beck, A. Bioremediation of Polycyclic Aromatic Hydrocarbon (PAH)-Contaminated Waste Using Composting Approaches. *Crit. Rev. Environ. Sci. Technol.* **2004**, *34*, 249–289. [CrossRef]

52. Orellana, R.; Cumsille, A.; Rojas, C.; Cabrera, P.; Seeger, M.; Cárdenas, F.; Stuardo, C.; González, M. Assessing technical and economic feasibility of complete bioremediation for soils chronically polluted with petroleum hydrocarbons. *J. Bioremediat. Biodegrad.* **2017**, *8*, 396. [CrossRef]

53. Méndez, V.; Fuentes, S.; Morgante, V.; Hernández, M.; González, M.; Moore, E.; Seeger, M. Novel hydrocarbonoclastic metal-tolerant Acinetobacter and Pseudomonas strains from Aconcagua river oil-polluted soil. *J. Soil Sci. Plant Nutr.* **2017**, *17*, 1074–1087. [CrossRef]

54. Macaya, C.C.; Méndez, V.; Durán, R.E.; Aguila-Torres, P.; Salvà-Serra, F.; Jaén-Luchoro, D.; Moore, E.R.B.; Seeger, M. Complete Genome Sequence of Hydrocarbon-Degrading Halotolerant Acinetobacter radioresistens DD78, Isolated from the Aconcagua River Mouth in Central Chile. *Microbiol. Resour. Announc.* **2019**, *8*, e00601-19. [CrossRef] [PubMed]

55. Forsyth, J.V.; Tsao, Y.M.; Bleam, R.D. *Bioremediation: When Is Augmentation Needed?* Battelle Press: Columbus, OH, USA, 1995; pp. 1–14.

56. Peters, M.; Timmerhaus, K.; West, R. *Plant Design and Economics for Chemical Engineers*; McGraw-Hill: New York, NY, USA, 1991; pp. 150–215.

57. Azubuike, C.C.; Chikere, C.B.; Okpokwasili, G.C. Bioremediation techniques–classification based on site of application: Principles, advantages, limitations and prospects. *World J. Microbiol. Biotechnol.* **2016**, *32*, 180. [CrossRef] [PubMed]

58. Sales da Silva, I.G.; Gomes de Almeida, F.C.; Padilha da Rocha e Silva, N.M.; Casazza, A.A.; Converti, A.; Asfora Sarubbo, L. Soil Bioremediation: Overview of Technologies and Trends. *Energies* **2020**, *13*, 4664. [CrossRef]

59. Nagkirti, P.; Shaikh, A.; Vasudevan, G.; Paliwal, V.; Dhakephalkar, P. Bioremediation of Terrestrial Oil Spills: Feasibility Assessment. In *Optimization and Applicability of Bioprocesses*; Purohit, H.J., Kalia, V.C., Vaidya, A.N., Khardenavis, A.A., Eds.; Springer: Singapore, 2017; pp. 141–173.

60. Loick, N.; Hobbs, P.J.; Hale, M.D.C.; Jones, D.L. Bioremediation of Poly-Aromatic Hydrocarbon (PAH)-Contaminated Soil by Composting. *Crit. Rev. Environ. Sci. Technol.* **2009**, *39*, 271–332. [CrossRef]

61. Diaz, L.F.; de Bertoldi, M. History of composting. In *Waste Management Series*; Diaz, L.F., de Bertoldi, M., Bidlingmaier, W., Stentiford, E., Eds.; Elsevier: Amsterdam, The Netherlands, 2007; Volume 8, pp. 7–24.

62. Shrestha, P.; Small, G.E.; Kay, A. Quantifying nutrient recovery efficiency and loss from compost-based urban agriculture. *PLoS ONE* **2020**, *15*, e0230996. [CrossRef]

63. Erhart, E.; Hartl, W.; Putz, B. Biowaste compost affects yield, nitrogen supply during the vegetation period and crop quality of agricultural crops. *Eur. J. Agron.* **2005**, *23*, 305–314. [CrossRef]

64. Ahmad, R.; Naveed, M.; Aslam, M.; Zahir, Z.A.; Arshad, M.; Jilani, G. Economizing the use of nitrogen fertilizer in wheat production through enriched compost. *Renew. Agric. Food Syst.* **2008**, *23*, 243–249. [CrossRef]

65. Machado, R.M.A.; Alves-Pereira, I.; Faty, Y.; Perdigão, S.; Ferreira, R. Influence of Nitrogen Sources Applied by Fertigation to an Enriched Soil with Organic Compost on Growth, Mineral Nutrition, and Phytochemicals Content of Coriander (*Coriandrum sativum* L.) in Two Successive Harvests. *Plants* **2021**, *11*, 22. [CrossRef]

66. Mehta, C.M.; Palni, U.; Franke-Whittle, I.H.; Sharma, A.K. Compost: Its role, mechanism and impact on reducing soil-borne plant diseases. *Waste Manag.* **2014**, *34*, 607–622. [CrossRef]

67. Litterick, A.M.; Harrier, L.; Wallace, P.; Watson, C.A.; Wood, M. The Role of Uncomposted Materials, Composts, Manures, and Compost Extracts in Reducing Pest and Disease Incidence and Severity in Sustainable Temperate Agricultural and Horticultural Crop Production—A Review. *Crit. Rev. Plant Sci.* **2004**, *23*, 453–479. [CrossRef]

68. Sánchez-Monedero, M.A.; Cayuela, M.L.; Sánchez-García, M.; Vandecasteele, B.; D'Hose, T.; López, G.; Martínez-Gaitán, C.; Kuikman, P.J.; Sinicco, T.; Mondini, C. Agronomic Evaluation of Biochar, Compost and Biochar-Blended Compost across Different Cropping Systems: Perspective from the European Project FERTIPLUS. *Agronomy* **2019**, *9*, 225. [CrossRef]

69. Dsouza, A.; Price, G.W.; Dixon, M.; Graham, T. A Conceptual Framework for Incorporation of Composting in Closed-Loop Urban Controlled Environment Agriculture. *Sustainability* **2021**, *13*, 2471. [CrossRef]

70. Adugna, G. A review on impact of compost on soil properties, water use and crop productivity. *Agric. Sci. Res. J.* **2018**, *4*, 93–104.

71. Hill, M.J.; Braaten, R.; McKeon, G.M. A scenario calculator for effects of grazing land management on carbon stocks in Australian rangelands. *Environ. Model. Softw.* **2003**, *18*, 627–644. [CrossRef]

72. Nikiema, J.; Brzezinski, R.; Heitz, M. Elimination of methane generated from landfills by biofiltration: A review. *Rev. Environ. Sci. Bio/Technol.* **2007**, *6*, 261–284. [CrossRef]

73. Lu, Y.; Zheng, G.; Zhou, W.; Wang, J.; Zhou, L. Bioleaching conditioning increased the bioavailability of polycyclic aromatic hydrocarbons to promote their removal during co-composting of industrial and municipal sewage sludges. *Sci. Total Environ.* **2019**, *665*, 1073–1082. [CrossRef] [PubMed]

74. Lin, C.; Cheruiyot, N.K.; Bui, X.-T.; Ngo, H.H. Composting and its application in bioremediation of organic contaminants. *Bioengineered* **2022**, *13*, 1073–1089. [CrossRef]

75. Cai, Q.-Y.; Mo, C.-H.; Wu, Q.-T.; Zeng, Q.-Y.; Katsoyiannis, A.; Férard, J.-F. Bioremediation of polycyclic aromatic hydrocarbons (PAHs)-contaminated sewage sludge by different composting processes. *J. Hazard. Mater.* **2007**, *142*, 535–542. [CrossRef]

76. Reyes-Bozo, L.; Godoy-Faundez, A.; Herrera-Urbina, R.; Higueras, P.; Salazar Navarrete, J.L.; Valdés-González, H.; Vyhmeister, E.; Antizar-Ladislao, B. Greening Chilean copper mining operations through industrial ecology strategies. *J. Clean. Prod.* **2014**, *84*, 671. [CrossRef]

77. D'Adamo, I.; Gastaldi, M.; Morone, P.; Rosa, P.; Sassanelli, C.; Settembre-Blundo, D.; Shen, Y. Bioeconomy of Sustainability: Drivers, Opportunities and Policy Implications. *Sustainability* **2022**, *14*, 200. [CrossRef]

78. EPA (United States Environmental Protection Agency). *Remediation Technology Cost Compendium [Electronic Resource]: Year 2000*; Office of Solid Waste and Emergency Response, Technology Innovation Office: Washington, DC, USA, 2001.

79. EPA (United States Environmental Protection Agency). Soil Vapor Extraction at the Hastings Groundwater Contamination Superfund Site, Well Number 3 Subsite, Hastings, Nebraska. Cost and Performance Report. 1995. Available online: https://frtr.gov/costperformance/profile.cfm?ID=104&CaseID=104 (accessed on 17 July 2022).

80. Leeson, A.P.; Graves, M.; Kramer, J. *Site-Specific Technical Report for Bioslurper Testing at Site ST-04, K.I. Sawyer AFB*; Battelle: Columbus, OH, USA, 1996.

81. EPA (United States Environmental Protection Agency). *Bioremediation Field Evaluation of Hill Air Force Base, Utah*; EPA: Washington, DC, USA, 1997.

82. EPA (United States Environmental Protection Agency). *Cost and Performance Summary Report Soil Vapor Extraction at the Intersil/Siemens Superfund Site Cupertino, California*; EPA: Washington, DC, USA, 1998.

83. EPA (United States Environmental Protection Agency). *Remediation Case Studies: In Situ Soil Treatment Technologies, (Soil Vapor Extraction, Thermal Processes) Volume 8*; EPA: Cincinatti, OH, USA, 1998.

84. EPA (United States Environmental Protection Agency). *Installation Restoration Program. LF-036 Groundwater Surface Water and Sediments*; Record of Decision Plattsburgh Air Force Base; EPA: Washington, DC, USA, 1995.

85. Baxter, L.; Dossey, R.; Eastty, B.; Lamb, R.E.; Lush, A.; McCain, S.; Myers, C.; Nowick, M.; Smith, S.; Stetson, J. *Final Environmental Assessment Addressing Construction of a Fitness Center at Beale Air Force Base, California*; HDR Environmental, Operations and Construction Inc.: Englewood, CO, USA, 2009.

86. Engineering Science, Inc. Part 1: Bioventing Pilot Test Work Plan for Installation Restoration Program Site 3, Fire Training Area, Battle Creek ANGB, Michigan. PART II: Draft Interim Pilot Test Results Report for Installation Restoration Program Site 3, Fire Training Area, Battle Creek ANGB, Michigan. 1992. Available online: https://apps.dtic.mil/sti/pdfs/ADA385759.pdf (accessed on 10 July 2022).

87. EPA (United States Environmental Protection Agency). *Soil Vapor Extraction at the Rocky Mountain Arsenal Superfund Site, Motor Pool Area (OU 18), Commerce City, Colorado*; EPA: Washington, DC, USA, 1995.

88. Wan, X.; Lei, M.; Chen, T. Cost–benefit calculation of phytoremediation technology for heavy-metal-contaminated soil. *Sci. Total Environ.* **2016**, *563–564*, 796–802. [CrossRef] [PubMed]

89. Line, M.A.; Garland, C.D.; Crowley, M. Evaluation of landfarm remediation of hydrocarbon-contaminated soil at the inveresk railyard, Launceston, Australia. *Waste Manag.* **1996**, *16*, 567–570. [CrossRef]

90. Parsons Engineering Science Inc. *Natural Attenuation of Chlorinated Solvents Performance and Cosr Results from Multiple Air Force Demonstration Sites*; Parsons Engineering Science Inc.: Denver, CO, USA, 1999.

91. Hinchee, R.; Downey, D.; Slaughter, J.; Selby, D.; Westray, M.; Long, G. Enhanced Bioreclamation of Jet Fuels: A Full-Scale Test at Eglin AFB, Florida. 1989, p. 164. Available online: https://www.researchgate.net/publication/235176012_Enhanced_Bioreclamation_of_Jet_Fuels_A_Full-Scale_Test_at_Eglin_AFB_Florida (accessed on 17 July 2022).

92. EPA (United States Environmental Protection Agency). *Use of Bioremediation at Superfund Sites*; EPA: Washington, DC, USA, 2001.

93. Wendell, S.J.; Johnson, K.; Sawyer, M.; Kelly, L.; Hellauer, K.; Schneider, R.L.; Gomez, C.; Perry, A.; Bates, S. *Final Cannon AFB Housing Privatization Environmental Assessment*; Geo-Marine Inc.: Plano, TX, USA, 2009.

94. Parsons Engineering Science Inc. *Intrinsic Remediation Engineering Evaluation/Cost Analysis for the Former Car Care Center, Bolling Air Force Base, Washington, District of Columbia*; Parsons Engineering Science Inc.: Denver, CO, USA, 1997.

95. EPA (United States Environmental Protection Agency). *Application, Performance, and Cost of Biotreatment Technologies for Contaminated Soils*; EPA: Washington, DC, USA, 2002.

96. UNITED STATES AIR FORCE. Joint Base Elmendorf-Richardson Community Involvenment Plan. 2011. Available online: https://www.jber.jb.mil/Portals/144/Services-Resources/environmental/restoration/Enviornmental-JBER-Community-Involvement-Plan-(2011).pdf (accessed on 10 July 2022).

97. EPA (United States Environmental Protection Agency). *In Situ Bioremediation at Vandenberg Air Force Base, Lompoc, California*; EPA: Washington, DC, USA, 2000.

98. EPA (United States Environmental Protection Agency). *Cleanup Activities Hanscom Field/Hanscom Air Force Base Bedford, MA*; EPA: Washington, DC, USA, 2009.
99. Department of the Air Force. Final Decision Document for no Further Action at B-58 Hustler Burial Site (Area of Concern 8). 2002. Available online: https://semspub.epa.gov/work/05/281571.pdf (accessed on 17 July 2022).
100. EPA (United States Environmental Protection Agency). *Remedial Action Plan for the Risk-Based. Air Force Center for Environmental Excellence Technology Transfer Division Brooks Air Force Base San Antonio, Texas*; EPA: Washington, DC, USA, 1996.
101. ESTCP. *Electronically induced Redox Barriers for Treatment of Groundwater at F.E. Warren Air Force Base, Wyoming*; ESTCP: Arlington, VA, USA, 2006.
102. State of Hawaii, Department of Health. Hazard Evaluation and Emergency Response Office Fiscal Year 1998 Activities (7/1/97-6/30/98). 1999. Available online: https://health.hawaii.gov/heer/ (accessed on 17 July 2022).
103. Parsons Engineering Science, Inc. *Confirmation Sampling and Analysis Plan for Installation Restoration Program Site ST35 Ordnance Testing Laboratory Oil Leak, Air Force Plant PJKS, Colorado*; Parsons Engineering Science Inc.: Denver, CO, USA, 1996.
104. Constantino, J.; Dominador, D. *Analysis of Camp Pendleton California Medical Treatment Facility Budget and Execution Process*; Naval Postgraduate School: Monterey, CA, USA, 2008.
105. Air Force Occupational and Environmental Health Laboratory. Hazardous Waste Staff Assistance Survey, Patrick AFB and Cape Canaveral AFS, Florida. 1997. Available online: https://www.osti.gov/biblio/6425972 (accessed on 17 July 2022).
106. Cost and Performance Report: Solar-Powered Remediation and pH Control. 2017. Available online: https://apps.dtic.mil/sti/citations/AD1036540 (accessed on 17 July 2022).
107. Parsons Engineering Science, Inc. *Bioventing Performance and Cost Results from Multiple Air Force Test Sites*; Parsons Engineering Science Inc.: Denver, CO, USA, 1996.
108. Density-Driven Groundwater Sparging at Amcor Precast Ogden, Utah. 1994. Available online: https://frtr.gov/costperformance/pdf/AmcorPrecastGWSparging.pdf (accessed on 17 July 2022).
109. Board, T.R.; National Academies of Sciences, Engineering, and Medicine. *Use and Potential Impacts of AFFF Containing PFASs at Airports*; The National Academies Press: Washington, DC, USA, 2017; p. 222.
110. Division of Environmental Remediation New York State Department of Environmental Conservation. Proposed Remedial Action Plan Fort Drum-Waste Disposal Areas. 2016. Available online: https://www.dec.ny.gov/data/DecDocs/623008/PRAP.HW.6230 08.2016-02-24.fianl%20PRAP%20PCE%20Plume.pdf (accessed on 17 July 2022).
111. Day, S.J.; Morse, G.K.; Lester, J.N. The cost effectiveness of contaminated land remediation stategies. *Sci. Total Environ.* **1997**, *201*, 125–136. [CrossRef]
112. EPA (United States Environmental Protection Agency). *Superfund Explanation of Significant Differences for the Record of Decision: Fairchild Air Force Base, Craig Road Landfill, WA 12/5/1994*; EPA: Washington, DC, USA, 1995.
113. EPA (United States Environmental Protection Agency). *Aerobic Degradation at Site 19, Edwards Air Force Base, California*; EPA: Washington, DC, USA, 2000.
114. EPA (United States Environmental Protection Agency). *Remediation Case Studies: Ex Situ Soil Treatment Technologies (Bioremediation, Solvent Extraction, Thermal Desorption) Volumen 7*; EPA: Cincinatti, OH, USA, 1998.
115. Parsons Engineering Science Inc. *Natural Attenuation of Fuel Hydrocarbons Performance and Cost Results from Multiple Air Force Demonstration Sites*; Parsons Engineering Science Inc.: Denver, CO, USA, 1999.
116. Carlton, G.N.; Smith, L.B. Exposures to Jet Fuel and Benzene During Aircraft Fuel Tank Repair in the U.S. Air Force. *Appl. Occup. Environ. Hyg.* **2000**, *15*, 485–491. [CrossRef] [PubMed]
117. EPA (United States Environmental Protection Agency). *Soil Vapor Extraction at the Verona Well Field Superfund Site, Thomas Solvent Raymond Road (OU-1), Battle Creek, Michigan: Cost and Performance Report*; EPA: Washington, DC, USA, 1995.
118. EPA (United States Environmental Protection Agency). *Cost and Performance Report Sand Creek Industrial Superfund Site, O.U. 1*; EPA: Washington, DC, USA, 1989.
119. Parsons Engineering Science, Inc. *Corrective Action Plan for Expanded Bioventing System Site SS-41, Former Building 93 (Fuel Pumping Station Number 3), Charleston Air Force Base, South Carolina*; Parsons Engineering Science Inc.: Denver, CO, USA, 1997.
120. Low Temperature Thermal Desorption at Longhorn Army Ammunition Plant. Burning Ground No. 3 Karnack, Texas; Cost and Performance Report; Karnack, Texas. 1999. Available online: https://frtr.gov/costperformance/profile.cfm?ID=138&CaseID=138 (accessed on 17 July 2022).
121. EPA (United States Environmental Protection Agency). *Vacuum-Enhanced, Low-Temperature Thermal Desorption at the FCX Washington Superfund Site Washington, North Carolina. Cost and Performance Report*; EPA: Washington, DC, USA, 2009.
122. EPA (United States Environmental Protection Agency). *Cost and Performance Report: Thermal Desorption at the T H Agriculture & Nutrition Company Superfund Site Albany, Georgia*; EPA: Washington, DC, USA, 1995.
123. EPA (United States Environmental Protection Agency). *Soil Vapor Extraction at the Fairchild Semiconductor Corporation Superfund Site San Jose, California. Cost and Performance Report*; EPA/Office of Solid Waste and Emergency Response, Technology Innovation Office: Washington, DC, USA, 1995.
124. EPA (United States Environmental Protection Agency). *Report of the Remediation System Evaluation, Site Visit Conducted at the Commencement Bay/South Tacoma Channel Well 12A Superfund Site*; EPA: Washington, DC, USA, 2001.
125. Parsons Engineering Science, Inc. *Intrinsic Remediation Treatibility Study for Site St-29 Patrick Air Force Base Florida*; Parsons Engineering Science Inc.: Denver, CO, USA, 1995.

126. EPA (United States Environmental Protection Agency). *Thermal Desorption at the Sand Creek Industrial Superfund Site, OU 5 Commerce City, Colorado*; EPA: Washington, DC, USA, 1999.

127. EPA (United States Environmental Protection Agency). *Thermal Desorption at the Metaltec Superfund Site, Franklin Borough, New Jersey*; EPA: Washington, DC, USA, 2000.

128. EPA (United States Environmental Protection Agency). *Thermal Desorption at the Sarney Farm Superfund Site, Amenia, New York*; EPA: Washington, DC, USA, 2001.

129. EPA (United States Environmental Protection Agency). *Superfund Record of Decision Garden State Cleaners, NJ. First Remedial Action*; EPA: Washington, DC, USA, 1991.

130. EPA (United States Environmental Protection Agency). Cost and Performance Report: Thermal Desorption at the McKin Company Superfund Site Gray, Maine. Available online: https://clu-in.org/products/costperf/THRMDESP/Mckin.htm (accessed on 17 July 2022).

131. Acharya, P.; Ives, P. Incineration at Bayou Bounfouca remediation project. *Waste Manag.* **1994**, *14*, 13–26. [CrossRef]

132. EPA (United States Environmental Protection Agency). *Remediation System Evaluation SMS Instruments Deer Park, New York*; EPA: Washington, DC, USA, 2003.

133. Parsons Engineering Science, Inc. *Site-Specific Technical Report for the Evaluation of Thermatrix GS Series Flameless Thermal Oxidizer for Off-Gas Treatment of Trichloroethene Vapors at Air Force Plant 4, Fort Worth, Texas*; Parsons Engineering Science Inc.: Denver, CO, USA, 1996.

134. EPA (United States Environmental Protection Agency). Cost and Performance Report: Thermal Desorption at the Outboard Marine Corporation Superfund Site Waukegan, Illinois. Available online: https://clu-in.org/products/costperf/THRMDESP/Omc.htm (accessed on 17 July 2022).

135. EPA (United States Environmental Protection Agency). *Final Record of Decision Operable Unit 1 Former Nebraska Ordnance Plant Site Mead, Nebraska*; EPA: Washington, DC, USA, 1995.

136. EEPA (United States Environmental Protection Agency). *On-Site Incineration at the Celanese Corporation Shelby Fiber Operations Superfund Site Shelby, North Carolina*; EPA: Washington, DC, USA, 1997.

137. EPA (United States Environmental Protection Agency). Cost and Performance Report: Thermal Desorption/Dehalogenation at the Wide Beach Development Superfund Site Brant, New York. Available online: https://clu-in.org/products/costperf/THRMDESP/Widebch.htm (accessed on 17 July 2022).

138. EPA (United States Environmental Protection Agency). Thermal Desorption at Port Moller Radio Relay Station Port Moller, Alaska. Cost and Performance Report. Available online: https://frtr.gov/costperformance/pdf/Port%20Moller.pdf (accessed on 17 July 2022).

139. EPA (United States Environmental Protection Agency). *Cost and Performance Summary Report Thermal Desorption at the Waldick Aerospace Devices Site Wall Township, New Jersey*; EPA: Washington, DC, USA, 1998.

140. EPA (United States Environmental Protection Agency). *Incineration at the MOTCO Superfund Site Texas City, Texas*; EPA: Washington, DC, USA, 1997.

 *sustainability*

Article

Sodium Hydroxide Hydrothermal Extraction of Lignin from Rice Straw Residue and Fermentation to Biomethane

Tawaf Ali Shah [1,*], Sabiha Khalid [2], Hiba-Allah Nafidi [3], Ahmad Mohammad Salamatullah [4] and Mohammed Bourhia [5]

1   National Institute for Biotechnology and Genetic Engineering (NIBGE), Jhang Road, Faisalabad 44000, Punjab, Pakistan
2   Department of Human Genetics and Molecular Biology, University of Health Sciences, Lahore 54600, Punjab, Pakistan
3   Department of Food Science, Faculty of Agricultural and Food Sciences, Laval University, Quebec City, QC G1V 0A6, Canada
4   Department of Food Science & Nutrition, College of Food and Agricultural Sciences, King Saud University, 11 P.O. Box 2460, Riyadh 11451, Saudi Arabia
5   Laboratory of Chemistry and Biochemistry, Faculty of Medicine and Pharmacy, Ibn Zohr University, Laayoune 70000, Morocco; mouhabourhi@gmail.com
*   Correspondence: tawafbiotech@yahoo.com

Citation: Shah, T.A.; Khalid, S.; Nafidi, H.-A.; Salamatullah, A.M.; Bourhia, M. Sodium Hydroxide Hydrothermal Extraction of Lignin from Rice Straw Residue and Fermentation to Biomethane. *Sustainability* **2023**, *15*, 8755. https://doi.org/10.3390/su15118755

Academic Editors: Idiano D'Adamo and Massimo Gastaldi

Received: 12 April 2023
Revised: 20 May 2023
Accepted: 23 May 2023
Published: 29 May 2023

Abstract: The purpose of the NaOH pretreatment of rice straw with a recycling strategy was to enhance the economic efficiency of producing biomethane. Anaerobic digestion is used for converting rice straw into biogas. In this work, 5% NaOH and rice straw mixed samples were autoclaved at 121 °C for 20 min for lignin removal. The NaOH black liquor was separated using filtration for the subsequent treatment cycle. The NaOH liquor was utilized in one more subsequent recycling procedure to test its ability to remove lignin from the rice straw. The 5% NaOH treatment results in a reduction in rice straw (RC) lignin of 73.6%. The lignin content of the recycled NaOH-filtrated rice straw samples (RCF1) was reduced by 55.5%. The 5% NaOH-treated rice straw sample yields a total cumulative biogas of 1452.4 mL/gVS, whereas the recycled NaOH-filtered (RCF1) samples generate 1125.2 mL/gVS after 30 days of incubation. However, after 30 days of incubation, the untreated rice straw (RCC) bottle produced a total of 285.5 mL/gVS of biogas. The total increase in methane output after NaOH treatment is 6–8 times greater, and the biogas yield improves by 80–124%. We show here that the recycled NaOH black solution has still the effectiveness to be used for successive pretreatment cycles to remove lignin and generate methane. In the meantime, the NaOH black solution contains useful materials (lignin, sugars, potassium, and nitrogen) that could be purified for commercial purposes, and more importantly recycling the NaOH solution decrease the chances of environmental pollution. Thus, recycling NaOH decreased chemical consumption, which would provide net benefits instead of using fresh NaOH solution, had a lower water consumption, and provided the prospect of producing an optimum yield of methane in anaerobic digestion. This method will decrease the chemical treatment costs for biomass pretreatment prior to anaerobic digestion. Recycling of NaOH solution and the integration of pretreatment reactors could be a novel bioprocessing addition to the current technology.

Keywords: lignin; straw; heating; NaOH; pretreatment; biogas

1. Introduction

One of the fundamental requirements for a sustainable way of life is energy. This need is primarily met by the energy generated from fossil fuel sources around the world. However, as these fossil fuel resources are depleted, there may be major energy shortages in the future [1]. The need for affordable, long-lasting energy is of the utmost importance. It must be both sustainable and cost-effective to meet the energy needs of human activity [2].

The primary problem is that the majority of the energy is derived from fossil fuels, which are expensive and generate greenhouse gases. Anaerobic digestion is a sustainable process for producing bioenergy from lignocellulosic waste biomass [3,4].

Rice straw production has increased as a result of rising energy consumption demands. The rice plant contains forty to sixty percent of its weight in rice straw, an agricultural waste. Rice straw contains 38–43% cellulose, 2–26% hemicellulose, and 15–20% lignin, depending on the variety and region [5,6]. Despite being one of the many crops that are commonly grown throughout the world, burning rice straw in open fields has a detrimental effect on the environment. The anaerobic digestion of rice straw to produce biogas is one possible method to use rice straw and minimize air pollution [7]. From anaerobic digestion, two useful products, biogas and digestate, are generated. Biogas can be used for lighting, electricity, and cooking. It can also be transformed into biomethane and used as a fuel for cars or injected into the pipeline to be transported as natural gas [8]. The composition of rice straw contains three major components: lignin, cellulose, and hemicellulose. Thus, rice straw's walls are tightly packed by layers of lignin, cellulose, and hemicellulose, and remain protected from enzymatic hydrolysis. In order to dissolve lignin and make cellulose and hemicellulose accessible to enzymatic action [9], a pretreatment step is therefore required before biogas production through anaerobic digestion [10]. Pretreatment aims to enhance surface area, promote lignocellulose structural opening, decrease lignin quantity, maintain hemicellulose, and decrease cellulose crystallinity. Additionally, biomass is processed to promote porosity in order to enhance enzyme access and assist the process of anaerobic digestion. When compared to acid or oxidative reagents, alkali pretreatment is the most efficient approach for rupturing the ester linkages between lignin, hemicellulose, and cellulose and limiting breakage of the hemicellulose polymers. According to previous studies, pretreatment with NaOH effectively increases the accessibility of enzymes to the cellulose while removing lignin and a small fraction of the hemicelluloses [11]. The majority of alkali pretreatment processes used in rice straw biogas production require heating and a substantial use of chemicals. Such pretreatment methods use a lot of water and produce harmful substances. Furthermore, treating alkali at high temperatures results in greater initial costs, increased treatment expenses, and most likely adverse environmental effects [12]. Biogas helps the economy by reducing poverty, offering affordable biofertilizer to increase soil fertility and reduce hunger, supporting new businesses, turning waste into usable products, and supporting global sustainability. Rice straw waste may be anaerobically digested to produce biogas, which small business owners can use to provide gas for cooking and farmers can use for electricity as well. Thus, the anaerobic digestion of waste biomass, such as rice straw, is a viable, sustainable, and profitable method for business owners to establish a biogas plant in an industrial area and supply energy to local populations [13]. It is estimated that crop waste production worldwide exceeds 9 billion tons. The huge quantity of crop waste can be efficiently managed with the advancement of anaerobic digestion technology. The production of biogas and its byproducts from crop waste can also help to solve challenges with food security, pollution in the environment, the cost of production, providing clean energy, and stability and sustainability in the economy [14].

To effectively pretreat crops and remove lignin, particularly rice straw, NaOH has shown a high rate of lignin removal efficiency compared to other alkalis, especially KOH and $Ca(OH)_2$ [15,16]. The treatment of rice straw with NaOH removes lignin and makes the cellulose component vulnerable to microbial hydrolysis. However, using a high concentration of alkalis not only increases the cost of the pretreatment process but also produces chemical inhibitors along with a high amount of volatile fatty acids (VFA), which stop the anaerobic digestion process [17]. Therefore, picking a suitable pretreatment method is a critical step for enhancing anaerobic digestion, but could also be economical for industrial applications [18].

NaOH black liquor can be reused for the pretreatment of biomass, rather than being wasted after initial treatment. Previously, NaOH with a high thermal heating process was applied and did not recycle this black liquor for further uses [15]. In such practices, the yield was improved, but the cost and environmental hazards were ignored. A few researchers filtered and recycled NaOH black liquor using extensive washing steps [19]. Because of such approaches, this important reagent is also unreasonably expensive for large-scale applications. The goal of this work was to reuse the black NaOH liquid repeatedly in order to identify a less expensive pretreatment procedure. In addition, the study focused on using NaOH in dilute concentrations with gentle heating conditions to remove lignin from rice straw as agribiomass. It was further noted that the effect of lignin removal on biogas yield and the anaerobic digestion process should be investigated.

2. Methodology

2.1. Materials and Methods

All essential chemicals, including NaOH, were bought from the Merck Group of Chemical Companies and Fisher Scientific. After being processed to a 40 mm mesh size, the rice straw was stored at room temperature in a polythene bag. The total solids (TS), volatile solids (VS), ash, moisture, lignin, hemicellulose, and cellulose were calculated using the normal NREL laboratory analytical process [20]. Using the oven-drying technique, the samples were dried in an air-drying oven for 24 h at 105 °C. This calculation Equation (1) was used to determine the moisture:

$$\%\text{moisture} = \frac{M_1 - M_2}{M} \times 100 \tag{1}$$

where M = initial sample weight/gm, M_1 = sample weight/gm. + container before drying, and M_2 = sample weight/gm + container after drying.

The moisture-free biomass was heated to 500 °C in a muffle furnace for 3 h, cooled to ambient temperature in a desiccator, and weighed to assess the samples' ash content. The ash was estimated as shown below in Equation (2).

$$\%\text{ash} = \frac{S_2 - S}{S_1} \times 100 \tag{2}$$

where S = burn dish weight, S_1 = sample without moisture, and S_2 = weight of sample plus dish after furnace ignition.

A clean crucible was heated for 24 h to 105–110 °C, and the substrate was weighted and dried in the oven set at 70 °C. The TS is calculated as shown in Equation (3):

$$\text{TS\%} = \frac{T - S}{S} \times 100 \tag{3}$$

where T = dried sample + dish and S = dish weight.

The volatile solids (VSs) were measured by heating the substrate to 450 °C for 1 h in a desiccator. The dry ash was collected from the muffle furnace, and VSs were calculated after cooling the sample as shown in Equation (4).

$$\text{VS\%} = \frac{V - D}{S} \times 100 \tag{4}$$

where V = substrate weight + dish and D = substrate weight + dish after ignition.

A 300 g dry sample was soaked in 3 mL of 72% sulfuric acid in a pressure tube for lignin measurement, and the sample was then incubated in a water bath at 30 °C for 60 min. A tiny glass rod was used to agitate the samples continuously. An amount of 84 mL of pure water was added to dilute the reaction to 4%. The reaction was stopped after full breakdown and cooled to ambient temperature. The soluble lignin was measured from the UV-visible spectrophotometer's absorbance value, and the insoluble lignin was measured from the weight of the ash after a rice straw sample that was oven-dried overnight was burned at 575 °C. The calculations were performed according to the NREL laboratory's analytical process [20]. The National Renewable Energy Laboratory's (NREL) established analytical method was used for the analysis of rice straw compositions of cellulose and hemicellulose sugars (hexose and pentose sugars) by using a high-pressure liquid chromatography (HPLC) system for rice straw samples.

2.2. Rice Straw Pretreatment

The experiment (Figure 1) was started using 5% NaOH for lignin degradation from rice straw (RC sample). All the rice straw samples (100 g/L) were immersed in 5% NaOH in a 100 mL solution in a 250 mL flask. The pH of NaOH black liquor was measured and it was 12.5 at the start of the reaction process. All the samples along with the control (rice straw dissolved in a 100 mL solution without NaOH) were then autoclaved at 121 °C for 20 min [21]. Under the same conditions, the control samples (RCC) were treated with distilled water. After heating, the black NaOH liquid was filtered and collected in a sterile flask. The solid rice straw residue was separated by filtration and rinsed with 100 mL of water. Before using it in an anaerobic digestion experiment, the washed solid residue was allowed to dry at room temperature. The 100 mL of water used in residue washing was added to the black liquor of NaOH. It was assumed that adding 100 mL of water to filtered black liquor of NaOH would reduce the starting concentration of 5% NaOH/100 mL by up to 2.5%. Again, 100 mL of filtered NaOH black liquid was added to a 250 mL flask containing 100 g/L of rice straw (RCF1) and autoclaved at 121 °C for 20 min. Under the same conditions, the control samples were treated with distilled water. The NaOH was collected in a sterile flask after being filtered, and the solid residue of the rice straw was separated and rinsed with 100 mL of water. With a spatula, the solid residue was separated and dried at room temperature.

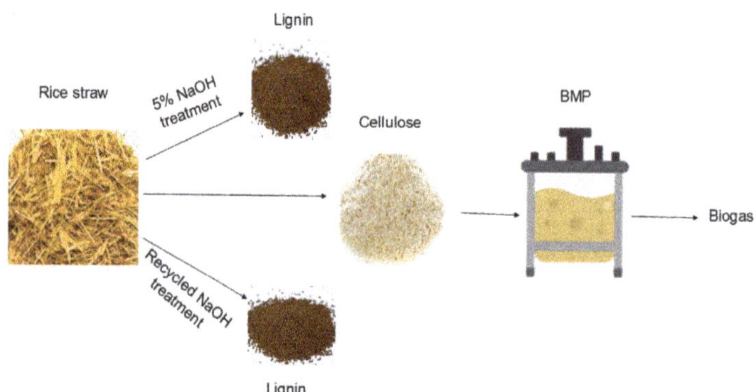

Figure 1. Experimental plan for 5% NaOH and recycled NaOH for lignin removal.

2.3. Scanning Electron Microscopy (SEM)

Scanning election microscope (SEM) machine was used to examine the pretreatment distortion and destructions on the structural surface of rice straw samples after and before treatment. The surface morphology of the untreated and pre-treated samples from both experimental groups were studied using a vacuum-desiccated SEM (S-3700 microscope Hitachi, manufacturer, Albany, CA, USA) with magnification ranges of 1000× to visualize the cracked and degraded tough structure of biomass samples as described previously [22].

2.4. Anaerobic Digestion Experiment

Anaerobic digestion experiments (ADE) were conducted on rice straw samples RC, RCF1, and RCC as a control sample. In each 250 mL serum bottle, the reactor volume was 100 mL. K_2HPO_4 (2.5 g/L), 3.5 g/L sodium bicarbonate solution, and 5 mL stock solutions of vitamin solutions thymine (0.01 g/L), folic acid (0.02 g/L), vitamin B12 (0.001 g/L), panthonic acid (0.05 g/L), riboflavin (0.05 g/L), and lipoic acid (0.05 g/L) were used as fermentation media. The fermentation media were autoclaved at 121 °C for 20 min. An inoculum of the full-scale anaerobic digester was enriched successively on fermentation media supplemented with 5 g/L alkali lignin, 5 g/L glucose, 5 g/L cellulose, and 5 g/L methanol under N_2-flushed conditions. This culture was used as an inoculum for biogas experiment. Based on the volatile solid (VS) content as food to microorganism ratio (F/M ratio), a 0.28 (S/I ratio) was determined for the ADE experiment. After flushing the serum bottles with N_2 gas for 6 min, the bottles were incubated at 40 °C and pH 8. There were control samples of inoculum without rice straw and rice straw without inoculum. Both untreated and pretreated rice straw sample fermentation tests were started in parallel to compare biogas and methane yield. The daily biogas production was assessed using the water displacement method, and serum vials were physically shaken on a daily basis. The GFM406 multichannel portable gas analyzer was used to determine the CH_4 and CO_2 composition of the biogas. The biogas data were examined using OBA, (https://biotransformers.shinyapps.io/oba1/, accessed on 20 March 2019). This biogas software package (R package) estimated the total biogas, methane, daily methane, daily biogas, rate of biogas, and methane. The R software derives all of these characteristics from daily biogas data for each bottle based on biogas volume and percent CH_4 [23]. The software makes use of the R function to estimate the theoretical and expected yield given a sample of raw data. Three groups make up the web-based interface known as the OBA Online Biogas App: (1) fundamental vectorized functions for typical conversions and calculations, (2) two data processing functions that compute BMP or comparable results from laboratory measurements, and (3) a function for predicting CH_4 production: basic biogas package operations. The oxygen demand, composition, molar mass, biogas volume, gas volume to moles, cumulative biogas, and methane production were all calculated using this software.

For the daily methane and cumulative methane yields for each of the examined samples, RCC, RC, and RCF1, a one-way ANOVA in Excel data analysis toolpak was carried out. Using Graphpad Prism 9.0.0, the standard deviation (SD), mean values, and significance between the data were determined.

3. Results

3.1. Pretreatment and Composition of Rice Straw

For each experimental analysis, the composition of the evaluated biomass must be found, because the chemical and biological processing generated changes in the physical properties, chemical structures, and chemical compositions of rice straw (RC). The selected rice straw for this study has 41.3, 22.2, and 20.3% cellulose, hemicellulose, and lignin, respectively. Total solids (TS), volatile solids (VS), and ash were 89.5, 80.2, and 9.6%, respectively, Table 1. The composition of the rice straw used in this study is comparable with that employed in an earlier article on rice straw that evaluated the effects of NaOH on lignin [24].

Table 1. The composition of rice straw.

Factor	Value
Moister (%)	9.6. ± 0.2
Volatile solid (VS %)	80.2 ± 2
Total solid (TS %)	89.5 ± 3
Ash (%)	7.2 ± 1
Hemicellulose	22.2 ± 2
Cellulose	41.3 ± 3
Acid-insoluble lignin	6.8 ± 2
Acid-soluble lignin	20.3 ± 1

The compositional value of rice straw, as indicated in Table 1, was estimated using a conventional NREL laboratory analytical process before NaOH treatment and was used to determine the level of delignification [20]. The samples treated with 5% NaOH showed 73.6% delignification for the RC samples. RCF1 samples, on the other hand, showed 55.5%

delignification. This was due to the addition of water to the NaOH black liquor for the following cycle after filtering. It was observed that the lignin value reduced with NaOH treatment in both RC and RCF1 samples. In contrast, no significant reduction in lignin removal was observed in untreated RCC samples (Figure 2).

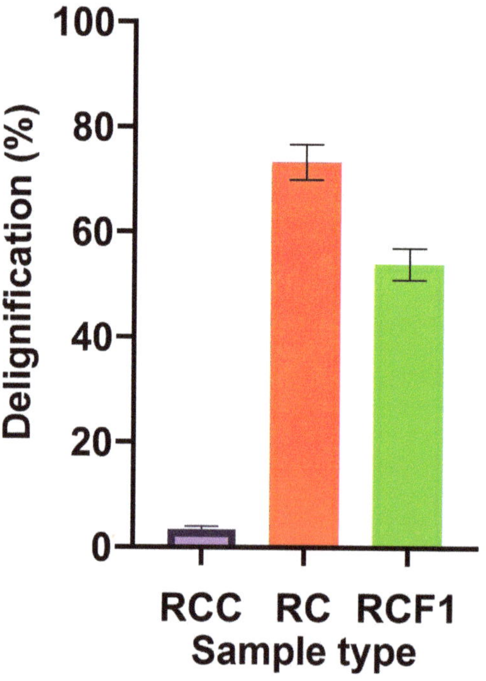

Figure 2. The percent value of delignification of rice straw with 5% NaOH treatment: RCC is the control sample, RC is the 5% NaOH-treated sample, and RCF1 represents the filtered sample.

3.2. SEM Analysis

SEM micrographs were used to compare the surface features of the untreated (RCC), 5% NaOH-treated (RC), and recycled filtered (RCF1) samples. The RCC images had a clean and smooth structure and were exceedingly compact. The dense structure of the rice straw was broken after NaOH pretreatment, and degradation was more visible on the surface of rice straw. The SEM micrographs of the autoclaved RCC samples without NaOH revealed that the hydrothermal pre-treatment without NaOH had a less damaging effect on their surface compared to the treated samples (Figure 3). The surface degradation effect was more intense and clearly visible in the case of the sample of the 5% NaOH-treated (RC) and recycled filtered samples (RCF1). In Figure 3, the deterioration effect was harsh for the 5% NaOH-treated (RC) sample, and gradually diminished in evident distortion on the surface of the recycled filtered sample (RCF1). The less severe effect in the SEM pictures of the recycled filtered sample (RCF1) supports the delignification rate and the relation of NaOH with lignin removal from the rice straw residue. Overall, the SEM micrographs clearly demonstrated ruptures in the silicon waxy structure, surface degradation, and morphological alterations in the rice straw as a result of heating and NaOH treatment, indicating that the fermentation process was accelerated.

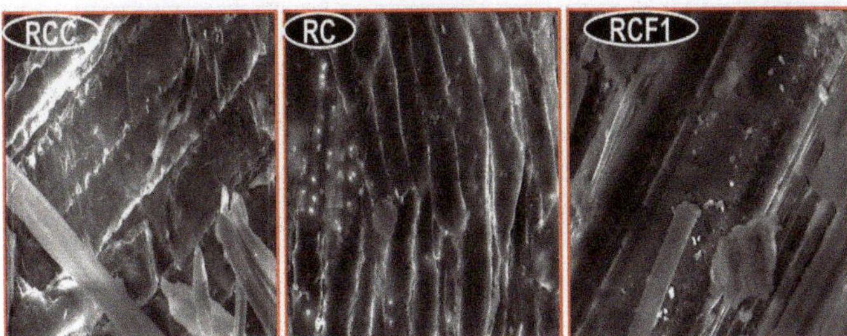

Figure 3. SEM images of the rice straw: RCC is control untreated sample, RC is 5% NaOH-treated sample, and RCF1 represents filtered sample of recycled NaOH treated with magnification of 1000× times.

3.3. Anaerobic Digestion for Biogas

Anaerobic digestion of the 5% NaOH-treated rice straw samples was compared to RCC samples that had not been treated. The improvement in biogas output in anaerobic digestion was investigated. The maximum daily biogas yields were 55.6 and 39.6 mL/gVS, respectively, from 5% NaOH-treated (RC) and recycled filtered sample (RCF1) bottles (Figure 4). The untreated RCC bottle yielded a lower biogas output at 14.7 mL/gVS. During the fermentation experiment using rice straw samples, it was observed that biogas production begins after 7 days of lag time in the case of untreated (RCC) bottles, but only 24 h in the case of treated RC and RCF1. Similarly, biogas generation continues until 27 days of anaerobic digestion for samples treated with 5% NaOH, whereas it stops after 24 days for untreated RCC bottles.

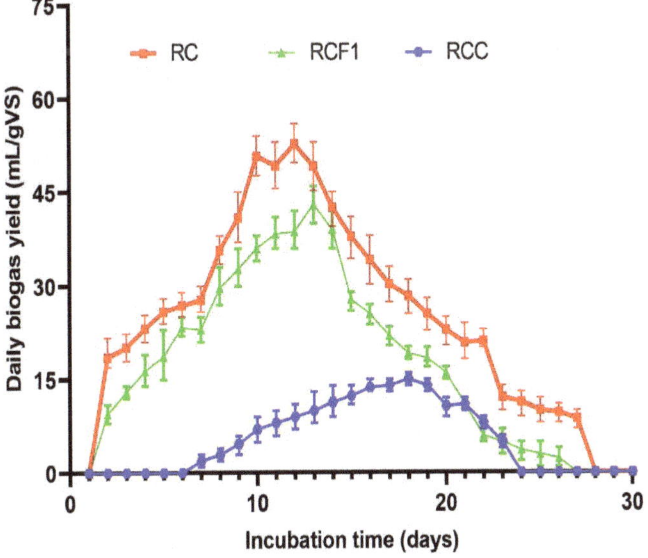

Figure 4. Daily biogas (mL/gVS): RCC is control untreated sample, RC is a 5% NaOH-treated sample, and RCF1 represents a filtered sample.

The daily methane output of the pretreated and untreated samples is shown in Figure 5. The greatest daily methane yield from RC and RCF1 bottles was 26.5 and 15.9 mL/gVS/d, respectively. The maximal daily methane output from the untreated RCC bottle was 4.5 mL/gVS/d.

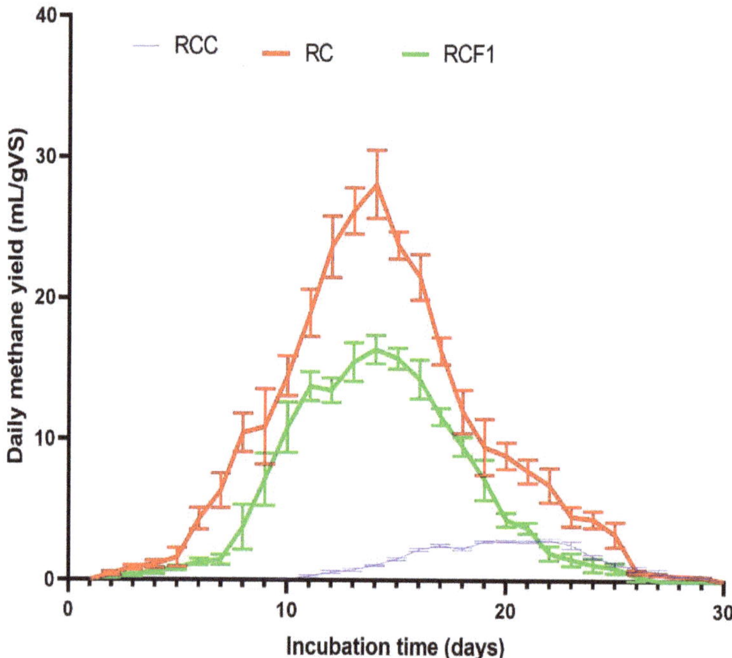

Figure 5. Daily methane (mL/gVS): RCC is control untreated sample, RC is a 5% NaOH-treated sample, and RCF1 represents a filtered sample.

After adding the inoculum to the reactor bottles to start the fermentation process, the daily production of biogas and methane began and after reaching its peak value, it continuously remained at a 30 mL per day yield for the next two weeks. The highest and optimum biogas production was observed from the 7th day until the 20th day of the fermentation process. The first two weeks of production showed the highest and maximum levels of biogas and methane. NaOH pretreatment enhanced the biodigestibility of RC and RCF-1 samples, facilitating anaerobic microbial utilization and reducing the time needed for digestion. This explains how the lignocellulose matrix was efficiently broken down by the NaOH pretreatment, supplying methanogens with enough organic materials to grow faster and produce more methane. However, in the untreated rice straw sample, the low biogas yield and low methane output provide indications that the lignin in the lignocellulose matrix of rice straw is not broken down, possibly leading to an inadequate supply of organic matter in the anaerobic reactor. As a result, during the lag phase of the anaerobic digestion time, the sample required additional time to generate biogas. As a result, the methanogens end up producing small amounts of methane and biogas.

The cumulative biogas from the fermented bottles of the treated and untreated samples is shown in (Figure 6). The maximum cumulative biogas yielded by the RC and RCF1 bottles was 1452.4 and 1125.2 mL/gVS, respectively. After 30 days of incubation, the maximum cumulative biogas from the untreated RCC bottle was only 285.5 mL/gVS.

Similarly, the cumulative methane yield in (Figure 7) for the treated and untreated samples shows a similar pattern. The RC and RCF1 bottles yielded the greatest cumulative methane levels of 396.7 and 274.4 mL/gVS, respectively. After 30 days of incubation, the total methane output from the untreated RCC bottle was only 30.2 mL/gVS.

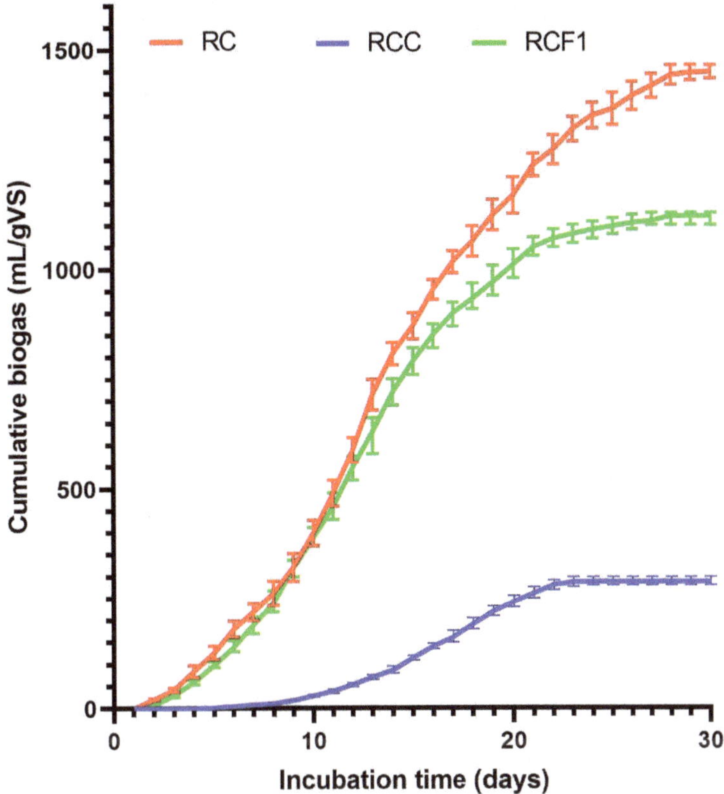

Figure 6. Cumulative biogas (mL/gVS): RCC is control untreated sample, RC is a 5% NaOH-treated sample, and RCF1 represents a filtered sample.

To observe a significant variance in the total methane production across all tests, a single-factor one-way ANOVA in the excel data analysis toolpak was used for the statistical difference, and the means of methane production among the three conditions—untreated, 5% NaOH-treated, and recycled-NaOH-solution-treated rice straw samples—were found. Table 2 displays the mean variance analysis (ANOVA) of single-factor results and multiple comparisons of the control and treatment samples. A total of 90 reaction samples and data from 30 of each sample were used for the substrate samples of untreated rice straw (RCC), treated rice straw (RC), and filtered NaOH-recovered liquor-treated rice straw (RCF1). According to Table 2, the mean methane yield for the RC and RCF1 samples at 95% probability is very different from that of the RCC sample. The average result consistently had a p-value of 1.33×10^{-8}, indicating that most of the variables for all the substrates under investigation—rice straw that had not been treated, rice straw that had been treated with recycled NaOH, and rice straw that had been treated with NaOH—were significant (RCF1). When RC was the substrate, followed by RCF1, the difference in methane output was significant, and it was possible to see a precise, significant difference in the value. The least amount of methane was produced by the RCC composed of rice straw that had not been processed. The cumulative methane yields from RC and RCF1 differ significantly from each other, as shown in Table 2 ($p > 0.05$). The yields of the treated samples and the RCC (control) samples are the same when compared to one another. The biogas yield was significantly affected by the NaOH treatment of the two RC and RCF1 substrates as a result. Increased lignin breakdown and cellulose digestibility, both of which encourage anaerobic digestion, are ascribed to this effect.

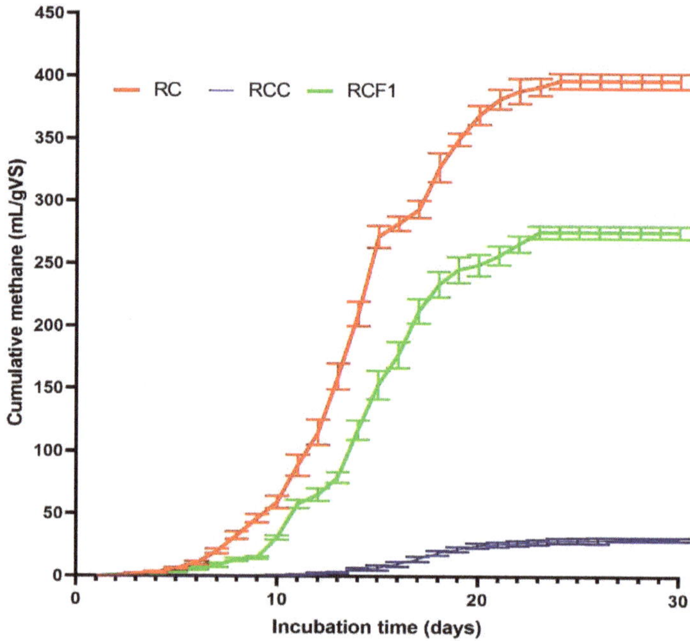

Figure 7. Cumulative methane (mL/gVS): RCC is control untreated sample, RC is a 5% NaOH-treated sample, and RCF1 represents a filtered sample.

Table 2. Comparison of cumulative methane yield of the RCC, RC, and RCF1 samples using one-way ANOVA.

ANOVA Summary				
Groups	Count	Sum	Average	Variance
RCC	30	420.758	14.02527	183.934
RC	30	6434.37	214.479	27,157.8
RCF1	30	4468.662	148.9554	14,449.81
ANOVA Results				

Source of Variation	SS	df	MS	F	P-value	F crit
Between Groups	626,811.8225	2	313,405.9	22.4978	1.33×10^{-8}	3.101296
Within Groups	1,211,954.77	87	13,930.51			
Total	1,838,766.592	89				

Table 3 displays the effects of several chemical pretreatments on certain straw biomasses. The table demonstrates that, although the same chemical pretreatment procedures were used, the effects of these pretreatments are not the same, e.g., 5% NaOH for rice straw and wheat straw. In a similar sense, it can be inferred that different procedures used for the same feedstock produced different outputs. It was found that after pretreatment with acids, NaOH, and biological and hydrothermal treatments, all reactions enhanced the biogas yield. In order to choose the appropriate chemical for the feedstock being used, this difference in yield increase can be further investigated using the same chemical concentration. In addition, it is necessary to identify the chemical concentration that can produce a yield increase compared to that of each substrate, due to differences in structural rigidity and responses to chemical effects. Our findings demonstrate that NaOH has a greater capacity to destroy lignin than other chemicals tested against all the substrates, wheat straw, corn straw, corn cob, rice straw, and sorghum stalk. It is also reported that the lignin-degrading bacteria also break down lignin and ultimately increase methane production, as shown in Table 3 for the batch fermentation method with *B. altitudinis* AN-2. The NaOH increased methane production in the treated samples compared to the untreated samples. This demonstrates that the choice of chemical pretreatment to be used will depend on the chemical, the treatment conditions, and the final products. The NaOH recycling pretreatment effect proved that it can be reused again and proved itself as a cost-effective chemical reagent compared to other chemical reagents used for lignin removal from complex biomass.

Table 3. Comparison of different chemical treatments and methane yields from biomass with the current study.

Substrate	Pretreatment Conditions	Methane Yield	Ref
Sugarcane bagasse	2% NaOH, Batch at 35 °C for 30 days,	222 mL CH_4/g VS	[25]
Wheat straw	3% NaOH, Batch at 37 °C for 30 days,	350 mL CH_4/g VS	[25]
Corn Straw	8% NaOH, 1L Erlenmeyer flask at 37 ± 1 °C	100.6 mL/g VS	[26]
Corn Straw	10% Ammonia, 1L Erlenmeyer flask at 37 °C	100.6 mL/g VS	[27]
Pinewood	Organosolv (150 °C, 1 h), at 39 ± 1 °C	387.4 mL/g VS	[28]
Rice straw	Hydrothermal 100 °C; 10 min	92.5 mL/g VS	[29]
Rice straw	Fungal treatment, Batch at 37 °C for 20 days	127 mL CH_4/g VS	[30]
Rice straw	NaOH+ hydrothermal	598.7 mLCH_4/g VS	
Sorghum stalk	H_2SO_4, 250 mL batch reactors at 37 °C	55% CH_4	[30]
Wheat straw	Acid steam, batch reactors at 37 °C	280 mL CH_4/g VS	[31]
Wheat straw	2% NaOH autoclaved, batch reactors at 37 °C	165.9 mL CH_4/g VS	[32]
Wheat straw	*Bacillus altitudinis* AN-2, batch fermentation	225.3 mL CH_4/g VS	[33]
Untreated straw	No treatment	78.5 mL CH_4/g VS	[33]
Corn cob	$Ca(OH)_2$ soaking (30 days), batch reactors	360.6 mL CH_4/g VS	[16]
Rice straw	5% NaOH autoclaved, batch reactors	430.8 mL CH_4/g VS	[15]
Rice straw	5% KaOH autoclaved, batch reactors	308.5 mL CH_4/g VS	[15]
Rice straw (RC)	5% NaOH autoclaved, batch reactors	396.7 mL CH_4/g VS	This study
Rice straw (RCF1)	5% NaOH autoclaved, batch reactors	274.4 mL CH_4/g VS	This study
Rice straw (RCC)	No treatment	30.2 mL CH_4/g VS	This study

4. Discussion

Anaerobic digestion is a wonderful and commonly used method for producing biogas from waste biomass and developing an effective waste management system. As a feeding substrate, animal waste such as manure and agricultural leftovers such as straw are commonly used. Rice straw is a crop leftover from rice, one of Pakistan's most important crops. Several factors, including lignin content, cellulose crystallinity, and particle size, limit the digestion of the hemicellulose and cellulose found in rice straws. Farmers burn the residual waste in the field, and there are no management mechanisms in the country either. If the correct management procedures are followed, this useful carbon source can be used for energy production. This research presented a recycled NaOH treatment of rice straw to remove lignin and expose hemicellulose and cellulose for effective digestion in the anaerobic fermentation process. The reprocessing of black NaOH liquid was utilized to test its effect in subsequent pretreatment and to ensure its reuse for biomass digestibility prior to anaerobic digestion. Initially, the rice straw's composition was measured, and the results of the contents were comparable to the composition of straw reported in earlier studies [34,35]. The 5% NaOH reduced the barrier lignin by 73.6%, which is comparable to the lignin removal potential of NaOH from rice straw residue found in prior work [15]. In prior research, we discovered that decreasing the alkali concentration under the same heating circumstances decreased the potential for lignin breakdown [15]. Similarly, in this study, a similar response for the lignin removal capacity of the recycled NaOH samples (RCF1) was observed. Its potential was reduced due to a decrease in NaOH concentration in the second reprocessing pretreatment. The effect of NaOH degradation on the surface of the rice straw samples also shows that the 5% NaOH samples caused higher degradation than the recycled samples. The RCC exhibits no significant structural degradation in the rice straw sample. Similarly, after pretreatment, there is damage and degradation on the surface of biomass straw [15,36]. During the rice straw fermentation experiment, it was discovered that biogas generation begins after 7 days of lag time in the case of untreated bottles, but just 24 h in the case of NaOH pretreatment bottles. This suggests that pretreatment reduces the time required for hydrolysis in the anaerobic fermentation process and aids in the start-up of biogas generation. Similarly, biogas generation continues for the samples treated with 5% NaOH until 27 days of anaerobic digestion, whereas biogas production stops after 23 days in the case of the untreated bottle. On the 10th day of the fermentation experiment, the concentrations of CH_4 and CO_2 were measured, and the methane concentration was 3%, rising to a maximum of 15.6% on the 20th day of the experiment in the untreated bottle. However, the concentration of CH_4 in the NaOH-treated bottles was 16% on the 6th day and rapidly grew to 57–67% on the 17–20th day of the fermentation experiment. Biogas and methane production ceased on the 27th and 28th days of the ADE experiment in the NaOH-treated bottles and on the 23rd day in the untreated bottle. The recycled RCF1 had a reduced biogas yield, which might be related to a decrease in the NaOH level during the filtration process for the second cycle, as well as the addition of an additional 100 mL of distilled water to adjust the volume for the following batch of the experiment. The explanation for the decrease in biogas output is obvious, because lignin removal and increasing biogas potential are directly related to the amount of lignin in lignocellulosic biomass [15]. Preferably, the biogas production and CH_4 percentage should be around 50–70% under standard temperature and pressure (STP) to be used for cooking, burning, and lighting. Only the NaOH-treated samples produced more than 65% raw biogas in this study, while the untreated samples produced just 15–17% CH_4 content. Similar observations of an increase in biogas have been reported previously in pretreatment with the alkalis $Ca(OH)_2$, KOH, and NaOH, depending on the concentration of the tested chemical and thermal heating from rice straw and wheat straw. The findings of this study and the related literature show that alkalis are a useful agent for delignifying biomasses, and that they should be utilized again rather than discarded. Lignin removal of waste biomass with NaOH pretreatment could be taken into consideration to proceed from lab research to an industrial-scale technique for less expensive biogas generation in order to enhance the anaerobic digestion process [37,38]. Such observations further highlighted the significance of pretreatment prior to the fermentation process in producing high-quality biogas. Even so, it is vital to remember that the pretreatment method must be affordable and may be adequate for pilot-scale waste biomass fermentation processing. The findings clearly show that processed rice straw enhances biogas and methane yields. When compared to the untreated sample, the NaOH samples produced roughly 6–8 times more biogas and methane. Overall, the NaOH-treated samples produced 85–124% more biogas than the untreated samples. The findings of this study are comparable to those of earlier studies [37,38]. The manipulated variables set for one-way ANOVA were RCC, RC, and RCF1 to compare the mean differences for daily methane yield and cumulative methane yield, and from the 90 data sets it is found that there is a significant improvement in the methane yield in both 5% NaOH-treated rice straw and 2.5% recycled-NaOH-treated rice straw samples. The P value of 0.0001 indicated that the results were significant. The results of this study demonstrated that pretreatment under mild circumstances improved the digestibility of the biomass straw and, as a result, increased output yield. It was also stated that NaOH black liquor should not be discarded. It should be recycled for the pretreatment purpose of disintegrating the lignin of lignocellulosic biomass before the fermentation process produces various bioproducts. To summarize this study, it was essential to reduce the cost of pretreatment for the efficient fermentation of rice stalks to methane. The rice straw pretreatment with 5% NaOH increased methane yield during initial testing. In the majority of instances, however, researchers discarded the NaOH solution after initial pretreatment, and recycling this black liquor solution as a treatment was not considered for lignin removal. We presumed that the reuse of alkaline pretreatment and the recovery of lignin are crucial issues for biorefinery in order to reduce pretreatment costs. For lignin removal from rice straw, an alternative strategy to the pretreatment procedure involving recycling the NaOH solution is presented in this study. We are able to demonstrate that the recycled NaOH solution removed lignin and increased methane yield. Relative to other pretreatment strategies, NaOH pretreatment is more effective at removing lignin and releases fewer inhibitors such as furfural and HMF. Through this method's design, not only was lignin removed, but the resulting NaOH solution was also recycled for the pretreatment of rice straw. Transformation of the recovered lignin revealed its potential in a variety of industrial fields. For instance, lignin has been identified as a highly valuable renewable polymer that can be used to manufacture phenolic resins and epoxies [39]. In addition, lignin can function as a binder and dispersant, such as in cement additives. During the pretreatment procedure, lignin and carbohydrates may be recovered for the ongoing utilization of biomass. In addition, no pollutant was discharged post alkali pretreatment. Improving alkaline pretreatment is both interesting and valuable because it concurrently reduces treatment costs and generates byproducts. In previous research, the recycling of a NaOH solution demonstrated an economic benefit, increased methane yield, and decreased pretreatment costs. According to the studies, black liquor recycling enhanced pretreatment efficacy and decreased chemical and water consumption by 40 to 60 percent [40]. Therefore, continuous-reactor studies are required to assess the economic and environmental benefits of recycling the black residue of NaOH, as well as the technical and financial viability and aggregate profitability of a biorefinery system. The results of this study demonstrated that pretreatment under mild circumstances improved the digestibility of the biomass straw and, as a result, increased output yield. NaOH black liquor should not be discarded, and should be recycled for the pretreatment purpose of disintegrating the lignin of lignocellulosic biomass before the fermentation process produces various bioproducts.

5. Conclusions

Two treatment options for lignin removal from rice straw were evaluated: NaOH and recycled NaOH. To accelerate the effect of NaOH on the rice straw substrate, mild thermal heating was utilized. The maximum dose examined in this study indicated that NaOH treatment considerably reduced lignin by up to 73.6%. The recycled filtered sample has a lignin removal percentage of 55.5%. This is due to NaOH concentration loss during the filtration process, which was neutralized by the addition of water in the second cycle. Additionally, no extra NaOH dose was utilized in the NaOH filtration process; therefore, the NaOH concentration fell in the subsequent heating cycles and the impact of treatment on lignin was reduced. The SEM findings clearly show that the NaOH treatment destroyed the surface morphology of the rice straw samples. The fermentation experiment findings showed a considerable improvement in biogas yield following the NaOH treatment when compared to the untreated sample. We concluded from this work that NaOH black liquor can be retested for lignin removal in agricultural waste biomass digestion to reduce the cost of chemical treatment for large-scale development.

Author Contributions: Conceptualization, writing and editing was done by T.A.S., S.K. prepared methodology, H.-A.N. did data curation, A.M.S. worked on and original draft preparation, review, project administration, and funding acquisition by M.B. All authors have read and agreed to the published version of the manuscript.

Funding: This work is financially supported by the Researchers Supporting Project number (RSP-2023R437), King Saud University, Riyadh, Saudi Arabia.

Institutional Review Board Statement: Not applicable.

Informed Consent Statement: Not applicable.

Data Availability Statement: The authors confirm that all data are included in this published article.

Acknowledgments: The authors would like to extend their sincere appreciation to the Researchers Supporting Project, King Saud University, Riyadh, Saudi Arabia for funding this work through the project number (RSP-2023R437). This work was supported by the research funding agency Higher Education Comission and International Research Support Initiative Programme (IRSIP) department, H9 Islamabad Pakistan.

Conflicts of Interest: The authors declare that they have no competing interest.

References

1. Arthur, P.M.A.; Konaté, Y.; Sawadogo, B.; Sagoe, G.; Dwumfour-Asare, B.; Ahmed, I.; Bayitse, R.; Ampomah-Benefo, K. Evaluating the Potential of Renewable Energy Sources in a Full-Scale Upflow Anaerobic Sludge Blanket Reactor Treating Municipal Wastewater in Ghana. *Sustainability* **2023**, *15*, 3743. [CrossRef]
2. Łysiak, G.; Kulig, R.; Al Aridhee, J.K. Toward New Value-Added Products Made from Anaerobic Digestate: Part 1—Study on the Effect of Moisture Content on the Densification of Solid Digestate. *Sustainability* **2023**, *15*, 4548. [CrossRef]
3. Kildegaard, G.; Balbi, M.D.P.; Salierno, G.; Cassanello, M.; De Blasio, C.; Galvagno, M. A cleaner delignification of urban leaf waste biomass for bioethanol production, optimised by experimental design. *Processes* **2022**, *10*, 943. [CrossRef]
4. Tsegaye, D.; Khan, M.M.; Leta, S. Optimization of Operating Parameters for Two-Phase Anaerobic Digestion Treating Slaughterhouse Wastewater for Biogas Production: Focus on Hydrolytic–Acidogenic Phase. *Sustainability* **2023**, *15*, 5544. [CrossRef]
5. Harun, S.N.; Hanafiah, M.M.; Noor, N.M. Rice straw utilisation for bioenergy production: A brief overview. *Energies* **2022**, *15*, 5542. [CrossRef]
6. Chen, C.; Deng, X.; Kong, W.; Qaseem, M.F.; Zhao, S.; Li, Y.; Wu, A.-M. Rice straws with different cell wall components differ on abilities of saccharification. *Front. Bioeng. Biotechnol.* **2021**, *8*, 624314. [CrossRef]
7. Shetty, D.J.; Kshirsagar, P.; Tapadia-Maheshwari, S.; Lanjekar, V.; Singh, S.K.; Dhakephalkar, P.K. Alkali pretreatment at ambient temperature: A promising method to enhance biomethanation of rice straw. *Bioresour. Technol.* **2017**, *226*, 80–88. [CrossRef]
8. Wang, Y.; Zhi, B.; Xiang, S.; Ren, G.; Feng, Y.; Yang, G.; Wang, X. China's Biogas Industry's Sustainable Transition to a Low-Carbon Plan—A Socio-Technical Perspective. *Sustainability* **2023**, *15*, 5299. [CrossRef]
9. Cui, C.; Yan, C.; Wang, A.; Chen, C.; Chen, D.; Liu, S.; Li, L.; Wu, Q.; Liu, Y.; Liu, Y.; et al. Understanding the Inhibition Mechanism of Lignin Adsorption to Cellulase in Terms of Changes in Composition and Conformation of Free Enzymes. *Sustainability* **2023**, *15*, 6057. [CrossRef]
10. Budiyono; Wicaksono, A.; Rahmawan, A.; Matin, H.H.A.; Wardani, L.G.K.; Kusworo, T.D.; Sumardiono, S. The effect of pretreatment using sodium hydroxide and acetic acid to biogas production from rice straw waste. *MATEC Web Conf.* **2017**, *101*, 02011. [CrossRef]
11. Anuradha, A.; Sampath, M.K. Optimization of alkali, acid and organic solvent pretreatment on rice husk and its techno economic analysis for efficient sugar production. *Prep. Biochem. Biotechnol.* **2022**, *53*, 279–287.
12. Syahri, S.N.K.M.; Abu Hasan, H.; Abdullah, S.R.S.; Othman, A.R.; Abdul, P.M.; Azmy, R.F.H.R.; Muhamad, M.H. Recent challenges of biogas production and its conversion to electrical energy. *J. Ecol. Eng.* **2022**, *23*, 251–269. [CrossRef]
13. Arshad, M.; Ansari, A.R.; Qadir, R.; Tahir, M.H.; Nadeem, A.; Mehmood, T.; Alhumade, H.; Khan, N. Green electricity generation from biogas of cattle manure: An assessment of potential and feasibility in Pakistan. *Front. Energy Res.* **2022**, *10*, 11485. [CrossRef]
14. Romero-Perdomo, F.; González-Curbelo, M.Á. Integrating Multi-Criteria Techniques in Life-Cycle Tools for the Circular Bioeconomy Transition of Agri-Food Waste Biomass: A Systematic Review. *Sustainability* **2023**, *15*, 5026. [CrossRef]
15. Shah, T.A.; Raheem, U.; Asifa, A.; Romana, T. Effect of Alkalis pretreatment on Lignocellulosic Waste Biomass for Biogas Production. *Int. J. Renew. Energy Res.* **2018**, *8*, 1318–1326.
16. Shah, T.A.; Tabassum, R. Enhancing Biogas Production from Lime Soaked Corn Cob Residue. *Int. J. Renew. Energy Res.* **2018**, *8*, 761–766.
17. Sumardiono, S.; Adisukmo, G.; Hanif, M.; Budiyono, B.; Cahyono, H. Effects of pretreatment and ratio of solid sago waste to rumen on biogas production through solid-state anaerobic digestion. *Sustainability* **2021**, *13*, 7491. [CrossRef]
18. Hidalgo-Sánchez, V.; Behmel, U.; Hofmann, J.; Borges, M.E. Enhancing Biogas Production of Co-Digested Cattle Manure with Grass Silage from a Local Farm in Landshut, Bavaria, through Chemical and Mechanical Pre-Treatment and Its Impact on Biogas Reactor Hydraulic Retention Time. *Sustainability* **2023**, *15*, 2582. [CrossRef]
19. Wang, W.; Tan, X.; Wang, Q.; Yu, Q.; Zhou, G.; Qi, W.; Zhuang, X.; Yuan, Z. Commentary on the Reuse of Black Liquor for the Enzymatic Hydrolysis and Ethanol Fermentation of Alkali-treated Sugarcane Bagasse. *J. Microbiol. Biotechnol.* **2017**, *6*, 228.
20. Sluiter, A.; Sluiter, J.; Wolfrum, E.J. *Methods for Biomass Compositional Analysis*; National Renewable Energy Laboratory (NREL): Golden, CO, USA, 2013.
21. Mosier, N.; Wyman, C.; Dale, B.; Elander, R.; Lee, Y.Y.; Holtzapple, M.; Ladisch, M. Features of promising technologies for pretreatment of lignocellulosic biomass. *Bioresour. Technol.* **2005**, *96*, 673–686. [CrossRef]
22. Qu, P.; Huang, H.; Zhao, Y.; Wu, G. Physicochemical changes in rice straw after composting and its effect on rice-straw-based composites. *J. Appl. Polym. Sci.* **2017**, *134*, 44878. [CrossRef]
23. Lou, X.F.; Nair, J.; Ho, G. Influence of food waste composition and volumetric water dilution on methane generation kinetics. *Int. J. Environ. Prot.* **2012**, *2*, 22–29.

24. Do, N.H.; Pham, H.H.; Le, T.M.; Lauwaert, J.; Diels, L.; Verberckmoes, A.; Tran, V.T.; Le, P.K. The novel method to reduce the silica content in lignin recovered from black liquor originating from rice straw. *Sci. Rep.* **2020**, *10*, 21263. [CrossRef] [PubMed]
25. Olatunji, K.O.; Ahmed, N.A.; Ogunkunle, O. Optimization of biogas yield from lignocellulosic materials with different pretreatment methods: A review. *Biotechnol. Biofuels* **2021**, *14*, 1–34. [CrossRef] [PubMed]
26. Song, Z.; Yang, G.; Guo, Y.; Zhang, T. Comparison of two chemical pretreatments of rice straw for biogas production by anaerobic digestion. *BioResources* **2012**, *7*, 3223–3236.
27. Song, Z.; Yang, G.; Liu, X.; Yan, Z.; Yuan, Y.; Liao, Y. Comparison of seven chemical pretreatments of corn straw for improving methane yield by anaerobic digestion. *PLoS ONE* **2014**, *9*, e93801. [CrossRef]
28. Arreola-Vargas, J.; Ojeda-Castillo, V.; Snell-Castro, R.; Corona-González, R.I.; Alatriste-Mondragón, F.; Méndez-Acosta, H.O. Methane production from acid hydrolysates of Agave tequilana bagasse: Evaluation of hydrolysis conditions and methane yield. *Bioresour. Technol.* **2015**, *181*, 191–199. [CrossRef]
29. Luo, T.; Huang, H.; Mei, Z.; Shen, F.; Ge, Y.; Hu, G.; Meng, X. Hydrothermal pretreatment of rice straw at relatively lower temperature to improve biogas production via anaerobic digestion. *Chin. Chem. Lett.* **2019**, *30*, 1219–1223. [CrossRef]
30. Dahunsi, S.; Adesulu-Dahunsi, A.; Osueke, C.; Lawal, A.; Olayanju, A.; Ojediran, J.; Izebere, J. Biogas generation from Sorghum bicolor stalk: Effect of pretreatment methods and economic feasibility. *Energy Rep.* **2019**, *5*, 584–593. [CrossRef]
31. Nkemka, V.N.; Murto, M. Biogas production from wheat straw in batch and UASB reactors: The roles of pretreatment and seaweed hydrolysate as a co-substrate. *Bioresour. Technol.* **2013**, *128*, 164–172. [CrossRef]
32. Chandra, R.; Takeuchi, H.; Hasegawa, T.; Kumar, R. Improving biodegradability and biogas production of wheat straw substrates using sodium hydroxide and hydrothermal pretreatments. *Energy* **2012**, *43*, 273–282. [CrossRef]
33. Shah, T.A.; Ali, S.; Afzal, A.; Tabassum, R. Simultaneous pretreatment and biohydrogen production from wheat straw by newly isolated ligninolytic Bacillus s strains with two-stage batch fermentation system. *BioEnergy Res.* **2018**, *11*, 835–849. [CrossRef]
34. Gustafson, V. Combined Production of Bioethanol and Biogas from Wheat Straw. Master's Thesis, Department of Chemical Engineering, Lund University, Lund, Sweden, 2015.
35. Cui, Z.; Shi, J.; Li, Y. Solid-state anaerobic digestion of spent wheat straw from horse stall. *Bioresour. Technol.* **2011**, *102*, 9432–9437. [CrossRef] [PubMed]
36. Zeng, J.; Singh, D.; Chen, S. Biological pretreatment of wheat straw by Phanerochaete chrysosporium supplemented with inorganic salts. *Bioresour. Technol.* **2011**, *102*, 3206–3214. [CrossRef]
37. Shah, T.A.; Ullah, R. Pretreatment of wheat straw with ligninolytic fungi for increased biogas productivity. *Int. J. Environ. Sci. Technol.* **2019**, *16*, 7497–7508. [CrossRef]
38. Shah, T.A.; Charles, C.; Orts, W.J.; Tabassum, R. Biological pretreatment of rice straw by ligninolytic Bacillus s strains for enhancing biogas production. *Environ. Prog. Sustain. Energy* **2018**, *38*, e13036. [CrossRef]
39. Chen, Y.; Peng, N.; Gao, Y.; Li, Q.; Wang, Z.; Yao, B.; Li, Y. Two-Stage Pretreatment of Jerusalem Artichoke Stalks with Wastewater Recycling and Lignin Recovery for the Biorefinery of Lignocellulosic Biomass. *Processes* **2023**, *11*, 127. [CrossRef]
40. D'adamo, I.; Gastaldi, M.; Morone, P.; Rosa, P.; Sassanelli, C.; Settembre-Blundo, D.; Shen, Y. Bioeconomy of sustainability: Drivers, opportunities and policy implications. *Sustainability* **2021**, *14*, 200. [CrossRef]

 *sustainability*

Perspective

Sustainalism: An Integrated Socio-Economic-Environmental Model to Address Sustainable Development and Sustainability

N. P. Hariram [1], K. B. Mekha [2], Vipinraj Suganthan [3] and K. Sudhakar [4,5,6,*]

1 Renewable Energy and Environmental Engg Focus Group, Universiti Malaysia Pahang, Paya Basar 26300, Malaysia
2 Integrated Centre for Green Development and Sustainability (ICFGS), Cuddalore 607001, India
3 Faculty of Engineering and Natural Sciences, Tampere University, P.O. Box 541, 33014 Tampere, Finland; vipinrajsugathan@tuni.fi
4 Faculty of Mechanical and Automotive Engineering Technology, Universiti Malaysia Pahang, Pekan 26600, Malaysia
5 Centre for Research in Advanced Fluid & Processes (Fluid Centre), Universiti Malaysia Pahang, Paya Basar 26300, Malaysia
6 Energy Centre, Maulana Azad National Institute of Technology, Bhopal 462003, India
* Correspondence: sudhakar@ump.edu.my

Abstract: This paper delves into the multifaceted concept of sustainability, covering its evolution, laws, principles, as well as the different domains and challenges related to achieving it in the modern world. Although capitalism, socialism, and communism have been utilized throughout history, their strengths and drawbacks have failed to address sustainable development comprehensively. Therefore, a holistic approach is necessary, which forms the basis for a new development model called sustainalism. This study proposes a new socio-economic theory of sustainalism that prioritizes quality of life, social equity, culture, world peace, social justice, and well-being. This paper outlines the six principles of sustainalism and identifies sustainalists as individuals who embrace these new concepts. This study also explores how to attain sustainalism in the modern world through a sustainable revolution, representing a step toward a sustainable era. In conclusion, this paper summarizes the key points and emphasizes the need for a new approach to sustainalism in the broader sense. The insights provided are valuable for further research on sustainalism and sustainability.

Keywords: sustainalist; sustainability; sustainable revolution; SDG; quality of life; sustainalism

Citation: Hariram, N.P.; Mekha, K.B.; Suganthan, V.; Sudhakar, K. Sustainalism: An Integrated Socio-Economic-Environmental Model to Address Sustainable Development and Sustainability. *Sustainability* **2023**, *15*, 10682. https://doi.org/10.3390/su151310682

Academic Editors: Idiano D'Adamo and Massimo Gastaldi

Received: 5 May 2023
Revised: 30 June 2023
Accepted: 3 July 2023
Published: 6 July 2023

1. Introduction

Evolution of Sustainable Development and Sustainability

Sustainability has been widely accepted since the olden days, especially in rural societies. The world's ancient cultures combine worship and religious convictions with environmental preservation, which calls on people to take care of the planet and keep it in good condition; this may be considered a demonstration of sustainability in the ancient ages. The "sustainability" term's origins can be found in the realm of hunting, wherein hunters and gatherers were eager to establish a stable means of subsistence. In old German, "sustenance" refers to provisions kept in reserve for emergencies. The verb "to sustain" or the phrase "sustainable" have both been "proven to be a derivation of the noun "sustenance" (literally retain, what one retains). Nowadays, the word "sustainable" still has the meaning of being "enduringly effective" in common usage [1].

Silent Spring by Racheal Carson, The Ecologist's *A Blueprint for Survival*, and *The Population Bomb* by Ehrlich are some early works that significantly impacted the world in the cradle stage of sustainable development during the 1960s. After that, within a short time, the words "sustainable" and "sustainability" were introduced

---

in the Oxford English Dictionary [2]. The word "sustainable" comes from the Latin word "sustinere". Thomas Malthu's postulates on the drastic consumption of natural resources and energy emerged, addressing the aftermaths of the population explosion. In his essay, he stated the principle of population, showing that population growth is not sustainable, which is not in proportion with the available resources and carrying capacity of the Earth [3].

The future and existence of humanity were described as "sustainability" in the British book *Blueprint for Survival* and a United Nations statement in 1978. Policy journals began using the word "sustainability" along with technical articles and studies around 1978. Most of these concentrated on the major sustainability domains and the environment. Soon, the World Bank started working to integrate sustainability into its organizational structure, operational procedures, and policy frameworks. Due to the fact that the term "sustainability" has roots in so many fundamentally different ideas, each with a compelling argument for its legitimacy, it seems futile to attempt to define it in a single sense [4]. The concept of sustainable development has acquired acceptance and significance theoretically. Its further development is frequently overlooked or minimized. While some people may think evolution is irrelevant, it may still be used to predict future trends and defects, which can be helpful now and in the future [5]. The unchecked economic expansion may cause the planet's carrying capacity to be exceeded and civilization to crumble. The ideas of sustainability and sustainable development, as a result, emerged [6].

The repercussions of anthropogenic activity and environmental devastation are becoming increasingly well-known, thanks to the media and publications. Works such as *Limits of Growth* or *Small is Beautiful* argued that economy-based development is unsustainable in this finite world of limited resources, and this started to question ongoing economic growth [7]. The early discourse was radical and demanded structural reform, arguing that capitalist economic development cannot be integrated with social and ecological development, which contradicts the concept of a sustainable world [8].

Reiterating the need for SD, the "World Commission on Environment and Development", headed by Gro Harlem Brundtland of Norway, produced the Brundtland Report titled "Our Common Future" in 1987. The report defined sustainable development as "the development that meets the demands of the current generation without compromising the ability of the future generation to meet their own needs", as was already mentioned. The Rio Earth Summit, also known as the UNCED or Rio Earth Summit, was inspired by the Brundtland Report in 1992 [9]. The main subject of discussion at the UNCED was the report's recommendations. The conference outcome document for the UNCED included Agenda 21 as one of the critical sustainable development outcomes. It urged that national policies be devised and implemented to address the economic, social, and environmental components of sustainable development after stating that sustainable development should become an essential item on the international community's agenda [10]. The World Summit on Sustainable Development (WSSD), also known as Rio+10, was convened in Johannesburg in 2002 to assess the status of putting the Rio Earth Summit's outcomes into practice. The World Summit on Sustainable Development (WSSD) introduced several multi-stakeholder partnerships for sustainable development and the Johannesburg Plan, an implementation plan for the measures outlined in Agenda 21 [11]. Figure 1 illustrates the sequence of activities associated with sustainable development.

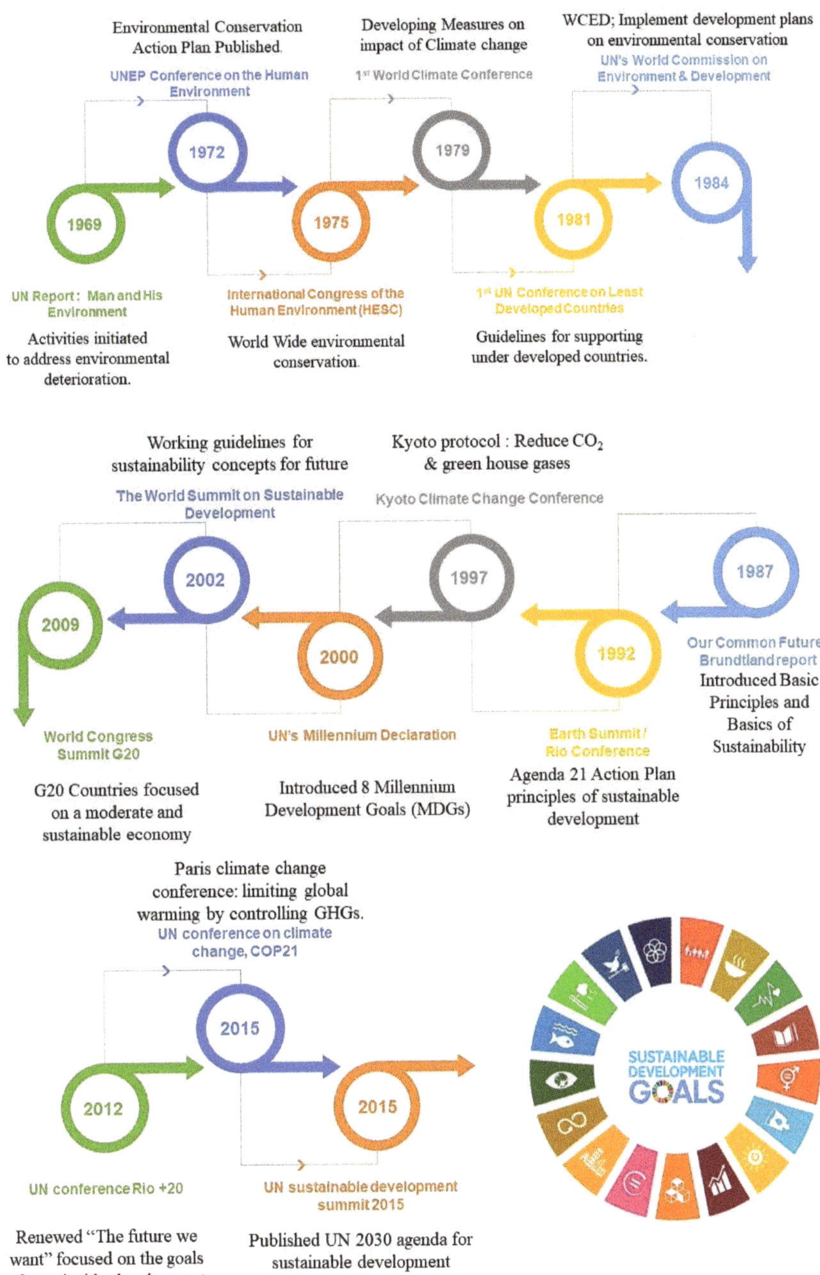

Figure 1. Overview of the various activities related to the concept of sustainable development till SDGs [12].

Rio+20, also known as the United Nations Conference on Sustainable Development (UNCSD), occurred in 2012, 20 years after the first Rio Earth Summit. The conference's two main sustainable development topics were the green economy and an institutional framework. The conference conclusion document "The Future We Want" placed a strong emphasis on sustainable development to the point that the term "sustainable development" was used [13]. The Rio+20 outcomes included a procedure for creating new SDGs, which would go into effect in 2015 and promote targeted action regarding sustainable development in all areas of the global development agenda. SD was thus one of the five main objectives of the United Nations in 2012 (UN), highlighting the important part that sustainable development (SD) should play in national and international development policies, programs, and agendas [14]. Table 1 provides an overview of the different definitions of sustainability and sustainable development.

Table 1. Different definitions of sustainability and sustainable development [15–18].

Source	Definition of Sustainability
The Brundtland Report	A process of change in which the exploitation of resources, the direction of investments, the orientation of technological development, and institutional changes are made consistent with future as well as present needs.
Earth Centre	Sustainability means that all living things on Earth have obligations to each other, the larger biosphere, and the subsequent generations.
NCARB	In sustainability, interrelated ecological, economic, and social systems succeed now without sacrificing their future prosperity.
UN	Sustainability is meeting the demands of the present without compromising the ability of future generations to satisfy their own needs.
Hannover principles	Sustainability is the conception and realization of ecologically, economically, and ethically sensitive as well as responsible expression as a part of the evolving matrix of nature.
Source	Definition of Sustainable Development
WCED	Sustainable development is the development or growth which meets the needs of the present without compromising the ability of future generations to meet their own needs.
Berke and Manta	Sustainable development is defined as a dynamic process connecting local and global concerns, as well as linking local social, economic, and ecological issues, to cater to the current and future generations' needs fairly.

2. Laws and Principles of Sustainability

The term "sustainability" is scrutinized by Albert Bartlett using different laws, hypotheses, observations, and predictions. They may not apply to small groups of people or to tribes living in primitive conditions as they are all based on populations and rates of resource and good consumption found in the world. The laws are more exacting than the hypotheses [19]. Though they are limited in many perspectives, some postulates may be proven correct by experience and given the status of laws. The observations may provide insight into the issues and possible solutions. These postulates can be classified into four categories: Population and consumption, Energy, Resource and Environment, and Human- and Society-centric (Table 2).

Table 2. Albert's laws on sustainability [20].

Population and Consumption	First Law: Population growth and increase in the rates of consumption of resources cannot be sustained. Second Law: The difficulty of transforming a society, with more growth in population and higher consumption of resources, into being sustainable is higher. Third Law: Population Momentum: The response time of populations to changes in the human fertility rate is the average length of a human life. Fourth Law: The size of the population that can be sustained (the carrying capacity) and the sustainable average standard of living of the people are inversely related. Eighth Law: Sustainability requires that the size of the population be less than or equal to the carrying capacity of the ecosystem for the desired standard of living. Ninth Law: The benefits of population growth and growth in the rates of consumption of resources accrue to a few; all of society bears the costs of population growth and growth in the consumption of resources. Seventeenth Law: If, for whatever reason, humans fail to stop population growth and growth in the rates of consumption of resources, nature will eliminate these growths.
Energy, Resource, and Environment	Tenth Law: Growth in the rate of consumption of a non-renewable resource, such as a fossil fuel, causes a dramatic decrease in the life expectancy of the resource. Eleventh Law: The time of expiration of non-renewable resources can be postponed, possibly for a very long time. Twelfth Law: When considerable efforts are made to improve the efficiency with which resources are used, the resulting savings are wholly and rapidly wiped out by the added resources consumed due to modest population increases. Thirteenth Law: The benefits of large efforts to preserve the environment are rapidly canceled by the added environmental demands resulting from small increases in the human population. Fourteenth Law: (Second Law of Thermodynamics) When rates of pollution exceed the natural cleansing capacity of the environment, it is easier to pollute than it is to clean up the environment.
Human—an Society-centric	Seventh Law: A society that has to import people to do daily work ("We can't find locals who will do the work") is not sustainable. Sixteenth Law: Humans will always be dependent on agriculture (This is the first of Malthus' two postulates). Eighteenth Law: In local situations within the State, creating jobs increases the number of people locally who are out of work. Nineteenth Law: Starving people do not care about sustainability.
Universal	Fifth Law: One cannot sustain a world in which some regions have high standards of living while others have low standards of living. Sixth Law: All countries cannot simultaneously be net importers of carrying capacity. Fifteenth Law: (Eric Sevareid's Law): solutions are the chief cause of problems. (Sevareid 1970) Twentieth Law: The addition of the word "sustainable" to our vocabulary, to our reports, programs, and papers, to the names of our academic institutes and research programs, and to our community initiatives is not sufficient to ensure that our society becomes sustainable. Twenty-first Law: Extinction is forever.

Principles of Sustainability

Sustainable development can only be realized if a few principles are followed. However, the economy, environment, and society are typically prioritized when discussing the basics of sustainable development [21]. Population control, human resource management, ecological and biodiversity preservation, production systems, the preservation of progressive culture, and public participation are among the issues addressed [22].

One of the tenets of sustainable development is the preservation of the environment. Since all life would end without the environment and biodiversity, they must be protected. The Earth's finite resources cannot meet the population's needs and means. Natural resource extraction must not exceed the Earth's capacity for sustainable development because resource depletion hurts the ecosystem [23]. This suggests that development activities need to consider the Earth's capabilities. Due to this, having renewable energy sources, such as solar, is essential, rather than relying too heavily on hydroelectricity and things made from petroleum. To accomplish sustainable development, population control is also crucial [24]. People can live by using the limited resources of the Earth. The growing population raises human needs, such as those for food, clothing, and housing, but there are limits to how much the world's resources can be expanded to provide. Therefore, population management and control are essential for sustainable development [25]. Effective human resource management is another integral component of sustainable development. The individuals are in charge of seeing that the principles are upheld. The environment must be used wisely and protected by humans. It is up to individuals to maintain peace on Mother Earth [13]. The argument is built on the premise that sustainable development cannot be accomplished solely through the efforts of one person or organization, which alludes to the system's theory. All individuals and relevant organizations must share this obligation. The concept of participation, which calls for optimistic attitudes from the populace in order to make real progress while accepting responsibility and accountability for stability, is the cornerstone of sustainable development [24]. In order to achieve genuine, long-lasting change, participation entails the combined effort of numerous people and organizations who are all working toward a shared vision of sustainability. We are more likely to succeed if we recognize the responsibility that is placed on each of us as well as the power that comes from working together for a common goal [26].

Sustainable development requires promoting socially progressive traditions, behaviors, and political cultures. In order to maintain social cohesion and support environmental appreciation and preservation for sustainable development, advanced traditional and political culture must be developed, nurtured, and expanded [22]. The systematic integration of ecological, social, and economic considerations into all areas of outcome across generations can be summed up as the central tenet of sustainable development. A socially progressive culture is crucial for sustainable development because it enables people to understand their obligations to society and the environment. Therefore, in order to achieve sustainable development, a progressive traditional and political culture must be created. This can be achieved by implementing measures such as encouraging grassroots organizations to increase public awareness and participation in sustainable development initiatives [27].

A systematic consensual and heuristic approach was used to arrive at sustainability principles based on studies of humans as a social species, the Laws of Thermodynamics, and the science behind them [28]. The lack of a comprehensive definition of "sustainability" and the recognition of the inherent issues involved in the current use of the term "sustainable development" served as the inspiration for this [29]. These pillars support several logical deductions about how social and ecological systems communicate. The guiding principles gradually evolved and were consented to after discussions with experts from the larger scientific community. A framework with logical guiding principles was used to apply the system conditions (Figure 2). These concepts effectively outline how the system parameters may be addressed using "back-casting".

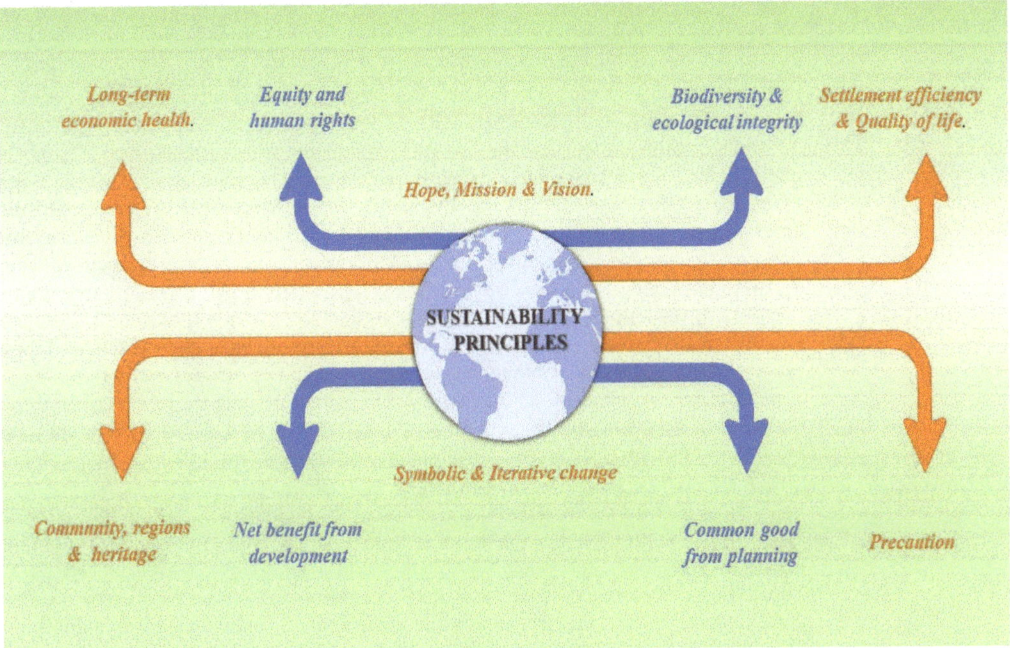

Figure 2. Sustainability principles.

In essence, they restate the definition of sustainability in a format that is applicable and relevant to all people, reclaiming it from the hazy "definition drift" observed for the term "sustainable development". As a result, they act as unyielding "custodians" of the sustainability idea [30].

3. Three Pillars and Domains of Sustainability

A sustainable structure is said to be built on three pillars. Three intersecting circles are a common visual depiction of sustainability and its dimensions [31]. Many research findings also use a nested approach with a particular dimension at the center. Present-day sustainable development is frequently represented by literal "pillars" supporting it. The hierarchy of the dimensions is highlighted in the schematic with the nested ellipses, with "environment" serving as the basis for the other two. Three interconnected "pillars", "dimensions", "components", "stool legs", "aspects", "perspectives", etc., are frequently used to describe sustainability and include economic, social, and environmental (or ecological) factors or "goals" [32]. It must be acknowledged that these conflicting terms are typically used synonymously, and our preference for "pillars" is largely arbitrary. The three intersecting circles of society, environment, and economy are frequently, though not always, used to represent this multi-stakeholder description, with sustainability situated at the crossroads. While frequently referred to as a "Venn diagram," this diagram frequently lacks the particularly emphasized attributes associated with such a construction. It describes "sustainability" in academic literature, policy documentation, business literature, and online [33].

Alternative ways of expressing the three concepts include using nested concentric circles or actual "pillars" to represent them visually and using them independently of visual aids to represent distinct categories of sustainability objectives or metrics [34]. While captivating due to their simplicity, the meaning these diagrams and the larger "pillar" concept themselves convey is frequently ambiguous, restricting their ability to be coherently operationalized. However, the conceptual underpinnings of this description and the time when it entered popular culture are unclear, and its precise meaning is up for debate. The three-pillar conception has undoubtedly attained widespread acceptance, but this should not obscure its flaws [7].

Critics argue that the sustainable development framework is not ambitious enough to address the scale and urgency of today's environmental and social challenges. Some also argue that focusing on economic growth and development can perpetuate the unsustainable use of natural resources and the unequal distribution of wealth and power. Some alternatives to traditional sustainable development have been proposed, including "Degrowth" and "Post-Development" [35,36]. Degrowth advocates for a reduction in consumption and production, particularly in wealthy countries, to reduce pressure on the environment and address social inequalities. Post-Development argues that the focus on economic growth has created a distorted view of development that prioritizes Western values and disregards the knowledge and values of marginalized communities [35,36]. Ultimately, the debate around sustainable development and its alternatives highlights the need for a more nuanced and inclusive approach to development that prioritizes both human well-being and environmental sustainability. In order to better navigate the turbulent and uncertain conditions that make up the post-Brundtland world, academics, development practitioners, environmental managers, sustainability advocates, and government planners must work together [37].

Since the Brundtland Report was released, mainstream sustainable development has advanced rapidly. The notion of sustainable development is firmly rooted in many government offices, corporate boardrooms, and the hallways of international NGOs and financial institutions, despite the risk of cooptation and abuse, frequently resulting in a scaling-back of its more progressive prescriptions for achieving sustainability [38]. At the very least, its willingness to offer some commonality for deliberations among various development and environmental sectors, which are frequently at odds, can be used to explain why sustainable development has endured. Strongest proponents of the idea, such as those in international environmental NGOs and intergovernmental organizations, are thus at ease advancing a concept that most effectively converts former opponents into social constructivism, contending that understanding the world invariably entails a series of mediations between human social relations and individual identities. Critics also tend to conduct qualitative research based on a case study methodology and emphasize the historical contingency of development processes. Perhaps most significantly, proponents of traditional sustainable development still view the policy-making process as a legitimate means of reform [39].

3.1. Domains of Sustainability

The framework assumes that three domains—the economy, the environment, and the social domain—should be considered when discussing sustainability [34]. These domains are claimed to connect as three separate spheres of life. Sustainability is foundational to public administration, policymaking, and political governance; variations of this three-domain framework can be found throughout all policy documents [40]. Despite this prevalence, the framework is infrequently subjected to sustained or in-depth stakeholder discussion. The economy (or profit), the environment (or ecology, the planet), and the social sector (or society, prosperity) are typically left as the framework for defining and operationalizing sustainability, with the economy typically leading the pack as the first between many "equals". This approach to sustainability is often seen as one that places too much emphasis on economic prosperity while not giving enough attention to social and environmental aspects. The idea of sustainability has become ubiquitous in the public sector, but its implementation and application often fail to capture the complex realities of social and environmental interdependencies [41]. In other words, economic factors have taken over almost all decision-making processes. They are viewed as fundamental to the human condition, defining and serving as a standard by which everything else is measured [42].

The three domains are considered separate actual activity spheres. Despite the justifications for why the three domains must be combined in an integrated assessment, they are still hypothesized as different spheres, pillars, or circles (Figure 3). This does not mean that various sustainability domains should not have their own integrity and measurement methods. However, in order to have it both ways, it is necessary to name different sustainability domains analytically and acknowledge that, in reality, they are dimensions of a whole rather than separate spheres that ought to be reconnected [43]. To do this, an integrated approach to sustainability needs to be adopted in which these different domains are conceptualized as parts of a system, not stand-alone spheres [44].

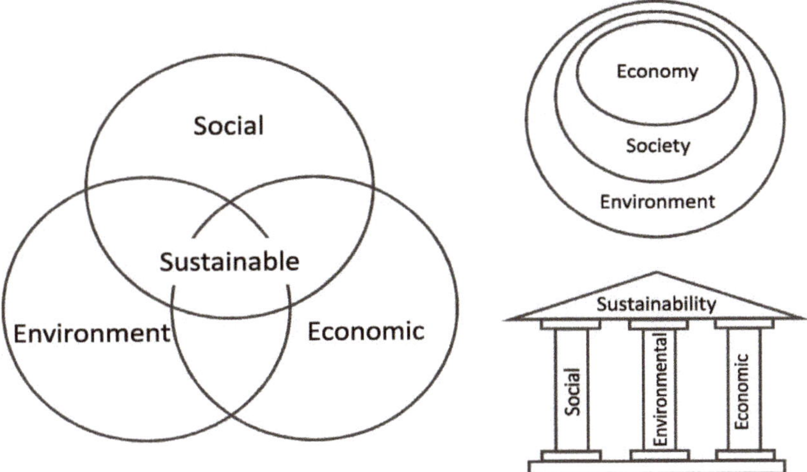

Figure 3. Three pillars and principles of sustainability.

One of the earliest critical public documents to use the three-domain model was *Our Common Future*, though this use is still largely implicit today. Ecology and economy were once distinct, conceptually and practically, but processes of globalization and growth have merged them. This merging of the two fields has created unique challenges in balancing human needs and environmental protection [45].

The demand for global interrelationships between the three domains has undoubtedly increased as expansion or development processes have ramped up. Ecologies, economies, and social relations have historically been intertwined in practice. Table 3 presents an overview of the criteria considered across different sustainability domains. The idea of sustainability as it relates to ecology, economics, politics, and culture—or even as it is used in one of those fields alone—is remarkably new [46].

Table 3. Domain-oriented principles and criteria.

Environmental Domain	Social Domain	Economical Domain
Protect the health of the ecosystem	Social justice and equity	Adequate funds for social growth
Avoid excess pollution	Social infrastructure	Create employment and fair trade
Shift to renewable resources	Engaged governance	Rise the income of the people
Intergenerational decisions	Social capital	High standard of living
Target welfare, not GDP	Community and culture	Free and sharing market
Restoration and conservation	N/A	Cost saving and green finance
N/A	N/A	Financial stability and security
N/A	N/A	Green and circular economy

The forty-year-old idea of sustainability is undergoing a paradigm shift in favor of The Dominant Domain Structure of Sustainability in the 21st century. We observe the same three-domain structure of sustainability in the various policy discourses linked to significant international organizations, with "the social" as a convenient term to group together those aspects of life that do not fall under the purview of the economy or the environment [47]. Since *Our Common Future*, three significant global initiatives have been launched. Each of these has impacted policy, administration, and sustainability governance, often directly influencing national and local engagement. The three major initiatives, the International Panel on Sustainable Development (IPCC), the Sustainable Development Goals (SDGs), and the United Nations Framework Convention on Climate Change (UNFCCC), are each instrumental in advancing global sustainability goals [48].

3.1.1. Political Domain

In order to envision and create a shared future, a deliberative political process is fundamentally required. In sustainability politics, resources are mobilized with an eye toward respect and harmony along multiple axes and over an extended period. Sustainability politics requires a unique approach to democratic deliberation that does not simply focus on immediate policy solutions but instead takes a broader and longer-term view of the impact of policies on society and the environment [44]. This approach enables us to identify and engage stakeholders, communities, and policy-makers in a shared vision of a sustainable future, allowing us to include all perspectives in the debate and reach equitable and effective decisions [44].

The topic of system regulation and governance identifies a fourth fundamental category of organization; the political sphere, which regulates the relations between (and within) the environmental sphere and the economic and social spheres. Establishing conventions, laws, and institutional frameworks for controlling society's social, economic, and, indirectly, environmental spheres creates the political sphere [49]. The political sphere serves as the "referee" who settles disputes involving the various, frequently incompatible claims made by participants in the social and economic spheres, both for themselves and concerning other spheres, such as the environment. This occurs through the political sphere as an intermediary, highlighting that, rather than direct environmental "governance", there is frequently an indirect connection between the political and environmental spheres [50]. The transition from politics to the environment may involve the "supply" of public policy meant to impact how environmental systems operate. Environmental–social then social–political integrations or environmental–economic then economic–political integrations are two ways to communicate societal demands "on behalf of the environment" [51]. Establishing conventions and procedures for regulating each sphere in relation to the others, to ensure the concurrent respect for quality/performance goals of all three spheres, constitutes the political or governance dimension of the organization [10]. This is the area of arbitrage between various principles and asserts of concern, obtained de facto or on purpose through coercion and institutional arrangements ranging from town and county councils to national government institutions to United Nations and other active international organizations. This type of regulation is a powerful force and creates a system of checks and balances that protects the interests of society and the environment. Such regulation works to ensure that all interests and needs, both those of individuals and larger groups, are taken into account to produce an optimal outcome for everyone involved [38].

3.1.2. Cultural Domain

In addition to art and literature, lifestyles, ways of interacting, value systems, traditions, and beliefs, culture is defined as the collection of unique spiritual, material, intellectual, and emotional characteristics of a society or social group. This leads to the perspectives and the character of a person as well as a society [52].

Along with sustainable development's environmental, social, and economic dimensions, culture has gained increasing attention in recent years. The protection of ecology and the environment is motivated by the environmental dimension. The economic dimension encourages the effective use of financial resources and aims for long-term benefits, whereas the social dimension concentrates on the needs of humans, both present and future [53]. Since culture was previously included in the social dimension of sustainability, it was not considered a separate dimension. The situation gradually changed, though culture is now acknowledged as crucial to achieving sustainable development [27]. A separate cultural dimension of sustainability has been established to focus on the protection of traditional values and lifestyles, as well as the preservation of tangible and intangible heritage [52]. This separation of culture from the social dimension of sustainability is significant due to globalization, which has led to increased displacement and destruction of traditional cultures. Furthermore, this cultural dimension of sustainability includes respect for cultural diversity and promoting intercultural dialogue [54].

3.2. SDGs/MDG Linkages with Sustainability Domains

The idea of sustainability has garnered attention on a global scale and has been thoroughly discussed by academics, professionals, and decision-makers [55].

Several development goals and policies have been established to follow a sustainable vision and mission with sustainable development plans for stakeholders and the correct direction for our future survival [56]. The Millennium Development Goals (MDGs) were first established; they include eight main goals and address environmental, social, and economic challenges. MDGs, particularly, improved mortality, public health, hunger, and poverty [57]. However, all over the world remain a significant number of problems and challenges [58].

The Millennium Development Goals (MDGs) are significant and affect the international mobilization strategy for addressing several crucial global socioeconomic concerns. They convey mass political concern about gender inequity, ecological degradation, hunger, malnutrition, poverty, and illness [59]. The MDGs strive to advance awareness and understanding, political responsibility, enhanced measures, excellent interpersonal communication, and social opinion by condensing these objectives into a manageable group of eight goals and setting measurable targets within a limited period [60]. Developing nations have made significant strides toward achieving the MDGs. However, advancement rates vary greatly between targets, countries, and areas. The likely gap in MDG achievement is grave, sad, and highly unpleasant for low-income people. Yet, there is a widespread feeling among policy leaders and civil society that progress against poverty, hunger, and disease is significant; that the MDGs have played an essential part in securing that progress [61]. The shortage results from operational errors affecting numerous parties in wealthy and poor nations. Rich countries, for instance, have not complied with their promises of formal development support [62].

The Sustainable Development Goals (SDGs) are a group of 17 worldwide objectives to change the world. They are a component of the 2030 Agenda for Sustainable Development, created as a "Structured Road map to drive into a more equitable and sustainable future of planet and humanity" [63]. Each of the 17 goals is interconnected and addresses domains of sustainability in the present scenario while simultaneously reducing challenges such as poverty and the effects of climate change worldwide. The SDGs are intended to "defend the earth and enhance the lives and aspirations of everyone worldwide", according to the UN [64].

The 17 SDGs generally aim to satisfy the following summarized visions and objectives.

- Healthy life without hunger, malnutrition, and poverty [65].
- Ensure that everyone has universal access and the opportunity to utilize vital amenities such as sustainable energy, water, and sanitation [66].
- Encourage the creation of development possibilities through equitable employment as well as quality and accessible education.
- Promote innovation and robust infrastructure to build towns and cities that produce and consume things sustainably.

- Lessen global inequalities, particularly those related to gender equality and discrimination. Supporting and protecting weaker sections all over the world [67].
- Protect the marine and terrestrial ecosystems while battling climate change to preserve ecological integrity and the survival of the planet [68].
- Encourage cooperation amongst various social actors to foster a peaceful atmosphere and assure ethical production, trade, and consumption.

The objectives can only be met if they are incorporated into every aspect of government. Due to the complementarities, achieving one aim may aid in attaining others at the same time. Consider how tackling climate change challenges could enhance energy security, human health, ecosystems, and marine health. The main characteristic of the SDGs is that they are not standalone goals. Most goals are interconnected and interdependent and are well-defined in their perspectives and their plan of action of applications. Interconnectedness implies that achieving one goal leads to supporting another; therefore, they should be seen as connecting frames of a holistic and harmonic prominent structure. This is the key feature of the 2030 Agenda for Sustainable Development, which was adopted in 2015 by the United Nations [69]. There are 17 SDGs and recommendations for collaborative relationships among and within people, policy-makers, and other stakeholders worldwide. They address social, economic, and environmental concerns and promote sustainable perspectives from a broader viewpoint [58]. Hence, to pursue a sustainable world, the focus will be global collaboration and reducing inequalities and discrimination inside the states. Accordingly, all countries must work together to create a unified plan for implementing these goals, which can only be accomplished if each country accepts accountability for its conduct [23].

The initiative that led to the creation of the future Global Goals was more comprehensive, with policy-making governments incorporating corporates, communities, NGOs, other stakeholders, and individuals right from the start. We all must move toward the same direction of sustainability to attain global goals. It will take an extraordinary effort from all facets of society to realize these goals, and business must play a significant part in that endeavor [70].

Evidence shows that all earlier objectives are closely related to environmental challenges and sustainability issues [71]. More specifically, environmental justice and sustainable development will only function to the extent of their weakest SDG, much like the adage "a chain is only as strong as its weakest link" [72]. As an illustration, significant changes in the water, energy, and food sectors will be necessary for climate change mitigation, which is also essential to safeguarding the welfare of people. In other words, environmental justice and sustainable development goals must all be addressed effectively in order to achieve the full range of objectives for global sustainability [73].

4. Sustainability Challenge and Nexus

The interpretation of definitions and concepts of sustainability and sustainable development is still incomplete, and the perspectives are different for stakeholders in various domains and spheres. Even though sustainability addresses the relation and harmony between environmental, economic, and social spheres more extensively, executing the concept in the present scenario must be very specific and introduce a holistic approach. The ecological and biological aspects must be considered to accomplish the sustainability goals, which have not yet been met. To get rid the planet of anthropogenic problems, an integrated approach is urgently and consistently needed [74]. Figure 4 illustrates the various sustainability challenges.

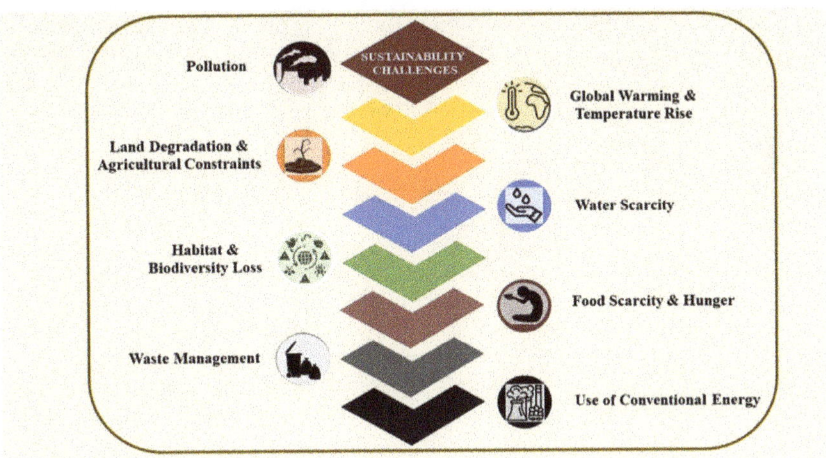

Figure 4. Sustainability challenges.

4.1. Pollution

The emergence of various pollutants over the past few decades as a result of human activity has had a negative impact on ecosystems. Due to rapid industrialization, rapid deforestation, and urbanization, all harming the ecosystems, conditions in developing countries are more dire [75]. Urban areas have developed into congestion hotspots, endangering mobility, air quality, water, and soil [76].

Due to the harm that plastics cause to ecosystems worldwide, it is a terrible ecological and environmental problem. Oil spills are another significant anthropogenic activity that harms the world's marine ecosystems [77].

Environmental pollution is a problem that affects both developed and developing nations, so it is a concern that is felt everywhere in the world. Researchers and decision-makers are eager to learn about the causes and effects of environmental pollution to develop potential remedies because it makes the Earth uninhabitable for living things and significantly contributes to global climate change. According to researchers, if environmental pollution persists, many regions will be covered by water while others will turn into deserts. In addition, extreme temperatures may exist everywhere [78]. The primary goal of the modern world is to combat environmental pollution by taking practical steps to safeguard those of us who live on the planet while being careful not to upset ecological balances. People must start reducing their waste and implementing environmentally friendly behaviors such as recycling and composting if they want the planet to remain habitable [44]. Additionally, people must avoid using too much energy or water from the environment [79]. Assisting citizens in lowering their carbon footprints, launching awareness campaigns, and implementing new laws that will lessen the effects of climate change are all things that governments should be implementing as well [77].

4.2. Global Warming and Temperature Rise

GHGs such as carbon dioxide (CO_2), methane (CH_4), nitrous oxide (N_2O), water vapor (H_2O), hydrofluorocarbons (HFCs), perfluorocarbons (PFCs), and sulfur hexafluoride (SF6) are among those contributing to the current climate change [80].

Extreme temperature changes over various parts of the Earth are predicted to happen sooner. With significant regional variations, the average global surface temperature has risen by 0.74 °C since the late 19th century and is projected to grow by 1.4–5.8 °C by 2100 [81]. A rise in sea level, changes in the distribution of plants and animals, increased environmental degradation, and natural disasters are all consequences of climate change. Other effects include hot weather, melting glaciers, polar warming, coral reef bleaching, extreme precipitation events, prolonged droughts, and dry periods [82]. Thus, it can be concluded that, when it comes to the adverse effects of human activity, global warming and climate change are at the top of the list and must be addressed right away to restore the balance. Biological alternatives and solutions can also play a significant role in addressing climate change [83].

4.3. Land Degradation and Agricultural Constraints

Land degradation is a problem that affects all life forms and is not just limited to the deterioration of soil quality. Climate change is a significant factor that contributes to land degradation, as well as to a decline in the fertility and productivity of agricultural lands [84].

As a result of erratic precipitation patterns and rising global temperatures, severe weather events such as droughts and floods have also become more frequent, aggravating wind or soil erosion [85].

Another threat that is affecting more and more parts of the world is drought. Frequent cycles of drought and flooding have been brought on by climate change, and long stretches of water scarcity have led to desertification. Droughts have increased significantly over the past 40 years, particularly in the tropics and subtropics. The world has been experiencing water stress and environmental problems since the Anthropocene era, primarily attributed to human activity. One of the most noticeable effects of drought is on nutrient uptake because water is the medium by which nutrients are transported in plants [86]. Due to the unevenly distributed C, N, and microbial diversity caused by these adverse effects, the soil becomes infertile over time. In addition to biodiversity loss, wildfires, and soil erosion, drought harms habitats.

With an estimated surface area of 1 billion ha, soil salinization is another significant factor in land degradation affecting most countries. The drought and salinization of lands are related. The main contributors to land salinization have been agricultural practices and ineffective irrigation techniques. Low-quality irrigation water causes salt to build up in the soil, and poor drainage only worsens the situation [87].

Floods degrade land quality, disrupt agroecosystem productivity, and disturb vegetation. Since the 1950s, floods have become more frequently related to climate change's effects [88]. Waterlogging-induced hypoxia in plants results in poor root development, which reduces nutrient and water uptake and stomatal conductance, which results in wilting and decreased productivity [89].

4.4. Habitat and Biodiversity Loss

The disappearance of biological diversity has grown into a complex and ongoing issue. Due to this, biological heterogeneity has decreased, which has led to an unprecedented decline in terrestrial and marine species, including flora and fauna, affecting the ecosystems' overall stability. The main issue is the extinction of plant species because they are crucial to keeping the ecosystem balanced and directly impact how it functions by providing a habitat for various other organisms [90]. Although extinction is a natural occurrence, there is no denying that a wide range of human activities also contributes to the loss of biological diversity. Over the past 40,000 years, extinctions due to human activity have increased. Within the next 240 years, the Earth will likely experience its sixth mass extinction if current trends hold [87]. In fact, some researchers contend that anthropogenic activities alone are to blame for the beginning of the sixth species mass extinction. Estimates indicate that artificial habitats have replaced natural vegetation on 43% of the Earth's land surface [88]. This century, the rate of extinction is predicted to rise by a factor of two; with accelerated climate change, this rate may grow even faster.

The main threats to biodiversity are habitat loss and fragmentation, which are mutually dependent. Due to the population explosion, there is an increased demand for resources, which has resulted in the degradation of natural habitats and a severe threat to the habitats of plants and animals [91]. Examples of this demand include the expansion of cattle ranching, mining, and building infrastructure. However, in addition to this, other anthropogenic activities are also significantly reducing the diversity of life on Earth.

4.5. Water Scarcity

Water resources are under pressure due to the ever-growing water demand brought on by population growth, economic development, and dietary changes. The World Economic Forum ranked the water supply crisis as the significantly higher risk facing our times [92].

It is crucial to comprehend water scarcity to create global, regional, national, and local policies. The Panta Rhei program set up a focused working group on "Water Scarcity Assessment: Methodology and Application" to create a cutting-edge methodology and assess water scarcity [93].

The northern hemisphere's middle-to-low latitudes generally have a high level of water scarcity, according to all the indicators. In almost all African nations, there is a severe problem with water scarcity, water poverty, and physical and economic water stress [94]. In order to retain objectivity and simplicity, all other water scarcity indicators created to date have been based solely on the physical quantity of water availability and use. Therefore, it is necessary to develop an integrated and consistent water scarcity assessment that simultaneously combines the physical, economic, and social aspects of water [95].

4.6. Food Scarcity and Hunger

The possibility of ending hunger by 2050 becomes doubtful with steady population growth. Hunger and malnutrition are primarily brought on by natural disasters, armed conflicts, population growth, and poverty [96]. The environment, the finite supply of food on the planet, and energy resources will not significantly impact these dynamics [97]. Therefore, while increased agricultural production and food preservation are essential to providing enough food for all, a more comprehensive approach is necessary to tackle the problem of global hunger [98].

4.7. Waste Management

Ecosystems and human health are seriously at risk due to the volume and complexity of the waste produced by the modern economy [99].

Waste management is also conducted to recover resources from the materials and lessen its environmental impact. Waste management can involve solid, liquid, or gaseous materials, and there are various techniques and procedures for each. Waste is managed using a variety of techniques, such as avoidance and reduction, energy recovery, recycling (physical and biological processing), and disposal (landfilling and incineration) [100]. The SDGs were created with this fundamental principle in mind. It is crucial to offer a comprehensive strategy based on sustainability as a concrete ideology that must be adopted as a way of life, a theory for formulating policies, and a sociopolitical concept of development to address the challenges the world is currently facing [101].

4.8. Industrialization and Sustainability Nexus

Industrialization refers to the economic and social change process that transforms a human group from an agrarian society into an industrial one. Regarding sustainability, industrialization has often been associated with negative impacts on the environment, natural resources, and human health and well-being. These negative impacts include pollution, deforestation, and the depletion of natural resources [102]. The nexus between industrialization and sustainability refers to the interconnected relationship between economic development, social well-being, and environmental protection. Sustainable development aims to achieve economic growth and improve living standards without compromising the ability of future generations to meet their own needs [103]. Industrialization can contribute to sustainable development by providing economic growth, jobs, and improved living standards. However, it also has the potential to harm the environment, deplete natural resources, and exacerbate social inequalities. Therefore, it is crucial to consider industrialization's environmental and social impacts and implement policies and practices that promote sustainable production and consumption [104].

4.9. Urbanization and Sustainability Nexus

Urbanization and sustainability are closely connected as urban areas are significant economic growth and development drivers, but they also have significant environmental and social impacts. Rapid urbanization can increase energy consumption, greenhouse gas emissions, environmental degradation, social inequality, and poverty. Urbanization can promote sustainable development by fostering compact, efficient, green cities [105]. Urbanization can impact the environment, including increased energy consumption and greenhouse gas emissions, water demand, and waste generation. However, sustainable urban development can help to mitigate these impacts through the promotion of energy-efficient buildings, the use of renewable energy sources, and the implementation of sustainable transportation systems. A nexus between urbanization and sustainability. The nexus between urbanization and sustainability refers to the interconnected relationship between the process of urbanization and the goal of sustainable development [79]. Urbanization can significantly impact the environment, economy, and society, and sustainable urban development is necessary to balance these impacts and promote liveable, resilient, and low-carbon cities. Urbanization can increase energy consumption, greenhouse gas emissions, and environmental degradation. It can also exacerbate social inequalities, poverty, and housing affordability issues. Sustainable urban development promotes social sustainability by addressing social equity, affordable housing, and access to services and job opportunities. It also supports the development of resilient cities that can adapt to the impacts of climate change and natural disasters. Sustainable urban planning is key to promoting sustainable urban development [106]. This includes promoting compact and efficient land use patterns, protecting natural areas and biodiversity, and integrating green spaces into the urban landscape. Additionally, sustainable transportation systems, such as public transportation, walking, and cycling, can reduce dependence on personal vehicles and improve air quality.

4.10. Globalization and Sustainability Nexus

Globalization and sustainability are related concepts that positively and negatively impact economic, social, and environmental development. On the one hand, globalization can lead to increased economic growth and improved living standards, greater access to goods and services, and enhanced communication and cultural exchange. However, it can also lead to adverse environmental impacts such as increased greenhouse gas emissions, biodiversity loss, and increased dependence on non-renewable resources [107]. Globalization can also exacerbate social inequalities, increasing poverty and marginalizing certain groups. The nexus between globalization and sustainability is complex, and it requires a holistic approach that considers the interrelated economic, social, and environmental aspects of global development. This can be achieved by implementing sustainable development policies, regulations, and initiatives that promote environmentally and socially responsible economic growth. One example of this nexus is sustainable trade, which supports economic growth while also addressing environmental and social concerns [108].

Globalization refers to the interconnectedness and interdependence of countries and economies by exchanging goods, services, information, and ideas. Globalization can lead to increased economic growth and improved living standards, but it can also contribute to environmental degradation and social inequality. On the other hand, sustainability aims to address these negative impacts by promoting environmentally and socially responsible economic development. Therefore, it is vital to find ways to balance the benefits of globalization with the need for sustainable development. This can include implementing policies and practices that promote sustainable production and consumption, protecting natural resources, and reducing inequality [107]. Globalization and sustainable development are related concepts that positively and negatively impact economic, social, and environmental aspects of development. On the one hand, globalization, which refers to the increased interconnectedness and interdependence of the world's economies, cultures, and populations, can lead to increased economic growth and improved living standards, greater access to goods and services, and enhanced communication and cultural exchange. Globalization can also lead to adverse environmental impacts such as increased greenhouse gas emissions, biodiversity loss, and dependence on non-renewable resources [109]. It can also exacerbate social inequalities, increasing poverty and marginalizing certain groups.

4.11. Climate Change and Sustainability Nexus

Climate change refers to the long-term changes in temperature, precipitation, wind patterns, and other measures of climate that occur over several decades or longer. It is primarily caused by burning fossil fuels, deforestation, and other human activities, which release greenhouse gases into the atmosphere, trapping heat and warming the planet. Addressing climate change is, therefore, a key component of sustainable development. This can include reducing greenhouse gas emissions by increasing the use of renewable energy, improving energy efficiency, and implementing carbon pricing [74]. Additionally, adaptation measures such as building sea walls or drought-resistant crops can help communities and ecosystems cope with the impacts of a changing climate. Therefore, it is essential to take action to reduce greenhouse gas emissions and to adapt to the changes that are already happening and those that are projected to occur in the future. Climate change can negatively impact the balance of global society by increasing poverty, reducing access to food and water, and exacerbating health problems [110]. On the other hand, addressing climate change through sustainable development practices such as renewable energy, energy efficiency, and sustainable land use can create economic opportunities and improve the well-being of communities.

The United Nations Framework Convention on Climate Change (UNFCCC) and the United Nations Sustainable Development Goals (SDGs) are closely related and mutually reinforcing. They provide a roadmap to achieve a better and more sustainable future for all. The Paris Agreement, adopted under the UNFCCC, aims to strengthen the ability of countries to address the impacts of climate change and to accelerate and intensify the actions and investments needed for a sustainable low carbon future. Sustainability plays a crucial role in addressing climate change. As previously mentioned, climate change is a significant threat to sustainable development, and addressing climate change is a critical component of sustainable development. Sustainability is vital in building resilience to climate change through disaster risk reduction and community preparedness [111]. Addressing climate change by promoting sustainable consumption and production patterns is desirable. This includes reducing waste, conserving natural resources, and promoting sustainable products and services [105]. To effectively face the challenges to sustainability due to climate change, a combination of mitigation and adaptation strategies, as well as international cooperation, access to finance and technology, and strengthened governance and institutions, are required.

4.12. Natural Disasters and Sustainability Nexus

Natural disasters can have a significant impact on sustainability. They can cause loss of life and injury, damage infrastructure and buildings, and disrupt economic activity. Additionally, natural disasters can exacerbate poverty, inequality, and vulnerability, particularly among marginalized communities. Climate change is projected to increase the frequency and intensity of many natural disasters, such as floods, droughts, and storms. This makes sustainability even more important as it can help to build resilience to these events [112].

Sustainability can help to reduce the risks and impacts of natural disasters in several ways [113]. Disaster risk reduction: By identifying and addressing the underlying risk factors that make communities and infrastructure vulnerable to natural disasters, sustainability can help to reduce the likelihood and severity of these events. Adaptation: Sustainability can help to build the resilience of communities and ecosystems to the impacts of natural disasters through strategies such as making sea walls, drought-resistant crops, and community-based adaptation programs. Sustainable land use: Sustainable land use can help to reduce the risk of natural disasters such as floods and landslides by preventing deforestation and protecting wetlands and other natural habitats. Sustainable infrastructure: Sustainable infrastructure, such as green buildings and resilient transportation systems, can help to reduce the impacts of natural disasters on communities and economies. Community-based approach: A community-based approach to sustainability can ensure that the most vulnerable communities are involved in the planning and implementation of disaster risk reduction and adaptation measures. In conclusion, natural disasters can significantly impact sustainability and climate change is projected to increase the frequency and intensity of many natural disasters [114]. Sustainability can help to reduce the risks and impacts of natural disasters by building resilience, protecting vulnerable communities, and promoting sustainable practices.

4.13. Population Rise and Sustainability Nexus

Population growth can significantly impact sustainability, both positively and negatively. On the one hand, a larger population can lead to increased economic growth, technological innovation, and cultural diversity [106]. On the other hand, a rapidly growing population can strain natural resources and lead to environmental degradation, urbanization, and increased demand for energy, food, and water. One of the main concerns regarding population growth is its impact on the environment [115]. As the population increases, so does food, water, and energy demand. This can lead to the overconsumption of natural resources, deforestation, and pollution. Additionally, population growth can lead to urbanization and land-use changes, which can cause a loss in biodiversity and ecosystem services.

It is essential to consider sustainability in population growth control and development policies to address these challenges. This can include family planning programs, education and healthcare access, and policies promoting sustainable consumption and production. Promoting sustainable urbanization, land-use planning, and investing in renewable energy and water conservation technologies is also essential. Population growth and sustainable development are closely linked, as controlling population growth can significantly impact the ability to achieve sustainable development goals [116]. To address these challenges of population growth, it is crucial to incorporate population dynamics into sustainable development policies. This can include family planning programs, education and healthcare access, and policies promoting sustainable consumption and production. Promoting sustainable urbanization, land-use planning, and investing in renewable energy and water conservation technologies can also be included [103].

5. Models and Principles of Socio-Economic Growth
Current Scenario

In the 21st century, the ideology of sustainability and sustainable development has been globally accepted and gained momentum. While our understanding of sustainability has significantly increased, development has become harder to define in many ways. However, studying sustainable development is insufficient because it is time to act.

In this post-Brundtland era, sustainability, through the Sustainable Development Goals, addresses challenges such as increasing conventional energy consumption, loss of biomass and land degradation, skepticism of science, financial disparities in life and opportunities, and a fragmented set of universal policies, institutional frameworks, and governance. Furthermore, due to several interconnected phenomena, the difficulties of sustainability and development are more complex today than they were in Brundtland's time. Adopting pluralistic and transdisciplinary approaches to sustainability analysis is a crucial strategy to face the challenges of the present scenario around the globe.

The contentious nature of sustainability as a dominant policy discourse has encouraged the formation of many public forums for discussion and engagement [117]. Though idealistic, the concepts and methods point to the fact that the deliberative democracy, which includes open discourse, open decision-making, holding decision-makers accountable, as well as reasoned and respectful debate, is essential to achieve green development in public spheres wherein the various sustainable development ideas can be discussed and improved upon and manage it socially, politically, and financially [118].

The task of defining sustainability and sustainable development has been a complex and ongoing endeavor for researchers. The lack of a universally accepted definition stems from the diverse interpretations attributed to the phrase, particularly in relation to its association with "economic growth". This has sparked a debate among scholars, as some argue that traditional notions of development, synonymous with continuous economic expansion, are incompatible with sustainability, given the finite resources of our planet [119]. Bringing together the diverse perspectives of sustainability under a unified framework has proven to be a challenging task. Over the years, the definitions of sustainability have evolved while retaining their core essence [105]. However, there is still a need for a comprehensive explanation that can encompass all the different domains of sustainability. Additionally, it is important to acknowledge that the concept of sustainability is influenced by the economic growth and political ideologies prevalent in different nations. Table 4 presents a comparative analysis of different socio-economic growth models.

Table 4. Comparison of capitalism, socialism, and communism.

	Pros	Cons
Capitalism	Most efficient and effective way to allocate resources	This leads to economic inequality, environmental degradation
	Create enormous wealth	Focus on profit over the well-being of people and communities
	Encourages innovation and hard work.	Different levels of government regulation and intervention
Socialism	More just and fair economic system	Lead to inefficiencies and a lack of financial incentives
	Reduce income inequality and provide a safety net for all	Government has a low level of control over the economy and the lives of its citizens
	Focus on the well-being of people and communities	It can limit individual freedom and personal responsibility
Communism	Meeting the basic needs of all members of society and maximizing the collective well-being	Economic inefficiency and widespread human rights abuses
	Seeks to eliminate the exploitation of one person by another	
	Create a society based on equality and cooperation	

Capitalism: Capitalism is an economic system in which the means of production, prices of commodities (goods and services), and distribution are privately owned and operated. The prices are determined by supply and demand in a competitive market [120]. In a capitalist economy, the goal is to make a profit, and businesses are free to operate and compete with one another [121].

Socialism: Socialism is an economic and political system in which the means of production and distribution are owned and controlled by the state community rather than private individuals [122]. In a socialist system, the wealth produced by the economy is shared more equally among the members of society, and there is often a strong emphasis on providing for the basic needs of all people, including healthcare, education, and social security [123].

Communism: Communism is a political and economic ideology that seeks to create a classless, stateless society in which the means of production and distribution are owned and controlled by the community as a whole [124]. Most generally, communism refers to community ownership of property, with the end goal being complete social equality via economic equality. Under communism, the goal is to create a society where everyone works according to their abilities and receives according to their needs [125]. Fundamentally, communism argues that all labor belongs to the individual laborer; no man can own another man's body, and therefore each man holds his work.

The universally accepted capitalistic GDP growth-driven economic model, which has prevailed for the last century, has proven to be a complete failure and is leading us toward a disastrous path. Reliant on fossil fuel burning, this model exacerbates climate change, pollution, biodiversity loss, and freshwater depletion while reinforcing unequal wealth distribution. Its promises of economic success and improved quality of life for citizens have come at the expense of the environment and social equity. In summary, these socio-economic models have their drawbacks and challenges, and there is no one-size-fits-all solution for achieving sustainable economic growth and development. Urgent and comprehensive revaluation is needed to shift toward alternative mechanisms prioritizing sustainable development, incorporating concepts such as the circular economy, nature-based solutions, social innovation, and responsible consumption and production patterns, to ensure a resilient and equitable future for all. A more holistic and inclusive approach must be identified and developed to create a more sustainable and equitable future.

6. Global Sustainability and Sustainalism: An Integrated Framework

The world's economic development model has become saturated with an excessive focus on increasing consumption. As a result, humanity is confronted with many significant threats, including climate change, health crises such as COVID-19, and economic instability. These crises have served as powerful reminders of the importance of cooperative actions and global solidarity. Therefore, the path to global sustainability lies in educating the masses and nurturing a knowledge-based economy and socially responsible society [126]. Sustainalism builds on the foundations laid by capitalism and socialism but takes the broader view that the challenges of today and tomorrow demand of us. The more considerable paradigm shift from capitalism, communism, and socialism is sustainalism. We need a paradigm shift from capitalism or moderated socialism [127]. The new model of social economy for the current generation is referred to as "sustainalism" [128].

6.1. Global Sustainability 6S Principles: A Tool to Achieve a Sustainable Economy

It is crucial to recognize that all life forms on Earth, including humans, animals, and plants, are intricately interconnected. In the era of globalization and rapid digitalization, countries worldwide, regardless of their size, wealth, or level of development, rely on each other in various aspects. This interconnectedness is fostered through economic, cultural, and social relations. The scientific community comprehends the inherent value of this interconnectedness and its implications for our collective well-being. Furthermore, the distribution of global resources is highly unequal, posing challenges to achieving sustainable development. A genuine appreciation for Mother Nature, the Earth, and its delicate ecosystems is at the core of sustainable actions. This sentiment serves as the foundation for sustainable practices. It drives individuals, organizations, and governments to adopt a holistic approach to development that integrates economic, social, and environmental considerations in every corner of the world.

A plan for a sustainable economy is presented in a simple equation format.

Sustainable Economy = 6S Principles of Global Sustainability
= Happiness + Well-being + Equality
= Regenerative Practices +Climate and Biodiversity Protection + Ecological Restoration

The "Global Sustainability Framework" is a novel comprehensive toolkit comprising 6S principles (Figure 5). This framework equips individuals, organizations, and governments with the tools to pursue global sustainability effectively. However, the responsibility for creating a sustainable future lies on the shoulders of global citizens, who must embrace this responsibility and work collectively to ensure a thriving and sustainable world for future generations.

Figure 5. Tool for sustainalism (6S principles).

1S—Sustainable Energy, Resource Efficiency, and Circular Economy:

1S principle is a compelling approach highlighting the urgency of transitioning to sustainable energy sources, promoting resource efficiency, and adopting a circular economy. Our reliance on finite energy sources, such as fossil fuels, poses significant risks for future generations [129]. By embracing sustainable energy, we can mitigate these risks and ensure a more secure and stable energy supply for the long term.

The need for a sustainable energy transition is evident. Investing in renewable energy infrastructure and promoting energy efficiency measures are crucial in reducing greenhouse gas emissions and combating climate change [130]. Technologies such as solar panels, wind turbines, and hydroelectric power plants offer promising avenues for generating clean and renewable energy [131].

In addition to transitioning to sustainable energy sources, resource efficiency plays a vital role in promoting sustainability. We can effectively utilize water, materials, and energy resources across various sectors by optimizing resource consumption, minimizing waste generation, and encouraging recycling and reuse [132]. Energy efficiency improvements, in particular, offer significant opportunities to reduce energy consumption and enhance sustainability [133].

The concept of a circular economy further strengthens the 1S principle. By designing out waste and pollution, extending the lifespan of products and materials, and regenerating natural systems, the circular economy offers a transformative approach to resource management (Figure 6). It shifts the focus from a linear take–make–dispose model to one that values waste as a resource and prioritizes reuse, remanufacturing, and recycling [134]. Embracing circular business models and strategies promotes sustainability, reduces resource depletion, and minimizes environmental impact [135].

Adopting a circular economic model is crucial for overcoming the challenges posed by climate change and resource depletion [136]. By fundamentally rethinking our approach to resource utilization, we can break free from the unsustainable practices of the past and build a resilient and sustainable economy. This shift benefits the environment, presents economic opportunities, and fosters innovation in sustainable technologies and practices [137].

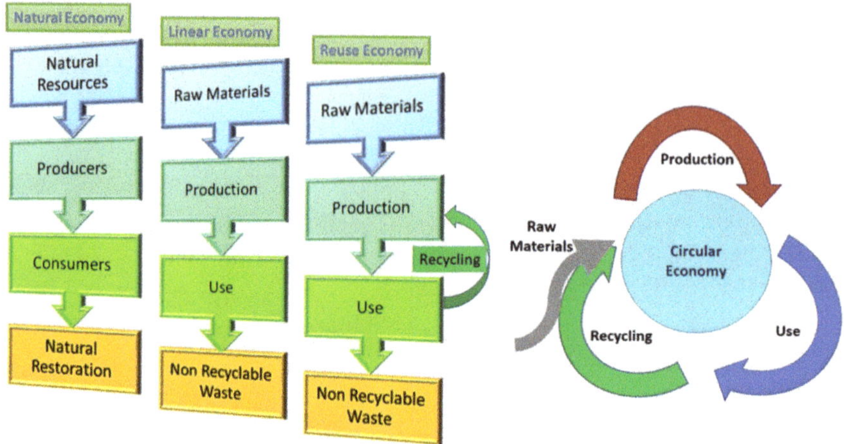

Figure 6. Concept of circular economy.

2S—Sustainable agriculture, agroforestry, and bioeconomy:

2S—Sustainable agriculture and agroforestry present a compelling solution to address food security challenges while promoting sustainability [138]. These practices go beyond conventional farming methods and integrate environmental, social, and economic considerations into agricultural systems [139].

Sustainable agriculture encompasses a range of practices prioritizing the conservation of natural resources, protection of the environment, and the well-being of farmers and communities. By adopting sustainable farming techniques, such as organic farming, precision agriculture, and crop rotation, farmers can minimize the use of harmful chemicals, preserve soil fertility, and reduce water consumption [140]. This approach ensures the long-term viability of agricultural land and safeguards the health of consumers and ecosystems.

Conversely, agroforestry combines agricultural activities with cultivating trees and woody plants in a mutually beneficial manner. This integrated approach offers numerous benefits for sustainable land management [141]. Trees and wood plants serve as natural allies, contributing to soil conservation by preventing erosion, enhancing soil health through increased organic matter, and improving water retention capabilities. They also play a vital role in conserving water resources by reducing evaporation and promoting water infiltration into the soil. Furthermore, the carbon sequestration potential of trees and woody plants helps mitigate climate change by absorbing atmospheric carbon dioxide [142].

Agroforestry systems also provide valuable habitats for wildlife, including birds, insects, and mammals [143]. Creating diverse and interconnected ecosystems contributes to biodiversity conservation and supports the preservation of valuable ecological services. Moreover, agroforestry allows farmers to diversify their income streams and improve economic stability by introducing a broader range of products and value-added opportunities [144].

Bioeconomy encompasses various production sectors, such as industrial and economic domains, which utilize biological resources and methods to produce bio-based goods and services [145]. The bioeconomy concept revolves around achieving sustainability by leveraging natural resources and processes in economic activities [146]. In doing so, it fosters the development of bio-based industries and generates employment opportunities [147]. A regenerative economy and bio-economy provide the opportunity to fulfill the needs of all individuals emphasizing the restoration, regenerative practices—embracing the potential of the bio-economy—and renewal of natural resources, thereby safeguarding the health and vitality of our planet.

We can address food security challenges by embracing sustainable agriculture and agroforestry while promoting environmental stewardship and social well-being [148]. These practices provide a holistic and resilient approach to agricultural production, ensuring the long-term availability of nutritious food, safeguarding natural resources, and supporting local communities.

3S—Sustainable building, health, and lifestyle:

Sustainable living is making environmentally, socially, and economically responsible choices to reduce one's impact on the planet and contribute to a more sustainable future [149]. This can involve making lifestyle changes, such as

Sustainable built environment: Incorporating sustainable transportation principles into urban planning and infrastructure development [150]. This includes designing cities and communities prioritizing walkability, cycling infrastructure, and efficient public transportation systems, reducing the need for long-distance commuting and promoting compact and sustainable development [150].

Green infrastructure/sustainable practices: Sustainable living often involves reducing one's consumption of resources, such as energy, water, and materials. This can be achieved through energy conservation, water conservation, and waste reduction [151]. It involves supporting businesses and organizations that adopt sustainable practices, such as using renewable energy, reducing waste, and protecting the environment [152]. Additionally, effective land use planning is vital in curbing and, ideally, preventing urban sprawl, contributing to the depletion of natural and agricultural lands. A sustainable lifestyle bestows positive effects on one's physical and mental health. The integration of sustainable building, health, and lifestyle creates a comprehensive approach to living in harmony with the environment while prioritizing personal well-being.

Protecting natural habitats: Sustainable living often involves protecting and preserving natural habitats, such as forests, wetlands, and oceans [153]. Implementing green infrastructure networks is crucial in filtering and purifying water and air while promoting energy and water efficiency through retrofitting measures in new and existing developments. By integrating human and environmental consciousness into our lives, we can make more sustainable choices that benefit both people and the planet [154,155].

4S—Sustainable mobility, transportation, and eco-tourism:

4S—Sustainable Mobility, Transportation, and Eco-Tourism offer a transformative approach to address sustainability challenges in transport and tourism. This comprehensive principle encompasses various elements that can revolutionize how we move and explore the world while minimizing environmental impact and fostering inclusive and responsible practices.

Sustainable Transportation Systems are crucial in reducing carbon emissions, alleviating traffic congestion, and improving air quality [156]. By promoting and prioritizing sustainable modes of transportation such as walking, cycling, public transit, and electric vehicles, we can create a more environmentally friendly and efficient transportation network [157].

Efficient and Integrated Transport Networks are essential for optimizing travel routes, enhancing connectivity, and reducing the overall environmental footprint of transportation activities [158]. By designing and implementing integrated transport systems that prioritize efficiency and sustainability, we can significantly improve the overall transportation experience [159].

Active and Shared Mobility encourages individuals to embrace operational modes of transportation such as walking and cycling while promoting shared mobility options such as carpooling and ride-sharing [160]. These initiatives help to reduce the reliance on private vehicles, decrease traffic congestion, and encourage sustainable travel choices.

Accessible and Inclusive Transport is vital to ensure that transportation systems cater to the diverse needs of individuals, including those with disabilities, the elderly, and those with limited mobility [161]. By designing infrastructure, vehicles, and services that are accessible and inclusive, we can create transportation systems that leave no one behind [162]. Moreover, fostering sustainable supply chains and collaboration across stakeholders is crucial. This involves promoting responsible sourcing, reducing carbon emissions in transportation, and minimizing environmental impacts throughout the supply chain.

Eco-Tourism and Sustainable Travel promote responsible tourism practices that minimize negative environmental impacts, support local communities, and preserve natural and cultural heritage [163]. Emphasizing eco-friendly behaviors, supporting local economies, and raising awareness about sustainable travel choices can lead to a more sustainable and enriching travel experience [164].

By integrating these elements within the 4S principle, we emphasize the importance of sustainable mobility, transportation systems, and eco-tourism in mitigating climate change, enhancing accessibility, and preserving our natural and cultural resources. It calls for prioritizing sustainable travel choices, embracing efficient transportation modes, and incorporating sustainability into urban planning and tourism practices for a more sustainable and inclusive future.

5S—Sustainable Education, Innovative Research, and Entrepreneurship:
5S—Sustainable Education, Innovative Research, and Entrepreneurship form a dynamic and transformative approach to addressing sustainability challenges. This principle highlights the critical need for an education system that actively prepares individuals to contribute to sustainable development, fosters innovative research, and nurtures entrepreneurial endeavors for a sustainable future [165].

Sustainable Education is the cornerstone of this principle, advocating for educational systems that integrate sustainability principles at all levels, from primary to higher education [166]. By infusing environmental awareness, social responsibility, and sustainable practices into the curriculum, we can equip students with the knowledge and skills necessary to navigate and thrive in a sustainable world. Hands-on experiences, experiential learning, and lifelong learning opportunities further empower individuals to adapt to evolving sustainability needs and foster a mindset of continuous growth and development [167]. Education for sustainable development emphasizes critical thinking, problem-solving, and global citizenship, preparing future generations to address sustainability challenges.

Education and Outreach initiatives are crucial in raising awareness about the importance of sustainable energy and its various applications. By implementing public information campaigns and sustainability education programs, we can engage and inspire individuals to embrace sustainable practices and promote the adoption of sustainable energy solutions [168]. These efforts contribute to a broader cultural shift toward sustainability and encourage active participation in creating a more sustainable future.

Knowledge Transfer and Collaboration are fundamental aspects of sustainable education, research, and entrepreneurship. Facilitating knowledge and technology exchange between academia, industry, and communities creates synergies that drive sustainable development forward [169]. Collaborative partnerships leverage expertise and resources, fostering innovation and enabling the practical application of research outcomes. By establishing strong ties with local organizations and businesses, educational institutions can provide students with real-world exposure to sustainability challenges and opportunities, empowering them to impact their community [170] positively.

Entrepreneurship and Social Innovation are integral to addressing sustainability challenges effectively [171]. We foster interdisciplinary collaboration and propel sustainable development by supporting research endeavors that tackle these challenges and contribute to innovative solutions. Cultivating an entrepreneurial culture encourages individuals to generate ideas and develop solutions aligned with sustainability goals [172]. Providing aspiring entrepreneurs in the sustainable sector with the necessary support, resources, and mentorship enables them to translate their ideas into impactful ventures.

In the digital age, incorporating technologies such as digitalization, AI, IoT, and intelligent and automated systems further amplifies the potential for sustainable education, research, and entrepreneurship. These tools enhance efficiency, optimize resource utilization, and enable more effective decision-making, contributing to sustainable practices and outcomes. This perspective empowers individuals to become agents of change, equipping them with the knowledge, skills, and entrepreneurial spirit needed to address sustainability challenges and create a more sustainable and prosperous future for all.

6S—Sustainable business, governance, and finance:

6S—Sustainable governance is a powerful approach that addresses sustainability challenges by promoting a culture of sustainability, ethics, and responsible decision-making within organizations [173]. This principle recognizes the significance of integrating social justice considerations into sustainable organizational culture and ethical governance practices [174]. We can leverage various approaches to achieve sustainable governance by adopting a holistic perspective.

A sustainable organizational structure is crucial for long-term stability and effectiveness while ensuring environmental, social, and economic sustainability. Organizations must proactively design systems that align with sustainable principles, enabling them to adapt to changing circumstances and prioritize sustainability in their operations. This involves considering the environmental impact of business practices, fostering social responsibility, and optimizing economic outcomes sustainably [175].

Ethical Leadership serves as a cornerstone of sustainable governance. Leaders at all levels of an organization must embody moral values, champion social justice, and address systemic inequalities. By embracing diversity and inclusion, ethical leaders create an environment that values the contributions of all individuals and promotes a sense of fairness and justice [176]. Ethical leadership fosters a culture where sustainable practices are embedded into decision-making processes and guides the organization toward long-term sustainability goals.

Social Justice values are integral to sustainable governance. Organizations must embed equity, fairness, and social justice principles into their organizational culture and decision-making processes [177]. This entails promoting inclusivity, embracing diversity, and providing equal opportunities for all organization members. By aligning all levels of the organization around shared values such as sustainability and social justice, we create a foundation for sustainable practices and facilitate collective efforts toward long-term sustainability.

Sustainable Policy and Stakeholder Engagement play a crucial role in sustainable governance. Policy interventions, such as tax credits, incentives for renewable energy, energy efficiency regulations, and funding for research and development, encourage adopting sustainable practices [178]. Moreover, engaging stakeholders, including employees, communities, customers, and other relevant actors, allows for a collaborative approach to decision-making [179]. By involving diverse perspectives, organizations can make more informed and sustainable decisions that consider the needs and interests of all stakeholders [180].

Collaborative Decision-Making is essential for sustainable governance. Encouraging collaboration and participation from all levels of the organization enables a diversity of perspectives and ideas to be considered. We can foster sustainable collaboration and drive positive change by establishing common agendas, engaging in participatory decision-making, and monitoring progress. This inclusive approach facilitates the identification of innovative solutions, ensures transparency in decision-making processes, and fosters a sense of ownership and commitment among stakeholders. By embracing collaborative decision-making, organizations can effectively address sustainability challenges and promote adopting sustainable practices. Figure 7 provides a concise overview of the essential components of the 6S principles, which serve as a roadmap for attaining global sustainability.

Figure 7. Global sustainability 6S principles: A tool to achieve sustainability.

By incorporating sustainable energy practices, resource efficiency, and circular economy principles (1S), we can reduce reliance on finite resources, mitigate climate change, and minimize waste. Sustainable agriculture and agroforestry (2S) contribute to food security, soil conservation, and habitat preservation. Sustainable mobility and transportation (3S) promote low-carbon options, reduce congestion, and enhance accessibility. Sustainable habitat and lifestyle (4S) focuses on sustainable urban development, healthy life style choices, physical and mental well-being, waste management, and responsible consumption. Sustainable education, innovative research, and entrepreneurship (5S) foster knowledge transfer, skills development, and solutions for sustainability challenges. Finally, sustainable governance (6S) promotes ethical leadership, social justice, and stakeholder collaboration.

By embracing the 6S—Sustainable governance principle, organizations can create a framework that fosters sustainability, ethics, and responsible decision-making. This approach integrates social justice considerations into organizational culture, promotes ethical leadership, engages stakeholders, embraces collaborative decision-making, and leads to a more sustainable and equitable future.

6.2. Concept of Sustainalism

The term sustainalism has not gained widespread usage. It is not well-defined or widely recognized in the literature. "Sustainalism" is a socio-political, economic, and environmental theory of global social organizations as a whole that advocates for the means of production, distribution, exchange, and symbiotic lifestyle to be owned or regulated by the worldwide community holistically [127]. Sustainalism is a social equity and inclusiveness theory built on the foundations laid by capitalism, communism, and socialism (Figure 8). Sustainalism is the new, inclusive, and more equitable socio-economic–environmental theory and practice model of the 21st century to meet the needs of the 10 billion people who will share a single planet in just a few decades from now. Sustainalism is an Earth-friendly conscious civilization that inculcates a socio-economic–environmental model to decarbonize the physical economy and embrace green products and services. Sustainalism is a term that has been used to describe a concept or approach that combines elements of sustainability with elements of traditionalism [126]. Sustainalism is a resource-efficient lifestyle wherein the materials are economically produced from 100% natural resources such as wood, plant fibers, etc.

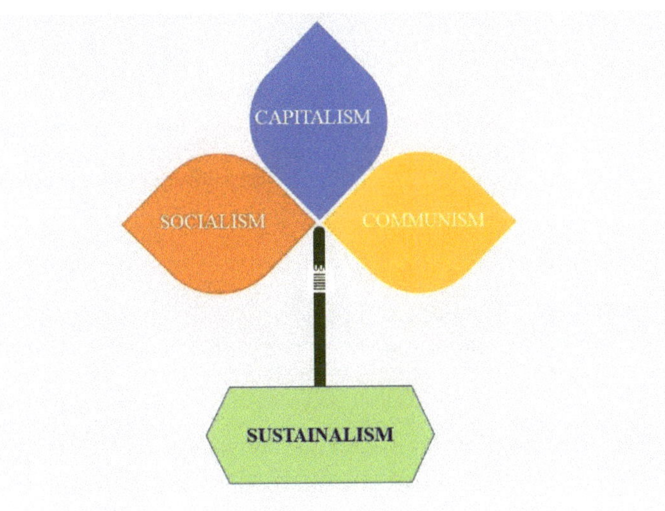

Figure 8. Concept of sustainalism.

Sustainalism is a new model of survival and a practice leading to a sustainable era. Sustainalism paves the path to a "Sustainable Revolution" which satisfies the needs of our future. Sustainalism is an art of social engineering toward a greener and more sustainable lifestyle while promoting harmonious relationships within the environmental, economic, social, political, and cultural domains (Figure 9). This includes adopting ecological principles, eco-lifestyle, decarbonizing the physical economy, as well as embracing green products and services [181]. Sustainalism is a collaborative practice of all the stakeholders, including governments, organizations, private sector, public sector, service sectors, corporates, entrepreneurs, investors, and individuals who are collectively responsible for shifting the global socio-economy from the current equilibrium, which is a status quo to a better, that is, cleaner, renewable, and sustainable planet [182].

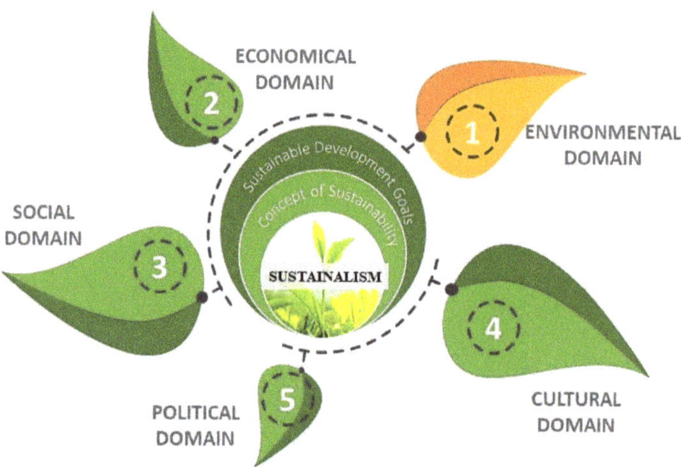

Figure 9. Domains of sustainalism.

6.3. Objective of Sustainalism

The objectives of sustainalism are as follows:

- To create economic growth and prosperity while protecting the environment and promoting social equality.
- To find ways to live and consume environmentally sustainably and maintain and preserve traditional cultural practices and values.
- To focus on local communities, self-sufficiency, and intergenerational equity.
- To complement education, leadership, and collective consciousness to sustain a quality life for society.
- To emphasize using nature-based solutions, such as green technology and carbon pricing, to address economic, environmental, and social problems.
- To advocate for creating new businesses, policies, and regulations that promote environmental, social, and economic sustainability in the short and long term.

6.4. Role of Individuals in Sustainalism: Sustainalist

"Sustainalist" is a term used to describe a person who prioritizes sustainability in all aspects of life, including environmental, economic, and social domains. It emphasizes the need to balance economic growth with environmental protection and social justice so that current needs can be met without compromising the ability of future generations to meet their own needs. The role of sustainalists in the future will be crucial in promoting and advocating for sustainable practices and policies that prioritize the well-being of current and future generations [183]. Sustainalists will work toward creating a more equitable and environmentally responsible world by promoting environmentally friendly technologies and practices, reducing carbon emissions, preserving natural resources, and practicing social justice and equality. They will also play a key role in raising public awareness about the importance of sustainalism, and influencing businesses, governments, and the public to adopt a new socio-economic–environmental model of sustainalism. By working toward a more sustainable future, sustainalists will help ensure that the planet remains habitable and that future generations can thrive. Becoming a sustainalist involves incorporating sustainable principles and practices into your daily life and advocating for policies and initiatives that prioritize sustainability [183].

Here are a few steps to becoming a sustainalist:

1. Educate yourself: Learn about the principles of sustainalism, including environmental, economic, and social domains, and the impact of human activities on the planet.
2. Reduce your carbon footprint: Start by reducing waste, conserving energy, and using environmentally friendly products.
3. Support sustainable businesses: Look for products and services that prioritize sustainability and support companies that have environmentally friendly practices.
4. Advocate for sustainable policies: Write to your local representatives, participate in environmental campaigns, and raise awareness about the importance of sustainalism.
5. Live sustainably: Incorporate sustainable practices into your daily life, such as cycling or taking public transportation instead of driving, eating a plant-based diet, and conserving water.
6. Lead by example: Encourage others to adopt sustainable practices by sharing your experiences and knowledge with family, friends, and colleagues.

By following these directions, we can take a step closer to realizing the principles of sustainalism.

6.5. Role of Society in Sustainalism

Sustainability, closely related to sustainalism, is a relatively new concept that has gained widespread recognition and support in recent years, and this trend will likely continue. However, the ideas and values associated with sustainalism will likely continue to be meaningful and influential. As the global population continues to grow and the impacts of human activity on the environment become more apparent, there will likely be increasing emphasis on finding ways to live and consume more sustainably. We have an exceptional opportunity for individuals, sectors, companies, and organizations to go down in history as the generations that changed the course of the world for the better [183]. This may involve a focus on local communities, self-sufficiency, intergenerational equity, and other values and practices associated with sustainalism. A sustainable lifestyle consists of promoting social equity, diversity, inclusion, social justice, fair labor practices, advocating for human rights and reducing inequality. Adopting a sustainable lifestyle requires a shift in our attitudes, values, and beliefs. It requires us to recognize our interconnectedness with each other and the natural world. It also requires us to take collective responsibility for our actions and their impact on the environment and society.

6.6. Role of Nations: Sustainable Revolution

Sustainable revolution refers to a significant transformation at the global level to achieve a more sustainable future [184]. It involves a shift in values, attitudes, and behaviors toward sustainalism and adopting sustainable practices and policies at all levels of society, from individuals to governments and various states [185]. The sustainable revolution aims to create a more equitable and environmentally responsible world wherein economic growth is balanced with environmental protection and social justice. "Sustainable Era" refers to a time in which sustainalism is the dominant paradigm, and sustainable practices and policies are widely adopted and implemented. A strong focus on environmental protection, resource conservation, and social equity and the widespread adoption of sustainable technologies and practices characterizes the sustainable era. In this era, economic growth is decoupled from environmental degradation, and the world operates in a way that prioritizes the well-being and survival of both current and future generations. The sustainable era aims to create a more livable and sustainable world for all.

7. Conclusions

In conclusion, sustainability has emerged as a crucial response to environmental degradation, social inequality, and economic instability. However, traditional approaches to sustainable development have proven inadequate in tackling these complex challenges, necessitating a more comprehensive and holistic approach.

To address these limitations and foster a paradigm shift toward a more sustainable and inclusive world, this paper proposes an integrated socio-economic and environmental model, the 6S principles of global sustainability. We have presented a novel perspective on achieving sustainable development goals through a social movement centered around sustainable education, sustainable living, peace, social justice, social equity, sustainable housing, sustainable networks (including mobility and health infrastructure), and sustainable energy. The 6S theoretical framework offers a clear roadmap toward achieving global sustainability and effectively tackles the challenges related to sustainability through an inclusive approach.

To enhance individual responsibility toward sustainable development, the concept of sustainalism is introduced. Building upon the principles of sustainalism, the Global Sustainability Framework recognizes the interconnectedness of different dimensions of sustainability and the diverse Sustainable Development Goals (SDGs). It highlights the importance of collective action, dedication, and collaboration among individuals, organizations, and governments for a fair and inclusive quality of life. Sustainalists adopt this new way of thinking and practice, recognizing the interdependence of all living beings and advocating for social and environmental justice.

Implementing the Global Sustainability Framework and embracing sustainalism necessitates a sustainable revolution—an unprecedented collective movement to reshape our societies, economies, and governance systems toward sustainability. The sustainable revolution offers a transformative pathway toward achieving an equitable world, marking a significant step toward a sustainable era.

By embracing the principles of sustainalism and adopting a sustainalist lifestyle, we can pave the way toward a more sustainable economy that balances humanity's and the environment's needs, benefiting everyone involved.

Author Contributions: Conceptualization, K.S.; formal analysis, K.B.M.; investigation, N.P.H. and V.S.; data collection, K.S. and K.B.M.; writing—original draft preparation, N.P.H. and K.S.; writing—review and editing, V.S. and N.P.H.; supervision, K.S. All of the authors contributed significantly to the completion of this review, conceiving and designing the study, and writing and improving the paper. All authors have read and agreed to the published version of the manuscript.

Funding: This research received no external funding.

Institutional Review Board Statement: Not Applicable.

Informed Consent Statement: Not Applicable.

Data Availability Statement: Not Applicable.

Acknowledgments: We would like to express our sincere gratitude to the editor, the anonymous reviewers for their invaluable support and constructive feedback on the manuscript. We also thank the ICFGS Community, a renowned knowledge think-tank on sustainability, for their valuable support throughout the process.

Conflicts of Interest: The authors declare no conflict of interest.

References

1. Boyer, R.H.; Peterson, N.D.; Arora, P.; Caldwell, K. Five Approaches to Social Sustainability and an Integrated Way Forward. *Sustainability* **2016**, *8*, 878. [CrossRef]
2. Brander, J.A. Viewpoint: Sustainability: Malthus revisited? *Can. J. Econ.* **2007**, *40*, 1–38. [CrossRef]
3. Keeble, B.R. The Brundtland Report: "Our Common Future". *Med. War* **1988**, *4*, 17–25. [CrossRef]
4. Kidd, C.V. The evolution of sustainability. *J. Agric. Environ. Ethics* **1992**, *5*, 1–26. [CrossRef]
5. Haferkamp, H.; Smelser, N.J. *Social Change and Modernity*; University of California Press: Berkeley, CA, USA, 1992.
6. Frick, K.T.; Weinzimmer, D.; Waddell, P. The politics of sustainable development opposition: State legislative efforts to stop the United Nation's Agenda 21 in the United States. *Urban Stud.* **2015**, *52*, 209–232. [CrossRef]
7. Purvis, B.; Mao, Y.; Robinson, D. Three pillars of sustainability: In search of conceptual origins. *Sustain. Sci.* **2019**, *14*, 681–695. [CrossRef]
8. Tulloch, L.; Neilson, D. The Neoliberalisation of Sustainability. *Citizsh. Soc. Econ. Educ.* **2014**, *13*, 26–38. [CrossRef]

9. Allen, C.; Metternicht, G.; Wiedmann, T. Initial progress in implementing the Sustainable Development Goals (SDGs): A review of evidence from countries. *Sustain. Sci.* **2018**, *13*, 1453–1467. [CrossRef]

10. Santander, P.; Sanchez, F.A.C.; Boudaoud, H.; Camargo, M. Social, political, and technological dimensions of the sustainability evaluation of a recycling network: A literature review. *Clean. Eng. Technol.* **2022**, *6*, 100397. [CrossRef]

11. Sen, S. Gender, environment and sustainability: The journey from 'silent spring' to 'staying alive'. *Int. J. Adv. Life Sci. Res.* **2020**, *3*, 11–22. [CrossRef]

12. Klarin, T. The Concept of Sustainable Development: From its Beginning to the Contemporary Issues. *Zagreb Int. Rev. Econ. Bus.* **2018**, *21*, 67–94. [CrossRef]

13. Wang, X.; Ren, H.; Wang, P.; Yang, R.; Luo, L.; Cheng, F. A Preliminary Study on Target 11.4 for UN Sustainable Development Goals. *Int. J. Geoherit. Park.* **2018**, *6*, 18–24. [CrossRef]

14. Weitz, N.; Carlsen, H.; Nilsson, M.; Skånberg, K. Towards systemic and contextual priority setting for implementing the 2030 Agenda. *Sustain. Sci.* **2018**, *13*, 531–548. [CrossRef]

15. Berke, P.; Manta, M. *Planning for Sustainable Development: Measuring Progress in Plans Lincoln Institute Product Code: WP99PB1*; Lincoln Institute of Land Policy: Cambridge, MA, USA, 1999.

16. Borgonovi, E.; Compagni, A. Sustaining Universal Health Coverage: The Interaction of Social, Political, and Economic Sustainability. *Value Health* **2013**, *16*, S34–S38. [CrossRef]

17. Kadir, S.A.; Jamaludin, M. Universal Design as a Significant Component for Sustainable Life and Social Development. *Procedia Soc. Behav. Sci.* **2013**, *85*, 179–190. [CrossRef]

18. Premalatha, M.; Tauseef, S.; Abbasi, T.; Abbasi, S. The promise and the performance of the world's first two zero carbon eco-cities. *Renew. Sustain. Energy Rev.* **2013**, *25*, 660–669. [CrossRef]

19. Bartlett, A.A. Reflections on Sustainability, Population Growth, and the Environment. *Popul. Environ.* **1994**, *16*, 5–35. [CrossRef]

20. Roeder, J.; Bartlett, A.A. Sponsored by the Association of Teachers in Independent Schools Affiliated with the Triangle Coalition for Science and Technology Education. In *The Meaning of Sustainability Background on Sustainability*; American Association of Physics Teachers: College Park, MD, USA, 2012; Volume 31.

21. Molinario, E.; Kruglanski, A.W.; Bonaiuto, F.; Bonnes, M.; Cicero, L.; Fornara, F.; Scopelliti, M.; Admiraal, J.; Beringer, A.; Dedeurwaerdere, T.; et al. Motivations to Act for the Protection of Nature Biodiversity and the Environment: A Matter of "Significance". *Environ. Behav.* **2020**, *52*, 1133–1163. [CrossRef]

22. Mensah, J.; Enu-Kwesi, F. Implications of environmental sanitation management for sustainable livelihoods in the catchment area of Benya Lagoon in Ghana. *J. Integr. Environ. Sci.* **2019**, *16*, 23–43. [CrossRef]

23. Biermann, F.; Kanie, N.; Kim, R.E. Global governance by goal-setting: The novel approach of the UN Sustainable Development Goals. *Curr. Opin. Environ. Sustain.* **2017**, *26*, 26–31. [CrossRef]

24. Taylor, C.D.; Gully, B.; Sánchez, A.N.; Rode, E.; Agarwal, A.S. Towards Materials Sustainability through Materials Stewardship. *Sustainability* **2016**, *8*, 1001. [CrossRef]

25. Collste, D.; Pedercini, M.; Cornell, S.E. Policy coherence to achieve the SDGs: Using integrated simulation models to assess effective policies. *Sustain. Sci.* **2017**, *12*, 921–931. [CrossRef] [PubMed]

26. Barbier, E.B. The Concept of Sustainable Economic Development. *Environ. Conserv.* **1987**, *14*, 101–110. [CrossRef]

27. Tjarve, B.; Zemīte, I. The Role of Cultural Activities in Community Development. *Acta Univ. Agric. Silvic. Mendel. Brun.* **2016**, *64*, 2151–2160. [CrossRef]

28. Hammond, G.P.; Winnett, A.B. The Influence of Thermodynamic Ideas on Ecological Economics: An Interdisciplinary Critique. *Sustainability* **2009**, *1*, 1195–1225. [CrossRef]

29. Redclift, M. The meaning of sustainable development. *Geoforum* **1992**, *23*, 395–403. [CrossRef]

30. Tyrrell, T.J.; Johnston, R.J. Tourism Sustainability, Resiliency and Dynamics: Towards a More Comprehensive Perspective. *Tour. Hosp. Res.* **2008**, *8*, 14–24. [CrossRef]

31. Ly, A.M.; Cope, M.R. New Conceptual Model of Social Sustainability: Review from Past Concepts and Ideas. *Int. J. Environ. Res. Public Health* **2023**, *20*, 5350. [CrossRef]

32. Fauré, E.; Arushanyan, Y.; Ekener, E.; Miliutenko, S.; Finnveden, G. Methods for assessing future scenarios from a sustainability perspective. *Eur. J. Futures Res.* **2017**, *5*, 17. [CrossRef]

33. Zijp, M.C.; Heijungs, R.; Van der Voet, E.; Van de Meent, D.; Huijbregts, M.A.J.; Hollander, A.; Posthuma, L. An Identification Key for Selecting Methods for Sustainability Assessments. *Sustainability* **2015**, *7*, 2490–2512. [CrossRef]

34. Sreenath, S.; Sudhakar, K.; Yusop, A. Sustainability at airports: Technologies and best practices from ASEAN countries. *J. Environ. Manag.* **2021**, *299*, 113639. [CrossRef]

35. Villamayor-tomas, S.; Muradian, R. *The Barcelona School of Ecological Economics and Political Ecology A Companion in Honour of Joan*; Springer Nature: Berlin/Heidelberg, Germany, 2023; ISBN 9783031225659.

36. Dunlap, A.; Ruelas, A.; Søyland, L. Debates in Post-development and degrowth. Tvergastein. Interdiscip. *J. Environ.* **2023**, *2*, 229.

37. Sneddon, C.; Howarth, R.B.; Norgaard, R.B. Sustainable development in a post-Brundtland world. *Ecol. Econ.* **2006**, *57*, 253–268. [CrossRef]

38. Fernando, R. Sustainable globalization and implications for strategic corporate and national sustainability. *Corp. Gov. Int. J. Bus. Soc.* **2012**, *12*, 579–589. [CrossRef]

39. Swart, R.; Raskin, P.; Robinson, J. The problem of the future: Sustainability science and scenario analysis. *Glob. Environ. Chang.* **2004**, *14*, 137–146. [CrossRef]

40. James, P.; Magee, L. Domains of Sustainability. In *Global Encyclopedia of Public Administration, Public Policy, and Governance*; Springer International Publishing: Berlin, Germany, 2016; pp. 1–17.

41. Wojewódzka-Wiewiórska, A.; Kłoczko-Gajewska, A.; Sulewski, P. Between the Social and Economic Dimensions of Sustainability in Rural Areas—In Search of Farmers' Quality of Life. *Sustainability* **2020**, *12*, 148. [CrossRef]

42. Hansmann, R.; Mieg, H.A.; Frischknecht, P. Principal sustainability components: Empirical analysis of synergies between the three pillars of sustainability. *Int. J. Sustain. Dev. World Ecol.* **2012**, *19*, 451–459. [CrossRef]

43. Babu, G.; Satya, S. Understanding the Inherent Interconnectedness and other Salient Characteristics of Nature crucial for Sustainability. *Environ. Dev. Sustain.* **2022**, 1–13. [CrossRef]

44. O'Connor, D.; Hou, D.; Ok, Y.S.; Song, Y.; Sarmah, A.K.; Li, X.; Tack, F.M. Sustainable in situ remediation of recalcitrant organic pollutants in groundwater with controlled release materials: A review. *J. Control. Release* **2018**, *283*, 200–213. [CrossRef]

45. Reddy, T.L.; Thomson, R.J.; Taryn, M.; Reddy, L. Environmental, social and economic sustainability: Implications for actuarial science. *Actuar. Inst.* **2015**, 23–27.

46. James, P.; Magee, L. *Global Encyclopedia of Public Administration, Public Policy, and Governance*; Springer: Cham, Switzerland, 2020. [CrossRef]

47. Bañon Gomis, A.J.; Guillén Parra, M.; Michael Hoffman, W.; Mcnulty, R.E.; Guillén, P.M.; McNulty, R.E. Rethinking the Concept of Sustainability. *Bus. Soc. Rev.* **2011**, *116*, 171–191. [CrossRef]

48. Boeske, J.; Murray, P.A. The Intellectual Domains of Sustainability Leadership in SMEs. *Sustainability* **2022**, *14*, 1978. [CrossRef]

49. Shrivastava, P.; Stafford Smith, M.; O'Brien, K.; Zsolnai, L. Transforming Sustainability Science to Generate Positive Social and Environmental Change Globally. *One Earth* **2020**, *2*, 329–340. [CrossRef] [PubMed]

50. Bouzarovski, S. Just Transitions: A Political Ecology Critique. *Antipode* **2022**, *54*, 1003–1020. [CrossRef]

51. Boas, I.; Biermann, F.; Kanie, N. Cross-sectoral strategies in global sustainability governance: Towards a nexus approach. *Int. Environ. Agreem. Politics Law Econ.* **2016**, *16*, 449–464. [CrossRef]

52. Lazar, N.; Chithra, K. Role of culture in sustainable development and sustainable built environment: A review. *Environ. Dev. Sustain.* **2022**, *24*, 5991–6031. [CrossRef]

53. Miska, C.; Szőcs, I.; Schiffinger, M. Culture's effects on corporate sustainability practices: A multi-domain and multi-level view. *J. World Bus.* **2018**, *53*, 263–279. [CrossRef]

54. Manitiu, D.N.; Pedrini, G. Urban smartness and sustainability in Europe. Anex anteassessment of environmental, social and cultural domains. *Eur. Plan. Stud.* **2016**, *24*, 1766–1787. [CrossRef]

55. D'adamo, I.; Gastaldi, M. Perspectives and Challenges on Sustainability: Drivers, Opportunities and Policy Implications in Universities. *Sustainability* **2023**, *15*, 3564. [CrossRef]

56. D'Adamo, I.; Gastaldi, M.; Morone, P. Economic sustainable development goals: Assessments and perspectives in Europe. *J. Clean. Prod.* **2022**, *354*, 131730. [CrossRef]

57. Opoku, A. SDG2030: A sustainable built environment's role in achieving the post-2015 United Nations Sustainable Development Goals. In Proceedings of the 32nd Annual ARCOM Conference, Manchester, UK, 5–7 September 2016; Volume 2.

58. Jucker, R.; Von Au, J. Improving Learning Inside by Enhancing Learning Outside: A Powerful Lever for Facilitating the Implementation of the UN SDGs. *Sustainability* **2019**, *12*, 104–108. [CrossRef]

59. Dhar, S. Gender and Sustainable Development Goals (SDGs). *Indian J. Gend. Stud.* **2018**, *25*, 47–78. [CrossRef]

60. Sørup, H.J.; Brudler, S.; Godskesen, B.; Dong, Y.; Lerer, S.M.; Rygaard, M.; Arnbjerg-Nielsen, K. Urban water management: Can UN SDG 6 be met within the Planetary Boundaries? *Environ. Sci. Policy* **2020**, *106*, 36–39. [CrossRef]

61. Costanza, R.; Daly, L.; Fioramonti, L.; Giovannini, E.; Kubiszewski, I.; Mortensen, L.F.; Pickett, K.E.; Ragnarsdottir, K.V.; De Vogli, R.; Wilkinson, R. Modelling and measuring sustainable wellbeing in connection with the UN Sustainable Development Goals. *Ecol. Econ.* **2016**, *130*, 350–355. [CrossRef]

62. Hassani, H.; Huang, X.; MacFeely, S.; Entezarian, M.R. Big Data and the United Nations Sustainable Development Goals (UN SDGs) at a Glance. *Big Data Cogn. Comput.* **2021**, *5*, 28. [CrossRef]

63. Miola, A.; Schiltz, F. Measuring sustainable development goals performance: How to monitor policy action in the 2030 Agenda implementation? *Ecol. Econ.* **2019**, *164*, 106373. [CrossRef]

64. Pedersen, C.S. The UN Sustainable Development Goals (SDGs) are a Great Gift to Business! *Procedia CIRP* **2018**, *69*, 21–24. [CrossRef]

65. Org, S.U. *Transforming Our World: The 2030 Agenda for Sustainable Development*; United Nations: New York, NY, USA, 2015.

66. Caiado, R.G.G.; Filho, W.L.; Quelhas, O.L.G.; Nascimento, D.L.d.M.; Ávila, L.V. A literature-based review on potentials and constraints in the implementation of the sustainable development goals. *J. Clean. Prod.* **2018**, *198*, 1276–1288. [CrossRef]

67. Salvia, A.L.; Leal Filho, W.; Brandli, L.L.; Griebeler, J.S. Assessing research trends related to Sustainable Development Goals: Local and global issues. *J. Clean. Prod.* **2019**, *208*, 841–849. [CrossRef]

68. Sharma, H.B.; Vanapalli, K.R.; Samal, B.; Cheela, V.S.; Dubey, B.K.; Bhattacharya, J. Circular economy approach in solid waste management system to achieve UN-SDGs: Solutions for post-COVID recovery. *Sci. Total Environ.* **2021**, *800*, 149605. [CrossRef]

69. Che, X.; Jiang, M.; Fan, C. Multidimensional Assessment and Alleviation of Global Energy Poverty Aligned with UN SDG 7. *Front. Energy Res.* **2021**, *9*, 777244. [CrossRef]

70. Duane, S.; Domegan, C.; Bunting, B. Partnering for UN SDG #17: A social marketing partnership model to scale up and accelerate change. *J. Soc. Mark.* **2022**, *12*, 49–75. [CrossRef]

71. Bexell, M.; Jönsson, K. Responsibility and the United Nations' Sustainable Development Goals. *Forum Dev. Stud.* **2017**, *44*, 13–29. [CrossRef]

72. Khaled, R.; Ali, H.; Mohamed, E.K. The Sustainable Development Goals and corporate sustainability performance: Mapping, extent and determinants. *J. Clean. Prod.* **2021**, *311*, 127599. [CrossRef]

73. van der Waal, J.W.; Thijssens, T. Corporate involvement in Sustainable Development Goals: Exploring the territory. *J. Clean. Prod.* **2020**, *252*, 119625. [CrossRef]

74. Agovino, M.; Casaccia, M.; Ciommi, M.; Ferrara, M.; Marchesano, K. Agriculture, climate change and sustainability: The case of EU-28. *Ecol. Indic.* **2019**, *105*, 525–543. [CrossRef]

75. Fuller, R.; Landrigan, P.J.; Balakrishnan, K.; Bathan, G.; Bose-O'Reilly, S.; Brauer, M.; Caravanos, J.; Chiles, T.; Cohen, A.; Corra, L.; et al. Pollution and health: A progress update. *Lancet Planet. Health* **2022**, *6*, e535–e547. [CrossRef]

76. Landrigan, P.J. Air pollution and health. *Lancet Public Health* **2017**, *2*, e4–e5. [CrossRef]

77. Rahman, F.A.; Aziz, M.M.A.; Saidur, R.; Abu Bakar, W.A.W.; Hainin, M.R.; Putrajaya, R.; Hassan, N.A. Pollution to solution: Capture and sequestration of carbon dioxide (CO_2) and its utilization as a renewable energy source for a sustainable future. *Renew. Sustain. Energy Rev.* **2017**, *71*, 112–126. [CrossRef]

78. Villarrubia-Gómez, P.; Cornell, S.E.; Fabres, J. Marine plastic pollution as a planetary boundary threat–The drifting piece in the sustainability puzzle. *Mar. Policy* **2018**, *96*, 213–220. [CrossRef]

79. Rana, M.P. Urbanization and sustainability: Challenges and strategies for sustainable urban development in Bangladesh. *Environ. Dev. Sustain.* **2011**, *13*, 237–256. [CrossRef]

80. Jang, S.M.; Hart, P.S. Polarized frames on "climate change" and "global warming" across countries and states: Evidence from Twitter big data. *Glob. Environ. Chang.* **2015**, *32*, 11–17. [CrossRef]

81. De_Richter, R.; Caillol, S. Fighting global warming: The potential of photocatalysis against CO_2, CH_4, N_2O, CFCs, tropospheric O3, BC and other major contributors to climate change. *J. Photochem. Photobiol. C Photochem. Rev.* **2011**, *12*, 1–19. [CrossRef]

82. Yoro, K.O.; Daramola, M.O. Chapter 1—CO_2 emission sources, greenhouse gases, and the global warming effect. In *Advances in Carbon Capture*; Rahimpour, M.R., Farsi, M., Makarem, M.A., Eds.; Elsevier BV: Amsterdam, The Netherlands, 2020; pp. 3–28.

83. Zandalinas, S.I.; Fritschi, F.B.; Mittler, R. Global Warming, Climate Change, and Environmental Pollution: Recipe for a Multifactorial Stress Combination Disaster. *Trends Plant Sci.* **2021**, *26*, 588–599. [CrossRef]

84. Vågen, T.-G.; Winowiecki, L.A.; Tondoh, J.E.; Desta, L.T.; Gumbricht, T. Mapping of soil properties and land degradation risk in Africa using MODIS reflectance. *Geoderma* **2016**, *263*, 216–225. [CrossRef]

85. Kelly, C.; Ferrara, A.; Wilson, G.A.; Ripullone, F.; Nolè, A.; Harmer, N.; Salvati, L. Community resilience and land degradation in forest and shrubland socio-ecological systems: Evidence from Gorgoglione, Basilicata, Italy. *Land Use Policy* **2015**, *46*, 11–20. [CrossRef]

86. Zhang, D.; Yan, M.; Niu, Y.; Liu, X.; van Zwieten, L.; Chen, D.; Bian, R.; Cheng, K.; Li, L.; Joseph, S.; et al. Is current biochar research addressing global soil constraints for sustainable agriculture? *Agric. Ecosyst. Environ.* **2016**, *226*, 25–32. [CrossRef]

87. McElwee, P.; Turnout, E.; Chiroleu-Assouline, M.; Clapp, J.; Isenhour, C.; Jackson, T.; Kelemen, E.; Miller, D.C.; Rusch, G.; Spangenberg, J.H.; et al. Ensuring a Post-COVID Economic Agenda Tackles Global Biodiversity Loss. *One Earth* **2020**, *3*, 448–461. [CrossRef]

88. Anderson, R.; Bayer, P.E.; Edwards, D. Climate change and the need for agricultural adaptation. *Curr. Opin. Plant Biol.* **2020**, *56*, 197–202. [CrossRef]

89. Mantyka-Pringle, C.S.; Visconti, P.; Di Marco, M.; Martin, T.G.; Rondinini, C.; Rhodes, J.R. Climate change modifies risk of global biodiversity loss due to land-cover change. *Biol. Conserv.* **2015**, *187*, 103–111. [CrossRef]

90. Kujala, H.; Whitehead, A.; Morris, W.; Wintle, B. Towards strategic offsetting of biodiversity loss using spatial prioritization concepts and tools: A case study on mining impacts in Australia. *Biol. Conserv.* **2015**, *192*, 513–521. [CrossRef]

91. Morand, S. Emerging diseases, livestock expansion and biodiversity loss are positively related at global scale. *Biol. Conserv.* **2020**, *248*, 108707. [CrossRef] [PubMed]

92. Liu, J.; Liu, Q.; Yang, H. Assessing water scarcity by simultaneously considering environmental flow requirements, water quantity, and water quality. *Ecol. Indic.* **2016**, *60*, 434–441. [CrossRef]

93. Vanham, D.; Hoekstra, A.Y.; Wada, Y.; Bouraoui, F.; de Roo, A.; Mekonnen, M.M.; van de Bund, W.J.; Batelaan, O.; Pavelic, P.; Bastiaanssen, W.G.M.; et al. Physical water scarcity metrics for monitoring progress towards SDG target 6.4: An evaluation of indicator 6.4.2 "Level of water stress". *Sci. Total Environ.* **2018**, *613–614*, 218–232. [CrossRef]

94. Wang, J.; Li, Y.; Huang, J.; Yan, T.; Sun, T. Growing water scarcity, food security and government responses in China. *Glob. Food Secur.* **2017**, *14*, 9–17. [CrossRef]

95. Pedro-Monzonís, M.; Solera, A.; Ferrer, J.; Estrela, T.; Paredes-Arquiola, J. A review of water scarcity and drought indexes in water resources planning and management. *J. Hydrol.* **2015**, *527*, 482–493. [CrossRef]

96. Folwarczny, M.; Christensen, J.D.; Li, N.P.; Sigurdsson, V.; Otterbring, T. Crisis communication, anticipated food scarcity, and food preferences: Preregistered evidence of the insurance hypothesis. *Food Qual. Prefer.* **2021**, *91*, 104213. [CrossRef]

97. Koyanagi, A.; Stubbs, B.; Oh, H.; Veronese, N.; Smith, L.; Haro, J.M.; Vancampfort, D. Food insecurity (hunger) and suicide attempts among 179,771 adolescents attending school from 9 high-income, 31 middle-income, and 4 low-income countries: A cross-sectional study. *J. Affect. Disord.* **2019**, *248*, 91–98. [CrossRef]

98. von Braun, J. Food insecurity, hunger and malnutrition: Necessary policy and technology changes. *N. Biotechnol.* **2010**, *27*, 449–452. [CrossRef]

99. Brunner, P.H.; Rechberger, H. Waste to energy—Key element for sustainable waste management. *Waste Manag.* **2015**, *37*, 3–12. [CrossRef]

100. Miezah, K.; Obiri-Danso, K.; Kádár, Z.; Fei-Baffoe, B.; Mensah, M.Y. Municipal solid waste characterization and quantification as a measure towards effective waste management in Ghana. *Waste Manag.* **2015**, *46*, 15–27. [CrossRef]

101. Pires, A.; Martinho, G. Waste hierarchy index for circular economy in waste management. *Waste Manag.* **2019**, *95*, 298–305. [CrossRef]

102. Kunkel, S.; Tyfield, D. Digitalisation, sustainable industrialisation and digital rebound—Asking the right questions for a strategic research agenda. *Energy Res. Soc. Sci.* **2021**, *82*, 102295. [CrossRef]

103. Nasrollahi, Z.; Hashemi, M.; Bameri, S. Environmental pollution, economic growth, population, industrialization, and technology in weak and strong sustainability: Using STIRPAT model. *Environ. Dev. Sustain.* **2020**, *22*, 1105–1122. [CrossRef]

104. Huang, Y.; Chen, C.; Su, D.; Wu, S. Comparison of leading-industrialisation and crossing-industrialisation economic growth patterns in the context of sustainable development: Lessons from China and India. *Sustain. Dev.* **2020**, *28*, 1077–1085. [CrossRef]

105. Cui, L.; Weng, S.; Nadeem, A.M.; Rafique, M.Z.; Shahzad, U. Exploring the role of renewable energy, urbanization and structural change for environmental sustainability: Comparative analysis for practical implications. *Renew. Energy* **2022**, *184*, 215–224. [CrossRef]

106. Arshad, Z.; Robaina, M.; Shahbaz, M.; Veloso, A.B. The effects of deforestation and urbanization on sustainable growth in Asian countries. *Environ. Sci. Pollut. Res.* **2020**, *27*, 10065–10086. [CrossRef]

107. Sharif, A.; Afshan, S.; Suki, N.M. Revisiting the Environmental Kuznets Curve in Malaysia: The role of globalization in sustainable environment. *J. Clean. Prod.* **2020**, *264*, 121669. [CrossRef]

108. Sarbu, R.; Alpopi, C.; Burlacu, S.; Diaconu, S. Sustainable Urban Development in the Context of Globalization and the Health Crisis Caused by the COVID-19 Pandemic. *SHS Web Conf.* **2021**, *92*, 01043. [CrossRef]

109. Umar, M.; Ji, X.; Kirikkaleli, D.; Shahbaz, M.; Zhou, X. Environmental cost of natural resources utilization and economic growth: Can China shift some burden through globalization for sustainable development? *Sustain. Dev.* **2020**, *28*, 1678–1688. [CrossRef]

110. Mi, Z.; Guan, D.; Liu, Z.; Liu, J.; Viguié, V.; Fromer, N.; Wang, Y. Cities: The core of climate change mitigation. *J. Clean. Prod.* **2019**, *207*, 582–589. [CrossRef]

111. Fedele, G.; Donatti, C.I.; Harvey, C.A.; Hannah, L.; Hole, D.G. Transformative adaptation to climate change for sustainable social-ecological systems. *Environ. Sci. Policy* **2019**, *101*, 116–125. [CrossRef]

112. Zhao, X.-X.; Zheng, M.; Fu, Q. How natural disasters affect energy innovation? The perspective of environmental sustainability. *Energy Econ.* **2022**, *109*, 105992. [CrossRef]

113. Peduzzi, P. The Disaster Risk, Global Change, and Sustainability Nexus. *Sustainability* **2019**, *11*, 957. [CrossRef]

114. Sarker, M.N.I.; Peng, Y.; Yiran, C.; Shouse, R.C. Disaster resilience through big data: Way to environmental sustainability. *Int. J. Disaster Risk Reduct.* **2020**, *51*, 101769. [CrossRef]

115. Arfanuzzaman; Dahiya, B. Sustainable urbanization in Southeast Asia and beyond: Challenges of population growth, land use change, and environmental health. *Growth Chang.* **2019**, *50*, 725–744. [CrossRef]

116. Marteleto, L.J.; Guedes, G.; Coutinho, R.Z.; Weitzman, A. Live Births and Fertility Amid the Zika Epidemic in Brazil. *Demography* **2020**, *57*, 843–872. [CrossRef]

117. Hammond, M. Sustainability as a cultural transformation: The role of deliberative democracy. *Environ. Politics* **2020**, *29*, 173–192. [CrossRef]

118. Kersting, N. Participatory Democracy and Sustainability. Deliberative Democratic Innovation and Its Acceptance by Citizens and German Local Councilors. *Sustainability* **2021**, *13*, 7214. [CrossRef]

119. Spaiser, V.; Ranganathan, S.; Swain, R.B.; Sumpter, D.J.T. The sustainable development oxymoron: Quantifying and modelling the incompatibility of sustainable development goals. *Int. J. Sustain. Dev. World Ecol.* **2016**, *24*, 457–470. [CrossRef]

120. Sweidan, O.D. State capitalism and energy democracy. *Geoforum* **2021**, *125*, 181–184. [CrossRef]

121. Postigo, J.C. Navigating capitalist expansion and climate change in pastoral social-ecological systems: Impacts, vulnerability and decision-making. *Curr. Opin. Environ. Sustain.* **2021**, *52*, 68–74. [CrossRef]

122. Yang, B. Confucianism, socialism, and capitalism: A comparison of cultural ideologies and implied managerial philosophies and practices in the P. R. China. *Hum. Resour. Manag. Rev.* **2012**, *22*, 165–178. [CrossRef]

123. Caeldries, F. On the sustainability of the capitalist order: Schumpeter's capitalism, socialism and democracyrevisited. *J. Socio. Econ.* **1993**, *22*, 163–185. [CrossRef]

124. Kautsky, J.H. Comparative communism versus comparative politics. *Stud. Comp. Communism* **1973**, *6*, 135–170. [CrossRef]

125. Johnson, S.; Kaufmann, D.; McMillan, J.; Woodruff, C. Why do firms hide? Bribes and unofficial activity after communism. *J. Public Econ.* **2000**, *76*, 495–520. [CrossRef]

126. Verhaar, H. The Age of Sustainalism: The Connected Lighting Revolution. Available online: https://www.ieta.org/resources/COP%2023/Side-Event-Presentations/SE19_RTCC-Philips%20Lighting%20-%20COP23%2010Nov2017.pdf (accessed on 15 January 2023).

127. Verhaar, H.; Affairs, G. X-Change Ignify. The Connected LED Lighting Revolution. 2018. Available online: https://www.a2ep.org.au/_files/ugd/c1ceb4_d270fb8ef8744ab2b594da727597c7f2.pdf?index=true (accessed on 15 January 2023).

128. Verhaar, H. The Age of Sustainalism: A New Growth Model for the 21st Century. UN Environment Programme. 18 January 2018. Available online: https://www.unenvironment.org/news-andstories/story/age-sustainalism-new-growth-model-21st-century (accessed on 15 January 2023).

129. Child, M.; Breyer, C. Transition and transformation: A review of the concept of change in the progress towards future sustainable energy systems. *Energy Policy* **2017**, *107*, 11–26. [CrossRef]

130. Sahoo, S.K. Renewable and sustainable energy reviews solar photovoltaic energy progress in India: A review. *Renew. Sustain. Energy Rev.* **2016**, *59*, 927–939. [CrossRef]

131. Kuzemko, C.; Bradshaw, M.; Bridge, G.; Goldthau, A.; Jewell, J.; Overland, I.; Scholten, D.; Van de Graaf, T.; Westphal, K. COVID-19 and the politics of sustainable energy transitions. *Energy Res. Soc. Sci.* **2020**, *68*, 101685. [CrossRef]

132. Tukker, A. Product services for a resource-efficient and circular economy—A review. *J. Clean. Prod.* **2015**, *97*, 76–91. [CrossRef]

133. Bach, V.; Berger, M.; Henßler, M.; Kirchner, M.; Leiser, S.; Mohr, L.; Rother, E.; Ruhland, K.; Schneider, L.; Tikana, L.; et al. Integrated method to assess resource efficiency—ESSENZ. *J. Clean. Prod.* **2016**, *137*, 118–130. [CrossRef]

134. Korhonen, J.; Honkasalo, A.; Seppälä, J. Circular Economy: The Concept and its Limitations. *Ecol. Econ.* **2018**, *143*, 37–46. [CrossRef]

135. Geissdoerfer, M.; Savaget, P.; Bocken, N.M.P.; Hultink, E.J. The circular economy—A new sustainability paradigm? *J. Clean. Prod.* **2017**, *143*, 757–768. [CrossRef]

136. Kirchherr, J.; Reike, D.; Hekkert, M. Conceptualizing the circular economy: An analysis of 114 definitions. *Resour. Conserv. Recycl.* **2017**, *127*, 221–232. [CrossRef]

137. Wei, G.; Zhang, J.; Usuelli, M.; Zhang, X.; Liu, B.; Mezzenga, R. Biomass vs inorganic and plastic-based aerogels: Structural design, functional tailoring, resource-efficient applications and sustainability analysis. *Prog. Mater. Sci.* **2022**, *125*, 100915. [CrossRef]

138. Umesha, S.; Manukumar, H.M.G.; Chandrasekhar, B. Chapter 3—Sustainable Agriculture and Food Security. In *Biotechnology for Sustainable Agriculture*; Singh, R.L., Mondal, S., Eds.; Woodhead Publishing: Cambridge, UK, 2018; pp. 67–92. [CrossRef]

139. Jayne, T.; Snapp, S.; Place, F.; Sitko, N. Sustainable agricultural intensification in an era of rural transformation in Africa. *Glob. Food Secur.* **2019**, *20*, 105–113. [CrossRef]

140. Singh, J.S.; Pandey, V.C.; Singh, D. Efficient soil microorganisms: A new dimension for sustainable agriculture and environmental development. *Agric. Ecosyst. Environ.* **2011**, *140*, 339–353. [CrossRef]

141. Wilson, M.H.; Lovell, S.T. Agroforestry—The Next Step in Sustainable and Resilient Agriculture. *Sustainability* **2016**, *8*, 574. [CrossRef]

142. Iiyama, M.; Neufeldt, H.; Dobie, P.; Njenga, M.; Ndegwa, G.; Jamnadass, R. The potential of agroforestry in the provision of sustainable woodfuel in sub-Saharan Africa. *Curr. Opin. Environ. Sustain.* **2014**, *6*, 138–147. [CrossRef]

143. Mbow, C.; Smith, P.; Skole, D.; Duguma, L.; Bustamante, M. Achieving mitigation and adaptation to climate change through sustainable agroforestry practices in Africa. *Curr. Opin. Environ. Sustain.* **2014**, *6*, 8–14. [CrossRef]

144. Krčmářová, J.; Kala, L.; Brendzová, A.; Chabada, T. Building Agroforestry Policy Bottom-Up: Knowledge of Czech Farmers on Trees in Farmland. *Land* **2021**, *10*, 278. [CrossRef]

145. D'adamo, I.; Gastaldi, M.; Morone, P.; Rosa, P.; Sassanelli, C.; Settembre-Blundo, D.; Shen, Y. Bioeconomy of Sustainability: Drivers, Opportunities and Policy Implications. *Sustainability* **2022**, *14*, 200. [CrossRef]

146. Aworunse, O.S.; Olorunsola, H.A.; Ahuekwe, E.F.; Obembe, O.O. Towards a sustainable bioeconomy in a post-oil era Nigeria. *Resour. Environ. Sustain.* **2023**, *11*, 100094. [CrossRef]

147. Stephenson, P.J.; Damerell, A. Bioeconomy and Circular Economy Approaches Need to Enhance the Focus on Biodiversity to Achieve Sustainability. *Sustainability* **2022**, *14*, 10643. [CrossRef]

148. Schaffer, C.; Eksvärd, K.; Björklund, J. Can Agroforestry Grow beyond Its Niche and Contribute to a Transition towards Sustainable Agriculture in Sweden? *Sustainability* **2019**, *11*, 3522. [CrossRef]

149. Zheng, H.W.; Shen, G.Q.; Wang, H. A review of recent studies on sustainable urban renewal. *Habitat Int.* **2014**, *41*, 272–279. [CrossRef]

150. Morris, R.; Alonso, I.; Jefferson, R.; Kirby, K. The creation of compensatory habitat—Can it secure sustainable development? *J. Nat. Conserv.* **2006**, *14*, 106–116. [CrossRef]

151. Kang, J.; Martinez, C.M.J.; Johnson, C. Minimalism as a sustainable lifestyle: Its behavioral representations and contributions to emotional well-being. *Sustain. Prod. Consum.* **2021**, *27*, 802–813. [CrossRef]

152. Vogel, S.M.; Vasudev, D.; Ogutu, J.O.; Taek, P.; Berti, E.; Goswami, V.R.; Kaelo, M.; Buitenwerf, R.; Munk, M.; Li, W.; et al. Identifying sustainable coexistence potential by integrating willingness-to-coexist with habitat suitability assessments. *Biol. Conserv.* **2023**, *279*, 109935. [CrossRef]

153. Shittu, O. Emerging sustainability concerns and policy implications of urban household consumption: A systematic literature review. *J. Clean. Prod.* **2020**, *246*, 119034. [CrossRef]

154. Witt, A.H.-D. Exploring worldviews and their relationships to sustainable lifestyles: Towards a new conceptual and methodological approach. *Ecol. Econ.* **2012**, *84*, 74–83. [CrossRef]
155. Woiwode, C.; Schäpke, N.; Bina, O.; Veciana, S.; Kunze, I.; Parodi, O.; Schweizer-Ries, P.; Wamsler, C. Inner transformation to sustainability as a deep leverage point: Fostering new avenues for change through dialogue and reflection. *Sustain. Sci.* **2021**, *16*, 841–858. [CrossRef]
156. Badassa, B.B.; Sun, B.; Qiao, L. Sustainable Transport Infrastructure and Economic Returns: A Bibliometric and Visualization Analysis. *Sustainability* **2020**, *12*, 2033. [CrossRef]
157. Moslem, S.; Ghorbanzadeh, O.; Blaschke, T.; Duleba, S. Analysing Stakeholder Consensus for a Sustainable Transport Development Decision by the Fuzzy AHP and Interval AHP. *Sustainability* **2019**, *11*, 3271. [CrossRef]
158. Bamwesigye, D.; Hlavackova, P. Analysis of Sustainable Transport for Smart Cities. *Sustainability* **2019**, *11*, 2140. [CrossRef]
159. Afrin, T.; Yodo, N. A Survey of Road Traffic Congestion Measures towards a Sustainable and Resilient Transportation System. *Sustainability* **2020**, *12*, 4660. [CrossRef]
160. Abduljabbar, R.L.; Liyanage, S.; Dia, H. The role of micro-mobility in shaping sustainable cities: A systematic literature review. *Transp. Res. Part D Transp. Environ.* **2021**, *92*, 102734. [CrossRef]
161. Banister, D. The sustainable mobility paradigm. *Transp. Policy* **2008**, *15*, 73–80. [CrossRef]
162. Holden, E.; Banister, D.; Gössling, S.; Gilpin, G.; Linnerud, K. Grand Narratives for sustainable mobility: A conceptual review. *Energy Res. Soc. Sci.* **2020**, *65*, 101454. [CrossRef]
163. Yogi, H.N. Eco-Tourism and Sustainability—Opportunities and Challenges in the Case of Nepal. 2010; pp. 1–60. Available online: https://docslib.org/doc/5102734/eco-tourism-and-sustainability-opportunities-and-challenges-in-the-case-of-nepal (accessed on 12 February 2023).
164. Mateoc-Sîrb, N.; Albu, S.; Rujescu, C.; Ciolac, R.; Țigan, E.; Brînzan, O.; Mănescu, C.; Mateoc, T.; Milin, I.A. Sustainable Tourism Development in the Protected Areas of Maramureș, Romania: Destinations with High Authenticity. *Sustainability* **2022**, *14*, 1763. [CrossRef]
165. Mohanty, A. Education for sustainable development: A conceptual model of sustainable education for India. *Int. J. Dev. Sustain.* **2019**, *7*, 2242–2255.
166. Burbules, N.C.; Fan, G.; Repp, P. Five trends of education and technology in a sustainable future. *Geogr. Sustain.* **2020**, *1*, 93–97. [CrossRef]
167. Ardoin, N.M.; Bowers, A.W. Early childhood environmental education: A systematic review of the research literature. *Educ. Res. Rev.* **2020**, *31*, 100353. [CrossRef]
168. Maqsood, A.; Abbas, J.; Rehman, G.; Mubeen, R. The paradigm shift for educational system continuance in the advent of COVID-19 pandemic: Mental health challenges and reflections. *Curr. Res. Behav. Sci.* **2021**, *2*, 100011. [CrossRef]
169. Nousheen, A.; Zai, S.A.Y.; Waseem, M.; Khan, S.A. Education for sustainable development (ESD): Effects of sustainability education on pre-service teachers' attitude towards sustainable development (SD). *J. Clean. Prod.* **2020**, *250*, 119537. [CrossRef]
170. Kopnina, H. Contesting 'Environment' Through the Lens of Sustainability: Examining Implications for Environmental Education (EE) and Education for Sustainable Development (ESD). *Cult. Unbound* **2014**, *6*, 931–947. [CrossRef]
171. Terán-Yépez, E.; Marín-Carrillo, G.M.; Casado-Belmonte, M.D.P.; Capobianco-Uriarte, M.D.L.M. Sustainable entrepreneurship: Review of its evolution and new trends. *J. Clean. Prod.* **2020**, *252*, 119742. [CrossRef]
172. Hummels, H.; Argyrou, A. Planetary demands: Redefining sustainable development and sustainable entrepreneurship. *J. Clean. Prod.* **2020**, *278*, 123804. [CrossRef]
173. Merad, M.; Dechy, N.; Marcel, F. A pragmatic way of achieving Highly Sustainable Organisation: Governance and organisational learning in action in the public French sector. *Saf. Sci.* **2014**, *69*, 18–28. [CrossRef]
174. Glass, L.-M.; Newig, J. Governance for achieving the Sustainable Development Goals: How important are participation, policy coherence, reflexivity, adaptation and democratic institutions? *Earth Syst. Gov.* **2019**, *2*, 100031. [CrossRef]
175. Visseren-Hamakers, I.J.; Razzaque, J.; McElwee, P.; Turnhout, E.; Kelemen, E.; Rusch, G.M.; Fernández-Llamazares, Á.; Chan, I.; Lim, M.; Islar, M.; et al. Transformative governance of biodiversity: Insights for sustainable development. *Curr. Opin. Environ. Sustain.* **2021**, *53*, 20–28. [CrossRef]
176. Chowdhury, M.H.; Quaddus, M.A. Supply chain sustainability practices and governance for mitigating sustainability risk and improving market performance: A dynamic capability perspective. *J. Clean. Prod.* **2021**, *278*, 123521. [CrossRef]
177. Lombardi, R.; Trequattrini, R.; Cuozzo, B.; Cano-Rubio, M. Corporate corruption prevention, sustainable governance and legislation: First exploratory evidence from the Italian scenario. *J. Clean. Prod.* **2019**, *217*, 666–675. [CrossRef]
178. Kern, F.; Rogge, K.S.; Howlett, M. Policy mixes for sustainability transitions: New approaches and insights through bridging innovation and policy studies. *Res. Policy* **2019**, *48*, 103832. [CrossRef]
179. Hörisch, J.; Schaltegger, S.; Freeman, R.E. Integrating stakeholder theory and sustainability accounting: A conceptual synthesis. *J. Clean. Prod.* **2020**, *275*, 124097. [CrossRef]
180. Silva, S.; Nuzum, A.-K.; Schaltegger, S. Stakeholder expectations on sustainability performance measurement and assessment. A systematic literature review. *J. Clean. Prod.* **2019**, *217*, 204–215. [CrossRef]
181. Cambridge Institute for Sustainability. *Developing the EU's 'Competitive Sustainability' for a Resilient Recovery and Dynamic Growth*; The University of Cambridge Institute for Sustainability: Cambridge, UK, 2020.

182. Feng, A.; Li, H. We Are All in the Same Boat: Cross-Border Spillovers of Climate Risk through International Trade and Supply Chain. *IMF Work. Pap.* **2021**, 1–57. [CrossRef]
183. Alam, M. Environmental Education and Non-governmental Organizations. In *The Palgrave Encyclopedia of Urban and Regional Futures*; Springer International Publishing: Cham, Switzerland, 2023; pp. 495–502.
184. Burns, T.R. The Sustainability Revolution: A Societal Paradigm Shift. *Sustainability* **2012**, *4*, 1118–1134. [CrossRef]
185. Ali, S.M.; Appolloni, A.; Cavallaro, F.; D'adamo, I.; Di Vaio, A.; Ferella, F.; Gastaldi, M.; Ikram, M.; Kumar, N.M.; Martin, M.A.; et al. Development Goals towards Sustainability. *Sustainability* **2023**, *15*, 9443. [CrossRef]

Review

Toward a Resilient Future: The Promise of Microbial Bioeconomy

Adenike Akinsemolu [1], Helen Onyeaka [2,*], Omololu Fagunwa [3] and Adewale Henry Adenuga [4]

[1] Institute of Advanced Studies, University of Birmingham, Birmingham B15 2TT, UK
[2] School of Chemical Engineering, University of Birmingham, Birmingham B15 2TT, UK
[3] School of Pharmacy, Queen's University Belfast, 97 Lisburn Road, Belfast BT9 7BL, UK
[4] Economics Research Branch, Agri-Food and Biosciences Institute AFBI Headquarters, 18a Newforge Lane, Belfast BT9 5PX, UK
[*] Correspondence: h.onyeaka@bham.ac.uk

Abstract: Naturally occurring resources, such as water, energy, minerals, and rare earth elements, are limited in availability, yet they are essential components for the survival and development of all life. The pressure on these finite resources is anthropogenic, arising from misuse, overuse, and overdependence, which causes a loss of biodiversity and climate change and poses great challenges to sustainable development. The focal points and principles of the bioeconomy border around ensuring the constant availability of these natural resources for both present and future generations. The rapid growth of the microbial bioeconomy is promising for the purpose of fostering a resilient and sustainable future. This highlights the economic opportunity of using microbial-based resources to substitute fossil fuels in novel products, processes, and services. The subsequent discussion delves into the essential principles required for implementing the microbial bioeconomy. There is a further exploration into the latest developments and innovations in this sub-field. The multi-sectoral applications include use in bio-based food and feed products, energy recovery, waste management, recycling, and cascading. In multi-output production chains, enhanced microbes can simultaneously produce multiple valuable and sustainable products. The review also examines the barriers and facilitators of bio-based approaches for a sustainable economy. Despite limited resources, microbial-based strategies demonstrate human ingenuity for sustaining the planet and economy. This review highlights the existing research and knowledge and paves the way for a further exploration of advancements in microbial knowledge and its potential applications in manufacturing, energy production, reduction in waste, hastened degradation of waste, and environmental conservation.

Keywords: sustainability; microorganisms; energy; biotechnology; biocatalysis; biotransformation; industrial applications; circular bioeconomy

Citation: Akinsemolu, A.; Onyeaka, H.; Fagunwa, O.; Adenuga, A.H. Toward a Resilient Future: The Promise of Microbial Bioeconomy. *Sustainability* **2023**, *15*, 7251. https://doi.org/10.3390/su15097251

Academic Editors: Idiano D'Adamo and Massimo Gastaldi

Received: 11 March 2023
Revised: 18 April 2023
Accepted: 19 April 2023
Published: 27 April 2023

1. Introduction

Human survival and development depend on natural resources, such as water, energy, and raw materials, which are unsteadily available. As human development increases and the global population rises, based on predictions, from 7.6 to 9.7 billion by 2050 and further to 11.2 billion by 2100 [1] the demand for available resources is increasing, making sustainability one of the major challenges with which the world's population is contending [2] The need for optimal utilisation of available scarce resources in a sustainable manner that guarantees their use for future generations can, therefore, not be overemphasised [3]. The sustainable use of available resources secures a continuous food supply, renewable energy, and a continuous raw materials supply, achieving zero hunger and other sustainable development goals [4,5] Guaranteeing the continuous availability of these resources for present and future needs is the core principle of bioeconomy [6].

The term "bioeconomy" was first coined by Zeman in the 1960s and is derived from the Greek words "bios", "oikos", and "nomos", which mean life, house, and law, respectively. The German Federal Government Bioeconomy Council also defines bioeconomy as the utilisation and production of biological resources (with knowledge inclusive) to provide services, processes, and products across all industry sectors and trade in a sustainable economy [7].

In the bioeconomy, resources are either sourced through nature or recycled after human usage. Besides land, water, air, biomass, technologies, and knowledge, microorganisms are also a crucial resource in the bioeconomy. Microorganisms are genetically altered to render them suitable for bioproduction processes [8] and serve as biocatalytic platforms for the microbial bioeconomy. The microbial bioeconomy involves utilising microorganisms to create sustainable and biodegradable products, replace non-renewable resources, and reduce environmental pollution. This concept aligns with the circular bioeconomy, which prioritises minimising waste and maximising resource efficiency [9].

The utilisation of microorganisms in the production of goods and services dates back to ancient civilisations with examples such as the Egyptians using yeast to bake leavened bread [10]. Recent developments in the microbial bioeconomy have focused on developing new and enhanced microorganisms as well as optimising production processes [11] For example, metabolic engineering and synthetic biology have allowed for the development of microorganisms with enhanced capabilities, such as the ability to produce biofuels from non-edible feedstock [12]. Additionally, advanced fermentation technologies have enabled the creation of high-value products, such as biodegradable plastics and specialty chemicals [13]. The microbial bioeconomy can promote economic sustainability and reduce the environmental impact of production processes [14]. By replacing non-renewable resources with renewable alternatives and using waste streams as feedstocks, the microbial bioeconomy can help reduce greenhouse gas emissions and other forms of pollution. For instance, using rice hull ash instead of cement in making concrete mix reduces carbon emissions [15]. Similarly, leveraging microorganisms can improve production process efficiency, save costs, and increase competitiveness [7].

By using microorganisms to produce biodegradable and sustainable products, the microbial bioeconomy offers a viable alternative to traditional production methods, which are based on the global economy principles. With continued advances and innovations, the microbial bioeconomy will likely become an increasingly important part of the global economy [7]. Genetically modified organisms (GMOs) are vital for a sustainable bioeconomy; they are employed in the creation of food and feed additives, pharmaceuticals, biomedicine, bioenergy, biofuels, bioplastics, environmental remediation and waste management, building and transportation system, forestry and agriculture, and other recyclable bio-based materials and products [7,16–21]. Wesseler et al. 2022. Some of these microorganisms and the products formed are shown in Table 1.

Table 1. Examples of microorganisms and the biodegradable and sustainable products they can produce.

Microorganism Type	Microorganism	Biodegradable and Sustainable Products	Industry/ Application	Role in Achieving Sustainability	References
Bacteria	*Escherichia coli*	Bioplastics	Packaging	Reduces reliance on fossil-fuel-based plastics, reduces waste	[22]
	Bacillus subtilis	Enzymes	Cleaning and detergent	Reduces environmental impact of cleaning products, promotes sustainable practices	[23]
	Saccharomyces cerevisiae	Bioethanol	Fuel	Provides sustainable alternative to fossil fuels	[24]
	Pseudomonas putida	Biodegradable polymers	Biodegradable materials	Reduces reliance on non-biodegradable materials, reduces waste	[25]

Table 1. *Cont.*

Microorganism Type	Microorganism	Biodegradable and Sustainable Products	Industry/ Application	Role in Achieving Sustainability	References
Fungi	*Candida albicans*	Biosurfactants	Cosmetics and personal care	Provides sustainable alternative to conventional surfactants	[26]
	Lactobacillus acidophilus	Probiotics	Food and beverage	Promotes sustainable agriculture practices, reduces food waste	[27]
	Trichoderma reesei	Cellulases	Paper and pulp	Promotes sustainable forestry practices, reduces waste	[28,29]
	Aspergillus niger	Organic acids	Food and beverage	Provides sustainable alternative to conventional food additives	[30]
	Rhizopus oryzae	Biodegradable plastics	Packaging	Reduces reliance on non-biodegradable materials, reduces waste	[31]

2. Methodology

This review discusses the microbial bioeconomy principles, current state, and recent advances and innovations. A systematic review was conducted to examine the applications of the microbial bioeconomy in various sectors, including food and beverage production, biotechnology, and environmental remediation. The review explores how microorganisms can help achieve economic sustainability and discusses some of the future trends in this field.

A keyword search of existing research, knowledge, and reports was performed using the keywords "microbial bioeconomy," "sustainability", "microorganisms", and "principles". The results were filtered for relevance, the author's authority, and timeliness and synthesised.

3. Applications of the Microbial Bioeconomy for Economic, Environmental, and Social Sustainability

3.1. Principles of the Microbial Bioeconomy

To achieve a sustainable and efficient microbial bioeconomy, adherence to certain guiding principles is crucial. Adhering to these principles can help create an environmentally friendly, economically sustainable, and socially responsible microbial economy.

Seven (7) Key Principles Central to the Microbial Bioeconomy

1. Use of renewable resources: The microbial bioeconomy relies on renewable resource utilisation, such as plant biomass, organic materials, and agricultural waste, as feedstocks for the production of products of high value. This approach helps to not only reduce reliance on non-renewable resources, such as fossil fuels, but also create a more sustainable and efficient economy [32]
2. Sustainable production: The microbial bioeconomy produces products and services that are environmentally friendly and sustainable. This involves using renewable resources, converting waste materials into valuable products, and optimising production processes to minimise waste and reduce greenhouse gas emissions [33].
3. Efficient utilisation of resources: The microbial bioeconomy seeks to use resources efficiently to minimise waste and reduce the environmental impact of production. This involves closed-loop systems where waste from one process is used as a resource [5].
4. Adaptability: The microbial bioeconomy technologies should be adaptable to different environments and conditions and should be able to be easily modified or scaled up as needed [34].
5. Innovation: The microbial bioeconomy should be driven by the development of new technologies and approaches along with a focus on innovation [35].

6. Collaboration: The development and implementation of the microbial bioeconomy technologies often require collaboration between researchers, industry, and government [36].

7. Transparency: The microbial bioeconomy should be transparent and open with information about processes and products and make this information readily available to stakeholders [33].

The seven principles are shown graphically in Figure 1.

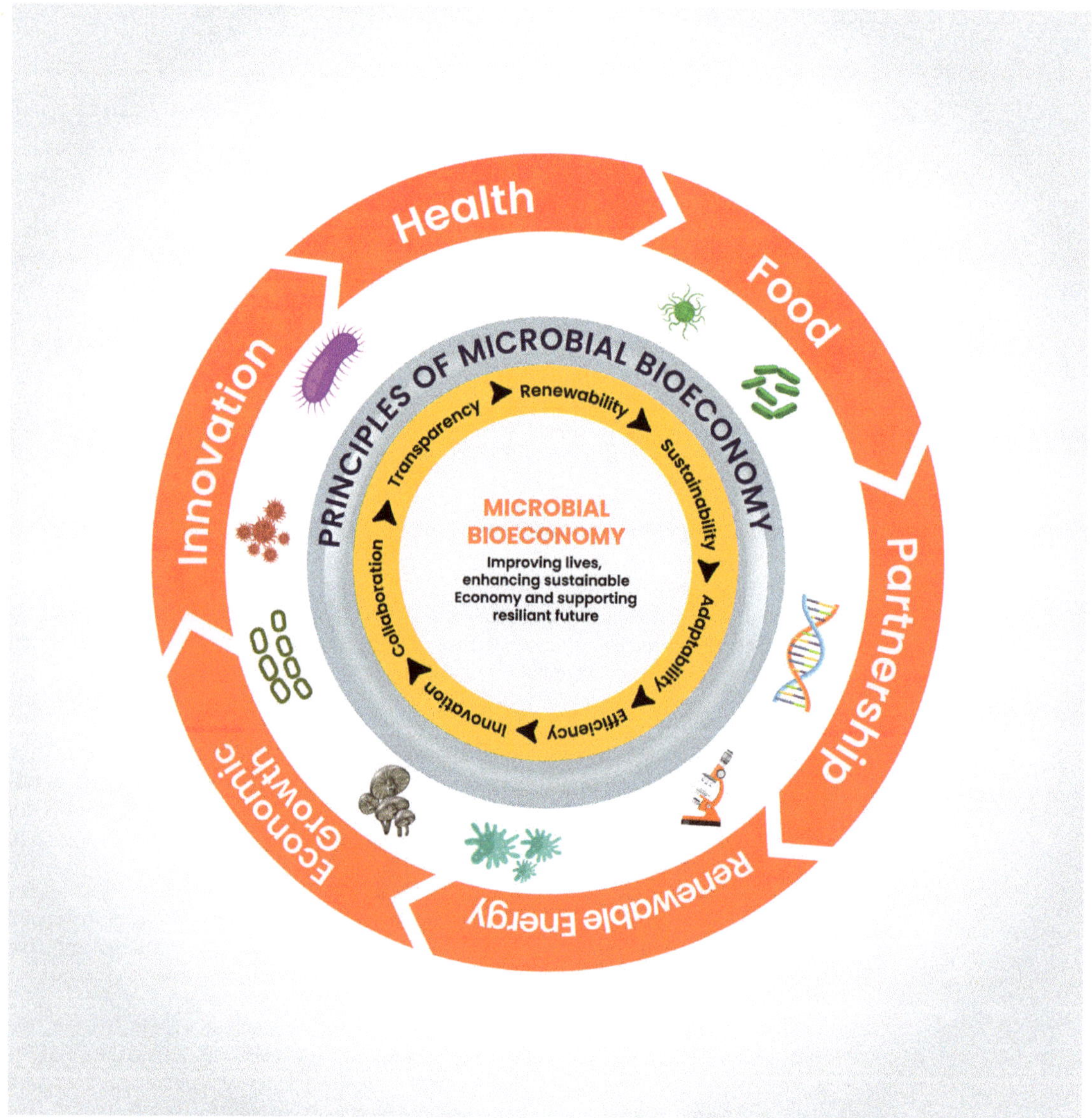

Figure 1. Principles of the Microbial Bioeconomy.

3.2. The Microbial Bioeconomy Current State

Based on a report by the [37] the global microbial market is worth €250 billion, with an annual growth of 5.6%. By 2030, it is predicted to reach €700 billion due to the rising demand for sustainable products, biotechnology advancements, and supportive policies. The European Union (EU) bioeconomy generated €2.29 trillion turnover in 2015 [38] However, the COVID-19 pandemic caused a 2% decrease in 2019 and a 0.4% decrease in 2020 in employment and gross value addition [38].

One of the key trends in the microbial bioeconomy is the growing demand for renewable and sustainable products and services driven by a growing awareness of the social and environmental impacts of non-renewable resources such as fossil fuels. Additionally, an improved public understanding of climate change and the devastations caused by non-renewable products have changed consumer preferences and resulted in more robust regulatory policies [37]. This trend creates new products and processes that rely on microorganisms, such as biofuels, biomaterials, and bioplastics [7].

Another significant trend is the increasing focus on waste management and resource recovery. Microorganisms are expected to play a critical role in this area by using anaerobic digestion and other processes to convert waste into valuable products, such as biogas, fertilisers, and biochemicals [39].

The microbial bioeconomy contributes to economic sustainability through biocatalysis, biotransformations, anaerobic digestion, and bioremediation processes. Biocatalysis has advantages over traditional chemical processes due to its improved selectivity and specificity [40,41]. Similarly, biotransformation is used in industries, such as fine chemicals, pharmaceuticals, food, and animal feed, to produce vitamins, flavour compounds, biofuels, and other biomaterials [40,42]. Utilisation of these bio-based processes can offer several advantages over traditional chemical processes, such as improved sustainability and lower production costs. For example, this biologically dependent process can use renewable resources, such as waste materials and biomass, as substrates either by decreasing or eliminating the use of chemicals and their environmental impacts. This can also lead to economic benefits, such as reducing raw material costs and developing new markets and products [43]. Anaerobic digestion and bioremediation processes can convert waste materials and pollutants into valuable products, such as biogas, fertilisers, and biochemicals, which can be used to produce other goods and services [44].

3.3. Recent Advances and Innovations in the Microbial Bioeconomy

Recent advances in biotechnology have led to significant progress, particularly in the development and application of microorganisms for producing goods and services. Advances in metabolic engineering and synthetic biology have made significant progress in creating new microbial strains with improved productivity and efficiency. The *Penicillium chrysogenum* X-1612 strain was genetically modified using X-ray mutagenesis for enhanced genetic expression [45] (These advances have also paved the way for developing sustainable products and processes, including the production of biofuels from waste materials and using microorganisms to remove pollutants from water and air [46,47].

In addition to technical advances, there have also been significant innovations in commercialising microbial products and services. For example, the rise of biotechnology start-ups and the increasing availability of venture capital have facilitated the development and growth of the microbial bioeconomy with greater availability in some countries than others. The most common way through which start-ups are commercialising microbial products and services is the repurposing of biomass from agricultural products, consultancy, and training [48].

3.4. Applications of the Microbial Bioeconomy in Various Sectors

The microbial bioeconomy is a fast-growing field with various commercial and industrial applications, bio-based product production, food and feed, energy recovery and composting, waste management, recycling, and cascading, as well as multi-output production chains [48].

3.4.1. Bio-Based Products, Food, and Feed

The use of microorganisms in producing bio-based products, including food and animal feed, has been widely acknowledged for a long time. One key advantage of using microorganisms to produce bio-based products is their capacity to utilise a variety of feedstocks. For instance, microorganisms, such as yeast and bacteria, can convert plant-based feedstocks, such as corn and wheat, into valuable products, such as lactic acid and ethanol [49] This renders the use of microorganisms a more cost-effective and efficient option to produce biofuels [50] Additionally, they can produce a variety of food products, such as fermented foods including yogurt, which is made using *Lactobacillus delbrueckii subsp. Bulgaricus* and *Streptococcus thermophilus* [51].

Similarly, *Salmonella typhimurium*, and *Bacillus subtilis* and *Escherichia coli* are well-studied to produce vitamins [52]. Microorganisms can be used for the production of high-quality proteins as animal feed, such as single-cell proteins (SCP) that include *Rhodopseudomonas faecalis* PA2 [53]. This can help reduce the reliance on limited and potentially environmentally damaging protein sources, such as fishmeal and soymeal, yielding better protein sources while promoting sustainability and environmental conservation.

3.4.2. Energy Recovery

Microorganisms are also used for energy recovery and composting by converting organic waste into a valuable soil amendment that improves its physical properties, such as drainage, water retention, structure, permeability, aeration, and water infiltration. There has been a growing interest recently in composting to recover energy from organic waste while promoting economic sustainability [54]. One of the main benefits of composting is that it decreases the quantity of waste sent to landfills, where organic waste decomposes anaerobically and produces methane, a potent greenhouse gas. In contrast, when organic waste is composted, it decomposes in the presence of oxygen and produces carbon dioxide, a less harmful greenhouse gas. By promoting anaerobic decomposition, certain microorganisms can convert food waste into biogas, providing a renewable energy source while promoting the sustainability of waste disposal processes by reducing the production of methane [55].

3.4.3. Waste Management, Recycling, and Cascading

Human activities have led to the proliferation of landfills and incineration facilities, which can be environmentally damaging and costly to maintain. Microorganisms can convert organic waste, such as food and agricultural wastes, into biofuels and biochemicals [46]. This reduces the amount of refuse that is sent to landfills and generates revenue from selling these products.

Another important component of the microbial bioeconomy in relation to waste management is composting; it involves the breakdown of organic matter in waste, converting them to compost. A diverse community of microorganisms, including bacteria, fungi, and protozoa, conducts the composting [11,56].

Composting provides economic benefits by reducing the cost of waste disposal and providing an alternative to landfill disposal [56]. It can also generate revenue through selling high-demand compost among farmers and gardeners, supporting local agriculture and food production for economic sustainability [57].

Several vital microorganisms play essential roles in composting; they include bacteria in the genera *Clostridium*, *Bacillus*, and *Pseudomonas*, and fungi from *the Aspergillus, Penicillium*, and *Trichoderma* genera [52].

Additionally, the microbial bioeconomy can be utilised in a cascading approach, where waste from one process is used as feedstock for another, allowing resources to be used efficiently and multiple valuable products to be produced from a single waste stream, which is a central principle of the circular economy approach [58]. For instance, *Saccharomyces cerevisiae* can convert waste wood into bioethanol [47] which can then be used as a feedstock for bioplastic production [59].

3.4.4. Integrated and Multi-Output Production Chains

The microbial bioeconomy can also support integrated and multi-output production chains, leading to cost-effective and efficient production. This method utilises microorganisms to simultaneously produce multiple products from a single feedstock. Studies have shown how effective this method is in producing biofuels and chemicals from a single microbial strain [60–63] and in producing biofuels, animal feed, and bioplastics using microalgae [64,65]. Such integrated production can increase the industrial process efficiency and sustainability. The discovery of CRISPR/Cas9 and other genetic engineering techniques has enabled the precise modification of various strains to produce targeted products. This flexibility has facilitated rapid and efficient responses to changes in market demand [66].

3.5. Role of the Microbial Bioeconomy in Achieving Sustainability in the Economy and Environment

The microbial bioeconomy plays a significant role in achieving both economic and environmental sustainability by using renewable resources and reducing waste and pollution. By using biomass and biological knowledge to provide food, feed, industrial products, bioenergy, and ecological services, the microbial bioeconomy aligns with several sustainable development goals, such as affordable and clean energy (Goal 7), sustainable cities and communities (Goal 11), and responsible consumption and production (Goal 12) [18,67,68]. By creating a balance between sustainability and economic aspirations, the microbial bioeconomy can help address global challenges, such as climate change mitigation, global food security, and sustainable resource management, leading to a more resilient and sustainable economy that benefits both people and the planet [69,70]. It creates a more resilient and sustainable economy that benefits both people and the planet by utilising the unique capabilities of microorganisms [18]. The contribution of the microbial bioeconomy to sustainable economic development may vary depending on the resources available in different regions and countries. However, the following sections outline potential ways that the microbial bioeconomy can support the achievement of the Sustainable Development Goals (SDGs).

3.5.1. Job Creation and Rural Development

The microbial-based bioeconomy can create jobs, drive economic growth, and contribute to achieving the Sustainable Development Goals of the United Nations, particularly No Poverty (SDG 1), Zero hunger (SDG 2), and Decent Work and Economic Growth (SDG 8) [71,72]. As biomass is widely available, the microbial bioeconomy can create modern jobs (biotechnologists and bioeconomists) in rural areas and promote social inclusion [69,73,74], For instance, in 2019, approximately 17.4 million people in the EU were working in the bioeconomy sectors, which was 8.3% of its total labour force. Bio-based employment can be generated from advances in the microbial bioeconomy as described in the following practical examples. *Spirulina,* a kind of blue-green algae, is an excellent source of protein and other nutrients, making it a potentially sustainable and nutritious food source. *Spirulina* cultivation can create jobs in the aquaculture and agriculture industries, as well as in the processing and packaging of spirulina-based products [72]. *Spirulina platensis* is also being explored for its potential use in animal feed production. Because of its high protein content, *spirulina* is used in animal feed, which can create jobs in the animal husbandry and feed manufacturing industries. Using microorganisms in waste treatment, biopesticide production, and bioremediation products can create jobs in the environmental industry and in research and development, engineering, and operations. Additionally, biopesticides made from microorganisms can help control agricultural pests and reduce the reliance on chemical pesticides, which have adverse effects on human health and the environment [75,76]. This can create jobs in the agricultural and biotechnology industries.

3.5.2. Climate Change Mitigation and Neutrality

Microorganisms play a vital role in mitigating and achieving climate neutrality, as they are used in various industrial processes that can help reduce greenhouse gas emissions and promote environmental sustainability. Using biological resources for food, feed, bio-based products, and bioenergy can align with the United Nations' Sustainable Development Goals (SDGs), such as Affordable and Clean Energy (SDG 7); Industry, Innovation, and Infrastructure (SDG 9); and Responsible Consumption and Production (SDG 12) [7]. Microorganisms, such as yeast and algae, can produce biofuels, such as ethanol and biodiesel, as alternatives for fossil fuels. The production of biofuels using microorganisms can reduce lifecycle greenhouse gas emissions, as biofuels have a lower net GHG emission compared to fossil fuels [77]. Additionally, microorganisms produce bioplastics from renewable materials, such as corn starch or sugarcane. Bioplastics reduce our dependence on fossil-based plastics, which emit greenhouse gases. They can be biodegraded by microorganisms, reducing plastic waste in the environment. The microbial bioeconomy can help create greener cities that operate on closed material and energy cycles, thereby, reducing emissions, waste, and losses [69]. For instance, in Hamburg, Germany, the world's first building with an algae façade made of glass bioreactors produces heat and biomass, and binds CO_2 through the green algae's photosynthesis. Model calculations suggest that the façade can convert approximately 48% of the incoming sunlight into usable bioenergy [69].

3.5.3. Ecosystem and Biodiversity Restoration

Microbes play a crucial role in ecosystem and biodiversity restoration in several ways. For example, certain bacteria can break down pollutants and waste products, making them less harmful to the environment [78]. This process, known as bioremediation, can help clean up contaminated soil and water, making it safer for plants, animals, and humans. One example of a bacterium that can be used in bioremediation is *Pseudomonas cepacia*. This bacterium secretes a bio-surfactant that cleans up hydrocarbon contamination [79]. Microorganisms, such as bacteria and fungi, can degrade organic waste and pollutants, aiding in wastewater treatment and soil remediation. Specific examples of microorganisms that are used in the bioremediation of crude oil include *Pseudomonas cepacian* [79,80]. *Bacillus cereus* [81], *Aspergillus oryzae* [82], *Bacillus coagulans* [80], *Citrobacter koseri* [80] and *Serratia ficariam* [83]. These microorganisms can degrade pollutant hydrocarbons, heavy metals, and pesticides and are also used in the bioremediation of dyes in textile industry wastewater. Specific examples of microorganisms effective in dye bioremediation include *Exiguobacterium indicum* [84] (*Exiguobacterium aurantiacums* [85] *Bacillus cereus* [86] and *Acinetobacter baumanii* [87].

Penicillium, which is known for breaking down cellulose, a major component of plant cell walls [88] can help release nutrients into the soil, making them more available to plants. Additionally, *Penicillium* can produce antibiotics that can kill or inhibit the growth of other microbes, which can help control soil-borne diseases. Other fungi involved in this process include *Trichoderma, Rhizopus,* and *Fusarium*.

Trichoderma is a diverse genus of fungi. Some can break down varieties of organic compounds, including lignin, cellulose, and hemicellulose [89]. This renders it an effective tool for improving soil health and supporting the growth of plants. *Rhizopus* is another fungus in which most group members can help decompose organic matter and release nutrients into the soil, and it is known for its ability to break down starch and other complex sugars. One of the most well-known is *Rhizopus stolonifer* (the common bread mould), making it an essential contributor to the soil nutrient cycle [90].

Furthermore, bacteria, such as *Microcystis aeruginosa*, for example, are responsible for converting carbon dioxide into organic compounds, which can be used as a source of energy by other organisms [91]. By participating in this process, microbes help maintain the balance of carbon in the atmosphere, which is essential for the planet's health. By cleaning up pollutants and improving soil health, microbes can help support crops and other vegetation growth, providing food and other resources.

3.6. Barriers to the Development of the Microbial Bioeconomy

Several limitations must be considered when using microorganisms in the bioeconomy [92]. One of the main limitations is the complexity of the microorganisms themselves. The genetic makeup of microorganisms can vary significantly, even within the same species, making it challenging to standardise the production process. Major advances are required, including genome sequencing and the creation of systems that can facilitate multitrophic and multi-layered production of microorganisms [93]. In addition, the process of genetically modifying microorganisms, an important component of the microbial bioeconomy, is complex and expensive, making it difficult to scale up production for commercial viability [93]. The use of these organisms raises ethical and safety concerns as they could escape into the environment and potentially cause harm to other organisms and could also spread to wild populations. Addressing these limitations is important for the successful implementation of the microbial bioeconomy.

To ensure sustainable production and prevent environmental harm, the microbial bioeconomy must consider resource availability and feedstocks. It involves developing microbial-based products and services while addressing bottlenecks through scientific policy and economic approaches. Policy recommendations include increasing research investments, incentivising microbial-based products, and facilitating public–private partnerships. Multi-disciplinary research from microbiologists, chemists, economists, and farmers is necessary, as in the case of biogas production, to evaluate the innovations' scientific and societal impacts. Clear communication and close engagement with society are also crucial [94].

Another barrier lies in combining biodiversity with synthetic biotechnology for industrial-scale CO_2 capture. The CO_2 emission from fossil fuels and increased global warming are major challenges that will have significant and lasting implications for future generations. The first indicators of artificial climate change are the rising frequency of droughts, wildfires, heat waves in southern nations, excessive rainfall, and flooding [95,96].

Microbial genomes hold valuable instructions for developing long-term CO_2 collection methods. There are seven distinct pathways involved in CO_2 fixation with the pathway used by cyanobacteria, green plants, algae, and related microbes being the most familiar. Additionally, the Wood–Ljungdahl pathway, an ancient evolutionary route, is well-preserved in acetogenic bacteria and methanogenic archaea that are commonly found in harsh environments [97,98].

The reverse tricarbonyl acid cycle, the 3-hydroxypropionate bicycle, the 3-hydroxypropionate/4-hydroxybutyrate cycle, the dicarboxylate/4-hydroxybutyrate cycle (DC/HB cycle), and a few more highly effective methods to fix CO_2 from the environment have also been developed by nature [99]. Thus, there can be an improvement in the methods for designing the artificial creatures and procedures required for industrial CO_2 capture by better understanding the molecular and structural principles [97].

4. Conclusions

The rapid growth of the microbial economy promises to advance sustainability through its positive implications on environmental conservation; reduced wastage, particularly in the food and agriculture industries; the repurposing of waste; and the development of new products using alternative cheaper and environmentally friendly raw materials. In the past decades, significant advancements in research and knowledge in the microbial economy have driven the increased application of the concept in the energy sector through the production of biofuels; the increased application of the concept in the manufacturing and processing sectors through the use of microorganisms to enhance, improve, and transform some processes; and entrepreneurship through the rising number of start-ups whose products and services capitalise on biological knowledge. The observed proliferation of the adoption of the microbial economy across several sectors of the economy has shown all indications of continuing in the foreseeable future. Future studies should explore the above-identified barriers to the mass adoption and application of the microbial economy in different industries, processes, and activities to drive better waste utilisation, new product and process innovation, environmental conservation, energy reclamation, soil and environmental conservation, and sustainability.

Author Contributions: Conceptualization, A.A. and H.O.; writing—original draft preparation, A.A.; writing—review and editing, A.A., H.O., O.F. and A.H.A.; visualization, A.A. and O.F.; supervision, H.O. All authors have read and agreed to the published version of the manuscript.

Funding: This research received no external funding.

Data Availability Statement: The data that support the findings of this study are available from the corresponding author upon reasonable request.

Conflicts of Interest: The authors declare that they have no known competing financial interests or personal relationships.

References

1. United Nations. *The World Population Prospects 2022: Summary of Results*; UN Department of Economic and Social Affairs, Population Division: New York, NY, USA, 2022. Available online: https://www.un.org/development/desa/pd/sites/www.un.org.development.desa.pd/files/undesa_pd_2022_wpp_key-messages.pdf (accessed on 8 December 2022).
2. D'Adamo, I.; Gastaldi, M.; Morone, P.; Rosa, P.; Sassanelli, C.; Settembre-Blundo, D.; Shen, Y. Bioeconomy of sustainability: Drivers, opportunities and policy implications. *Sustainability* **2021**, *14*, 200. [CrossRef]
3. Tien, N. Natural Resources Limitation and the Impact on Sustainable Development of Enterprises. *Int. J. Res. Financ. Manag.* **2020**, *3*, 80–84.
4. Karp, A.; Beale, M.H.; Beaudoin, F.; Eastmond, P.J.; Neal, A.L.; Shield, I.F.; Townsend, B.J.; Dobermann, A. Growing Innovations for the Bioeconomy. *Nat. Plants* **2015**, *1*, 15193. [CrossRef] [PubMed]
5. Antar, M.; Lyu, D.; Nazari, M.; Shah, A.; Zhou, X.; Smith, D.L. Biomass for a Sustainable Bioeconomy: An Overview of World Biomass Production and Utilization. *Renew. Sustain. Energy Rev.* **2021**, *139*, 110691. [CrossRef]
6. Albrecht, K.; Ettling, S. Bioeconomy Strategies across the Globe. *Int. J. Rural. Dev.* **2014**, *3*, 10–13.
7. Thrän, D. Introduction to the Bioeconomy System. In *The Bioeconomy System*; Thrän, D., Moesenfechtel, U., Eds.; Springer: Berlin/Heidelberg, Germany, 2022; pp. 1–19, ISBN 978-3-662-64414-0.
8. Onyeaka, H.; Ekwebelem, O.C. A Review of Recent Advances in Engineering Bacteria for Enhanced CO2 Capture and Utilization. *Int. J. Environ. Sci. Technol.* **2022**, *20*, 4635–4648. [CrossRef]
9. United States Environmental Protection Agency. What Is a Circular Economy? In *US EPA Publications and Reports*; United States Environmental Protection Agency: Washington, DC, USA, 2022.
10. Lonestar College. History of Biotechnology. 2022. Available online: https://www.lonestar.edu/history-of-biotechnology.htm (accessed on 8 December 2022).
11. Liu, T.; Klammsteiner, T.; Dregulo, A.M.; Kumar, V.; Zhou, Y.; Zhang, Z.; Awasthi, M.K. Black soldier fly larvae for organic manure recycling and its potential for a circular bioeconomy: A review. *Sci. Total Environ.* **2022**, *833*, 155122. [CrossRef]
12. Hanif, M.; Bhatti, I.A.; Zahid, M.; Shahid, M. Production of Biodiesel from Non-Edible Feedstocks Using Environment Friendly Nano-Magnetic Fe/SnO Catalyst. *Sci. Rep.* **2022**, *12*, 16705. [CrossRef]
13. Nduko, J.M.; Taguchi, S. Microbial Production of Biodegradable Lactate-Based Polymers and Oligomeric Building Blocks From Renewable and Waste Resources. *Front. Bioeng. Biotechnol.* **2021**, *8*, 618077. [CrossRef] [PubMed]
14. Bian, B.; Bajracharya, S.; Xu, J.; Pant, D.; Saikaly, P.E. Microbial Electrosynthesis from CO_2: Challenges, Opportunities and Perspectives in the Context of Circular Bioeconomy. *Bioresour. Technol.* **2020**, *302*, 122863. [CrossRef]
15. Romero-Perdomo, F.; González-Curbelo, M.Á. Integrating Multi-Criteria Techniques in Life-Cycle Tools for the Circular Bioeconomy Transition of Agri-Food Waste Biomass: A Systematic Review. *Sustainability* **2023**, *15*, 5026. [CrossRef]
16. Kircher, M. Economic trends in the transition into a circular bioeconomy. *J. Risk Financ. Manag.* **2022**, *15*, 44. [CrossRef]
17. Wesseler, J.; Kleter, G.; Meulenbroek, M.; Purnhagen, K.P. EU Regulation of Genetically Modified Microorganisms in Light of New Policy Developments: Possible Implications for EU Bioeconomy Investments. *Appl. Econ. Perspect. Policy* **2022**, 1–21. [CrossRef]
18. Bose, D.; Dey, A.; Banerjee, T. Aspects of Bioeconomy and Microbial Fuel Cell Technologies for Sustainable Development. *Sustainability* **2020**, *13*, 107–118. [CrossRef]
19. Gilbert, J.A.; Stephens, B. Microbiology of the built environment. *Nat. Rev. Microbiol.* **2018**, *16*, 661–670. [CrossRef]
20. Shi, T.-Q.; Peng, H.; Zeng, S.-Y.; Ji, R.-Y.; Shi, K.; Huang, H.; Ji, X.-J. Microbial Production of Plant Hormones: Opportunities and Challenges. *Bioengineered* **2017**, *8*, 124–128. [CrossRef]
21. Voidarou, C.; Antoniadou, M.; Rozos, G.; Tzora, A.; Skoufos, I.; Varzakas, T.; Lagiou, A.; Bezirtzoglou, E. Fermentative Foods: Microbiology, Biochemistry, Potential Human Health Benefits and Public Health Issues. *Foods* **2020**, *10*, 69. [CrossRef]
22. Atiwesh, G.; Mikhael, A.; Parrish, C.C.; Banoub, J.; Le, T.A.T. Environmental impact of bioplastic use: A review. *Heliyon* **2021**, *7*, e07918. [CrossRef]

23. Bhange, K.; Chaturvedi, V.; Bhatt, R. Simultaneous production of detergent stable keratinolytic protease, amylase and biosurfactant by Bacillus subtilis PF1 using agro industrial waste. *Biotechnol. Rep.* **2016**, *10*, 94–104. [CrossRef]

24. Azhar, S.H.M.; Abdulla, R.; Jambo, S.A.; Marbawi, H.; Gansau, J.A.; Faik, A.A.M.; Rodrigues, K.F. Yeasts in sustainable bioethanol production: A review. *Biochem. Biophys. Rep.* **2017**, *10*, 52–61. [CrossRef]

25. Ene, N.; Soare Vladu, M.G.; Lupescu, I.; Ionescu, A.D.; Vamanu, E. The Production of Biodegradable Polymers-medium-chain-length Polyhydroxyalkanoates (mcl-PHA) in *Pseudomonas putida* for Biomedical Engineering Applications. *Curr. Pharm. Biotechnol.* **2022**, *23*, 1109–1117. [CrossRef]

26. Gupta, P.L.; Rajput, M.; Oza, T.; Trivedi, U.; Sanghvi, G. Eminence of microbial products in cosmetic industry. *Nat. Prod. Bioprospecting* **2019**, *9*, 267–278. [CrossRef] [PubMed]

27. Iseppi, R.; Zurlini, C.; Cigognini, I.M.; Cannavacciuolo, M.; Sabia, C.; Messi, P. Eco-friendly edible packaging systems based on live-*Lactobacillus kefiri* MM5 for the control of Listeria monocytogenes in fresh vegetables. *Foods* **2022**, *11*, 2632. [CrossRef] [PubMed]

28. Peciulyte, A.; Anasontzis, G.E.; Karlström, K.; Larsson, P.T.; Olsson, L. Morphology and enzyme production of *Trichoderma reesei* Rut C-30 are affected by the physical and structural characteristics of cellulosic substrates. *Fungal Genet. Biol.* **2014**, *72*, 64–72. [CrossRef] [PubMed]

29. Harman, G.E.; Kubicek, C.P. *Trichoderma and Gliocladium. Volume 1: Basic Biology, Taxonomy and Genetics*, 1st ed.; CRC Press: London, UK, 1998; ISBN 978-0-429-07893-4.

30. Rasoulnia, P.; Mousavi, S.M. Maximization of organic acids production by *Aspergillus niger* in a bubble column bioreactor for V and Ni recovery enhancement from power plant residual ash in spent-medium bioleaching experiments. *Bioresour. Technol.* **2016**, *216*, 729–736. [CrossRef] [PubMed]

31. Birania, S.; Kumar, S.; Kumar, N.; Attkan, A.K.; Panghal, A.; Rohilla, P.; Kumar, R. Advances in development of biodegradable food packaging material from agricultural and agro-industry waste. *J. Food Process Eng.* **2021**, *45*, e13930. [CrossRef]

32. Chandel, A.K.; Garlapati, V.K.; Kumar, S.P.J.; Hans, M.; Singh, A.K.; Kumar, S. The Role of Renewable Chemicals and Biofuels in Building a Bioeconomy. *Biofuels Bioprod. Biorefining* **2020**, *14*, 830–844. [CrossRef]

33. Issa, I.; Delbrück, S.; Hamm, U. Bioeconomy from Experts' Perspectives—Results of a Global Expert Survey. *PLoS ONE* **2019**, *14*, e0215917. [CrossRef]

34. Jadhav, D.A.; Mungray, A.K.; Arkatkar, A.; Kumar, S.S. Recent Advancement in Scaling-up Applications of Microbial Fuel Cells: From Reality to Practicability. *Sustain. Energy Technol. Assess.* **2021**, *45*, 101226. [CrossRef]

35. Schütte, G. What Kind of Innovation Policy Does the Bioeconomy Need? *New Biotechnol.* **2018**, *40*, 82–86. [CrossRef] [PubMed]

36. Lokko, Y.; Heijde, M.; Schebesta, K.; Scholtès, P.; Van Montagu, M.; Giacca, M. Biotechnology and the Bioeconomy—Towards Inclusive and Sustainable Industrial Development. *New Biotechnol.* **2018**, *40*, 5–10. [CrossRef] [PubMed]

37. European Commission, Directorate-General for Research and Innovation. A Sustainable Bioeconomy for Europe: Strengthening the Connection between Economy, Society and the Environment: Updated Bioeconomy Strategy, Publications Office. 2018. Available online: https://data.europa.eu/doi/10.2777/792130 (accessed on 8 December 2022).

38. Piotrowski, S.; Carus, M.; und Carrez, D. European Bioeconomy in Figures. 2016. Available online: https://biconsortium.eu/sites/biconsortium.eu/files/documents/European%20Bioeconomy%20in%20Figures%202008%20-%202016_0.pdf (accessed on 8 December 2022).

39. Joshi, S.; Robles, A.; Aguiar, S.; Delgado, A.G. The Occurrence and Ecology of Microbial Chain Elongation of Carboxylates in Soils. *ISME J.* **2021**, *15*, 1907–1918. [CrossRef]

40. Singh, R. Microbial Biotransformation: A Process for Chemical Alterations. *J. Bacteriol. Mycol. Open Access* **2017**, *4*, 47–51. [CrossRef]

41. Marulanda, V.A.; Gutierrez, C.D.B.; Alzate, C.A.C. Thermochemical, Biological, Biochemical, and Hybrid Conversion Methods of Bio-Derived Molecules into Renewable Fuels. In *Advanced Bioprocessing for Alternative Fuels, Biobased Chemicals, and Bioproducts*; Elsevier: Amsterdam, The Netherlands, 2019; pp. 59–81. ISBN 978-0-12-817941-3.

42. Bell, E.L.; Finnigan, W.; France, S.P.; Green, A.P.; Hayes, M.A.; Hepworth, L.J.; Lovelock, S.L.; Niikura, H.; Osuna, S.; Romero, E.; et al. Biocatalysis. *Nat. Rev. Methods Primers* **2021**, *1*, 46. [CrossRef]

43. Ahmed, I.; Zia, M.A.; Afzal, H.; Ahmed, S.; Ahmad, M.; Akram, Z.; Sher, F.; Iqbal, H.M.N. Socio-Economic and Environmental Impacts of Biomass Valorisation: A Strategic Drive for Sustainable Bioeconomy. *Sustainability* **2021**, *13*, 4200. [CrossRef]

44. Nattassha, R.; Handayati, Y.; Simatupang, T.M.; Siallagan, M. Understanding Circular Economy Implementation in the Agri-Food Supply Chain: The Case of an Indonesian Organic Fertiliser Producer. *Agric. Food Secur.* **2020**, *9*, 10. [CrossRef]

45. Adrio, J.L.; Demain, A.L. Genetic Improvement of Processes Yielding Microbial Products. *FEMS Microbiol. Rev.* **2006**, *30*, 187–214. [CrossRef]

46. Karmee, S.K. Liquid Biofuels from Food Waste: Current Trends, Prospect and Limitation. *Renew. Sustain. Energy Rev.* **2016**, *53*, 945–953. [CrossRef]

47. Cunha, M.; Romaní, A.; Carvalho, M.; Domingues, L. Boosting Bioethanol Production from Eucalyptus Wood by Whey Incorporation. *Bioresour. Technol.* **2018**, *250*, 256–264. [CrossRef]

48. Donner, M.; de Vries, H. Innovative Business Models for a Sustainable Circular Bioeconomy in the French Agrifood Domain. *Sustainability* **2023**, *15*, 5499. [CrossRef]

49. Lin, Y.; Tanaka, S. Ethanol Fermentation from Biomass Resources: Current State and Prospects. *Appl. Microbiol. Biotechnol.* **2006**, *69*, 627–642. [CrossRef] [PubMed]
50. Singh, R.; Langyan, S.; Rohtagi, B.; Darjee, S.; Khandelwal, A.; Shrivastava, M.; Singh, A. Production of biofuels options by contribution of effective and suitable enzymes: Technological developments and challenges. *Mater. Sci. Energy Technol.* **2022**, *5*, 294–310. [CrossRef]
51. Nagaoka, S. Yogurt Production. In *Lactic Acid Bacteria*; Kanauchi, M., Ed.; Methods in Molecular Biology; Springer: New York, NY, USA, 2019; Volume 1887, pp. 45–54. ISBN 978-1-4939-8906-5.
52. Wang, Y.; Liu, L.; Jin, Z.; Zhang, D. Microbial Cell Factories for Green Production of Vitamins. *Front. Bioeng. Biotechnol.* **2021**, *9*, 661562. [CrossRef]
53. Patthawaro, S.; Saejung, C. Production of Single Cell Protein from Manure as Animal Feed by Using Photosynthetic Bacteria. *Microbiol. Open* **2019**, *8*, e913. [CrossRef]
54. Saradha Devi, G.; Vaishnavi, S.; Srinath, S.; Dutt, B.; Rajmohan, K.S. Chapter 19—Energy Recovery from Biomass Using Gasification. In *Current Developments in Biotechnology and Bioengineering*; Varjani, S., Pandey, A., Gnansounou, E., Khanal, S.K., Raveendran, S., Eds.; Elsevier: Amsterdam, The Netherlands, 2020; pp. 363–382. [CrossRef]
55. National Space Agency. Electrified Bacteria Clean Wastewater, Generate Power. NASA Spin-off 2019. Available online: https://spinoff.nasa.gov/Spinoff2019/ee_1.html (accessed on 8 December 2022).
56. Bernal, M.P.; Sommer, S.G.; Chadwick, D.; Qing, C.; Guoxue, L.; Michel, F.C., Jr. Current Approaches and Future trends in Compost Quality Criteria for Agronomic, Environmental, and Human Health Benefits. *Adv. Agron.* **2017**, *144*, 143–233.
57. Panoutsou, C.; Alexopoulou, E. Costs and Profitability of Crops for Bioeconomy in the EU. *Energies* **2020**, *13*, 1222. [CrossRef]
58. Capozzi, V.; Fragasso, M.; Bimbo, F. Microbial Resources, Fermentation and Reduction of Negative Externalities in Food Systems: Patterns toward Sustainability and Resilience. *Fermentation* **2021**, *7*, 54. [CrossRef]
59. Knowledge of Wharton Staff. The Brazilian Bioplastics Revolution. 2009. Available online: https://knowledge.wharton.upenn.edu/article/the-brazilian-bioplastics-revolution (accessed on 8 December 2022).
60. Jang, Y.S.; Lee, J.Y.; Lee, J.; Park, J.H.; Im, J.A.; Eom, M.H.; Lee, J.; Lee, S.H.; Song, H.; Cho, J.H.; et al. Enhanced butanol production obtained by reinforcing the direct butanol-forming route in Clostridium acetobutylicum. *mBio* **2012**, *3*, e00314-12. [CrossRef]
61. Aindrila, M. Tolerance engineering in bacteria for the production of advanced biofuels and chemicals. *Trends Microbiol.* **2015**, *23*, 498–508. [CrossRef]
62. Adegboye, M.F.; Ojuederie, O.B.; Talia, P.M.; Babalola, O.O. Bioprospecting of microbial strains for biofuel production: Metabolic engineering, applications, and challenges. *Biotechnol. Biofuels* **2021**, *14*, 5. [CrossRef] [PubMed]
63. Nyyssölä, A.; Ojala, L.S.; Wuokko, M.; Peddinti, G.; Tamminen, A.; Tsitko, I.; Nordlund, E.; Lienemann, M. Production of Endotoxin-Free Microbial Biomass for Food Applications by Gas Fermentation of Gram-Positive H2-Oxidizing Bacteria. *ACS Food Sci. Technol.* **2021**, *1*, 470–479. [CrossRef]
64. Benemann, J. Microalgae for Biofuels and Animal Feeds. *Energies* **2013**, *6*, 5869–5886. [CrossRef]
65. Onen Cinar, S.; Chong, Z.K.; Kucuker, M.A.; Wieczorek, N.; Cengiz, U.; Kuchta, K. Bioplastic Production from Microalgae: A Review. *Int. J. Environ. Res. Public Health* **2020**, *17*, 3842. [CrossRef] [PubMed]
66. Guo, N.; Liu, J.-B.; Li, W.; Ma, Y.-S.; Fu, D. The Power and the Promise of CRISPR/Cas9 Genome Editing for Clinical Application with Gene Therapy. *J. Adv. Res.* **2022**, *40*, 135–152. [CrossRef] [PubMed]
67. United Nations. *The Sustainable Development Goals Report 2022*. United Nations Publications; United Nations: New York, NY, USA, 2022. Available online: https://unstats.un.org/sdgs/report/2022/ (accessed on 8 December 2022).
68. Tan, E.C.D.; Lamers, P. Circular Bioeconomy Concepts—A Perspective. *Front. Sustain.* **2021**, *2*, 701509. [CrossRef]
69. Wesseler, J.; von Braun, J. Measuring the bioeconomy: Economics and policies. *Annu. Rev. Resour. Econ.* **2017**, *9*, 275–298. [CrossRef]
70. Staffas, L.; Gustavsson, M.; McCormick, K. Strategies and Policies for the Bioeconomy and Bio-Based Economy: An Analysis of Official National Approaches. *Sustainability* **2013**, *5*, 2751–2769. [CrossRef]
71. Gomez, J.A.; Höffner, K.; Barton, P.I. From Sugars to Biodiesel Using Microalgae and Yeast. *Green Chem.* **2016**, *18*, 461–475. [CrossRef]
72. Jung, F.; Krüger-Genge, A.; Waldeck, P.; Küpper, J.-H. Spirulina Platensis, a Super Food? *J. Cell. Biotechnol.* **2019**, *5*, 43–54. [CrossRef]
73. Azevedo, S.G.; Sequeira, T.; Santos, M.; Mendes, L. Biomass-related sustainability: A review of the literature and interpretive structural modeling. *Energy. Elsevier* **2019**, *171*, 1107–1125.
74. Demirbas, A. Political, Economic and Environmental Impacts of Biofuels: A Review. *Appl. Energy* **2009**, *86*, S108–S117. [CrossRef]
75. Behera, L.; Datta, D.; Kumar, S.; Kumar, S.; Sravani, B.; Chandra, R. Role of Microbial Consortia in Remediation of Soil, Water and Environmental Pollution Caused by Indiscriminate Use of Chemicals in Agriculture: Opportunities and Challenges. In *New and Future Developments in Microbial Biotechnology and Bioengineering*; Elsevier: Amsterdam, The Netherlands, 2022; pp. 399–418. ISBN 978-0-323-85577-8.
76. Lin, F.; Mao, Y.; Zhao, F.; Idris, A.L.; Liu, Q.; Zou, S.; Guan, X.; Huang, T. Towards Sustainable Green Adjuvants for Microbial Pesticides: Recent Progress, Upcoming Challenges, and Future Perspectives. *Microorganisms* **2023**, *11*, 364. [CrossRef] [PubMed]
77. United States Environmental Protection Agency. Economics of Biofuels. In *US EPA Publications and Reports*; United States Environmental Protection Agency: Washington, DC, USA, 2022.

78. Rabbani, A.; Zainith, S.; Deb, V.K.; Das, P.; Bharti, P.; Rawat, D.S.; Kumar, N.; Saxena, G. Microbial Technologies for Environmental Remediation: Potential Issues, Challenges, and Future Prospects. In *Microbe Mediated Remediation of Environmental Contaminants*; Elsevier: Amsterdam, The Netherlands, 2021; pp. 271–286. ISBN 978-0-12-821199-1.

79. Soares da Silva, R.D.C.F.; Luna, J.M.; Rufino, R.D.; Sarubbo, L.A. Ecotoxicity of the Formulated Biosurfactant from *Pseudomonas cepacia* CCT 6659 and Application in the Bioremediation of Terrestrial and Aquatic Environments Impacted by Oil Spills. *Process Saf. Environ. Prot.* **2021**, *154*, 338–347. [CrossRef]

80. Fagbemi, O.K.; Sanusi, A.I. Chromosomal and Plasmid Mediated Degradation of Crude Oil by *Bacillus coagulans*, *Citrobacter koseri* and *Serratia ficaria* Isolated from the Soil. *Afr. J. Biotechnol.* **2017**, *16*, 1242–1253. [CrossRef]

81. Deng, Z.; Jiang, Y.; Chen, K.; Gao, F.; Liu, X. Petroleum Depletion Property and Microbial Community Shift After Bioremediation Using *Bacillus halotolerans* T-04 and *Bacillus cereus* 1-1. *Front. Microbiol.* **2020**, *11*, 353. [CrossRef]

82. Singh, R.; Kumar, M.; Mittal, A.; Mehta, P.K. Microbial Enzymes: Industrial Progress in 21st Century. *3 Biotech* **2016**, *6*, 174. [CrossRef]

83. Dos Santos, R.A.; Rodríguez, D.M.; da Silva, L.A.R.; de Almeida, S.M.; de Campos-Takaki, G.M.; de Lima, M.A.B. Enhanced production of prodigiosin by Serratia marcescens UCP 1549 using agrosubstrates in solid-state fermentation. *Arch Microbiol.* **2021**, *203*, 4091–4100. [CrossRef]

84. Solís, M.; Solís, A.; Pérez, H.I.; Manjarrez, N.; Flores, M. Microbial Decolouration of Azo Dyes: A Review. *Process Biochem.* **2012**, *47*, 1723–1748. [CrossRef]

85. Palanivelan, R.; Sakthi Thesai, A.; Ramya, S.; Ayyasamy, P.M. Effect of Multiple Factors on Azo Dye Decolorization using a Moderate Halophilic Bacterium *Exiguobacterium aurantiacum* (ESL52). *Glob. Sci.* **2019**, *22*, 206–216.

86. Emadi, Z.; Sadeghi, R.; Forouzandeh, S.; Mohammadi-Moghadam, F.; Sadeghi, R.; Sadeghi, M. Simultaneous Anaerobic Decolorization/Degradation of Reactive Black-5 Azo Dye and Chromium (VI) Removal by *Bacillus cereus* Strain MS038EH Followed by UV-C/H2O2 Post-Treatment for Detoxification of Biotransformed Products. *Arch. Microbiol.* **2021**, *203*, 4993–5009. [CrossRef]

87. Sreedharan, V.; Saha, P.; Rao, K.V.B. Dye Degradation Potential of *Acinetobacter baumannii* Strain VITVB against Commercial Azo Dyes. *Bioremediation J.* **2021**, *25*, 347–368. [CrossRef]

88. Todero Ritter, C.E.; Camassola, M.; Zampieri, D.; Silveira, M.M.; Dillon, A.J.P. Cellulase and Xylanase Production by *Penicillium echinulatum* in Submerged Media Containing Cellulose Amended with Sorbitol. *Enzym. Res.* **2013**, *2013*, 240219. [CrossRef] [PubMed]

89. Do Vale, L.H.F.; Filho, E.X.F.; Miller, R.N.G.; Ricart, C.A.O.; de Sousa, M.V. Cellulase Systems in Trichoderma. In *Biotechnology and Biology of Trichoderma*; Elsevier: Amsterdam, The Netherlands, 2014; pp. 229–244, ISBN 978-0-444-59576-8.

90. Briggs, G.M. *Inanimate Life*; Milne Open Textbooks: Geneseo, NY, USA, 2021; ISBN 978-1-942341-82-6.

91. Straub, C.; Quillardet, P.; Vergalli, J.; de Marsac, N.T.; Humbert, J.-F. A Day in the Life of Microcystis Aeruginosa Strain PCC 7806 as Revealed by a Transcriptomic Analysis. *PLoS ONE* **2011**, *6*, e16208. [CrossRef]

92. Talwar, N.; Holden, N.M. The Limitations of Bioeconomy LCA Studies for Understanding the Transition to Sustainable Bioeconomy. *Int. J. Life Cycle Assess.* **2022**, *27*, 680–703. [CrossRef] [PubMed]

93. Yarnold, J.; Karan, H.; Oey, M.; Hankamer, B. Microalgal Aquafeeds as Part of a Circular Bioeconomy. *Trends Plant Sci.* **2019**, *24*, 959–970. [CrossRef]

94. United Kingdom Department of Business, Energy and Industrial Strategy. *Growing the Economy: A National Strategy to 2030*; United Kingdom Department of Business, Energy and Industrial Strategy: London, UK, 2018. Available online: https://www.gov.uk/government/publications/bioeconomy-strategy-2018-to-2030/uk-bioeconomy-strategy-background-analytical-note (accessed on 8 December 2022).

95. Karl, T.R.; Trenberth, K.E. Modern global climate change. *Science* **2003**, *302*, 1719–1723. [CrossRef] [PubMed]

96. IPCC. Climate Change 2014. Synthesis Report. 2014. Available online: https://www.ipcc.ch/assessment-report/ar5/ (accessed on 8 December 2022).

97. Antranikian, G.; Streit, W.R. Microorganisms Harbor Keys to a Circular Bioeconomy Making Them Useful Tools in Fighting Plastic Pollution and Rising CO_2 Levels. *Extremophiles* **2022**, *26*, 10. [CrossRef]

98. Schuchmann, K.; Müller, V. Autotrophy at the thermodynamic limit of life: A model for energy conservation in acetogenic bacteria. *Nat. Rev. Microbiol.* **2014**, *12*, 809–821. [CrossRef]

99. Steffens, L.; Pettinato, E.; Steiner, T.M.; Mall, A.; König, S.; Eisenreich, W.; Berg, I.A. High CO_2 levels drive the TCA cycle backwards towards autotrophy. *Nature* **2021**, *592*, 784–788. [CrossRef]

MDPI

St. Alban-Anlage 66

4052 Basel

Switzerland

Tel. +41 61 683 77 34

Fax +41 61 302 89 18

www.mdpi.com

Sustainability Editorial Office

E-mail: sustainability@mdpi.com

www.mdpi.com/journal/sustainability